Lecture Notes in Computer Science 4425

Commenced Publication in 1973
Founding and Former Series Editors:
Gerhard Goos, Juris Hartmanis, and Jan van Leeuwen

Giambattista Amati   Claudio Carpineto
Giovanni Romano (Eds.)

Advances in Information Retrieval

29th European Conference on IR Research, ECIR 2007
Rome, Italy, April 2-5, 2007
Proceedings

Springer

Volume Editors

Giambattista Amati
Claudio Carpineto
Giovanni Romano
Fondazione Ugo Bordoni
Via Baldassarre Castiglione 59, 00142 Rome, Italy
E-mail:{gba, carpinet, romano}@fub.it

Library of Congress Control Number: 2007923290

CR Subject Classification (1998): H.3, H.2, I.2.3, I.2.6-7, H.4, H.5.4, I.7

LNCS Sublibrary: SL 3 – Information Systems and Application, incl. Internet/Web and HCI

ISSN	0302-9743
ISBN-10	3-540-71494-4 Springer Berlin Heidelberg New York
ISBN-13	978-3-540-71494-1 Springer Berlin Heidelberg New York

Springer is a part of Springer Science+Business Media

springer.com

Typesetting: Camera-ready by author, data conversion by Scientific Publishing Services, Chennai, India
Printed on acid-free paper SPIN: 12038824 06/3142 5 4 3 2 1 0

Preface

This volume contains the papers presented at ECIR 2007, the 29th European Conference on Information Retrieval. The conference was organized by the Fondazione Ugo Bordoni (FUB), in cooperation with the Information Retrieval Specialist Group of the British Computer Society (BCS-IRSG) and the ACM Special Interest Group on Information Retrieval (ACM SIGIR). It was held during April 2–5, 2007 in Rome, hosted by the National Research Council of Italy (CNR).

The conference was initially established by BCS-IRSG in the United Kingdom under the name "Annual Colloquium on Information Retrieval Research". When the involvement of Continental Europe became more important, it was decided to alternate the conference venue between the UK and Continental Europe. Starting from 2001, the conference was renamed ECIR. In the last few years, ECIR has grown steadily, becoming the major European forum for research on information retrieval.

ECIR 2007 has dramatically confirmed the growth trend, with a high record number of submissions. In four years the number of full-paper submissions to ECIR has nearly tripled, going from 88 (2004, Sunderland) to 124 (2005, Santiago de Compostela), then to 175 (2006, London), and now to 220 (2007, Rome). ECIR 2007 has also attracted more and more people from outside Europe, thus making the conference a large, truly international event. Furthermore, in spite of these major changes, the traditional strong focus on students has been well preserved, for their participation has been massive.

The large number of sponsors of ECIR 2007 both reflects the increasing importance of the conference and is a key to its success. The sponsorship contribution was primarily used to support student attendance and to invite three internationally renowned researchers to give a keynote speech, namely, Andrei Broder (Yahoo! Research), Stephen Robertson (Microsoft Research Cambridge and City University London), and Marco Gori (University of Siena).

Turning to the reviewing process, the Program Committee (PC) of ECIR 2007 was formed by 147 members, 41 of whom were non-European. To increase consistency of refereeing, the PC was split in subgroups of homogeneous expertise with three members each, and each subgroup was then assigned a bunch of related papers.

In response to the call for papers, 220 submissions were received. Each submission was normally reviewed by three PC members. In case of a strong disagreement, the reviewers were given a chance to change their reviews before the PC meeting. The reviews were then thoroughly discussed by the 19 participants in the PC meeting, which was held at FUB in Rome. Decisions were based on the scores assigned by the reviewers and on the reviews themselves. When the reviews of a paper were not sufficient to make a decision, the paper went through an extra review at the PC meeting. Out of the 220 submissions,

42 were accepted for presentation at the conference. Interestingly, 23 of these had a full-time student as the primary author.

In addition, at the PC meeting, it was decided to create a short-paper session because there were many papers with good scores that would have been rejected due to high competition and limited capacity for oral presentation at the conference. Nineteen papers were accepted as short papers. Short papers were given 8 pages in the proceeding (instead of 12), with the short-paper session consisting of a brief oral presentation of all short papers followed by a poster-style exhibition.

There was also a separate call for posters. Each of the 72 poster submissions was normally reviewed by three PC members, as with paper submissions. Twenty-one posters were accepted for presentation.

The topics covered by the accepted papers span well-established as well as emerging research areas, with a concentration on indexing techniques, distributed information retrieval, and query processing. The 42 full papers have been grouped in the following way: Theory and Design (5), Efficiency (5), Peer-to-Peer Networks (4), Result Merging (2), Queries (4), Relevance Feedback (2), Evaluation (2), Classification and Clustering (4), Filtering (4), Topic Identification (2), Expert Finding (2), XML (2), Web IR (2), Multimedia IR (2).

The success of ECIR 2007 was due to a huge number of people and several organizations who were involved in the various stages of the whole process. We thank the researchers who submitted their results, the members of the Program Committee for reviewing many papers in a short time, the members of the local Organizing Committee for their hard work over many months, the members of the Award Committees for choosing the best paper and the the best student paper, the keynote speakers for accepting our invitation, the Italian National Research Council for providing the conference venue, the sponsoring organizations for providing the money. We are deeply indebted to all of them. We would also like to thank the members of BCS-IRSG and the organizers of ECIR 2006 for their useful help. Finally, a special thanks to Fondazione Ugo Bordoni, whose support was invaluable for running the whole conference.

We would like to conclude this preface on a more personal note. Although the amount of work required was definitely harder than expected, we now feel that it was worth it! Organizing and running ECIR 2007 was a great experience for us. We are very happy especially because of our own historical involvement in ECIR. We have seen how the conference has grown over time and we are proud to have contributed to making it a key event in the information retrieval field.

Enjoy the papers!

January 2007

Giambattista Amati
Claudio Carpineto
Giovanni Romano

Organization

ECIR 2007 was organized by the Fondazione Ugo Bordoni (FUB), in collaboration with the Information Retrieval Specialist Group of the British Computer Society (BCS-IRSG) and the ACM Special Interest Group on Information Retrieval (ACM-SIGIR).

General and Program Chairs

Giambattista Amati, Fondazione Ugo Bordoni, Rome, Italy
Claudio Carpineto, Fondazione Ugo Bordoni, Rome, Italy
Giovanni Romano, Fondazione Ugo Bordoni, Rome, Italy

Local Organization Committee

Marilena Carletti, Fondazione Ugo Bordoni
Annalisa Filardo, Fondazione Ugo Bordoni
Sara Saverione, Fondazione Ugo Bordoni
Stefania Vinci, Fondazione Ugo Bordoni
Guido Zampilloni, Fondazione Ugo Bordoni

Program Committee

Maristella Agosti, University of Padova, Italy
James Allan, University of Massachusetts, Amherst, USA
Massih-Reza Amini, Université Pierre et Marie Curie, France
Einat Amitay, IBM Research Lab, Haifa, Israel
Giuseppe Attardi, University of Pisa, Italy
Leif Azzopardi, University of Strathclyde, UK
Ricardo Baeza-Yates, Yahoo! Research, Barcelona, Spain
Alex Bailey, Google, UK
Álvaro Barreiro, Universidade da Coruña, Spain
Roberto Basili, University of Tor Vergata, Roma, Italy
Holger Bast, Max Planck Institute for Informatics, Germany
Micheline Beaulieu, University of Sheffield, UK
Michel Beigbeder, École Nationale Supérieure, Saint-Etienne, France
Nicholas Belkin, Rutgers University, USA
Gloria Bordogna, IDPA CNR, Italy
Theo Bothma, University of Pretoria, South Africa

Mohand Boughanem, University Paul Sabatier, France
Giorgio Brajnik, Università di Udine, Italy
Martin Braschler, Zurich University of Applied Sciences, Switzerland
Peter Bruza, Queensland University of Technology, Australia
Stefan Büttcher, University of Waterloo, Canada
Wray Buntine, Helsinki Institute of Information Technology, Finland
Fidel Cacheda, Universidade da Coruña, Spain
Jamie Callan, Carnegie Mellon University, USA
Caterina Caracciolo, FAO, Italy
David Carmel, IBM Research Lab, Haifa, Israel
Carlos Castillo, Yahoo! Research, Spain
Yves Chiaramella, IMAG, Grenoble, France
Paul-Alexandru Chirita, L3S Research Center, Hannover, Germany
Charles L. A. Clarke, University of Waterloo, Canada
Paul Clough, University of Sheffield, UK
Gordon Cormack, University of Waterloo, Canada
Nick Craswell, Microsoft, USA
Fabio Crestani, University of Strathclyde, UK
Bruce Croft, University of Massachusetts, Amherst, USA
Pablo De la Fuente, Universidad de Valladolid, Spain
Renato De Mori, University of Avignon, France
Maarten de Rijke, University of Amsterdam, The Netherlands
Arjen de Vries, CWI, The Netherlands
Marcello Federico, Istituto Trentino di Cultura, Italy
Ronen Feldman, Bar-Ilan University, Israel
Juan Manuel Fernández Luna, University of Granada, Spain
Paolo Ferragina, University of Pisa, Italy
Nicola Ferro, University of Padova, Italy
Edward Fox, Virginia Tech, USA
Johannes Fürnkranz, TU Darmstadt, Germany
Norbert Fuhr, University of Duisburg-Essen, Germany
Patrick Gallinari, LIP6, Université Pierre et Marie Curie, France
Éric Gaussier, Xerox Research Centre Europe, France
C. Lee Giles, Pennsylvania State University, USA
Mark Girolami, University of Glasgow, UK
Ayse Goker, Robert Gordon University, UK
Julio Gonzalo, UNED, Spain
Margaret Graham, Northumbria University, UK
Warren Greiff, The MITRE Corporation, USA
David Grossman, Illinois Institute of Technology, USA
Antonio Gulli, ASK, Italy
Cathal Gurrin, Dublin City University, Ireland
Preben Hansen, SICS, Sweden
David Hawking, CSIRO ICT Centre, Australia

Ben He, University of Glasgow, UK
William Hersh, Oregon Health Sciences University, USA
Djoerd Hiemstra, University of Twente, The Netherlands
Eduard Hoenkamp, University of Nijmegen, The Netherlands
Andreas Hotho, Universitt Kassel, Germany
Xiaohua Hu, Drexel University, USA
Theo Huibers, University of Twente, The Netherlands
Peter Ingwersen, Royal School of Library and Information Science, Denmark
Mario J. Gaspar da Silva, University of Lisbon, Portugal
Kalervo Jarvelin, University of Tampere, Finland
Gareth Jones, Dublin City University, Ireland
Joemon Jose, University of Glasgow, UK
Jaap Kamps, University of Amsterdam, The Netherlands
Jussi Karlgren, SICS, Sweden
Gabriella Kazai, Microsoft, Cambridge, UK
Manolis Koubarakis, Technical University of Crete, Greece
Wessel Kraaij, TNO TPD, The Netherlands
S. Ravi Kumar, Yahoo! Research, USA
Kui-Lam Kwok, Queens College, City University of New York, USA
Mounia Lalmas, Queen Mary, University of London, UK
Monica Landoni, University of Strathclyde, UK
Birger Larsen, Royal School of Library and Information Science, Denmark
Mun-Kew Leong, Laboratories of Information Technology, Singapore
David Lewis, David D. Lewis Consulting, USA
Xuelong Li, School of Computer Science and Information Systems, London, UK
Yang Lingpeng, Institute for Infocomm Research, Singapore
Christina Lioma, University of Glasgow, UK
David Losada, University of Santiago de Compostela, Spain
Craig Macdonald, University of Glasgow, UK
Andrew MacFarlane, City University, London, UK
Marco Maggini, University of Siena, Italy
Bernardo Magnini, Istituto Trentino di Cultura, Italy
Massimo Melucci, University of Padova, Italy
Alessandro Micarelli, University of Roma Tre, Italy
Stefano Mizzaro, University of Udine, Italy
Dunja Mladenic, Jozef Stefan Institute, Slovenia
Marie-Francine Moens, Katholieke Universiteit Leuven, Belgium
Alistair Moffat, University of Melbourne, Australia
Josiane Mothe, IRIT, France
Gheorghe Muresan, Rutgers University, USA
Jian-Yun Nie, University of Montreal, Canada
Michael Oakes, University of Sunderland, UK
Stanislaw Osinski, Poznan Supercomputing and Networking Center, Poland
Iadh Ounis, University of Glasgow, UK

Marius Pasca, Google, USA
Gabriella Pasi, University of Milano - Bicocca, Italy
Jan O. Pedersen, Yahoo!, USA
Nils Pharo, Oslo University College, Norway
Vassilis Plachouras, University of Glasgow, UK
Jay Ponte, Google, USA
Victor Poznanski, Sharp, UK
Andreas Rauber, Vienna University of Technology, Austria
Stephen Robertson, Microsoft Research, UK
Thomas Rolleke, Queen Mary, University of London, UK
Stefan Rüger, Imperial College, UK
Song Ruihua, Microsoft Research Asia, China
Ian Ruthven, University of Strathclyde, UK
Dominich Sandor, University of Veszprem, Hungary
Ralf Schenkel, Max-Planck-Institut für Informatik, Germany
Hinrich Schuetze, University of Stuttgart, Germany
Giovanni Semeraro, University of Bari, Italy
Stefan Siersdorfer, Max Planck Institute for Computer Science, Germany
Fabrizio Silvestri, ISTI CNR, Italy
Alan Smeaton, Dublin City University, Ireland
Barry Smyth, University College Dublin, Ireland
Vaclav Snasel, VSB-Technical University Ostrava, Czech Republic
Eero Sormunen, University of Tampere, Finland
Amanda Spink, Queensland University of Technology, Australia
John Tait, University of Sunderland, UK
Martin Theobald, Max Planck Institute for Informatics, Germany
Ulrich Thiel, Fraunhofer IPSI, Germany
Anastasios Tombros, Queen Mary, University of London, UK
Andrew Tomkins, Yahoo!, USA
Stephen Tomlinson, Hummingbird, USA
Pertti Vakkari, University of Tampere, Finland
Keith van Rijsbergen, University of Glasgow, UK
Olga Vechtomova, University of Waterloo, Canada
Dawid Weiss, Poznan University of Technology, Poland
Ross Wilkinson, CSIRO ICT Centre, Australia
Wensi Xi, Google, USA
Tao Yang, ASK, USA
Elad Yom-Tov, IBM Research Lab, Haifa, Israel
Hugo Zaragoza, Yahoo! Research, Barcelona, Spain
Hua-Jun Zeng, Microsoft Research Asia, China
ChengXiang Zhai, University of Illinois at Urbana-Champaign, USA
Dell Zhang, University of London, UK
Justin Zobel, RMIT, Australia

Paper Awards (Sponsored by Yahoo! Research)

Best Paper Award Committee

Fabio Crestani, University of Strathclyde, UK (Chair)
David Losada, University of Santiago de Compostela, Spain
Hinrich Schuetze, University of Stuttgart, Germany

Best Student Paper Award Committee

Maristella Agosti, University of Padova, Italy (Chair)
David Lewis, David D. Lewis Consulting, USA
Giovanni Semeraro, University of Bari, Italy

Additional Reviewers

Eija Airio, University of Tampere, Finland
Mark Baillie, University of Strathclyde, UK
Nuno Cardoso, University of Lisboa, Portugal
Mauro Cettolo, Istituto Trentino di Cultura, Italy
Marco Degemmis, University of Bari, Italy
Gianna Del Corso, University of Pisa, Italy
Chris H.Q. Ding, Lawrence Berkeley National Laboratory, USA
Nicola Fanizzi, University of Bari, Italy
Stefano Ferilli, University of Bari, Italy
Claudio Giuliano, Istituto Trentino di Cultura, Italy
Milen Kouylekov, Istituto Trentino di Cultura, Italy
Jie Lu, Carnegie Mellon University, USA
Pasquale Lops, University of Bari, Italy
Joao Magalhaes, Imperial College, UK
Inderjeet Mani, The MITRE Corporation, USA
Vanessa Murdock, Yahoo! Research, Barcelona, Spain
Matteo Negri, Istituto Trentino di Cultura, Italy
Domenico Redavid, University of Bari, Italy
James Thom, RMIT, Australia

Table of Contents

Keynote Talks

Theory and Design

Efficiency

Peer-to-Peer Networks (In Memory of Henrik Nottelmann)

Result Merging

Queries

Relevance Feedback

Evaluation

Classification and Clustering

Filtering

Topic Identification

Expert Finding

XML IR

Web IR

Multimedia IR

Short Papers

Posters

The Next Generation Web Search and the Demise of the Classic IR Model

Andrei Broder

Yahoo! Research, USA
broder@yahoo-inc.com

Abstract. The classic IR model assumes a human engaged in activity that generates an "information need". This need is verbalized and then expressed as a query to search engine over a defined corpus. In the past decade, Web search engines have evolved from a first generation based on classic IR algorithms scaled to web size and thus supporting only informational queries, to a second generation supporting navigational queries using web specific information (primarily link analysis), to a third generation enabling transactional and other "semantic" queries based on a variety of technologies aimed to directly satisfy the unexpressed "user intent", thus moving further and further away from the classic model.

What is coming next? In this talk, we identify two trends, both representing "short-circuits" of the model: The first is the trend towards context driven Information Supply (IS), that is, the goal of Web IR will widen to include the supply of relevant information from multiple sources without requiring the user to make an explicit query. The information supply concept greatly precedes information retrieval; what is new in the web framework, is the ability to supply relevant information specific to a given activity and a given user, while the activity is being performed. Thus the entire verbalization and query-formation phase are eliminated. The second trend is "social search" driven by the fact that the Web has evolved to being simultaneously a huge repository of knowledge and a vast social environment. As such, it is often more effective to ask the members of a given web milieu rather than construct elaborate queries. This short-circuits only the query formulation, but allows information finding activities such as opinion elicitation and discovery of social norms, that are not expressible at all as queries against a fixed corpus.

G. Amati, C. Carpineto, and G. Romano (Eds.): ECIR 2007, LNCS 4425, p. 1, 2007.

The Last Half-Century: A Perspective on Experimentation in Information Retrieval

Stephen Robertson

Microsoft Research Cambridge and City University London, UK
ser@microsoft.com

Abstract. The experimental evaluation of information retrieval systems has a venerable history. Long before the current notion of a search engine, in fact before search by computer was even feasible, people in the library and information science community were beginning to tackle the evaluation issue. Sometimes it feels as though evaluation methodology has become fixed (stable or frozen, according to your viewpoint). However, this is far from the case. Interest in methodological questions is as great now as it ever was, and new ideas are continuing to develop. This talk will be a personal take on the field.

G. Amati, C. Carpineto, and G. Romano (Eds.): ECIR 2007, LNCS 4425, p. 2, 2007.

Learning in Hyperlinked Environments

Marco Gori

Dipartimento di Ingegneria dell'Informazione, University of Siena, Italy
marco@dii.unisi.it

Abstract. A remarkable number of important problems in different domains (e.g. web mining, pattern recognition, biology ...) are naturally modeled by functions defined on graphical domains, rather than on traditional vector spaces. Following the recent developments in statistical relational learning, in this talk, I introduce Diffusion Learning Machines (DLM) whose computation is very much related to Web ranking schemes based on link analysis. Using arguments from function approximation theory, I argue that, as a matter of fact, DLM can compute any conceivable ranking function on the Web. The learning is based on a human supervision scheme that takes into account both the content and the links of the pages. I give very promising experimental results on artificial tasks and on the learning of functions used in link analysys, like PageRank, HITS, and TrustRank. Interestingly, the proposed learning mechanism is proven to be effective also when the rank depends jointly on the page content and on the links. Finally, I argue that the propagation of the relationships expressed by the links reduces dramatically the sample complexity with respect to traditional learning machines operating on vector spaces, thus making it reasonable the application to real-world problems on the Web, like spam detection and page classification.

G. Amati, C. Carpineto, and G. Romano (Eds.): ECIR 2007, LNCS 4425, p. 3, 2007.

A Parameterised Search System

Roberto Cornacchia and Arjen P. de Vries

CWI, Kruislaan 413, 1098SJ, Amsterdam, The Netherlands
{roberto,arjen}@cwi.nl

Abstract. This paper introduces the concept of a Parameterised Search System (PSS), which allows flexibility in user queries, and, more importantly, allows system engineers to easily define customised search strategies. Putting this idea into practise requires a carefully designed system architecture that supports a declarative abstraction language for the specification of search strategies. These specifications should stay as close as possible to the problem definition (i.e., the retrieval model to be used in the search application), abstracting away the details of the physical organisation of data and content. We show how extending an existing XML retrieval system with an abstraction mechanism based on array databases meets this requirement.

1 Introduction

For many years, information retrieval (IR) systems could be adequately described as software that assign an estimate of relevancy to a pair of document and query, each represented as a 'bag-of-words'. The implementation of such search systems has been relatively straightforward, and most engineers code the retrieval model directly on top of an inverted file structure.

Trends in research and industry motivate however a reconsideration of the above characterisation of IR. First, modern retrieval systems have become more complex, as they exploit far more than 'just' the text. For example, the ranking function combines query and document text with other types of evidence, derived from, e.g., document markup, link structure, or various types of 'context information'. Also, work tasks supported by search have become diverse. Within organisations, *enterprise search* refers to intranet search, but also search over collections of e-mail, finding expertise, etc. [1]. People use web search indeed for the goal of 'finding information about', but also to book a hotel, find a job, hunt for a house, just to name a few. Companies are targeting these niche markets with specialised search engines (known as *vertical search*).

Today, the development of such specialised applications is the job of information retrieval specialists. We expect however that, very soon, *any* software developer should be able to develop applications involving search. Actually, Hawking has stated that *'an obvious reason for poor enterprise search is that a high performing text retrieval algorithm developed in the laboratory cannot be applied without extensive engineering to the enterprise search problem, because of the complexity of typical enterprise information spaces'* [1]. Simplifying the process of tailoring search to a specific work task and user context should therefore be an important goal of IR research!

G. Amati, C. Carpineto, and G. Romano (Eds.): ECIR 2007, LNCS 4425, pp. 4–15, 2007.

This paper proposes that the engineering of search systems may proceed analogous to the development of office automation applications using relational database management systems – define the 'universe of discourse'; design a conceptual schema; express the user application in terms of this schema; and, design the user interface. So, software developers of a search application should have access to a high-level declarative language to specify collection resources *and retrieval model*. The search engine can then be parameterised for optimal effectiveness: adapted to the work task and user context, optimised for specific types of content in the collection, and specialised to exploit domain knowledge.

1.1 Anatomy of a Parameterised Search System

We refer to this new generation of information retrieval systems as *parameterised search engines* (PSSs). Fig. 1 illustrates how a PSS differs from the traditional search engine, in the so-called *abstraction language*. Its main purpose is to decouple search strategies from algorithms and data structures, bringing what the database field calls data independence to IR systems. This abstraction language should enable the search system developer to specify search strategies declaratively, ideally without any consideration of the physical representation of document structure and content.

Compare this to current practise using IR software like Lucene, Lemur, or Terrier. These systems provide a variety of well-known ranking functions, implemented on top of the physical document representation. The modular design of the IR toolkit allows application developers to select one of these ranking functions suited for their search problem. Design decisions about how to rank, e.g., weighting various types of documents differently with respect to their expected contribution to relevancy, will however be part of the application code.

Summarising, a PSS provides a declarative IR language. Search application developers specify the desired search strategy for a specific user and task context in this language. The system translates expressions in this language into operations on its internal data structures to perform the actual retrieval.

1.2 Approach, Contributions and Outline

This raises many open questions, the obvious one of course what the abstraction language could look like. The main contribution of this work is to demonstrate feasibility of the idea of a PSS through a (preliminary) prototype implementation. We propose a very specific instantiation of a PSS, by extending an existing XML IR system (PF/Tijah) with a formal language for the specification of retrieval models (Matrix Framework for IR). Another contribution is to operationalise this theoretical framework for IR on an array database system (SRAM), such that it can be deployed in practical search system implementation.

The remainder is organised as follows. Section 2 describes the layered architecture of the XML retrieval system PF/Tijah. Section 3 details how we turn PF/Tijah into a PSS named Spiegle. We introduce its two main ingredients, the Matrix Framework for IR formalism and the array data-model abstraction of

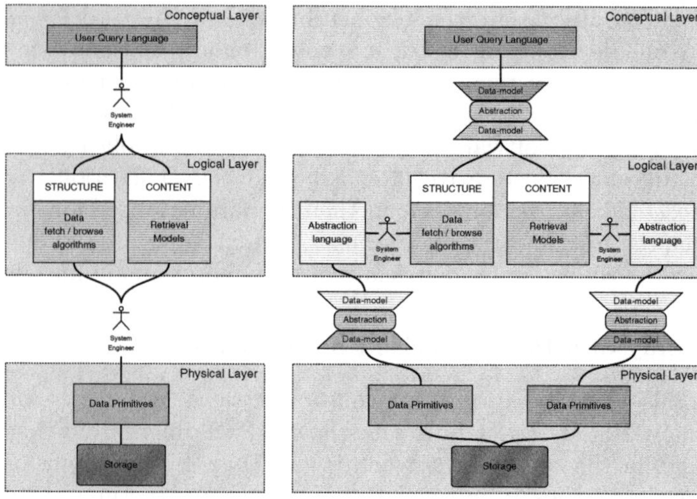

(a) Traditional (b) Parameterised

Fig. 1. Comparison of Search System architectures

SRAM, and explain how these are integrated. Section 4 provides implementation details. Related research is discussed in Section 5, before we conclude and outline future work in Section 6.

2 Querying Content and Structure in PF/Tijah

A preliminary requirement of implementing a PSS is that it allows to express search strategies that refer to structure *and* content. Spiegle meets this requirement by building upon PF/Tijah [2], a flexible environment for setting up XML search systems.

PF/Tijah integrates XQuery and NEXI: XQuery to query and transform the *structure* of XML documents, NEXI (Narrowed Extended XPath) [3] to rank XML elements by their *content*. The resulting query language provides a powerful way to customise (mainly) the structural aspects of a retrieval strategy. Its layered system architecture, depicted in Fig. 2(a), uses the PathFinder (PF) XQuery system [4] to query by structure, as well as to construct the preferred result presentation. PathFinder translates the XQuery (non-ranked) part of a query into a relational query plan, independently from the ranking component of the system.

The Tijah XML retrieval system provides the XML IR support, by processing the NEXI expressions in a query. NEXI is a subset of XPath (it allows only descendant and self axis steps) that allows an additional `about()` clause, which ranks the selected XML elements by their *content*. The following example retrieves, from a database of scientific publications, *The title of documents written by John Smith, where the abstract is about "IR DB integration"*:

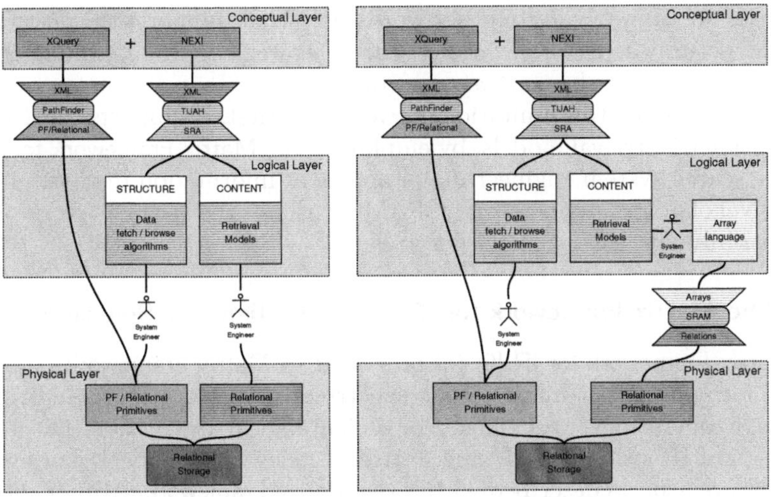

(a) PF/Tijah (b) Spiegle

Fig. 2. Architectures of PF/Tijah and Spiegle

```
let $c := doc("papers.xml")//DOC[author = "John Smith"]
let $q := "//text[about(.//Abstract, IR DB integration)];"
for $res in tijah-query($c, $q)
return $res/title/text()
```

PF/Tijah translates the NEXI expression `$q` into a logical query plan consisting of SRA (Score Region Algebra) operators [5].The SRA algebra includes operators that perform the following tasks: *(i)* selection of the XML elements that are to be ranked; *(ii)* a parameterised *element score computation*, implementing the `about()` clause according to the desired retrieval model; *(iii)* a parameterised *element score combination*, i.e. compute the score of `AND` and `OR` combinations of `about()` clauses; *(iv)* a parameterised *element score propagation*, needed when scores need to be propagated to a common ancestor before being combined. As depicted in Fig. 2(a), structured information retrieval queries involve operations that act on *structure* (like *(i)*) and on *content* (e.g. *(ii)*). Each SRA operator implementation is given in terms of relational database operators (by a system engineer). Several retrieval models are supported out-of-the-box, selected by an extra 'options' parameter to the `tijah-query()` function.

3 Spiegle: Turning an XML Retrieval System into a PSS

While PF/Tijah provides a powerful query language to embed search functionality in data processing, it does *not* support customisation of the retrieval model. Although advanced users may in principle modify the pre-defined mapping of SRA operators to relational query plans to implement new ranking functions, doing so is far from trivial. In other words, PF/Tijah supports the customisation of the structural aspects of various search strategies, but it is inflexible with respect to modifying the *content* aspects.

Spiegle overcomes this limitation in two steps. First, it supports the declarative specification of retrieval models, by employing the Matrix Framework for IR. This way, the search system engineer may implement information retrieval models at a high level of abstraction. Second, SRAM translates the specified retrieval model automatically into a relational query plan.

3.1 The Matrix Framework for IR: A Formalism for Search Strategies

The Matrix Framework for IR [6] (abbreviated to Matrix Framework) is a mathematical formalism that maps a wide spectrum of IR concepts to matrix spaces and matrix operations, providing a convenient logical abstraction that facilitates the design of IR systems. Indexing, retrieval, relevance feedback and evaluation measures are aspects described within the Matrix Framework. Also, it establishes a consistent notation for frequencies in event spaces, readily available as building blocks for IR applications in common matrix operation libraries.

We introduce only the part of the Matrix Framework that is focused on indexing and retrieval. First, we define three vectors, one for each of the dimensions used in the framework: $D = [w_d]_{N\times 1}$ for *documents*, $T = [w_t]_{S\times 1}$ for *terms* and $L = [w_l]_{R\times 1}$ for *locations*, with $1 \leq d \leq N$, $1 \leq t \leq S$ and $1 \leq l \leq R$. The quantities $w_d \geq 0$, $w_t \geq 0$ and $w_l \geq 0$ are the weight of document d, term t and location l, respectively. In the simplest case, these weights are boolean values that denote presence/absence in the collection. The term "location" is rather generic and covers concepts indicating document components of varying granularity, such as section, paragraph, position, or XML element.

The content and the structure of each data collection are entirely described by the two boolean matrices $LD_{L\times D}$ (location-document) and $LT_{L\times T}$ (location-term), whereas each query is described by a vector $Q_{T\times 1}$. As defined in (1), each value $LD(l,d)$ tells whether location l belongs to document d and each value $LT(l,t)$ encodes the occurrence of term t at location l of the collection. Finally, each query Q is described as a bit-vector of all collection-terms.

$$LD(l,d) = \begin{cases} 0, & \text{if } l \notin d \\ 1, & \text{if } l \in d \end{cases}, \quad LT(l,t) = \begin{cases} 0, & \text{if } t \notin l \\ 1, & \text{if } t \in l \end{cases}, \quad Q(t) = \begin{cases} 0, & \text{if } t \notin Q \\ 1, & \text{if } t \in Q \end{cases} \tag{1}$$

Standard IR statistics are defined as simple operations on matrices LD and LT:

DT_{nl}, DT – **#term occurrences and term presence**

$$DT_{nl} = LD^T \cdot LT, \quad DT = min(DT_{nl}, Ones[|D| \times |T|]),$$

D_{nl}, T_{nl} – **#per-document and #per-term locations**

$$D_{nl} = LD^T \cdot L, \quad T_{nl} = LT^T \cdot L, \tag{2}$$

D_{nt}, T_{nd} – **#terms per document and #documents per term**

$$D_{nt} = DT \cdot T, \quad T_{nd} = DT^T \cdot D,$$

where $Ones[A \times B]$ defines a matrix of size $A \times B$, filled with 1's.

A number of standard IR frequencies are easily derived from the quantities defined above:

DT_f – **within-document term frequency**

$$DT_f(d,t) = \frac{DT_{nl}(d,t)}{D_{nl}(d)},$$

D_{tf}, D_{itf} – **term frequency and inverse term frequency of a document**

$$D_{tf}(d) = \frac{D_{nt}(d)}{|T|}, \quad D_{itf}(d) = -logD_{tf}(d), \tag{3}$$

T_{df}, T_{idf} – **document frequency and inverse document frequency of a term**

$$T_{df}(t) = \frac{T_{nd}(t)}{|D|}, \quad T_{idf}(t) = -logT_{df}(t),$$

T_{nl} – **collection frequency of a term**

$$T_f(t) = \frac{T_{nl}(t)}{|L|}.$$

Finally, it is possible to define a number of retrieval models using the quantities and the frequencies described above. The *tf.idf* approach is specified as $RSV_{tf.idf} = DT_f \cdot diag\,(T_{idf}) \cdot Q$. Refer to [6] for many other retrieval models.

The Matrix Framework expresses indexing and retrieval tasks declaratively, starting from a simple matrix-based representation of collection and queries, and following a set of generic definitions that can be stored as a well-structured system library. This facilitates the engineering of extensible retrieval systems, as it allows a programming interface based on the array data-model (suited for IR). Physical implementation is delegated to another, dedicated layer.

3.2 SRAM: An Array Data-Model Implementation

The SRAM system is a prototype tool for mapping sparse arrays to relations and array operations to relational expressions. While SRAM syntax allows to express array operations on an element by element basis, the system translates such element-at-a-time operations to collection-oriented database queries, suited for (potentially more efficient) bulk processing.

The SRAM language defines operations over arrays declaratively in comprehension syntax [7], which allows to declare arrays by means of the following construct: A = [<array-cell value> | <array axes>]. The section <array axes> specifies the *shape* $\mathcal{S}_A$ of result array A, i.e., its number of dimensions and their domain, denoted as: $\mathtt{i}_0 < \mathtt{N}_0, \ldots, \mathtt{i}_{n-1} < \mathtt{N}_{n-1}$. The value of each dimension variable $\mathtt{i}_j$ ranges from 0 to $\mathtt{N}_j - 1$. The section <array-cell value> assigns a value to each cell indexed by the index values enumerated by the <array axes> section. For example, the expression Tidf = [-log(Tdf(t)) | t < nTerms] defines a new array $T_{idf}[nTerms]$, where the value of each cell is computed by applying the function $-\log$ to corresponding cells of array T_{df}. The explicit domain specification may be omitted when its range is clear from the context, e.g., the more convenient [-log(Tdf(t)) | t] is equivalent to [-log(Tdf(t)) | t < nTerms].

The language supports aggregations over any array dimension (sum, prod, min, max). For example, summing per document the term frequency counts (in a document-term matrix DT_f) of query terms (in binary vector Q) is expressed as: P = [sum([DTf(d,t) * Q(t) | t]) | d]. The shape $\mathcal{S}_P$ of array P is determined by the rightmost axis d.

Retrieving the *top-N* values is allowed for one-dimensional arrays only. The result consists of the *positions* (in order) of the values in the original array. So, `T = topN(P, N, <ASC|DESC>)` returns a *dense* array T with $\mathcal{S}_T = [N]$. The actual values can then be fetched by dereferencing the original array, `S = P(T)`.

The SRAM syntax allows definitions and assignments. Definitions are indicated by the symbol "=", as in all the expressions seen above. They are expanded symbolically by a preprocessor at every occurrence in the array expression. Assignments, indicated by the symbol ":=", translate to database expressions whose result is stored permanently in tables, named as indicated by the left part of the assignment expression: `<array name> := <comprehension expression>`.

3.3 How Spiegle Applies the Array Data-Model

Fig. 2 shows how Spiegle inherits PF/Tijah's logical layer, which provides an algebraic abstraction (SRA) between the conceptual and physical layers. The Spiegle architecture takes the point of logical abstraction further, by exploiting the clean distinction between *structure* and *content* operations offered by SRA. In the following, we first explain how the Matrix Framework can handle XML documents. Next, we see how to bootstrap, index and rank XML collections using the array data-model. The SRAM scripts shown are translated on-the-fly into database queries executed by the backend.

The Matrix Framework and XML data. The formalism of the Matrix Framework is not restricted to flat text search. Mapping the XML structure to the three dimensions location, document and term is all that is needed to apply the Matrix Framework to semi-structured collections. Each XML file can be seen as a collection of documents, where each distinct XML element represents a different one. The main difference from modelling plain text collections is that locations are now shared among different (nested) elements. Therefore, locations are defined as *term and XML tag* position in the file. Consider the following excerpt of the `papers.xml` collection of scientific publications, where location, document and term identifiers are annotated next to each piece of text for the reader's convenience, with `l`, `d` and `t` prefixes respectively:

```
<PAPERS>[l0,d0]
 <DOC>[l1,d1]
  <text>[l2,d2]
   <Abstract>[l3,d3] IR[l4,t0] as[l5,t1] simple[l6,t2] as[l7,t1] cakes[l8,t3]</Abstract>[l9]
   ...
  </text>[l40]
 </DOC>[l41]
 <DOC>[l42,d15] ..... </DOC>[l120] ...
</PAPERS>[l30000000]
```

Notice that *(i)* the two `DOC` elements have different identifiers; *(ii)* the two `as` occurrences have the same term identifier; *(ii)* locations 3 to 9 belong to documents 0,1,2 and 3.

Bootstrapping XML collections. A Matrix Framework representation of a collection is obtained from vectors L, D and T and matrices LD and LT.

For the example collection papers.xml, this corresponds to a SRAM script with the following definitions.

papers.ram :
```
// global information for collection "papers.xml"
nLocs=30000000 , nDocs=2000000 , nTerms=400000
L  = [ 1 | l < nLocs ]
D  = [ 1 | d < nDocs ]
T  = [ 1 | t < nTerms ]
LD = [nLocs,nDocs],  bool, sparse("0"), "LD_table"
LT = [nLocs,nTerms], bool, sparse("0"), "LT_table"
```

First, the length of each dimension is defined as nLocs, nDocs, nTerms. Then, vectors L, D and T are defined, using a constant weight 1. Finally, matrices LD and LT are declared as persistently stored in the database, as they are supposed to be the outcome of an earlier parsing and pre-processing phase on the XML file. For each persistent matrix, the dimensionality, the element type, the information on sparsity (LD and LT are both sparse with common value 0) and the name of the corresponding relational table are specified. Section 4 gives further details on the automatic creation of matrices LD and LT.

A script file similar to papers.ram is created automatically for each collection and it is included in every SRAM script that uses that particular collection. Notice that the script file above only contains SRAM *definitions* and no assignments (see Section 3.2), which result in simple in-memory declarations.

Array System Libraries. The uniform approach to IR tasks like indexing and retrieval that the Matrix Framework provides is easily operationalised by collecting the formulae presented in Section 3.1 in a system library composed by SRAM expressions (automatically made collection-dependent using a unique prefix for the collection's stored arrays). Fig. 3 shows an excerpt of such a library, that is used in the subsequent paragraphs about indexing and retrieval of XML collections. One can observe that the SRAM expressions are an almost direct transcription of mathematical formulae to ASCII characters, which demonstrates the intuitiveness of array comprehensions as an IR query language.

Indexing XML collections. The indexing phase creates statistical information about a given collection. As described in Section 3.1, this entails the computation of the matrices defined in (2) and (3). The definition of such matrices in SRAM syntax is given in Fig. 3, file MF_Indexing.ram. Notice that this file contains array *assignments*, that create persistent arrays in the database. The SRAM script for indexing the example papers.xml collection becomes straightforward: load the collection's global definitions, followed by the generic index-creation assignments:

```
#include "papers.ram"       // global definitions for papers.xml
#include "MF_Indexing.ram"  // create index arrays
```

Ranking XML collections. Recall the example query of Section 2, where NEXI expression //text[about(.//Abstract, IR DB integration)] ranks the *"text" elements that contain an "Abstract" element about "IR DB integration"*. This structured IR part of the query is translated to SRA algebra. Selection of text and their containing Abstract elements is performed by *structure*

```
#include "MF_DocContext.ram"
bm25(d,k1,b) = sum( [ w(d,t,k1,b) * Q(t) | t ] )
w(d,t,k1,b)  = log( nDocs / Tdf(t) ) * (k1+1) * DTf(d,t)
               / ( DTf(d,t) + k1 * (1 - b + b * Dnl(d) / avgDnl) )

langmod(d,a) = sum( [ lp(d,t,a) * Q(t) | t ] )
lp(d,t,a)    = log( a*DTf(d,t) + (1-a)*Tf(t) )
```

MF_RetrievalModels.ram

```
#include "LinearAlgebra.ram"
DTnl := mxMult(mxTrnsp(LT), LD)
Dnl  := mvMult(mxTrnsp(LD), L)
Tnl  := mvMult(mxTrnsp(LT), L)
DT   := [ min(DTnl(d,t),1) | d,t ]
DTf  := [ DTnl(d,t)/Dnl(d) | d,t ]
Tf   := [ Tnl(t)/nLocs | t ]
```

MF_Indexing.ram

```
DTnl = [ DTnl(d,t) * DX(d) | d,t ]
Dnl  = [ Dnl(d)    * DX(d) |   d ]
DT   = [ DT(d,t)   * DX(d) | d,t ]
DTf  = [ DTf(d,t)  * DX(d) | d,t ]
```

MF_DocContext.ram

```
mxTrnsp(A)   = [ A(j,i) | i,j ]
mxMult(A,B) = [ sum([ A(i,k) * B(k,j) | k ]) | i,j ]
mvMult(A,V) = [ sum([ A(i,j) * V(j) | j ]) | i ]
```

LinearAlgebra.ram

Fig. 3. SRAM libraries (excerpts)

operations implemented using PathFinder primitives. We call the resulting node-set the *ranking-context*, represented in the Matrix Framework by a binary vector DX.

Ranking the `Abstract` elements of the *ranking-context* is performed by a *content* SRA operation, implemented in Spiegle by a function with signature:

```
Function rank(collection, rankingContext, queryTerms, N,
              retrievalModel, param1, param2, ...) := rankedContext
```

This function turns the *ranking-context* of a given query into a top-N *ranked-context* against the specified query terms, by applying the desired retrieval model. Its body executes a SRAM script, customised at each call with the value of the current function parameters. The script of the example NEXI query above, with parameters `collection`="`papers`", `retrievalModel`="`langmod`" and N=20, corresponds to:

```
// global definitions for collection papers.xml
#include "papers.ram"
// ranking-context and query terms
DX = [nDocs],  bool, sparse("0"), "DX_table"
Q  = [nTerms], bool, sparse("0"), "Q_table"
// include retrieval model definitions
#include "MF_RetrievalModels.ram"
// retrieve top N documents using "langmod" (Language Modelling)
S = [ langmod(param1,param2,...) | d ]
D20 := topN( [S(d) | d<20], DESC )
S20 := S(D20)
```

First, global definitions for the collection `papers.xml` are loaded. Then, the *ranking-context* and the query terms are declared as persistent arrays, previously stored in the database by the system. Definitions of the available retrieval models are loaded (file `MF_RetrievalModels.ram`) and the selected one used to rank documents. Finally, the top 20 document identifiers and scores are computed and used together as the *ranked-context* returned by the SRA operator.

Fig. 3, file `MF_RetrievalModels.ram` shows an excerpt of the available retrieval model expressions. These get customised to the current querying scenario by declaring the two arrays DX (*ranking-context*) and Q (*query-terms*) before each query is executed. The first line includes library file `MF_DocContext.ram`. The included script limits the document axes of the index arrays to the current *ranking-context* by multiplying the document axes with the binary vector representing the *ranking-context* (the result of this multiplication corresponds to precisely those portions of index arrays that contain information about the node-set to be ranked).

4 Implementation Details

Spiegle uses PathFinder's efficient 'document shredder' to turn any XML data file into database relations (also from remote sources). PF/Tijah extends the standard PathFinder data-model by indexing text words (terms) in addition to XML nodes. XML elements are encoded as *regions*. Regions are stored as tuples `<start,end,id,type>`, where `id` is the identifier associated with the XML element (a node tag or a term) and `type` can be either `node` or `term`. The `start` and `end` positions identify the text region in the original text file (each term is a text region of length 1). An additional relation is created for efficient term access (representing an inverted file structure). A more detailed description of PathFinder and PF/Tijah indexing schemes can be found in [4,2].

To use the Matrix Framework as described above, matrices LD and LT should be prepared during this indexing phase. Clearly, all required information is available in the tables created by PF/Tijah. Positions delimited by `start` and `end` values become "locations", whereas the combination of `id` and `type` values become the document and term identifiers for matrices LD and LT.

Many storage schemes have been proposed for sparse multi-dimensional arrays, with strong emphasis on the special case of two-dimensional arrays [8]. SRAM represents an n-dimensional sparse array A as relation R_A, with ε_A denoting its default value. Each tuple encodes a vector $\mathbf{i} = (i_0, \ldots, i_{n-1})$ of index values and a cell value $A(\mathbf{i})$; only those cells for which $A(\mathbf{i}) \neq \varepsilon_A$ are stored:

$$A \mapsto \mathsf{R}_A(\mathsf{i}_0, \ldots, \mathsf{i}_{n-1}, \mathsf{v}) = \{(i_0, \ldots, i_{n-1}, A(\mathbf{i})) \mid A(\mathbf{i}) \neq \varepsilon_A\}.$$

This storage scheme naturally extends to dense arrays, for which all tuples are stored physically. Data access patterns are optimised by creating standard relational indexing structures on top of such relations, or by explicit tuple clustering/sorting (the index columns form the relation's primary key).

The life-cycle of array queries through the SRAM architecture can be summarised by the following sequence of transformations: Array comprehension $\mapsto$

Array algebra $\mapsto$ Relational algebra $\mapsto$ Relational plan. The first translation step generates an array-algebra tree, which represents the sequence of operations performed on the stored arrays (reshaping, selection, aggregation, or function application). A cost-based optimiser normalises and rewrites the array-algebra tree. No physical details are involved in the process yet: only arrays' size and density are taken into account. The next step maps the array data-model to the relational data-model, by means of translation rules that take into account the storage details: although common values of a sparse arrays are not stored, they may affect the final result. Standard relational optimisation techniques finalise the relational expression. The last step of the translation process maps relational algebra expressions to physical query plans for the database engine. Data access paths and algorithms are chosen for the physical implementation of relational operators in the query tree. For more in-depth details about the array-algebra and the mapping and optimisation rules used in the translation process, see [9].

5 Related Work

This work addresses some of the main issues about IR&DB integration [10,11]. We are not aware of previous work that provides a declarative language for the definition of the retrieval model as part of an XML retrieval system. However, Inquery's [12] (now Indri) query language can be considered to provide some early version of a parameterised search system, and it has often been used for precisely this flexibility (for example, in cross-language IR).

Already ten years ago, Fuhr argued in favour of data independence in IR [13], pointing out how this would reduce problems in plain text search with noun phrase search and treatment of compound words, and (semi-)structured data types to capture attributes like author, journal title or publication year.

De Vries defined the notion of 'content independence' [14] to refer to the decoupling between search strategies and content representation. His definition has been refined by Mihajlovic into the two related concepts 'retrieval model independence' and 'content description independence' [5]. Wen et al. use 'media independence' for a similar separation of concerns as Mihajlovic's content description independence [15]. Yet, none of these authors has proposed a declarative language that actually achieves the goal of separating the retrieval model from its actual implementation.

6 Conclusions and Future Work

This paper has argued that modern IR application requirements force us to reconsider the design characteristics of search systems. We promote an innovation in the search system engineering process, by introducing more flexibility in the IR system's architecture. The increased flexibility aims to reduce the effort of adapting search functionalities to work task and user context.

We defined the architectural requirements of so-called *Parameterised Search Systems*: *(i)* a layered architecture that allows structural *and* content information to be exploited for the search task; *(ii)* a convenient abstraction from the physical details that discloses the retrieval engine's capabilities to the unique needs of each particular combination of collection characteristics, user preferences, and search strategies.

We indicated the architecture of the PF/Tijah XML IR system as a possible foundation to build upon, and the Matrix Framework for IR as a very well-suited abstraction to express retrieval strategies. Finally, we showed how SRAM can operationalise the Matrix Framework on top of a database system. The result of this effort is Spiegle: the first prototype of parameterised XML search system.

Future work includes further exploiting the array abstraction for the implementation of structure operations. This will allow for simpler inclusion of structural information in the retrieval strategy and better opportunities for optimising the final relational query plan. Early results of [9] have demonstrated that SRAM's multi-stage query translation process can give excellent run-time performance on a large collection of web-data, given an efficient back-end system. Integrating this new back-end into the Spiegle architecture is on-going work.

References

1. Hawking, D.: Challenges in enterprise search. In: Proc. ADC. (2004) 15–26
2. Hiemstra, D., Rode, H., van Os, R., Flokstra, J.: PFTijah: text search in an XML database system. In: Proc. OSIR. (2006)
3. R.A.O'Keefe, Trotman, A.: The simplest query language that could possibly work. In: Proc. INEX. (2004)
4. Boncz, P., Grust, T., van Keulen, M., Manegold, S., Rittinger, J., Teubner, J.: MonetDB/XQuery. In: Proc. SIGMOD. (2006) 479–490
5. Mihajlovic, V.: Score Region Algebra. A Flexible Framework for Structured Information Retrieval. PhD thesis, University of Twente (2006)
6. Rölleke, T., Tsikrika, T., Kazai, G.: A General Matrix Framework for Modelling Information Retrieval. IP&M **42**(1) (2005) 4–30
7. Buneman, P., Libkin, L., Suciu, D., Tannen, V., Wong, L.: Comprehension syntax. SIGMOD Record **23**(1) (1994) 87–96
8. Demmel, J., Dongarra, J., Ruhe, A., van der Vorst, H.: Templates for the solution of algebraic eigenvalue problems: a practical guide. SIAM (2000)
9. Cornacchia, R., Héman, S., Zukowski, M., de Vries, A., Boncz, P.: Flexible and efficient IR using Array Databases. Technical Report INS-E0701, CWI (2007)
10. Chaudhuri, S., Ramakrishnan, R., Weikum, G.: Integrating DB and IR Technologies: What is the Sound of One Hand Clapping? In: Proc. CIDR, Asilomar, CA, USA (2005) 1–12
11. Amer-Yahia, S., Case, P., Rölleke, T., Shanmugasundaram, J., Weikum, G.: Report on the DB/IR Panel at Sigmod 2005. SIGMOD Record **34**(4) (2005) 71–74
12. Callan, J.P., Croft, W.B., Harding, S.M.: The INQUERY retrieval system. In: Proc. DEXA. (1992) 78–83
13. Fuhr, N.: Object-oriented and database concepts for the design of networked information retrieval systems. In: Proc. CIKM. (1996) 164–172
14. de Vries, A.: Content independence in multimedia databases. JASIST **52**(11) (2001) 954–960
15. Wen, J., Li, Q., Ma, W., Zhang, H.: A multi-paradigm querying approach for a generic multimedia database management system. SIGMOD Record **32**(1) (2003) 26–34

Similarity Measures for Short Segments of Text

Donald Metzler[1], Susan Dumais[2], and Christopher Meek[2]

[1] University of Massachusetts
Amherst, MA
[2] Microsoft Research
Redmond, WA

Abstract. Measuring the similarity between documents and queries has been extensively studied in information retrieval. However, there are a growing number of tasks that require computing the similarity between two very short segments of text. These tasks include query reformulation, sponsored search, and image retrieval. Standard text similarity measures perform poorly on such tasks because of data sparseness and the lack of context. In this work, we study this problem from an information retrieval perspective, focusing on text representations and similarity measures. We examine a range of similarity measures, including purely lexical measures, stemming, and language modeling-based measures. We formally evaluate and analyze the methods on a query-query similarity task using 363,822 queries from a web search log. Our analysis provides insights into the strengths and weaknesses of each method, including important tradeoffs between effectiveness and efficiency.

1 Introduction

Retrieving documents in response to a user query is the most common text retrieval task. For this reason, most of the text similarity measures that have been developed take as input a query and retrieve matching documents. However, a growing number of tasks, especially those related to web search technologies, rely on accurately computing the similarity between two very short segments of text. Example tasks include query reformulation (query-query similarity), sponsored search (query/ad keyword similarity), and image retrieval (query-image caption similarity).

Unfortunately, standard text similarity measures fail when directly applied to these tasks. Such measures rely heavily on terms occurring in both the query and the document. If the query and document do not have any terms in common, then they receive a very low similarity score, regardless of how topically related they actually are. This is well-known as the vocabulary mismatch problem. This problem is only exacerbated if we attempt to use these measures to compute the similarity of two short segments of text. For example, "UAE" and "United Arab Emirates" are semantically equivalent, yet share no terms in common.

Context is another problem when measuring the similarity between two short segments of text. While a document provides a reasonable amount of text to infer the contextual meaning of a term, a short segment of text only provides a limited context. For example, "Apple computer" and "apple pie" share the term apple, but are topically distinct. Despite this, standard text similarity measures would say that these

G. Amati, C. Carpineto, and G. Romano (Eds.): ECIR 2007, LNCS 4425, pp. 16 – 27, 2007.

two short segments of text are very similar. However, computing the similarity between the query "Apple computer" and a full document about "apple pie" will produce a low similarity score since the document contains proportionally less text that is relevant to the query, especially compared to a full document about "Apple business news".

In this paper, we explore the problem of measuring similarity between short segments of text from an information retrieval perspective. Studies in the past have investigated the problem from a machine learning point of view and provided few, if any comparisons to standard text similarity measures. In this work, we describe a set of similarity measures that can be used to tackle the problem. These measures include simple lexical matching, stemming, and text representations that are enriched using web search results within a language modeling framework. In addition, we formally evaluate the measures for the query-query similarity task using a collection of 363,822 popular web queries. Our analysis provides a better understanding of the strengths and weaknesses of the various measures and shows an interesting tradeoff between effectiveness and efficiency.

The remainder of this paper is laid out as follows. First, Section 2 provides an overview of related work. We then describe the various ways to represent short segments of text in Section 3. Section 4 follows up this discussion by describing the similarity measures we investigated. Section 5 provides the details of our experimental evaluation on the query-query similarity task. Finally, in Section 6 we wrap up and provide conclusions and directions of future work.

2 Related Work

Many techniques have been proposed to overcome the vocabulary mismatch problem, including stemming [5,9], LSI [3], translation models [1], and query expansion [6,14]. This section describes several of these techniques that are most related to our work. The task we focus on is a query-query similarity task, in which we compare short text segments, such as "Apple computer", "apple pie", "MAC OS X", and "iMAC".

Translation models, in a monolingual setting, have been used for document retrieval [1], question answering [8], and detecting text reuse [7]. The goal is to measure the likelihood that some candidate document or sentence is a translation (or transformation) of the query. However, such models are less likely to be effective on very short segments of texts, such as queries, due to the difficulty involved in estimating reliable translation probabilities for such pieces of text.

Query expansion is a common technique used to convert an initial, typically short, query into a richer representation of the information need [6,10,14]. This is accomplished by adding terms that are likely to appear in relevant or pseudo-relevant documents to the original query representation. In our query-query matching work, we explore expanding both the original and candidate query representations.

Sahami and Heilman proposed a method of enriching short text representations that can be construed as a form of query expansion [11]. Their proposed method expands short segments of text using web search results. The similarity between two short segments of text can then computed in the expanded representation space. The expanded representation and DenseProb similarity measure that we present in

Sections 3 and 4 are similar to this approach. However, we estimate term weights differently and analyze how such expansion approaches compare, in terms of efficiency and effectiveness, to other standard information retrieval measures.

Finally, since we evaluate our techniques on a query-query similarity task, it should be noted that this problem, and the related problem of suggesting and identifying query-query reformulations has been investigated from a number of angles, ranging from machine learning approaches [4] to query session log analysis[2]. These techniques are complimentary to the core representational and similarity ideas that we explore in our work.

3 Text Representations

Text representations are an important part of any similarity measure. In this section, we describe three different ways of representing text. Although these representations can be applied to text of any length, we are primarily interested in using them to represent short segments of text.

3.1 Surface Representation

The most basic representation of a short segment of text is the surface representation (i.e. the text itself). Such a representation is very sparse. However, it is very high quality because no automatic or manual transformations (such as stemming) have been done to alter it. While it is possible that such transformations enhance the representation, it is also possible that they introduce noise.

3.2 Stemmed Representation

Stemming is one of the most obvious ways to generalize (normalize) text. For this reason, stemming is commonly used in information retrieval systems as a rudimentary device to overcome the vocabulary mismatch problem. Various stemmers exist, including rule-based stemmers [9] and statistical stemmers [5].

Although stemming can significantly improve matching coverage, it also introduces noise, which can lead to poor matches. Using the Porter stemmer, both "marine vegetation" and "marinated vegetables" stem to "marin veget", which is undesirable. Overall, however, the number of meaningful matches introduced typically outweighs the number of spurious matches.

Throughout the remainder of this paper, we use the Porter stemmer to generate all of our stemmed representations.

3.3 Expanded Representation

Although stemming helps overcome the vocabulary mismatch problem to a certain extent, it does not handle the contextual problem. It fails to discern the difference between the meaning of "bank" in "Bank of America" and "river bank". Therefore, it is desirable to build representations for the short text segments that include contextually relevant information.

<query>*apple pie***</query>**

<title>Applie pie – Wikipedia, the free encyclopedia**</title>**
<snippet>In cooking, an apple pie is a fruit pie (or tart) in which the principal filling ingredient is apples . Pastry is generally used top-and-bottom, making a double-crust pie, the upper crust of which ...**</snippet>**
<url>en.wikipedia.org/wiki/Apple_ pie**</url>**

<title>All About Food – Apple Pies**</title>**
<snippet>Apple Pie. Recipes. All-American Apple Pie. American Apple Pie. Amish Apple Pie . Apple Cream Pie. Apple Crumble Pie. Apple Pie . Apple Pie in a Brown Bag. Best Apple Pie**</snippet>**
<url>fp.enter.net/~rburk/pies/ applepie/applepie.htm**</url>**

<title>Apple Pie Recipe**</title>**
<snippet>Apple Pie Recipe using apple peeler corer slicer ... Apple Pie Recipe. From Scratch to Oven in 20-Minutes. Start by preheating the oven. By the time it's ...**</snippet>**
<url>applesource.com/applepierecipe.htm**</url>**
...

Fig. 1. Example expanded representation for the text "apple pie." The expanded representation is the concatenation of the title and snippet elements.

One approach is to enrich the representation using an external source of information related to the query terms. Possible sources of such information include web (or other) search results returned by issuing the short text segment as a query, relevant Wikipedia articles, and, if the short text segment is a query, query reformulation logs. Each of these sources provides a set of contextual text that can be used to expand the original sparse text representation.

In our experiments, we use web search results to expand our short text representations. For each short segment of text, we run the query against a commercial search engine's index and retrieve the top 200 results. The titles and snippets associated with these results are then concatenated and used as our expanded representation. In Figure 1, we show a portion of the expanded representation for the short text segment "apple pie". As we see, this expanded representation contains a number of contextually relevant terms, such as "recipe", "food", and "cooking" that are not present in the surface representation. We note that this expanded representation is similar to the one proposed in [11].

4 Similarity Measures

In this section we describe three methods for measuring the similarity between short segments of text. These measures are motivated by, and make use of, the representations described in the previous section. We also propose a hybrid method of combining the ranking of the various similarity measures in order to exploit the strengths and weaknesses of each.

4.1 Lexical

The most basic similarity measures are purely lexical. That is, they rely solely on matching the terms present in the surface representations. Given two short segments of text, Q and C, treating Q as the query and C as the candidate we wish to measure the similarity of, we define the following lexical matching criteria:

- Exact – Q and C are lexically equivalent. (Q: "seattle mariners tickets", C: "seattle mariners tickets")
- Phrase – C is a substring of Q. (Q: "seattle mariners tickets", C: "seattle mariners")
- Subset – The terms in C are a subset of the terms in Q. (Q: "seattle mariners tickets", C: "tickets seattle")

These measures are binary. That is, two segments of text either match (are deemed 'similar') or they do not. There is no graded score associated with the match. However, if necessary, it is possible to impose such a score by looking at various characteristics of the match such as the length of Q and C, or the frequency of the terms in some collection.

It should also be noted that exact matches $\subseteq$ phrase matches $\subseteq$ subset matches. Exact matches are very high precision (excellent matches), yet very low recall since they miss a lot of relevant material. At the other extreme, subset matches are lower precision, but have higher recall. Any candidate C that contains a term that does not appear in the query Q will not match under any of these rules, which is very undesirable. Therefore, we expect that matches generated using these lexical rules will be have high precision but poor recall.

4.2 Probabilistic

As we just described, lexical matching alone is not enough to produce a large number of relevant matches. In order to improve recall, we must make use of the expanded text representations. To do so, we use the language modeling framework to model query and candidate texts.

To utilize the framework, we must estimate unigram language models for the query (θ_Q) and each candidate (θ_C). For ranking purposes, we use the negative KL-divergence between the query and candidate model, which is commonly used in the language modeling framework [14]. This results in the following ranking function:

$$\begin{aligned} -KL(\theta_Q, \theta_C) &= H(\theta_Q) - CE(\theta_Q, \theta_C) \\ &\equiv \sum_{w \in V} P(w \mid \theta_Q) \log P(w \mid \theta_C) \end{aligned} \tag{1}$$

where V is the vocabulary, H is entropy, CE is cross entropy, and $\equiv$ denotes rank equivalence.

The critical part of the ranking function is how the query and candidate language models are estimated. Different estimates can lead to radically different rankings. We now describe how we estimate these models using the representations available to us.

We begin with the query model. The most straightforward way of estimating a query model is to use the surface representation. This is estimated as:

$$P(w \mid \theta_Q) = \frac{tf_{w,QS}}{\mid QS \mid} \tag{2}$$

where QS denotes the query surface representation, $tf_{w,QS}$ is the number of times w occurs in the representation, and $|QS|$ is the total number of terms in QS. This estimate will be very sparse since we are using the surface representation. This allows Equation 1 to be computed very efficiently since most terms in the summation ($w \in V$) will be zero.

We also consider the case when we use the expanded representation of the query, as described in Section 3.3. The estimate, which is analogous to the unexpanded case, is:

$$P(w \mid \theta_Q) = \frac{tf_{w,QE} + \mu_Q P(w \mid C)}{|QE| + \mu_Q} \qquad (3)$$

where QE is the query expanded representation, and μ_Q is a smoothing parameter. This type of estimation is commonly used in the language modeling community and is often referred to as Dirichlet or Bayesian smoothing [13]. Since this estimate is much more dense than the unexpanded estimate, it is more time consuming to evaluate Equation 1. Due to the amount of data we work with in our experiments, we truncate this distribution by only keeping the 20 most likely terms and setting the remaining probabilities to 0. Pruning similar to this was done in [11] for the same reason.

Finally, we describe how the candidate model is estimated. Rather than exploring both estimates using both unexpanded and expanded representations, we restrict ourselves to expanded representations. Therefore, we get the following estimate:

$$P(w \mid \theta_C) = \frac{tf_{w,CE} + \mu_C P(w \mid C)}{|CE| + \mu_C} \qquad (4)$$

where CE is the candidate expanded representation, and μ_C is a smoothing parameter. Unlike the expanded query model, we do not truncate this distribution in any way.

Finally, it is important to recall that expanded representations may be created using any number of external sources. Our use of the web was simply a matter of convenience. However, we can use this same general framework with expanded representations generated using any possible external text source.

4.3 Hybrid

We are often interested in taking the matches generated by several different similarity measures and combining them. We call these *hybrid* techniques. Given two or more lists of matches, we stack the lists according to some pre-defined ordering (denoted ">") of the lists, to form a combined list. For example, given match lists A and B, and ordering A > B, we form the hybrid list AB, which is list B appended to the end of list A. Since the same match may occur in more than one set of results, we must

Table 1. Overview of query representation, candidate representation, and similarity measure used for each matching method

Method Name	*Query Representation*	*Candidate Representation*	*Similarity Measure*
Lexical	Surface	Surface	Hybrid (Exact > Phrase > Subset)
Stemming	Stemmed	Stemmed	Hybrid (Lexical > Exact Stems)
SparseProb	Surface	Expanded	Probabilistic
DenseProb	Expanded	Expanded	Probabilistic
Backoff	Various	Various	Hybrid (Exact > Exact Stems > DenseProb)

remove duplicates from the combined list. Our deduplication policy states that we keep the highest ranked match and remove all others. Although this combination scheme is naïve, it has the advantage that there are no combination parameters to learn.

4.4 Summary of Methods Evaluated

Table 1 summarizes the methods we evaluate in the next section. For each method, we include the query and candidate representations and the similarity measure used.

The Lexical method, which considers the surface forms of the query and candidate, makes use of a hybrid technique that ranks exact matches first, then phrase matches, and finally subset matches. The Stemming method also uses a hybrid technique that first ranks matches using the Lexical method just described and then ranks any exact matches that result after stemming both the query and the candidate. We refer to these types of matches as "exact stems" matches.

The SparseProb method is the first of the two probabilistic methods. It uses the unexpanded query representation, the expanded candidate representation, and ranks using the negative KL-divergence, whereas the DenseProb method uses expanded representations for both the query and the candidate and also ranks using the negative KL-divergence.

Finally, the Backoff method is a hybrid method that ranks exact matches, exact stems matches, and then DenseProb matches. The goal here is to see what benefit, if any, is achieved by replacing the phrase and subset matches from the Stemming method with DenseProb matches. We hypothesize that the DenseProb matches will be better than the often poor phrase and subset matches.

Many other query/candidate representation combinations are possible beyond those listed in Table 1. For example, it may be reasonable to use an expanded query form and a surface candidate form. However, in order to maintain a reasonable scope, we constrain ourselves to the methods described in this section.

Table 2. Examples matches taken from our test collection for the query "seattle mariners". The Seattle Mariners are a baseball team from Seattle. For each method, we show the 10 matches with the highest similarity score.

Query: "seattle mariners"				
Lexical	*Stemming*	*SparseProb*	*DenseProb*	*Backoff*
seattle mariners	seattle mariners	seattle mariners tickets	seattle mariners tickets	seattle mariners
seattle	seattle	mariners tickets	mariners tickets	seattle mariner
mariners	mariners	seattle mariners	seattle mariners baseball	seattle mariners tickets
	seattle mariner	seattle mariners baseball	seattle mariners	mariners tickets
		seattle mariners schedule	seattle mariners schedule	seattle mariners baseball
		mariners baseball	mariners baseball	seattle mariners schedule
		seattle baseball	seattle baseball	mariners baseball
		mariners	red sox mariners tickets	seattle baseball
		mariners schedule	mariners schedule	red sox mariners tickets
		seattle mariner	cheap mariners tickets	mariners schedule

5 Experimental Evaluation

In this section we evaluate the similarity measures proposed in Section 4. We begin by showing some illustrative examples of matches generated using our algorithms. We then formally evaluate the methods in the context of a query-query similarity task.

5.1 Illustrative Examples

Table 2 provides illustrative matches returned using the various matching techniques described in Section 4. Although many of these results look reasonable, it is difficult to quantify how much better any one method is by simply looking at these results. Therefore, in the next section we formally evaluate the different match types.

5.2 Query-Query Similarity

We now describe our query-query similarity experiments. Here, we are interested in evaluating how well the various methods we described in Section 4 can be used to find queries that are similar to some target query. This task is a general task that is widely applicable. For example, such a query-query similarity system could be used to recommend alternative queries to users of a web search engine or for session boundary detection in query log analysis.

5.2.1 Data

The following data resources were used in our experimental evaluation. A sample of 363,822 popular queries drawn from a 2005 MSN Search query log was used as our

Table 3. Description of the relevance judgment scale

Judgment	*Description*	*Examples (Query / Candidate)*
Excellent	The candidate is *semantically equivalent* to the user query.	atlanta ga / atlanta georgia
Good	The candidate is related to (but not identical to) the query intent and it is likely the *user would be interested in the candidate.*	seattle mariners / seattle baseball tickets
Fair	The candidate is related to the query intent, but in an overly vague or specific manner that results in the *user having little, if any, interest in the candidate.*	hyundia azera / new york car show
Bad	The candidate is *unrelated* to the query intent.	web visitor count / coin counter

candidate pool of queries to match against. For each query, we generated an expanded representation, as described in Section 3.3. In our experiments, we set μ_Q to 0 and μ_C to 2500. To handle this amount of data, we built an index out of the expanded representations using the Indri search system [12].

We also randomly sampled a set of 120 queries from the same log to use as target queries. These target queries were then matched against the full set of 363k queries. For each of these target queries, we ran the methods described in Section 4 and pooled the results down to a depth of 25 per method. A single human assessor then judged the relevance of each candidate result with respect to the target query using a 4-point judgment scale. Table 3 provides a description and examples of each type of judgment.

The result of this assessment was 5231 judged target/candidate pairs. Of these judgments, 317 (6%) were Excellent, 600 (11%) were Good, 2537 (49%) were Fair, and 1777 (34%) were Bad. In order to determine the reliability of the judgments, four assessors judged 10 target queries. The inter-annotator agreement was then computed for these queries and was found to be 60%. However, when Excellent and Good judgments were binned and Fair and Bad judgments were binned, the agreement increased to 80%. This indicates the boundary between Fair and Bad is interpreted differently among users. For this reason, we will primarily focus our attention on the boundary between Excellent and Good and between Good and Fair. In addition, the Excellent and Good matches are the most interesting for many practical applications including query suggestion and sponsored search.

5.2.2 Evaluation

We are interested in understanding how our matching methods compare to each other across various relevance criteria. Since we are interested in using standard information retrieval metrics, such as precision and recall, we must binarize the relevance judgments. For each experiment, we state the relevance criteria used.

We first evaluate the methods using precision-recall graphs using two different relevance criteria. The results are given in Figure 2. For the case when Excellent matches are considered relevant (left panel), we see that the Lexical and Stemming methods outperform the probabilistic methods, especially at lower recall levels.

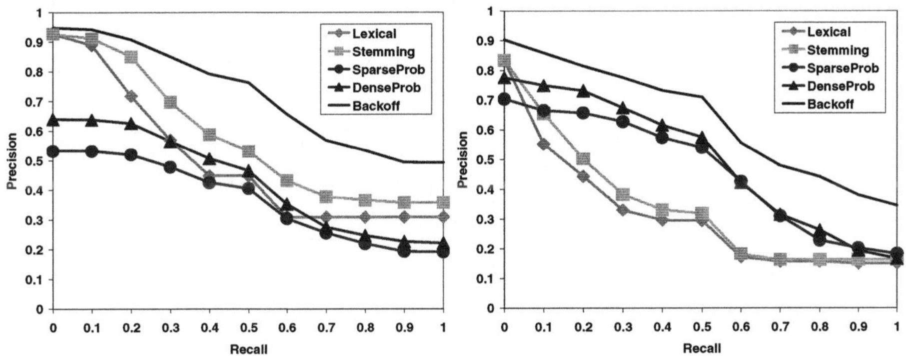

Fig. 2. Interpolated, 11-point precision-recall curves for the five matching methods described in Section 4. On the left, candidates judged 'Excellent' are considered relevant. On the right, candidates judged 'Excellent' or 'Good' are considered relevant.

This is not surprising, since we expect lexical matches to easily find most of the Excellent matches. In addition, we see that Stemming consistently outperforms the Lexical method. However, the Backoff method dominates the other methods at all recall levels. This results from backing off from stricter matches to less strict matches. For example, for the query "atlanta ga", the Lexical method will match "atlanta ga", but neither the Lexical nor the Stemming methods will match "atlanta georgia", which is actually an Excellent match that is found using the DenseProb method.

When we relax the relevance criteria and consider both Excellent and Good judgments to be relevant (right panel), we see an interesting shift in the graph. Here, the probabilistic methods, SparseProb and DenseProb, outperform the Lexical and Stemming methods at all recall levels, except very low levels. This suggests that the Lexical and Stemming methods are good at finding Excellent matches, but that they are worse at finding Good matches compared to the probabilistic methods. We further test this hypothesis later in this section. However, once again, we see that the Backoff method outperforms all of the methods at all recall levels.

One reason why the Backoff method is superior to the non-hybrid probabilistic methods is the fact that the SparseProb and DenseProb methods often fail to return exact matches high in the ranked list. This is caused by truncating the expanded query distribution before computing the KL divergence. Since exact matches account for a majority of the Excellent judgments, this causes the entire curve to be shifted down. By forcing the exact and exact stems matches to occur first, we are 'stacking the deck' and promoting matches that are likely to be high precision. This, combined with the high recall of the DenseProb method, results in a superior matching method.

It is clear that exact matches are very likely to result in Excellent matches. However, it is not clear how phrase and subset lexical matches compare to stemming and probabilistic matches. To measure this, we compute the precision at k for the Lexical and Backoff methods, where k is the number of results returned by the Lexical method. This evaluation allows us to quantify the improvement achieved by replacing the low precision phrase and subset matches with the high precision exact stems matches and high recall DenseProb matches.

Table 4. Precision at *k*, where *k* is the number of matches returned using the Lexical method. In this table, the evaluation set of queries was stratified according to *k*. Queries indicates the the number of queries associated with each *k* . Only values of *k* associated with 10 or more queries are shown.

		R = {Excellent}		R = {Excellent, Good}	
k	Queries	Lexical	Backoff	Lexical	Backoff
1	40	0.7500	0.8125	0.7500	0.8125
2	38	0.3235	0.4853	0.3382	0.5882
3	31	0.2688	0.4194	0.3978	0.5914

The results are presented in Table 4 for two relevance criteria. We stratify the queries with respect to *k*, the number of Lexical method matches for the query, and compute precision at depth *k* over these queries. We only include values of *k* associated with 10 or more queries, since it misleading to compute and compare means over smaller samples. As the results show, the Backoff method is superior in every case. This suggests that the stemming and probabilistic matches (used in the Backoff method) are considerably better at finding both Excellent and Good matches compared to the phrase and subset matches (used in the Lexical method).

5.2.3 Effectiveness vs. Efficiency

One important practical aspect of the techniques developed is efficiency. Generating lexical and stemming matches is very efficient. The probabilistic methods are slower, but not unreasonable. Generating matches against our collection of 363,822 candidates using a modern single CPU machine takes 0.15 seconds per query using the SparseProb method and 3 seconds per query using the DenseProb method.

The DenseProb method requires, *a priori*, an index of expanded representations for both the candidates and the incoming queries. If we are asked to generate DenseProb matches for a query that is not in our index, then we must generate this representation on the fly. However, the SparseProb method does not exhibit this behavior and can be used to efficiently generate matches for *any* incoming query.

Therefore, SparseProb is the the best choice in terms of speed and coverage. However, if speed is not an issue, and high quality results are important, then DenseProb is the better choice.

6 Conclusions and Future Work

In this paper we studied the problem of measuring the similarity between short segments of text. We looked at various types of text representations, including surface, stemmed, and expanded. We showed how web search results can be used to form expanded representations of short text segments. We then described several similarity measures based on these representations, including lexical matching and probabilistic measures based on language models estimated from unexpanded and expanded representations. We then formally evaluated and compared these measures in the context of a query-query similarity task over a large collection of popular web search queries. Our results showed that lexical matching is good for finding semantically identical matches and that the probabilistic methods are better at finding interesting topically related matches. It was shown that a simple hybrid technique that combines lexical, stemmed, and probabilistic matches results in far superior performance than any method alone.

The probabilistic framework presented in this paper provides a general method for measuring the similarity between two short segments of text. Although we chose to use web search results as the basis of our expanded representation in this work, an interesting direction of future work would be to use a variety of other sources of external text, such as query reformulation logs, queries that result in similar click patterns, and Wikipedia. It would also be worthwhile to evaluate these techniques in an end-to-end application, such as a query-query reformulation system, in order to see what impact they have in a more practical setting.

References

[1] Berger, A. and Lafferty, J. Information retrieval as statistical translation. In *Proceedings of SIGIR '99*, pages 222-229, 1999.

[2] Cucerzan, S. and Brill, E. Extracting semantically related queries by exploiting user session information. Technical Report, Microsoft Research, 2005.

[3] Deerwester, S., Dumais, S., Landauer, T., Furnas, G. and Harshman, R. Indexing by latent semantic analysis. In *JASIST*, 41(6), pages 391-407, 1990.

[4] Jones, R. Generating query substitutions. In *Proceedings of WWW 2006*, pages 387-396, 2006.

[5] Krovetz, R. Viewing morphology as an inference process. In *Proceedings of SIGIR '93*, pages 191-202, 1993.

[6] Lavrenko, V. and Croft, W.B. Relevance based language models. In *Proceedings of SIGIR '01*, pages 120-127, 2001.

[7] Metzler, D., Bernstein, Y., Croft, W.B., Moffat, A., and Zobel, J. Similarity measures for tracking information flow. In *Proceedings of CIKM '05*, pages 517-524, 2005.

[8] Murdock, V. and Croft, W.B. A Translation Model for Sentence Retrieval. In *Proceedings of HLT/EMNLP '05*, pages 684-691, 2005.

[9] Porter, M. F. An algorithm for suffix stripping. *Program*, 14(3), pages 130-137, 1980.

[10] Rocchio, J. J. *Relevance Feedback in Information Retrieval*, pages 313-323. Prentice-Hall, 1971.

[11] Sahami, M. and Heilman, T. A web-based kernel function for measuring the similarity of short text snippets. In *Proceedings of WWW 2006*, pages 377-386, 2006.

[12] Strohman, T., Metzler, D., Turtle, H., Croft, W. B. Indri: A language model-based search engine for complex queries. In *Proceedings of the International Conference on Intelligence Analysis*, 2005.

[13] Zhai, C. and Lafferty, J. A study of smoothing methods for language models applied to ad hoc information retrieval. In *Proceedings of SIGIR '01*, pages 334-342, 2001.

[14] Zhai, C. and Lafferty, J. Model-based feedback in the language modeling approach to information retrieval. In *Proceedings of CIKM '01*, pages 403-410, 2001.

Multinomial Randomness Models for Retrieval with Document Fields

Vassilis Plachouras[1] and Iadh Ounis[2]

[1] Yahoo! Research, Barcelona, Spain
[2] University of Glasgow, Glasgow, UK
vassilis@yahoo-inc.com, ounis@dcs.gla.ac.uk

Abstract. Document fields, such as the title or the headings of a document, offer a way to consider the structure of documents for retrieval. Most of the proposed approaches in the literature employ either a linear combination of scores assigned to different fields, or a linear combination of frequencies in the term frequency normalisation component. In the context of the Divergence From Randomness framework, we have a sound opportunity to integrate document fields in the probabilistic randomness model. This paper introduces novel probabilistic models for incorporating fields in the retrieval process using a multinomial randomness model and its information theoretic approximation. The evaluation results from experiments conducted with a standard TREC Web test collection show that the proposed models perform as well as a state-of-the-art field-based weighting model, while at the same time, they are theoretically founded and more extensible than current field-based models.

1 Introduction

Document fields provide a way to incorporate the structure of a document in Information Retrieval (IR) models. In the context of HTML documents, the document fields may correspond to the contents of particular HTML tags, such as the title, or the heading tags. The anchor text of the incoming hyperlinks can also be seen as a document field. In the case of email documents, the fields may correspond to the contents of the email's subject, date, or to the email address of the sender [9]. It has been shown that using document fields for Web retrieval improves the retrieval effectiveness [17,7].

The text and the distribution of terms in a particular field depend on the function of that field. For example, the title field provides a concise and short description for the whole document, and terms are likely to appear once or twice in a given title [6]. The anchor text field also provides a concise description of the document, but the number of terms depends on the number of incoming hyperlinks of the document. In addition, anchor texts are not always written by the author of a document, and hence, they may enrich the document representation with alternative terms.

The combination of evidence from the different fields in a retrieval model requires special attention. Robertson et al. [14] pointed out that the linear combination of scores, which has been the approach mostly used for the combination of fields, is difficult to interpret due to the non-linear relation between the assigned scores and the term frequencies in each of the fields. Hawking et al. [5] showed that the term frequency

G. Amati, C. Carpineto, and G. Romano (Eds.): ECIR 2007, LNCS 4425, pp. 28–39, 2007.

normalisation applied to each field depends on the nature of the corresponding field. Zaragoza et al. [17] introduced a field-based version of BM25, called BM25F, which applies term frequency normalisation and weighting of the fields independently. Macdonald et al. [7] also introduced *normalisation 2F* in the Divergence From Randomness (DFR) framework [1] for performing independent term frequency normalisation and weighting of fields. In both cases of BM25F and the DFR models that employ normalisation 2F, there is the assumption that the occurrences of terms in the fields follow the same distribution, because the combination of fields takes place in the term frequency normalisation component, and not in the probabilistic weighting model.

In this work, we introduce weighting models, where the combination of evidence from the different fields does not take place in the term frequency normalisation part of the model, but instead, it constitutes an integral part of the probabilistic randomness model. We propose two DFR weighting models that combine the evidence from the different fields using a multinomial distribution, and its information theoretic approximation. We evaluate the performance of the introduced weighting models using the standard .Gov TREC Web test collection. We show that the models perform as well as the state-of-the-art model field-based PL2F, while at the same time, they employ a theoretically founded and more extensible combination of evidence from fields.

The remainder of this paper is structured as follows. Section 2 provides a description of the DFR framework, as well as the related field-based weighting models. Section 3 introduces the proposed multinomial DFR weighting models. Section 4 presents the evaluation of the proposed weighting models with a standard Web test collection. Sections 5 and 6 close the paper with a discussion related to the proposed models and the obtained results, and some concluding remarks drawn from this work, respectively.

2 Divergence from Randomness Framework and Document Fields

The Divergence From Randomness (DFR) framework [1] generates a family of probabilistic weighting models for IR. It provides a great extent of flexibility in the sense that the generated models are modular, allowing for the evaluation of new assumptions in a principled way. The remainder of this section provides a description of the DFR framework (Section 2.1), as well as a brief description of the combination of evidence from different document fields in the context of the DFR framework (Section 2.2).

2.1 DFR Models

The weighting models of the Divergence From Randomness framework are based on combinations of three components: a randomness model $\mathcal{RM}$; an information gain model $\mathcal{GM}$; and a term frequency normalisation model.

Given a collection D of documents, the randomness model $\mathcal{RM}$ estimates the probability $P_{\mathcal{RM}}(t \in d|D)$ of having tf occurrences of a term t in a document d, and the importance of t in d corresponds to the informative content $-\log_2(P_{\mathcal{RM}}(t \in d|D))$. Assuming that the sampling of terms corresponds to a sequence of independent Bernoulli trials, the randomness model $\mathcal{RM}$ is the binomial distribution:

$$P_{\mathcal{B}}(t \in d|D) = \binom{TF}{tf} p^{tf}(1-p)^{TF-tf} \tag{1}$$

where TF is the frequency of t in the collection D, $p = \frac{1}{N}$ is a uniform prior probability that the term t appears in the document d, and N is the number of documents in the collection D. A limiting form of the binomial distribution is the Poisson distribution $\mathcal{P}$:

$$P_{\mathcal{B}}(t \in d|D) \approx P_{\mathcal{P}}(t \in d|D) = \frac{\lambda^{tf}}{tf!} e^{-\lambda} \quad \text{where} \quad \lambda = TF \cdot p = \frac{TF}{N} \tag{2}$$

The information gain model $\mathcal{GM}$ estimates the informative content $1 - P_{risk}$ of the probability P_{risk} that a term t is a good descriptor for a document. When a term t appears many times in a document, then there is very low risk in assuming that t describes the document. The information gain, however, from any future occurrences of t in d is lower. For example, the term 'evaluation' is likely to have a high frequency in a document about the evaluation of IR systems. After the first few occurrences of the term, however, each additional occurrence of the term 'evaluation' provides a diminishing additional amount of information. One model to compute the probability P_{risk} is the Laplace after-effect model:

$$P_{risk} = \frac{tf}{tf + 1} \tag{3}$$

P_{risk} estimates the probability of having one more occurrence of a term in a document, after having seen tf occurrences already.

The third component of the DFR framework is the term frequency normalisation model, which adjusts the frequency tf of the term t in d, given the length l of d and the average document length $\bar{l}$ in D. *Normalisation 2* assumes a decreasing density function of the normalised term frequency with respect to the document length l. The normalised term frequency tfn is given as follows:

$$tfn = tf \cdot \log_2(1 + c \cdot \frac{\bar{l}}{l}) \tag{4}$$

where c is a hyperparameter, i.e. a tunable parameter. Normalisation 2 is employed in the framework by replacing tf in Equations (2) and (3) with tfn.

The relevance score $w_{d,q}$ of a document d for a query q is given by:

$$w_{d,q} = \sum_{t \in q} qtw \cdot w_{d,t} \quad \text{where} \quad w_{d,t} = (1 - P_{risk}) \cdot (-\log_2 P_{\mathcal{RM}}) \tag{5}$$

where $w_{d,t}$ is the weight of the term t in document d, $qtw = \frac{qtf}{qtf_{max}}$, qtf is the frequency of t in the query q, and qtf_{max} is the maximum qtf in q. If $P_{\mathcal{RM}}$ is estimated using the Poisson randomness model, P_{risk} is estimated using the Laplace after-effect model, and tfn is computed according to normalisation 2, then the resulting weighting model is denoted by PL2. The factorial is approximated using Stirling's formula: $tf! = \sqrt{2\pi} \cdot tf^{tf+0.5} e^{-tf}$.

The DFR framework generates a wide range of weighting models by using different randomness models, information gain models, or term frequency normalisation models. For example, the next section describes how normalisation 2 is extended to handle the normalisation and weighting of term frequencies for different document fields.

2.2 DFR Models for Document Fields

The DFR framework has been extended to handle multiple document fields, and to apply per-field term frequency normalisation and weighting. This is achieved by extending normalisation 2, and introducing *normalisation 2F* [7], which is explained below.

Suppose that a document has k fields. Each occurrence of a term can be assigned to exactly one field. The frequency tf_i of term t in the i-th field is normalised and weighted independently of the other fields. Then, the normalised and weighted term frequencies are combined into one pseudo-frequency tfn_{2F}:

$$tfn_{2F} = \sum_{i=1}^{k} w_i \cdot tf_i \log_2 \left(1 + c_i \cdot \frac{\overline{l_i}}{l_i}\right) \tag{6}$$

where w_i is the relative importance or weight of the i-th field, tf_i is the frequency of t in the i-th field of document d, l_i is the length of the i-th field in d, $\overline{l_i}$ is the average length of the i-th field in the collection D, and c_i is a hyperparameter for the i-th field. The above formula corresponds to normalisation 2F. The weighting model PL2F corresponds to PL2 using tfn_{2F} as given in Equation (6). The well-known BM25 weighting model has also been extended in a similar way to BM25F [17].

3 Multinomial Randomness Models

This section introduces DFR models which, instead of extending the term frequency normalisation component, as described in the previous section, use document fields as part of the randomness model. While the weighting model PL2F has been shown to perform particularly well [7,8], the document fields are not an integral part of the randomness weighting model. Indeed, the combination of evidence from the different fields takes place as a linear combination of normalised frequencies in the term frequency normalisation component. This implies that the term frequencies are drawn from the same distribution, even though the nature of each field may be different.

We propose two weighting models, which, instead of assuming that term frequencies in fields are drawn from the same distribution, use multinomial distributions to incorporate document fields in a theoretically driven way. The first one is based on the multinomial distribution (Section 3.1), and the second one is based on an information theoretic approximation of the multinomial distribution (Section 3.2).

3.1 Multinomial Distribution

We employ the multinomial distribution to compute the probability that a term appears a given number of times in each of the fields of a document. The formula of the weighting model is derived as follows. Suppose that a document d has k fields. The probability that a term occurs tf_i times in the i-th field f_i, is given as follows:

$$P_{\mathcal{M}}(t \in d|D) = \binom{TF}{tf_1 \quad tf_2 \ldots tf_k \quad tf'} p_1^{tf_1} p_2^{tf_2} \ldots p_k^{tf_k} p'^{tf'} \tag{7}$$

In the above equation, TF is the frequency of term t in the collection, $p_i = \frac{1}{k \cdot N}$ is the prior probability that a term occurs in a particular field of document d, and N is the number of documents in the collection D. The frequency $tf' = TF - \sum_{i=1}^{k} tf_i$ corresponds to the number of occurrences of t in other documents than d. The probability $p' = 1 - k\frac{1}{k \cdot N} = \frac{N-1}{N}$ corresponds to the probability that t does not appear in any of the fields of d.

The DFR weighting model is generated using the multinomial distribution from Equation (7) as a randomness model, the Laplace after-effect from Equation (3), and replacing tf_i with the normalised term frequency tfn_i, obtained by applying normalisation 2 from Equation (4). The relevance score of a document d for a query q is computed as follows:

$$w_{d,q} = \sum_{t \in q} qtw \cdot w_{d,t} = \sum_{t \in q} qtw \cdot (1 - P_{risk}) \cdot \Big(- \log_2(P_{\mathcal{M}}(t \in d|D)\Big)$$

$$= \sum_{t \in q} \frac{qtw}{\sum_{i=1}^{k} tfn_i + 1} \cdot \Big(- \log_2(TF!) + \sum_{i=1}^{k} \Big(\log_2(tfn_i!) - tfn_i \log_2(p_i)\Big)$$

$$+ \log_2(tfn'!) - tfn' \log_2(p')\Big) \tag{8}$$

where qtw is the weight of a term t in query q, $tfn' = TF - \sum_{i=1}^{k} tfn_i$, $tfn_i = tf_i \cdot \log_2(1 + c_i \cdot \frac{\overline{l_i}}{l_i})$ for the i-th field, and c_i is the hyperparameter of normalisation 2 for the i-th field. The weighting model introduced in the above equation is denoted by ML2, where M stands for the multinomial randomness model, L stands for the Laplace after-effect model, and 2 stands for normalisation 2.

Before continuing, it is interesting to note two issues related to the introduced weighting model ML2, namely setting the relative importance, or weight, of fields in the document representation, and the computation of factorials.

Weights of fields. In Equation (8), there are two different ways to incorporate weights for the fields of documents. The first one is to multiply each of the normalised term frequencies tfn_i with a constant w_i, in a similar way to normalisation 2F (see Equation (6)): $tfn_i := w_i \cdot tfn_i$. The second way is to adjust the prior probabilities p_i of fields, in order to increase the scores assigned to terms occurring in fields with low prior probabilities: $p_i := \frac{p_i}{w_i}$. Indeed, the assigned score to a query term occurring in a field with low probability is high, due to the factor $-tfn_i \log_2(p_i)$ in Equation (8).

Computing factorials. As mentioned in Section 2.1, the factorial in the weighting model PL2 is approximated using Stirling's formula. A different method to approximate the factorial is to use the approximation of Lanczos to the Γ function [12, p. 213], which has a lower approximation error than Stirling's formula. Indeed, preliminary experimentation with ML2 has shown that using Stirling's formula affects the performance of the weighting model, due to the accumulation of the approximation error from computing the factorial $k + 2$ times (k is the number of fields). This is not the case for the Poisson-based weighting models PL2 and PL2F, where there is only one factorial computation for each query term (see Equation (2)). Hence, the computation of factorials in Equation (8) is performed using the approximation of Lanczos to the Γ function.

3.2 Approximation to the Multinomial Distribution

The DFR framework generates different models by replacing the binomial randomness model with its limiting forms, such as the Poisson randomness model. In this section, we introduce a new weighting model by replacing the multinomial randomness model in ML2 with the following information theoretic approximation [13]:

$$\frac{TF!}{tf_1!tf_2!\cdots tf_k!tf'!}{p_1}^{tf_1}{p_2}^{tf_2}\cdots{p_k}^{tf_k}{p'}^{tf'} \approx \frac{1}{\sqrt{2\pi TF}^k}\frac{2^{-TF\cdot D\left(\frac{tf_i}{TF},p_i\right)}}{\sqrt{p_{t1}p_{t2}\cdots p_{tk}p'_t}} \tag{9}$$

$D\left(\frac{tf_i}{TF},p_i\right)$ corresponds to the information theoretic divergence of the probability $p_{ti} = \frac{tf_i}{TF}$ that a term occurs in a field, from the prior probability p_i of the field:

$$D\left(\frac{tf_i}{TF},p_i\right) = \sum_{i=1}^{k}\left(\frac{tf_i}{TF}\log_2\frac{tf_i}{TF\cdot p_i}\right) + \frac{tf'}{TF}\log_2\frac{tf'}{TF\cdot p'} \tag{10}$$

where $tf' = TF - \sum_{i=1}^{k} tf_i$. Hence, the multinomial randomness model $\mathcal{M}$ in the weighting model ML2 can be replaced by its approximation from Equation (9):

$$w_{d,q} = \sum_{t\in q} qtw \cdot \frac{\frac{k}{2}\log_2(2\pi TF)}{\sum_{i=1}^{k} tfn_i + 1} \cdot \Big(\sum_{i=1}^{k}\Big(tfn_i\log_2\frac{tfn_i/TF}{p_i} + \frac{1}{2}\log_2\frac{tfn_i}{TF}\Big) + tfn'\log_2\frac{tfn'/TF}{p'} + \frac{1}{2}\log_2\frac{tfn'}{TF}\Big) \tag{11}$$

The above model is denoted by $\mathrm{M_DL2}$. The definitions of the variables involved in the above equation have been introduced in Section 3.1.

It should be noted that the information theoretic divergence $D\left(\frac{tf_i}{TF},p_i\right)$ is defined only when $tf_i > 0$ for $1 \leq i \leq k$. In other words, $D\left(\frac{tf_i}{TF},p_i\right)$ is defined only when there is at least one occurrence of a query term in all the fields. This is not always the case, because a Web document may contain all the query terms in its body, but it may contain only some of the query terms in its title. To overcome this issue, the weight of a query term t in a document is computed by considering only the fields in which the term t appears.

The weights of different fields can be defined in the same way as in the case of the weighting model ML2, as described in Section 3.1. In more detail, the weighting of fields can be achieved by either multiplying the frequency of a term in a field by a constant, or by adjusting the prior probability of the corresponding field.

An advantage of the weighting model $\mathrm{M_DL2}$ is that, because it approximates the multinomial distribution, there is no need to compute factorials. Hence, it is likely to provide a sufficiently accurate approximation to the multinomial distribution, and it may lead to improved retrieval effectiveness compared to ML2, due to the lower accumulated numerical errors. The experimental results in Section 4.2 will indeed confirm this advantage of $\mathrm{M_DL2}$.

4 Experimental Evaluation

In this section, we evaluate the proposed multinomial DFR models ML2 and M_DL2, and compare their performance to that of PL2F, which has been shown to be particularly effective [7,8]. A comparison of the retrieval effectiveness of PL2F and BM25F has shown that the two models perform equally well on various search tasks and test collections [11], including those employed in this work. Hence, we experiment only with the multinomial models and PL2F. Section 4.1 describes the experimental setting, and Section 4.2 presents the evaluation results.

4.1 Experimental Setting

The evaluation of the proposed models is conducted with the .Gov TREC Web test collection, a crawl of approximately 1.25 million documents from the .gov domain. The .Gov collection has been used in the TREC Web tracks between 2002 and 2004 [2,3,4]. In this work, we employ the tasks from the Web tracks of TREC 2003 and 2004, because they include both informational tasks, such as the topic distillation (td2003 and td2004, respectively), as well as navigational tasks, such as named page finding (np2003 and np2004, respectively) and home page finding (hp2003 and hp2004, respectively). More specifically, we train and test for each type of task independently, in order to get insight on the performance of the proposed models [15]. We employ each of the tasks from the TREC 2003 Web track for training the hyperparameters of the proposed models. Then, we evaluate the models on the corresponding tasks from the TREC 2004 Web track.

In the reported set of experiments, we employ $k = 3$ document fields: the contents of the <BODY> tag of Web documents (b), the anchor text associated with incoming hyperlinks (a), and the contents of the <TITLE> tag (t). More fields can be defined for other types of fields, such as the contents of the heading tags <H1> for example. It has been shown, however, that the body, title and anchor text fields are particularly effective for the considered search tasks [11]. The collection of documents is indexed after removing stopwords and applying Porter's stemming algorithm. We perform the experiments in this work using the Terrier IR platform [10].

The proposed models ML2 and M_DL2, as well as PL2F, have a range of hyperparameters, the setting of which can affect the retrieval effectiveness. More specifically, all three weighting models have two hyperparameters for each employed document field: one related to the term frequency normalisation, and a second one related to the weight of that field. As described in Sections 3.1 and 3.2, there are two ways to define the weights of fields for the weighting models ML2 and M_DL2: (i) multiplying the normalised frequency of a term in a field; (ii) adjusting the prior probability p_i of the i-th field. The field weights in the case of PL2F are only defined in terms of multiplying the normalised term frequency by a constant w_i, as shown in Equation (6).

In this work, we consider only the term frequency normalisation hyperparameters, and we set all the weights of fields to 1, in order to avoid having one extra parameter in the discussion of the performance of the weighting models. We set the involved hyperparameters c_b, c_a, and c_t, for the body, anchor text, and title fields, respectively, by directly optimising mean average precision (MAP) on the training tasks from the Web track of TREC 2003. We perform a 3-dimensional optimisation to set the values of the hyperparameters. The optimisation process is the following. Initially, we apply a simulated annealing algorithm, and then, we use the resulting hyperparameter values as a starting point for a second optimisation algorithm [16], to increase the likelihood of detecting a global maximum. For each of the three training tasks, we apply the above optimisation process three times, and we select the hyperparameter values that result in the highest MAP. We employ the above optimisation process to increase the likelihood that the hyperparameters values result in a global maximum for MAP. Figure 1 shows the MAP obtained by ML2 on the TREC 2003 home page finding topics, for each iteration of the optimisation process. Table 1 reports the hyperparameter values that resulted in the highest MAP for each of the training tasks, and that are used for the experiments in this work.

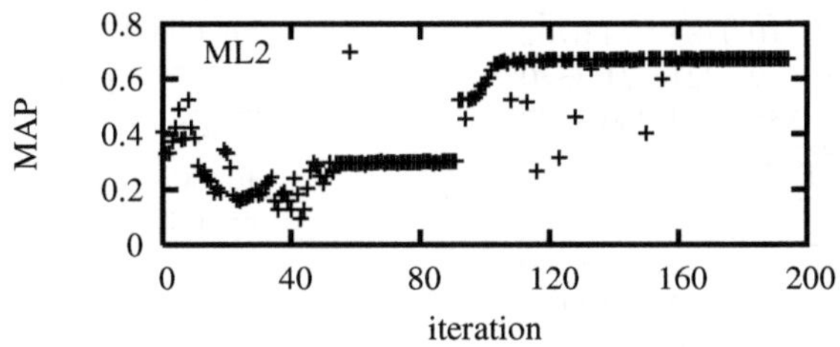

Fig. 1. The MAP obtained by ML2 on the TREC 2003 home page finding topics, during the optimisation of the term frequency normalisation hyperparameters

The evaluation results from the Web tracks of TREC 2003 [3] and 2004 [4] have shown that employing evidence from the URLs of Web documents results in important improvements in retrieval effectiveness for the topic distillation and home page finding tasks, where relevant documents are home pages of relevant Web sites. In order to provide a more complete evaluation of the proposed models for these two types of Web search tasks, we also employ the length in characters of the URL path, denoted by $URLpathlen$, using the following formula to transform it to a relevance score [17]:

$$w_{d,q} := w_{d,q} + \omega \cdot \frac{\kappa}{\kappa + URLpathlen} \tag{12}$$

where $w_{d,q}$ is the relevance score of a document. The parameters ω and κ are set by performing a 2-dimensional optimisation as described for the case of the hyperparameters c_i. The resulting values for ω and κ are shown in Table 2.

4.2 Evaluation Results

After setting the hyperparameter values of the proposed models, we evaluate the models with the search tasks from TREC 2004 Web track [4]. We report the official TREC evaluation measures for each search task: mean average precision (MAP) for the topic distillation task (td2004), and mean reciprocal rank (MRR) of the first correct answer for both named page finding (np2004) and home page finding (hp2004) tasks.

Table 1. The values of the hyperparameters c_b, c_a, and c_t, for the body, anchor text and title fields, respectively, which resulted in the highest MAP on the training tasks of TREC 2003 Web track

ML2			
Task	c_b	c_a	c_t
td2003	0.0738	4.3268	10.8220
np2003	0.1802	4.7057	8.4074
hp2003	0.1926	310.3289	624.3673
M_DL2			
Task	c_b	c_a	c_t
td2003	0.2562	10.0383	24.6762
np2003	1.0216	9.2321	21.3330
hp2003	0.4093	355.2554	966.3637
PL2F			
Task	c_b	c_a	c_t
td2003	0.1400	5.0527	4.3749
np2003	1.0153	11.9652	9.1145
hp2003	0.2785	406.1059	414.7778

Table 2. The values of the hyperparameters ω and κ, which resulted in the highest MAP on the training topic distillation (td2003) and home page finding (hp2003) tasks of TREC 2003 Web track

ML2		
Task	ω	κ
td2003	8.8095	14.8852
hp2003	10.6684	9.8822
M_DL2		
Task	ω	κ
td2003	7.6974	12.4616
hp2003	27.0678	67.3153
PL2F		
Task	ω	κ
td2003	7.3638	8.2178
hp2003	13.3476	28.3669

Table 3 presents the evaluation results for the proposed models ML2, M_DL2, and the weighting model PL2F, as well as their combination with evidence from the URLs of documents (denoted by appending U to the weighting model's name). When only the document fields are employed, the multinomial weighting models have similar performance compared to the weighting model PL2F. The weighting models PL2F and M_DL2 outperform ML2 for both topic distillation and home page finding tasks. For the named page finding task, ML2 results in higher MRR than M_DL2 and PL2F.

Using the Wilcoxon signed rank test, we tested the significance of the differences in MAP and MRR between the proposed new multinomial models and PL2F. In the case of the topic distillation task td2004, PL2F and M_DL2 were found to perform statistically significantly better than ML2, with $p < 0.001$ in both cases. There was no statistically significant difference between PL2F and M_DL2. Regarding the named page finding task np2004, there is no statistically significant difference between any of the three proposed models. For the home page finding task hp2004, only the difference between ML2 and PL2F was found to be statistically significant ($p = 0.020$).

Regarding the combination of the weighting models with the evidence from the URLs of Web documents, Table 3 shows that PL2FU and M_DL2U outperform ML2U for td2004. The differences in performance are statistically significant, with $p = 0.002$ and $p = 0.012$, respectively, but there is no significant difference in the retrieval effectiveness between PL2FU and M_DL2U. When considering hp2004, we can see that PL2F outperforms the multinomial weighting models. The only statistically significant difference in MRR was found between PL2FU and M_DL2FU ($p = 0.012$).

Table 3. Evaluation results for the weighting models ML2, M_DL2, and PL2F on the TREC 2004 Web track topic distillation (td2004), named page finding (np2004), and home page finding (hp2004) tasks. ML2U, M_DL2U, and PL2FU correspond to the combination of each weighting model with evidence from the URL of documents. The table reports mean average precision (MAP) for the topic distillation task, and mean reciprocal rank (MRR) of the first correct answer for the named page finding and home page finding tasks. ML2U, M_DL2U and PL2FU are evaluated only for td2004 and hp2004, where the relevant documents are home pages (see Section 4.1).

Task	ML2	M_DL2	PL2F
		MAP	
td2004	0.1241	0.1391	0.1390
		MRR	
np2004	0.6986	0.6856	0.6878
hp2004	0.6075	0.6213	0.6270
Task	ML2U	M_DL2U	PL2FU
		MAP	
td2004	0.1916	0.2012	0.2045
		MRR	
hp2004	0.6364	0.6220	0.6464

A comparison of the evaluation results with the best performing runs submitted to the Web track of TREC 2004 [4] shows that the combination of the proposed models with the evidence from the URLs performs better than the best performing run of the topic distillation task in TREC 2004, which achieved MAP 0.179. The performance of the proposed models is comparable to that of the most effective method for the named page finding task (MRR 0.731). Regarding the home page finding task, the difference is greater between the performance of the proposed models with evidence from the URLs, and the best performing methods in the same track (MRR 0.749). This can be explained in two ways. First, the over-fitting of the parameters ω and κ on the training task may result in lower performance for the test task. Second, using field weights may be more effective for the home page finding task, which is a high precision task, where the correct answers to the queries are documents of a very specific type.

From the results in Table 3, it can be seen that the model M_DL2, which employs the information theoretic approximation to the multinomial distribution, significantly outperforms the model ML2, which employs the multinomial distribution, for the topic distillation task. As discussed in Section 3.2, this may suggest that approximating the multinomial distribution is more effective than directly computing it, because of the number of computations involved, and the accumulated small approximation errors from the computation of the factorial. The difference in performance may be greater if more document fields are considered.

Overall, the evaluation results show that the proposed multinomial models ML2 and M_DL2 have a very similar performance to that of PL2F for the tested search tasks. None of the models outperforms the others consistently for all three tested tasks, and the weighting models M_DL2 and PL2F achieve similar levels of retrieval effectiveness. The next section discusses some points related to the new multinomial models.

5 Discussion

This section discusses (i) the advantages of the proposed multinomial models compared to the existing field-based weighting models, and (ii) the use of normalisation 2 (or normalisation 2F) for weighting fields in any of the field-based DFR weighting models.

The proposed models result in similar retrieval effectiveness to that of PL2F (Equation (6)), and also provide a new approach to the combination of evidence from the fields, compared to PL2F or BM25F [17], where a weighted sum aggregates term frequencies. Indeed, by employing multinomial distributions, the combination of fields takes place in the probabilistic weighting model. Hence, the weight of a term in a document depends explicitly on the distribution of term frequencies in the different fields.

A second advantage of the multinomial models over PL2F or BM25F is that they allow for a more principled approach to the weighting of fields, rather than just multiplying term frequencies by a constant. As suggested earlier, in the case of ML2 and M_DL2, the prior probability of each field can be used as a weight for that field. The same approach cannot be applied to PL2F, because the randomness model does not consider document fields.

Normalisation 2 is primarily used for normalising the frequency of terms in a document, or in the document fields. In addition, it can also be used to weight the document fields, possibly avoiding the introduction of additional hyperparameters. Indeed, from the equation of normalisation 2: $tfn_i = tf_i \cdot \log_2 \left(1 + c_i \cdot (\overline{l_i}/l_i)\right)$, where $c_i \in (0, +\infty)$, it can be seen that applying a very high value for a particular document field, such as the title field, results in weak term frequency normalisation, and also multiplies the original term frequency. In this way, it may not be necessary to employ separate hyperparameters for field weights, thus reducing the imposed training overhead.

Overall, the proposed multinomial models offer a novel and effective way to combine document fields in a theoretically driven approach. Their introduction in the DFR framework can also generate a family of new weighting models, by combining different information gain or term frequency normalisation models.

6 Conclusions

In this work, we have introduced two new weighting models that combine document fields for Information Retrieval. While field-based weighting models, such as PL2F [7], or BM25F [17], combine evidence from fields in the term frequency normalisation component, we take a different approach. In the context of the DFR framework [1], we employ multinomial randomness models, and model the document fields in the probabilistic retrieval model. The first model, ML2, employs directly the multinomial distribution to assign a relevance score to documents, and the second model, M_DL2, uses an information theoretic approximation of the multinomial distribution.

We have performed experiments in the context of the .Gov TREC Web test collection. The evaluation results show that the new models perform as well as PL2F for a range of Web search tasks, such as topic distillation, named page finding and home page finding. In particular, for the topic distillation task, the model M_DL2 performs as well as PL2F, and significantly outperforms ML2, suggesting that it is more effective to approximate the multinomial distribution, than to compute it directly.

The proposed multinomial models represent a novel and effective approach to the combination of document fields, which is achieved in a principled way within a probabilistic framework. As a result, one of their advantages is that, for example, they allow for the investigation of the weighting of fields in terms of the prior probabilities of each field.

References

1. Amati, G., van Rijsbergen, C.J.: Probabilistic models of information retrieval based on measuring divergence from randomness. ACM TOIS **20** (2002) 357–389
2. Craswell, N., Hawking, D.: Overview of TREC-2002 web track. In: Proceedings of TREC-2002, Gaithersburg, MD, USA (2002)
3. Craswell, N., Hawking, D., Wilkinson, R., Wu, M.: Overview of the TREC-2003 web track. In: Proceedings of TREC-2003, Gaithersburg, MD, USA (2003)
4. Craswell, N., Hawking, D.: Overview of TREC-2004 web track. In: Proceedings of TREC-2004, Gaithersburg, MD, USA (2004)
5. Hawking, D., Upstill, T., Craswell, N.: Toward better weighting of anchors. In: Proceedings of the 27th annual international ACM SIGIR conference on Research and Development in Information Retrieval, ACM Press (2004) 512–513
6. Jin, R., Hauptmann, A.G., Zhai, C.X.: Title language model for information retrieval. In: Proceedings of the 25th annual international ACM SIGIR conference on Research and Development in Information Retrieval, ACM Press (2002) 42–48
7. Macdonald, C., Plachouras, V., He, B., Lioma, C., Ounis, I.: University of Glasgow at WebCLEF 2005: Experiments in per-field normalisation and language specific stemming. In: Proceedings of the Cross Language Evaluation Forum (CLEF) 2005. (2005)
8. Macdonald, C., He, B., Plachouras, V., Ounis, I.: University of Glasgow at TREC 2005: Experiments in Terabyte and Enterprise Tracks with Terrier. In: Proceedings of TREC-2005, Gaithersburg, Maryland USA (2005)
9. Macdonald, C., Ounis, I.: Combining fields in known-item email search. In: SIGIR '06: Proceedings of the 29th annual international ACM SIGIR conference on Research and Development in Information Retrieval, ACM Press (2006) 675–676
10. Ounis, I., Amati, G., Plachouras, V., He, B., Macdonald, C., Lioma, C.: Terrier: A High Performance and Scalable Information Retrieval Platform. In: Proceedings of ACM SIGIR'06 Workshop on Open Source Information Retrieval (OSIR) (2006)
11. Plachouras, V.: Selective Web Information Retrieval. PhD thesis, Department of Computing Science, University of Glasgow (2006)
12. Press, W., Teukolsky, S., Vetterling, W., Flannery, B.: Numerical Recipes in C: The Art of Scientific Computing. Cambridge University Press (1992)
13. Rényi, A.: Foundations of probability. Holden-Day (1970)
14. Robertson, S., Zaragoza, H., Taylor, M.: Simple BM25 extension to multiple weighted fields. In: Proceedings of the 13th ACM Conference on Information and Knowledge Management (CIKM'04), ACM Press (2004) 42–49
15. Soboroff, I.: On evaluating web search with very few relevant documents. In: Proceedings of the 27th annual international ACM SIGIR conference on Research and Development in Information Retrieval, ACM Press (2004) 530–531
16. Yuret, D.: From Genetic Algorithms To Efficient Optimization. Master Thesis, MIT, A.I. Technical Report No. 1569. (1994)
17. Zaragoza, H., Craswell, N., Taylor, M., Saria, S., Robertson, S.: Microsoft Cambridge at TREC-13: Web and HARD tracks. In: Proceedings of TREC-2004, Gaithersburg, MD, USA (2004)

On Score Distributions and Relevance

Stephen Robertson

Microsoft Research, 7 JJ Thomson Avenue, Cambridge CB3 0FB, UK
ser@microsoft.com

Abstract. We discuss the idea of modelling the statistical distributions of scores of documents, classified as relevant or non-relevant. Various specific combinations of standard statistical distributions have been used for this purpose. Some theoretical considerations indicate problems with some of the choices of pairs of distributions. Specifically, we revisit a generalisation of the well-known inverse relationship between recall and precision: some choices of pairs of distributions violate this generalised relationship. We identify the choices and the violations, and explore some of the consequences of this theoretical view.

1 Introduction

The idea of modelling the distributions of scores of relevant and non-relevant documents in an information retrieval system has been around for a long time (see Swets [1]), but in recent years has taken a new lease of life [2,3,4]. Various combinations of statistical distributions have been proposed, for example two normal distributions of equal variance [1], two unequal variance normals or two exponentials [5], two Poisson distributions [6], two gamma distributions [2], an exponential for non-relevant and a normal for relevant [3,4,7], an exponential and a gamma [4].

Clearly a strong argument for choosing any particular combination of distributions is that it gives a good fit to some set of empirical data, and some of the above authors address this question in various ways. However, we do not attempt in this paper any such empirical analysis. Nor does it claim any fundamentally new theoretical results. Rather, it revisits old work [8,9] in order to consider some theoretical properties which might be desirable for such distributions. The primary argument of the paper is that, putting aside considerations of empirical fit, some combinations of distributions exhibit undesirable or anomalous features which reduce their theoretical value. This argument generalises a point made by Bookstein [6] about the Swets unequal variance model. Some of these considerations were also aired in [10] in the context of an analysis of the relation between performance and collection size. The contribution of the paper is to bring together and clarify the theoretical issues, and to connect them with the recent work on score distributions.

Note that many other authors model or analyse score distributions without reference to relevance. This work is not discussed here. Also, the work depends on an assumption of the binary nature of relevance; a different approach would have

G. Amati, C. Carpineto, and G. Romano (Eds.): ECIR 2007, LNCS 4425, pp. 40–51, 2007.

to be taken to take account of degrees or grades or ranks of assessed relevance. The link between relevance and ranking is assumed to be the probability ranking principle [11], which asserts that a search system should rank output in order of probability of (assumed binary) relevance.

In the next section, we introduce the main theoretical argument of the paper. In Sect. 3 we analyse in detail one of the early suggestions, the case of two normal distributions of unequal variance. Then in Sect. 4 we define a simple test and apply it to five different sets of distributional assumptions that have been suggested in the literature. Finally we discuss some further issues and conclude.

2 Recall and Fallout

We consider the output of a retrieval system, as a result of a search query, to be a list of documents ranked by score or retrieval status value, and the user action to involve reading down the list until some stopping point. This stopping point then corresponds, explicitly or implicitly, to a threshold on the score: everything above this threshold has been retrieved, i.e. seen by the user; everything below has not. We further model the situation in terms of the distributions of scores in the populations of relevant and non-relevant documents: a signal detection (SD) theory view of retrieval. In this case, the two natural parameters for evaluation are recall, which corresponds to the proportion of the relevant distribution exceeding the threshold, and fallout (the same for the non-relevant distribution). We interpret these parameters in a probabilistic fashion as the values of the respective cumulative distribution functions, cumulated from the right (i.e. from the high-score end). In this case, the natural performance graph to consider is a graph of recall against fallout, referred to in the SD context as the receiver operating characteristic or ROC curve.

The recall-fallout graph is not normally used for real retrieval experiments, partly because real fallout values are typically so small, but also so unevenly distributed, that it is difficult to display such graphs in a reasonable way. One solution is to transform fallout in some way, e.g. by using a log scale. However, for the purpose of considering some theoretical characteristics, it is appropriate to think in terms of a recall-fallout graph on linear scales. We also present all such graphs with fallout on the x-axis and recall on the y-axis. All such curves may be presumed to pass through (0,0) (very high threshold, nothing retrieved) and (1,1) (very low threshold, everything retrieved). Going down the ranking from the top to the bottom, i.e. lowering the threshold, corresponds to traversing the curve from bottom left (0,0) to top right (1,1).

An example of an idealised smooth curve is shown in Fig. 1 (a detailed derivation of this curve is given in Section 3). We can also see in this figure two other properties of the recall-fallout graph on linear scales. Assuming that this curve represents a single request, the slope of the line OA from the origin to a point A on the curve is a monotonic function of the precision at point A. Also the slope of the tangent at A represents the 'instantaneous' precision – that is, the probability that a document having exactly that score is relevant. A mathematical

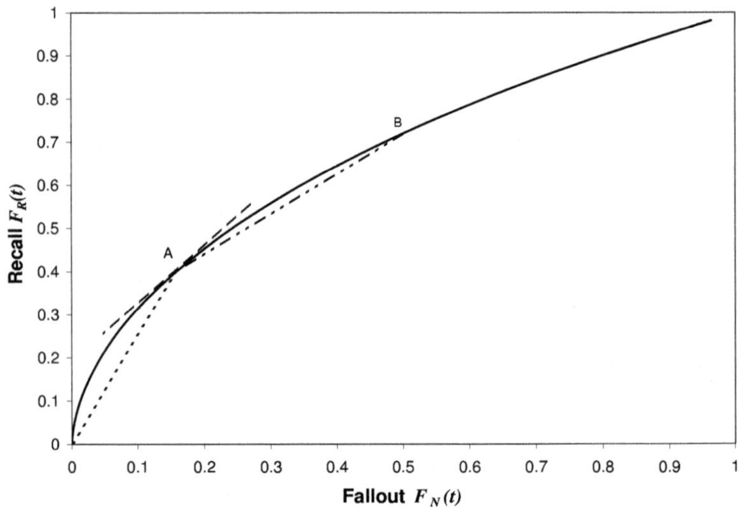

Fig. 1. Receiver Operating Characteristic (ROC) curve for the distributions discussed in Section 3

explanation of these points is given in Section 3. First, we formulate the Convexity Hypothesis, which provides a strong expectation on the shape of the curve.

2.1 The Convex Curve

The straight line on the linear recall-fallout graph, from point (0,0) to (1,1), represents a random ordering of the document collection – identical relevant and non-relevant score distributions [12]. Other straight lines may also be interpreted as random orderings of sets. For example, suppose that we have two points A, B on the recall-fallout graph, corresponding to two score thresholds t_A and t_B, with $t_A > t_B$. Then the straight line from A to B corresponds to retrieval of all documents at A (those whose scores exceed t_A), followed by a random ordering of the documents scoring between t_A and t_B.

It follows that we would in general expect the recall-fallout curve to be convex, when viewed from the top left (0,1). If we found a scoring function which generated a curve containing a concavity, we could improve upon it simply by means of a randomisation process on that section of the ranked list corresponding to the concavity in the curve – this operation would replace the concave section by a straight line, thus raising this part of the curve. (Actually, we could do better than this: the concave part represents scores which tell us something about likely relevance, but in the reverse order – a suitable re-ordering of score values would get us as far above the straight line as the concavity is below it.) Thus even if it is not always the case that the curve is convex, we would certainly expect it of a good system, because a departure from convexity implies that the system can be very easily improved. We may therefore state the following hypothesis, with the support of the above arguments:

Convexity hypothesis. For all good systems, the recall-fallout curve (seen from the ideal point of recall=1, fallout=0) is convex.

This result is related to, but somewhat stronger than, the usual inverse relationship between recall and precision – that is, the R-P relationship follows from convexity [12,8]. We see the convexity hypothesis as a generalisation of the hypothesis of the inverse R-P relationship.

The same hypothesis can also be formulated as a condition on the instantaneous precision, or the probability of relevance of a document at an exact score. The condition is that this should be a monotonic increasing function of the score – the higher the score the higher the probability of relevance. This condition is assumed in [4]; their use of it will be discussed further below.

The convexity hypothesis is the basis for the theoretical arguments of this paper.

3 Score Distributions: Details and an Example

Consider a pair of score distributions for relevant and non-relevant documents. In Fig. 2, we see an example of a pair of normal distributions. The normal is used as example only, but we will generally be using continuous distributions, although it is likely that the scoring function has some degree of granularity, and also we are dealing with finite collections of documents. These distributions are shown in the form of density functions (the usual bell curve). The x-axis is the score or retrieval status value, denoted v; the two distributions are denoted $f_R(v)$ and $f_N(v)$ for relevant and non-relevant documents respectively. All the equations in this section apply to any pair of continuous distributions, but the diagrams relate to the pair of normals.

As indicated above, we turn them into cumulative distributions *from the right* – see Fig. 3. These functions are defined as follows:

$$F_R(t) = \int_{v=t}^{\infty} f_R(v)dv$$

and similarly for F_N.

At any given cut-off or threshold t (examples shown in the form of vertical lines), the cumulative distributions give the probability of retrieving a relevant or non-relevant document respectively at or above that threshold score. These two probabilities may be equated with the traditional measures of recall and fallout respectively. That is, the probabilities can be used as *definitions* of recall and fallout, and observed recall and fallout values are then estimates of these measures:

$$\begin{aligned}
\text{Recall at threshold } t &= Pr(d \text{ retrieved at or above threshold } t|d \text{ relevant}) \\
&= Pr(v(d) \geq t|d \text{ relevant}) \\
&= F_R(t) \qquad (1)
\end{aligned}$$

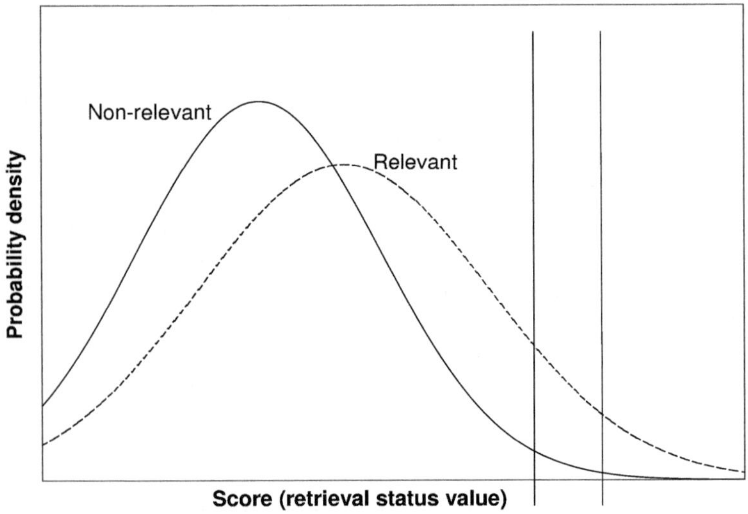

Fig. 2. SD model, normal distributions unequal variance: relevant mean 2.5 variance 1.2; non-relevant mean 1.8 variance 1

(where d is a random document), and similarly for fallout and F_N. We can similarly define precision P, and identify it as a function of recall, fallout and generality G[1], as follows:

$$\begin{aligned} P &= Pr(d \text{ relevant} | v(d) \geq t) \\ &= \frac{GF_R(t)}{GF_R(t) + (1-G)F_N(t)} \end{aligned}$$

We reformulate this as odds:

$$\frac{P}{1-P} = \frac{G}{1-G} \frac{F_R(t)}{F_N(t)} \tag{2}$$

which gives us the monotonic relation between precision and the slope of the straight line OA of Fig. 1. Similarly we can define the odds that a document has a score between two limits:

$$Odds(d \text{ relevant} | t_1 \leq v(d) \leq t_2) = \frac{G}{1-G} \frac{F_R(t_1) - F_R(t_2)}{F_N(t_1) - F_N(t_2)} \tag{3}$$

which gives us the corresponding relation for the line AB. Furthermore, letting $(t_2 - t_1) \to 0$ gives the instantaneous precision result.

We now treat the score v as defining parametrically a relation between recall and fallout, and draw the ROC curve for these two parameters. The curve already presented in Fig. 1 is based on the distributions used here. It does not actually reach (1,1) because it was plotted only down to a threshold of zero; the assumed

[1] Generality is the proportion of documents in the collection that are relevant.

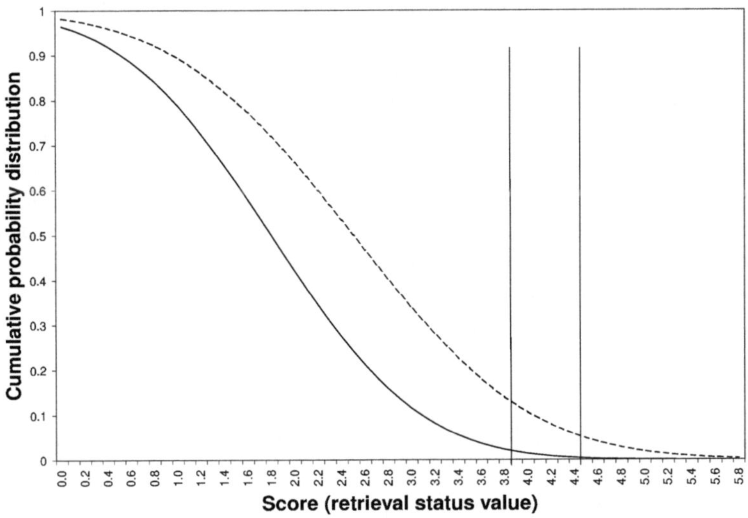

Fig. 3. SD model, normal distributions unequal variance, as Fig. 2, cumulative form

normal distributions both go below zero. The curve does indeed appear to be convex. However, in the full curve there would actually be a small concavity, at the right-hand end, invisible on the scale on which the graph is shown. This is because the relevant document distribution assumed, with a larger variance than the non-relevant, predicts a slightly larger number of documents with significant

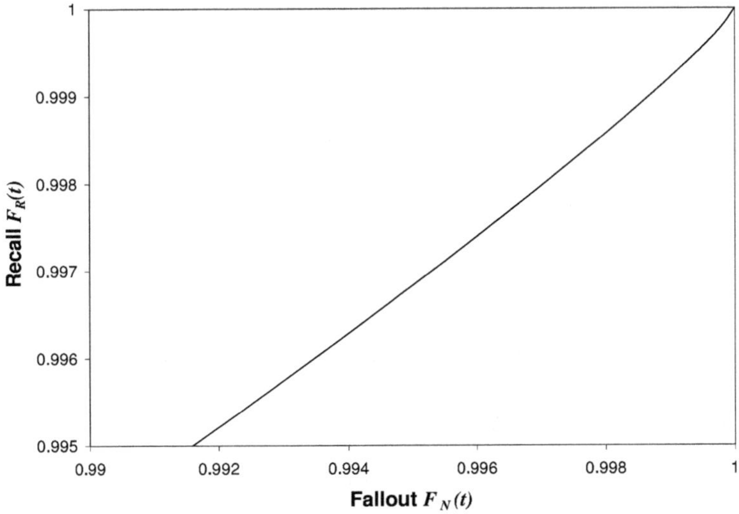

Fig. 4. Top end of the ROC curve for the distributions used in Figs 2 and 3

negative scores than the non-relevant. The curve is extended to (1,1) and the top right corner is blown up in Fig. 4; now the concavity is clearly seen.

In this case we may take this to be an artifact of the model, and of no practical significance whatever, because (a) the system probably does not calculate negative scores anyway, and (b) the number of documents in that range predicted by the distributions is probably measured in very small fractions of a document. It could be that the two normal model gives a fair approximation to real score distributions, and this theoretical anomaly is of no concern. However, the conclusion must be that the 2-normal (unequal variance) model is theoretically flawed, irrespective of its practical usefulness. It therefore seems useful to investigate the conditions under which a pair of distributions will violate the convexity hypothesis.

4 Convexity Condition and Distributional Assumptions

The convexity condition is given in [8] as:

$$\frac{d^2(\text{Recall})}{dt^2} < \frac{d^2(\text{Fallout})}{dt^2} \frac{\frac{d(\text{Recall})}{dt}}{\frac{d(\text{Fallout})}{dt}} \tag{4}$$

for some controlling variable t. As above, we identify recall and fallout with $F_R(t)$ and $F_N(t)$ respectively. We note that

$$\frac{dF_R(t)}{dt} = -f_R(t)$$

As the density function f is always positive, this expression is negative. The condition can be expressed as:

$$\frac{1}{f_R(t)} \frac{df_R(t)}{dt} > \frac{1}{f_N(t)} \frac{df_N(t)}{dt} \tag{5}$$

throughout the range of t. We can now test this condition on a number of the pairs of distributions that have been proposed for modelling scores. For each distribution, we need the function

$$g(t) = \frac{1}{f(t)} \frac{df(t)}{dt} \tag{6}$$

derived from its density function f, and then we can compare $g_R(t)$ and $g_N(t)$.

4.1 Two Exponential Distributions

The case of two exponential distributions (one of the models suggested [5]) is simple. The exponential density function is

$$f(t) = \frac{1}{\mu} \exp\left(-\frac{t}{\mu}\right)$$

where μ is the mean. Thus

$$g(t) = -\frac{1}{\mu}.$$

Since μ_R would be larger than μ_N, the convexity condition holds for all t.

4.2 Two Normal Distributions

Here

$$f(t) = \frac{1}{\sigma\sqrt{2\pi}} \exp\left(\frac{-(t-\mu)^2}{2\sigma^2}\right)$$

where μ is the mean and σ^2 is the variance. Thus

$$g(t) = \frac{1}{\sigma^2}(\mu - t)$$

Again, we expect μ_R to be larger than μ_N. If the two variances are equal (the first model proposed in [1]), then the convexity condition always holds. But if $\sigma_R^2 > \sigma_N^2$ (as in the example above), there will be some small value of t (or perhaps a large negative value) below which the condition is not satisfied: the reverse is the case. If $\sigma_R^2 < \sigma_N^2$, the departure from convexity occurs at the other end.

4.3 Two Poisson Distributions

This combination was suggested in [6], specifically in response to the kind of anomaly just observed. The Poisson distribution is discrete, so the analysis above based on continuous distributions does not apply. However, we can define a function analogous to $g(t)$ above, as follows:

$$g(k) = \frac{P(k+1) - P(k)}{P(k)}$$

for each integral threshold $k = 0, 1, ...$, where P is the probability of observation k. The probability function for the Poisson distribution is:

$$P(k) = \frac{\lambda^k \exp(-k)}{k!}$$

where λ is the Poisson mean, from which

$$g(k) = \frac{\lambda}{k+1} - 1$$

Once more, we expect λ_R to be larger than λ_N, so the condition is always satisfied. This is consistent with the argument in [6].

4.4 Two Gamma Distributions

This configuration is used in [2]. The density is:

$$f(t) = \left(\frac{t}{b}\right)^{c-1} \frac{1}{b\,\Gamma(c)} \exp\left(-\frac{t}{b}\right)$$

where b is the scale parameter and c is the shape parameter; the mean is bc. Thus

$$g(t) = \frac{c-1}{t} - \frac{1}{b}$$

Here if either c or b is the same for the two distributions, but the other varies in the way we would expect (higher mean for relevants), the condition is satisfied. The range of variations for which the condition is satisfied is in fact quite wide, although one could certainly construct examples which violate the condition for some t.

In fact Baumgarten's model is slightly more complex, involving shifted gamma distributions (i.e. shifted along the t-axis by a small amount).

4.5 Exponential Non-relevants and Normal Relevants

This combination is used in [4], [3] and [7], making it currently the most popular model. If we examine the formulae for $g(t)$ in sections 4.2 and 4.1, we see that in the exponential case $g(t)$ is constant, while in the normal case it declines linearly with t. Therefore there is always a t above which the condition is not satisfied. This affects the bottom left end of the recall-fallout graph, whatever the parameter values.

There is also a problem at the top right end (low t). Because the exponential is defined over the positive real numbers only, but the normal necessarily extends over the negatives as well, the curve hits the fallout=1 line below the recall=1 level. Thereafter it climbs straight up the fallout=1 line to the point (1,1). Thus this end also violates the convexity condition, again for all parameter values.

5 Discussion

5.1 Score Range

In practice, score distributions may be truncated. It is common, for example, for scores to be constrained to be positive, either as a mathematical consequence of the scoring formula or as a matter of practical convenience. Indeed, most of the above theoretical distributions are also confined to the positive real numbers, although the normal is not. Many scoring systems also, however, constrain the scores below a maximum. For example, some produce scores that are normalized to the range [0,1]. All the above theoretical distributions extend to infinity in the positive direction. This fact produces its own theoretical problems:

Should the fitted distribution be a truncated version of the theoretical one, i.e. normalised so that its integral over the truncated range is unity? This is potentially problematic, because it affects such statistics as the mean. Many authors ignore this issue – e.g. [4] considers a scoring system which produces scores in the range [0,1], but does not worry about the implied truncation.

5.2 Non-convexity

This truncation might have the side-effect of resolving the non-convexity problem, by putting the non-convex part of the curve out of effective scoring range. In the case of [4], however, the non-convexity of the normal-exponential model does affect them, and they recognise it as a problem, at least at the high-threshold end (the non-convexity at the other end is avoided by the truncation at zero). They observe that for some of their topics, the non-convexity at the high t end falls within the scoring range [0,1]. In their terms, the probability of relevance as a function of score is no longer monotonic in these cases: after a certain point it declines. They resolve the problem by redefining the probability of relevance: when the predicted function starts declining, they replace it with a straight line from the maximum reached to the point (1,1) (that is, score=1 and probability of relevance=1). They do not give any justification for this procedure, other than that one would expect probability of relevance to be a monotonic function of score.

On the basis of the above analysis of the recall-fallout graph, one could devise an alternative procedure. Since a straight line on a recall-fallout graph represents a random ordering of some set of documents, we could perform a procedure similar to that of [4] but on the recall-fallout graph. We illustrate the procedure in Fig. 5. Replacing the concave section of the curve with the straight line is equivalent to randomly reordering all documents which score in that range. This is thus a well-founded form of extrapolation.

5.3 Monotonic Transformations of the Score

One characteristic of all the above analysis is that it assigns a status to scores which they might not possess. Systems produce scores in order to rank documents, and care not at all about the scale or shape of the scoring function. Thus any monotonic transformation of a score produces a new score which is indistinguishable from the old, in terms of ranking. Factors which do not affect the rank order may be arbitrarily included or removed at any stage. This fact is often used to simplify scoring functions, or their calculation.

Thus for example some scoring functions produce numbers that are restricted to the range [0,1] because they are intended to model probabilities. Independence assumptions lead one to multiply multiple probabilities; the result is another probability. On the other hand, it is often easier to use log-probabilities (or log-odds) and add them rather than multiplying them. The resulting logarithmic (or logistic) scale looks quite different, and belongs to the range $(-\infty, 0)$ (or to $(-\infty, \infty)$). But a system using such a scale might then decide to normalise

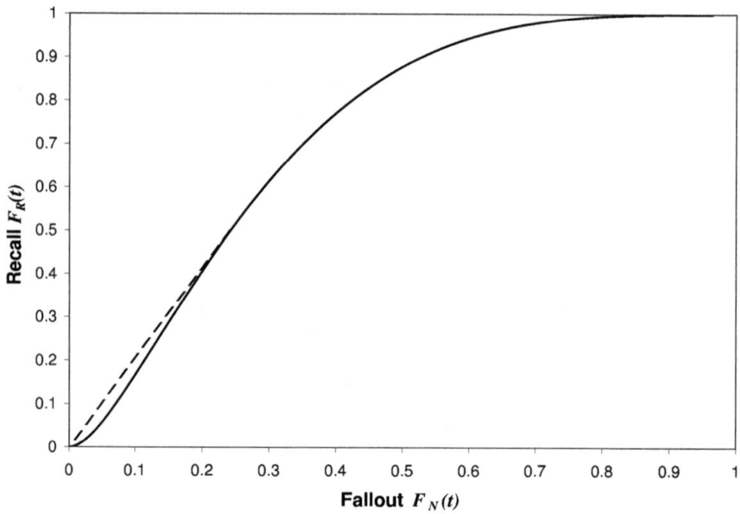

Fig. 5. Concavity at the high-threshold end

back to [0,1] *linearly*, by taking account of the observed maximum and minimum values, rather than non-linearly, by reversing the logarithmic or logistic transformation.

All such operations will drastically affect the distributions of scores, while not at all affecting the resulting ranked output or any performance curve. Thus observed distributions might depend on, in effect, accidental characteristics of the system.

6 Conclusion

We have seen that we would normally expect the recall-fallout curve to be convex in the sense defined above. That is, if we find a system which violates this condition, then the system can be improved merely by adding a randomisation process. Therefore we would at least expect good systems to satisfy the condition already.

We have seen that under some models of the distributions of relevant and non-relevant scores. models which have been proposed and/or used by researchers, this convexity condition is violated. While the violation may relate to some part of the score range which is not normally encountered, any violation seems at least to raise questions about the general validity of the distributional model under consideration.

Specifically, the model that appears to be most frequently used at present, the normal/exponential mixture, *always* violates the convexity condition at both ends of the range of theoretically possible scores. While this result does not invalidate the model as a reasonable approximation to the true distributions, it does put into question its general validity.

References

1. Swets, J.A.: Information retrieval systems. Science **141**(3577) (July 1963) 245–250
2. Baumgarten, C.: A probabilistic solution to collection fusion problem in distributed information retrieval. In Hearst, M., Gey, F., Tong, R., eds.: SIGIR'99: Proceedings of the 22nd Annual International ACM SIGIR Conference on Research and Development in Information Retrieval, New York, ACM Press (1999) 246–253
3. Arampatzis, A., van Hameren, A.: The score-distributional threshold optimization for adaptive binary classification tasks. In Croft, W.B., Harper, D.J., Kraft, D.H., Zobel, J., eds.: SIGIR 2001: Proceedings of the 24th Annual International ACM SIGIR Conference on Research and Development in Information Retrieval, New York, ACM Press (2001) 285–293
4. Manmatha, R., Rath, T., Feng, F.: Modelling score distributions for combining the outputs of search engines. In Croft, W.B., Harper, D.J., Kraft, D.H., Zobel, J., eds.: SIGIR 2001: Proceedings of the 24th Annual International ACM SIGIR Conference on Research and Development in Information Retrieval, New York, ACM Press (2001) 267–275
5. Swets, J.A.: Effectiveness of information retrieval methods. American Documentation **20** (1969) 72–89
6. Bookstein, A.: When the most 'pertinent' document should not be retrieved – an analysis of the Swets model. Information Processing and Management **13** (1977) 377–383
7. Collins-Thompson, K., Ogilvie, P., Zhang, Y., Callan, J.: Information filtering, novelty detection and named page finding. In Voorhees, E.M., Harman, D.K., eds.: The Eleventh Text REtrieval Conference, TREC 2002. NIST Special Publication 500-251, Gaithersburg, MD: NIST (2003) 107–118
8. Robertson, S.E.: Explicit and implicit variables in information retrieval systems. Journal of the American Society for Information Science **26**(4) (1975) 214–222
9. van Rijsbergen, C.J.: Retrieval effectiveness. In Voigt, M.J., Hanneman, G.J., eds.: Progress in communication sciences. Volume 1., Ablex Publishing (1979) 91–118
10. Hawking, D., Robertson, S.: On collection size and retrieval effectiveness. Information Retrieval **6** (2003) 99–150
11. Robertson, S.E.: The probability ranking principle in information retrieval. Journal of Documentation **33** (1977) 294–304
12. Robertson, S.E.: The parametric description of retrieval tests. part 1: The basic parameters. Journal of Documentation **25**(1) (1969) 1–27

Modeling Term Associations for Ad-Hoc Retrieval Performance Within Language Modeling Framework

Xing Wei and W. Bruce Croft

Center for Intelligent Information Retrieval
University of Massachusetts Amherst
140 Governors Drive
Amherst, MA 01003
{xwei,croft}@cs.umass.edu

Abstract. Previous research has shown that using term associations could improve the effectiveness of information retrieval (IR) systems. However, most of the existing approaches focus on query reformulation. Document reformulation has just begun to be studied recently. In this paper, we study how to utilize term association measures to do document modeling, and what types of measures are effective in document language models. We propose a probabilistic term association measure, compare it to some traditional methods, such as the similarity co-efficient and window-based methods, in the language modeling (LM) framework, and show that significant improvements over query likelihood (QL) retrieval can be obtained. We also compare the method with state-of-the-art document modeling techniques based on latent mixture models.

Keywords: Information Retrieval, Language Model, Term/Word Associations/ Relationships, Term/Word similarity, Document Model, Topic Model.

1 Introduction

Modeling term associations is important to Information Retrieval (IR) systems. It is well-known that ranking algorithms solely based on matching the literal words that are present in queries and documents will fail to retrieve much relevant information. For example, matching only the word "fruit" will miss the documents containing "apple" that are also relevant to "fruit". For this reason, term associations, which are also called "term relationships" or "word similarity" in literature, have been introduced to add new terms to the query/document representations that are related to the original terms. There can be associations between two single terms (term-term association); or between two groups of terms (term group association).

There has been much research in IR to associate related terms for queries and/or documents. Manual techniques such as using hand-crafted thesauri and automatic techniques such as clustering all attempt to provide a solution, with varying degrees of success. Although manual processing can usually provide precise and useful information with relatively less noise, an automatic method is expected to be more effective due to many problems related with manual processing [15], such as labor intensiveness, inconsistencies and ambiguity. Most automatic approaches to modeling term associations are based on term co-occurrence or grammatical analysis.

G. Amati, C. Carpineto, and G. Romano (Eds.): ECIR 2007, LNCS 4425, pp. 52 – 63, 2007.

Grammatical analysis is provides very specific knowledge about term relationships, but it is not as robust as using term co-occurrence [12]. Accurate but limited knowledge that provides few related terms is unlikely to substantially improve the retrieval output. Term co-occurrence has been widely used in term-association studies based on the intuition that co-occurring words are more likely to be similar, such as in term-term association models (e.g., measuring term similarity with co-efficient of two term-document vectors, which was widely used in earlier work such as term clustering [15, 23, 21] and Hyperspace Analogue to Language (HAL) [4]) and term group association models (e.g., measuring document similarity with co-efficient of two document-term vectors in document clustering [16] and Latent Dirichlet Allocation [2]). After term associations are constructed by these methods, some post-processing techniques can be applied to further improve the associations such as in [6], or to make the results compatible with systems using clustering such as in [15, 16].

With the term associations derived from previous methods, texts are reformulated (i.e. usually expanded) to improve the retrieval effectiveness. Some reformulations are not as explicit as replacing query terms with new terms, but instead the reformulation process is implicit, such as in the spreading activation techniques [22, 7, 8], in which the expansion is actually acquired during the process of following links between nodes that represent terms or documents. Both query and document reformulation processes have been investigated.

Query reformulation has been extensively studied with many term-association models in various IR frameworks [10, 21, 14, 26, 19] (In the works that phrases are considered, such as [14], we view a phrase as one term in this paper). The well-known pseudo-relevance feedback process, which expands the initial query vocabulary by adding terms contained in previously retrieved documents, is one of the best query expansion techniques in terms of retrieval performance [19]. Most relevance feedback models do term group association to find terms related to the entire query, which contains more information than individual words and thus can produce better results [21, 14]. Some query expansion techniques based on term-term associations such as [1] do post-processing to generate associations with the entire query. These query-based expansion processes have to be done online, in that they require an extra search for each query, which negatively affects query response time. Also, the efficiency of an IR system depends heavily on the number of terms of the query submitted to the system; query expansion therefore has its disadvantages in spite of the generally good retrieval results.

Document reformulation can be done offline without query inputs, thus being transparent to users and more efficient in terms of query response time. Offline processing, however, can be time-consuming and memory-expensive because it needs to process the associations of every term in every document of the entire collection, which is one of the reasons that document expansion was not popular until recent years. Two types of term associations have been applied to document reformulation: (i) Term group associations for document reformulation are usually based on documents. In the cluster-based document model [16], related documents are grouped and used to expand documents; in the LDA-based document model [24], documents are associated with related terms. Improvements have been obtained on several TREC collections with both of these two models, but they are both very expensive and difficult to apply to large collections, and parameter tuning for these models makes them even more expensive. (ii) Simple term-term association has significant advantages over term group association considering the offline efficiency of document reformulation. Cao et al. reformulate documents within the language modeling framework using term associations extracted both from a manually built thesaurus (WordNet) and from a co-occurrence based automatic technique, which considers term co-occurrence in a fixed-sized window. They achieve significant improvements over a baseline query likelihood system on some TREC collections [5], and obtain better results by further processing the original term-term associations with Markov chains [6]. The window-based approach, however, always requires an appropriate setting for the window size, and the improvements using only the automatic model are not as impressive.

Cao et al.'s work sheds light on the effectiveness of integrating term-term associations into the language modeling framework, which has been confirmed by a number of groups to be a theoretically attractive and potentially very effective probabilistic framework for studying information retrieval problems [20]. On the other hand, the assumption of the term independence ("bag of words") of the unigram language model is well known to be inappropriate for natural language. This has led many language model researchers to study term associations.

As a summary, we are interested in an automatic term-association method based on term co-occurrence in the language modeling framework, especially for dealing with document reformulation. Although term-association models have been studied for decades, none of the association processes have been performed within the language modeling framework, even that some integration processes of term associations are carried out with language models and some association processes like the window-based co-occurrence model are probabilistic methods. In this paper we study the traditional term co-occurrence based automatic term-association methods in the document reformulation task, and propose a new and simple method, which is based on the language modeling approach and thus fits within this framework naturally, to model term associations for retrieval operations.

2 Related Work

The history of examining term associations to improve retrieval effectiveness is almost as long as the history of IR itself. Since the binary term matching model, IR researchers have been trying to expand the matching of literal terms to include the matching of many other related words.

2.1 Hand-Crafted Thesauri

The earliest method of detecting and using term associations in IR was by building hand-crafted thesauri. This approach still attracts considerable interest from the IR community and open resources like WordNet and the open directory project[1] have been studied extensively [5, 6, 9].

Manual indexing has often been viewed as a gold standard and a thesaurus as a "correct" way of incorporating new words or phrases, but building and maintaining a

[1] http://www.dmoz.com/

thesaurus is very labor-intensive and it is very difficult to get people to agree on the semantic classifications involved. Inconsistencies and ambiguity in the use of these thesauri have produced poor results when they are used for retrieval experiments. Also, it is a fact that human beings tend to stick to obvious principles of classification. It is easy for human beings to group such words as "fruit" and "apple" together, but it may be difficult for them to find out that "boundary", "layer" and "flow" are related by their combined use in aerodynamic contexts [15]. Therefore, an automatic, instead of a manual approach, is expected to be more effective for improving retrieval.

2.2 Similarity Coefficient

A variety of similarity coefficients have been developed and applied to measure term associations in IR environments, such as the Cosine metric, weighted and unweighted Tamimoto [15], etc. The coefficient used in Qiu & Frei's Concept Based Query Expansion is one example [21]. They built a term-document matrix and computed the similarity between any two terms as follows;

$$SIM(t_i, t_j) = \sum_{k=1}^{n} d_{ik} \cdot d_{jk}, \tag{1}$$

$$d_{ik} = \frac{(0.5 + 0.5 \times \frac{ff(d_k, t_i)}{\max ff(t_i)} iif(d_k)}{\sqrt{\sum_{j=1}^{n} ((0.5 + 0.5 \times \frac{ff(d_j, t_i)}{\max ff(t_i)} iif(d_j))^2}}, \tag{2}$$

where $ff(d_k, t_i)$ is the frequency of term t_i in document d_k, $iff(d_k)$=log(m/| d_k|), m is the number of terms in the collections and $|d_k|$ is the number of different terms in document d_k. max $ff(t_i)$ is the maximum frequency of term t_i in all documents. The d_{ik}'s and d_{jk}'s signify feature weights of the indexing features (documents). Then, the similarity between a term and a query is defined as the weighted sum of the similarity values between the term and individual terms in the query. To expand a query, terms with the highest similarity to the query are added and the weight of each added term takes its similarity value with the original query. Significant improvements in retrieval effectiveness were reported in their paper [21].

Although many techniques in this area have been tested and some interesting results were obtained, most of the techniques have been used to do query expansion. Few studies on document modeling with term similarity coefficients have been conducted.

2.3 Co-occurrence in Windows

Another important group of term association measures estimates the conditional probability of a term given another term. Van Rijsbergen [23] and Cao et al. [5] compute the conditional probability using co-occurrence samples. To compute the conditional probability of two terms by their co-occurrence in a window is a practical method for both its simplicity and effectiveness. A fixed-sized window is applied to measure the co-occurrence in [5] and a sliding-window method (Hyperspace

Analogue to Language, HAL) is described in [4]. A typical computation of the co-occurrence probability (the strength of term association) is as follows:

$$P(t_j \mid t_i) = f(t_i, t_j) / \sum_k f(t_i, t_k) , \quad (3)$$

where $f(t_i, t_j)$ is the frequency of co-occurrences of t_i and t_j.

2.3.1 Fixed-Sized Window

A fixed-sized window is often used to measure the co-occurrence of two terms. In this window-based method, two words are considered as co-occurring once when the distance between them is less than the window size. For instance, Xu & Croft developed a metric used for query expansion based on the fixed-sized window method and achieved excellent performance [25, 26]; Cao et al. applied fixed windows in document modeling in combination with WordNet [5] and obtained significant improvements on two TREC collections.

2.3.2 Sliding Window

In addition to setting a threshold to judge the co-occurrence of terms as in the fixed-sized window method, the distance between two words are also taken into account in some term-association models, such as in [4, 11, 17, 1]. Sliding window method is one of the examples, which is also called HAL Space (Hyperspace Analogue to Language) [4, 17]. By moving a window across the text, an accumulated co-occurrence matrix for all terms is produced. Compared to the fixed-sized window method, the sliding window method takes accumulated co-occurrence in all possible fixed-sized windows and in this way, the strength of association between two words is inversely proportional to their distance. Some interesting results with the sliding window method are obtained in previous works, including query expansion tasks in the language modeling framework [1, 4, 17]. However, its effectiveness on document modeling tasks is still unknown.

In both the fixed-sized window and the sliding window methods, the size of the window is a parameter that needs to be determined.

2.4 Latent Mixture Models

Because of the success of statistical approaches to representing text, IR has the potential of benefiting from recent advances in the fields of statistical modeling and machine learning. Research in these fields has led to new mathematical models that effectively represent documents through latent mixture modeling techniques. Some of these models have also been studied in IR research with interesting results, such as the mixture of unigrams model [18] and (probabilistic) Latent Semantic Indexing ((p)LSI) [12]. The Latent Dirichlet Allocation (LDA) model [2], which possesses fully generative semantics and overcomes the drawbacks of previous latent mixture models such as pLSI, has quickly become one of the most popular probabilistic text modeling techniques in machine learning. LDA has recently been shown to outperform both the unigram document model and the cluster-based document model in the language modeling framework for IR [24].

However, latent mixture models are usually very expensive and difficult to apply on large collections. There is often no exact inference techniques for these models and approximation techniques have to be adopted to iteratively approach the solution. Parameter tuning for these complicated models makes them even more expensive. Furthermore, they require a new training process for each new collection; in contrast, term-term associations can often be used across collections.

3 Modeling Term Associations by Joint Probability

3.1 Term-Association Models

Previous research described in Section 1 and Section 2 has shown the effectiveness of modeling and integrating term associations into information retrieval processes. Especially, constructing term-term associations and integrating them into document models is an attractive way considering both of its online efficiency and large-collection feasibility. Also, the recently developed language modeling framework has opened up new ways of thinking about retrieval problems. Its solid theoretical setting and promising experimental results provide and motivate new directions of the construction and integration process of term associations. In this section, we present an approach in the language modeling framework to estimating the conditional probability of terms by joint probability through Bayesian rule, and the joint probability will be computed by unigram document models.

To get a sense of the association or closeness between two terms, w and t, we consider $P(w|t)$, which is the probability of observing w when t is given. By Bayesian rule, we have

$$P(w \mid t) = P(wt) / P(t) \ , \tag{4}$$

To estimate the join probability of observing the word w and the term t, instead of counting co-occurrence samples in windows, we assume that w and t are identical and independent samples from a unigram document model D. Then the total probability of observing w together with t is:

$$P(wt) = \sum_{D \in \Pi} P(D)P(wt \mid D) = \sum_{D \in \Pi} P(D)P(w \mid D)P(t \mid D) \ , \tag{5}$$

where $\prod$ represent some finite universe of unigram document models. We choose to use unigram priors $P(D)$ and limit the universe $\prod$ to the collection we test on. Then,

$$P(w \mid t) = \frac{\sum_{D} P(w \mid D)P(t \mid D)}{\sum_{w}\sum_{D} P(w \mid D)P(t \mid D)} \ . \tag{6}$$

Thus, for each term t, there is a list of words w with the probability $P(w|t)$ representing the association of w and t. We can view this probability as the association/closeness between w and t.

3.2 Document Language Models with Term Associations

The basic approach for using language models for IR is the query likelihood model where each document is scored by the likelihood of its model generating a query Q.

$$P(Q \mid D) = \prod_{q \in Q} P(q \mid D) \, , \tag{7}$$

where D is a document model, Q is the query and q is a query term in Q. $P(Q|D)$ is the likelihood of the document model generating the query terms under the 'bag-of-words' assumption that terms are independent given the documents. And $P(q|D)$ is specified by the document model with Dirichlet smoothing [27],

$$P_U(w \mid D) = \frac{N_d}{N_d + \mu} P_{ML}(w \mid D) + (1 - \frac{N_d}{N_d + \mu}) P_{ML}(w \mid coll) \, , \tag{8}$$

where $P_{ML}(w|D)$ is the maximum likelihood estimate of word w in the document D, and $P_{ML}(w|coll)$ is the maximum likelihood estimate of word w in the entire collection. N_d is document length. μ is the Dirichlet prior, and in our experiments we used a fixed value with μ=1000.

In the original query likelihood model, documents are estimated by the independence assumption, which is not appropriate to natural language that is much more complicated than simple "bags of words". Modeling term associations is a straightforward way to integrate related words into text models. To integrate the association information into document models, we first compute the word distribution in documents through the probabilistic association measure (Eqn (9)), and then combine it with the original term model by linear combination:

$$P_T(w \mid D) = \sum_{t \in D} P(w \mid t) P(t \mid D) \, . \tag{9}$$

It is similar to the retrieval methodology using translation models proposed by Berger and Lafferty to incorporate term associations into document language models [3]. With the translation model, the document model becomes

$$P_{TR}(w \mid D) = \sum_{t} tr(w \mid t) P(t \mid D) \, , \tag{10}$$

where $tr(w|t)$ is the translation model for mapping a document term t to an arbitrary term w. The translation probability $tr(w|t)$ describes the degree of link between a term w and the document term t. If we set $tr(w|t)$ to be $P(w|t)$, then Eqn (9) and Eqn (10) will be same.

The linear combination method is widely used in integrating related words into document models, such as in [16, 5, 24]. The final document model would be

$$\begin{aligned} P(w \mid D) &= \lambda P_U(w \mid D) + (1 - \lambda) P_T(w \mid D) \\ &= \lambda (\frac{N_d}{N_d + \mu} P_{ML}(w \mid D) + (1 - \frac{N_d}{N_d + \mu}) P_{ML}(w \mid coll)) \, . \\ &\quad + (1 - \lambda) \sum_{t \in D} P(w \mid t) P(t \mid D) \end{aligned} \tag{11}$$

where λ is the integration co-efficient. This is the only parameter to our model, and is also one of the parameters to the other models we compare to in Section 4.

In this paper we try several association measures to model $P(w|t)$ in Eqn (11), including the similarity co-efficient, the fixed-sized window method, the sliding window method, and the joint probability method we propose. In the similarity co-efficient method, we normalize its co-efficient to be consistent with the probabilistic application as following:

$$P(t_j \mid t_i) = SIM(t_i, t_j) / \sum_k SIM(t_i, t_k) \cdot \qquad (12)$$

4 Experiments and Results

4.1 Data

We conduct experiments on five data sets taken from TREC: the Associated Press Newswire (AP) 1988-90 with queries 51-150, Wall Street Journal (WSJ) 1987-92 with queries 51-100 and 151-200, Financial Times (FT) 1991-94 with queries 301-400, San Jose Mercury News (SJMN) 1991 with queries 51-150, and LA Times (LA) with queries 301-400. Queries are taken from the "title" field of TREC topics. Queries that have no relevant documents in the judged pool for a specific collection have been removed from the query set for that collection.

4.2 Parameters

There are several parameters that need to be decided in our experiments. For the retrieval experiments, the proportion of the term-association part in the linear combination must be specified (λ in (11)). For the similarity measure, the window sizes need to be determined. We use the AP collection as our training collection to estimate the parameters. The WSJ, FT, SJMN, and LA collections are used for testing whether the parameters optimized on AP can be used consistently on other collections. At the current stage of our work, the parameters are selected through exhaustive search or manually hill-climbing search. All parameter values are tuned based on mean average precision (MAP).

4.3 Experimental Results

In all experiments, both the queries and documents are stemmed, and stopwords are removed.

4.3.1 Other Term-Associating Methods

We test the effectiveness of some traditional term-term associating methods that we discussed in Section 2 in language document models, and present the retrieval results in Table 1.

Similarity co-efficient: With the parameter setting λ=0.8, which was obtained by training on the AP collection, we run experiments with the similarity co-efficient based document models (SCDM) on other collections. Some improvements, including significant improvements on one of the five collections, are achieved over query likelihood retrieval by integrating the similarity co-efficient into document models.

Fixed-sized window: With λ=0.7 and window size W=30, which were obtained by training on the AP collection, we run experiments with the fixed-sized window based document models (FWDM) on other collections. Significant improvements on two of the five collections are obtained over query likelihood retrieval.

Sliding window: Retrieval results of the document models based on the sliding window method, with λ=0.6 and W=50, are shown in Table 1. Significant improvements on two of the five collections over the query likelihood retrieval are achieved. Table 1 also shows that the sliding window performs better than the fixed-sized window, which was adopted in [5] and [6] as an automatic term associating method to be integrated into language document models.

Table 1. Comparison of query likelihood retrieval (QL) and retrieval with document models based on similarity coefficient (SCDM), fixed-sized window method (FWDM), or sliding window method (SWDM). The evaluation measure is average precision. %chg denotes the percentage change in average precision. Stars indicate statistically significant differences with a 95% confidence according to the Wilcoxon test.

Collection	QL	SCDM	%chg over QL	FWDM	%chg over QL	SWDM	%chg over QL	%chg over FWDM
AP	0.2161	0.232	+7.62*	0.2381	+10.15*	0.2375	+9.88*	-0.25
FT	0.2558	0.2652	+3.68	0.2640	+3.22	0.2690	+5.14	+1.86*
SJMN	0.1985	0.2068	+4.18	0.2118	+6.67*	0.2142	+7.86*	+1.12
LA	0.2290	0.2305	+0.62	0.2362	+3.12	0.2485	+8.48	+5.20*
WSJ	0.2908	0.2866	-1.44*	0.2827	-2.79	0.2905	-0.10	+2.76*

4.3.2 Term Associations by Joint Probability

We test document models based on the term-associating method by joint probability (JPDM) that we present, and show the retrieval results in Table 2. λ=0.6 for these experiments, and we process only the top 400 related terms of each term. On four of the five collections JPDM retrieval achieves significant improvements over query likelihood retrieval. On the WSJ collection, no improvements are achieved with λ=0.6, and then we especially tuned λ for it and obtained improvement with λ=0.2 as shown in the last line of Table 2.

In previous experiments, we build term associations for each collection respectively. To test the easy applicability of the term-associating method we present, we also run experiments with the term associations constructed only from the AP collection (JPDM-ap), or all of the five collections (JPDM-all). Results of JPDM-ap are presented in Table 2 and JPDM-all in Table 3.

JPDM-all achieves the best performance among JPDM, JPDM-all and JPDM-ap. This shows that more training data lead to higher performance, because more data can imply more knowledge about the term associations. At the same time, term associations trained only on the AP collection are also effective on other collections.

Table 2. Comparison of query likelihood retrieval (QL) and retrievals with JPDM and JPDM-ap

Collection	QL	JPDM	%chg over QL	JPDM-ap	%chg over QL	%chg over JPDM
AP	0.2161	0.2400	+11.03*	0.2400	+11.03*	0
FT	0.2558	0.2754	+7.66*	0.2636	+3.05	-4.28
SJMN	0.1985	0.2180	+9.80*	0.2139	+7.74*	-1.88
LA	0.2290	0.2516	+9.85*	0.2426	+5.91	-3.59
WSJ	0.2908	0.2870	-1.32	0.2884	-0.83	+0.49
WSJ (λ=0.2)	0.2908	0.2971	+2.15	N/A	N/A	N/A

Table 3. Comparison of query likelihood retrieval (QL) and retrievals with LBDM, JPDM, and JPDM-all

Collection	QL	LBDM	JPDM-all	%chg over QL	%chg over JPDM	%chg over LBDM
AP	0.2161	0.2629	0.2422	+12.05*	+0.92*	-7.91*
FT	0.2558	0.2795	0.2842	+11.10	+3.20	+1.68
SJMN	0.1985	0.2279	0.2186	+10.10*	+0.27*	-4.06*
LA	0.2290	0.2563	0.2547	+11.21*	+1.24	-0.63
WSJ	0.2908	0.3244	0.2910	+0.07	+1.41*	-10.30*

So, the term associations built by joint probability do not have to be trained on the specific collection of experiments.

Table 3 shows the comparison of JPDM-all and LDA-based document models (LBDM) [24]. The LBDM achieves better performance than the term association model we propose. However, based on our experiments, the term association modeling is much faster than the LDA model estimation. Also, we have shown that it is very easy and effective to apply the term associations trained on other collections, which is impossible for the LDA model training.

5 Conclusions and Future Work

We have proposed a probabilistic term association model in the language modeling framework, which measures term associations through their joint probability, and a document retrieval model that integrates term associations into document models through linear combination. We did experiments and compared the model we proposed with other popular term-association methods on ad-hoc retrieval tasks.

The experimental results showed that modeling term associations through joint probability was effective in the language modeling framework. Document models that include term associations outperformed the query likelihood model, and term associations constructed by joint probability achieved better performance than other term-association models, such as window co-occurrence methods, in the language modeling framework. Comparing the two window co-occurrence methods, the sliding window method performs better than the fixed-sized window method on the retrieval tasks. We also showed that term associations trained on other collections were effective in our model, and more training data leads to better performance.

Although the retrieval with term-associating model did not obtain improvements over the LDA-based document models [24], the results are interesting and encouraging considering the cost of LDA training.

For future work, we plan to investigate whether several association measures can be combined in one document modeling. We will also combine the term-association based document models with latent mixture model based document models and test the effectiveness of this combination. In addition, studying post-processing with the probabilistic term associations obtained from this paper would also be interesting.

Acknowledgments

This work was supported in part by the Center for Intelligent Information Retrieval and in part by NSF grant #CNS-0454018. Any opinions, findings and conclusions or recommendations expressed in this material are the authors' and do not necessarily reflect those of the sponsor.

References

1. Bai, J., Song, D., Bruza, P., Nie, J.-Y. and Cao, G.: Query expansion using term relationships in language models for information retrieval. In Fourteenth International Conference on Information and Knowledge Management (CIKM 2005).
2. Blei, D. M., Ng, A. Y., and Jordan, M. J.: Latent Dirichlet allocation. In Journal of Machine Learning Research, 3, 993-1022 (2003).
3. Berger, A. and Lafferty, J.: Information retrieval as statistical translation. In Proceedings of the 22nd annual international ACM SIGIR conference on Research and development in information retrieval, 222-229, August 15-19, 1999, Berkeley, California, United States.
4. Burgess, C., Livesay, K., and Lund, K., Explorations in Context Space: Words, Sentences, Discourse. Discourse Processes, 25(2&3), 211-257 (1998).
5. Cao, G., Nie, J.-Y., and Bai, J.: Integrating word relationships into language models. In Proceedings of SIGIR 2005, 298-305.
6. Cao, G., Nie, J.-Y., and Bai, J.: Constructing Better Document and Query Models with Markov Chains. In Proceedings of the ACM 15th Conference on Information and Knowledge Management (CIKM), November 2006, Arlington, USA.
7. Croft, W.B., Lucia, T.J., Cringean, J., and Willett, P.: Retrieving Documents By Plausible Inference: An Experimental Study. Information Processing and Management, 25, 599-614 (1989).
8. Croft, W.B. and Thompson, R.: I3R : A New Approach to the Design of Document Retrieval Systems. Journal of the American Society for Information Science, 38(6), 389-404, (1987).
9. Croft, W.B. and Wei, X.: Context-Based Topic Models for Query Modification. CIIR Technical Report, IR-424 (2005).
10. Fang, H. and Zhai, C.: Semantic Term Matching in Axiomatic Approaches to Information Retrieval. In Proceedings ACM SIGIR 2006, 115-122.

11. Gao, J.F., Nie, J.-Y., Zhang, J., Xun, E., Zhou, M. and Huang, C.: Improving Query Translation for CLIR using Statistical Models. In Proceedings of the 24th ACM SIGIR Conference on Research and Development in IR, pp. 96-104 (2001).
12. Manning, C.D., Raghavan, P., and Schütze, H.: Introduction to Information Retrieval, Cambridge University Press (2007).
13. Hofmann, T.: Probabilistic latent semantic indexing. In Proceedings of SIGIR 1999, Berkeley, CA, USA.
14. Jing, Y. and Croft, W.B.: An Association Thesaurus for Information Retrieval, In Proceedings RIAO-94, 146-160 (1994).
15. Jones, K. S.: Automatic Keyword Classification for Information Retrieval. London: Butterworths (1971).
16. Liu, X. and Croft, W.B. Liu, X., and Croft, W. B.: Cluster-based retrieval using language models, in Proceedings of SIGIR 2004, 186-193.
17. Lund, K. and Burgess, C.: Producing High-dimensional Semantic Spaces from Lexical Co-occurrence. Behavior Research Methods, Instruments,& Computers, 28(2), 203-208 (1996).
18. McCallum, A.: Multi-label text classification with a mixture model trained by EM. In AAAI workshop on Text Learning (1999).
19. Lavrenko, V. and Croft, W.B.: Relevance-based language models. In Research and Development in Information Retrieval, 120-127 (2001).
20. Ponte, J. and Croft, W.B. : A language modeling approach to information retrieval. In Proceedings of ACM SIGIR 1998 275-281.
21. Qui, Y. and Frei, H., Concept based query expansion, In Proceedings of ACM SIGIR 1993, 160-169.
22. Salton G. and Buckley, C.: On the Use of Spreading Activation Methods in Automatic Information Retrieval. In Technical Report 88-907, Department of Computer Science, Cornell University.
23. Van Rijsbergen, C. J.: Automatic Classification. In: Information Retrieval. 2nd edn. Chapter 3. London: Butterworths. (1979). http://citeseer.ist.psu.edu/vanrijsbergen79 information.html
24. Wei, X. and Croft, W.B. LDA-based Document Models for Ad-hoc Retrieval. In Proceedings of SIGIR 2006, 178-185.
25. Xu, J.: Solving the Word Mismatch Problem Through Automatic Text Analysis. Ph.D. Dissertation. Department of Computer Science, University of Massachusetts (1997).
26. Xu, J. and Croft, W.B.: Query expansion using local and global document analysis. In Proceedings of the 1996 ACM SIGIR Conference on Research and Development in Information Retrieval.
27. Zhai, C. and Lafferty, J.: A study of smoothing methods for language models applied to ad hoc information retrieval. In Proceedings of ACM SIGIR 2001, 334-342.

Static Pruning of Terms in Inverted Files

Roi Blanco and Álvaro Barreiro

IRLab, Computer Science Department
University of Coruña, Spain
rblanco@udc.es, barreiro@udc.es

Abstract. This paper addresses the problem of identifying collection dependent stop-words in order to reduce the size of inverted files. We present four methods to automatically recognise stop-words, analyse the tradeoff between efficiency and effectiveness, and compare them with a previous pruning approach. The experiments allow us to conclude that in some situations stop-words pruning is competitive with respect to other inverted file reduction techniques.

1 Introduction

Inverted files are the data structures employed by most modern retrieval systems [14] to associate index terms (words, stems, phrases, bigrams, etc...) with document occurrences. Indexes are organised into posting lists containing several pointers which carry the correspondence information. Fast query evaluation is normally done by repeatedly accessing the on-disk index file and fetching the information for every query term. Disk accessing times are the bottleneck for most retrieval systems, and there had been many solutions to improve query evaluation times without affecting retrieval effectiveness, such as lossless compression techniques [7]. More recently, a new family of lossy compression algorithms, namely *pruning*, has emerged to try to improve the efficiency while retaining high effectiveness values. Pruning techniques aim at removing unnecessary information by determining a set of non relevant pointers in each posting list and ruling them out of the retrieval. If the pointer set is dependent on each query, it is called *dynamic pruning* [13], whereas if the pruning can be made off-line it is said to be *static*. Recent works demonstrated that static pruning can produce very compact indices whilst not suffering from an unacceptable precision loss [2]. Also, this technique has been applied in web retrieval [4].

This paper presents several techniques for reducing the size of the inverted file by identifying a stop-words set dependent on the collection. The main difference between this method and the one described in [2] (hereinafter *Carmel's* method) is that the whole term is removed from the index instead of deleting single occurrences. We introduce several techniques based on the terms' *informativeness value*, in particular inverse document frequency (*idf*) and residual inverse document frequency (*ridf*), and a novel method based on the *term discriminative value.* Discarding a whole term determines that the index term is not useful in every possible

G. Amati, C. Carpineto, and G. Romano (Eds.): ECIR 2007, LNCS 4425, pp. 64–75, 2007.

context (query). Although this claim may seem too aggressive (or naive), except for a predetermined and well-known set of function words, we found out that in some scenarios these algorithms prove to be competitive or even better than the methods based on the pruning of term-document occurrences. Other works ([2],[4]) size the amount of pruning as the percentage of pointers removed from the inverted file, and in [2] Carmel et al. advanced that it is not known how static pruning would behave in conjunction with the traditional lossless compression methods, and that further research was needed in order to clarify it. This paper also presents the experiments and results assessing the relationship between the amount of pointers and the real space savings, for five well known coding algorithms. We advance a good and stable behaviour of the static pruning methods for every coding scheme tested. Experiments also report on query times in a real retrieval platform.

The rest of the paper is organised as follows: section 2 describes Carmel's method, section 3 introduces the term pruning methods, the experiments and results are presented in section 4 and the paper ends with a conclusions and further work section.

2 Static Index Pruning of Posting Entries

Carmel et al. in [2] proposed and successfully tested a method for removing information from an inverted file. The algorithm operates in a per-term basis, selecting the *less necessary* information from every single posting list in order to reduce the total index size.

There are two parameters involved in the so-called top-k pruning algorithm: k and ϵ. The procedure to select which postings are removed from the index is as follows. First, for every term in the lexicon, the algorithm computes the contribution of every document occurrence to the final score using the score function of the retrieval system. Then it retrieves the k-th highest score z_t and sets a threshold $\tau_t = \epsilon * z_t$. Finally, every document occurrence which score is lower than τ_t is dropped out from the posting list.

It is worth to point out that this is an *idealised pruning algorithm*, as the top k documents scores for a query with less than $\frac{1}{\epsilon}$ terms are guaranteed to be the same, within an error of ϵ, when the original or pruned inverted file is used. However, the algorithm has the problem of obtaining negligible pruning levels. In order to obtain any significant index reduction it is necessary to shift every document occurrence score in the term lists, by subtracting a global minimum score to every document score. The real procedure is to apply the pruning algorithm after this ad-hoc modification of the inverted file. This accomplishes excellent results but the aforementioned property is not proved to hold. As well, there is another variation of the algorithm, namely δ-top answers, that consists of keeping the entries whose score value under a query q is at least δ times the highest score of all the documents under q. The implementation we employed here considered the BM25 score [11] instead of Smart's tf-idf (used in [2]) and we decided to skip any shifting implying that higher pruning levels were obtained by setting a higher ϵ value.

3 Static Index Pruning of Term Posting Lists

Traditionally, stop word removal aims at identifying noisy terms that may hurt precision, and to the best of our knowledge it has not been used for efficiency purposes.

It is clear that removing high-frequency terms from an uncompressed inverted file may lead to substantial space savings, as they tend to engross most of the occurrences (according to Zipf's law). How this may affect to compressed inverted files is disccussed in [14]. The claim is that the higher the frequency of the word, the better a parametrised compression model such as Golomb will adapt to it, so the less space it will consume in a compressed form. In general, it is a commonly accepted idea that stop-words should be in the inverted file since removing high-frequency words would result in very small space savings. However, we believe that if it is possible to obtain a good ranking of terms according to their *importance*, it would be interesting to establish the tradeoff between retrieval accuracy and the index reduction implied by the removal of the *less important* terms. In fact, some authors [10] report that building a manual extended stop-list speeds searches. We propose to study this effect with techniques that obtain *informativeness* (3.1) and *discriminative* (3.2) rankings.

3.1 Stop-Words List Based on idf and ridf

The inverse document frequency is a term informativeness measure, therefore it can be used to produce a ranking of bad terms (those with lower idf values). We used a common *idf* normalisation introduced by Robertson and Sparck-Jones in [9] that performed well for identifying dynamic stop-words in [6]. If D is the total number of documents in the collection, and df the number of documents the term t appears in (document frequency), then the *idf* for term t is:

$$idf = \log\left(\frac{D - df - 0.5}{df + 0.5}\right) \tag{1}$$

Residual idf is defined in [3] as the difference between the observed idf (IDF) and the idf expected under the assumption that the terms follow an independence model, such as Poisson ($I\hat{D}F$). To the best of our knowledge it has not been used for identifying collection-dependent stop-words, although in [8] it is employed successfully for named entity recognition. If tf is the total number of tokens for a term t, then the ridf devised by a Poisson distribution is

$$RIDF = IDF - I\hat{D}F = -\log(\frac{df}{D}) + \log(1 - e^{-\frac{tf}{D}}) \tag{2}$$

Church and Gale [3] claim that the more a term deviates from Poisson, the more dependent on hidden variables, and more useful the term is to discriminate between documents containing it on the basis of the hidden dependencies. In order to compute the *idf* and *ridf* values for every term appearing in the collection, it is only necessary to traverse the lexicon file once.

3.2 Stop-Words List Based on Salton's Term Discrimination Model

Salton's Term Discrimination Model (TDM) [12] is one of the first computationally attractive attempts to find an effective ranking of words, based on the analysis of the *Discriminative Value* (DV) of a term and it was used for automatic indexing. The model is embodied into the vector-space framework for Information Retrieval and its use has been limited to small collections (Cranfield, Medlars, Time). However, the usefulness of the model has not been clearly stated in the following years, nor it has been applied in large TREC collections. This paper proposes to revisit the original model and to determine to which extend it may be worthy as a tool for finding stop-words.

The Term Discrimination Model measures the importance of every index term based on the influence it has on a document space. The main assumption is that a document space with distant vectors is preferable for retrieval. A good document space is one that maximises the average separation between every pair of vectors, because it would be easier to distinguish among the retrieved documents. Under this claim, and given that terms act as dimensions of the document space, it is possible to rank the index terms according to how much each term affects the *density* of the vector space, i.e. how good as *discriminators* they are. The DV of a term t is defined as how much the removal of t from the vector space decreases the total space density.

Let $\{t_1 \ldots t_T\}$ and $\{d_1 \ldots d_D\}$ be the term and the document set respectively, where every document d_i is represented by a term frequency component vector $\langle tf_{i1}, tf_{i2} \ldots tf_{iT}\rangle$. The calculation of every document-to-document distance as a measure of the space density is computationally unaffordable for very large collections. One possible variation could be a definition of the density measure related to documents-to-centroid distances. In this case, the DV for a term t_k is

$$DV_k = \sum_{i=1}^{D} \text{distance}(d_i^k, c^k) - \sum_{i=1}^{D} \text{distance}(d_i, c) = Q_k - Q, \tag{3}$$

where Q is the space density, Q_k is the space density after the term t_k is removed, d_i^k is the document obtained after removing the term t_k from d_i, c is the document centroid and c^k is the document centroid resulting after the removal of the term t_k.

A straight implementation of eq. 3 is very time consuming. For every term, it requires the computation of the similarities between every document and the centroid, forcing to traverse T times a direct file of D documents. Next it follows a reformulation of eq. 3 that allows to save most of the operations by storing some data in main memory and reducing drastically the total computation time. First, let TF_j^k (TF_j) be the j-th component of the centroid c^k (c):

$$TF_j = \tfrac{1}{D}\textstyle\sum_{i=1}^{D} tf_{ij};\ TF_j^k = TF_j \text{ if } j \neq k;\ TF_j^k = 0 \text{ if } j = k.$$

Equation 3 can be rewritten as follows, where tf_{ij}^k is the j-th component of d_i^k.

$$DV_k = \sum_{i=1}^{D}\sum_{j=1}^{T} \frac{tf_{ij}^k \times TF_j^k}{|d_i^k| \times |c^k|} - \sum_{i=1}^{D}\sum_{j=1}^{T} \frac{tf_{ij} \times TF_j}{|d_i| \times |c|}, \tag{4}$$

Let $w_i = \sum_{j=1}^{T} tf_{ij} TF_j$, which is a value that can be precomputed for each d_i, then

$$\sum_{j=1}^{T} tf_{ij}^k \times TF_j^k = \begin{cases} w_i & \text{if } t_k \notin d_i \\ w_i - tf_{ik} TF_k & \text{if } t_k \in d_i, \end{cases} \tag{5}$$

and Q_k can be expressed as

$$Q_k = \sum_{i \backslash t_k \in d_i}^{D} \frac{w_i - tf_{ik} TF_k}{|c^k| \times |d_i^k|} + \sum_{i \backslash t_k \notin d_i}^{D} \frac{w_i}{|c^k| \times |d_i^k|} \tag{6}$$

Taking into account that $|d_i^k| = |d_i|$ if $t_k \notin d_i$, and that $\sum_{i \backslash t_k \notin d_i} \frac{w_i}{|d_i|} = \sum_{i=1}^{D} \frac{w_i}{|d_i|} - \sum_{i \backslash t_k \in d_i} \frac{w_i}{|d_i|}$, then Q_k can be finally rewritten as:

$$Q_k = \frac{1}{|c^k|} \left(\sum_{i \backslash t_k \in d_i}^{D} \left(\frac{w_i - tf_{ik} TF_k}{|d_i^k|} - \frac{w_i}{|d_i|} \right) + \sum_{i=1}^{D} \frac{w_i}{|d_i|} \right) \tag{7}$$

Since Q is constant, the Q_k values will suffice to compute the rank produced by the TDM. The reformulation of Q_k introduced in eq. 7 allows the computation of this rank with just one single pass to a direct file to calculate the w_i and $|d_i|$ values, and another one to the inverted file to recalculate every single term contribution. If we use the cosine normalisation, then $|d_i| = \sqrt{\sum_{j=1}^{T} tf_{ij}^2}$, $|c| = \sqrt{\sum_{j=1}^{T} TF_j^2}$, implying that $|d_i^k| = \sqrt{|d_i|^2 - t_{ik}^2}$, $|c^k| = \sqrt{|c|^2 - TF_k^2}$. Finally we propose another last modification to this model, in which the contribution $\frac{1}{|c^k|} \sum_{i=1}^{D} \frac{w_i}{|d_i|}$ is dropped out from eq. 7. This factor is dominant in the final value of Q_k and very dependent on the $|c^k|$ value. This is a problem in large collections because the method is too biased for high frequency terms (concretely on the factor TF_j appearing on $|c^k|$), ranking them higher.

This efficient implementation of the Term Discrimination Model requires $2|D| + |T|$ extra pointers to store the document lengths, the w_i (for each document) and the TF_j (for each term) values. Considering 16-byte double precision floats, these amounts sum up to approximately 12 MB for the 2 Gigabyte TREC web collection.

The approach described here will be referred as *tdm1* and we denote as *tdm2* another variation that employs a term frequency normalisation factor in the fashion of BM25 [11]:

$$\hat{tf_{ij}} = \frac{(k_1 + 1) tf_{ij}}{tf_{ij} + k_1 \left((1 - b) + b \frac{len(d_i)}{avglen} \right)} \tag{8}$$

In equation 8, $len(d_i)$ stands for the number of tokens in the document d_i, $avglen$ is the average document length in the collection and we used the recommended values for $k_1 = 1.2$ and for $b = 0.75$. In the implementation of *tdm2* we considered the simplification of not recomputing the average document length every time a term is removed from the collections. Once the term frequencies are computed according to eq. 8 the process follows as described for *tdm1*.

4 Experiments and Results

4.1 Experimental Setting

We report our empirical findings using the five pruning methods described in sections 2 and 3. The evaluation tries to assess how the mean average precision (MAP) and precision at ten (P@10) vary as the number of deleted occurrences from the inverted file increases. Intentionally, we chose settings that devise high precision values in order to measure the decrease in precision when augmenting the pruning level. We used Porter's algorithm for stemming. BM25 (eq. 9) was selected as the scoring function for every method, as it has proved to be robust in the IR literature:

$$score(d,Q) = \sum_{t \in Q} \log_2 \left(\frac{D - df_t + 0.5}{df_t + 0.5} \right) \frac{(k_1+1)tf}{K+tf} \frac{(k_3+1)qtf}{k_3+qtf} \tag{9}$$

where qtf is the frequency of the term in the query, $K = k_1((1-b)+b\frac{d_l}{avgl})$, and d_l and $avgl$ are the document and average document length respectively. The recommended values [11] are: $k_1 = 1.2$, $k_3 = 1000$ and $b = 0.75$.

We experimented with TREC topics from 401 to 450 in the LATimes and WT2g collections, short queries (title) and long queries (title plus description). Note that the narrative field was discarded as it hurts precision using these settings. Regarding to Carmel's method, the k value was set to 10, and the different pruning levels were obtained by modifying ϵ.

For the TDM-based methods, another condition was taken into account in order to smooth the correlation between the frequency range and the discrimination value. We introduced a document frequency threshold based on the size the collection: only terms with document frequency in the collection greater than 400(2000) where pruned for the LATimes(WT2g) collection.

A second class of experiments try to assess the real tradeoff between the pruning level and the disk space occupied by the inverted file, using different posting-list compression methods. We experimented with five different coding algorithms [7] for the document pointers: three non-parametrised methods (γ, δ, *variable byte*), a local parametrised method (*Golomb coding*), and a context-sensitive method (*interpolative coding*). Within-document term frequencies were coded with unary code, except for the case of variable byte where they were coded with variable bytes as well.

Finally, a third experiment measured the real query time performance of the system for one term-based method (*ridf*) and Carmel's method, to try to determine the final speedup effect of pruning on a retrieval platform.

Indexing and retrieval was carried out using the Terrier IR platform[1] v1.0.0, developed at the University of Glasgow. The pruning and compression program suite was implemented on top of it.

[1] http://ir.dcs.gla.uk/terrier

4.2 Precision vs. Pruning

Figures 1 to 4 show MAP and P@10 results for the LATimes collection, for both short and long queries. The precision curves end when the number of terms deleted forces any query to be empty. In general, all the methods that prune terms are able to increase initial MAP and P@10 values. Overall, *tdm1* achieved the highest values in precision with a pruning level around 20%-30%. If Fox's stop-list [5] is applied the results are: MAP 0.2695(0.2524) and P@10 0.2933(0.2911) for long(short) queries at a 26.7% pruning level. The best values achieved with the *tdm1*, MAP 0.2839(0.2544) and P@10 0.3224 (0.3022), are better than those attained by Fox's stop-list. Term pruning methods present a good behaviour at certain levels, being ridf remarkably stable and smooth and *tdm1* very good at increasing precision, although at the cost of being too aggressive. The other two methods, *tdm2* and *idf* are very correlated and perform slightly worse than ridf for most of the cases.

Tables 1 and 2 summarise the results for the WT2g collection. Results are analogous to the ones obtained in the LATimes, although short queries benefit more from precision gains. It is remarkable that Carmel's method is able to improve P@10 values in the WT2g collection at very high pruning levels (short queries only).

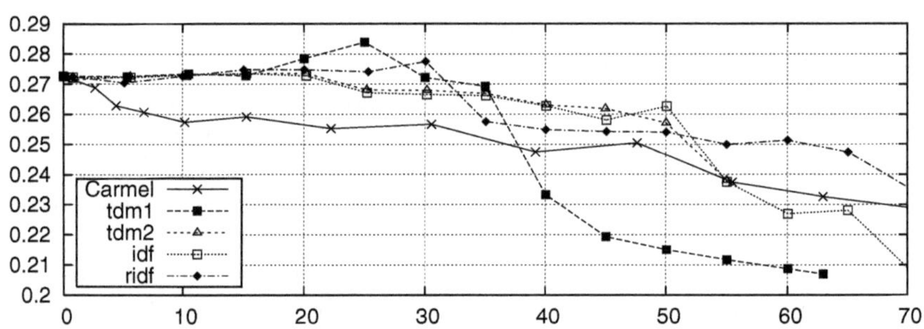

Fig. 1. MAP vs. %pruning for LATimes & long queries

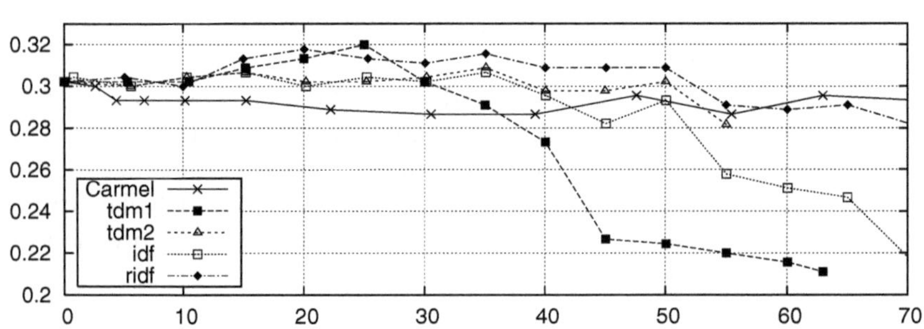

Fig. 2. P@10 vs. %pruning for LATimes & long queries

Table 1. Precision vs. %pruning WT2g & long queries

		pruning									
		0%	10%	15%	20%	25%	30%	40%	50%	60%	65%
tdm1	MAP	0.2966	**0.3006**	**0.3062**	**0.3074**	0.2892	0.2704	0.2473	0.2151	0.2054	–
	P@10	0.4780	0.4780	0.4780	**0.4860**	**0.4660**	0.4460	0.3980	0.3440	0.3143	–
tdm2	MAP	0.2966	0.2985	0.2942	0.2925	0.2755	0.2741	0.2602	0.2436	0.2166	0.2054
	P@10	0.4780	0.4620	0.4600	0.4680	0.4360	0.4400	0.4080	0.3780	0.3583	0.3208
idf	MAP	0.2966	0.2987	0.2945	0.2928	0.2749	0.2733	0.2599	0.2410	0.2163	–
	P@10	0.4780	0.4640	0.4620	0.4700	0.4380	0.4380	0.4100	0.3760	0.3204	–
ridf	MAP	0.2966	0.3000	0.3050	0.2970	**0.2922**	**0.2962**	**0.2881**	**0.2625**	**0.2325**	**0.2322**
	P@10	0.4780	**0.4800**	**0.4880**	0.4640	0.4600	**0.4640**	**0.4560**	**0.4320**	**0.3760**	**0.3653**

		pruning									
		0%	9.3%	14.0 %	19.1%	24.2%	30.3%	39.4%	52.1%	58.1 %	66.0%
Carmel	MAP	0.2966	0.2789	0.2779	0.2712	0.2634	0.2591	0.2606	0.2405	0.2283	0.2188
	P@10	0.4780	0.4480	0.4460	0.4540	0.4440	0.4480	0.4400	0.4280	0.4220	0.4200

Every method presented needs to set some threshold in order to *stop pruning*, be it the ϵ parameter (Carmel's method) or the percentage of pruning (term pruning methods). We carried out a third experiment in order to find an automatic threshold using Fox's stop-list as *relevance* information, i.e. good stop-words. The procedure is as follows: the list of terms is sorted according to a first measure and split into several intervals bounded by the *relevant* (trusted) stop-words. For every term and using a second measure, its informativeness value v_1 and the value of the lower bound of its corresponding interval v_2 are compared. If $v_2 \geq v_1$ the term is pruned. Combining the *ridf* (first) and *tdm2* (second) measures this approach gives, for long(short) queries, MAP values of 0.2685(0.2490) and P@10 values of 0.3044(0.2889) at a 56% pruning level in the LATimes collection. These precision values are obtained automatically and comparable with the ones obtained by Fox's stop-list alone, but at a higher pruning level.

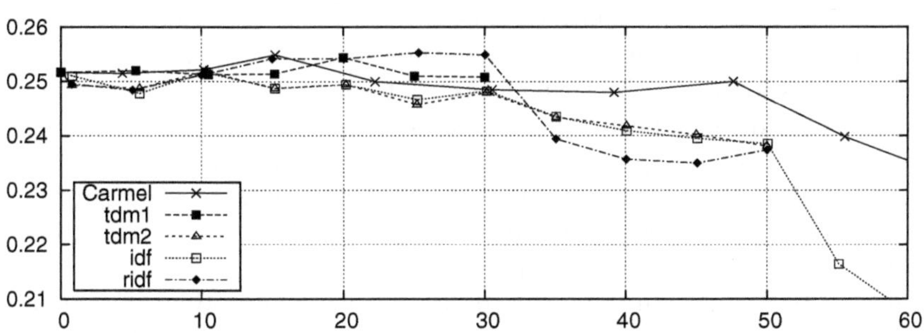

Fig. 3. MAP vs. %pruning LATimes & short queries

4.3 Index Compression vs. Index Pruning

Figure 5 shows the real tradeoff between pruning level and disk space usage (WT2g collection). The graphs reflect how the inverted file size decreases when the number of pruned pointers increases using different coding methods. Sizes are

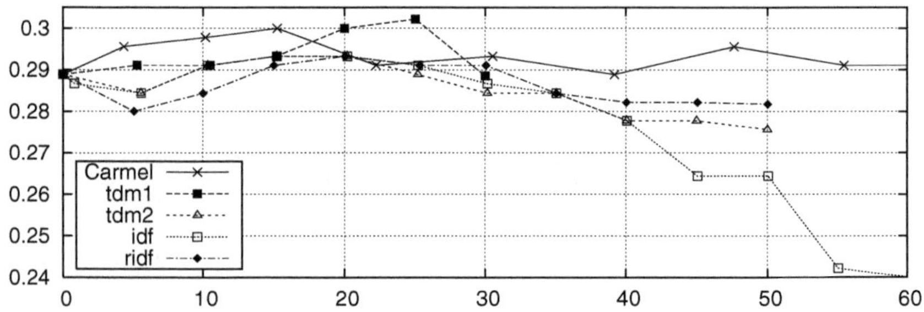

Fig. 4. P@10 vs. %pruning LATimes & short queries

Table 2. Precision vs. %pruning WT2g & short queries

		pruning									
		0%	10%	15%	20%	25%	30%	35%	40%	50%	55%
tdm1	MAP	0.2540	**0.2688**	**0.2719**	**0.2661**	0.2524	0.2470	–	–	–	–
	P@10	0.4180	**0.4540**	**0.4560**	**0.4620**	**0.4480**	**0.4271**	–	–	–	–
tdm2	MAP	0.2540	0.2635	0.2641	0.2600	0.2498	0.2490	0.2393	0.2351	0.2172	–
	P@10	0.4180	0.4360	0.4360	0.4300	0.4040	0.4060	0.3800	0.3620	0.3553	–
idf	MAP	0.2540	0.2635	0.2644	0.2602	0.2503	0.2495	0.2408	0.2351	0.2172	0.2109
	P@10	0.4180	0.4380	0.4380	0.4300	0.4080	0.4040	0.3780	0.3600	0.3480	0.3163
ridf	MAP	0.2540	0.2619	0.2640	0.2636	**0.2594**	**0.2572**	**0.2524**	**0.2509**	**0.2333**	**0.2254**
	P@10	0.4180	0.4400	0.4360	0.4204	0.4143	0.4204	**0.4020**	**0.3939**	**0.3633**	**0.3653**
		pruning									
		0%	9.27%	14.0%	19.1%	24.2%	31.0%	39.6%	45.2%	52.1%	58.9%
Carmel	MAP	0.2540	0.2634	0.2632	0.2622	0.2606	0.2558	0.2548	0.2526	0.2397	0.2301
	P@10	0.4180	0.4360	0.4360	0.4360	0.4380	0.4420	0.4400	0.4500	0.4580	0.4500

relative with respect to the original inverted index except in the last graph, where the size is absolute. Only the posting list file is considered since the space reduction due to the lexicon file is not significant. The behaviour is stable for every compression algorithm, which proves that measuring the pruning level as the number of deleted occurrences is a valid indicator of the final compressed file, despite of the coding method used. The best reduction is obtained for the method based on *ridf* although with minor differences. The final figure shows the relative performance of the different coding algorithms, measured in megabytes (pruning values obtained with *ridf*).

It is possible to explain the values in figure 5 as follows. Real coding of posting lists is based on document gaps. A document gap is the difference between two consecutive document identifiers in the same list. For a given term with consecutive document identifiers a, b, c the cost of coding its postings would be $\phi(b-a)+\phi(c-b)$ and for bit-based coding methods $\phi(x) = O\left(\log(x)\right)$. Carmel's method may prune the document occurrence with identifier b resulting in a coded posting list reduction from $\log(b-a)+\log(c-b)$ to $\log(c-a)$. Methods that prune every term occurrence do not leave this $\log(c-a)$ gap in the posting list when they operate, as they remove the whole list, thus they may yield less average bits per gap values. The first slope in the graphs is due to the fact that the

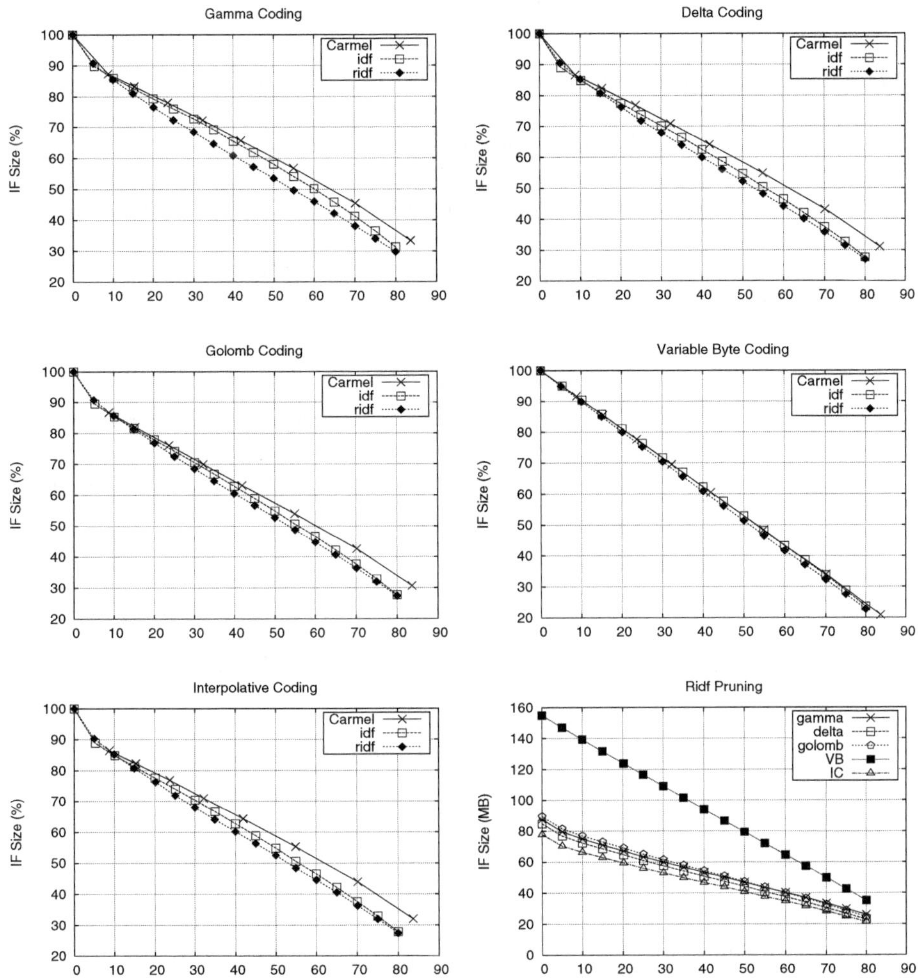

Fig. 5. Effect on Inverted File size vs. %pruning

first terms in being pruned are the ones with highest document frequency, which happen to be the ones with the highest within-document term frequencies. When those frequencies are coded in unary ($\phi(x) = x$) the space saved when they are removed is more noticeably. In fact, if the frequencies are coded with gamma, the slope softens. It is interesting to notice that Carmel's method follows this behaviour too, which indicates that if ϵ is low, it is only able to delete occurrences of terms with high document frequency.

In the case of variable byte coding, 90% of the pointers require just one byte and therefore there is no noticeable difference among the methods. Variable byte is clearly the worst method with respect to inverted file size, although it is interesting because of its faster decompression times.

4.4 Query Times vs. Pruning

Figure 6 reports on average query times for *ridf* and Carmel's method on the LATimes collection with fifty queries (topics from 401 to 450). There is a query processing time reduction which is more important in the case of long queries. The different behaviour between the methods is due to the number of disk accesses, main bottleneck for query evaluation in retrieval systems. Every query term is processed if the inverted file is pruned with Carmel's method, and this is the reason why query processing time varies smoothly with respect to the pruning level. In the *ridf*-based pruning method, query processing times can be drastically reduced at pruning levels that maintain or even improve the precision values.

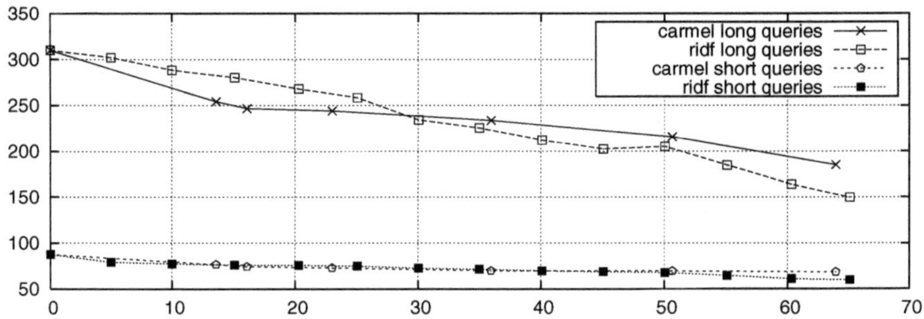

Fig. 6. Average query processing time (ms) vs. %pruning

5 Conclusions and Future Work

We implemented several pruning techniques based on the *informativeness* and *discriminative value* of terms. We also evaluated the behaviour of precision with respect to pruning, and the final effect in index file reduction and query processing times. Those methods have been compared with the well-known pruning method introduced by Carmel et al. [2]. We found out that *tdm1* is good if only high values of precision are desired, although it is very aggressive, and *ridf* is easy to implement and very stable. In general, pruning whole terms is better for maintaining or improving MAP, and it keeps precision values at high pruning levels with long queries, whereas pruning pointers is better with respect to P@10. In particular, Carmel's method behaved very well for P@10 and short queries in the WT2g collection. Therefore, methods that prune terms could be useful in applications such as indexing collections for PDAs and mobile devices, and desktop search.

One future research line is to design a pointer-based pruning method that operates selectively over posting lists, driven by a global term rank. Another topic of research is to address the problem of pruning while allowing for phrasal queries. None of the methods presented here is appropriate for processing phrasal queries. To tackle these problems it is necessary to develop an explicit pruning method for this purpose [4] or to combine a pruned inverted file with a next-word index [1].

Acknowledgements. The work reported here was co-funded by SEUI and FEDER under project MEC TIN2005-08521-C02 and "Xunta de Galicia" under project PGIDIT06PXIC10501PN. Roi Blanco is supported by a grant of DXID of the "Xunta de Galicia". We also thank the support of the "Galician Network of NLP&IR" (2006/03).

References

1. D. Bahle, H. Williams, and J. Zobel. Efficient phrase querying with an auxiliary index. In *Proc. of ACM SIGIR* 2002, pages 215–221.
2. D. Carmel, D. Cohen, R. Fagin, E. Farchi, M. Herscovici, Y. Maarek, and A. Soffer. Static index pruning for information retrieval systems. In *Proc. of ACM SIGIR* 2001, pages 43–50.
3. K. Church and W. Gale. Poisson mixtures. *Natural Language Engineering*, 2(1):163–190, 1995.
4. E. S. de Moura, C. F. dos Santos, D. R. Fernandes, A. S. Silva, P. Calado, and M. A. Nascimento. Improving web search efficiency via a locality based static pruning method. In *Proc. of WWW* 2005, pages 235–244.
5. C. Fox. A stop list for general text. *SIGIR Forum*, 24(1-2):19–21, 1990.
6. R.T.W. Lo, B. He, and I. Ounis. Automatically building a stopword list for an information retrieval system. In *Proc. of DIR'05*, Utrecht, Netherlands, 2005.
7. A. Moffat and A. Turpin. *Compression and Coding Algorithms.* Kluwer Academic Publishers, Norwell, MA, USA, 2002.
8. J. D. M. Rennie and T. Jaakkola. Using term informativeness for named entity detection. In *Proc. of ACM SIGIR* 2005, pages 353–360.
9. S. Robertson and K. Sparck Jones. Relevance weighting of search terms. *JASIS*, 27:129–146, 1976.
10. S. E. Robertson and S. Walker. Okapi/Keenbow at TREC-8. In *Text REtrieval Conference*, pages 151–162, 2000.
11. S. E. Robertson, S. Walker, M. Hancock-Beaulieu, A. Gull, and M. Lau. Okapi at TREC-4. In *Text REtrieval Conference*, pages 21–30, 1996.
12. G. Salton, C. S. Yang, and C. T. Yu. A theory of term importance in automatic text analysis. *JASIS*, 26(1):33–44, 1975.
13. H. Turtle and J. Flood. Query evaluation: Strategies and optimizations. *IP&M*, 31(6):831–850, 1995.
14. I. H. Witten, A. Moffat, and T. C. Bell. *Managing Gigabytes: Compressing and Indexing Documents and Images.* Morgan Kaufmann Publishers, San Francisco, CA, 1999.

Efficient Indexing of Versioned Document Sequences

Michael Herscovici[1], Ronny Lempel[2], and Sivan Yogev[2]

[1] Google Inc., Haifa, Israel (work done while at IBM)
[2] IBM Haifa Research Lab, Israel

Abstract. Many information systems keep multiple versions of documents. Examples include content management systems, version control systems (e.g. ClearCase, CVS), Wikis, and backup and archiving solutions. Often, it is desired to enable free-text search over such repositories, i.e. to enable submitting queries that may match any version of any document. We propose an indexing method that takes advantage of the inherent redundancy present in versioned documents by solving a variant of the multiple sequence alignment problem. The scheme produces an index that is much more compact than a standard index that treats each version independently. In experiments over publicly available versioned data, our method achieved compaction ratios of 81% as compared with standard indexing, while supporting the same retrieval capabilities.

1 Introduction

In many business applications, information systems keep multiple versions of documents. Examples include content management systems, version control systems (e.g. ClearCase and CVS), Wikis, and backup and archiving solutions. Email, where each reply or forward operation in a thread often repeats some previously sent content, can also be seen as having evolving document versions. Often, it is desired to enable free-text search over such repositories, i.e. to enable submitting queries that may match any version of any document. A straightforward way to support free-text search over corpora of versioned documents is to index each version of each document separately, essentially treating the versions as independent entities. However, intuition suggests that when each version is not significantly different than its predecessor, the redundancy of the data can be exploited to index the data in a more compact manner, while still preserving the retrieval functionality supported by the full index.

This paper proposes an indexing method that eliminates much of the redundancy present in versioned documents and produces an index that is much more compact than a standard index that treats each version independently. The scheme involves first solving a unique variant of the multiple sequence alignment problem, and then indexing just once every token of the alignment that is common to a sequence of versions. Experiments over publicly available versioned data showed that our compact index is considerably smaller than the corresponding standard index, while supporting the same retrieval capabilities.

G. Amati, C. Carpineto, and G. Romano (Eds.): ECIR 2007, LNCS 4425, pp. 76–87, 2007.

The rest of this paper is organized as follows: Section 2 surveys related work. Section 3 assumes that an alignment of multiple sequences is given, and shows how to index those sequences efficiently while supporting search operations over the resulting index. The ensuing analysis yields an optimality criterion for the given alignment, and Section 4 presents a polynomial time algorithm that, under certain assumptions, produces an optimal alignment according to that criterion. Section 5 reports on experiments demonstrating the savings in index space that our scheme achieves on real-life versioned data. Section 6 addresses implementation details that were omitted in previous sections. We conclude in Section 7.

2 Related Work

2.1 The Inverted Index Data Structure

The inverted index (sometimes called inverted file) is the data structure of choice for full-text indexing in search engines [1, 2]. Many papers have described algorithms for building inverted indices, e.g. [2, 3, 4, 5]. An inverted index contains a *postings list* for each unique term that appears in the corpus. Each postings list consists of *posting elements*, with each element corresponding to a single appearance of the term in the corpus. A simple abstraction of a posting element is as a pair $\langle d, o \rangle$ that represents the *location* - the document id and the offset within the document - of the occurrence[1]. The elements within each postings list are typically sorted by increasing locations. Another important structure in an inverted index is the *lexicon*, or *dictionary*, which is a lookup table that for each term t in the corpus, points to the postings list corresponding to t.

The description above implies that the number of posting elements in the index equals the total number of terms appearing in the corpus. Our goal is to reduce this factor when there is inherent redundancy in the set of indexed documents. This problem was addressed in 1992 by Anick and Flynn [6], who identified corporate helpdesk corpora as a domain with evolving data. In the proposed indexing solution, changes to the indexed documents are saved in time-stamped "delta entries" consisting of the added/deleted text. This enables performing "historical queries", where the index is rolled back to a certain point in time and the query is evaluated for that time. The complexity of historical query evaluation depends on the number of rollback actions, and therefore performing a query on the entire history of the index could be inefficient. Recently, Broder et al. [7] showed how to index certain redundant content in an efficient manner, by identifying content that can be indexed just once and logically *shared* among several documents. Their method was applied to two domains - the indexing of near-duplicate Web pages (e.g. as identified by the shingling method described in [8]) and the indexing of threaded email and newsgroup discussions. For example, in the latter case, the model in [7] assumes that when replying to or forwarding an email m, the entire content of m is kept in an uninterrupted fashion in the reply. Under that somewhat limiting assumption, the content of each message in the

[1] Indexing the offsets within the document is required for efficient support of phrase queries and for proximity-based ranking of search results.

thread is thus contained in all "downstream" messages (replies and forwards that include it), and can be shared with those messages. In a sense, this work extends [7] by indexing redundant content in a compact manner without any restrictions or limiting assumptions on the structure of the corpus. We note that in a different domain, Ferragina et al. [9] proposed a compression scheme for XML that preserves the ability to efficiently search and navigation the data.

2.2 Sequence Comparison

Methods to compare sequences of symbols (e.g. the Levenshtein edit distance [10] and others) are of great importance in textual applications (spell checking, data compression, coding theory), molecular biology (DNA, RNA and protein sequences analysis), and many other fields. A widely studied problem is the *longest common subsequence* (LCS) problem (see [11] for a review), where given two strings, we look for a subsequence of both with maximal length. An efficient algorithm for this problem, proposed by Myers [12], is implemented by the UNIX *diff* command [13], and included in GNU diff 1.15. For strings of lengths m and n, its complexity is $O((n+m)D)$, where D is the minimal number of insertion/deletion operations required to transform one string into the other. As with many comparison problems, the general LCS problem is NP-Hard [14].

3 Efficient Indexing of Aligned Sequences

This section shows how to efficiently index sequence of versioned documents. We begin by defining an alignment of a single sequence of versions, and transforming the alignment into an inverted index. Subsequently, 3.1 shows that the index supports the basic operators that are required in free-text search engines. Subsection 3.2 analyzes the factors that influence the size of the index, and derives an optimality criterion on the alignment matrix that will guide us to an alignment algorithm in Section 4.

Definition 1. *Let $s_1, s_2, \ldots, s_n$ be n sequences over alphabet Σ. An $(n+1) \times L$ matrix M is an exclusive alignment of $s_1, \ldots, s_n$ if:*

1. *The first row of M (row 0) is a supersequence of length L of $s_1, \ldots, s_n$.*
2. *All other entries of M are binary.*
3. *For each row $i = 1, \ldots, n$, one can reconstruct s_i by concatenating the symbols in $M_{0,j}$ s.t. $M_{i,j} = 1$.*

For example, Equation 1 shows an exclusive alignment matrix of the sequences

$$s_1 = \text{ABCDEF},\ s_2 = \text{ABXEFY},\ s_3 = \text{XCDEFY},\ s_4 = \text{ZBXCDFY}$$

$$M = \begin{pmatrix} Z & A & B & X & C & D & E & F & Y \\ 0 & 1 & 1 & 0 & 1 & 1 & 1 & 1 & 0 \\ 0 & 1 & 1 & 1 & 0 & 0 & 1 & 1 & 1 \\ 0 & 0 & 0 & 1 & 1 & 1 & 1 & 1 & 1 \\ 1 & 0 & 1 & 1 & 1 & 1 & 0 & 1 & 1 \end{pmatrix} \qquad (1)$$

Let there be a sequence of n versioned documents $d_1, \ldots, d_n$, where each document is represented as a sequence of words (tokens). Let M denote an alignment matrix of the documents, where each column corresponds to a token (Section 4 will explain how to derive M).

From the alignment we derive $\binom{n+1}{2}$ *virtual* documents $\{v_{j,i}, 1 \le i \le j \le n\}$ as follows: each virtual document $v_{j,i}$ will contain the tokens in row 0 of M, corresponding to columns having a maximal run of 1s that starts at row i and ends at row j. Furthermore, the virtual documents will be ordered by lexicographic ordering of the pair $\langle j, i\rangle$, i.e. primarily by increasing values of the end of the run of 1s, and within all runs ending at a particular index j, by increasing index of the beginning of the run. To exemplify this process, Equation 2 depicts the transformation of the alignment shown in Equation 1 into 10 virtual documents (consider each letter to be a token):

$$\begin{pmatrix} Z & A & B & X & C & D & E & F & Y \\ 0 & 1 & 1 & 0 & 1 & 1 & 1 & 1 & 0 \\ 0 & 1 & 1 & 1 & 0 & 0 & 1 & 1 & 1 \\ 0 & 0 & 0 & 1 & 1 & 1 & 1 & 1 & 1 \\ 1 & 0 & 1 & 1 & 1 & 1 & 0 & 1 & 1 \end{pmatrix} \Longrightarrow \begin{array}{ll} 1.\ v_{1,1} = \text{CD} & 6.\ \ v_{3,3} = \text{(empty)} \\ 2.\ v_{2,1} = \text{AB} & 7.\ \ v_{4,1} = \text{F} \\ 3.\ v_{2,2} = \text{(empty)} & 8.\ \ v_{4,2} = \text{XY} \\ 4.\ v_{3,1} = \text{E} & 9.\ \ v_{4,3} = \text{CD} \\ 5.\ v_{3,2} = \text{(empty)} & 10.\ v_{4,4} = \text{ZB} \end{array} \quad (2)$$

The process above transformed a single group of 4 (physical) versioned documents into 10 virtual documents. Given k sequences of versioned documents

$$d_1^1, \ldots, d_{n_1}^1,\ \ d_1^2, \ldots, d_{n_2}^2,\ \ \ldots,\ \ d_1^k, \ldots, d_{n_k}^k$$

we construct $N \triangleq \sum_{i=1}^{k} \binom{n_i+1}{2}$ virtual documents, and order them as follows:

$$v_{1,1}^1, \ldots, v_{n_1,n_1}^1,\ \ v_{1,1}^2, \ldots, v_{n_2,n_2}^2,\ \ \ldots\ ,\ \ v_{1,1}^k, \ldots, v_{n_k,n_k}^k$$

We now simply build the inverted index that corresponds to the N virtual documents, assigning them documents identifiers (docids) $1, \ldots, N$. In addition,we also require four predicates per virtual document $X = \text{docid}(v_{j,i}^k)$ in the index:

$$\text{from}(X) = i,\ \ \text{to}(X) = j,\ \ \text{root}(X) = \text{docid}(v_{1,1}^k),\ \ \text{last}(X) = \text{docid}(v_{n_k,n_k}^k)$$

3.1 Supporting the Various Search Operators

Given an inverted index with N virtual documents, this section presents an efficient document-at-a-time algorithm [15] to support basic Boolean search queries, consisting of required and/or forbidden terms[2]. The algorithm returns a list of virtual documents, corresponding to physical documents that contain all the required terms and none of the forbidden terms. Using the predicates *root*, *from* and *to*, we will map the resulting virtual documents to the corresponding

[2] For ease of presentation, this section deals only with Boolean queries. Section 6 presents extensions for supporting TF/IDF scoring.

original versioned documents. To simplify our algorithm, we swap every forbidden term t with a virtual required term (denoted by *neg(t)*) that "virtually" appears in all the documents in which the forbidden term does not appear, and only in those. Formally then, a query Q is a set of size $|Q|$ of required terms (real and virtual), $t_1, \ldots, t_{|Q|}$.

We denote by p_t the current position within the postings list of the term t; p_t is often called the *cursor* of term t in the IR literature. Next, we define a few primitive functions used in our algorithm. First, the primitive *next*(p_t, docid) sets p_t to the first virtual document in t's postings list whose id is greater than *docid* (or to ∞ if no such document exists) and returns that document id. Figure 3 shows how to implement the next function for the virtual term *neg(t)* using t's cursor p_t. Second, the function *location*(root, from, to) returns the id of the virtual document corresponding to the range [*from*, *to*], given the id of the virtual *root* document (corresponding to the range [1,1]) of a group of versioned documents. This can simply be calculated as follows:

$$\text{location(root, from, to)} = \text{root+(from-1)+}\binom{\text{to}}{2}$$

Third, the function *overlapDocsRange*(docid1, docid2) returns the id of the virtual document that corresponds to the intersection of the ranges represented by *docid1* and *docid2*, or ∞ if the two ranges do not intersect.

Our algorithm uses a modification of the *zig-zag join* [16] technique, in which the cursors of all required terms (real or virtual) are advanced in alternating order, until they align at some document id. That document then contains all of the terms, i.e. matches the query. At each step of a zig-zag join, a cursor that lags behind the most advanced cursor is chosen, and is advanced using the *next* operator to a point at or beyond the most advanced cursor. In our case, we slightly modify the classic zig-zag join, since cursor positions do not necessarily need to align at some virtual document, but rather on a set of virtual documents whose ranges intersect. Figure 1 presents the standard outer shell document-at-a-time evaluation. The *nextCandidate* function, depicted in Figure 2, performs the zig-zag join and returns the virtual document id representing the next range

```
function search(Query Q)
   // we assume that the cursors of all physical terms are initialized to position 0
   candidate ← 0
   while candidate ≠ ∞ do
      // Find a virtual document containing all required (real or virtual) terms
      candidate ← nextCandidate(candidate, Q)
      output candidate
   end while
end function
```

Fig. 1. Enumerating all documents that match the query Q

```
function nextCandidate(docid, Query Q)
  // advance t_1 beyond the last document in docid's range
  nextd ← next(p_t1, location(root(docid), to(docid), to(docid)) )
  align ← 2
  // perform a zig-zag join on ranges of virtual documents
  while (align ≠ |Q| + 1) ∧ (nextd ≠ ∞) do
    // advance term t_align to or beyond the beginning of nextd's range
    temp ← next(p_t_align, location(root(nextd), 1, from(nextd)) − 1)
    // surely now to(temp)≥from(nextd)
    if (root(temp) == root(nextd)) ∧ (from(temp) ≤ to(nextd)) then
      nextd ← overlapDocsRange(nextd, temp)
      align ← align + 1
    else
      nextd ← next(p_t1, location(root(temp), 1, from(temp)) − 1)
      align ← 2
    end if
  end while
  return nextd
end function
```

Fig. 2. Zig-zag join beyond a given start-point argument

on which all cursors intersect. Finally, Figure 3 presents the implementation of the *next* function of the virtual cursor that corresponds to a negative term. For ease of presentation, an easily eliminated tail recursion is used there.

Theorem 1. *The algorithm shown in Figure 1 outputs a virtual document if and only if the range of physical documents corresponding to it match the query, i.e. contain all the required terms and none of the forbidden terms.*

The proof follows in the spirit of [7], with some additional interval algebra over the virtual document identifiers (e.g. the *overlapDocsRange* function).

3.2 Implications on Alignment Optimality

The size of a classical inverted index depends on three main factors. First, the size of the lexicon. Since the vocabulary of the index does not change with our approach, the lexicon's size does not vary between standard indexing and our proposed scheme. Second, the total number of posting elements in the various postings lists. This factor is reduced by our scheme, since instead of having a posting element per token per document, the number of posting elements in our scheme is equal to the total number of runs of 1 in the columns of the alignment matrix. The third factor is the compressibility of each postings list. Postings lists are typically compressed by some form of *delta encoding*, i.e. compression of the numeric gaps between successive document identifiers and offsets therein [2, 3]. Since our scheme enlarges the space of document identifiers by transforming

```
function next(p_{t=neg(w)}, docid) // Invariant: from(docid) always equals to(docid)
   if docid ≥ p_w then
      p_w ← next(p_w, docid)
   end if
   target ← docid + 1
   // we now know that to(p_w) is at or beyond to(target)
   if (p_w = ∞) ∨ (root(p_w) > root(target)) then
      // return the id corresponding to the range that starts at to(target)
      // and continues until the end of target's version group
      p_t ← location(root(target), to(target), to(last(target)))
      return p_t
   end if
   // here we know that p_w and target share the same root
   if from(p_w) > to(target) then
      // return the id corresponding to the range [to(target),from(p_w)-1]
      p_t ← location(root(target), to(target), from(p_w) − 1)
      return p_t
   end if
   // the range of p_w immediately follows docid; we therefore apply tail recursion
   p_t ← next(p_t, location(root(target), to(p_w), to(p_w)))
end function
```

Fig. 3. Advancing a virtual cursor of a negative term

each group of n versioned documents into $\binom{n+1}{2}$ virtual documents, the postings lists on virtual documents should not compress as well as postings lists on the physical documents, introducing some overhead into our scheme. In addition to the above factors, our scheme requires additional overhead external to the index due to the four predicates required for all virtual documents.

In this paper, we do not attempt to adapt the compression of the postings lists to better fit our indexing scheme. Thus, the reduction of index space will depend on our ability to align sequences of versioned documents in a manner that minimizes the number of runs of 1 in the columns of the alignment matrix.

4 Indexing-Optimal Alignment of Multiple Sequences

As argued in Section 3.2, the degree of freedom in our scheme that affects the size of the resulting inverted index is the number of runs of 1 in the alignment matrix. Thus, we aim to find an alignment matrix for a set of versioned documents that has as few runs of 1 in its columns as possible. For two strings, finding an alignment with a minimal number of runs of 1 is identical to the LCS problem described in Section 2.2, since each alignment column includes a single run of 1, and the optimal LCS solution minimizes the number of columns. As with LCS, the generalization to aligning N strings is NPH, since even if all rows of the alignment matrix are given (which, in our case, they are not), finding the permutation of the rows that minimizes the number of runs of 1 in the columns is a known NP-Hard problem called *Consecutive Blocks Minimization* [14].

We thus limit ourselves to cases where the document versions evolve in a linear fashion, i.e. the versions do not branch. In such cases, it is natural to assume that each version is relatively close to the previous one, and so we heuristically align the documents (and order the corresponding rows of the matrix) according to the temporal order of the documents' creation. We hereby prove that by restricting the space of possible alignment matrices to those where row i corresponds to version i, an alignment that minimizes the number of runs of 1 can be found by a greedy polynomial-time algorithm.

Theorem 2. *Let $S = (s_1, s_2, \ldots, s_n)$ be an ordered set of sequences, and denote the length of s_i by l_i. Let the* **cost** *of an exclusive alignment M ($c(M)$) be the number of runs of* 1 *in M, and assume that M^* is an exclusive alignment of S with minimal cost. Then:*

$$c(M^*) = \sum_{i=1}^{n} l_i - \sum_{i=2}^{n} lcs(s_{i-1}, s_i)$$

where $lcs(s_i, s_j)$ denotes the length of the LCS of sequences s_i, s_j.

Proof: The number of runs of 1 is the total number of 1 in the matrix, minus the number of occurrences of 1 below another 1. The total number of 1 in every exclusive alignment of S is $\sum_{i=1}^{n} l_i$. Assume by contradiction that

$$c(M^*) < \sum_{i=1}^{n} l_i - \sum_{i=2}^{n} lcs(s_{i-1}, s_i)$$

This means that the number of occurrences of 1 below another 1 is greater than $\sum_{i=2}^{n} lcs(s_{i-1}, s_i)$, and so there exists an index $2 \leq i \leq n$ such that the number of 1s in row i of M^* which occur below 1s in row $i-1$ is greater than $lcs(s_{i-1}, s_i)$. This contradicts the definition of LCS, and therefore

$$c(M^*) \geq \sum_{i=1}^{n} l_i - \sum_{i=2}^{n} lcs(s_{i-1}, s_i)$$

To show that $c(M^*) = \sum_{i=1}^{n} l_i - \sum_{i=2}^{n} lcs(s_{i-1}, s_i)$, we construct an exclusive alignment matrix M' of $(s_1, \ldots, s_n)$ as follows: initialize M' with s_1 as the supersequence in row-0, and add an all-1 row. Now, iteratively for $i = 2, \ldots, n$, add row i according to the pairwise LCS alignment of s_{i-1} and s_i. This involves adding columns to M' corresponding to indices where s_i contains a symbol that is not part of the LCS. That symbol is added to the supersequence in row-0, and below that we add 0 for rows $1, \ldots, i-1$ and a 1 in row i. This process ensures that the number of runs of 1 starting in row i is $l_i - lcs(s_{i-1}, s_i)$, and so

$$c(M') = \sum_{i=1}^{n} l_i - \sum_{i=2}^{n} lcs(s_{i-1}, s_i),$$

completing the proof. □

According to Theorem 2, given an ordered set of versioned documents, a greedy algorithm constructs an exclusive alignment matrix whose number of runs of 1 is minimal. Furthermore, the complexity of finding M is polynomial, and is equivalent to solving $n-1$ pairwise LCS problems. Finally, we state a criterion on $\{s_1, s_2, \ldots, s_n\}$ where, when met, guarantees that aligning them in natural order is indeed optimal:

Theorem 3. *Let $s_1, \ldots, s_n$ be strings satisfying that for each $i = 2, \ldots, n$,*

$$lcs(s_i, s_{i-1}) \geq lcs(s_i, s_{i-2}) \geq \ldots lcs(s_i, s_1) .$$

Then, there exists an exclusive alignment of $s_1, \ldots, s_n$ whose cost is no larger than that of any exclusive alignment on any permutation of those strings.

5 Experiments

This section presents experimental results for publicly available text repositories where document versions evolve in a linear fashion. The repositories tested were (1) 4323 documents corresponding to versions of 222 Wikipedia entries of countries, and (2) 2055 documents corresponding to versions of 142 MediaWiki PHP source files. In both cases, at most 20 versions of each specific resource (Wikipedia entry or PHP source file) were crawled.

Our indexing code was written in Java, using the LCS algorithm of the Diff class [17] and version 1.9.1 of the Apache Lucene search library[3]. The Original documents were tokenized using Lucene's StandardTokenizer; for each set of versions, the token sequences were exclusively aligned by the algorithm described in Section 4 and virtual documents were constructed from the resulting alignment. We processed the sequence of virtual documents through one additional change of discarding all empty virtual documents (to which the alignment assigned no tokens). This required adding two predicates for translating between the non empty documents' identifiers and the full virtual identifier space, and performing slight changes to the functions in Section 3.1 . Next, the original documents and the non-empty virtual documents were indexed separately by Lucene.

To optimize the compressibility of the Lucene index over the original documents, we added those documents to the Lucene index by version sets, and within each version group by increasing version numbers. Since Lucene sorts its postings lists by document insertion order, this process naturally clusters terms in consecutive documents, improving the compressibility of the postings lists.

For each of the two repositories, we measured two compression ratios:

Alignment ratio: the number of tokens in the set of virtual documents, divided by the number of tokens in the set of original documents. This ratio examines the reduction in the number of posting elements in the version-aware index.

Index ratio: The size (in bytes) of the Lucene index resulting from the virtual documents plus the overhead needed for the required predicates, divided by the size of the index of the original versioned documents.

[3] `http://lucene.apache.org/`

Table 1. Experimental setup - input documents and resulting compression ratios

Repository	No. original version sets	No. original documents	No. virtual documents (non-empty)	Alignment ratio	Index ratio
Wikipedia countries	222	4323	45138 (7019)	7.46%	12%
MediaWiki source files	142	2055	19144 (4240)	8.63%	18.5%

Table 2. Indexing time measurements

Repository	Indexing time orig. docs	Alignment time orig. docs	Indexing time virt. docs	Overall virt. docs
Wikipedia countries	125 sec	295 sec	61 sec	356 sec
MediaWiki source files	33 sec	116 sec	32 sec	148 sec

The results of the experiments on the different repositories are given in table 1. Note the savings of over 81% in actual index space are achieved in both cases.

Next, we turned to measure indexing time. In order to assess the affect of producing and indexing virtual documents on the index building time, Table 2 reports the following per each repository: (1) the time to index original texts (each version separately), (2) the time to align the original texts, and (3) the time to index the resulting virtual documents. The sum of the latter two represents the total time needed to process the versioned texts into a version-aware compact index. As expected, the overall time for version-aware indexing exceeds the original indexing time by far, and is dominated by the alignment time which is expected to remain dominant even if further optimized. However, it seems that as the collection size grows, the actual indexing time of the non-empty virtual documents somewhat compensates for the time spent performing the alignments. To fully characterize this behavior, more experiments and analysis are needed.

We did not perform extensive experiments comparing the query-time performance of the version-aware index to that of the standard index over the physical documents. However, since the algorithmic overhead of our scheme beyond the standard zig-zag join involves very few operations (see Section 3.1) while the amount of required I/O is greatly reduced due to the much smaller index, we anticipate that the runtime performance of our index scheme will be on par with or better than that of a standard Lucene index over all physical documents.

6 Implementation Issues

Scoring documents: Section 3.1 showed that our indexing scheme supports Boolean queries, i.e. can identify documents that match Boolean predicates on their contents. However, in order to rank the matching documents by relevance, search systems must enumerate the occurrences of all query terms in each matching document. To support such ranking, whenever a virtual document

$v^k_{to,from}$ representing the range [from,to] of version group k is returned from the *nextCandidate* function of Figure 2, we score the to-$from$+1 "real" documents represented by that range. We stream through the postings lists of all positive query terms, starting from virtual document $v^k_{from,1}$ and ending at $v^k_{to,to}$, attributing each term occurrence within those virtual documents to the corresponding "real" documents.

Suppoting exact-phrase queries: The indexing scheme described in Section 3 aligns documents according to the words they contain; the alignment then governs the distribution of the words among several virtual documents. This process does not preserve word co-occurrence patterns - words that appear next to each other in a certain version may be assigned to different virtual documents. Thus, the indexing scheme as presented does not support exact-phrase queries or proximity-based ranking considerations. To overcome this limitation, one may align the versioned documents by sentences: in tokenized documents, each sentence is hashed into an integer value, transforming documents into integer sequences. Those integers are then aligned, assigned to virtual documents, and transformed back to sentences. This variant of our scheme keeps sentences intact, enabling to support exact-phrase queries (and proximity based scoring) within sentence boundaries.

Note that indexing documents that were aligned by sentences will result in lesser space savings as compared to documents aligned by individual words, since any change in a sentence between version i and $i+1$ causes the entire sentence to be associated with another virtual documents. Consequently, when computing the exclusive alignment by sentences, the cost of starting a new run of 1 in some column equals the number of words in the sentence associated with that column. Therefore, the optimality criterion of the alignment becomes the weighted sum of runs of 1, where the weight of each run in column j is the number of tokens in sentence j. Our row-alignment algorithm adapts to this by applying the Needleman-Wunsch algorithm [18] instead of the LCS algorithm.

7 Conclusions and Future Work

This paper presented a scheme to efficiently index corpora consisting of sequences of document versions. Such sequences naturally arise in content management systems, in email, in code repositories and in Wikis. The indexing scheme avoids re-indexing certain units that repeat within the versions but still supports search operations as if the entire text of all versions was indexed. Using a variant of the multiple sequence alignment problem, we showed a greedy, polynomial-time algorithm that optimally solves the problem under some natural assumptions. We validated our approach on two real-life corpora consisting of sequences of versioned documents, where we demonstrated savings of 81% in index space as compared to a standard approach that indexes all versions of each document. The limitations of our scheme are that (1) it increases indexing time, and therefore is only applicable in cases where such an increase can be tolerated, and (2) as in [7], it assumes that the index over the versions is built in batch mode, i.e the indexing process accepts a batch of version sequences and builds an index for searching over them.

For future work, we intend to explore supporting tree-like evolution patterns of versioned documents (i.e. to support branches). Also, while batched indexing may suffice for many applications (e.g. archiving), we intend to investigate *incremental* indexing schemes for versioned data.

References

[1] Arasu, A., Cho, J., Garcia-Molina, H., Paepcke, A., Raghavan, S.: Searching the web. ACM Transactions on Internet Technology **1**(1) (2001) 2–43

[2] Witten, I., Moffat, A., Bell, T.: Managing Gigabytes. second edn. Morgan Kaufmann Publishers, Inc., San Francisco, CA (1999)

[3] Baeza-Yates, R., Ribeiro-Neto, B.: Modern Information Retrieval. ACM Press / Addison Wesley, New York, NY (1999)

[4] Heinz, S., Zobel, J.: Efficient single-pass index construction for text databases. JASIST **54**(8) (2003) 713–729

[5] Melnik, S., Raghavan, S., Yang, B., Garcia-Molina, H.: Building a distributed full-text index for the web. In: Proc. 10th International World Wide Web Conference (WWW 2001), ACM Press (2001) 396–406

[6] Anick, P.G., Flynn, R.A.: Versioning a full-text information retrieval system. In: Proc. 15th Annual International ACM SIGIR Conference on Research and Development in Information Retrieval. (1992) 98–111

[7] Broder, A.Z., Eiron, N., Fontoura, M., Herscovici, M., Lempel, R., McPherson, J., Qi, R., Shekita, E.J.: Indexing of shared content in information retrieval systems. In: Proc. 10th International EDBT Conference. (2006) 313–330

[8] Broder, A.Z., Glassman, S.C., Manasse, M.S.: Syntactic clustering of the web. In: Proc. 6th International WWW Conference. (1997)

[9] Ferragina, P., Luccio, F., Manzini, G., Muthukrishnan, S.: Compressing and searching xml data via two zips. In: Proc. 15th International World Wide Web Conference (WWW'2006). (2006) 751–760

[10] Levenshtein, V.I.: Binary codes capable of correcting deletions, insertions and reversals. Soviet Physics Doklady **10**(8) (1966) 707–710

[11] Apostolico, A.: String editing and longest common subsequences. In Rozenberg, G., Salomaa, A., eds.: Handbook of Formal Languages. Volume 2 Linear Modeling: Background and Application., Springer-Verlag, Berlin (1997) 361–398

[12] Myers, E.W.: An o(ND) difference algorithm and its variations. Algorithmica **1**(2) (1986) 251–266

[13] Miller, W., Myers, E.W.: A file comparison program. Software – Practice and Experience **15**(11) (1985) 1025–1040

[14] Garey, M.R., Johnson, D.S.: Computers and Intractability, A Guide to the Theory of NP-Completeness. W.H. Freeman and Company, New York (1979)

[15] Turtle, H., Flood, J.: Query evaluation: strategies and optimizations. Inf. Process. Manage. **31**(6) (1995)

[16] Garcia-Molina, H., Ullman, J., Widom, J.: Database System Implementation. Prentice Hall (2000)

[17] Gathman, S.D.: Diff java class (2003) `http://www.bmsi.com/java/Diff.java`.

[18] Needleman, S., Wunsch, C.: A general method applicable to the search for similarities in the amino acid sequence of two proteins. J. Molecular Biology **48**(3) (1970) 443–453

Light Syntactically-Based Index Pruning for Information Retrieval

Christina Lioma and Iadh Ounis

University of Glasgow, G12 8QQ, UK
{xristina,ounis}@dcs.gla.ac.uk

Abstract. Most index pruning techniques eliminate terms from an index on the basis of the contribution of those terms to the content of the documents. We present a novel syntactically-based index pruning technique, which uses exclusively shallow syntactic evidence to decide upon which terms to prune. This type of evidence is document-independent, and is based on the assumption that, in a general collection of documents, there exists an approximately proportional relation between the frequency and content of 'blocks of parts of speech' (*POS blocks*) [5]. POS blocks are fixed-length sequences of nouns, verbs, and other parts of speech, extracted from a corpus. We remove from the index, terms that correspond to low-frequency POS blocks, using two different strategies: (i) considering that low-frequency POS blocks correspond to sequences of content-poor words, and (ii) considering that low-frequency POS blocks, which also contain 'non content-bearing parts of speech', such as prepositions for example, correspond to sequences of content-poor words. We experiment with two TREC test collections and two statistically different weighting models. Using full indices as our baseline, we show that syntactically-based index pruning overall enhances retrieval performance, in terms of both average and early precision, for light pruning levels, while also reducing the size of the index. Our novel low-cost technique performs at least similarly to other related work, even though it does not consider document-specific information, and as such it is more general.

1 Introduction

The field of Information Retrieval (IR) addresses the general problem of how to retrieve information, which is relevant to a user need, from a given repository of information, such as a document collection. Information in the document collection is represented in the form of an index, which contains statistics on term frequencies in each document and in the whole collection. An integral part of the index is the postings file, which records information on which terms appear in which documents and the term frequency statistics of these terms [7]. Usually, terms are associated with individual weights, which capture the importance of the terms to the content of each document. Term weights can be computed using various term weighting schemes. A matching function then estimates the likely

G. Amati, C. Carpineto, and G. Romano (Eds.): ECIR 2007, LNCS 4425, pp. 88–100, 2007.

relevance of a document to a query, on the basis of term weights, and the most relevant documents are identified and retrieved [11].

Very often, stopwords are removed from the index. Stopword removal is a way of pruning terms that harm retrieval performance. Generally, *index pruning* consists of replacing a full index by a smaller index, with the aim of improving system efficiency, without harming significantly retrieval performance [7]. System efficiency relates to issues such as the computational costs associated with storing large indices, or the time needed to query large indices. Retrieval performance relates to how relevant the returned results are (precision), or how many of the relevant results are returned (recall). Typically, system efficiency benefits from index pruning, because pruned indices tend to be more economical to store, and less time-consuming to query. Overall retrieval performance tends to decrease after index pruning, although early precision can be enhanced, for moderate or light pruning [4,10].

We present a light index pruning technique, which uses shallow syntactic evidence to prune low-content terms from the index, at indexing time. Specifically, this shallow syntactic evidence consists of blocks of part of speech (*POS blocks*), which are induced from a corpus. Any general corpus, such as a test collection, can be used. Firstly, a POS tagger maps every term in the corpus to a part of speech. We define a *POS block* as a fixed-length block of parts of speech, which we extract from text in a recurrent and overlapping way. For example, for a given sentence ABCDEFGH, where parts of speech are denoted by the single letters A, B, C, D, E, F, G, H, and where POS block length = 4, the POS blocks extracted are ABCD, BCDE, CDEF, DEFG, and EFGH. We extract all possible POS blocks from the corpus, without considering the order in which POS blocks occur, or the documents in which they occur. It has been shown that low-frequency POS blocks correspond to low-content words [5], unlike the case of individual words, where low-frequency words tend to be high in content [6]. On this basis, we hypothesise that pruning the words corresponding to low-frequency POS blocks from an index corresponds to eliminating content-poor words, and may enhance retrieval performance. In order to test the validity of this hypothesis, we take the following steps. Firstly, we POS tag a corpus and extract POS blocks from it. Secondly, we set a cutoff threshold θ, which controls which POS blocks from the corpus are used for pruning the index. Thirdly, we POS tag the collection to be indexed, and remove from it the words which correspond to POS blocks bounded by θ. With regards to which POS blocks from the corpus are used for index pruning, we test two different strategies, which we call *Rank A* and *Rank B*, respectively. *Rank A* considers the raw frequency of a POS block as indicative of the content salience of the words corresponding to that POS block. More simply, it assumes that low-frequency POS blocks correspond to low-content words. Using this strategy, POS blocks are frequency-sorted, and the θ least frequent POS blocks are used for index pruning. *Rank B* considers both the frequency of a POS block, and the part of speech classification (*POS class*) of its components, as indicative of the content salience of the words corresponding to that POS block. More simply, it assumes that low-frequency POS blocks that contain non content-bearing parts of speech, such as prepositions for example, correspond to low-content words. Using this strategy, POS blocks are sorted according to their frequency and member parts of speech, and the θ least frequent POS blocks are used for index pruning. Note that this is a low-cost approach, because it simply requires running the collection through the POS tagger once at indexing time, and is not born down by query or document-centric parameters.

This paper is organised as follows. Section 2 presents related work. Section 3 presents in details our proposed syntactically-based index pruning technique. Section 4 presents and discusses our experiments using *Rank A* and *Rank B* strategies. Section 5 summarises our findings and states intended future work.

2 Related Studies

Index pruning is used in IR to improve system efficiency, without harming significantly retrieval performance [4,7]. Typically, the data pruned from the index is estimated to be the least important to retrieval performance, according to some relevance criteria [2,4,7,10]. Index pruning is uniform when it is applied to all the documents in the same way, regardless of document- or term-specific criteria. A detailed overview of index pruning methods is given in [4]. In the same study, Carmel et al. investigate uniform and term-based index pruning methods, and report that early precision is not affected by moderate pruning, unlike average precision, which seems to decrease approximately linearly with the amount of data pruned. An alternative to pruning terms from an index, is replacing the documents in the index by their respective summaries [2,10]. Brandow et al. show that summary indexing improves precision at the cost of a large loss in recall [2]. This claim is also supported by Sakai and Sparck Jones, who report that moderate summary indexing does not affect early precision [10]. Overall, the consensus seems to be that light or moderate index pruning does not decrease significantly early precision, but may decrease average precision. As long as retrieval performance is not significantly hurt by index pruning, pruning techniques are applied, driven primarily by system efficiency gains [7].

The syntactically-based index pruning technique we present, differs from the above work in two ways. Firstly, our aim is not to mainly improve system efficiency, but also to enhance retrieval performance. Hence, we only and solely apply light pruning, wishing to validate solely the applicability of our novel syntactically-based pruning technique, and not its detailed effect upon general system efficiency, even though we do report on the index compression resulting from our pruning techniques, as compression in index is typically related to gains in efficiency. Secondly, the pruning criteria we use are not lexical, but exclusively shallow syntactic. More simply, our pruning criteria do not relate to words, but parts of speech. Also, our pruning technique does not use relevance weight metrics, or other document-specific criteria, to decide which terms to prune. The only two criteria used are the frequency of a POS block in a corpus, and the POS class of the members of a POS block, namely whether they are nouns, verb, prepositions, and so on. In this respect, our technique is novel, generic, and document-independent. Note that, in literature, restrictions are usually

Table 1. Open class parts of speech and their % frequency in WT10G

Part of Speech (POS)	POS Class	% in WT10G
Noun	open	38
Adjective		8
Verb		7
Participle		4

introduced in the pruning strategies. For example, in [4], terms are pruned only from the postings list, while in [10], there is a minimum length for a summary. On the opposite, our application includes no such restrictions. We prune terms from all the data structures of the index, and also allow for documents to have all of their terms pruned. Applying such restrictions may further refine our technique, and lead to further improving our reported results.

3 Syntactically-Based Index Pruning

We present the steps taken in order to test the hypothesis that shallow syntactic evidence can indicate low-content terms, whose elimination from the index can enhance retrieval performance, while reducing the index size.

All words in language are syntactically classified as either open or closed class words. The open class contains nouns, verbs, and generally content-rich words, while the closed class contains prepositions, conjunctions, and generally content-poor words that mainly perform linguistic well-formedness functions, instead of bearing content. These syntactic categories of words collectively constitute the *parts of speech*. Following from [5], we consider a shallow categorisation of parts of speech, namely one that only distinguishes between 14 parts of speech, as it is enough to distinguish between content words and stop words. Table 1 displays the 4 open class parts of speech we use, out of all 14. Using a POS tagger, we extract POS blocks from a corpus[1]. Section 1 presented and illustrated how POS blocks are extracted from text. At the end of this stage, we have a list of all the POS blocks induced from the corpus. In order to turn the list of POS blocks extracted from the corpus into evidence that can be used to indicate low-content terms, we test two different strategies, namely *Rank A* and *Rank B*.

Rank A. We consider the raw frequency of a POS block in the corpus as indicative of the content salience of the words potentially associated to that POS block [5]. We sort all the POS blocks extracted from the corpus in order of raw frequency, and assume that $low-frequency \approx low-content$.

Rank B. We consider both the frequency of a POS block, and the POS class of its components, as indicative of the content salience of the words corresponding to that POS block. The open class parts of speech we use are displayed in

[1] We use WT10G, but generally, any corpus can be used.

Table 1. In order to represent this combination of frequency and POS class in a quantitative way, we introduce an estimator of content for POS blocks. This estimator approximates the potential amount of content of the words corresponding to a POS block on the basis of: (i) the POS class of the POS block components, and (ii) the length of the POS block. This content score estimator is based on two assumptions, namely that: (i) only open class parts of speech correspond to content-bearing words (see Table 1); (ii) nouns are slightly more content-bearing than adjectives, verbs, and participles. Both of these assumptions are based on linguistic intuition. Specifically, in this paper, the content score $cs_{POSblock}$ of a POS block is estimated as follows:

$$cs_{POSblock} = \frac{C_N + C_{AVP} \cdot \varrho}{l_{POSblock}} \qquad (1)$$

where C_N = number of nouns in the POS block, C_{AVP} = number of adjectives and/or verbs and/or participles in the POS block, $l_{POSblock}$ = POS block length, and ϱ = penalising parameter. ϱ is a penalising parameter applied to adjectives, verbs, and participles, following from the intuition that they are slightly less content-bearing than nouns. Using the statistics found in Table 1 as a guide, we set $\varrho = 0.17$, as follows. Adjectives, verbs and participles occur in the corpus approximately 19% (=8% + 7% + 4%) of the times, while nouns occur approximately 38% of the times. We estimate $\varrho = \frac{19/3}{38} \approx 0.17$. Using Equation (1), the content score of a POS block can be between 0 and 1, where 0 and 1 denote no content and the maximum amount of content, respectively. For example, using the here-proposed content score estimator for POS blocks, the POS blocks *noun + noun + noun + noun* and *adjective + noun + preposition + adverb* score $cs = 1$ and $cs = 0.29$, respectively.

Having established a quantitative estimator of content for POS blocks, we come back to the second strategy used to test our hypothesis. Specifically, we multiply the raw frequency and content score of POS blocks, and sort the POS blocks extracted from the corpus, according to the product of this multiplication. Multiplication is a simple way of combining frequency to content score, which practially implements our assumption, and hence is suitable for our experimentation. Other linear or log-scale combination approaches may be also used.

Strategies *Rank A* and *Rank B* are used for index pruning as follows. Firstly, identically to Section 1, we POS tag the collection to be indexed, only that this time we retain information on which POS blocks correspond to which terms in the collection. Secondly, we define a cutoff threshold θ to control the number of POS blocks used for index pruning. Then, only for those POS blocks bounded by θ, we remove from the collection the terms corresponding to those POS blocks. Terms are removed from all the data structures at indexing time. Note that the value of θ does not correspond to the actual frequency (for *Rank A*), or product of frequency and content score (for *Rank B*) of a POS block, but to the number of POS blocks which are to be used for index pruning. Note also, that, for the POS blocks selected for pruning to be low-content, we start counting θ POS blocks from the the lowest ranking to the higher ranking. For example, let us assume that *Rank A* contains 10 POS blocks, in decreasing order of frequency. Then, setting $\theta = 3$, means that the words corresponding to the 10nth, 9nth, and 8th POS blocks are pruned from the index.

4 Evaluation

4.1 Experimental Settings

We evaluate our hypothesis on WT2G (2GB) and WT10G (10GB), from the 1999, 2000 and 2001 Small Web, Web, and Adhoc tracks of the TREC Web Track, respectively, using topics 401-550 from the corresponding tasks[2]. We experiment with Title-only queries, as such queries are more representative of real user queries on the Web. During indexing, we apply stopword removal and Porter's full stemming. We select the largest of the two collections, namely WT10G, as the corpus from which we extract POS blocks, and POS tag it using the Tree-Tagger [3]. We set POS block length to $l_{POSblock} = 4$[4]. We the BM25 [9] and PL2 [1] weighting models. For all our experiments, we use the Terrier IR platform [8]. We use the default values of all weighting model parameters: (i) for BM25, $k1 = 1.2$, $k3 = 1000$, and $b = 0.75$ [9]; (ii) for PL2, $c = 10.99$ with WT2G, and $c = 13.13$ with WT10G[5]. We use default values, instead of tuning these parameters, because our focus is to test our index pruning hypothesis, and not to optimise retrieval performance. If the said parameters are optimised, retrieval performance may be further improved. We use mean average precision (MAP) and precision at 10 (P@10) to evaluate the impact of pruning on retrieval performance, using the full index as a baseline. We use a metric of similarity for the top k retrieved results, namely the *symmetric difference* [4] between the full and pruned indices, to evaluate the impact of pruning on early precision. We set k to 10, in accordance with P@10. The maximum and minimum symmetric difference scores of 1 and 0 occur when the top k results of the two indices are the same or disjoint, respectively, without considering the order of the results. In addition, we report the compression in index resulting from our pruning technique, as such compression is typically associated with gains in system efficiency.

4.2 Results and Discussion

We conduct experiments to test the hypothesis that pruning words which correspond to low-frequency POS blocks from the index, can enhance retrieval performance, at low pruning levels, using strategies *Rank A* and *Rank B*. By pruning levels, we denote the amount of data pruned from the full index.

---

[2] Information on the TREC datasets is found at: http://trec.nist.org/

[3] http://www.ims.uni-stuttgart.de/projekte/corplex/TreeTagger/

[4] We have also experimented with $l_{POSblock} = 3$ and 5, and observed little variation in retrieval performance. Generally, POS block length may vary, as long as it is kept within a range that appropriately models the adjacency of the terms in the sentence. This point is discussed in [5].

[5] Default settings for PL2 are suggested at: http://ir.dcs.gla.ac.uk/terrier/doc/dfr_description.html

Table 2. Collection statistics after pruning. *POS blocks* θ = number of POS blocks used (in multiples of 1,000). *Rank A* = POS blocks sorted according to their frequency only. *Rank B* = POS blocks sorted according to the product of their frequency and content. *tokens* = individual words pruned (% from full index). *terms* = unique terms pruned (% from full index). *postings* = document pointers pruned from the postings list (% from full index).

POS blocks θ (1,000)	% pruned from full index						collection
	Rank A			**Rank B**			
	tokens	terms	postings	tokens	terms	postings	
23	18.39	5.07	14.56	20.37	14.09	15.42	WT2G
22	12.13	3.55	9.46	14.35	12.09	10.52	
21	8.47	2.16	6.53	10.99	10.78	7.77	
20	6.00	1.46	4.57	8.78	10.31	5.98	
19	4.38	1.10	3.31	7.36	9.97	4.82	
18	3.24	0.82	2.44	6.36	9.77	4.02	
17	2.44	0.65	1.83	5.70	9.61	3.49	
16	1.84	0.50	1.37	5.24	9.50	3.13	
15	1.41	0.39	1.04	4.90	9.42	2.86	
14	1.10	0.31	0.80	4.65	9.35	2.67	
13	0.82	0.24	0.61	4.46	9.30	2.52	
23	16.57	4.68	14.30	19.93	14.69	15.23	WT10G
22	11.16	3.23	9.38	14.57	12.72	10.56	
21	7.90	1.96	6.56	11.56	11.41	7.87	
20	5.70	1.39	4.65	9.62	11.00	6.12	
19	4.24	1.08	3.41	8.27	10.64	4.96	
18	3.33	0.84	2.55	7.37	10.47	4.17	
17	2.43	0.66	1.93	6.73	10.29	3.63	
16	1.86	0.53	1.46	6.28	10.17	3.26	
15	1.44	0.40	1.11	5.96	10.10	2.99	
14	1.13	0.31	0.86	5.71	10.04	2.78	
13	0.86	0.25	0.66	5.52	9.99	2.63	

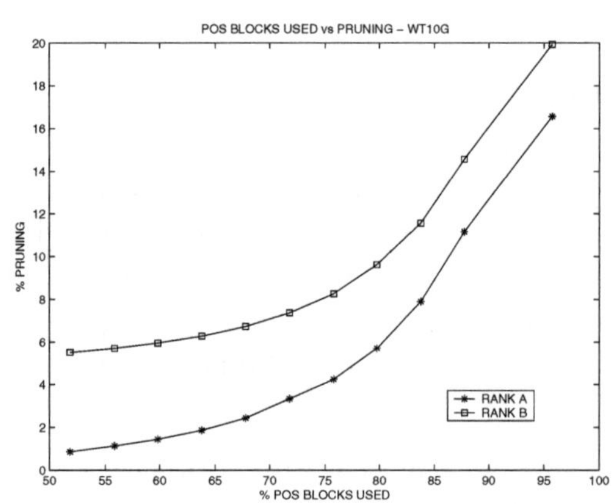

Fig. 1. % POS blocks used vs % pruning for WT10G

Table 2 displays statistics relating to the effect of each of our two pruning strategies on WT2G and WT10G, during indexing, separately for tokens, unique terms, and postings. Column *POS blocks θ (1,000)* contains the number of POS blocks used for pruning, in multiples of 1,000. We clearly see that pruning terms from more POS blocks (increasing θ) leads to more index compression (more terms being removed from the index). We also see that *Rank B* is more effective than *Rank A*, in the sense that it leads to more index compression, throughout. This is due to the fact that *Rank A* is a raw frequency sort of POS blocks, while *Rank B* is a sorting of a combination of POS block frequency and POS class information. More simply, the lowest ranked POS block in *Rank B* is not necessarily the least frequent POS block in the corpus, but a POS block that combines both (i) very low frequency, and (ii) very low content score. Very low content score practically translates to a POS block which does not contain any nouns, adjectives, or verbs. The fact that *Rank B* is better than *Rank A* is graphically illustrated in Figure 1, which plots various pruning levels against the % of POS blocks used. This % is the proportion of POS blocks used for pruning WT10G, out of all the POS blocks extracted from WT10G. We clearly observe that using the same number of POS blocks with *Rank A* and *Rank B* results in more index compression for the latter, than for the former. In Figure 1, we also observe that, even pruning the terms corresponding to 95% of all the POS blocks extracted from WT10G, only results in reducing the WT10G full index by 16-20%. Since we only use the lowest frequent POS blocks for pruning, this seems to indicate that there exists a very large number of POS blocks of low frequency, which is one of the properties of a power law distribution[6].

Table 3 displays retrieval performance scores at different index compression levels, separately for *Rank A* and *Rank A*, and for each collection. Pruning levels are reported in % reduction of postings, similarly to [4]. We see that light pruning leads to an overall improvement in MAP and P@10 over the full index, which is sometimes statistically significant. Two important observations are drawn from this table. Firstly, at no point does pruning hurt significantly retrieval. This point is very encouraging, considering that our techniques uses no document-specific criteria. Secondly, light pruning can improve both MAP and P@10. In fact, the best obtained MAP and P@10 scores for WT2G, namely MAP = 0.320 and P@10 = 0.468, are not given by the full index, but by pruning 2.86% and 1.37% of the index, respectively. Both of these scores are statistically very significant ($p << 0.01$). Similarly for WT10G, the best overall MAP and P@10 scores, namely MAP = 0.210 and P@10 = 0.328, are not given by the full index, but by pruning 0.66% and 2.99% of the index. The best overall retrieval scores are separately displayed in Table 4. Finally, in Table 3 we observe that PL2 performs better than BM25, which could be due to the default parameter settings used. Even so, both PL2 and BM25 outperform scores reported in [3,4], using TF·IDF and the same settings.

[6] Indeed we can report that the distribution of POS blocks in WT10G follows a Zipfian distribution.

Table 3. Pruning using POS blocks from Rank A and Rank B. *Prune* (%) = reduction in postings from full index. Grey-shaded = full index. Boldface = equal to or better than the full index. * and ** = stat. significance at $p < 0.05$ and $p < 0.01$ (Wilcoxon matched-pairs signed-ranks test), respectively.

Rank A					Rank B					
Prune (%)	MAP		P@10		Prune (%)	MAP		P@10		Collection
	BM25	PL2	BM25	PL2		BM25	PL2	BM25	PL2	
14.56	0.243**	0.298**	**0.432****	0.454**	15.42	0.246*	0.301**	**0.430***	0.454**	WT2G
9.46	0.250	0.303	**0.426**	**0.456**	10.52	0.254	0.308	**0.438**	**0.456**	
6.53	0.253	0.306	**0.428**	**0.466**	7.77	**0.258**	0.310	**0.432**	**0.466**	
4.57	**0.257**	0.310	**0.442**	**0.462**	5.98	0.257	0.311	**0.436**	**0.466**	
3.31	**0.259**	0.313	**0.442**	**0.460**	4.82	**0.259**	0.312	**0.434**	**0.464**	
2.44	**0.259**	0.313	**0.440**	**0.462**	4.02	**0.260**	**0.317**	**0.440**	**0.462**	
1.83	**0.260**	0.315	**0.442**	**0.464**	3.49	**0.259**	**0.319****	**0.432**	**0.462****	
1.37	**0.261***	**0.317****	**0.438***	**0.468****	3.13	**0.260**	**0.319****	**0.434**	**0.460****	
1.04	**0.261***	**0.318***	**0.438***	**0.462***	2.86	**0.258**	**0.320****	**0.434**	**0.456****	
0.80	**0.260**	**0.318**	**0.434**	**0.462**	2.67	**0.258**	**0.318**	**0.430**	0.454	
0.61	**0.260***	**0.318***	**0.436***	**0.460***	2.52	0.257*	**0.318**	**0.432***	**0.456**	
0.00	0.258	0.317	0.426	0.456	0.00	0.258	0.317	0.426	0.456	
14.30	0.175**	0.195**	0.293**	0.298**	15.23	0.175**	0.199**	0.295**	0.307**	WT10G
9.38	0.182*	0.203	**0.304***	0.307	10.56	0.179**	0.204	**0.300****	0.307	
6.56	0.185*	0.206**	**0.302***	0.316**	7.87	0.182*	0.207	**0.303***	0.313	
4.65	**0.187**	0.207	**0.300**	0.317	6.12	0.184	0.207	**0.306**	0.316	
3.41	0.186	0.206	**0.301**	0.312	4.96	0.185	0.207	**0.302**	0.325	
2.55	**0.187**	0.208*	0.298	0.319*	4.17	0.185	0.208	**0.303**	0.325	
1.93	**0.187**	**0.209**	**0.300**	0.323	3.63	0.185	**0.209**	**0.301**	**0.326**	
1.46	0.186	**0.209**	**0.302**	0.324	3.26	0.186	0.208	**0.303**	**0.326**	
1.11	**0.187**	**0.209**	**0.302**	0.324	2.99	0.186*	**0.209**	**0.303***	**0.328**	
0.86	**0.187**	**0.209**	**0.302**	**0.326**	2.78	0.186	**0.209**	**0.301**	**0.328**	
0.66	**0.188**	**0.210**	**0.303**	**0.326**	2.63	0.186*	**0.209***	**0.301***	**0.328***	
0.00	0.187	0.209	0.300	0.326	0.00	0.187	0.209	0.300	0.326	

Figure 2 plots MAP and P@10 versus index pruning. For both WT2G and WT10G, and for index compression more than roughly 6% of the full index, the relation between average precision and pruning becomes practically decreasing linear, where as pruning increases, average precision decreases, for both *Rank A* and *Rank B*. For index compression less than 6%, varying pruning leads to variations in average precision, which can be either increasing or decreasing, but only slightly. Using *Rank A* results in more variations for this pruning range (<6%), than using *Rank B*, for both collections and with both weighting models. This seems to suggest that *Rank B* is more stable. With regards to precision at 10, index compression roughly less than 8-10% of the full index seems to generally increase precision, with the exception of using PL2 on the WT10G collection. For index compression more than 10% of the full index, there is a slight degradation in early precision. Overall, *Rank A* and *Rank B* perform very similarly, in terms of retrieval performance. BM25 and PL2 also behave very similarly, indicating that our conclusions drawn from these runs are consistent across two statistically different weighting models, hence they are general.

Figure 3 plots the similarity of the top 10 results to the full index versus pruning, using a symmetric difference [4] estimation. We observe that the early precision obtained by *Rank B* approximates the full index more closely than that obtained by *Rank A*. This observation, which is consistent for both collections

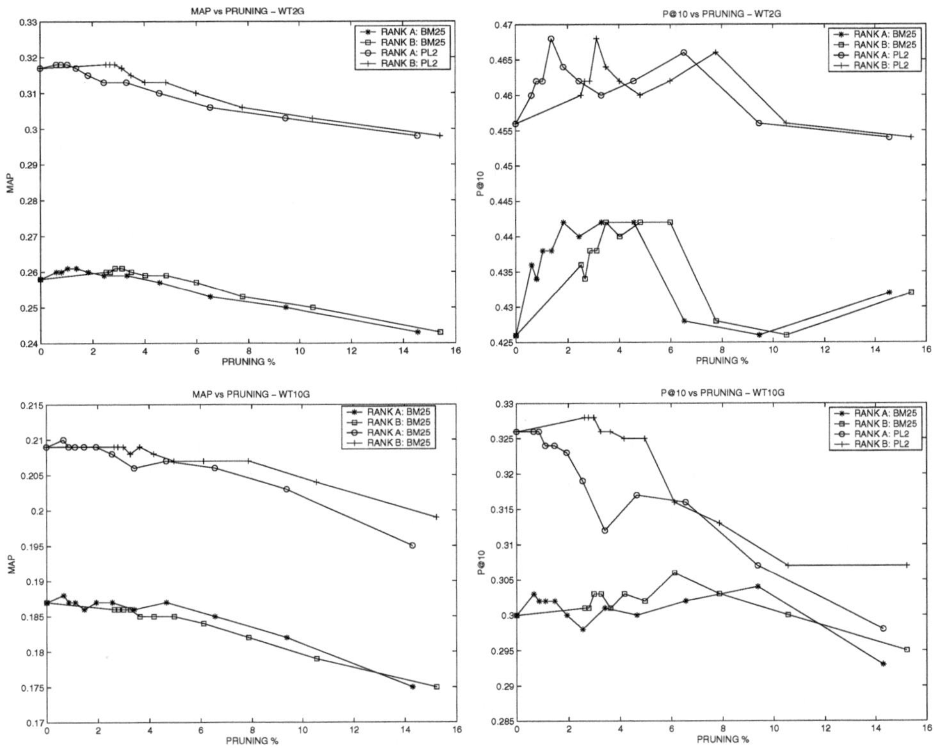

Fig. 2. Precision vs Pruning (% postings)

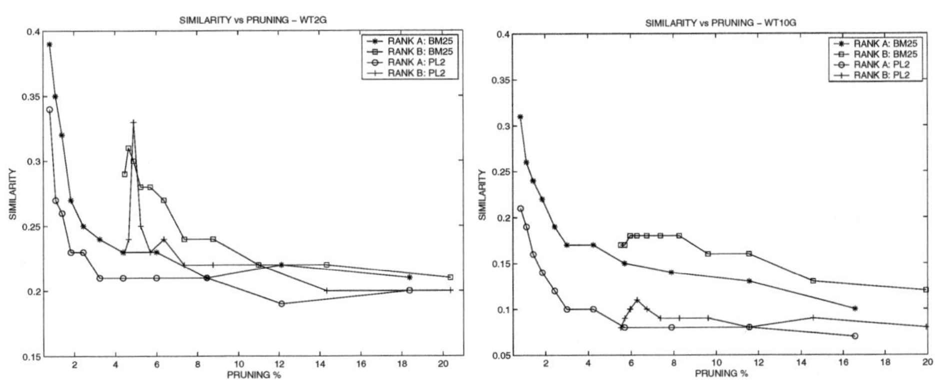

Fig. 3. Top 10 similarity at varying levels of pruning

and both weighting models, indicates that introducing the content score (Equation 1) of a POS block into the frequency ranking of POS blocks is a better pruning strategy.

Table 4. Best MAP and P@10 scores. $\Delta_F\%$ MAP & $\Delta_F\%$ $P@10$ = % difference from full index in MAP & P@10, respectively. *Prun.%* = % pruning in postings from full index. *strategy* = pruning strategy and weighting model. ** = stat. significance at $p < 0.01$ (Wilcoxon matched-pairs signed-ranks test).

best MAP	Δ_F% MAP	Prun. %	strategy	best P@10	Δ_F % P@10	Prun. %	strategy	coll
0.320**	+1.0	2.86	Rank B: PL2	0.468**	+2.6	1.37	Rank A: PL2	WT2G
0.210	+0.5	0.66	Rank A: PL2	0.328	+0.6	2.99	Rank B: PL2	WT10G

Table 5 compares the performance of syntactically-based index pruning to other index pruning work [3,4]. We compare with reported experiments that use the same collections, topics, and similar pruning levels to ours (under column *experimental settings*). We see that, both *Rank A* and *Rank B* pruning strategies are at least comparable to the uniform pruning strategy of Carmel et al. [4], (marked ⊕). Note that we prune terms using document-independent syntactic evidence (Section 1), and from the whole index, while [4] prune terms according to their contribution to the relevance score of a document, and only from the postings lists. On the basis of these two key-differences, we consider the fact that our technique is comparable to that of [4], as very promising. Table 5 also includes the results of pruning reported in [3], (marked ⊗), whereby, in addition to using document-specific information and pruning the postings lists only, a term-based strategy is used. Our equivalent run applies a uniform, as opposed to term-based, technique, which is generally considered less effective [3,4]. Still, we observe no significant difference between the two runs, a fact which is an additional credit to our technique. When we repeat this run using the exact 50 topics used in [3], (marked ⋄), we observe that our technique outperforms that of [3] in P@10, with a slight decrease in MAP. We consider this performance notable, considering how much more refined is the pruning approach applied in [3], as already discussed.

Table 5. Comparison of our runs to other index pruning work (grey-shaded). $\Delta_F\%$ MAP & $\Delta_F\%$ $P@10$ = % difference in MAP & P@10 from full index. *Prun.%* = % pruning in postings from full index. ⊕ = run described in [4]. ⊗ = run described in [3]. ⋄ repeats the run of the preceeding row using 50, instead of 100, topics. Major differences appear in boldface.

Prun. %	Δ_F% MAP	Δ_F% $P@10$	experimental settings
15.4	-4.7	+0.9	WT2G, 401-450 (Title), Rank B, uniform prun. from all index
14.6	-5.8	+1.4	WT2G, 401-450 (Title), Rank A, uniform prun. from all index
⊕13.2	-4.0	+2.5	WT2G, 401-450 (Title), uniform prun. from **postings**
10.5	-1.6	+2.8	WT2G, 401-450 (Title), Rank B, uniform prun. from all index
⊗10.7	-1.9	0.0	WT10G, 501-550 (Title), **term-based** prun. from **postings**
10.6	-2.4	0.0	WT10G, 451-550 (Title), Rank B, uniform prun. from all index
⋄10.6	-2.9	+4.3	WT10G, 501-550 (Title), Rank B, uniform prun. from all index

5 Conclusion

We proposed a novel, low-cost, unsupervised statistical technique for index pruning, with uses shallow syntactic evidence to reduce noise from the index. We hypothesised that pruning the words corresponding to low-frequency POS blocks from an index corresponds to eliminating content-poor words, and may enhance retrieval performance. We presented POS blocks, as fixed-length blocks of parts of speech, and assumed that low-frequency POS blocks correspond to low-content words, following from [5]. On the basis of this, we tested two pruning strategies: Firstly, terms corresponding to θ low-frequency POS blocks, were pruned from the index (*Rank A*). Secondly, terms corresponding to θ low-frequency POS blocks which were also estimated to contain 'non content-bearing parts of speech', such as prepositions for example, were pruned from the index. We experimented with various values of θ, and reported on the associated effect on pruning levels and retrieval performance, while also making a note of the associated gain in index compression. Both strategies behaved similarly, with *Rank B* providing results closer to the full index, for early precision. Overall, by compressing the index up to a maximum of roughly 14-15% (see Table 3), our proposed syntactically-based pruned indices outperfomed the full indices, in terms of MAP and P@10, for both collections. Additionally, for similar index compression levels, our syntactically-based technique was shown to be comparable to [3,4], which used more refined document-specific and term-based pruning approaches. In the future, we wish to experiment with higher index compression levels, and also applying our pruning technique more intelligently, for example on a per-document basis.

References

1. Amati, G.: Probabilistic Models for Information Retrieval based on Divergence from Randomness. Phd thesis. Department of Computing Science, University of Glasgow (2003)
2. Brandow, R., Mitze, K., Rau, L.: Automatic Condensation of Electronic Publications by Sentence Selection. Information Processing and Management, 31(5). (1995) 675-685
3. Carmel, D., Amitay, E., Herscovici, M., Maarek, Y., Petruschka, Y., and Soffer, A.: Juru at TREC 10 - Experiments with Index Pruning. In: Text REtrieval Evaluation Conference (TREC 2001) 228-265
4. Carmel, D., Cohen, D., Fagin, R., Farchi, E., Herscovici, M., Maarek, Y., and Soffer, A.: Static Index Pruning for Information Retrieval Systems. In: ACM Conference on Research and Development in Information Retrieval (SIGIR 2001) 43-50
5. Lioma, C., Ounis, I.: Examining the Content Load of Part of Speech Blocks for Information Retrieval. In: Proceedings of the International Committee on Computational Linguistics and the Association for Computational Linguistics (COLING/ACL 2006)
6. Luhn, H., P.: The Automatic Creation of Literature Abstracts. (1958) 159-165
7. Witten, I. H., Moffat, A., Bell, T. C.: Managing Gigabytes: Compressing and Indexing Documents and Images. 2nd edn. Morgan Kaufmann, San Francisco (1999)

8. Ounis, I., Amati, G., Plachouras, V., He, B., Macdonald, C., Lioma, C.: Terrier: A High Performance and Scalable Information Retrieval Platform. In: ACM Conference on Research and Development in Information Retrieval Workshop on Open Source Information Retrieval (OSIR 2006)
9. Robertson, S., Walker, S.: Some Simple Approximations to the 2-Poisson Model for Probabilistic Weighted Retrieval. In: ACM Conference on Research and Development in Information Retrieval (SIGIR 1994) 232-241
10. Sakai, T., Sparck Jones, K.: Generic Summaries for Indexing in Information Retrieval. In: ACM Conference on Research and Development in Information Retrieval (SIGIR 2001) 190-198
11. van Rijsbergen, C., J.: Information Retrieval. Butterworths, London (1979)

Sorting Out the Document Identifier Assignment Problem

Fabrizio Silvestri

Institute for Information Science and Technologies
ISTI - CNR, via Moruzzi, 1, 56126 Pisa, Italy
fabrizio.silvestri@isti.cnr.it

Abstract. The compression of Inverted File indexes in Web Search Engines has received a lot of attention in these last years. Compressing the index not only reduces space occupancy but also improves the overall retrieval performance since it allows a better exploitation of the memory hierarchy. In this paper we are going to empirically show that in the case of collections of Web Documents we can enhance the performance of compression algorithms by simply assigning identifiers to documents according to the lexicographical ordering of the URLs. We will validate this assumption by comparing several assignment techniques and several compression algorithms on a quite large document collection composed by about six million documents. The results are very encouraging since we can improve the compression ratio up to 40% using an algorithm that takes about ninety seconds to finish using only 100 MB of main memory.

1 Introduction

Indexes in Web Search Engines (WSEs) are usually represented using the popular *Inverted File* (IF) data structure [15]. Given a set of documents, an IF is composed by two distinct sets: the *Lexicon* and the *Posting Lists*. The *Lexicon* represents the set of terms that can be found within the whole document set. To each term of the lexicon a *Posting List* is associated containing information (the so-called *posting*) on all the documents containing that term. For example, the index entry $< t_1; 5; 3, 4, 10, 20, 23 >$ states that term t_1 (stored within the Lexicon) appears in five documents, namely 3, 4, 10, 20, and 23. The set containing all these lists is stored within the Posting Lists section.

One of the main reasons why IFs (or one of their variations) are usually adopted in real world WSEs, is that they can be easily compressed to reduce memory occupancy. Compressing indexes in WSEs has been also proved to enhance efficiency of the retrieval process [2,11,14]. A reduction in space occupancy, in fact, usually corresponds to a better utilization of the memory hierarchy.

The majority of the techniques adopted for compressing IFs are based on their d-gapped representation [15]. Posting lists are usually scanned sequentially. For this reason, it is possible to represent those lists by taking differences among successive identifiers (with the obvious exception of the first one). This way, the

G. Amati, C. Carpineto, and G. Romano (Eds.): ECIR 2007, LNCS 4425, pp. 101–112, 2007.

previous list would be represented as $< t_1; 5; 3, 1, 6, 10, 3 >$. Encoding each integer with a technique requiring few bits for smaller values will result in a reduction of the utilized space. Variable-length encoding schemata allow IF indexes to be represented concisely since small *d*-gaps are much more frequent than large ones. This feature of posting lists, usually called *Clustering property* [7] is passively exploited by compression algorithms. However, by mapping DocIDs in a way that increases the frequency of small *d*-gaps, it is very likely that we can enhance the effectiveness of any variable-length encoding scheme [13]. It has been shown that an effective way to compute such a mapping is clustering the collection of documents and computing the mapping by considering the way documents are grouped within the clusters.

Indeed, clustering is an expensive operation especially in a highly dimensional domain like textual documents. Even if a number of scalable clustering techniques are known, a cheaper approach would be desirable.

As often happens when dealing with large scale WSEs, the simplest solution usually results to be the most effective one. What we are going to empirically validate in this paper is, in fact, that a very effective mapping can be obtained by considering the sorted list of URLs referencing the web documents of the collection. Furthermore, it is very likely that this lexicographical ordering of the URLs is already used by link databases; we will show, then, that it would introduce benefits also for assigning numerical identifiers to documents in Web Search Engines indexes.

As already pointed out in another paper [10], numbering the URLs following their lexicographical order improves the performance of web graph compression algorithms. In the following we are going to show that using this kind of numbering scheme we can also obtain better compression ratios by spending just a few seconds even on a very large number of documents. In fact, sorting a list of 5.9 million URLs would take about 90 seconds against about 45 minutes needed by our previous clustering solutions.

While the motivations presented in [10] clearly explain why sorting URLs can improve compression effectiveness for Web graphs, it is not so easy how to argument why sorting URLs can improve index compression too.

Luhn's hypothesis [9] can help in finding a reasonable motivation to this phenomenon. Luhn's hypothesis states that the significance of a word can be effectively measured by its frequency of occurrence. In particular, Luhn stated that words occurring less (more) than a given cut-off value are not significant as they are too rare (too common).

It is reasonable to say that lists referring to common terms are highly compressible since they contain relatively small *d*-gaps. On the other hand, rare terms occur only once or twice and do not impact on compression performance.

The lists we should care about are instead those whose length falls between the two Luhn's cut-offs. These lists, in fact, are likely to be referring to terms that are either highly correlated (e.g. "new" and "york"), or highly uncorrelated (e.g. "loudspeakers", and "octopussy").

At least in principle, by placing documents containing correlated terms closer and by separating documents containing uncorrelated terms, we should observe a gain in compressibility of postings. Unfortunately this problem has been shown to be NP-complete [3], and several heuristics have been proposed in the past [12,5,13,4].

Our hypothesis is that *documents sharing correlated and discriminant terms are very likely to be hosted by the same site thus will also share a large prefix of their URLs.*

To partially evaluate the validity of our hypothesis we performed a simple experiment. We sorted the URLs identifying the documents of our test collection, and we took the first 35,000 documents. Then, between all the possible pairs of documents we measured the similarities using the Jaccard measure, and the similarity among their relative URLs by counting the number of tokens in common (i.e. the similarity among http://www.aaa.bbb.cc/people/n1.surname1 and http://www.aaa.bb.cc/people/n2.surname2 is 2 since they share the server part and the first subdirectory). Among those pairs of URLs with similarity 0, the large majority (about 89.7%) have also Jaccard similarity equal to 0. The document similarity increased by considering URLs with similarity equal to 1 and 2. In the former case the 18.1% of documents are at distance 0.1, in the latter the large majority of document pairs (52.1%) have similarity equal to 0.9 meaning that they share a large number of terms. This simple experiment empirically shown that the two measures might be linked in some manner.

In our opinion, the most important strength point of this paper is the simplicity of the algorithm employed. Instead of thinking of sophisticated metric, or algorithms, to measure the distance between documents, in the case of Web collections we simply need to sort the URL list lexicographically, and assign identifiers to documents accordingly.

The paper is organized as follows. Section 2 reviews the state-of-the-art on the assignment problem and analyzes pros and cons when compared against our solution. Section 3 briefly recalls the assignment problem definition and the main definitions that will be used throughout this paper. Section 4 shows the results of our empirical evaluation of the URL sorting assignment heuristics. Finally Section 5 concludes the paper, and present some issues that we are going to face in the future.

2 Related Work

In recent years several works have discussed approaches to the encoding of lists of integer values. These techniques have usually been applied to the compression of full-text indexes represented by means of *d*-gapped inverted lists [15]. Only recently, though, some works have been done in order to enhance the compressibility of indexes through a clever assignment of identifiers to documents [12,5,13,4].

Shieh *et al.* [12] proposed a DocID reassignment algorithm adopting a Travelling Salesman Problem (TSP) heuristic. A *similarity* graph is built by considering each document of the collection as a vertex and by inserting an edge between any pair of vertices whose associated documents share at least one term.

Moreover, edges are weighted by the number of terms shared by the two documents. The TSP heuristic algorithm is then used to find a cycle in the *similarity* graph having maximal weight and traversing each vertex exactly once. The sub-optimal cycle found is finally broken at some point and the DocIDs are reassigned to the documents according to the ordering established. The rationale is that since the cycle preferably traverses edges connecting documents sharing a lot of terms, if we assign close DocIDs to these documents, we should expect a reduction in the average value of *d*-gaps and thus in the size of the compressed IF index. The experiments conducted demonstrated a good improvement in the compression ratio achieved. Unfortunately, this technique requires to store the whole graph in the main memory, and is too expensive to be used for real Web collections: the authors reported that reordering a collection of approximately 132,000 documents required about 23 hours and 2.17 GBytes of main memory.

Blelloch and Blandford [5] also proposed an algorithm (hereinafter called *B&B*) that permutes the document identifiers in order to enhance the clustering property of posting lists. Starting from a previously built IF index, a similarity graph G is considered where the vertices correspond to documents, and the edges are weighted with the *cosine similarity* [8] measure between each pair of documents. The *B&B* algorithm recursively splits G into smaller subgraphs $G_{l,i} = (V_{l,i}, E_{l,i})$ (where l is the level, and i is the position of the subgraph within the level), representing smaller subsets of the collection. Recursive splitting proceeds until all subgraphs become singleton. The DocIDs are then reassigned according to a *depth-first* visit of the resulting tree. The main drawback of this approach is its high cost both in time and space: similarly to [12] it requires to store the whole graph G in the main memory. Moreover, the graph splitting operation is expensive, although the authors proposed some effective sampling heuristics aimed to reduce its cost. In [5] the results of experiments conducted with the TREC-8 ad hoc track collection are reported. The enhancement of the compression ratio obtained is significant, but execution times reported refer to tests conducted on a sub-collection of only 32,000 documents. The paper addresses relevant issues, but due to its cost, the *B&B* algorithm also seems unfeasible for real Web collections. Several transactional-model-based solution have been compared against *B&B* in [13] and they showed that there is actually room for a lot of improvements to their method. In particular, not from an effectiveness point of view, but more from a scalability perspective.

In [4], Blanco and Barreiro studied the effect of dimensionality reduction on reassignment algorithms based on the Greedy-NN TSP algorithm. Basically, they first reduce dimensionality of the input matrix through a Singular Value Decomposition (SVD) transformation and then they apply the Greedy-NN TSP algorithm. They also tested the effect of *blocking* (i.e., the division of the dataset in subsets) on the effectiveness of the algorithm. Results are very good and they were able to obtain good compression ratios with low running times. Even thought the results are good, using the SVD transformation on the matrix might spoil the scalability of the method. Indeed, the block partitioning approach proposed seems to reduce this effect but costs quite a lot in terms of effectiveness degradation.

In our opinion, another drawback of all the previous approaches is that they focus on *reassigning* DocIDs appearing in a previously built IF index. The strength point of this work, however, is that DocIDs are assigned on the fly, before (and not either during, or after) the inversion of the document collection. In order to compute efficiently and effectively a good assignment, a new model to represent the collection of documents is needed. In a previous work [13], there have been presented some results relative to four assignment algorithms based on clustering. The clustering approach resulted to be scalable and space-effective. This means that it can be used to assign DocIDs even before the spidered collection will be processed by the Indexer. Thus, when the index will be actually committed on disk, the new DocID assignment will be already computed. Conversely, the other methods proposed so far require that the IF index has already been computed in advance. They also proposed a new model that allowed the assignment algorithm to be placed into the typical spidering-indexing life-cycle of a WSE. The *Transactional Model* was based on the popular *bag-of-words* model. This work remain valid for generic textual document collections. Though, the main concern about this kind of techniques within Web collections is that usually document identifiers are used also for other purposes, like for instance addressing snippets, retrieve ranking information, and so on. Renumbering the collection, thus, may not be feasible in practice for large scale WSEs.

The solution of considering the sorted list of URLs has already been used in [10] to enhance the compressibility of Web Graphs. Web graphs may be represented by using adjacency lists. They are basically lists that contain, for each vertex v of the graph, the list of vertices directly reachable from v. It has been observed that almost 80% of all links are local, that is, point to pages of the same site. Starting from this observation, it is obvious that assigning closer identifiers to URLs referring to the same site will result in adjacency lists that will contain the around 80% of ids very close among them. Representing these lists using a d-gapped representation will thus lead to d-gapped adjacency lists having long runs of 1's. Starting from this assumption, in [10] and [6], authors show that exploiting the lexicographical ordering of URLs leads to an enhancement in performance of Web graphs encoding algorithms.

The main matter of this paper is an *evaluation of the compression algorithms efficiency, when an inverted index is built over a collection whose numerical identifiers are assigned according to lexicographically sorted URLs.*

3 Assignment of DocIds

Basically, the previously proposed clustering-based (re)assignment algorithms were trying to reduce the average gap value by clustering together documents having a number of terms in common. The distance measure used in clustering was thus based on this concept of number of shared terms and the complexity of clustering algorithms depended on the number of distance computations[1]. The

[1] The distance measure used was the *Jaccard* distance that depends on the cardinality of the intersection between the set of terms contained within two documents A, and B, and on the cardinality of their union. $d(A,B) = 1 - \frac{|A \cap B|}{|A \cup B|}$.

complexity is generally linear ($O(|D|)$) in the case of k-means based clustering, or superlinear ($O(|D| \, log \, |D|)$) in the case of hierarchical clustering methods (like *B&B*). Indeed, this complexity results consider distance computations as a constant complexity ($O(1)$) operation. Computing the distance between two documents, in fact, means finding the intersection among two sets of term, and this is clearly not a cheap operation.

The algorithm we are going to present instead is indeed trivial, since it just consists of sorting the list of URLs, yet very efficient, since it does not require any set intersection operations. The computational complexity of this approach is $O(|D| \, log \, |D|)$. Differently from the clustering methods, the complexity is expressed as the number of string comparisons instead of number of set intersections. Furthermore, sorting the list of URLs is very effective from the compression ratio point of view. Another non-trivial advantage of this solution with respect to the previously proposed ones, is that a lot of scalable external memory string sorting algorithms exists while, currently, no assignment (or even re-assignment) algorithms have been proposed using external memory based techniques. One could obviously use one of the many external memory clustering solutions that exists in literature, but these will still require a lot of time to complete their operations.

Throughout the rest of this paper we will compare our assignment strategy based on URLs sorting against the previously proposed k-scan algorithm [13].

4 Experiments

We experimented our solution on an index built upon the WBR99 collection. WBR99 consists of 5,939,061 documents, about 22 GB uncompressed, representing a snapshot of the Brazilian Web (domains .br) as spidered by www.todobr.com.br. As the hardware platform, we used a Pentium IV 3.2GHz, with 1GB of RAM, local disk, and Linux as the operating system.

The tests we perform are aimed at showing the superiority of our approach with respect to the clustering approach. We will show the improvements in both compression ratio, reordering time, and space consumed of various encoding algorithms.

We considered several encoding schemata in our experiments:

- Elias' Gamma (GAMMA);
- Elias' Delta (DELTA);
- Golomb Code (GOLOMB);
- Variable Byte (VB);
- Interpolative Coding (INTERP);
- Simple9 (S9).

Simple9, hereinafter S9, is the encoding scheme described in [1]. Gamma, Delta, Golomb, Interpolative (INTERP) and Variable Byte (VB) are five popular encoding schemata described in [15].

In order to assess the validity of our method we performed our experiments by comparing not only different encoding methods but also different orderings for assigning document IDs. The orderings compared are the following:

- A random ordering (*Random*). For each document an ID is assigned u.a.r. by considering the set of previously unassigned IDs.
- The original numbering of documents given to documents in the index (*Original*).
- Block, and transactional-model based k-scan clustering (*Clustering*). In this case k, the number of clusters, has been set equal to 900, and the block size has been set equal to $900,000$ documents. Since this kind of clustering is sensitive to initial ordering [13], we fed the algorithm with the ordering found in the original index.
- URL-based sorting (*Url sorting*).
- k-scan clustering of the document collection again using $k = 900$, and $900,000$ documents for each block (*Clustering + Url Sorting*). The sorted list of URLs has been taken as the initial ordering of the documents.

4.1 Results

Regarding the enhancements in compression ratios, Table 1 and Figure 1 show comparisons between six different compression ratios and five different orderings.

Table 1. Results of various assignment algorithms using different encoding schemata. In bold are represented the best results obtained for each encoding schemata.

	VB	GAMMA	DELTA	S9	INTERP	GOLOMB
Random	11.4	12.72	12.71	15.41	11.13	11.31
Original	11.25	12.34	12.32	15.2	10.94	11.12
Url sorting	**9.72**	**7.72**	**7.69**	14.34	7.48	**8.23**
Clustering	9.81	8.82	8.8	**14.03**	**7.26**	8.63
Clustering + Url sorting	10.03	8.96	8.95	14.15	7.31	8.9

In Table 1, the best compression ratio achieved by each methods and for each ordering schema has been represented in bold. As we expected the compression ratio in the case of identifiers assigned by using our URL sorting method performs better than clustering in almost all the cases except for S9, and INTERP. Anyway, the enhancements in terms of compression ratio of the URL sorting method against the Clustering one is visible only in the case of Gamma and Delta. Here the URL sorting has produced a posting list section that is 13% smaller than the Clustering ordering. In the remaining cases Clustering and URL-ordering are comparable.

Since Clustering is sensitive to the initial ordering, we also tested a hybrid solution consisting in performing Clustering on the list of documents ordered according to their URLs. Contrarily to what we expected this method did not perform better neither than URL sorting, nor than Clustering (except for S9 and INTERP). We have not found a good explanation of the reason why this happens, we reserve to better investigate this issue in the near future. Another positive result is that differently from what was observed in [13] VB can be improved.

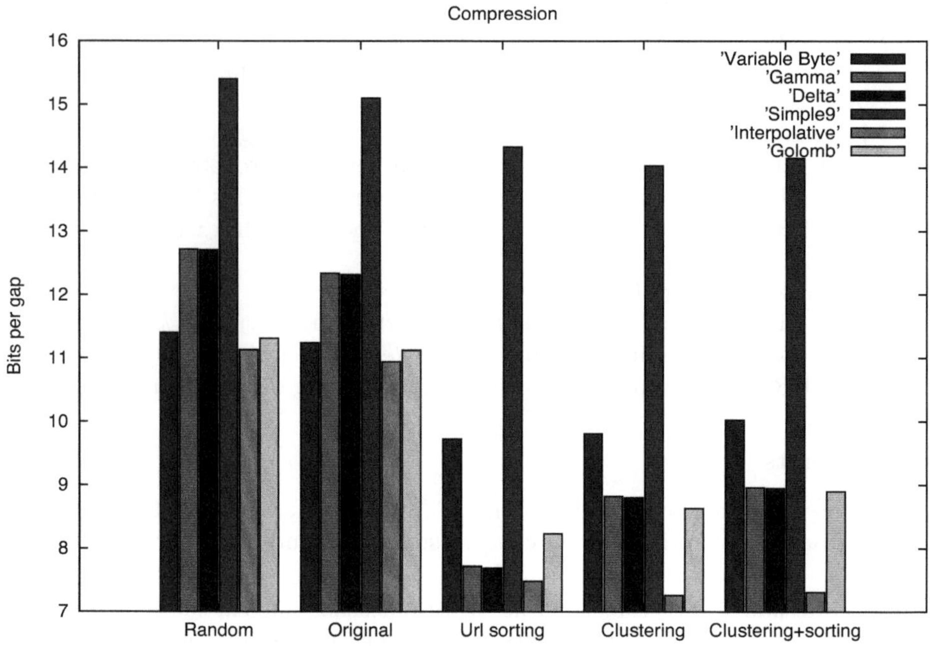

Fig. 1. Compression ratios of six different encoding techniques when applied to five different orderings

To further confirm our results we also measured the distribution of d-gaps within three different ordering: original, clustering and URL sorting. Figure 2 reports the three distributions.

As the histogram shows, the number of very small d-gaps dramatically increases in clustering and URL sorting based ordering with respect to the original ordering. In particular the number of d-gaps equal to 1 and 2 increases in the cases of URL sorting from around $75,000,000$ to $325,000,000$. The URL sorting successfully increases the number of small d-gaps up to the d-gap equal to seven and decreases the other. The clustering schema, instead, successfully increases the small d-gaps up to the d-gap equal to fifteen. This is the main reason why URL sorting is slightly better than clustering. This observation also confirms our hypothesis made in Section 3. *When we reduce the average gap, the resulting IF results smaller almost independently from the encoding method actually adopted.*

As we have seen, compression efficiency gains are comparable in most cases to those obtained by clustering. The main improvement, though, is in the resources consumed by our novel assignment algorithm compared to those needed by solutions based on clustering. As already said sorting a list of about six million URLs took just ninety seconds on the testing platform to complete and occupied just 95MB of main memory. On the other hand, clustering the same collection of documents required the partitioning of the collection into seven blocks of

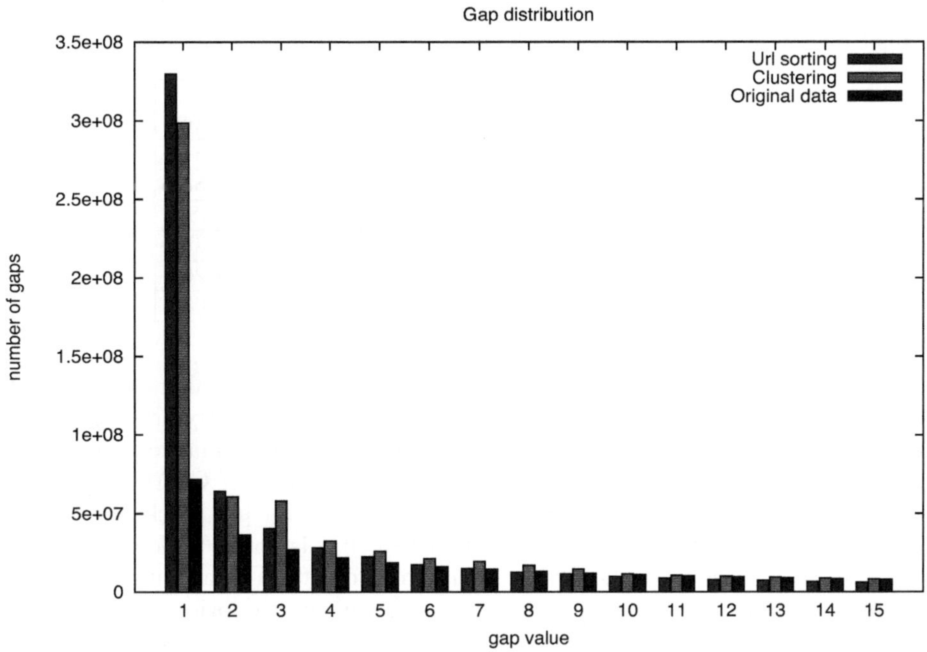

Fig. 2. Distribution of the *d*-gaps within the index organized according to the three different orderings: original, clustering and URL sorting

Table 2. Time (in seconds) spent for clustering each block of documents

	Block Size				
	100,000	300,000	500,000	700,000	900,000
Time (s)	39	119	197	274	352

900,000 documents each (except for the last block composed by around 600,000 documents), each block took about 352 seconds, for a total of about 45 minutes of CPU-time. The memory occupancy in the case of clustering has been around 1.2GB. It is thus clear that, while the ordering algorithm can scale to billions of documents without any particular problems, Clustering solutions cannot afford to achieve the same performance figures. In fact, as Table 2 shows, reducing the block size to 100,000 documents will result in a slightly reduction of total completion time around 100 seconds less than the case with blocks of 900,000 documents. In all the experiments the number of total clusters has been kept equal to 1,000.

On the other hand a reduction of block size will impact in compression performance. Table 3 shows the relation among block size and compression ratio.

Table 3. Bits per gap of various encoding schemata when the collection is reordered using Clustering, and by varying block size

	Block Size				
	100,000	300,000	500,000	700,000	900,000
VB	11.24	10.9	10.56	9.98	9.81
GAMMA	12.32	12.18	10.83	9.03	8.82
DELTA	12.32	12.17	10.81	9.02	8.8
S9	15.01	14.91	14.46	14.11	14.03
INTERP	10.89	9.23	8.01	7.62	7.26
GOLOMB	11.07	10.79	9.98	9.11	8.63

As it is possible to observe in the table, by reducing the number of documents for each block, the compression ratio dramatically decreases. By halving the size of the blocks, in fact, almost all of the methods loose performance between one and two bits per each posting. For example, in the case of GAMMA, passing from 900,000 documents per block to 500,000 documents per block, the number of bits per posting increases from 8.82 to 10.83 that is about 23% worse.

To be more precise, we should have performed our tests comparing the sorting technique against a clustering technique not requiring the prior partitioning of the collection into blocks. To date, there have not been proposed any external memory document assignment algorithm. Actually one could think about using one of the many out-of-core clustering methods that exist in the literature. Anyway, these clustering methods will be even slower than the blocked solution and is not really clear whether the compression ratios will increase further.

5 Conclusions and Future Works

We have shown that a simple sorting of the list of URLs associated to a collection of Web documents is very effective and results to be very fast. In all the experiments performed the compression ratio has increased by numbering the documents according to the sorted list of URLs.

This fully confirms our initial claim. *Ordering of the URLs used by link databases introduces benefits also if the same ordering would be used for assigning numerical identifiers to documents in Web Search Engines indexes.*

The benefits of sorting, when compared to clustering, are multiple. Compression ratios are about 5% better. The time needed to compute the assignment is two orders of magnitude smaller than the time needed by clustering. The space occupied by the clustering algorithms is dramatically bigger than the space needed to sort a list of URLs. In our tests, clustering used around one KB per each document assigned, requiring the prior subdivision of the collection in blocks. Furthermore, the URL-ordering method is more scalable to collections of even billion of URLs.

Nevertheless, due to the very limited time consumed by the sorting algorithm, one may think about placing the assignment module before the indexing phase will actually took place.

In conclusion, the URL sorting technique is the most efficient technique for assigning docIDs in the case of Web Search Engines when considering the classic time-space trade-off. In the other cases, for instance desktop search, enterprise search, email search systems, where URL information are not available, we might use folder's names or e-mail threads. Anyway, if none of this information is available clustering is still viable.

Some points remain to be investigated further. An important issue, when dealing with encoding methods is the time spent in decoding lists, and thus in resolving queries. So far, methods based on the Variable Byte schema (i.e. byte-aligned methods) have been shown to be the most effective, offering the best trade-off between decoding speed, and space occupancy. From what it can be seen in Figure 1, Gamma, and Delta scheme, now, has compression ratios far better than those of Variable-Byte. This could mean that the Delta schema may become that of reference for compressing posting lists in Web Search Engines. In the near future, we are going to evaluate the decoding speed of the various encoding scheme in light of this new ordering schema on a real IR system. It should be pointed out, in fact, that in real IR systems IF indices contain information about term frequencies and term positions. This data obviously affects the performance (in terms of retrieval time) of a retrieval system and can be only measured experimentally.

References

1. Vo Ngoc Anh and Alistair Moffat. Inverted index compression using word-aligned binary codes. *Inf. Retr.*, 8(1):151–166, 2005.
2. Vo Ngoc Anh and Alistair Moffat. Simplified similarity scoring using term ranks. In *SIGIR '05: Proceedings of the 28th annual international ACM SIGIR conference on research and development in information retrieval*, pages 226–233, New York, NY, USA, 2005. ACM Press.
3. Roi Blanco and Alvaro Barreiro. Characterization of a simple case of the reassignment of document identifiers as a pattern sequencing problem. In *SIGIR '05: Proceedings of the 28th annual international ACM SIGIR conference on research and development in information retrieval*, pages 587–588, New York, NY, USA, 2005. ACM Press.
4. Roi Blanco and Alvaro Barreiro. Document Identifier Reassignment Through Dimensionality Reduction. In *Advances in Information Retrieval: 27th European Conference on IR research, ECIR 2005, Santiago de Compostela, Spain, March 21-23, 2005. Proceedings*, pages 375 – 387, 2005.
5. Dan Blandford and Guy Blelloch. Index compression through document reordering. In *Proceedings of the Data Compression Conference (DCC'02)*, pages 342–351, Washington, DC, USA, 2002. IEEE Computer Society.
6. P. Boldi and S. Vigna. The webgraph framework i: compression techniques. In *WWW '04: Proceedings of the 13th international conference on World Wide Web*, pages 595–602, New York, NY, USA, 2004. ACM Press.
7. A. Bookstein, S. T. Klein, and T. Raita. Modeling word occurrences for the compression of concordances. *ACM Trans. Inf. Syst.*, 15(3):254–290, 1997.
8. Chris Buckley. Implementation of the smart information retrieval system. Technical Report TR85–686, Cornell University, Computer Science Department, May 1985.

9. H. P. Luhn. The Automatic Creation of Literature Abstracts. *IBM Journal of Research Development*, 2(2):159–165, 1958.
10. Keith H. Randall, Raymie Stata, Janet L. Wiener, and Rajiv G. Wickremesinghe. The link database: Fast access to graphs of the web. In *DCC '02: Proceedings of the Data Compression Conference (DCC '02)*, page 122, Washington, DC, USA, 2002. IEEE Computer Society.
11. Falk Scholer, Hugh E. Williams, John Yiannis, and Justin Zobel. Compression of inverted indexes for fast query evaluation. In *SIGIR '02: Proceedings of the 25th annual international ACM SIGIR conference on research and development in information retrieval*, pages 222–229, New York, NY, USA, 2002. ACM Press.
12. Wann-Yun Shieh, Tien-Fu Chen, Jean Jyh-Jiun Shann, and Chung-Ping Chung. Inverted file compression through document identifier reassignment. *Information Processing and Management*, 39(1):117–131, January 2003.
13. Fabrizio Silvestri, Salvatore Orlando, and Raffaele Perego. Assigning identifiers to documents to enhance the clustering property of fulltext indexes. In *SIGIR '04: Proceedings of the 27th annual international ACM SIGIR conference on research and development in information retrieval*, pages 305–312, New York, NY, USA, 2004. ACM Press.
14. A. Trotman. Compressing inverted files. *Information Retrieval*, 6(1):5–19, January 2003.
15. Ian H. Witten, Alistair Moffat, and Timothy C. Bell. *Managing Gigabytes – Compressing and Indexing Documents and Images.* Morgan Kaufmann Publishing, San Francisco, second edition edition, 1999.

Efficient Construction of FM-index Using Overlapping Block Processing for Large Scale Texts*

Di Zhang[1], Yunquan Zhang[1,2], and Jing Chen[3]

[1] Institute of Software, Chinese Academy of Sciences
[2] State Key Laboratory of Computer Science
{zhangdi,zyq}@mail.rdcps.ac.cn
[3] Microsoft Research Asia
jingchen@microsoft.com

Abstract. In previous implementations of FM-index, the construction algorithms usually need several times larger memory than text size. Sometimes the memory requirement prevents the FM-index from being employed in processing large scale texts. In this paper, we design an approach to constructing FM-index based on overlapping block processing. It can build the FM-index in linear time and constant temporary memory space, especially suitable for large scale texts. Instead of loading and indexing text as a whole, the new approach splits the text into blocks of fixed size, and then indexes them respectively. To assure the correctness and effectiveness of query operation, before indexing, we further append certain length of succeeding characters to the end of each block. The experimental results show that, with a slight loss on the compression ratio and query performance, our implementation provides a faster and more flexible solution for the problem of construction efficiency.

Keywords: FM-index, Self-index, Block processing.

1 Motivation

In recent years, the amount of digital information is growing rapidly, and a large part of these data is in the form of text. This situation has made the space consumption of indices on texts become a serious problem. With the attempt to reduce the size of indices, a powerful concept of *self-index* has emerged[13]. Self-index is an index that provides fast search functionality using space proportional to the k-th order entropy[12] of the text. Besides search functionality, it also contains enough information to efficiently reproduce any text substring. A self-index can therefore replace the text. The exciting concept has triggered much interest on this issue and produced surprising results in very few years.

* This work was supported in partial by the National Natural Science Foundation of China under contract No.60303020 and No.60533020, the National 863 Plan of China under contract No.2006AA01A125 and No. 2006AA01A102, and Foundation of CAS under contract No. KSH1-02.

G. Amati, C. Carpineto, and G. Romano (Eds.): ECIR 2007, LNCS 4425, pp. 113–123, 2007.

FM-index is the first self-index designed by Ferragina and Manzini [2]. They realized the potential of the relationship between text compression and text indexing, in particular regarding the correspondence between the entropy of a text and the regularities in some widely used indexing data structures. The FM-index is designed to occupy space close to that of the best existing compression techniques, and provide search and text recovering functionality with almost optimal time complexity [5].

Many other researches follow the idea of FM-index, such as Huff-FMI[6], RL-FMI[9], AT-FMI[4]. Despite their efficiency in compression and searching, another important concern is the construction efficiency. In practical applications, the construction phase of FM-index is either too slow, or require too much working space[8]. Some studies have notice the problem. In paper [7], the construction algorithm can build the FM-index in $O(n \log\log |\Sigma|)$ time and $O(n \log |\Sigma|)$-bit working space.

Apart from full-text indices, we observed that many compression tools such as bzip2[14], usually split their input data into blocks of equal size before compression. This idea also could be employed by FM-index construction.

This paper is organized as follows. First, we introduce some background knowledge on FM-index, including the basic algorithm and construction procedure. Then, we discuss possible ways of implementing a block version of the FM-index, and give a complete solution of block construction. Finally, a series of experiments on various kinds of texts will be performed to validate the improvements we have made.

2 Background

2.1 FM-index

The algorithm of FM-index is based upon the relationship between the Burrows-Wheeler compression algorithm [1] and the suffix array data structure [11]. FM-index is a sort of compressed suffix array that takes advantage of the compressibility of the indexed data in order to achieve space occupancy close to the information theoretical minimum. Precisely, given text $T[1, n]$ to be indexed, the FM-index occupies at most $5nH_k(T) + o(n)$ bits of storage, where $H_k(T)$is the k-th order entropy of T, and allows the search for the *occ* occurrences of a pattern $P[1, p]$ within T in $O(p + occ \log^{1+\varepsilon} n)$ time, where $\varepsilon > 0$ is an arbitrary constant fixed in advance[2].

The Burrows-Wheeler transform produces a permutation of the original text, denoted by $T^{bwt} = bwt(T)$. String T^{bwt} is the result of the following steps: (1) append a special end marker #, which is the smallest character lexicographically, to the end of T; (2) form a conceptual matrix $\mathcal{M}$ whose rows are the cyclic shifts of string $T\#$ sorted in lexicographic order; (3) construct T^{bwt} by taking the last column of $\mathcal{M}$. The first column is denoted by F[2].

The suffix array $\mathcal{A}$ of text T is represented implicitly by T^{bwt}: $A[i] = j$ iff the i-th row of $\mathcal{M}$ contains string $t_j t_{j+1} \cdots t_n \# t_1 \cdots t_{j-1}$.[6] The novel idea of the FM-index is to store T^{bwt} in compressed form, and then simulate the search in the suffix array. The FM-index can also locate the text positions where P occurs, and display any length of substrings around the occurrences. The details of search algorithm are referred to [2] and [6].

Paper [3] introduced the first implementation of FM-index, now it has been upgraded to the version 2[1]. It provides a group of standard APIs of self-index, it is easy for later researchers to modify or extend its function.

2.2 Procedure of Construction

In above implementation of FM-index, the program constructs the index for a text as follow steps:

1. Load the whole text into main memory;
2. Scan the text to find a special character that never occurs. If any, insert it to the source data homogeneously, otherwise (this is not common), every position in text should be marked; build suffix array by sorting, also get all positions for characters in sequence L simultaneously;
3. Calculate Occ for every character, and build the mapping from original alphabet of ASCII code to new alphabet; generate sequence L from suffix array built in step 2, and transfer it to new alphabet Σ according to the mapping in step 4;
4. Record the positions of the special character in input file according to the order in L. The positions can be denoted as $loc\_occ[i] = j$, which means the i-th special character in L corresponding to input file is located in position j;
5. Compress the buckets and generate auxiliary information, and save them to disk.

In the above procedures, memory consumption mainly includes the following parts: load whole text to memory(step 1); build suffix array(step 2); BWT (step 3); and some other auxiliary data structures. All these usually add up to several times larger memory requirement than the original text size. Detailed experimental results about memory consumption can be found in Section 4.2.

3 Block FM-index

3.1 Basic Block FM-index

The original construction method requires enough memory to load an input text and build the index as a whole, which is infeasible for large texts like DNA sequences, whose size is often in the orders of Gigabytes.

There are some possible alternatives when dealing with large amounts of data. One is to optimize the program to enhance the ceiling of input data size. Although it permits larger files to be indexed than before, there will still be even larger files that could not fit into main memory. Another choice is to use secondary memory when main memory is not enough. This will make the construction much costly because excessive increments in I/O operations[13].

[1] FM-index Version 2, http://roquefort.di.unipi.it/˜ferrax/fmindexV2/index.html

Having considered the above alternatives, we find another way to overcome the memory problem. Our solution is to naturally split the data into independent blocks of a predetermined size before the data is indexed. These blocks are then processed through the FM-index algorithm in order, and their indexes are concatenated back again to form the final output file. The format of new output file is not compatible to the original FM-index, but it is just an index container actually. It is composed of two parts, including header and body. The header contains some necessary information to recover the text, such as length of input file, block size, etc.; and the body contains separated indexes ordered by the position of blocks in input file.

Besides building, the process of query operation should also be modified. The program executes a given query on all indexed blocks respectively, collects the results and submits them to the user.

3.2 Overlapping Block FM-index

The above block model has a problem that can't be ignored. That is, if an occurrence of the pattern crosses through two adjacent blocks, it will be lost in results set because either block contains only part of the pattern. In other words, in text $T[1,n]$, given an occurrence $occ[h, h+p-1]$ of pattern $P[1,p]$, if $occ[h, h+p-1]$ lies on the boundary of block B_i and block B_{i+1}, it will be divided into two parts: $occ[h, h+m]$ and $occ[h+m+1, h+p-1]$. In this instance, neither of the blocks will contain the full length of $occ[h, h+p-1]$. It will be lost in query results.

To solve this problem, the above model should be adjusted. The first solution is to search across blocks from end to start using the backward method[2]. Although this method is intuitive, but it has the following shortcomings: (1) the implementation of block version tied closely to the concrete implementation of original FM-index. This limits the application scope of the block model that may be applied on other self-indices, and the software will be hard to reuse if FM-index upgrades its version; (2) the query process is not independent of each block, which is not conducive to the future parallelization of query process in SMP or clusters for very large data sets.

In aware of these shortcomings, we propose another method named *overlapping block model*. It is based on an obvious fact that, in common query task, the pattern is far shorter than target file, actually there are always several orders of magnitude difference. Thus, if we estimate the maximum length r of possible query patterns, we could concatenate each block and its succeeding characters of length r to generate a new block, then process it in building procedure.

This overlapping block model is shown as Figure 1. Assumes the block size is s and overlapping length is r, the number of blocks will be $b = \lceil (n-r)/(s-r) \rceil$. In practice, since the parameters s and r are both predetermined, b can be denoted as: $b = \Theta(n)$. For a given text $T[1,n]$ with a constant alphabet Σ, the construction time and space for each block can be presumed as constants CT and CM in theory. Thus, the construction time of $T[1,n]$ is $O(b) = O(n)$. Noticed that the memory can be reused in serial processing of blocks, the construction space of $T[1,n]$ is CM, which is a constant number. The above analysis results can be

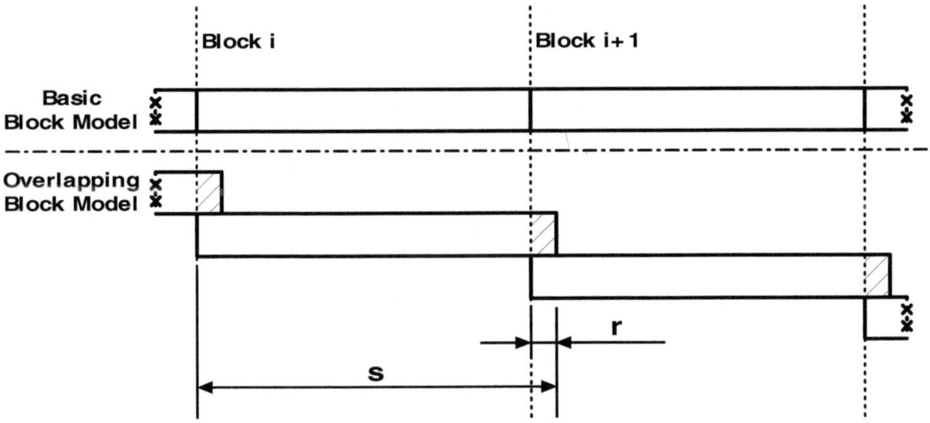

Fig. 1. Overlapping block model for FM-index

confirmed by experiments in Section 4.2. On other important aspects of performance, such as storage occupancy and query performance, since the explicit conclusions can't be made from general theoretical analysis, we incline to use the experiments to illustrate the impacts of block model. Related experimental results will be given in Section 4.1.

There are also some drawbacks in overlapping model. First, it produces some redundancy. In fact, compared to the size of the entire document, the redundant part of index, which is usually negligible, will not impact the compression ratio substantively. Second, there should be a solution to deal with very long search patterns such as $P[1,p](p > r)$, which still have possible loss of occurrences. Here is a solution for this situation. First we break the long patterns $P[1,p]$ into short sub-patterns $P'_1[1,r], P'_2[r+1,2r] \cdots P'_k[(k-1)r+1,p]$ with equal length $\lceil p/r \rceil$. All the sub-patterns can be searched correctly in block model. And then we use them to query the block index respectively, merge the nearby items in different result sets according to the sequence of sub-patterns in the original long pattern. That means, given a result set R for pattern P, and sub-result sets R'_i for sub-patterns $P'_i(1 \leq i \leq k)$: $occ[h, h+p-1] \in R$, iff $occ'_1[h, h+r-1] \in R'_1 \wedge \cdots \wedge occ'_k[h+(k-1)r, h+p-1] \in R'_k$. Although this solution is somewhat inefficient, in practice we can set r big enough to avoid splitting pattern frequently.

4 Experimental Results

The overlapping block FM-index contains two key parameters: the block size s and overlapping length r. In this section we will perform a series of experiments to show the effects of these parameters on the compression and query performance. Furthermore, we demonstrated the performance enhancements of the block model in contrast to the original version.

Our experiments ran on a Pentium 4 processor at 3.06 GHz, with 1 MB L2 cache and 512MB of DDR2 RAM. The operating system was Fedora core 5.

Table 1. Files used in the experiments. To understand the features of the files, we also provide alphabet size and compression ratio under bzip2 (level 9).

Collection	Size (bytes)	Content	Alphabet Size	Bzip2-9 comp. ratio
sources	210,866,607	C/Java source code	230	18.68%
dna	403,927,746	gene DNA sequences	16	25.76%
english	2,210,395,553	English text files	239	28.35%
xml	294,724,056	XML format text	97	11.39%

We compiled the code with gcc 3.4.4 using optimization option "-O9". The texts from Pizza&Chili Corpus[2], which is a special test suite for compressed index, are used for experiments in this paper. They are listed in Table 1. In the following sections, we use file-n to denote the first n Megabytes of file.

In addition, we use the default settings of FM-index in the following experiments. These settings are: superbucket size = 16; bucket size = 512; frequency of marked characters = 0.02.

4.1 The Effects of Parameters

Block Size. From the point of view of construction, of course the smaller the block size, the less space and time the program consumes. But on the other side, the block size should be big enough to guarantee a better compression ratio through BWT. Furthermore, more blocks will bring more auxiliary information, thus the compression ratio would be decreased. However, because the bucket size is fixed, the enhancement of compression ratio is limited in certain scope. It can not be exceeded whatever large block size is set to.

To obtain more comprehensive results, we selected a number of truncated texts of 100MB as test suites. The results in Figure 2 validated the above analysis. For a specific text, when the block size is larger than a critical value, the growth of compression ratio is not obvious.

In practical applications, we can choose a suitable block size to emphasize on compression ratio or construction time. In the following experiments, we will set 10MB as our default block size.

Overlapping Length. Because the overlapping length is very small compared to the block size, its effect on compression ratio can be ignored. Now we only experiment the effects of overlapping length on query performance. We randomly select substrings from *english-100*, and group them by length. Each group has 100 members. Then these groups of substrings are queried as patterns on the indices.

As shown in Figure 3, the block model has brought some reduction on query performance, but the loss is generally acceptable. Only in the situation that needs to split the very long patterns into short ones would lead to substantial decline in performance. Anyhow, the overlapping length can be adjusted for specific query tasks. To ensure that there is no need to split patterns in most tasks, we set the overlapping length to 1KB in the following experiments.

[2] Pizza&Chili Corpus, http://pizzachili.di.unipi.it/index.html

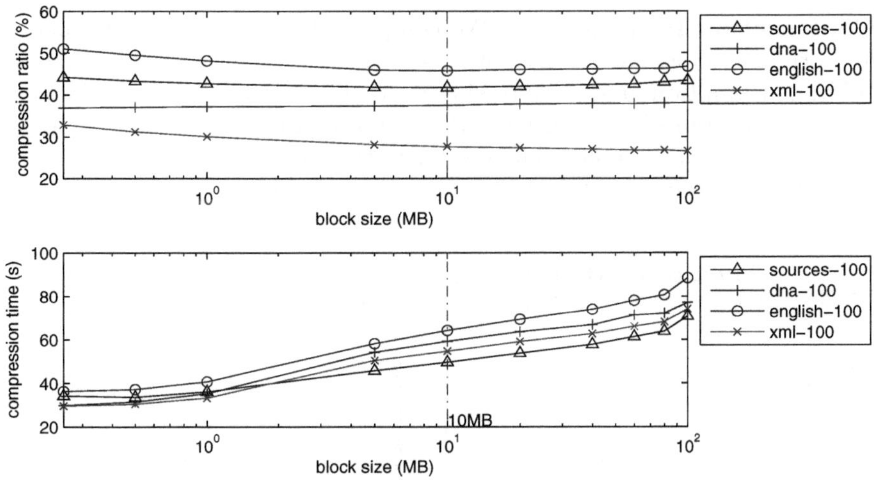

Fig. 2. Compression ratio (percentage), and construction time (seconds) as a function of the block size (MB). The block FM-index was built for the first 100MB of the listed four files, with a 64 bytes overlapping length.

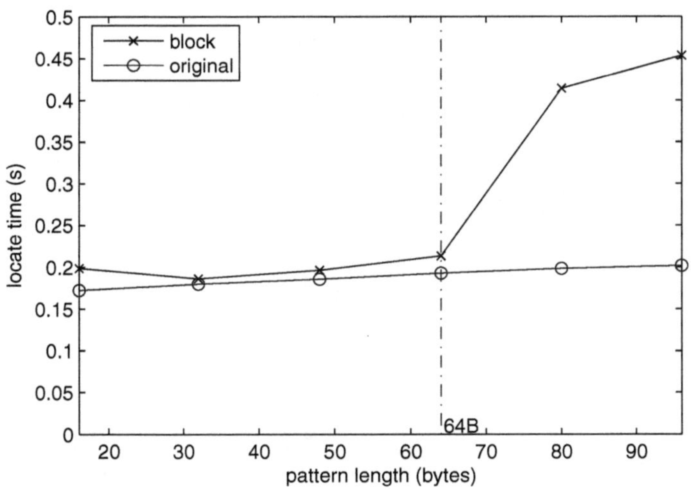

Fig. 3. The average time (seconds) for a locate operation as a function of pattern length (bytes), compared with block FM-index and original FM-index. The substrings of certain length from *english-100* files, are selected at random as query patterns. We set the overlapping length to 64B.

4.2 Comparison Between Block FM-index and FM-index

Some comparisons between the FM-index and other self-index tools have been made in [3] and [10]. Here we focus on comparisons of performance between block FM-index and original FM-index.

Construction Performance. We compare the performance of block version and original version as the size of *english-n* grows, including construction time, RAM consumption, and compression ratio. The results are shown in Figure 4.

From the results, we confirmed the perviously mentioned enhancement of performance. The block model achieved linear time and constant space performance, while the compression ratio loss is not obvious when dealing with large texts.

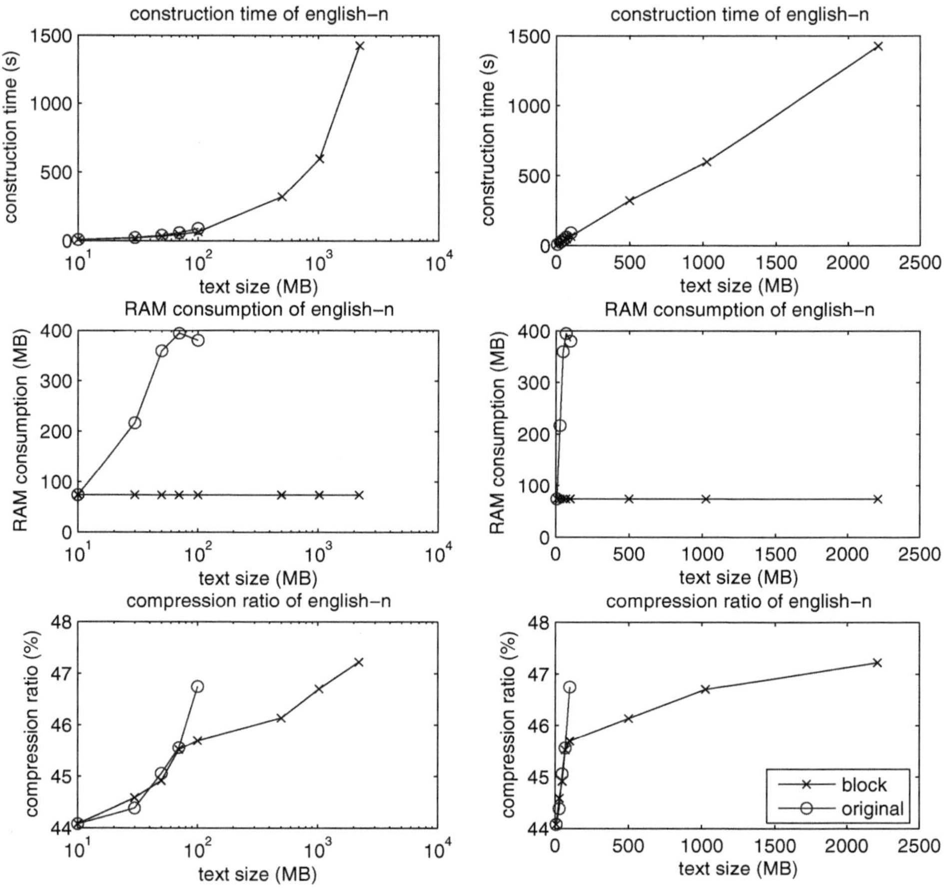

Fig. 4. Construction time (second), RAM consumption (MB), compression ratio (percentage) as a function of the size of truncated version of file *english* (MB). In every row, we use two coordinate systems to illustrate one graphics: the left column uses a semi-logarithmic coordinates, while the right column uses the ordinary coordinates. We set parameters as mentioned above: block size = 10MB, overlapping length = 1KB. When we built the FM-index for the text file that larger than 100MB, the OS would prompt there is not enough memory to allocate in test machine, thus no data are collected under such situation. In addition, at the point of 100MB, there is some decline on RAM requirement, instead of keeping on increasing. We estimate that it is due to the virtual memory management strategy of OS.

In our experiments, not only the entire *english* of over 2GB has been built into FM-index successfully, but also the progress of building is visible by counting blocks that have been indexed. The comparison of performance is shown in Figure 4.

Query Performance. Figure 5 shows a comparison of query performance between the block FM-index and the FM-index.

The experimental results show that for data of Gigabytes, the performance of locate, display, and extract operations are closer to the original version, but the performance loss of count is more significant than others. For count operation, because the occurrences in overlapping zone may be counted twice, we can merely get an approximate result in count operation. If the accurate result of a count operation is c, the overlapping model will give an approximate result, which is a

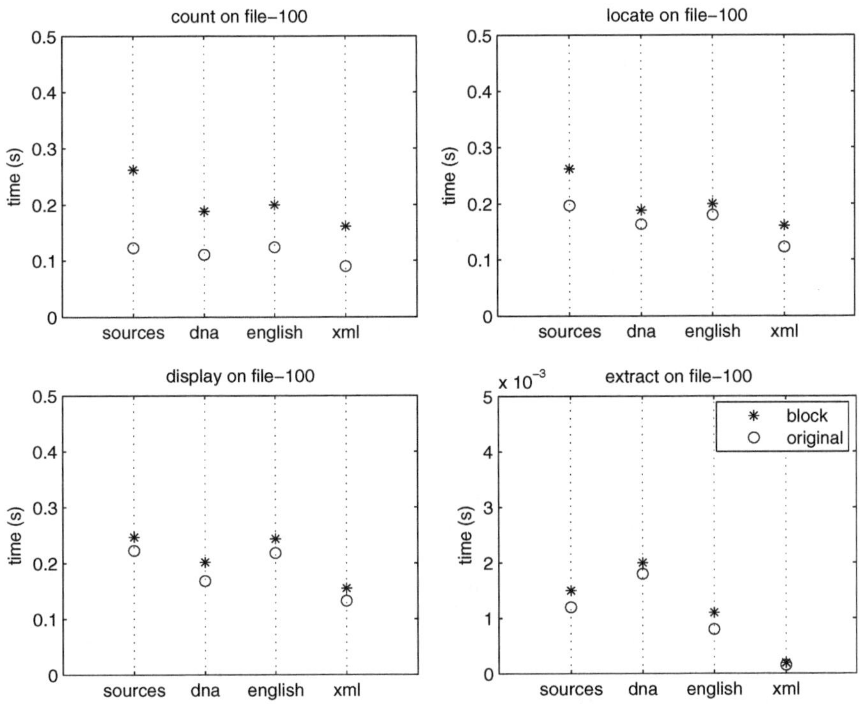

Fig. 5. Comparison on query performance of block FM-index and original FM-index, including the count, locate, display and extract operations. We built index for the first 100MB of the listed files, using original FM-index and block FM-index respectively. The file size is restricted to 100MB because the original FM-index can not index larger texts on our experimental platform. We haven't compared the RAM use of query operations because they are negligible in practice. The query patterns are selected randomly from corresponding files, with the average length of 50 bytes. In experiment, display length = 10, and length of extract = 50.

range between $\max(\lceil c/b \rceil, c - b)$ and c. Thus, in block FM-index we use locate operation to replace count operation to ensure the accuracy of the results.

5 Conclusion

This paper focuses on a practical method for constructing the FM-index for large scale texts efficiently. We proposed an overlapping block model with its implementation, and performed a series of experiments on it. It can construct the FM-index in $O(n)$ time and $O(1)$ space, without significant slowdown on query operations . Its implementation, which is an extension to the original self-index tools, can build indexes for texts of Gigabytes, just using a commodity PC instead of an expensive server.

In the future, based upon the block FM-index, we will try to parallelize the FM-index algorithm both for construction process and query operations. By parallel processing, the performance will hopefully be further improved for large scale texts.

Acknowledgements

Special thanks to Prof. Jiachang Sun for his indispensable guidance[15].

References

[1] Burrows, M., Wheeler, D.: A block sorting lossless data compression algorithm. Technical Report 124(1994). Digital Equipment Corporation.
[2] Ferragina, P. and Manzini, G.: Opportunistic data structures with applications. In Proc. 41st IEEE Symposium on Foundations of Computer Science (FOCS) (2000), pp. 390-398.
[3] Ferragina, P. and Manzini, G.: An experimental study of an opportunistic index. In Proc. 12th Annual ACM-SIAM Symposium on Discrete Algorithms (SODA) (2001), pp. 269-278.
[4] Ferragina, P., Manzini, G., Makinen, V., and Navarro, G.: An alphabet-friendly FM-index. In Proc. 11th International Symposium on String Processing and Information Retrieval (SPIRE), LNCS v. 3246 (2004), pp. 150-160.
[5] Ferragina, P., Manzini, G., Mäkinen, V., and Navarro, G.: Compressed representation of sequences and full-text indexes. Technical Report 2004-05 (Dec.) (2004), Technische Fakultät, Universität Bielefeld, Germany.
[6] Grabowski, S., Mäkinen, V., and Navarro, G.: First Huffman, then Burrows-Wheeler: an alphabet-independent FM-index. In Proc. 11th International Symposium on String Processing and Information Retrieval (SPIRE), LNCS v. 3246 (2004), pp. 210-211. Short paper. Full version as Technical Report TR/DCC-2004-4, Department of Computer Science, University of Chile, July 2004.
[7] Hon, W.-K., Sädakane, K., and Sung, W.-K.: Breaking a time-and-space barrier in constructing full-text indices. In Proc. 44th IEEE Symposium on Foundations of Computer Science (FOCS) (2003), pp. 251-260.
[8] Hon, W.-K., Lam, T.-W., Sung, W.-K., Tse, W.-L, Wong, C.-K., and Yiu, S.-M.: Practical aspects of compressed suffix arrays and FM-index in searching DNA sequences. In Proceedings of the 6th Workshop on Algorithm Engineering and Experiments. SIAM Press, Philadelphia, Pa. (2004), 31-38.

[9] Mäkinen, V. and Navarro, G.: New search algorithms and time/space tradeoffs for succinct suffix arrays. Technical Report C-2004-20 (April) (2004), University of Helsinki, Finland.
[10] Mäkinen, V. and Navarro, G.: Succinct suffix arrays based on run-length encoding. In Proc. 16th Annual Symposium on Combinatorial Pattern Matching (CPM), LNCS v. 3537 (2005), pp. 45-56.
[11] Manber, U., Myers, G.: Suffix arrays: new method for on-line string searches. SIAM Journal on Computing, 22(5) (1993): 935-948.
[12] Manzini, G. An analysis of the Burrows-Wheeler transform. Journal of the ACM 48, 3 (2001), 407-430.
[13] Navarro, G., Mäkinen, V.: Compressed Full-Text Indexes. Technical Report TR/DCC-2005-7 (2005), Dept. of Computer Science, University of Chile, June 2005.
[14] Seward, J.R.: Bzip2 and libbzip2: a program and library for data compression. http://sources.redhat.com/bzip2/ (1998)
[15] Sun, J.C.: Matrix analysis to additive schwarz methods. Journal of Computational Mathematics, Vol.13, No.4 (1995), 325-336.

Performance Comparison of Clustered and Replicated Information Retrieval Systems

Fidel Cacheda[1], Victor Carneiro[1], Vassilis Plachouras[2], and Iadh Ounis[3]

[1] Department of Information and Communication Technologies, University of A Coruña
Facultad de Informática, Campus de Elviña s/n, 15071 A Coruña, Spain
{fidel, viccar}@udc.es
[2] Yahoo! Research
Ocata 1, 1st floor, 08003 Barcelona, Spain
vassilis@yahoo-inc.com
[3] Department of Computing Science, University of Glasgow
Glasgow, G12 8QQ, UK
ounis@dcs.gla.ac.uk

Abstract. The amount of information available over the Internet is increasing daily as well as the importance and magnitude of Web search engines. Systems based on a single centralised index present several problems (such as lack of scalability), which lead to the use of distributed information retrieval systems to effectively search for and locate the required information. A distributed retrieval system can be clustered and/or replicated. In this paper, using simulations, we present a detailed performance analysis, both in terms of throughput and response time, of a clustered system compared to a replicated system. In addition, we consider the effect of changes in the query topics over time. We show that the performance obtained for a clustered system does not improve the performance obtained by the best replicated system. Indeed, the main advantage of a clustered system is the reduction of network traffic. However, the use of a switched network eliminates the bottleneck in the network, markedly improving the performance of the replicated systems. Moreover, we illustrate the negative performance effect of the changes over time in the query topics when a distributed clustered system is used. On the contrary, the performance of a distributed replicated system is query independent.

Keywords: distributed information retrieval, performance, simulation.

1 Introduction

The information available over the Internet has increased spectacularly in the last years, and we can expect that it will continue growing at the same rate, at least in the short term. Simultaneously, Web search engines have grown in importance as the users need to find, recover, and filter all the information available in this environment. Therefore, Web search engines must manage a large amount of information, and make it possible for users to locate the information required in a very short time, while simultaneously dealing with a large number of queries.

Information Retrieval (IR) systems based on a single centralised index present several problems, such as the lack of scalability, or server overloading and failures [11],

G. Amati, C. Carpineto, and G. Romano (Eds.): ECIR 2007, LNCS 4425, pp. 124 – 135, 2007.

which make them unsuitable for highly loaded systems, such as Web search engines. For this reason, the methods based on the distribution of the documents index for searching and storage are widely used. For example, the Google web search service is based on a distributed and replicated IR architecture [1].

A distributed IR system is made up of two components: the brokers (dispatchers or receptionists) and the query servers. The brokers receive the queries from the users, distribute them to the query servers, and send back the final results to the user. The query servers hold the distributed index, process the queries, and send their partial results back to the brokers for the final merging.

An inverted index can be distributed over a collection of servers following two main strategies: global inverted files (term partitioning), or local inverted files (document partitioning). In this work, we will focus on the local inverted file strategy as it has been found to be more efficient than the global inverted file strategy [17] [21].

The index distribution is necessary to deal with a high volume of data (probably because it cannot be indexed by a single machine) and to keep the response times low. When the volume of queries handled by the system must be increased (e.g. because multiple users simultaneously send queries to a Web search engine), then the IR system must be parallelised to process multiple queries concurrently. Two main types of systems can be defined to increase the query throughput: replicated and clustered.

A replicated system is composed of one or more distributed IR systems. Each distributed system indexes the whole collection, and all the distributed systems that have been replicated have the same number of query servers. The brokers, in this case, must decide initially which replica will process the query, and then broadcast the query to all the query servers in the replica. The brokers must balance the load through all the replicas to obtain an optimal performance.

A clustered system is divided into groups of computers (or clusters), where each group operates as an autonomous distributed IR system. Each cluster can be composed of a different number of query servers. Each cluster is responsible for one disjoint part of the whole collection of documents, and each cluster could use distribution and replication to store its respective index. In this case, a broker must determine the appropriate cluster for each query and then submit the query to it. A clustered system must be configured a-priori based on the distribution of the queries that the IR system expects to receive. For example, if 40% of the queries submitted to the IR system are related to "*Entertainment*", we may assign 40% of our resources (e.g. query servers) to the "*Entertainment*" cluster in order to improve its response time and throughput. This implies that a change in the queries distribution may affect the overall system performance. For example, if the number of "*Entertainment*" queries drops to 20%, this cluster may improve its performance, probably at the expense of other clusters.

In this paper, we present a detailed performance analysis, based on simulations, of a clustered system compared to a replicated system. We also study the effect of changes in the query topics over time, based on work by Spink, Jansen, Wolfram and Saracevic [22].

The performance analysis in distributed IR is used to study different configurations and measure different parameters, usually considering a fixed set of resources. Two parameters are usually considered in the performance analysis of a distributed IR system: response time and throughput [19]. In the former, we are interested in the average time to answer a query when the system is idle. In the latter, we are interested in the maximum query processing rate that the system is able to achieve. This is especially interesting when designing a large-scale IR system (e.g. hundreds of computers) in order to determine the optimal configuration for some fixed benchmarks[1], and to detect and prevent possible bottlenecks.

Many previous articles have studied different performance parameters of pure distributed IR systems, such as [8], [10], [14] or [21], to name but a few. On the other side, several previous articles examined the effects of different parallelisation techniques in a distributed IR system. Tomasic and Garcia-Molina [23] simulated a small group of servers and studied the effect of multiprogramming on the throughput using various inverted index organisations. Frieder and Siegelmann [9] studied the organisation of the data to improve the performance of parallel IR systems using multiprocessor computers. Lu and McKinley [16] analysed the effects of partial replication to improve the performance in a collection of 1TB. Moffat, Webber, Zobel and Baeza-Yates [18] presented a replication technique for a pipelined term distributed system, which significantly improves the throughput over a basic term distributed system.

In [5] and [6], the authors analysed the performance of a distributed, replicated and clustered system using a simple network model. They identified two main bottlenecks: the brokers and the network. The high load on the brokers was due to the number of local answer sets to be sorted. The network bottleneck was due to the high number of query servers and the continuous data interchange with the brokers, especially in a replicated IR system. The analysis of the clustered systems indicated that the best throughput was achieved when a great number of query servers was used, outperforming a replicated system. However, the clustered systems must be configured a-priori based on the queries distribution that the IR system is expected to receive.

In [4], a more realistic network simulation model is presented, and the authors described some solutions for the main bottlenecks of a distributed IR system. They showed that the use of a switched network reduces the saturation of the interconnection network. They also showed that the brokers' bottleneck can be improved by reducing the number of partial results sent by the query servers (with a negligible probability of changing the system's precision and recall retrieval performances), or by using a hierarchical distributed broker model.

The main objective of this paper is to compare the performance of a replicated and clustered IR system, both in terms of throughput and response time, using the extended simulation model introduced in [4], and to compare the obtained results with those previously reported in [5] and [6].

The paper is organised as follows. The simulation model is described in Section 2. Section 3 describes the simulations performed for the clustered and replicated systems and the results obtained. A discussion of the results obtained is presented in Section 4. The main conclusions of the work and possible future research directions are presented in Section 5.

2 Simulation Model

The simulation model of a distributed IR system used in this work is based on the work described in [4], where the authors implemented a discrete event-oriented

[1] An example of fixed benchmarks is that the maximum response time should be one second per query and the minimum throughput should be twenty queries per second.

simulator using the JavaSim simulation environment [15]. The defined simulation model represents a local inverted file strategy (see Section 1). All the queries are stored in a global queue, which is controlled by one or more central brokers. Each broker will take one query and will send it to all the query servers through a network [21]. Each query server then processes the whole query locally, obtains the answer set for that query, ranks the documents, selects a certain number of documents from the top of the ranking and returns them to the broker. The broker collects all the local answer sets and combines them into a global and final ranked set of documents.

From the brokers' point of view, the time to process the *ith* query (named t_i) is divided into three phases: the time to process the query in the query servers (P_1), the time to receive all the partial answers from the query servers to the broker (P_2) and the merging and ranking of the final results by the broker (P_3). Therefore, the processing time for a query q_i is given by:

$$t_i = \overbrace{\max(t_{i,j})}^{P_1} + \overbrace{\max(ra_{i,j})}^{P_2} + \overbrace{\sum_j tc(tr_{i,j})}^{P_3}$$

where the following parameters are introduced:

- q_i: vector of keywords for the *ith* query.
- $t_{i,j}$: total time (in milliseconds) to complete the processing of query q_i at query server *j*.
- $ra_{i,j}$: time to receive the local answer set for query q_i from the query server *j*.
- $tr_{i,j}$: number of documents from the top ranking in query q_i returned as the local answer set for query server *j*, where $tr_{i,j} \leq tr_{max}$, and tr_{max} is the maximum number of top ranked documents in the local answer (we consider the top 1000 documents only).
- $tc(n)$: time to merge and sort *n* documents, which is computed following the logarithmic model: $tc(n) = tc_0 + tc_1 \times n + tc_2 \times \ln(n)$, as described in [4].

From the query servers' point of view, the time to process the *ith* query by the *jth* query server (named $t_{i,j}$) is divided into five phases: the time to receive the query from the broker ($P_{1,1}$), the initialisation time ($P_{1,2}$), the seek time ($P_{1,3}$), the reading time from disk ($P_{1,4}$), and the ranking of the partial results ($P_{1,5}$).

$$t_{i,j} = \overbrace{rq_{i,j}}^{P_{1,1}} + \overbrace{ti}^{P_{1,2}} + \overbrace{k_i \times ts}^{P_{1,3}} + \overbrace{\sum_{k \in q_i} d_{k,j} \times tr}^{P_{1,4}} + \overbrace{tc(r_{i,j})}^{P_{1,5}}$$

where these new parameters are introduced:

- $rq_{i,j}$: time to receive the query q_i for the query server *j*.
- *ti*: initialisation time, including memory allocation and output display, if necessary.
- k_i: number of keywords in query q_i.
- *ts*: average seek time for a single disk.
- *tr*: average time to read the information about one document in an inverted list and to do its processing (seek time is excluded).
- $d_{k,j}$: number of documents of the inverted list for keyword *k* on query server *j*.
- $r_{i,j}$: number of results obtained for query q_i on query server *j*.

The Terrier[2] IR system described in [20] is used to estimate the parameters for the analytical model, obtaining the following values: ti = 62.335ms, ts = 0.03ms, tr = 1.15μs, tc_0 = -470, tc_1 = 0.0, tc_2 = 62 [4]. The document model parameters ($d_{k,j}$ and $r_{i,j}$) are simulated from the SPIRIT collection, which consists of 94,552,870 documents and 1 terabyte (TB) of text [13]. Each query is generated as a sequence of K terms ($t_1,\ldots,t_k$), independently and identically distributed, following the skewed query model [12]. The skewed query model sets the probability of a term occurring in a query proportional to its frequency in the vocabulary, and provides more realistic queries than the uniform query model [6].

The network parameters ($rq_{i,j}$ and $ra_{i,j}$) that determine the transmission times among the hosts cannot be estimated using an analytical model, as they depend directly on the network load of each moment. Therefore, a network simulation model is defined.

In [5] and [6], the network simulation model was based on a shared access local area network (LAN), where the transmission media is shared out among all the hosts, which must compete to access the media and send their transmissions. This network simulation model had certain limitations (e.g. not considering the maximum number of hosts connected to the LAN or the maximum size of the network) that reduced the capacities of the simulated IR systems.

With the aim of improving the limitations of this initial network model, a new model was defined in [4], equivalent to a switched network FastEthernet 100BASE-T at 100Mbps. The switched LAN is the evolution of the shared access networking technology and it is based on a device named switch, which centralises the communication among the hosts. In this way, the switch will reduce the transmission conflicts, because a host only has to compete with other hosts that want to communicate with the same destination, increasing the effective network speed.

Using this new network model, a more extensive and realistic simulation model is defined, where the hosts are interconnected via one or more switches, depending on the number of hosts to be interconnected (assuming that each switch has a capacity for 64 hosts). Moreover, the overhead estimation is carried out exhaustively, taking into account the different headers of the communication protocols, IP fragmentation, and even the propagation delay [4]. The design of this new network model has also extended the capacity to represent multicast messages. The multicast messages allow sending one message to multiple recipients, instead of sending one message to each recipient (unicast messages). In a distributed IR system based on local inverted files, multicast messages are especially useful to reduce the number of messages required to distribute the queries to the query servers from the brokers.

In [4], an extended description of the switched network simulation model can be found, along with a detailed comparison of the real IR system with the simulation model, confirming their correspondence. A brief description of the network simulation model can also be found in the short article [3]. In all the experiments reported in this paper, this new switched network simulation model is used in order to obtain realistic conclusions when simulating and comparing the clustered and replicated systems.

---

[2] A core version of the Terrier system can be downloaded from http://ir.dcs.gla.ac.uk/terrier

3 Experiments

The objective of the experiments in this paper is to compare the performance of a replicated and clustered IR system, both in terms of throughput and response time, using a realistic setting based on the switched network simulation model described in the previous section. In [5] and [6], the main conclusions showed that a clustered system will outperform a replicated system if a notable number of query servers is used (e.g. 1024). These experiments, however, were based on a shared access network, which produced the saturation of the network in the replicated system. Moreover, only four replicas were defined in the considered replicated system.

In the new experiments conducted in this paper, we provide a detailed comparison between a replicated system and a clustered system using 1024 query servers and a switched network supporting multicast. In addition, we consider replicated systems with up to 32 replicas.

3.1 Experimental Setting

For the replicated system we examine different configurations for the 1024 query servers: *1, 2, 4, 8, 16 and 32* replicas (with 1024, 512, 256, 128, 64 and 32 query servers per replica, respectively). The optimal number of brokers required for the replicated system is calculated as *3R*, where *R* is the number of replicas, as shown in [4]. The optimal number of brokers is the minimum number of brokers necessary to obtain the best throughput and response time performance (there are no improvements by further increasing this number).

As mentioned in Section 1, a clustered system must be configured a-priori based on the distribution of queries that the IR system is likely to receive. For the configuration of the clustered system in the experiments reported in this section, we used the work by Spink et al. [22], where a set of real Web queries is categorised into 11 different topics considering three different years: 1997, 1999 and 2001. Table 1 provides a summary of the 11 topics and the percentage of queries through the different years.

We assume that each topic is indexed in a different cluster. The SPIRIT collection [13] is divided into 11 sub-collections with an inverted file of approximately the same size that is 8.5 million documents in each sub-collection. Therefore, the 11 defined clusters will index the same number of documents, although using a different number of servers. This setting is selected because we are more interested in the configuration of each cluster, rather than the distribution of the topics. Hence, the configurations of the clusters will fit the same throughput curve, generating a simpler simulation model.

In the reported simulations, the number of queries is fixed to 200 and the queries will retrieve 3 million documents on average. The base sub-collection of 8.5 million documents has been distributed over *N* query servers using a switched network and three brokers, where *N* = *1, 2, 4, 8, 16, 32, 64, 128, 256 and 512*. In Table 1, the column *Configuration* describes the query servers assigned to each topic. The first number represents the number of the distributed query servers, and the second, the number of replicas in each cluster.

Table 1. Distribution of queries across general topic categories, and the used configurations for the simulated clustered system. The column Configuration describes the query servers assigned to each topic. The first number represents the number of the distributed query servers, and the second one represents the number of replicas in each cluster.

Topics	1997	1999	2001	Configuration
Entertainment	19.64 %	7.73 %	6.65 %	67 * 3
Pornography	16.54 %	7.73 %	8.56 %	56 * 3
Commerce	13.03 %	24.73 %	24.76 %	66 * 2
Computers	12.24 %	11.13 %	9.65 %	63 * 2
Sciences	9.24 %	8.02 %	7.55 %	48 * 2
People	6.43 %	20.53 %	19.75 %	66 * 1
Society	5.44 %	4.43 %	3.96 %	56 * 1
Education	5.33 %	5.52 %	4.55 %	55 * 1
Arts	5.14 %	1.33 %	1.16 %	53 * 1
Non-English	3.84 %	7.03 %	11.36 %	39 * 1
Government	3.13 %	1.82 %	2.05 %	32 * 1

The clustered system is configured in accordance with the distribution of the topics of the year 1997. The replications for the most popular topics are maximised, but the number of query servers in each replica is kept as close as possible to 64 in order to obtain an appropriate response time. Indeed, in [4], the authors studied the improvement obtained with a switched network and the figures showed that with less than 64 query servers the performance of the system decreases importantly. The number of brokers is selected taking into account the sum of the replicas in each cluster (i.e. $R=18$ replicas), and calculating the optimal number of brokers as $3R$, as described in [4]. For completeness, we also report results with the optimal number of brokers $2R + 1$, suggested in [5] and [6].

In all our experiments, as stressed in Section 1, the performance is measured using the throughput and the response time. The throughput is measured considering that the system is operating in batch mode and that there is a processing queue of 200 queries. The response time is measured assuming that the queries will arrive to the IR system following an Exponential distribution [7], with mean 500 milliseconds and simulating 200 queries.

3.2 Replicated and Clustered System Comparison

The results obtained for the simulated clustered and replicated systems are presented in Table 2. The column *Replicated* describes the configurations for the studied replicated system. The first number represents the number of replicas, and the second one represents the number of the distributed query servers. In all the replicated system configurations, the optimal number of brokers is defined as $3R$. The shaded cells represent the optimal configurations for the replicated and clustered systems. In this case, the optimal configuration is the one that achieves the best trade-off between the minimal response time and the maximal throughput.

Regarding the replicated systems, we observe that the throughput increases as the number of replicas increases due to the higher level of parallelism in the system. At the same time, the response time of the system is decreasing as the replication increases, except for the last configuration (*32x32*), where the reduced distribution of the index in each replication (only 32 query servers) increases the response time to 2658 milliseconds per query (17% more than the optimal configuration).

On the other hand, Table 2 shows that the throughput of the clustered system is maximised if *3R* brokers are used. This configuration outperforms the *2R + 1* configuration in terms of throughput, extending the conclusions obtained in [4] for the replicated system to the clustered system. The number of brokers does not have an important repercussion on the response time as the queries are processed nearly sequentially, which leads to minimal parallelisation. In a clustered system, the number of replicas (*R*) is calculated as the sum of all the replicas through all the clusters (e.g. 18 replicas, obtained from the configuration in Table 1).

Comparing the two types of systems, the results show that a replicated system with 16 replicas will achieve a better throughput and response time than the clustered system defined. In both cases the level of parallelism achieved is quite similar (with 16 and 18 parallel groups, respectively). The main benefit achieved with the clustered system is a reduction in the network traffic, which is crucial if the network is the main bottleneck of the system. However, the switched network has solved this problem improving markedly the performance of the replicated system, which is able to outperform the clustered system.

This result suggests that the distribution must be used to reduce the response times and the replication must be used to increase the query throughput of the system. For example, this is the approach used by the Google web search service [1].

The main conclusion of the above set of experiments is that the performance of a clustered system (both in throughput and response time) does not improve the performance obtained by the best replicated system. This result is related to the use of a switched network. The switched network has eliminated the bottleneck in the network, markedly improving the performance of the replicated systems. On the other hand, the main advantage of a clustered system is the reduction of network traffic, which is less relevant when a switched network model is used.

Table 2. Throughput (queries/second) and response time (milliseconds) for the clustered and replicated systems (with the optimal number of brokers as *3R*), using a switched network supporting multicast

Replicated	Throughput	Response Time	*Clustered* Brokers	*Clustered* Year	Throughput	Response Time
1x1024	0.70	4247.83	*3R*	*1997*	7.60	2404.11
2x512	1.38	4257.67	*3R*	*1999*	3.23	2828.11
4x256	2.69	3231.22	*3R*	*2001*	3.59	2960.87
8x128	5.03	2354.92	*2R+1*	*1997*	7.17	2380.20
16x64	8.47	2274.09	*2R+1*	*1999*	3.11	3165.59
32x32	12.92	2658.93	*2R+1*	*2001*	3.43	2863.65

3.3 Query Topics Change

As we described in the introduction, a clustered system must be configured a-priori based on the distribution of the queries that the IR system expects to receive. In our experiments above, the clustered system was configured based on the queries distribution for the year 1997 (Table 1, second column).

In this section, we study the effect of changes in the topics distribution over time in the performance of clustered systems. Obviously, the performance of a replicated system is query independent. Therefore, the performance values obtained for a replicated system do not change: 8.47 queries per second and 2.27 seconds per query on average, for the optimal configuration (See Table 2, row *16x64*). However, the performance of a clustered system when the queries distribution varies can be severely affected, as we will discuss below.

Indeed, in the experiments reported in Table 2, we also used the query distribution for the years 1999 and 2001 (Table 1, third and forth column respectively) to simulate the queries in the clustered system. The results (see Table 2) show more than 50% throughput reduction in the *3R* configuration (from 7.60 queries per second to 3.23 and 3.59 queries per second for the years 1999 and 2001, respectively) and in the *2R + 1* configuration (from 7.17 queries per second to 3.11 and 3.43 queries per second for the years 1999 and 2001, respectively), and an increase in the response time between 17% and 33%, in both configurations, for the years 1999 and 2001.

The negative effect on the response time is less marked, because each cluster has been configured with enough query servers per replica to obtain an appropriate response time. On the other hand, the changes in the popular topics imply that smaller clusters will receive more queries, decreasing their throughput, while larger clusters will receive fewer queries and may have idle periods, decreasing the overall throughput of the system.

In [5] and [6], the throughput worsening due to the changes in the topics distribution was also confirmed, although the distribution for the year 2001 was the baseline and the reduction in the performance was mainly manifested in the year 1997, with nearly no changes with the year 1999.

The important repercussion of the changes in the queries distribution in the performance of the clustered systems entails a permanent monitoring of the queries submitted to the IR system and, if necessary, the modification of the clusters configuration. This point raises new research concerns, such as the estimation of the threshold in the topics change where it is worth modifying the configuration, looking for a balance between the performance improvement and the cost of modifying the cluster configuration of the IR system.

4 Discussion

We have shown that the performance obtained (both in terms of throughput and response time) for a clustered system does not improve over the one obtained by the best replicated system.

This result, which in some way contradicts the conclusions in [5] and [6], is related to the use of a switched network. The main benefit of a clustered system is the reduction of network traffic. However, the switched network has eliminated the bottleneck in the network, markedly improving the performance of the replicated systems, which exceeds that of the clustered systems. It is interesting to state that the clustered system results represent a best-case performance as we have assumed that documents can be split in non-overlapping clusters and queries can be assigned univocally to one cluster.

Moreover, we have illustrated the negative performance effect of the changes over time in the topics distribution, when a clustered system is used, as opposed to the performance of a replicated system, which is query independent. As a consequence, the configuration of a clustered system needs to be modified according to the topics distribution and their drift over time. This may prove to be a practical problem in the Web context, where the information needs of the users and their interest may markedly vary over time, for example, in reaction to contemporary events and concerns [2].

When building a large-scale IR system, the index distribution is necessary to deal with a high volume of data (and to keep the response times low) and the parallelisation is necessary to process multiple queries concurrently. The results obtained in this work suggest that the replicated systems are the best option for the parallelisation in terms of performance (throughput and response time) and stability through the time (as it is query independent).

Finally, it is important to mention that there could be other factors not represented in these simulations that could improve the performance of the clustered systems. For example, one benefit of the clustered systems is the reduction in the size of the collection indexed by each cluster. This may produce a reduction in the number of relevant results obtained by the IR system, since the final results must be associated with the relevant cluster for the query. In all the experiments reported in this paper, both types of systems retrieve exactly the same number of relevant documents for each query. Nevertheless, in a real clustered system, some documents associated with other less relevant clusters will not be included and this could reduce the overall response time. It is however difficult to precisely model this as it depends on factors such as the size of the cluster, the length of the query and its type. In addition, the number of returned documents may affect precision and recall. As a consequence, in comparing the clustered and replicated systems, we simulated systems that return the same number of documents making the comparison fair in terms of precision and recall measures.

5 Conclusions

In this work, we have presented a detailed study of a clustered system and several replicated systems, comparing their performance in terms of throughput and response time. Using the extended simulation network model introduced in [4] to represent a switched network, we can perform a more accurate and realistic evaluation of the two types of distributed IR systems.

We showed that the throughput and response time for a clustered system do not improve the values obtained by the best replicated system. Moreover, the performance of a replicated system is independent of the changes over time in the query topics whilst there is a negative effect on the clustered system performance. This implies that the configuration of a clustered system needs to be adapted dynamically to the topics distribution. A possible future work would be to investigate how to automatically define the optimal configuration of a clustered system over time.

It is also important to consider that there are other factors that have not been taken into account, such as the reduction in the number of relevant documents, which may improve the performance of clustered systems. Another future work would be to study this effect on the response time and its repercussion in terms of precision and recall.

This work suggests that the replicated IR systems should be used to obtain a better degree of parallelism and throughput. In this sense, in our future work, we would like to study different solutions to reduce data interchange through the interconnection network and the workload on the brokers for the replicated and distributed IR systems.

Acknowledgements

The work of the first and second authors has been partially supported by the Spanish government under project TSI2005-07730.

The work of the third and fourth authors is funded by a UK Engineering and Physical Sciences Research Council (EPSRC) project grant, number GR/R90543/01. The project funds the development of the Terrier Information Retrieval framework (url: http://ir.dcs.gla.ac.uk/terrier).

We would also like to thank Mark Sanderson and Hideo Joho for giving us access to the 1TB dataset used for the SPIRIT Project.

References

1. Barroso, L.A., Dean, J., Hölzle, U.: Web search for a planet: The Google cluster architecture. *IEEE Micro*, 23(2), (2003) 22-28.
2. Beitzel, S.M., Jensen, E.C., Chowdhury, A., Grossman, D., Frieder, O.: Hourly Analysis of a Very Large Topically Categorized Web Query Log. In Proc. of the 27th Conf. on Research and Development in Information Retrieval. New York: ACM Press. (2004) 321-328
3. Cacheda, F., Carneiro, V., Plachouras, V., Ounis, I.: Performance Network Analysis for Distributed Information Retrieval Architectures. In Proc. of 27th European Conf. on Information Retrieval Research (ECIR'05), LNCS Vol, 3408, (2005) 527-529.
4. Cacheda, F., Carneiro, V., Plachouras, V., Ounis, I.: Performance Network Analysis for Distributed Information Retrieval Architectures. Information Processing and Management Journal, published on-line (2006)
5. Cacheda, F., Plachouras, V,. Ounis, I.: Performance Analysis of Distributed Architectures to Index One Terabyte of Text. In Proc. of 26th European Conf. on Information Retrieval Research, LNCS Vol. 2997, (2004) 394-408.
6. Cacheda, F., Plachouras, V., Ounis, I.: A Case Study of Distributed Information Retrieval Architectures to Index One Terabyte of Text. Information Processing and Management Journal, 41(5), (2005) 1141-1161
7. Cacheda, F., Viña, A.: Experiences retrieving information in the World Wide Web. In Proc. of the 6th IEEE Symposium on Computers and Communications. IEEE Computer Society, (2001) 72-79
8. Cahoon, B., McKinley, K.S.: Performance evaluation of a distributed architecture for information retrieval. In Proc. of 19th ACM-SIGIR International Conf. on Research and Development in Information Retrieval, New York: ACM Press. (1996) 110-118.
9. Frieder, O., Siegelmann, H. T.: On the Allocation of Documents in Multiprocessor Information Retrieval Systems. In Proc. of the 14th ACM-SIGIR Conf. on Research and Development in Information Retrieval. New York: ACM Press. (1991) 230-239

10. Hawking, D.: Scalable text retrieval for large digital libraries. Lecture Notes in Computer Science, Vol. 1324. (1997) 127-146
11. Hawking, D., Thistlewaite, P.: Methods for Information Server Selection. ACM Transactions on Information Systems, Vol. 17(1). (1999) 40-76
12. Jeong, B., Omiecinski, E.: Inverted File Partitioning Schemes in Multiple Disk Systems. IEEE Transactions on Parallel and Distributed Systems, Vol. 6(2). (1995) 142-153
13. Jones, C. B., Purves, R., Ruas, A., Sanderson, M., Sester, M., van Kreveld, M., Weibel, R.: Spatial information retrieval and geographical ontologies an overview of the SPIRIT project. In Proc. of the 25th ACM-SIGIR Conf. on Research and Development in Information Retrieval. New York: ACM Press. (2002) 387-388
14. Lin, Z., Zhou, S.: Parallelizing I/O intensive applications for a workstation cluster: a case study. ACM SIGARCH Computer Architecture News, Vol. 21 (5). (1993) 15-22
15. Little, M. C.: *JavaSim User's Guide. Public Release 0.3, Version 1.0.* University of Newcastle upon Tyne. Retrieved 1 June, 2003. http://javasim.ncl.ac.uk/manual/javasim.pdf
16. Lu, Z., McKinley, K.: Partial collection replication versus caching for information retrieval systems. In Proc. of the 25th ACM-SIGIR Conf. on Research and Development in Information Retrieval. New York: ACM Press. (2000) 248-255
17. Moffat, A., Webber, W., Zobel, J.: Load Balancing for Term-Distributed Parallel Retrieval. In Proc. of the 29th ACM-SIGIR Conf. on Research and Development in Information Retrieval. New York: ACM Press. (2006) 348-355
18. Moffat, A., Webber, W., Zobel, J., Baeza-Yates, R.: A pipelined architecture for distributed text query evaluation. *Information Retrieval*, published on-line. (2006)
19. Moffat, A., Zobel, J.: What does it mean to "measure performance"? In Proc. of the 5th International Conf. on Web Information Systems, LNCS Vol. 3306. (2004) 1-12
20. Ounis, I., Amati, G., Plachouras, V., He, B., Macdonald, C., Lioma, C.: Terrier: A High Performance and Scalable Information Retrieval Platform. In Proc. of ACM SIGIR'06 Workshop on Open Source Information Retrieval. (2006)
21. Ribeiro-Neto, B., Barbosa, R.: Query performance for tightly coupled distributed digital libraries. Proc. 3rd ACM Conf. on Digital Libraries. New York: ACM Press. (1998) 182-190
22. Spink, A., Jansen, B. J., Wolfram, D., Saracevic, T.: From e-sex to e-commerce: Web search changes. IEEE Computer Vol. 35(3). (1998) 107-109
23. Tomasic, A., Garcia-Molina, H.: Performance of inverted indices in shared-nothing distributed text document information retrieval systems. In Proc. 2nd Inter. Conf. on Parallel and Distributed Info. Systems. San Diego, California: IEEE Computer Society. (1993) 8-17

A Study of a Weighting Scheme for Information Retrieval in Hierarchical Peer-to-Peer Networks

Massimo Melucci and Alberto Poggiani

University of Padova
Department of Information Engineering
{massimo.melucci,alberto.poggiani}@dei.unipd.it

Abstract. The experimental results show that the proposed simple weighting scheme helps retrieve a significant proportion of relevant data after traversing only a small portion of a peer-to-peer hierarchical peer network in a depth-first manner. A real, large, highly heterogeneous test collection searched by very short, ambiguous queries was used for supporting the results. The efficiency and the effectiveness would suggest the implementation, for instance, in audio-video information retrieval systems, digital libraries or personal archives.

1 Introduction

A Web search engine (SE) is the main instrument for finding the documents relevant to user's need [1]. When some SEs are inter-connected to each other in an anarchic way, the stores of information may become much large without big investments. Peer-to-Peer (P2P) Information Retrieval (IR) can succeed in accessing much larger stores of information. P2P-IR systems perform content-based search technique across a network of peers. This paradigm favors scalability, increases system resilience, avoids unique points of failure, distributes indexing and query processing tasks to multiple computing nodes, re-uses processing power and memory at low-cost from shared, under utilized resources. In a digital library system, for instance, a network of peers can support the federation of heterogeneous, independent information sources without cumbersome organization and coordination overhead.

The peers have little knowledge about each other, therefore the evidence about document relevance is rather limited and can definitely be less readily collected than the evidence collected from centralized or distributed systems. These networks can also be very large, yet the bandwidth is rather small. Therefore query flooding is impracticable and a small portion of the network has to be selected in order to contact the peers that potentially store relevant documents. What is needed is a model which ranks documents and peers in a way that the probability of reaching all and only relevant data is maximized by minimizing network traversal. Clearly, then, retrieval modeling in P2P network becomes crucial in order to improve recall-oriented task completion. In particular, it was found that term weighting schemes can be tuned in order for improving IR effectiveness [2].

G. Amati, C. Carpineto, and G. Romano (Eds.): ECIR 2007, LNCS 4425, pp. 136–147, 2007.

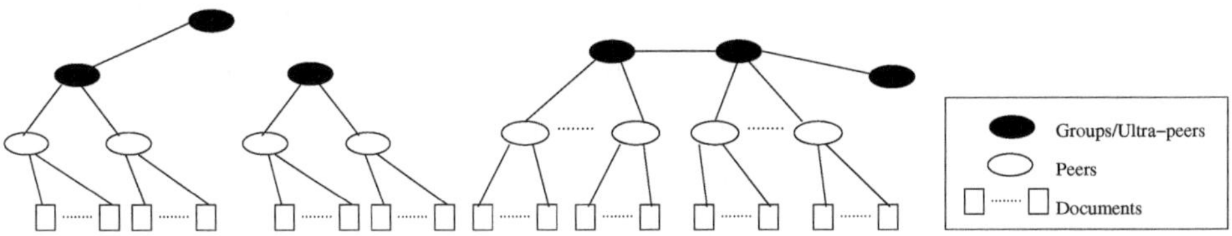

Fig. 1. A pictorial description of the P2P architecture

This paper reports on the evaluation of a simple weighting scheme for IR across P2P networks. The weighting scheme was proposed in [3,4] and evaluated by using the 1988 Associated Press test collection and the TREC-3, TREC-4 and TREC-5 topics. Although the evaluation in [3,4] gave useful insights about the effectiveness of the weighting scheme, its size was quite limited and partially comparable to the realistic collections. While some research works employed artificial relevance judgments, the evaluation reported in this paper in contrast employed a real, large and well-known test collection. The results help draw more stable conclusions about the effectiveness and the efficiency of the weighting scheme than the similar conclusions drawn after using medium-size test collections.

Because of the shortage of space, the illustration of an exhaustive background of this subject is impossible. Some references about motivations, architectures and approaches to P2P-IR and to distributed IR are, for example, [5,6,7,8,9,10,11,12,13].

2 The Weighting Scheme

The P2P network is hierarchically organized as three levels – groups, peers and documents – such that an element of every level belongs to the element of the upper level. A pictorial description is provided in Fig. 1. How documents, peers, groups belonging to these levels are scored with respect to a query Q is illustrated in the following. The peer sends the query to the ultra-peer of its group. The ultra-peer routes the query to the top ranked peers of its own group, to itself and to the top-ranked connected neighboring ultra-peers. Every selected peer ranks its documents with respect to the query and sends them back to the ultra-peer with the statistics needed to merge the result lists coming from the other peers. The top-ranked connected neighboring ultra-peers recursively replicate the process. Time-To-Live (TTL) is decreased by 1 for every query forward. If an ultra-peer cannot answer the query, the peers of its group are not searched, and the query is in this case forwarded to every neighboring ultra-peer. When a peer or an ultra-peer contact a peer, the connected peers are weighted and ranked.

As top-ranked documents, peers and ultra-peers are needed, a ranking scheme is defined as follows. The ultra-peers u's are ranked by

$$w_u^{(3)} = \sum_{i \in Q} w_{i,u}^{(3)} \tag{1}$$

where $w_{i,u}^{(3)}$ is the weight of keyword i in the group led by u — the superscript identifies the level of the network. Moreover, $w_{i,u}^{(3)} = \sum_{p \in u} w_{i,p}^{(2)}$, provided $w_{i,p}^{(2)}$ is the weight of i in the peer $p \in u$. The latter is computed as $w_{i,p}^{(2)} = \left(\sum_{j \in p} w_{i,j}^{(1)}\right) \mathrm{irf}_{i,u}^{(2)}$ where $\mathrm{irf}_{i,u}^{(2)} = \log(N_u^{(2)} + 0.5)/n_{i,u}^{(2)}$, $N_u^{(2)}$ is the number of peers p in the group led by ultra-peer u, $n_{i,u}^{(2)}$ is the number of peers p whose documents are indexed by i and $w_{i,j}^{(1)}$ is the weight of i in the document j of peer p. The peers of a group have been ranked by

$$w_p^{(2)} = \sum_{i \in Q} w_{i,p}^{(2)} \,. \tag{2}$$

Lastly, $w_{i,j}^{(1)} = \mathrm{TF}_{i,j} \cdot \mathrm{IDF}_{i,p}$ where $\mathrm{TF}_{i,j} = \log\left(f_{i,j,p} + 1\right) / \sum_{i \in j} \log\left(f_{i,j,p} + 1\right) \cdot U_{j,p}/(1 + 0.0115 \cdot U_{j,p})$ and $\mathrm{IDF}_{i,p} = \log\left(N_p^{(1)} - n_{i,p}^{(1)}\right)/n_{i,p}^{(1)}$, $U_{j,p}$ is the number of unique terms in the document j, $N_p^{(1)}$ is the total number of documents in peer p, $n_{i,p}^{(1)}$ is the number of documents in peer p indexed by i, as reported in [14]. Therefore the score of document j in peer p for the given query Q will be

$$\sum_{i \in Q} w_{i,j}^{(1)} \,. \tag{3}$$

3 Experiments

The experiments aimed at measuring the proportion of recall as the portion of visited network increased. The portion of visited network was function of the first ultra-peer contacted, the number of times the query is routed to an ultra-peer, that is the TTL, the number of ultra-peers contacted by an ultra-peer, and the number of peers to which a query was routed. The baseline experiment consisted of a local search such that every document has been stored into a single peer, as it were a traditional centralized IR system. The experiments have been performed by using the MySQL fulltext capabilities [15].

Test Collection. A P2P network was simulated by using a large test collection in a laboratory setting. Previous studies used different experimental testbeds. It was decided to use and organize a well-known test collection so as to have a realistic P2P network. The choice fell back on DLLC (Digital Libraries Lu and Callan), i.e. the testbed created at Carnegie Mellon University for evaluating the performance of content-based retrieval in P2P networks. DLLC is based on the TREC WT10g Web test collection of the Text Retrieval Conference (TREC), which is a 10GB, 1.69 million document set. Every document in this collection contains the HTML code of a webpage. One index on fulltext webpage body and title was built by our experimental system. For full documentation and applications, see [10,12,16,17,18,19,20].

Network Topology. Some important remarks and findings about recent snapshots (Apr. 2004 – Feb. 2005) of Gnutella network reported in [21,22,23] have been in

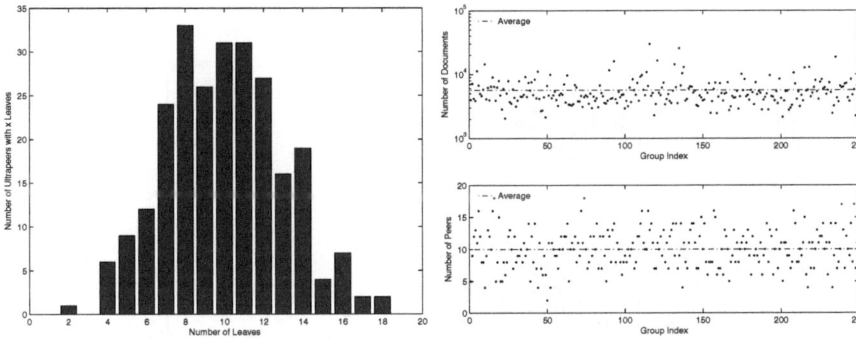

Fig. 2. (a) Distribution of Group Sizes. (b) Distribution of Peers and Documents over Groups.

this paper taken into account, that is: The node degree in the top-level overlay does not exhibit a power-law distribution — it has rather a significant spike around 30, which is the default maximum degree in the most popular Gnutella implementation (LimeWire and BearShare). There are much fewer peers with a degree greater than 30. The number of peers with degree lower than 30 is very significant, and the distribution of degree is quite uniform between 0 and 30. A significant minority of ultra-peers are connected to less than 30 leaves, which indicates the availability of open slots for new leaves for joining the overlay. This clearly contrasts with the power-law degree distribution reported in previous studies [24].

The 250 peers with the highest number of documents have been elected as ultra-peers. The ultra-peers have not been assumed to have greater computational power than peers. For creating the groups, every peer was assigned to one and only to one randomly chosen ultra-peer. This reflects peers' behavior in real Gnutella 0.6 environment — a peer does not select by content for connecting to the group, but joins the first group selected from a list which answers peer's request of connections. Doing so a Gaussian distribution of the number of peers over the groups was achieved, as depicted in Fig. 2(a), while Fig. 2(b) shows the distribution of documents and of peers over the groups. The connections between ultra-peers have been generated randomly. Every ultra-peer has a number of connection uniformly distributed between 1 and 5; every ultra-peer has on average three neighbors ultra-peer. In this way the topology resembles the topology described in [21,22,23].

Queries. Experiments used the topics of the TREC-9 and TREC-2001 Web tracks [25]. As in [4,10] only the title field of the topics was used for generating the queries. The reason was that, in real P2P settings, queries are on average very short [26] and the title meets this condition. On average each topic in TREC 9 has 47.7 relevant documents while in TREC 2001 has 62.2 relevant documents as depicted in Fig. 3(a). Fig. 3(b) shows the fraction of the total number of peers that a query needs to be forwarded to, in order to achieve 100%

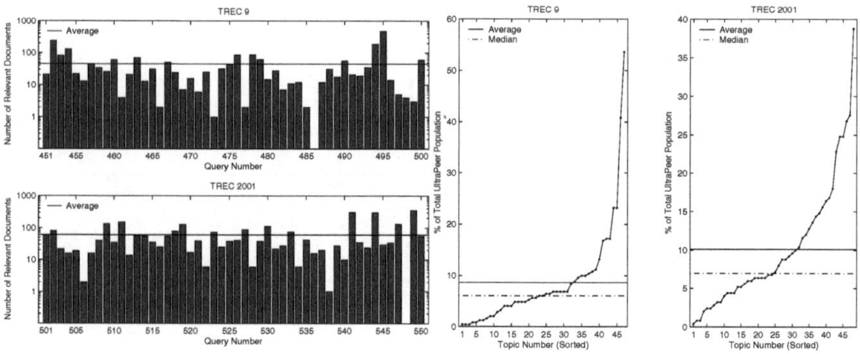

Fig. 3. (a) Distribution of Relevant Documents over the Topics. (b) Number of peer collections needed for 100% recall (sorted per topic figure).

recall. In this case 70% of the queries can fully be answered after contacting only less than 1% of the total peer population, and 90% needs for contacting not more than 2%. Of course, these percentages are valid when the peers containing the relevant documents are known, and should not be confused with the percentages given in the following. These figures give an idea of the difficulty of the task studied. Due to the "right" peers are unknown, peer and ultra-peer selection is a hard task and should work very well so as to make the P2P-IR system effective.

It should also be noted that "real" queries, and not "artificial" queries such as those employed in other studies, have been used in our experiments. Therefore "real" relevance assessments have been used thus making the task much more difficult and results less "good" than the task based on "artificial" queries and assessments: Our interest was indeed in realistic IR across P2P networks.

Evaluation Settings. The following parameters have been used for testing the retrieval algorithm: TTL is the number of times a query is routed to an ultra-peer. If TTL $= 1$ only the ultra-peer leading the group of the peer to which the query was submitted is contacted. If TTL $= 2$ the first ultra-peer and its m top-ranked neighbors are contacted, and so on. Since the contacted ultra-peers are ranked by Equation 1, the m top-ranked ultra-peers are selected. In the experiments $m \in \{1, 2, 3\}$. The contacted peers are ranked by Equation 2 and the k peers with the highest $w_p^{(2)}$ values have been matched against the query, and then selected. In the experiments $k \in \{1, 3, 5, 7, 10\}$. The retrieved documents are ranked by using Equation 3 and the n top-ranked documents have been given back to the ultra-peer. In the experiments it was supposed $n = 20, 50$ are the most common values set by an end user.

For each chosen combination of values of TTL, m, k, n two runs have been performed: one for TREC-9 topics and TREC-2001 topics. The statistics have been averaged over 40 different starting ultra-peers, i.e. each run was repeated for every starting ultra-peer, with different out-degrees and capability of reaching other ultra-peers. The starting ultra-peers have been randomly selected and are

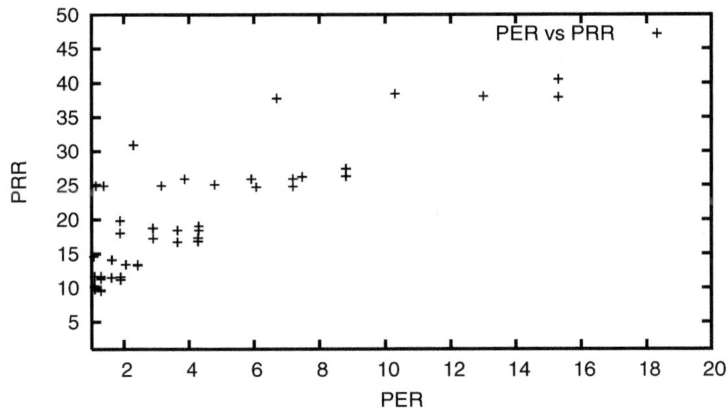

Fig. 4. The relationship between PER and PRR

always the same in every simulation. In this way our result are independent of the peer which originated the queries.

Three quantities have been measured and averaged over the starting ultra-peers for measuring to what extent the network can efficiently and effectively retrieve relevant documents: *Peer Exploration Ratio* (PER) is the ratio between the average number of selected peers and the total number of peers the network is composed of, i.e. 2500. *Ultra-Peer Exploration Ratio* (UPER) is the ratio between the average number of selected *ultra*-peers and the total number of ultra-peers the network is composed of, i.e. 250. *Peer Recall Ratio* (PRR) is the ratio between the number of relevant documents retrieved by visiting the network and the number of relevant documents retrieved by the baseline run. That is, if N documents are retrieved by visiting the network, the corresponding baseline run retrieved N documents from the centralized collection.

The relationship between PER and PRR is depicted in Figure 4 on the basis of the result tables reported in the following. The plot shows an almost-linear relationship between PER and PRR, although there are some variations which are explained in the rest of the section.

Weighting Scheme Evaluation Results. The following questions have been answered: (i) How much should TTL be? (ii) Is ultra-peer ranking effective? (iii) Is peer ranking effective? The experimental results are reported in the following tables for the TREC-9 topic set — the results for TREC-2001 were similar and have not been reported here due to the shortage of space. Every table is organized by TTL, m, k and each row includes the average number of actual contacted peers $\bar{k}$, the number of documents retrieved by every peer (n), the average number of actual contacted ultra-peers ($\bar{m}$), the number of documents retrieved across the network ($\tilde{N}$), and the number of relevant documents retrieved across the network (RDN). For example, the first row of Table 1(a) tells that when (i) $m = 1$ ultra-peers are contacted since the query is issued by a peer, (ii) the ultra-peer does not contact any other ultra-peer (TTL = 1) and (iii) the query

Table 1. TTL = 1, 2, $\bar{m} = 1$

k	n	$\bar{k}$	$\bar{N}$	RDN	PER	PRR
1	20	0.9	617.9	3.3	0.04%	7.4%
3	20	2.5	1458.7	4.4	0.10%	5.5%
5	20	3.7	2031.9	4.6	0.15%	4.8%
7	20	4.7	2422.6	4.6	0.19%	4.5%
10	20	5.5	2771.0	4.7	0.22%	4.4%
10	50	5.5	4625.8	5.5	0.22%	4.0%

(a) TTL=1, $m = 1$

k	n	$\bar{k}$	$\bar{N}$	RDN	PER	PRR
1	20	1.8	1332.7	7.9	0.07%	9.8%
3	20	5.0	3114.6	10.9	0.20%	8.2%
5	20	7.6	4310.0	11.3	0.31%	7.2%
7	20	9.6	5110.3	11.4	0.38%	6.9%
10	20	11.4	5804.9	11.7	0.45%	6.8%
10	50	11.4	9862.1	13.8	0.45%	5.9%

(b) TTL=2, $m = 1$

Table 2. TTL = 2, $m = 2, 3$

k	n	$\bar{k}$	$\bar{N}$	RDN	PER	PRR
1	20	2.6	1821.4	10.2	0.10%	9.8%
3	20	7.0	4287.7	14.0	0.38%	8.7%
5	20	10.6	5956.5	14.6	0.42%	7.7%
7	20	13.3	7063.6	14.9	0.53%	7.3%
10	20	15.8	8017.5	15.3	0.63%	7.1%
10	50	15.8	13552.0	17.9	0.63%	6.3%

(a) TTL=2, $m = 2$

k	n	$\bar{k}$	$\bar{N}$	RDN	PER	PRR
1	20	3.1	2159.6	11.3	0.12%	10.3%
3	20	8.4	5106.1	15.9	0.34%	8.6%
5	20	12.8	7110.8	16.6	0.51%	7.8%
7	20	16.1	8441.3	16.9	0.64%	7.3%
10	20	19.0	9577.1	17.4	0.76%	7.0%
10	50	19.0	16147.3	20.1	0.76%	6.6%

(b) TTL=2, $m = 3$

is routed to $k = 1$ top ranked peer of the group led by the ultra-peer, then, 617.9 documents are on average retrieved — the average has been computed over 40 starting ultra-peers; due to $n = 20$ are selected from the peer and the ultra-peer, 3.3 documents are relevant, thus exploring 0.04% of the network and achieving 7.4% of the recall which would be achieved if every document was collected in one centralized server.

How much should TTL be? Let us concentrate on a given PER and observe PRR for different TTL's (Tables 1 and 2). It can be noted that PRR varies when the value of PER is reached at different TTL's; for example, about PER = 2.0% of peers have been visited by only accessing ultra-peers with $m = 3, k = 10, \text{TTL} = 3$ and PRR was about 11%. A comparable PER was observed with $m = 3, k = 5, \text{TTL} = 4$, but PRR is significantly higher. The same pattern can be observed for different TTL's as reported in Tables 2, 3 and 4. It can be concluded that TTL is affecting effectiveness, provided PER. This suggests that the "best" strategy would increase the number of "hops" between the ultra-peers, i.e. TTL. In other words, the values of PRR with similar PER values suggest that a "depth-first search" (big TTL, small m and k) is better than a "breadth-first search" (small TTL and big k). Nevertheless, this pattern probably depends on the dataset and the actual content distribution between different ultra-peers. It was found, indeed, that different datasets and distributions of relevant data across the peers lead to opposite conclusions [4]. If the relevant documents were concentrated in a very few peers, then a depth-first strategy would be less effective than

Table 3. TTL = 3, 4 and TREC-9 topic set

k	n	$\bar{k}$	$\tilde{N}$	RDN	PER	PRR
$m = 1$, $\bar{m} = 2.85$, UPER = 1.14						
1	20	2.8	1979.6	12.9	0.11%	11.9%
3	20	7.5	4592.5	17.4	0.30%	9.9%
5	20	11.3	6326.6	17.9	0.45%	9.0%
7	20	14.1	7480.6	18.1	0.57%	8.4%
10	20	16.7	8444.6	18.5	0.67%	8.1%
10	50	16.7	14303.5	22.2	0.67%	7.5%
$m = 2$, $\bar{m} = 5.33$, UPER = 2.13						
1	20	5.2	3767.5	20.7	0.21%	12.2%
3	20	14.4	8822.4	28.7	0.58%	11.1%
5	20	21.8	12212.5	30.1	0.87%	10.0%
7	20	27.4	14477.7	30.7	1.10%	9.8%
10	20	32.3	16356.9	31.2	1.29%	9.5%
10	50	32.3	27596.5	37.5	1.29%	9.7%
$m = 3$, $\bar{m} = 7.73$, UPER = 3.09						
1	20	7.6	5447.0	27.6	0.31%	13.1%
3	20	21.1	12842.0	39.1	0.84%	12.2%
5	20	32.1	17867.0	41.0	1.29%	11.7%
7	20	40.4	21240.6	42.3	1.62%	11.5%
10	20	47.5	23984.3	42.9	1.90%	11.2%
10	50	47.5	40302.2	50.7	1.90%	11.6%

(a) TTL=3

k	n	$\bar{k}$	$\tilde{N}$	RDN	PER	PRR
$m = 1$, $\bar{m} = 3.77$, UPER = 1.51						
1	20	3.7	2631.6	18.0	0.15%	13.1%
3	20	9.9	6080.1	24.2	0.40%	11.1%
5	20	15.0	8383.7	25.0	0.60%	9.9%
7	20	18.7	9929.6	25.3	0.75%	9.3%
10	20	22.1	11226.3	25.8	0.88%	9.2%
10	50	22.1	19050.1	30.6	0.88%	9.0%
$m = 2$, $\bar{m} = 9.83$, UPER = 3.93						
1	20	9.7	7009.1	37.6	0.39%	15.2%
3	20	26.8	16464.7	52.6	1.07%	14.6%
5	20	40.8	22899.9	55.3	1.63%	14.1%
7	20	51.4	27237.8	56.6	2.06%	13.4%
10	20	60.6	30808.3	57.6	2.42%	13.3%
10	50	60.6	52030.4	68.3	2.42%	13.4%
$m = 3$, $\bar{m} = 17.05$, UPER = 6.82						
1	20	16.9	12125.9	57.4	0.68%	17.1%
3	20	47.2	28806.6	82.3	1.89%	18.0%
5	20	72.2	40273.3	87.4	2.89%	17.2%
7	20	91.1	48008.6	89.9	3.64%	16.7%
10	20	107.3	54265.9	91.5	4.29%	16.8%
10	50	107.2	91298.8	107.0	4.29%	17.3%

(b) TTL=4

the strategy adopted in these experiments. While clustering would help increase the concentration of relevant documents in a very few peers and the knowledge of the location of these relevant documents [7], this technique is infeasible in realistic or large experimental settings.

Is Ultra-Peer Ranking Effective? The weighting scheme proposed in this paper aims at reducing the portion of network visited for achieving a high proportion of recall. Therefore, ultra-peer ranking selects the ultra-peers which lead to the peers which are rich of relevant documents. If TTL = 1, ultra-peer ranking is ignored due to only one ultra-peer, i.e. the starting ultra-peer, is contacted independently of m. When TTL > 1, every contacted ultra-peer is called for choosing the "best" ultra-peer out of m connected ultra-peers and ranking is needed.

Let us consider TTL = 2 in Tables 2(a) and 2(b). PRR little increases from $m = 1$ to $m = 2$ and from the latter to $m = 3$ for a given k; for example, if $k = 1$, PRR increases from 9.8% to 10.3% when $m = 2$ increases to $m = 3$, respectively. If TTL = 4, and for $k = 1$, something similar happens, as reported in Table 3(b) — PRR increases from 13.1% ($m = 1$), to 15.2% ($m = 2$), and to 17.1% ($m = 3$). This small increase signifies that considering the second or the third ranked ultra-peer other than the top-ranked ultra-peer little affects effectiveness. Our explanation is that the single ultra-peer selected with $m = 1$

Table 4. TTL = 5, 6 and TREC-9 topic set

k	n	$\bar{k}$	$\tilde{N}$	RDN	PER	PRR
$m = 1$, $\bar{m} = 4.68$, UPER = 1.87						
1	20	4.9	3248.5	22.4	0.18%	13.6%
3	20	12.3	7483.0	30.1	0.49%	12.2%
5	20	18.4	10317.9	31.3	0.74%	11.0%
7	20	23.1	12229.1	31.8	0.92%	10.5%
10	20	27.3	13824.6	32.4	1.09%	10.3%
10	50	27.3	23493.5	38.2	1.09%	10.1%
$m = 2$, $\bar{m} = 17.22$, UPER = 6.89						
1	20	17.1	12346.0	64.6	0.68%	19.0%
3	20	47.3	29109.6	93.0	1.89%	19.8%
5	20	72.1	40572.6	98.4	2.88%	18.7%
7	20	91.0	48350.9	101.0	3.64%	18.4%
10	20	107.7	54799.5	102.7	4.31%	18.4%
10	50	107.7	92637.6	121.7	4.31%	19.0%
$m = 3$, $\bar{m} = 34.64$, UPER = 13.85						
1	20	34.5	24855.1	113.2	1.38%	24.9%
3	20	96.7	59182.8	164.4	3.87%	25.9%
5	20	148.1	82810.4	175.9	5.92%	25.9%
7	20	186.9	98779.2	181.6	7.48%	26.2%
10	20	220.3	111784.4	185.0	8.81%	26.3%
10	50	220.3	188350.0	219.8	8.81%	27.4%

(a) TTL=5

k	n	$\bar{k}$	$\tilde{N}$	RDN	PER	PRR
$m = 1$, $\bar{m} = 5.56$, UPER = 2.23						
1	20	5.5	3840.6	26.5	0.22%	14.4%
3	20	14.5	8832.1	36.1	0.58%	13.5%
5	20	21.8	12172.7	37.4	0.87%	12.0%
7	20	27.3	14425.7	38.2	1.09%	11.7%
10	20	32.2	16318.3	39.0	1.29%	11.5%
10	50	32.2	27706.4	45.6	1.29%	11.3%
$m = 2$, $\bar{m} = 28.39$, UPER = 11.36						
1	20	28.3	20533.9	104.6	1.13%	24.9%
3	20	78.6	48538.5	151.1	3.14%	24.9%
5	20	120.0	67704.0	160.1	4.80%	25.1%
7	20	151.8	80672.3	164.9	6.07%	24.7%
10	20	180.0	91507.1	168.0	7.20%	24.8%
10	50	180.0	154478.6	198.7	7.20%	25.9%
$m = 3$, $\bar{m} = 59.94$, UPER = 23.97						
1	20	59.8	43126.5	192.9	2.39%	30.9%
3	20	168.0	102958.4	284.0	6.72%	37.7%
5	20	257.5	144294.2	303.8	10.30%	38.4%
7	20	325.4	172316.9	313.5	13.02%	38.0%
10	20	384.1	195307.3	320.6	15.37%	37.9%
10	50	384.1	329435.7	380.9	15.37%	40.5%

(b) TTL=6

accounts for the largest proportion of PRR and that the second or third top-ranked ultra-peer is important but to a much less extent than the top-ranked ultra-peer. Something similar happens for the other TTL's and for the other k's, as reported in Tables 3(a), 4(a) and 4(b).

In general, the first top-ranked ultra-peer accounts for the largest proportion of PRR. This outcome is an evidence supporting the hypothesis that the weighting scheme is capable for selecting the "best" ultra-peers.

Is Peer Ranking Effective? Once the best ultra-peers have been selected, the best peers belonging to those groups had to be chosen. The values of PRR achieved with different k values for the same TTL and m have been computed for evaluating peer ranking effectiveness. In this way to what extent PRR varies by varying the number of select peers is measured. The previously cited tables report the PRR values achieved varying TTL, for each value of k and a fixed value of m.

A different pattern can be observed when the values of PRR are compared by varying k. The results bear evidence of the initial contribution to PRR of the one or three top-ranked peers. As the number of top-ranked peers increases, the values of PRR decreases. This result suggests that the selection of the first top-ranked peer accounts for a very significant proportion of PRR and that additional peers are little useful, if not disadvantageous, for increasing the quantity of relevant documents.

The negative trend of PRR when k increase is reverted once the exploration of the network becomes larger due to larger TTL's and/or m's. Indeed, Tables 3(b), 4(a) and 4(b) report that PRR increases if the number of peers increases, yet the contribution of the k-top ranked peers, $k > 1$, decreases as k increases — a similar pattern was observed as far concerned the ultra-peers.

One explanation can be provided considering the distribution of relevant documents, which are concentrated in a very small subset of the peers. If TTL and m are small, the event that the ultra-peer leading to a group of peers storing many relevant documents can be reached is very improbable. The query is little likely to have been routed by the contacted ultra-peer(s) to "relevant" peers. Therefore increasing k causes an increase of retrieved documents without an increase of relevant documents. On the contrary, if TTL and m are not small, the contacted ultra-peers may lead to the group of peers storing the relevant documents. The query is likely to have been routed to the "relevant" peers because of the higher number of contacted ultra-peers. As the relevant documents may be concentrated in some of the contacted peers, one group is likely to include all these peers. Therefore increasing k causes an increase of retrieved relevant documents.

Conclusions about Effectiveness and Efficiency. In general, the results suggests that a partial exploration of the network produces levels of recall which should be regarded as acceptable if the difficulty of the test collection is considered. Indeed, after contacting about 16% of the peers, a system based on the proposed weighting scheme can retrieve about 40% of the relevant documents that can be retrieved by a centralized system, as reported in Table 4(b).

Provided the characteristics of the test collection and of the network topology, it is our belief this is an encouraging results. Although the adoption of a similarity-based algorithm for organizing leaf nodes by topic as performed in DLLC could improve the routing step, this adoption is not affordable nor is it realistic with our system — a peer is required the statistics of the full network for connecting to the right group.

One effective strategy might be: (*i*) Decide for a "depth-first visit" strategy. (*ii*) Two or three at most top-ranked ultra-peers are selected by the ultra-peer that first received the query. (*iii*) The contacted ultra-peers contact very few peers, perhaps only the first top-ranked peer. The contacted peers retrieve in parallel some documents which are returned to the calling ultra-peer. Because k is low, the network bandwidth used is relatively small.

Peers and ultra-peers need to communicate to each other some data about local indexes for helping rank peer and ultra-peer. Because the network is a hierarchy — every peer connects to one ultra-peer — a peer periodically communicates a summary of its own index to one ultra-peer. The summary is a straightforward list of the term weights computed for every peer — this information is directly gathered from the local indexes. Term weight list update is required only if a new document is added to the peer and the latter decides to make the document as publicly available. The ultra-peers are connected to a few others. Every ultra-peer transmits a summary of the summaries received from its own peers to the other ultra-peers. The summary of the summaries is a list of term weights for every group. For both lists compression algorithms can be performed thus making synchronization quite fast.

4 Conclusions

This paper covers an empirical study using standard collections in a simulated P2P testbed. The basic rationale taken for the particular formulas used is the idea to select ultra-peers, then peers and lastly documents using the same type of formula. The particular parameters chosen for studying have been consistent with the characteristics of the P2P networks. In the future non-uniform parameters will be investigated. The approach is exploration rather than hypothesis testing. Although this study is mostly suggestive, it is a reasonable approach at this stage of the development and may provide results that can immediately be adopted.

One of the main points made in this paper concerning the efficacy of the ranking scheme is that the first peer or ultra-peer visited gives a quite large proportion of recall and that subsequent ultra-peers little contribute. It is thought that this is an evidence of good ranking, yet there may be obviously many more relevant peers or ultra-peers, as recall is very low after one ultra-peer is visited — therefore the addition of the next high-ranking peers should also contribute to recall; these are not contributions as large as the first, but still positive. This is true, but one should have to do with efficiency because PER would increase more quickly than PRR, namely, the increase of PER is not compensated by a significant increase of PRR — such an issue should be considered.

References

1. A. Broder. A taxonomy of Web search. *SIGIR Forum*, 36(2):3–10, 2002.
2. G. Salton and C. Buckley. Term weighting approaches in automatic text retrieval. *IPM*, 24(5):513–523, 1988.
3. M. Melucci and R. Castiglion. A weighing framework for information retrieval in peer-to-peer networks. In *Proc. of DEXA Workshop*, pages 374–378, Copenaghen, August 22-26 2005. IEEE Press.
4. M. Melucci and R. Castiglion. An evaluation of a recursive weighing scheme for information retrieval in peer-to-peer networks. In *Proc. of CIKM Workshop on IR in P2P Networks*, pages 9–16, Bremen, Germany, November 4 2005. ACM Press.
5. M. Bawa, G. S. Manku, and P. Raghavan. Sets: Search enhanced by topic-segmentation. In *Proc. of SIGIR*. ACM Press, 2003.
6. S. Chernov. *Result Merging in a Peer-to-Peer Web Search Engine*. PhD thesis, University of Saarland, February 2005.
7. I.A. Klampanos, V. Poznanski, J. Jose, and P. Dickman. A Suite of Testbeds for the Realistic Evaluation of Peer-to-Peer Information Retrieval Systems. *Proc. of ECIR*, volume 3408 of *LNCS*. Springer-Verlag.
8. J. Callan. Distributed information retrieval. In W. B. Croft, editor, *Advances in information retrieval*, chapter 5, pages 127–150. Kluwer Academic Publishers, 2000.

9. L. Gravano, K. Chang, A. Paepcke, and H. Garcia-Molina. STARTS: Stanford proposal for internet retrieval and search. Technical Report SIDL-WP-1996-0043, Computer Science Department, Stanford University, 1996.
10. J. Lu and J. Callan. Federated search of text-based digital libraries in hierarchical peer-to-peer networks. In *Proc. of SIGIR*, Sheffield, UK, 2004. ACM Press.
11. L. S. Larkey, M. E. Connell, and J. P. Callan. Collection selection and results merging with topically organized U.S. patents and TREC data. In *Proc. of CIKM*, pages 282–289. ACM Press, McLean, Virginia, US 2000.
12. J. Lu and J. Callan. Merging retrieval results in hierarchical peer-to-peer networks. In *Proc. of SIGIR*, Sheffield, UK, 2004. ACM Press.
13. L.Si and J. Callan. A semi-supervised learning method to merge search engine results. *ACM TOIS*, 21(4):457–491, October 2003.
14. A. Singhal, C. Buckley, and M. Mitra. Pivoted document length normalization. In *Proc. of SIGIR*, pages 21–29, Zurich, Switzerland, 1996. ACM Press.
15. P. Gulutzan. MySQL's full-text formulas, January 2006. `http://www.databasejournal.com/features/mysql/article.php/3512461`.
16. J. Lu and J. Callan. Content-based retrieval in hybrid peer-to-peer networks. In *Proc. of CIKM*, 2003.
17. J. Lu and J. Callan. Peer-to-peer testbed definitions: trecwt10g-2500-bysource-v1 and trecwt10g-query-bydoc-v1, January 2006. `http://hartford.lti.cs.cmu.edu/callan/Data`.
18. D. Hawking. Overview of the TREC-9 Web track. In E. M. Voorhes and D. K. Harman, editors, *Proc. of TREC*, pages 87–101, Gaithersburg, Maryland, September 2001. Department of Commerce, NIST.
19. P. Bailey, N. Craswell, and D. Hawking. Engineering a multi-purpose test collection for Web retrieval experiments. *IPM*, 39(6):853–871, November 2003.
20. H. Nottelmann and N. Fuhr. Comparing different architectures for query routing in peer-to-peer networks. In *Proc. of ECIR*, LNCS, London, UK, 2006. Springer.
21. D. Stutzbach, R. Rejaie, and S. Sen. Characterizing unstructured overlay topologies in modern P2P file-sharing systems. In *Proc. of IMC*, pages 49–62, 2005.
22. S. Zhao, D. Stutzbach, and R. Rejaie. Characterizing files in the modern Gnutella network: A measurement study. In *Proc. of MMCN*, San Jose, CA, January 2006.
23. D. Stutzbach and R. Rejaie. Characterizing the two-tier Gnutella topology. In *Proc. of SIGMETRICS*, pages 402–403, Banff, Alberta, Canada, 2005. ACM Press.
24. Q. Lv, P. Cao, E. Cohen, K. Li, and S. Shenker. Search and replication in unstructured peer-to-peer networks. In *Proc. of ICS*, pages 84–95. ACM Press, 2002.
25. TREC. Text REtrieval Conference, January 2006. `http://trec.nist.gov`.
26. S. Kwok. P2P searching trends: 2002-2004. *IPM*, 42(1):237–247, January 2006.

A Decision-Theoretic Model for Decentralised Query Routing in Hierarchical Peer-to-Peer Networks

Henrik Nottelmann and Norbert Fuhr

Department of Informatics, University of Duisburg-Essen,
47048 Duisburg, Germany
norbert.fuhr@uni-due.de

Abstract. Efficient and effective routing of content-based queries is an emerging problem in peer-to-peer networks, and can be seen as an extension of the traditional "resource selection" problem. The decision-theoretic framework for resource selection aims, in contrast to other approaches, at minimising overall costs including e.g. monetary costs, time and retrieval quality. A variant of this framework has been successfully applied to hierarchical peer-to-peer networks (where peers are partitioned into DL peers and hubs), but that approach considers retrieval quality only. This paper proposes a new model which is capable of considering also the time costs of hubs (i.e., the number of hops in subsequent steps). The evaluation on a large test-bed shows that this approach dramatically reduces the overall retrieval costs.

1 Introduction

Peer-to-peer (P2P) networks have emerged recently as an alternative to centralised architectures. The major problem in such networks is query routing, i.e. deciding to which other peers the query has to be sent for high efficiency and effectiveness. In contrast to the traditional resource selection problem, this process is inherently decentralised in peer-to-peer networks and based on local knowledge.

The decision-theoretic framework (DTF) [7, 3] computes an optimum selection based on cost estimations. These cost estimations include several important factors like retrieval quality, time or money. A user can weight these cost sources for specifying her own selection policy, e.g. preferring cheap digital libraries (DLs), or high quality DLs. Resource descriptions, i.e. statistical aggregation of the DLs, are employed for estimating costs, in particular to approximate the retrieval quality.

[8] presents a heuristic extension of the DTF for hierarchical peer-to-peer networks. In such a P2P topology, peers are partitioned into low-end DL peers hosting the documents, and hubs which act as directory peers. Only hubs are responsible for routing; a DL receiving a query only returns result documents but does not forward the query to other peers. Costs for hubs are computed by simply aggregating the resource descriptions of all DLs in a certain neighbourhood of that hub, by assuming that these DLs are merged into a single virtual collection. This approach, however, does not allow to estimate time costs properly: Those costs depend on the peers a selected hub itself selects, and thus cannot be estimated via simple aggregated statistics.

G. Amati, C. Carpineto, and G. Romano (Eds.): ECIR 2007, LNCS 4425, pp. 148–159, 2007.

In this paper, we present the first theoretical model for decentralised query routing in hierarchical P2P networks which considers time costs, and we give experimental results demonstrating the validity of this model. The basic idea is to estimate the costs of the DLs a neighbour hub would select in subsequent phases. This estimation is based on statistical aggregations of the DLs' content, as well as the distances of the DLs to the hub. The advantages are two-fold: Hub costs are only based on those DLs the hub potentially selects (i.e., can contribute to the final result). Second, this approach allows us to take the distances of selected DLs (and, thus, the associated time costs) into account.

This paper is structured as follows: Section 2 briefly describes DTF. An overview over important aspects of resource selection in peer-to-peer networks is given in section 3. Then, section 4 presents a new approach for estimating hub costs which inherently also considers the number of hubs associated with the selection of a hub. An evaluation of the proposed approach is shown in section 5. Section 6 summarises work related to resource selection in distributed IR.

2 The Decision-Theoretic Framework for Resource Selection

Most resource selection approaches (e.g. CORI) only consider retrieval quality. As an approximation, they compute a similarity score of each DL to the query, and select a fixed number of top-ranked libraries. Other aspects like execution time of the DLs are neglected.

In contrast, the decision-theoretic framework (DTF) [7, 3] is capable of dealing with different selection criteria, which are unified under the notion of "costs". As the actual costs are unknown in advance, expected costs (for digital library DL_i when s_i documents are retrieved for query q) are regarded instead.

Different sources can be considered:

Effectiveness: Probably most important, a user is interested in getting many relevant documents. In a simple model, effectiveness costs are based on the expected number $s_i - E[r_i(s_i,q)]$ of non-relevant documents, where $E[r_i(s_i,q)]$ denotes the expected number of relevant documents among the s_i top-ranked documents.

Time: We assume uniform costs for transmitting a result document, thus these costs can be neglected for selection. As a consequence, the expected costs $EC_i^t(s_i)$ are based on the initial costs for contacting a DL.

Money: Monetary costs are important for some applications, but can be neglected in the context of this paper.

These cost sources are weighted by user-specific parameters c^e (for effectiveness) and c^t (time) parameters. They allow a user to specify her own selection policy, e.g. good results vs. fast results. Thus, the expected costs (for digital library DL_i when s_i documents are retrieved for query q) are computed as:

$$EC_i(s_i,q) := c^e \cdot [s_i - E[r_i(s_i,q)]] + c^t \cdot EC_i^t(s_i) . \tag{1}$$

A user also specifies the total number n of documents to be retrieved out of m libraries, and the task is to compute an optimum solution (employing the algorithm presented in [3]):

$$s := \operatorname*{argmin}_{\sum_{i=1}^{m} s_i = n} \sum_{i=1}^{m} EC_i(s_i, q).$$

Relevance costs are computed in two steps:

1. First, the expected number $E(\text{rel}|q,DL)$ of relevant documents in the library is computed based on statistical aggregations (called "resource description") of the DL.
2. Then, a linearly decreasing approximation of the recall-precision function is used for computing the expected number $E[r_i(s_i,q)]$ of relevant retrieved documents.

For the first step, the resource descriptions store the DL size $|DL|$ and the average (expectation) $\mu_t = E[w(d,t)|d \in DL]$ of the indexing weights $w(d,t)$ (for document d and term t). For a query with term weights $a(q,t)$ (summing up to one) and a linear retrieval model, the expected number $E(\text{rel}|q,DL)$ of relevant documents in DL w. r. t. query q can be estimated as:

$$\begin{aligned} E(\text{rel}|q,DL) &= \sum_{d \in DL} Pr(\text{rel}|q,d) \approx \sum_{d \in DL} \sum_{t \in q} a(q,t) \cdot w(d,t) \\ &= |DL| \cdot \sum_{t \in q} a(q,t) \cdot \sum_{d \in DL} \frac{w(d,t)}{|DL|} \\ &= |DL| \cdot \sum_{t \in q} a(q,t) \cdot \mu_t \,, \end{aligned} \tag{2}$$

where $Pr(\text{rel}|q,d)$ denotes the probability that document d is relevant.

In a second step, $E(\text{rel}|q,DL)$ is mapped onto the expected number $E[r_i(s_i,q)]$ of relevant retrieved documents. Assuming a linearly decreasing recall-precision function $P: [0,1] \to [0,1]$, $P(R) := 1 - R$, with expected precision $E[r_i(s_i,q)]/s_i$ and expected recall $E[r_i(s_i,q)]/E(\text{rel}|q,DL_i)$, we can estimate the number of relevant documents when retrieving s_i documents:

$$E[r_i(s_i,q)] := \frac{E(\text{rel}|q,DL_i) \cdot s_i}{E(\text{rel}|q,DL_i) + s_i}. \tag{3}$$

For DTF, the libraries have to return the probabilities of relevance of the result documents, thus no further normalisation step is required.

3 Resource Selection in Peer-to-Peer Networks

Query routing (i.e., resource selection) is a crucial task in peer-to-peer networks, as contacting all connected peers does not scale [9]. This section introduces several competing approaches for resource selection in peer-to-peer networks.

3.1 Network Topologies

A direct neighbour $P \in nb(P')$ of a peer P' is another peer P if and only if there is a (direct) connection link. The distance between two peers is the minimum number of hops (i.e., links) required to go from one peer to the other.

In this paper, we regard hierarchical peer-to-peer topologies, which are based on a partition of peers into DL peers (sometimes also called “leaves”) and hubs. DLs are typically end-user machines which answer but do not forward queries, while hubs are responsible for routing and, thus, high-bandwidth computers which are nearly permanently online. Each DL peer is connected to at least one hub but not to other DL peers, which reduces the number of messages during query routing (i.e., resource selection). This results in a simple yet reasonable and efficient topology, called hierarchical peer-to-peer networks.

In this paper, we focus on HyperCube graphs (HyperCubes for short) [11] . A (binary) HyperCube is a regular d-dimensional structure, where each peer is connected to exactly d other peers (one per dimension). Messages arriving via a connection on dimension $k \in \{0, 1, \ldots, d-1\}$ can only be forwarded to peers on strictly higher dimensions $k' > k$. A consequence is that the dimensions define (starting from an arbitrary peer) a spanning tree on the network, which ensures that there is exactly one path from one peer to another peer. It also corresponds to a clearly defined partition of the whole network.

Positions of missing peers are filled with “virtual peers” (see [11] for details), which are then replaced by “shortcuts” to all original peers which can be contacted through virtual peers only. As a consequence, peers can be connected to more or less than d neighbours.

In this paper, we also ensure that each DL is connected to exactly one hub, so that (given the HyperCube) there is exactly one path from any hub to any DL in the network (i.e., cycles do not occur).

3.2 Centralised and Decentralised Selection

A simple selection strategy is to use the P2P network for a “cost estimation collection phase”, where the query is flooded in a Gnutella-like way through the hub network. Each hub computes cost estimations of its neighbour DLs, and sends them to the hub starting the routing process. Then, a single central selection is computed, and the selected DL peers are notified directly. This centralised selection strategy yields a global optimum, but is rather inefficient (see [9] for Gnutella, and section 5.2 for hierarchical networks). HyperCubes can improve the cost estimation collection phase as each hub is connected exactly once (and not multiple times).

In contrast, decentralised selection computes a local optimum selection on every hub receiving the query, by considering locally available descriptions of all DLs and hubs in a predefined distance. Thus, a hub decides locally how many documents should be retrieved from neighbour DLs, and how many documents are to be delivered by neighbour hubs (which itself compute a local selection). This decentralised selection method produces an overhead as a cost estimation and selection has to be performed on every hub. On the other hand, this method cuts down the number of hubs that are traversed, and thus saves time and bandwidth. In HyperCubes, hub descriptions are based on disjoint sets of DLs, which should improve the selection accuracy. The following Section describes how resource descriptions for hubs can be computed and employed for decentralised selection.

4 Cost Estimation for Hubs

A hub description is a representative of the neighbourhood of a hub. Basically, its statistical characteristics are defined by combining the resource descriptions of a set of DL peers.[1]

A naive approach is to combine the documents of all DLs in a neighbourhood in a large "virtual" collection, and use the description of that collection as the hub description [8]. This, however, has two drawbacks: A hub is not a monolithic DL, its selection results in further selection steps which ignores most of the DLs in the neighbourhood. Additionally, time costs (i.e., the number of hops) are not considered in such a setting.

The basic idea presented in this paper is to approximate the selection step in a selected hub: When we estimate the costs of a hub, we assume that the hub selects the best DLs in the neighbourhood (but no hubs), and only consider these selected DLs in the cost estimation. This approach also allows us to estimate the time costs associated with selecting a hub.

4.1 Hub Resource Descriptions

In the traditional decision-theoretic framework (see section 2), the resource description contains the average indexing weight $\mu_t = E[w(d,t)|d \in DL]$ for each index term t. Time costs can easily be added for DLs by using a constant value (e.g., one) for the one hop to the neighbour DL, by setting $EC_i^t(s_i) = 1$ iff $s_i > 0$, and $= 0$, otherwise.

However, a hub is a representative of a sub-network (a "neighbourhood"), and its selection results in contacting further peers (DLs and hubs) with additional costs for those hops. In addition, a term t can occur in DL peers in different distances for a neighbour hub H, so constant time costs are not sufficient for hubs. In the following, we show how the content of a resource description is modified for hubs, so that time costs can be considered as well.

We start with a simple scenario, where a hub H is connected to m libraries $DL_1,\ldots,$ $DL_m \in nb(H)$. Neighbour hubs $H' \in nb(H)$ are not considered so far, the approach is extended to this case in section 4.3. We further assume that resource descriptions are given for all DL_i, i.e. the average indexing weights $\mu_{t,i} = E(t \leftarrow d|d \in DL_i)$.

Each term t can be regarded as a single-term query. Then, the number $R_i(t) = E(rel|t,DL_i)$ of relevant documents in each DL_i can be easily estimated according to equation (2) as:

$$R_{t,i} = E(rel|t,DL_i) = |DL_i| \cdot \mu_{t,i} \,.$$

The results are rounded to natural numbers, to ease the further processing steps.

For the hub H under consideration, the discrete empirical distribution $Pr(R_t)$ of the number of relevant documents is computed over all neighbour DLs. Let us assume a term t_1 for which 4 of the 6 neighbour DLs have 3 relevant documents for t_1 and the 2 other neighbour DLs have 5 relevant documents. Then, the resulting distribution is defined by $Pr(R_{t_1} = 3) = 4/6$ and $Pr(R_{t_1} = 5) = 2/6$.

[1] In peer-to-peer networks, co-operative peers can be assumed, so query-based sampling is not considered here; each DL returns its description upon request.

This distribution $Pr(R_t)$ forms a compact representation of the content of the hub's neighbourhood, and is used as the resource description of hub H. In other words, for each term t a function $Pr(R_t = \cdot)$ is stored.

4.2 Cost Estimation

At query time, the resource description of hub H is employed for estimating retrieval costs. Costs for hubs are estimated in 6 subsequent phases, incorporating only neighbour DLs:

1. For the query, the distribution of relevant documents in all DLs is computed.
2. For all DLs, the number of relevant documents is estimated based on the distribution.
3. The DLs are ranked w. r. t. their number of relevant documents, and the best DLs are selected.
4. Costs are estimated based on the best selected DLs.
5. Minimum costs are computed.

In the first phase, the distribution $Pr(R_q)$ of the relevant documents $R_q = E(\text{rel}|q,DL)$ (for a query q and a randomly chosen library DL) is computed. Remember that with a linear retrieval function, we have:

$$R_q = E(\text{rel}|q,DL) = |DL| \cdot \sum_{t \in q} a(q,t) \cdot \mu_t = \sum_{t \in q} a(q,t) \cdot R_t \,.$$

Thus, the random variable R_q can be considered as the linear combination of the random variables R_t. The distribution $Pr(R_q)$ can thus be computed via convolution:

$$Pr(R_q) = \sum_{R_t:\, R_q = \sum_{t \in q} a(q,t) \cdot R_t} \prod_{t \in q} Pr(R_t).$$

Here, basically, the probabilities of all possible cases for the R_t which lead to a fixed value of R_q are summed up, assuming independence of the R_t. (Since the distributions $Pr(R_t)$ are very sparse, the convolution can be computed reasonably efficiently.) As an example, assume—in addition to the distribution for term t_1 (see section 4.1)—a second term t_2 with $Pr(R_{t_2} = 1) = Pr(R_{t_2} = 3) = 1/2$, and further assume $a_{t_1} = a_{t_2} = 1/2$. Then, the case $R_q = 2$ can only be caused by $R_{t_1} = 3$ and $R_{t_2} = 1$ with $Pr(R_q = 2) = 2/3 \cdot 1/2 = 1/3$. Similarly, the case $R_q = 3$ can be caused by either $R_{t_1} = R_{t_2} = 3$ or by $R_{t_1} = 5$ and $R_{t_2} = 1$, thus $Pr(R_q = 3) = 2/3 \cdot 1/2 + 1/3 \cdot 1/2 = 1/2$. Finally, $Pr(R_q = 4) = 2/3 \cdot 1/2 = 1/6$ is caused by the case $R_{t_1} = 5$ and $R_{t_2} = 3$.

In a second step, the frequencies $Pr(R_q)$ and the number m of neighbour DLs are used for estimating the number of relevant documents for the DLs. E.g., for $m = 6$ DLs in total, 2 DLs have 2 relevant document, 3 DL have 3 relevant documents, and the sixth DL contains 4 relevant documents. Interpolation is used for computing a value for each DL in cases where the probability for a R_q value does not correspond to a natural number of occurrences.

Third, the DLs are ranked w. r. t. their number of relevant documents, i.e. $R_1 \geq R_2 \geq \ldots \geq R_m$. For $1 \leq l \leq m$, $R(l) := \sum_{i=1}^{l} R_i$ denotes the sum of the number of relevant documents in the top l DLs.

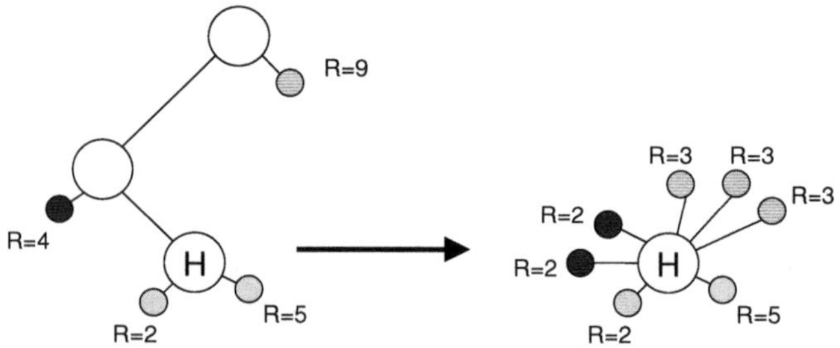

Fig. 1. Hub costs in larger neighbourhoods

In a fourth step, we assume that a hub is the combination of all l selected DLs, i.e. a hub is considered as a single DL. With the recall-precision function and equation (3), we can compute the number $r(s,R(l))$ of relevant documents in the result set when retrieving s documents from the union of the l selected DLs. Following equation (1), the costs $EC(s,l,R(l))$ when l neighbour DLs are selected can be computed as:

$$EC_H(s,l,R(l)) := c^e \cdot [s - r(s,R(l))]] + c^t \cdot (l+1) .$$

Note that the number of hops equals the number of selected neighbour DLs (each one can be reached via one hop in a later phase) plus 1 hop for reaching the hub itself.

The final cost estimations can be easily computed in a sixth step:

$$\begin{aligned} EC_H(s) &= \min\{EC_H(s,l,R(l)) | 1 \leq l \leq m\} \\ &= \min\{c^e \cdot [s - r(s,R(l))]] + c^t \cdot (l+1) | 1 \leq l \leq m\} . \end{aligned}$$

These cost estimations can be used in the usual selection process.

4.3 Considering a Larger Neighbourhood

So far, only neighbour DLs are considered for computing the resource description of a hub. However, hubs typically are connected to other (neighbour) hubs, which themselves have DLs (and, potentially, other hubs) attached. We apply a trick and conceptually replace hubs and their attached DLs by new virtual DLs directly connected to the hub. Thus, the network structure is "flattened", and costs can be estimated for all DLs in the same way, regardless of their distance.

The horizon h defines the maximum distance between the hub H and the DLs to be considered for its hub description. In section 4.2, only neighbour DLs are considered, which equals to a horizon $h = 1$. For a horizon $h > 1$, the neighbourhood function nb is extended as follows:

$$\begin{aligned} nb^1(H) &:= nb(H), \\ nb^h(H) &:= \bigcup_{H' \in nb(H)} nb^{h-1}(H') . \end{aligned}$$

In other words, $nb^h(H)$ describes all peers in a distance of exactly h hops (from H).

The key idea for considering remote DLs is the following: Costs remain untouched if such a library DL_i' is replaced by two "virtual" DLs connected directly to H, where the relevant documents are uniformly distributed over both DLs (i.e., with $R/2$ relevant documents): To obtain R relevant documents, two hops (one for each new virtual DL) are required. For implementing this scheme, the neighbouring hub H' has to send its cumulated statistics about $DL_i' \in nb(H')$ to hub H. Costs are then estimated as described in section 4.2, without a need for caring about the distance of DL peers. A similar approach is used for $h > 2$, were a DL in distance of $h' \leq h$ is replaced by h' DLs with R/h' relevant documents.

An example is shown in figure 1. Here, the DLs directly connected to the hub H remain untouched. The DL connected to the direct neighbour hub of H (with $R = 4$ relevant documents) is replaced by two virtual hubs with $R = 4/2 = 2$, while the DL with $R = 9$ connected to the neighbour hub of the neighbour hub (i.e., the DLs in $nb^3(H)$) is replaced by three virtual DLs with $R = 9/3 = 3$.

5 Evaluation

The proposed approach has been evaluated on a large test-bed. This sections describes the setup of the experiments and results in terms of efficiency, effectiveness and costs.

5.1 Experimental Setup

The WT10g collection is used as a basis for our experiments. The topology "cmu" (taken from [5]) employes a hierarchical P2P network, where the WT10g collection is divided into 11,485 collections according to the document URLs; 2,500 collections (containing in total 1,421,088 documents) were chosen randomly, each of them forming one DL peer. Hubs are formed by similarity of the DL peers, each hub is connected to 13-1,013 DLs (379.8 on average). Neighbour hubs are selected randomly so that each hub has 1-7 hub neighbours (3.8 on average).

The topology "hc-1" regarded in this paper is a HyperCube derived from the "cmu" topology in the following way: it consists of 25 hubs as in "cmu" (with dimension $d = 5$); each hub is connected to 4–14 other hubs (5.8 on average). Each DL is connected to exactly one hub, randomly chosen out of the "cmu" connections, so that each hub is connected to 3–254 DLs (100 on average). Thus, "hc-1" completely avoids cycles and yields disjoint hub neighbourhoods.

The WT10g collection only provides 100 topics with relevance judgements. For large P2P networks, more queries are required. Thus, we use 1,000 random queries (from the test set) generated from title fields of documents. Each query contains up to 6 terms, the average is 2.9 terms per query. In all experiments, $n = 50$ documents are requested. As these queries were created artificially, no relevance judgements are available. Pseudo-relevance judgements were obtained by combining all 2,500 collections into one centralised collection (using system-wide IDF values), and marking the 50 top-ranked documents[2] for each query as "relevant". Thus, the experiments

[2] Documents are ranked using the same indexing weights and retrieval functions as all DLs.

measure how well distributed retrieval approximates the centralised collection. Document indexing weights are computed based on the BM25 formula [10].

For resource selection, we set $c^e = 1$ in all cases (for effectiveness costs, i.e. the number of non-relevant documents), and vary the parameter c^t to simulate different user preferences. Similarly, for computing costs after retrieval (the actual costs connected to the query) we set $c^e = 1$ and use varying parameters for the time component. Please note that for $c^t = c^e = 1$, one hop corresponds to one non-relevant document. For the same number of documents, selecting an additional DL thus can only be compensated if that DL returns an additional relevant document. Similarly, for $c^t = \frac{c^e}{4} = 0.25$, four hops correspond to one relevant document.

For result merging, we assume that a hub propagates the hub-local idf values[3] to its directly connected DLs, and then merge the returned ranking lists according to descending RSVs.

Table 1. Results for centralised and decentralised resource selection

(a) $h = 1$

	Dec., $c^t = 0$	Cent., $c^t = 0$	Dec., $c^t = 0.1$	Dec., $c^t = 0.25$	Dec., $c^t = 0.5$	Dec., $c^t = 1$
P@10	0.4385 / 0.0%	0.6586 / 50.2%	0.4241 / -3.3%	0.4044 / -7.8%	0.3527 / -19.6%	0.2981 / -32.0%
P@30	0.2087 / 0.0%	0.3774 / 80.8%	0.2171 / 4.0%	0.2206 / 5.7%	0.2072 / -0.7%	0.1856 / -11.1%
MAP	0.1307 / 0.0%	0.2565 / 96.3%	0.1387 / 6.1%	0.1450 / 10.9%	0.1399 / 7.0%	0.1299 / -0.6%
Precision	0.1571 / 0.0%	0.2688 / 71.1%	0.1677 / 6.7%	0.1751 / 11.4%	0.1700 / 8.2%	0.1597 / 1.6%
Recall	0.1354 / 0.0%	0.2576 / 90.3%	0.1435 / 6.0%	0.1493 / 10.3%	0.1438 / 6.2%	0.1325 / -2.1%
#Hops	40.9 / 0.0%	55.5 / 35.5%	23.6 / -42.4%	14.9 / -63.6%	9.8 / -76.0%	6.6 / -83.8%
Costs	35.78 / 0.0%	34.52 / -3.5%	37.54 / 4.9%	38.63 / 8.0%	40.00 / 11.8%	41.98 / 17.3%

(b) $h = 2$

	Dec., $c^t = 0$	Cent., $c^t = 0$	Dec., $c^t = 0.1$	Dec., $c^t = 0.25$	Dec., $c^t = 0.5$	Dec., $c^t = 1$
P@10	0.5694 / 0.0%	0.6586 / 15.7%	0.5560 / -2.4%	0.5278 / -7.3%	0.4797 / -15.8%	0.4198 / -26.3%
P@30	0.3141 / 0.0%	0.3774 / 20.2%	0.3300 / 5.1%	0.3301 / 5.1%	0.3167 / 0.8%	0.2920 / -7.0%
MAP	0.2066 / 0.0%	0.2565 / 24.2%	0.2206 / 6.8%	0.2232 / 8.0%	0.2194 / 6.2%	0.2079 / 0.6%
Precision	0.2360 / 0.0%	0.2688 / 13.9%	0.2520 / 6.8%	0.2559 / 8.4%	0.2536 / 7.4%	0.2448 / 3.7%
Recall	0.2139 / 0.0%	0.2576 / 20.5%	0.2278 / 6.5%	0.2295 / 7.3%	0.2255 / 5.4%	0.2136 / -0.1%
#Hops	45.2 / 0.0%	55.5 / 22.7%	23.3 / -48.4%	14.9 / -67.1%	10.1 / -77.7%	7.2 / -84.2%
Costs	34.10 / 0.0%	34.52 / 1.3%	35.73 / 4.8%	36.84 / 8.1%	38.24 / 12.2%	40.61 / 19.1%

(c) $h = 3$

	Dec., $c^t = 0$	Cent., $c^t = 0$	Dec., $c^t = 0.1$	Dec., $c^t = 0.25$	Dec., $c^t = 0.5$	Dec., $c^t = 1$
P@10	0.6307 / 0.0%	0.6586 / 4.4%	0.6284 / -0.4%	0.6048 / -4.1%	0.5618 / -10.9%	0.4987 / -20.9%
P@30	0.3757 / 0.0%	0.3774 / 0.5%	0.4012 / 6.8%	0.4045 / 7.7%	0.3939 / 4.8%	0.3652 / -2.8%
MAP	0.2541 / 0.0%	0.2565 / 0.9%	0.2765 / 8.8%	0.2822 / 11.1%	0.2816 / 10.8%	0.2688 / 5.8%
Precision	0.2810 / 0.0%	0.2688 / -4.4%	0.3056 / 8.8%	0.3125 / 11.2%	0.3142 / 11.8%	0.3036 / 8.0%
Recall	0.2625 / 0.0%	0.2576 / -1.8%	0.2844 / 8.3%	0.2891 / 10.1%	0.2892 / 10.2%	0.2752 / 4.8%
#Hops	45.0 / 0.0%	55.5 / 23.3%	22.0 / -51.0%	14.4 / -68.1%	10.1 / -77.6%	7.6 / -83.2%
Costs	33.08 / 0.0%	34.52 / 4.4%	34.13 / 3.2%	35.17 / 6.3%	36.62 / 10.7%	39.64 / 19.8%

[3] In order to reduce the experimental effort, we used system-wide idf values in our experiments, since earlier experiments [8] showed that the difference between system-wide and hub-local idf values is negligible.

5.2 Results

Table 1 depicts the results for our HyperCube topology. First, the tables show that centralised selection ("cent.") outperforms decentralised variants ("dec.") in terms of effectiveness. Compared to $c^t = 0$ (ignoring time costs), precision in the top ranks decreases with increasing time costs c^t (for a single hop). Precision in lower ranks, mean average precision (MAP) as well as set-based precision and recall, however, increase up to $c^t = 0.5$ (for $h = 1$) or $c^t = 1$ (for $h = 3$ and the set-based values), before these values decrease again. As intended, efficiency dramatically increases (i.e., less hubs and DLs are selected) with increasing c^t. Both effects nearly balance so that the overall costs only slightly increase.

The table also reveals that for any fixed time cost user parameter c^t, effectiveness increases with a larger horizon h. In other words, a larger hub neighbourhood (with more DLs considered) improves the cost estimation process. This fact shows that our model makes good use of the knowledge provided. As also can be seen from these figures, decentralised selection with time costs considered outperforms the two other approaches.

As a summary, incorporating time costs in the selection process dramatically reduces the final costs w. r. t. the user's preference. Moreover, the approach is capable of adjusting to increasing time costs per hop c^t so that the final costs increase only marginally. In addition, broadening the horizon leads to increased retrieval quality and marginally lower costs.

6 Related Work

In contrast to the decision-theoretic framework (DTF) employed in this paper, most of the other selection algorithms compute a score for every library. Then, the top-ranked documents of the top-ranked libraries are retrieved and merged in a data fusion step.

The GlOSS system [4] is based on the vector space model and – thus – does not refer to the concept of relevance. For each library, a goodness measure is computed which is the sum of all scores (in the experiments reported, SMART scores) of all documents in this library w. r. t. the current query. Libraries are ranked according to the goodness values.

The state-of-the-art system CORI [1] uses the INQUERY retrieval system which is based on inference networks. The resource selection task is reduced to a document retrieval task, where a "document" is the concatenation of all documents of one library. The indexing weighting scheme is quite similar to one employed in DTF, but applied to libraries instead of documents. Thus, term frequencies are replaced by document frequencies, and document frequencies by collection frequencies. CORI also covers the data fusion problem, where the library score is used to normalise the document score. Experiments showed that CORI outperforms GlOSS [2].

Another ranking approach is based on language models [13]. Basically, the language model of the collection is smoothed with a collection-independent (system-wide) language model, and KL-divergence is used for ranking the DLs. The final document ranking is computed in a result merging step by using the original (collection-biased) document probabilities, the DL scores, a smoothing factor, and Bayesian inversion. The quality of this approach is slightly better than CORI.

The language model approach has been extended towards hierarchical peer-to-peer networks in [6] for ranking neighbour peers (leaves and hubs). Hubs are described by neighbourhood (which is not limited to the directly connected DLs), where the influence of term frequencies of distant DLs is exponentially decreased. DLs and hubs are selected separately, as DL and hub descriptions are not in the same order of magnitude. A fixed number of hubs is selected, while a modified version of the semi-supervised learning algorithm [12] is employed for computing a threshold for the number of selected leaves.

The decision-theoretic framework has been extended towards peer-to-peer networks in [8]. There, an extensive discussion of resource selection architectures for peer-to-peer networks is presented. The architectures are classified based on the underlying resource selection approach (DTF and CORI as a baseline), design choices like the locality of knowledge (e.g. IDF values) and selections (centralised vs. decentralised), as well as the network topology (hierarchical networks with DLs and hubs, distributed hash tables and HyperCubes). Time costs, however, are not regarded there. The evaluation shows that DTF slightly outperforms CORI in peer-to-peer networks. Centralised selection has higher effectiveness than decentralised selection, but has an expensive cost estimation collection phase. Distributed hash tables [14] and HyperCubes can reduce that effort.

7 Conclusion and Outlook

This paper presents the first theoretical model for decentralised query routing in hierarchical peer-to-peer networks, which also incorporates time costs (in addition to traditional retrieval quality measures). For this, the decision-theoretic framework has been extended to estimate the costs of DLs a neighbour hub would select in subsequent phases. This estimation is based on statistical aggregations of the DLs' content, as well as the distance of the DLs to the hub. The advantages are two-fold: Hub costs are only based on those DLs which the hub potentially selects (and, thus, can contribute to the final result). Second, this approach allows us to take the distance of selected DLs (and, thus, the associated time costs) into account.

The evaluation shows that the new P2P variant of the decision-theoretic framework is capable to optimise the selection quality when time costs are considered. The final costs (w. r. t. the user's preference) are dramatically reduced. Moreover, the approach can adjust to increasing time costs c^t per hop so that the final costs increase only marginally. Furthermore, broadening the horizon leads to increased retrieval quality and marginally lower costs.

Here we have tested our model under optimum conditions, in order to demonstrate its general validity. Future work will concentrate on the development of approximations for less favourable settings. First, we will replace the empirical term distributions $Pr(R_t)$ by appropriate theoretical distributions. Another issue is the reduction of the required knowledge about the neighbourhood when constructing resource descriptions. Currently, each hub has to provide separate statistics of all DLs in distance 1, for all

DLs in distance 2, and so on. As an alternative, approximate aggregated descriptions will be investigated. In a similar way, we will work on modifications of the approach for effectively dealing with cycles in the network.

References

[1] J. P. Callan, Z. Lu, and W. B. Croft. Searching distributed collections with inference networks. In E. A. Fox, P. Ingwersen, and R. Fidel, editors, *Proceedings of the 18th Annual International ACM SIGIR Conference on Research and Development in Information Retrieval*, pages 21–29, New York, 1995. ACM. ISBN 0-89791-714-6.

[2] J. French, A. Powell, J. Callan, C. Viles, T. Emmitt, K. Prey, and Y. Mou. Comparing the performance of database selection algorithms. In *Proceedings of the 22nd International Conference on Research and Development in Information Retrieval*, pages 238–245, New York, 1999. ACM.

[3] N. Fuhr. A decision-theoretic approach to database selection in networked IR. *ACM Transactions on Information Systems*, 17(3):229–249, 1999.

[4] L. Gravano and H. Garcia-Molina. Generalizing GIOSS to vector-space databases and broker hierarchies. In U. Dayal, P. M. D. Gray, and S. Nishio, editors, *VLDB'95, Proceedings of 21th International Conference on Very Large Data Bases*, pages 78–89, Los Altos, California, 1995. Morgan Kaufman.

[5] J. Lu and J. Callan. Content-based retrieval in hybrid peer-to-peer networks. In D. Kraft, O. Frieder, J. Hammer, S. Qureshi, and L. Seligman, editors, *Proceedings of the 12th International Conference on Information and Knowledge Management*, New York, 2003. ACM.

[6] J. Lu and J. Callan. Federated search of text-based digital libraries in hierarchical peer-to-peer networks. In J. Callan, N. Fuhr, and W. Nejdl, editors, *SIGIR Workshop on Peer-to-Peer Information Retrieval*, 2004.

[7] H. Nottelmann and N. Fuhr. Evaluating different methods of estimating retrieval quality for resource selection. In J. Callan, G. Cormack, C. Clarke, D. Hawking, and A. Smeaton, editors, *Proceedings of the 26th Annual International ACM SIGIR Conference on Research and Development in Information Retrieval*, New York, 2003. ACM.

[8] H. Nottelmann and N. Fuhr. Comparing different architectures for query routing in peer-to-peer networks. In *ECIR*, pages 253–264. Springer, 2006.

[9] J. Ritter. Why Gnutella can't scale. No, really., 2001. `http://www.darkridge.com/~jpr5/doc/gnutella.html`.

[10] S. E. Robertson, S. Walker, M. Hancock-Beaulieu, A. Gull, and M. Lau. Okapi at TREC. In *Text REtrieval Conference*, pages 21–30, 1992.

[11] M. Schlosser, M. Sintek, S. Decker, and W. Nejdl. Digital libraries. In *1st Workshop on Agents and P2P Computing*, 2005.

[12] L. Si and J. Callan. A semi-supervised learning method to merge search engine results. *ACM Transactions on Information Systems*, 24:457–49, 2003.

[13] L. Si, R. Jin, J. Callan, and P. Ogilvie. Language modeling framework for resource selection and results merging. In C. Nicholas, D. Grossman, K. Kalpakis, S. Qureshi, H. van Dissel, and L. Seligman, editors, *Proceedings of the 11th International Conference on Information and Knowledge Management*, New York, 2002. ACM. `http://www-2.cs.cmu.edu/~callan/Papers/cikm02-lsi.pdf`.

[14] I. Stoica, R. Morris, D. Karger, F. Kaashoek, and H. Balakrishnan. Chord: A scalable peer-to-peer lookup service for internet applications. In *ACM SIGCOMM*, 2001. `http://www.acm.org/sigcomm/sigcomm2001/p12-stoica.pdf`.

Central-Rank-Based Collection Selection in Uncooperative Distributed Information Retrieval

Milad Shokouhi

School of Computer Science and Information Technology
RMIT University, Melbourne 3001, Australia
milad@cs.rmit.edu.au

Abstract. Collection selection is one of the key problems in distributed information retrieval. Due to resource constraints it is not usually feasible to search all collections in response to a query. Therefore, the central component (broker) selects a limited number of collections to be searched for the submitted queries. During the past decade, several collection selection algorithms have been introduced. However, their performance varies on different testbeds. We propose a new collection-selection method based on the ranking of downloaded sample documents. We test our method on six testbeds and show that our technique can significantly outperform other state-of-the-art algorithms in most cases. We also introduce a new testbed based on the TREC GOV2 documents.

1 Introduction

Distributed information retrieval (DIR) has attracted considerable research interest during recent years. Centralized search engines are not capable of indexing the *hidden web* [Raghavan and Garcia-Molina, 2001]. In addition, it is not feasible to crawl and index the web documents with the same rate that they change. DIR has been introduced as a solution to these deficiencies. DIR techniques provide a search service over non-crawlable pages. They also return the recent version of webpages without consuming costly resources for crawling. Distributed search can be divided into three major steps; Firstly suitable collections are selected for a query. Secondly, the query is sent to the selected collections and they search their documents for suitable answers. Finally, the results from selected collections are returned to the broker that then merges them for presentation to the user.

In this paper, we focus on the collection selection stage. In DIR systems, each collection is represented by a set of documents and vocabularies usually known as collection summaries or representation sets. In *cooperative* environments, collections provide the broker with their term statistics and summaries [Gravano et al., 1997]. However, in real-life environments such as the web, collections may be *uncooperative*. In this case, collections are not willing to share their information and the broker should gather small summary sets for each collection by sampling [Callan and Connell, 2001]. Random queries are sent to each collection and results are downloaded and stored as collection representation sets.

G. Amati, C. Carpineto, and G. Romano (Eds.): ECIR 2007, LNCS 4425, pp. 160–172, 2007.

We propose a novel collection selection algorithm that can be used in both cooperative and uncooperative environments. However, we focus on the latter scenario as it is more similar to the practical situations. For each entered query, our method ranks collections according to the ranking of their centrally held sampled documents. The sampled documents from all collections are gathered in a single central index. Each query is executed on this index and the collection weights are calculated according to the ranking of their sampled documents. In the next sections we show that this simple idea can outperform the state-of-the-art methods on many testbeds including our new testbed created from the TREC GOV2 documents.

2 Collection Selection

For a given query, the broker ranks available collections based on the computed similarity values for their representation sets. It is not usually feasible to search all collections for a query. Therefore, the broker selects a few collections that are more likely to return relevant documents. For this purpose, the broker evaluates the similarity of the entered query with the collection summaries. These summaries vary from term statistics in cooperative situations [Callan et al., 1995; Gravano et al., 1997] to a limited number of documents downloaded by sampling for uncooperative environments [Callan and Connell, 2001; Craswell et al., 2000]. The broker chooses those collections that have a summary similar to the query.

Introduced by Gravano et al. [1999], GLOSS uses the document frequency and weight of each term to select suitable collections. However, GLOSS uses unrealistic assumptions such as uniform term weight distribution across all collections. In CVV [Yuwono and Lee, 1997], the fraction of documents inside each collection that contain the query terms is used for collection selection.

CORI [Callan et al., 1995] applies inference networks for collection selection. It has been reported as the most effective method in many papers [Craswell et al., 2000; Powell and French, 2003], but there are question marks over its effectiveness [D'Souza et al., 2004b].

Nottelmann and Fuhr [2003] suggest a decision-theoretic framework (DTF) that selects the best collections while reducing the overall costs such as time and money. Despite having a theoretical foundation, the reported performance of DTF is worse than CORI for short queries.

During recent years, new collection selection algorithms have been claimed to produce better results than CORI [D'Souza etal.,2004a; Si et al., 2002: Si and Callan, 2003a: 2004]. Si et al. [2002] developed a language modeling framework for distributed information retrieval. Their system slightly outperforms CORI in some cases. REDDE [Si and Callan, 2003a] ranks the collections based on the estimated number of relevant documents they contain. REDDE has been shown to be very effective on some testbeds. However, as we show in the next sections, it produces poor results for some other testbeds.

Si and Callan [2004] presented their Unified Utility Maximization (UUM) framework for collection selection. UUM performs slightly better than REDDE on some testbeds. However, it uses training data and logistic models that require human relevance judgments.

We use CORI and REDDE as the benchmarks of our experiments because they are the two well-known algorithms that do not require training data for collection selection, and they both can work on uncooperative environments. In the next section, we introduce a novel approach for selecting collections in DIR systems.

3 Central-Rank-Based Collection Selection

In uncooperative environments, collection summaries usually consist of a limited number of documents downloaded by query-based sampling [Callan and Connell, 2001]. The query is compared to each of these summaries and collections are selected based on the similarity of their summaries with the query [Callan et al., 1995] or according to the estimated number of relevant documents they contain [Si and Callan, 2003a]. The ranking of sampled documents contains useful information that could be applied for selecting suitable collections. The set of all collection summaries together approximates a global index of documents in all collections and the ranking of summary documents could be seen as the result of a query against this index.

We propose *central-rank-based collection selection* (CRCS) as a new method that ranks collections according to the ranking of their summary documents and show that it can effectively select suitable collections.

In CRCS, an effective retrieval model is applied to the index of all sampled documents from collections. We refer to this model as the *central sample search engine* (CSSE). For each query, CSSE ranks the downloaded documents from collections. Then, the weight of each collection is calculated based on the ranks of its sampled documents that are in the top γ results. We have arbitrarily set γ to 50 for our experiments. How to identify an optimum value for this number is left as future work. The documents ranked after γ are less likely to be relevant and should have no impact on collection weights.

The top γ documents are ranked according to their probabilities of relevance. Intuitively, the weight of a collection with a sampled document at rank one should be incremented more than another collection with a sampled document at rank say 40.

The impact of a sampled document D on the weight of its original collection c is computed according to the position of that document in the top γ results. In the simplest form, this can be computed linearly as below:

$$R(D_j) = \begin{cases} \gamma - j & \textit{if } j < \gamma \\ 0 & \textit{otherwise} \end{cases} \tag{1}$$

where $R(D_j)$ represents the impact of document D at the jth rank of results returned by CSSE. The impact of documents decreases linearly according to their ranks. However, previous studies [Joachims et al., 2005; Manmatha et al., 2001] suggest that the importance of documents for users and their probabilities of relevance have a negative exponential relation with the document ranks. Therefore, it might be more suitable to assign document scores in a negative exponential manner as follows:

$$R(D_j) = \alpha \exp(-\beta \times j) \qquad \textit{if } D_j \in S_i \tag{2}$$

Here, j is the rank of document D. Coefficient parameters α and β are two constants respectively set to 1.2 and 2.8 in our experiments according to the suggested figures by Joachims et al. [2005]. We use CRCS(l) when the impact values are computed linearly and CRCS(e) when Eq. (2) is applied.

The size of a collection is also important for calculating its final weight. If two collections contribute the same number of documents in the top-ranked results, the larger collection is likely to contain more relevant documents due to its greater size. Collection size is an important factor that has to be considered during collection selection. KL-divergence [Si et al., 2002], REDDE [Si and Callan, 2003a] and UUM [Si and Callan, 2004] all consider a collection size parameter in their calculations. Since the exact values for collection sizes are not usually available in uncooperative environments, we recommend that the size of each collection be estimated using the capture-history method [Shokouhi et al., 2006b].

In our experiments, we assume that the size of all collection summaries is the same (300 documents). In practice, collection summaries might be different in size. Larger summaries are more likely to contain the query terms, which means that collections with larger summaries are more likely to be selected for any given query. To overcome this bias, CRCS divides the weight of each collection by the size of its representation set.

Putting this together, CRCS calculates the weight of each collection as below:

$$C_i = \frac{CH_i}{CH_{max} \times |S_i|} \times \sum_{D \in S_i} R(D_j) \tag{3}$$

where, CH_i is the size of collection i estimated by the capture-history method [Shokouhi et al., 2006b]. We normalize the collection sizes by dividing the size of each collection by the size of the largest collection involved (CH_{max}). $|S_i|$ is the size of the representation set for collection i that is the number of documents downloaded by query-based sampling [Callan and Connell, 2001] from that collection. The weight of each collection is calculated by summing up the impact values for its summary documents. In summary, CRCS computes the final weights of collections as below:

- CSSE runs the query and ranks the sampled documents on the broker.
- The top γ documents returned by CSSE are selected and their impact values on the weights of their corresponding collections are calculated
- Collections are ranked according to the impact values of their sampled documents and their estimated sizes.

A similar technique is suggested by Craswell et al. [2000] for collection selection. In their approach, the broker sends a number of training multi-term probe queries to collections. The top results from each collection are downloaded and gathered together in a single index. Then, the broker applies an effective retrieval model to rank the downloaded documents for the initial training queries. The search effectiveness of collection are computed according to their contribution to the top n (they suggested $n = 20$) results when the query is executed on the downloaded documents. Our approach is different to their technique in several ways; they calculate an effectiveness score for each collection according to its performance for the training queries. The final weight of a collection is computed by adding its effectiveness factor to the score calculated by a standard collection selection algorithms such as CORI. Unlike our approach, their suggested technique has not been used independently without relying on other collection selection methods. Also, while CRCS can select effective collections online, the suggested method by [Craswell et al., 2000] cannot capture this without a sufficient number of offline training queries.

REDDE [Si and Callan, 2003a] also uses a similar strategy for collection selection. However, it ignores the rank difference of documents in the central sample index and concentrates on selecting collections with the highest number of relevant documents (high recall). In contrast, our novel CRCS method focuses on selecting collections with high-quality documents (high precision).

In the following sections we compare the performance of CRCS with the other state-of-the-art methods on different testbeds.

4 Experimental Testbeds and Evaluation Metrics

Several testbeds have been developed for DIR experiments. These testbeds may be useful for evaluating DIR methods for specific applications such as enterprise search. However, most of them are not suitable for evaluating DIR techniques on the web because:

- The proposed testbeds are usually much smaller than the web collections and cannot be considered as good instances of the current web servers. The largest testbed reported so far is about 18 GB used by Hawking and Thomas [2005]. The other common testbeds are at most one sixth of this size[Powell and French, 2003: Si and Callan, 2003b:a].
- Collections are generated artificially by allocating each document to a collection according to a predefined criteria. This might be the author name or the year of publication[Powell and French, 2003; Si and Callan, 2003b;a].
- Collections usually only contain documents from the TREC newswire data. Considering the diversity of topics on the web, testbeds containing only news-related documents are not sufficient for unbiased evaluations.

In addition to the testbed deficiencies, there is a common defect with related work in this area; proposed algorithms are usually only tested on a few testbeds and their robustness has not been evaluated on different test data. In Section 5, we show that the performance of DIR methods vary substantially on different testbeds. This is consistent with previous work by D'Souza et al.

Table 1. Testbed statistics

		Number of Documents ×1000			Size (MB)		
Testbed	Size (GB)	Min	Avg	Max	Min	Avg	Max
trec123-100col-bysource	3.2	0.7	10.8	39.7	28	32	42
trec4-kmeans	2.0	0.3	5.7	82.7	4	20	249
100-col-GOV2	110.0	32.6	155.0	717.3	105	1126	3891

[2004a] that investigates the impact of testbed characteristics on collection selection algorithms. An algorithm that produces the best results on one dataset does not necessarily perform as well on another. We also introduce a new testbed based on the TREC GOV2 data. Our new testbed is more than six times larger than the largest DIR testbed reported so far; it is about 36 times larger than trec123-100col-bysource [Powell and French, 2003; Si and Callan, 2003b;a; 2004; Si et al., 2002], and 55 times larger than trec4-kmeans [Si and Callan, 2003b; Xu and Croft, 1999]. We test our method on six testbeds, so that we can fairly compare its performance and robustness with other available approaches. Table 1 includes information about the three major testbeds that have been used in our experiments. The other three testbeds are all generated from trec123-100col-bysource, thus we do not present them in the table.

- **trec4-kmeans (trec4)**: One hundred collections created from the TREC4 data. A k-means clustering algorithm have been used to organize the collections by topic [Xu and Croft, 1999]. TREC topics 201–250 (description) and their relevance judgments are used for performance evaluations. Collections in this testbed are small and the average query length is 7.2 words.
- **trec123-100col-bysource (uniform)**: One hundred collections are created by organizing the documents from the TREC disks 1, 2, and 3 by source and publication date. It is one of the most popular DIR testbeds and has been used in many papers [Powell and French, 2003; Si and Callan, 2003b;a; 2004; Si et al., 2002]. TREC topics 51–100 (title) are used as queries with the average length of 3 words per query.
- **100-col-GOV2 testbed (GOV2)**: In this new testbed, documents from the largest 100 servers in the TREC GOV2 data—in terms of the number of crawled pages—have been extracted and located in one hundred separate collections. TREC topics 701–750 (title) were used as queries. On average there are 3.2 words per query. The documents in all collections are from crawled webpages and the size of this testbed is many times larger than the current alternatives.

The other three testbeds are generated artificially from the uniform testbed [Powell and French, 2003; Si and Callan, 2003a; 2004].

- **trec123-AP-WSJ-60col (relevant)**: 24 Associate Press collections in the uniform testbed are collapsed into a single large APress collection. The same process is applied to 16 Wall Street Journal collections and they create a large

WSJournal collection. The other collections remain unchanged. Two large collections have higher density of relevant documents for the TREC queries.
- **trec123-2ldb-60col (representative)**: Collections in the uniform testbed are sorted by their name. Every fifth collection starting with the first is merged into a large "representative" collection. The same process is applied to every fifth collection starting from the second collection and they form another large representative collection. The other collections are unchanged.
- **trec123-FR-DOE-81col (nonrelevant)**: The 13 Federal Register and 6 Department of Energy collections in the uniform testbed are collapsed in two large collections respectively called FR and DOE. The rest of collections remain as were before. The larger collections have a lower density of relevant documents.

Collection selection algorithms are often compared using a recall metric R_k [Powell and French, 2003; Si and Callan, 2003a; 2004]:

$$R_k = \frac{\sum_{i=1}^{k} E_i}{\sum_{i=1}^{k} B_i} \tag{4}$$

Here, E is the collection selection ranking (CORI, REDDE, CRCS). B is the baseline ranking that is the relevance based ranking (RBR) [Powell and French, 2003] in our experiments. E_i and B_i are respectively the number of relevant documents in the ith ranked collection of E and B.

Choosing collections with a greater number of relevant documents (high recall) does not always lead to higher search effectiveness in DIR experiments [Si and Callan, 2003a; 2004]. Therefore, we also assess the effectiveness of algorithms on each testbed using relevance judgments. In all testbeds, we download 300 documents by query-based sampling[Callan and Connell, 2001] for each collection. Although it has been argued that using static summary sizes for collections is not always recommended [Baillie et al., 2006; Shokouhi et al., 2006a], we use this number to make our results comparable to other published work in this area [Callan and Connell, 2001; Craswell et al., 2000; Si and Callan, 2003a; 2004]. Each collection returns at most 100 answers to the broker for the entered query. We used SSL [Si and Callan, 2003b] algorithm to merge the returned results from the selected collections. The next section discusses the experimental results.

5 Results

Figure 1 depicts the performance of different collection selection algorithms on the trec4 and uniform testbeds. The horizontal axis in these figures shows the cutoff values, which are the number of collections that are selected for each query. The goal of collection selection methods is to select a few collections that contain the best answers. Therefore, we only show the R_k values for cutoffs smaller than 20. This number is consistent with DIR experiments that are reported elsewhere [Si and Callan, 2003a; 2004]. On the trec4 testbed, all methods produce almost

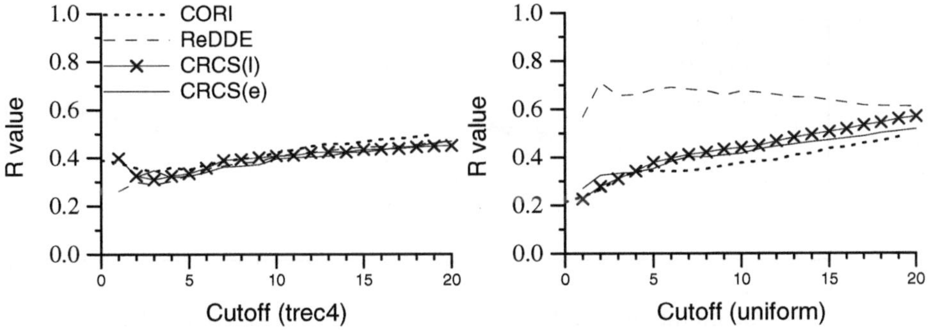

Fig. 1. R values for the CORI, REDDE and CRCS algorithms on the trec4-kmeans (left) and trec123-100col-bysource (right) testbeds

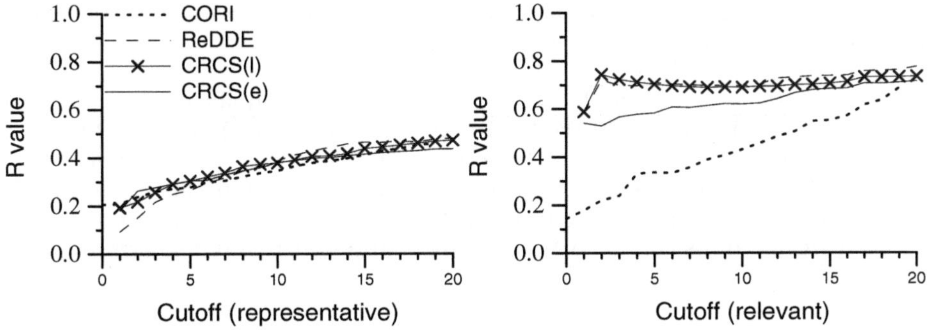

Fig. 2. R values for the CORI, REDDE and CRCS algorithms on the trec123-2ldb-60col (left) and trec123-AP-WSJ-60col (right) testbeds

the same R_k values for different cutoff points. For the uniform testbed, however, REDDE has a clear advantage while other methods have similar performance. We show later in this section that the advantage in recall does not have any significant impact on the final search effectiveness.

The R_k values for the representative and relevant testbeds are illustrated in Fig. 2. On the representative testbed, the difference between methods is negligible. On the relevant testbed, CORI produces far worse results than the other approaches. REDDE and CRCS(l) show similar outputs and they both work better than CRCS(e) for smaller cutoff points.

Figure 3 shows the R_k values produced by different methods on the GOV2 and non-relevant testbeds. The difference between methods on the non-relevant dataset is negligible for all cutoff values. On the GOV2 testbed, CORI selects collections with more relevant documents for smaller cutoff points. For larger cutoffs, all methods show similar outputs.

Higher recall values in the collection selection stage do not always lead to high precision in the final results [Si and Callan, 2003a; 2004]. Therefore, we

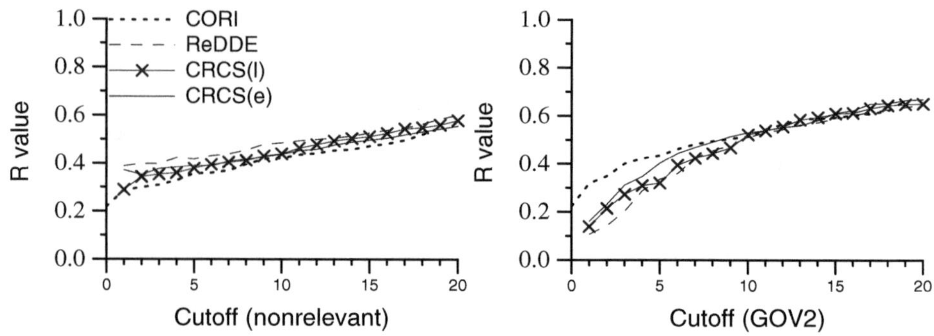

Fig. 3. R values for the CORI, REDDE and CRCS algorithms on the trec123-FR-DOE-81col (left) and 100-col-GOV2 (right) testbeds

Table 2. Performance of different methods for the Trec4 (trec4-kmeans) testbed. TREC topics 201–250 (long) were used as queries.

	Cutoff=1				Cutoff=5			
	P@5	**P@10**	**P@15**	**P@20**	**P@5**	**P@10**	**P@15**	**P@20**
CORI	0.3000	0.2380	0.2133†	0.1910†	0.3480	0.2980	0.2587	0.2380
ReDDE	0.2160	0.1620	0.1373	0.1210	0.3480	0.2860	0.2467	0.2190
CRCS(l)	0.2960	0.2260	0.2013	0.1810†	0.3520	0.2920	0.2533	0.2310
CRCS(e)	0.3080	0.2400	0.2173†	0.1910†	0.3880	0.3160	0.2680	0.2510

evaluate the search effectiveness of algorithms using the TREC queries and relevance judgments. We only report the results for cutoff=1 and cutoff=5. The former shows the system outputs when only the best collection is selected while for the latter, the best five collections are chosen to get searched for the query. We do not report the results for larger cutoff values because cutoff=5 has shown to be a reasonable threshold for DIR experiments on the real web collections [Avrahami et al., 2006]. The $P@x$ values show the calculated precision on the top x results.

We select REDDE as the baseline as it does not require training queries and its effectiveness is found to be higher than CORI and older alternatives [Si and callan, 2003a]. The following tables compare the performance of discussed methods on different testbeds. We used the t-test to calculate the statistical significance of difference between approaches. For each table, † and ‡ respectively indicate significant difference at the 99% and 99.9% confidence intervals between the performance of REDDE and other methods.

Results in Table 2 show that on the trec4 testbed, methods produce similar precision values when five collections are selected per query. The numbers also suggest that REDDE is not successful in selecting the best collection. It produces poorer results than the other methods and the difference is usually significant for P@15 and P@20.

Table 3. Performance of collection selection methods for the uniform (trec123-100col-bysource) testbed. TREC topics 51–100 (short) were used as queries.

	Cutoff=1				Cutoff=5			
	P@5	**P@10**	**P@15**	**P@20**	**P@5**	**P@10**	**P@15**	**P@20**
CORI	0.2520	0.2140	0.1960	0.1710	0.3080	0.3060	0.2867	0.2730
ReDDE	0.1920	0.1660	0.1413	0.1280	0.2960	0.2820	0.2653	0.2510
CRCS(l)	0.2120	0.1760	0.1520	0.1330	0.3440	0.3240	0.3067	0.2860
CRCS(e)	$0.3800^{\ddagger}$	$0.3060^{\ddagger}$	$0.2613^{\ddagger}$	$0.2260^{\dagger}$	0.3960	$0.3700^{\dagger}$	$0.3480^{\dagger}$	$0.3310^{\dagger}$

Table 4. Performance of collection selection methods for the representative (trec123-2ldb-60col) testbed. TREC topics 51–100 (short) were used as queries.

	Cutoff=1				Cutoff=5			
	P@5	**P@10**	**P@15**	**P@20**	**P@5**	**P@10**	**P@15**	**P@20**
CORI	*0.2160*	*0.2040*	*0.1773*	*0.1730*	0.3520	0.3500	0.3347	0.3070
ReDDE	0.3320	0.3080	0.2960	0.2850	0.3480	0.3220	0.3147	0.3010
CRCS(l)	0.3160	0.2980	0.2867	0.2740	0.3160	0.3160	0.2973	0.2810
CRCS(e)	0.2960	0.2760	0.2467	0.2340	0.3400	0.3500	0.3333	0.3090

On the uniform testbed (Table 3), CRCS(e) significantly outperforms the other alternatives for both cutoff values. CORI, REDDE, and CRCS(l) show similar performance on this testbed. Comparing the results with the R_k values in Figure 1 confirms our previous statement that selecting collections with a high number of relevant documents does not necessarily lead to an effective retrieval.

For the representative testbed as shown in Table 4, there is no significant difference between methods for cutoff=5. CRCS(l), CRCS(e) and REDDE produce comparable performance when only the best collection is selected. The precision values for CORI when cutoff=1 are shown in italic to indicate that they are significantly worse than REDDE at the 99% confidence interval.

On the relevant testbed (Table 5), all precision values for CORI are significantly inferior to that of REDDE for both cutoff values. REDDE in general produces higher precision values than CRCS methods. However, none of the gaps are detected statistically significant by the t-test at the 99% confidence interval.

Table 5. Performance of collection selection methods for the relevant (trec123-AP-WSJ-60col) testbed. TREC topics 51–100 (short) were used as queries.

	Cutoff=1				Cutoff=5			
	P@5	**P@10**	**P@15**	**P@20**	**P@5**	**P@10**	**P@15**	**P@20**
CORI	*0.1440*	*0.1280*	*0.1160*	*0.1090*	*0.2440*	*0.2340*	*0.2333*	*0.2210*
ReDDE	0.3960	0.3660	0.3360	0.3270	0.3920	0.3900	0.3640	0.3490
CRCS(l)	0.3840	0.3580	0.3293	0.3120	0.3800	0.3640	0.3467	0.3250
CRCS(e)	0.3080	0.2860	0.2813	0.2680	0.3480	0.3420	0.3280	0.3170

Table 6. Performance of collection selection methods for the non-relevant (trec123-FR-DOE-81col) testbed. TREC topics 51–100 (short) were used as queries.

	Cutoff=1				Cutoff=5			
	P@5	**P@10**	**P@15**	**P@20**	**P@5**	**P@10**	**P@15**	**P@20**
CORI	0.2520	0.2100	0.1867	0.1690	0.3200	0.2980	0.2707	0.2670
ReDDE	0.2240	0.1900	0.1813	0.1750	0.3480	0.2980	0.2707	0.2610
CRCS(l)	0.2040	0.1860	0.1813	0.1800	0.3360	0.3220	0.2973	0.2860
CRCS(e)	$0.3200^{\dagger}$	$0.2820^{\dagger}$	$0.2533^{\dagger}$	0.2240	0.3880	0.3600	$0.3387^{\dagger}$	$0.3210^{\dagger}$

Table 7. Performance of collection selection methods for the GOV2 (100-col-GOV2) testbed. TREC topics 701–750 (short) were used as queries.

	Cutoff=1				Cutoff=5			
	P@5	**P@10**	**P@15**	**P@20**	**P@5**	**P@10**	**P@15**	**P@20**
CORI	$0.1592^{\dagger}$	$0.1347^{\dagger}$	$0.1143^{\dagger}$	$0.0969^{\dagger}$	0.2735	0.2347	0.2041	0.1827
ReDDE	0.0490	0.0327	0.0286	0.0235	0.2163	0.1837	0.1687	0.1551
CRCS(l)	0.0980	0.0755	0.0667	0.0531	0.1959	0.1510	0.1442	0.1286
CRCS(e)	0.0857	0.0714	0.0748	0.0643	0.2776	0.2469	0.2272	0.2122

The results for CRCS(l), REDDE and CORI are comparable on the non-relevant testbed (Table 6). CRCS(e) significantly outperforms the other methods in most cases. On the GOV2 testbed (Table 7), CORI produces the best results when cutoff=1 while in the other scenarios there is no significant difference between the methods.

Overall, we can conclude that CRCS(e) selects better collections and its high performance remains robust. In none of the reported experiments, the precision values for CRCS(e) were significantly poorer than any other method at the 99% confidence interval. However, in many cases, the performance of CRCS(e) was significantly better than the second best method at the 99% or 99.9% confidence intervals. CORI and REDDE showed variable performance on different testbeds, each outperforming the other on some datasets.

6 Conclusions

We have introduced a new collection selection method for uncooperative DIR environments. We have shown that our proposed CRCS method can outperform the current state-of-the-art techniques. We investigated the robustness of different collection selection algorithms and showed that the performance of REDDE and CORI changes significantly on different testbeds while CRCS produces robust results. We also introduced a new testbed for DIR experiments based on the TREC GOV2 dataset. Our proposed testbed is about 36 times larger than the most well-known DIR testbeds and more than 6 times larger than the largest DIR testbed ever reported. Moreover, unlike traditional DIR testbeds that documents are assigned artificially into collections, collections in this testbed contain downloaded web documents arranged by server.

Experiments reported in this paper are based on the assumption that all collections are using the same retrieval model with equal effectiveness. However, in practice, collections often use different retrieval models and have different effectiveness. We plan to extend our experiments on collections with different retrieval models. Finally, the most proper values for α, β and γ have not been investigated and will be explored in our future research.

Acknowledgment

I am grateful to Justin Zobel, for his valuable comments on this work.

References

T. Avrahami, L. Yau, Luo Si, and Jamie Callan. The FedLemur: federated search in the real world. *Journal of the American Society for Information Science and Technology*, 57(3):347–358, 2006.

M. Baillie, L. Azzopardi, and F. Crestani. Adaptive query-based sampling of distributed collections. In *SPIRE String Processing and Information Retrieval Symposium*, pages 316–328, Glasgow, UK, 2006.

J. Callan and M. Connell. Query-based sampling of text databases. *ACM Transactions on Information Systems*, 19(2):97–130, 2001.

J. Callan, Z. Lu, and W. B. Croft. Searching distributed collections with inference networks. In *Proc. ACM SIGIR Conf.*, pages 21–28, Seattle, Washington, 1995.

N. Craswell, P. Bailey, and D. Hawking. Server selection on the World Wide Web. In *Proc. ACM Conf. on Digital Libraries*, pages 37–46, San Antonio, Texas, 2000.

D. D'Souza, J. Thom, and J. Zobel. Collection selection for managed distributed document databases. *Information Processing and Management*, 40(3):527–546, 2004a.

D. D'Souza, J. Zobel, and J. Thom. Is CORI effective for collection selection? an exploration of parameters, queries, and data. In *Proc. Australian Document Computing Symposium*, pages 41–46, Melbourne, Australia, 2004b.

L. Gravano, C. K. Chang, H. Garcia-Molina, and A. Paepcke. STARTS: Stanford proposal for Internet meta-searching. In *Proc. ACM SIGMOD Conf.*, pages 207–218, Tucson, Arizona, 1997.

L. Gravano, H. Garcia-Molina, and A. Tomasic. GlOSS: text-source discovery over the Internet. *ACM Transactions on Database Systems*, 24(2):229–264, 1999.

D. Hawking and P. Thomas. Server selection methods in hybrid portal search. In *Proc. ACM SIGIR Conf.*, pages 75–82, Salvador, Brazil, 2005.

T. Joachims, L. Granka, B. Pan, H. Hembrooke, and G. Gay. Accurately interpreting clickthrough data as implicit feedback. In *Proc. ACM SIGIR Conf.*, pages 154–161, Salvador, Brazil, 2005.

R. Manmatha, T. Rath, and F. Feng. Modeling score distributions for combining the outputs of search engines. In *Proc. ACM SIGIR Conf.*, pages 267–275, New Orleans, Louisiana, 2001.

H. Nottelmann and N. Fuhr. Evaluating different methods of estimating retrieval quality for resource selection. In *Proc. ACM SIGIR Conf.*, pages 290–297, Toronto, Canada, 2003.

A. L. Powell and J. French. Comparing the performance of collection selection algorithms. *ACM Transactions on Information Systems*, 21(4):412–456, 2003.

S. Raghavan and H. Garcia-Molina. Crawling the hidden web. In *Proc. 27th Int. Conf. on Very Large Data Bases*, pages 129–138, Roma, Italy, 2001. Morgan Kaufmann Publishers Inc.

M. Shokouhi, F. Scholer, and J. Zobel. Sample sizes for query probing in uncooperative distributed information retrieval. In *Proc. Asia Pacific Web Conf.*, pages 63–75, Harbin, China, 2006a.

M. Shokouhi, J. Zobel, F. Scholer, and S.M.M. Tahaghoghi. Capturing collection size for distributed non-cooperative retrieval. In *Proc. ACM SIGIR Conf.*, pages 316–323, Seattle, Washington, 2006b.

L. Si and J. Callan. Unified utility maximization framework for resource selection. In *Proc. ACM CIKM Conf.*, pages 32–41, Washington, 2004.

L. Si and J. Callan. Relevant document distribution estimation method for resource selection. In *Proc. ACM SIGIR Conf.*, pages 298–305, Toronto, Canada, 2003a.

L. Si and J. Callan. A semisupervised learning method to merge search engine results. *ACM Transactions on Information Systems*, 21(4):457–491, 2003b.

L. Si, R. Jin, J. Callan, and P. Ogilvie. A language modeling framework for resource selection and results merging. In *Proc. ACM CIKM Conf.*, pages 391–397, McLean, Virginia, 2002.

J. Xu and B. Croft. Cluster-based language models for distributed retrieval. In *Proc. ACM SIGIR Conf.*, pages 254–261, Berkeley, California, United States, 1999.

B. Yuwono and D. L. Lee. Server ranking for distributed text retrieval systems on the Internet. In *Proc. Conf. on Database Systems for Advanced Applications*, pages 41–50, Melbourne, Australia, 1997.

Results Merging Algorithm Using Multiple Regression Models

George Paltoglou[1], Michail Salampasis[2], and Maria Satratzemi[1]

[1] University of Macedonia, Egnatias 156, 54006 Thessaloniki, Greece
[2] Technological Educational Institute of Thessaloniki, P.O. BOX 141, 57400 Thessaloniki, Greece
{gpalt,cs1msa}@it.teithe.gr, maya@uom.gr

Abstract. This paper describes a new algorithm for merging the results of remote collections in a distributed information retrieval environment. The algorithm makes use only of the ranks of the returned documents, thus making it very efficient in environments where the remote collections provide the minimum of cooperation. Assuming that the correlation between the ranks and the relevancy scores can be expressed through a logistic function and using sampled documents from the remote collections the algorithm assigns local scores to the returned ranked documents. Subsequently, using a centralized sample collection and through linear regression, it assigns global scores, thus producing a final merged document list for the user. The algorithm's effectiveness is measured against two state-of-the-art results merging algorithms and its performance is found to be superior to them in environments where the remote collections do not provide relevancy scores.

Keywords: Distributed Information Retrieval, Results Merging, Algorithms.

1 Introduction

With the proliferation of the Web, it has become increasingly difficult for users to find relevant information to satisfy their information needs. Often, they are faced with the decision of choosing amongst several information sources in order to find the most appropriate that will provide the necessary information. A solution to this problem is provided by *search engines*, which give users a starting point in finding the information they need.

General purpose search engines nonetheless, only offer a limited solution. They cannot index the whole of the WWW because of its prohibitive size and rate of growth and thus index only parts of it. In addition to that, a large number of web sites, collectively known as *invisible web* [2, 14], are either not reachable by search engines or do not allow their content to be indexed by them and offer their own search capabilities. Thus, a user posing a query to general purpose search engines may be missing on relevant and highly qualitative information.

Distributed Information Retrieval (DIR [3]) offers a possible solution to the above problems by offering users the capability of simultaneously searching remote document collections (i.e. search engines or specific sites) through a single interface.

The challenge posed by DIR is on how to combine the results from multiple, independent, heterogeneous document collections into a single merged result in such

G. Amati, C. Carpineto, and G. Romano (Eds.): ECIR 2007, LNCS 4425, pp. 173 – 184, 2007.

a fashion that the effectiveness of the combination approximates or even surpasses the effectiveness of searching the entire set of documents as a single collection, if one was possible. This process can be perceived as four separate but often interleaved sub-processes. The *source representation* stage [4, 16], in which surrogates of the available remote collections are created. The *source selection* [6, 12, 16] stage, in which a subset of the available information collections is chosen to process the query. The *query submission* stage, in which the query is submitted to the selected collections. The *results merging* [8, 15] stage, in which the separate results are combined into a single merged result list which is returned to the user.

The focus of this paper is on the last part of distributed information retrieval process. Research [3, 6, 8, 15] has shown that the results merging phase is one of the most important elements in DIR, because it directly affects the response that the users get in return to their information need. Even if the most appropriate information sources have been chosen in the previous stages, if the merging isn't effective the overall quality of the retrieval process will deteriorate.

The rest of the paper is divided as follows. Section 2 reports on prior work. Section 3 describes the new methodology proposed. Section 4 describes the setup of the experiments. Section 5 reports and discusses the results and section 6 concludes the paper, summarizing the findings and presenting ideas for further development.

2 Prior Work

Significant research has been made in distributed information retrieval in recent years. Most of the focus has been on source selection, but significant progress has also been made in results merging, too.

The STARTS [9] initiative is an attempt to facilitate the task of querying multiple document sources through a commonly agreed protocol. It provides a solution for acquiring resource descriptions in a cooperative environment and thus facilitates source selection and results merging.

When cooperation from collections is not available (i.e. *isolated environments*), techniques have been developed that allow for the estimation of their contents. Query-based sampling [4] is such a technique that creates samples of the collections through multiple one-term queries. Through the sample, important statistics concerning the contents of the collection (such as terms, term frequencies, document frequencies etc) can be inferred.

Merging the result lists from individual collection is a complex problem not only because of the variety of retrieval engines that may be used by the individual collections, but also because of the diversity of collection statistics. Some of the results merging algorithms that have been proposed in recent years make use only of the ranked lists of documents returned by the individual collections, while others assume that the remote collections also return relevancy scores.

One of the first experiments in results merging was made by Voorhees in [17]. In that work two approaches where tested: one simple interleaving algorithm and a probabilistic biased c-faced die algorithm. The interleaving approach is based on the assumption that all chosen collections have the same number of relevant documents and works by simply interleaving their results one by one. It was found to be highly ineffective since this assumption is rather improbable in most environments. The c-faced die approach produced better results and was considered the most sophisticated technique that could be adopted in isolated environments in the absence of both sample collections and relevancy scores. The probabilistic nature of the algorithm was later re-examined in [18], and various deterministic approaches were presented.

In environments where the remote collections return not only ranked lists of documents but also relevancy scores, a variety of approaches have been proposed. *Raw score merging* merges the results as they are returned from the remote collections in a descending order, but it was found to be inefficient since for it to function properly, it required that the relevancy scores be comparable (i.e. between 0 and 1). The problem of incomparable scores was overcome by *normalizing* the returned scores at a common range. This approach produced better results, but the problem of different corpus statistics, eventually resulted in incomparable scores.

Weighted scores merging overcomes the above issue by assigning each document a score which is based both on the relevancy of the document itself and the relevancy of the collection where it belongs. The CORI results merging algorithm [3, 6] is based on a heuristic weighted scores merging algorithm and is considered state-of-the-art. The final score of each document is calculated as shown below:

$$C`_i = (C_i - C_{min}) / (C_{max} - C_{min}) \quad (1)$$

$$D` = (D - D_{min}) / (D_{max} - D_{min}) \quad (2)$$

$$D`` = (D` + 0.4 * D` * C`_i) / 1.4 \quad (3)$$

Equations (1) and (2) are used to normalize the collection and document scores to a range of 0 to 1 while equation (3) assigns the final weighted document relevancy score.

The work presented in this paper is influenced by the work in [15]. The algorithm presented in that paper, named *semi-supervised learning (SSL)*, makes use of a centralized index, comprised of all the sampled documents from the remote collections. The algorithm takes advantage of the common documents between the centralized index and the remote collections and their corresponding relevancy scores to estimate a linear regression model between the two scores.

The work presented in this paper differs from that work considerably in that it does not rely on relevancy scores returned from the remote collections, only ranked lists of documents. It is therefore much more efficient in environments where the remote collection provide minimum cooperation.

The results merging problem is often confused with the *metasearch* problem, where a number of information retrieval algorithms pose a query to a single document collection or multiple similar collections [10]. Most of the approaches under that context are based on the concept that the more a document appears at the returned lists of the individual algorithms, the more relevant it is. The work presented in this paper assumes that the remote collections have no documents in common (i.e. are non-intersecting), thus making the utilization of the above algorithms inappropriate.

Last but not least, a number of approaches download "on the fly", partially or fully, the returned documents in order to produce a final ranking [8]. The advantage of these methods is that they can estimate "first hand" the relevancy of a document. The disadvantages are that they have a significant time overhead and increased bandwidth requirements even if the download is only partial.

3 Results Merging Algorithm Using Multiple Regression Models

Most successful results merging algorithms [6, 15] rely on the concept that the remote collections return relevancy scores along with their ranked lists. However, in most modern environments, that is not the case. More often, remote collections return only ranked lists, relying on the fact that the average user has no need for relevancy scores, since they cannot be directly interpreted. Unfortunately, much less information is conveyed in rankings. Even if a collection is minimally relevant to a query, and the returned documents are only remotely relevant themselves, the rankings are the same as those provided by a very relevant collection returning very relevant documents.

3.1 Maximizing Usage of Local Resources

The motivation behind the results merging algorithm presented in this paper is to function effectively in environments where the remote collections provide the minimum of cooperation. By minimum, it is assumed that the remote collections are able to run queries and return ranked lists of documents, without scores. In order to achieve the goal of effectiveness, it is important for the algorithm to maximize the usage of available local resources. The sample collections that are created through query-sampling are readily available to the algorithm since they are stored locally and are under the control of the local authority. Their primarily use is for source selection and are usually discarded afterwards. It was not until the work in [15] that they were also utilized in later stages of the distributed information retrieval process.

In that work, all the sample documents from the individual collections were indexed into one single *centralized index*. The purpose of that index was to function as a representative of a single global index that would be created if all the documents were available for indexing. Specifically, the document and term frequency patterns were expected to resemble those that would be available under a centralized system.

The present work goes a step further and exploits this idea even more by regarding the sample collection of each remote collection as its representative. Under that notion, it is assumed that important statistics between the two collections (sample – remote) share common patterns. It is not assumed, nor is it necessary, that the sample collection be regarded as a complete representative of the remote collection, only that the most prominent features of the remote would still be valid in the sample.

3.2 Lack of Relevancy Scores

Previous work [1] attempted to make up for the lack of relevancy scores wherever that it was encountered by assuming that there is a linear mapping between the ranks of documents and the relevancy scores. Specifically, in cases where the remote collections did not return relevancy scores, artificial scores were assigned to the returned documents, giving a score of 0.6 to the 1st ranked document and decreasing at a steady rate until assigning a score of 0.4 to the last.

It has been shown in [7] none the less, that the decrease in relevancy is not linearly connected to the ranking. Specifically, it was shown that a logistic function, with $b<0$, would provide a better mapping. According to that work, the probability that document D_i is relevant given its rank x_i, is given by the equation:

$$\text{Prob}\left[D_i \text{ is } \text{Rel}/x_i\right] = \frac{e^{a+bx_i}}{1+e^{a+bx_i}} \quad (4)$$

Figure 1 demonstrates the expected correlation between relevancy and ranking.

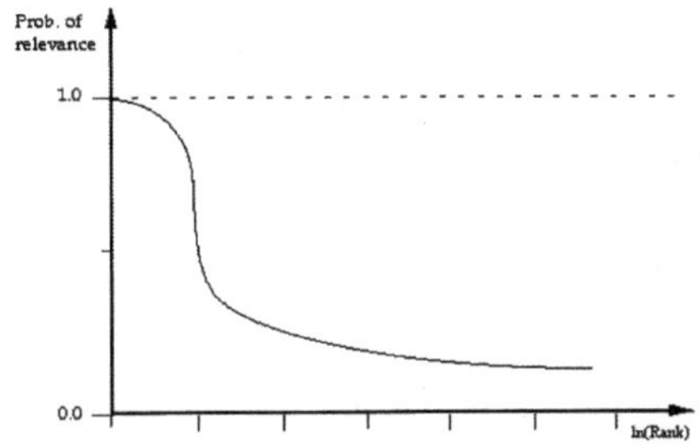

Fig. 1. Graph demonstrating the expected correlation between rank *(x)* and probability of relevance *(y)*. A logistic function with $b < 0$ was used to generate the S-curve.

Based on the above, the present work moves away from assigning artificial scores linearly and attempts to estimate the actual graph for each individual collection in order to produce accurate relevancy scores.

3.3 Estimating Local Relevancy Scores from Rankings

Following the source selection stage, the query is executed at the remote collections and in parallel on the locally stored samples of the selected collections and on the centralized index. We have used the inquery retrieval algorithm to query the local collections and the centralized index but any effective retrieval algorithm would do (kl divergence [19], okapi [13] etc). For each collection, two lists of documents are returned, one from the remote collection, containing only a ranked list of documents and one from the local sample, containing relevancy scores. The result list from the centralized index is disregarded for the time being, as it is incorporated into the algorithm at a latter stage. For each collection, the two document lists are compared and all the common documents are stored along with the rank that they obtained at the remote collection and the relevancy score at the local sample.

3.3.1 Estimating the S-Curve for Each Collection

Given the common documents found between the remote and the sampled collections, the algorithm estimates the S-curve for each collection, thus assigning local scores to the unseen documents returned from the remote collections. Influenced by the work in [7], we hypothesize that the correlation between the rank X of a document and the relevancy score Y is given by a logistic function:

$$Y = \frac{e^{a+b*X}}{1+e^{a+b*X}} \quad (5)$$

Applying the following transformations, we are able to modify the above equation into a linear one:

$$Y/(1-Y) = e^{a+b*X} \tag{6}$$

$$\ln[Y/(1-Y)]=a + b*X \tag{7}$$

$$\text{logit}[Y]=a + b*X \tag{8}$$

We need to estimate the parameters a, b of the above model in order to estimate the S-curve for each collection. Since equation (8) is a linear one, that estimation can be accomplished through linear regression.

3.3.2 Linear Regression

A linear regression model can be formally stated as:

$$y=a + b*x + e \tag{9}$$

where x is the independent variable (in our case, the rank of the document at the remote collection), y is the dependent variable (the similarity score of the document in the sample collection), a and b are the parameters of the model and e is the error (see below).

The observations used for the estimation of the model are pairs (x_i, y_i) i=1,…,n where x_i is the rank of the ith common document, y_i is the relevancy score of the document in the sample collection and n is the number of common documents found. Under that context, the aim of the model is to estimate parameters a and b that minimize the error e which represents the difference between the observed values of y and the ones estimated through the model. The best way to accomplish this is through least-squares regression analysis. In particular, the algorithm aims at minimizing the sum of squared residuals S:

$$S=\sum_{i=1}^{n}[y_i-\hat{y}_i]^2 \tag{10}$$

where y_i is the observed relevancy score of the ith common document and $\hat{y}_i$ is the one estimated by the model.

The problem can be formalized using matrix terminology as follows:

$$Y=X*B + e \tag{11}$$

where

$$Y=\begin{bmatrix} y_1 \\ y_2 \\ .... \\ y_n \end{bmatrix},\ X=\begin{bmatrix} 1\ x_1 \\ 1\ x_2 \\ ...... \\ 1\ x_n \end{bmatrix},\ B=\begin{bmatrix} a \\ b \end{bmatrix},\ e=\begin{bmatrix} e_1 \\ e_2 \\ .... \\ e_n \end{bmatrix}$$

where n is the number of common documents found. The optimal solution for parameters a and b is the one that minimizes S in equation (10) and is given by:

$$B=(X`X)^{-1}X`Y \tag{12}$$

In the experiments that were conducted we used only the first 10 common documents and disregarded the rest. Also, in order to better simulate the decline at the end of the graph (Fig.1), we inserted a "fake" common document at rank 3000, with score 0.001. Although these two adjustments were not necessary, they were found to increase the effectiveness of the algorithm.

The above process is repeated for each collection chosen by the source selection algorithm. Applying equation (5) with the estimated parameters for each remote collection, the algorithm assigns local relevancy scores to all the documents returned.

3.4 Estimating Global Relevancy Scores

Having estimated a local score for each document returned from the remote collections, the algorithm moves to the second phase, which is to estimate global scores for the returned documents. It is at this stage that the result list returned from the centralized index comes into use. As previously, the algorithm locates the common documents returned from the sample collections and the centralized index and stores their respective scores and fits a linear regression model to the data.

One might question whether linear regression is the best choice, or whether a polynomial (non-linear) regression would fit the data in a better way. Various reasons suggested the above decision. First of all, linear regression is consistent with the work done in [15], where a linear regression model is used to fit the relevancy scores attained from common documents at the remote collection and at the centralized index. Also, extensive early experiments showed that the benefit from going from a linear to a non-linear model would be minimal.

The pairs (x_i, y_i) i=1,…,m that will serve in the estimation of the parameters a and b in this case are x_i (the score assigned to the ith common document from the sample collection) and y_i (the score assigned to the ith common document from the centralized index). Again, the preferred methodology for estimating the parameters is the least-squares regression analysis, for which the optimal solution is given by equation (12).

Having estimated parameters a and b for each collection, the algorithm applies equation (9) to all the documents returned from the remote collections, using as independent variable the local score (x) that was attributed to them during the first phase of the algorithm and producing a final global score (y).

4 Experiment Setup

We used the TREC123 testbed, divided into 100 collections [12], that has been used extensively in DIR experiments. More information is provided below:

Trec123-100col-bysource: The documents in TREC 1, 2, 3 CDs are divided in 100 non-intersecting collections, organized by source and publication date. The contents of the collections are somewhat heterogeneous.

Table 1. Statistics about the collection

Name	Number of Collections	Size in GBs	Number of Documents		
			Min	Max	Avg
Trec123-100col	100	3.2	752	39713	10782

For queries, we used the title field from TREC topics 51-150. The average length of the queries is 3.1, which is typical for web queries.

Table 2. Statistics about the queries

Name	Number of Queries	TREC Topic Set	TREC Topic Field	Av. Length in Words
Trec123-100col	100	51-150	Title	3.1

For the experiments that were conducted, we used query-based sampling [4] for source representation, sending one-word queries and downloading the first four documents until a total of 300 documents per collection had been indexed. For source selection, we used CORI [6], as it is one of the best performing algorithms:

$$T=df / (df+50+150*cw/avg\_cw) \tag{13}$$

$$I=\log[(|DB|+0.5)/cf] / \log[|DB|+1.0] \tag{14}$$

$$p(r_k/C_i)=b + (1-b)*T*I \tag{15}$$

where df is the number of documents in collection C_i that contain term r_k, cf is the number of collection that contain term r_k, cw is the number of terms in C_i, avg_cw is the average cw, |DB| is the number of available remote collections and b is the default belief, set to the default value of 0.4.

An important factor that had to be examined was the retrieval algorithm that would be used at the remote collections. Two strategies were possible; we could either assume that all the remote collections operate on the same information retrieval algorithm, or that those differ. In order to make the experiments more realistic, the second approach was adopted. Three retrieval algorithms were implemented (inquery [5], kl-divergence language model [19] and the popular okapi algorithm [13]) and they were assigned to the remote collections in a round robin fashion. All the algorithms, including our own, were implemented using the Lemur Toolkit [11].

We also had to carefully consider the range of artificial scores that CORI and SSL would use. In [1] it was suggested that a good range would be 0.6–0.4. Indeed, after testing various ranges, it was found that although the differences weren't significant that range produced slightly better results.

5 Results

5.1 Number of Common Documents

For the algorithm to function properly, a minimum number of common documents need to be found between the sample and the remote collections and between the sample collections and the centralized index. Effectively, since we only utilize the first 10 common documents between the sample and the remote collections, any more than 10 are not needed. We report on the number of common documents found in both phases in a variety of settings below.

5.1.1 Remote and Sample Collections

Figure 2 shows the number of common documents that were found between the sample and the remote collections for queries 51-150 when 10 collections were selected to return 1000 and 300 documents each (left and right graph respectively). In the first case, the number of common documents is between 20 and 40 in the majority of queries, but there are 33 occasions out of 1000 (100 queries*10 collections) in which less than 3 documents were located. For these collections, although regression would be possible, a "fall-back" strategy was adopted and manufactured scores were assigned to the returned documents, as in [1].

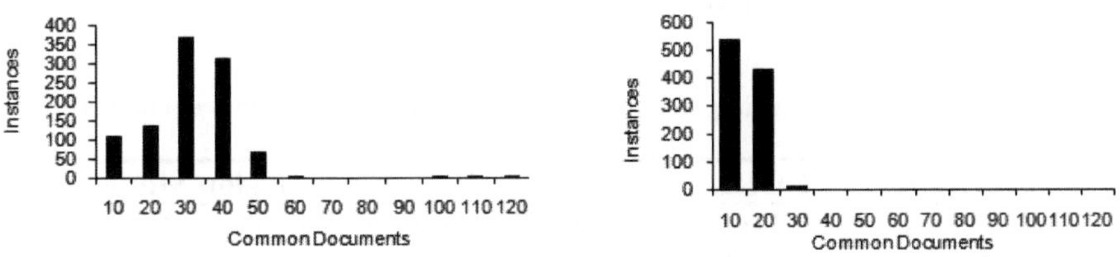

Fig. 2. Number of common documents found between the sample and the remote collections, when 1000 and 300 documents were returned from each (left and right graph respectively). The instances *(y axis)* refer to the number of times that the number of common documents found between a sample and a remote collection is as the corresponding interval at the *x axis*.

When 300 documents are requested, the number of common documents is, as expected, decreased. In 541 instances out of 1000, 0–10 common documents are found. The number of occasions where less than 3 documents were located and a "fall-back" strategy had to be adopted increased to 52, which although higher than in the previous setting, is still only a small fraction (5.2%).

5.1.2 Sample Collections and Centralized Index

A second question raised is whether enough common documents are found between the sample collections and the centralized index. Since both the sample collections and the centralized index are under our control, we have chosen to always request the maximum number of returned documents. Results are shown on figure 3.

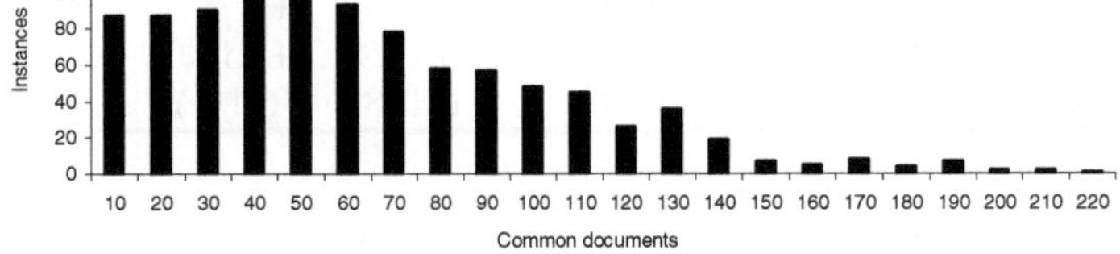

Fig. 3. Number of common documents between the samples and the centralized index

Although there is still a small number of instances (35 out of 1000) where there are less than 3 common documents, in the majority of cases enough documents are found to perform the second stage regression.

5.2 Precision

In distributed information retrieval environments it is usually inefficient to retrieve all the relevant documents scattered in the remote collections. Especially, when the focus of the retrieval is on the results merging part of the process, the focus is on precision.

Tables 3 to 5 report the results on precision at rankings 5 to 30 in a variety of settings. The percentages in parentheses report the difference between the new algorithm and CORI (left parenthesis) and SSL (right parenthesis) respectively.

Table 3. Precision using 10 collections, each one returning 1000 documents

Precision	CORI	SSL	Multiple Regression Models
At 5 docs:	0.1460	0.1560	0.1620 (+10.9%) (+3.8%)
At 10 docs:	0.1370	0.1350	0.1440 (+5.1%) (+6.6%)
At 15 docs:	0.1260	0.1267	0.1327 (+5.3%) (+4.7%)
At 20 docs:	0.1180	0.1205	0.1260 (+6.7%) (+4.5%)
At 30 docs:	0.1113	0.1103	0.1133 (+1.7%) (+2.7%)

Table 4. Precision using 10 collections, each one returning 300 documents

Precision	CORI	SSL	Multiple Regression Models
At 5 docs:	0.1340	0.1600	0.1780 (+32.8%) (+11.2%)
At 10 docs:	0.1340	0.1470	0.1540 (+14.9%) (+4.7%)
At 15 docs:	0.1333	0.1373	0.1427 (+7.0%) (+3.9%)
At 20 docs:	0.1300	0.1290	0.1335 (+2.7%) (+2.6%)
At 30 docs:	0.1260	0.1177	0.1177 (-6.5%) (+0.0%)

Table 5. Precision using 3 collections, each one returning 300 documents

Precision	CORI	SSL	Multiple Regression Models
At 5 docs:	0.1360	0.1580	0.1740 (+27.9%) (+10.1%)
At 10 docs:	0.1360	0.1300	0.1480 (+8.8%) (+13.9%)
At 15 docs:	0.1293	0.1187	0.1380 (+6.7%) (+16.2%)
At 20 docs:	0.1205	0.1075	0.1255 (+4.1%) (+16.7%)
At 30 docs:	0.1153	0.0947	0.1083 (-6.1%) (+14.4%)

It can be seen that the new results merging algorithm outperforms both CORI and SSL in most settings, often by a nontrivial margin. Even in settings where the remote collections return only a limited number of documents (i.e. 300) the new algorithm still manages to keep the precision at high levels.

6 Conclusions and Future Work

In this paper a new results merging algorithm was presented. It was designed explicitly to function effectively in settings where the remote collections return only ranked lists of documents, without relevancy scores. The effectiveness of the algorithm was tested against two state-of-the-art algorithms and was found to be superior to them.

Even more, the algorithm could be implemented to make use of relevancy scores whenever they are available, combining rank and score in producing the final document list.

Lastly, a methodology was provided on how to assign scores in environments where only ranked lists are returned using sampled documents. Although the particular methodology wasn't explicitly tested, the performance of the results merging algorithm indicate that there is at least some merit in it.

Acknowledgements

This work was supported by PENED, measure 8.3, action 8.3.1 under, project number 404, named "Collection fusion algorithms".

References

1. Avrahami, T. T., Yau, L., Luo Si, Callan, J.: The Fedlemur Project: Federated Search in the Real World. J. Am. Soc. Inf. Sci. Technol. 57, no. 3 (2006) 347-58
2. Bergman, M.: The deep web: surfacing the hidden value http://www.brightplanet.com/resources/details/deepweb.html BrightPlanet. (2001)
3. Callan J.: Distributed Information Retrieval, In W.B. Croft, editor, Advances in information retrieval, Kluwer Academic Publishers, chapter 5, 127-150
4. Callan, J., Connell M.: Query-based Sampling of Text Databases. ACM Trans. Inf. Syst. 19, no. 2 (2001) 97-130
5. Callan, J. P., Croft, W., B., Harding, St., M.: Inquery Retrieval System. 3rd International Conference on Database and Expert Systems Applications (1992) 78-83.
6. Callan, J., Zhihong, L. U., Croft, W. B.: Searching Distributed Collections With Inference Networks. SIGIR '95 (1995) 21-28
7. Calve, A. Le, Savoy, J.: Database Merging Strategy Based on Logistic Regression. Inf. Process. Manage. 36, no. 3 (2000) 341-59
8. Craswell, N., Hawking D., Thistlewaite P. B.: Merging Results from Isolated Search Engines. Australasian Database Conference (1999) 189-200
9. Gravano, L., Chang, C., Garcia-Molina, H., Paepcke A.: STARTS: Stanford proposal for internet meta-searching. 20th SIGMOD (1997) 207-218.
10. Lee, J., H.: Analyses of multiple evidence combination. (1997) 267–276
11. Lemur Toolkit http://www.lemurproject.org
12. Powell, A., L., French, J., C., Callan J., Connell, M., Viles, C., L.: The Impact of Database Selection on Distributed Searching. SIGIR '00 (2000) 232-239
13. Robertson, S., E., Walker, S., Hancock-Beaulieu M., Gatford M.: Okapi at Trec-3. TREC-3 (1994) 109-126.

14. Sherman, C.: Search for the invisible web. Guardian Unlimited (2001)
15. Si L., Callan J.: A Semisupervised Learning Method to Merge Search Engine Results. ACM Trans. Inf. Syst. 21, no. 4 (2003) 457-91
16. Si L., Callan, J.: Relevant Document Distribution Estimation Method for Resource Selection. SIGIR '03 (2003) 298-305
17. Voorhees, E. M., Gupta, N. K., Johnson-Laird, B.: The Collection Fusion Problem. TREC-3 (1994) 500-725
18. Yager, R., R., Rybalov, A.: On the Fusion of Documents From Multiple Collection Information Retrieval Systems. J. Am. Soc. Inf. Sci. 49, no. 13 (1998) 77-84
19. Zhai, C., Lafferty, J.: A study of smoothing methods for language models applied to ad hoc information retrieval, SIGIR'01 (2001) 334-342.

Segmentation of Search Engine Results for Effective Data-Fusion

Milad Shokouhi

School of Computer Science and Information Technology
RMIT University, Melbourne 3001, Australia
milad@cs.rmit.edu.au

Abstract. Metasearch and data-fusion techniques combine the rank lists of multiple document retrieval systems with the aim of improving search coverage and precision.

We propose a new fusion method that partitions the rank lists of document retrieval systems into chunks. The size of chunks grows exponentially in the rank list. Using a small number of training queries, the probabilities of relevance of documents in different chunks are approximated for each search system. The estimated probabilities and normalized document scores are used to compute the final document ranks in the merged list. We show that our proposed method produces higher average precision values than previous systems across a range of testbeds.

1 Introduction

In traditional information retrieval, a centralized search system is responsible for crawling and indexing the documents and ranking them for user queries.

In the practice, there are many cases that combining the outputs of multiple retrieval systems may produce better results than using a single monolithic system. Metasearch [Meng et al., 2002], distributed information retrieval (DIR) [Callan, 2000] and data-fusion [Croft, 2000] techniques all combine the results of different retrieval models to produce a *richer* rank list.

Compared to centralized search engines, metasearch engines have several advantages. Some of these benefits, as noted by other researchers [Aslam and Montague, 2001; Meng et al., 2002; Vogt, 1999] are listed below:

High recall: Compared to centralized search engines, metasearch engines can provide a broader search coverage over the internet documents. Commercial search engines index varying portions of the web graph and the amount of overlap between their sets of indexed documents is small. For example, the amount of overlap between the indexed documents by Google[1] and Yahoo![2] is estimated to be less than 45% [Bar-Yossef and Gurevich, 2006].

[1] www.google.com

[2] www.search.yahoo.com

G. Amati, C. Carpineto, and G. Romano (Eds.): ECIR 2007, LNCS 4425, pp. 185–197, 2007.

High precision: A document that is returned by multiple search engines for a query is more likely to be relevant than another document that is returned by only a single search engine. A previous study suggests that search systems are not likely to return the same nonrelevant documents [Lee, 1997] while they may return the same relevant documents. Therefore, the search precision improves by giving high ranks to documents that are returned by multiple search engines. This is also known as the *chorus effect* [Vogt, 1999].

Typically, the top-ranked answers returned by search engines have higher density of relevant documents compared to the other results. Therefore, a result list that is generated from the top-ranked documents of different search engines is more likely to contain relevant documents than the outputs of a single search engine. This is also known as the *skimming effect* [Vogt, 1999].

Removal of spam pages: Spam pages can be avoided by combining the results of multiple search engines. It is unlikely that several search engines return a particular spam page for a given query [Dwork et al., 2001].

Easy updates: The contents of many web documents such as news pages change frequently. While centralized search engines cannot update their index with the same rate, metasearch techniques can be useful for providing a search service over the most recent documents [Rasolofo et al., 2003].

These benefits were originally suggested for metasearch engines but most of them are also valid for DIR and data-fusion techniques.

Data-fusion methods merge the results of different ranking functions that are applied on a single text collection [Aslam and Montague, 2001; Croft, 2000; Fox and Shaw, 1993; Lee, 1997; Lillis et al., 2006; Vogt, 1999]. In metasearch—also known as collection fusion—the query is sent to different search engines that may have different rates of overlap. In other words, data-fusion is a special form of metasearch where the overlap between search engines is 100% [Vogt, 1999].

When the amount of overlap between search systems is negligible, metasearch can be classified under the DIR category. Therefore, DIR can be considered as a special form of metasearch where the overlap between search systems in either none or negligible [Si and Callan, 2003].

In this paper, we propose a novel data-fusion technique that can effectively merge the results of different ranking functions that are applied to a single collection. Our algorithm partitions the rank lists returned by search engines into separate chunks. The probability of relevance for documents in each chunk is computed for all search engines. The final score of a document is calculated according to its probabilities of relevance in multiple result sets. We show that the merged lists produced by our method have higher mean average precision compared to the alternatives.

2 Data Fusion

Perhaps, the simplest data-fusion method is the round-robin strategy [Savoy et al., 1996] or *interleaving* [Voorhees et al., 1994], where documents returned by all search systems are treated equally and are merged according to their ranks. The final merged list produced by the round-robin strategy contains the top-ranked documents from all search systems followed by the second top-ranked documents and so forth. Since the search effectiveness of different ranking functions is ignored by round-robin, it is not suitable for environments where both poor and effective search systems are involved. A more sophisticated model of interleaving is suggested by Voorhees et al. [1995]. In this approach, the effectiveness of retrieval models is approximated by training queries. The number of documents that are included in the final merged list from each collection depends on the approximated effectiveness of that model in the training phase.

Several combination methods are suggested by Fox and Shaw [1993; 1994] to compute the score of documents that are returned by more than a single search system. CombSUM and CombMNZ are the most successful of these methods and are used in many metasearch engines such as Metacrawler [Selberg and Etzioni, 1997] and SavvySearch [Dreilinger and Howe, 1997]. In CombSUM, when a document d is returned by multiple retrieval models, all scores are added together to produce the final score. In CombMIN and CombMAX, respectively the minimum and maximum values among the scores reported by different search systems are used as the final score. CombMNZ adds all the reported scores for a document and multiplies the sum value to the number of retrieval models that have returned that document (d) as below:

$$CombMNZ_d = \sum_{c}^{N} D^c \times |D^c > 0| \tag{1}$$

Here, N is the number of input rank lists for data-fusion and $|D^c > 0|$ is the number of rank lists that contain the document d. D^c is the normalized score of document d in the rank list c and is computed as:

$$D^c = \frac{S_d - D^c_{min}}{D^c_{max} - D^c_{min}} \tag{2}$$

Where S_d is the score of document d in the rank list c before normalization. D^c_{min} and D^c_{max} are the minimum and maximum document scores available in the rank list. We use CombMNZ as one of our baselines because it is a common benchmark for the data-fusion experiments [Lee, 1997; Aslam and Montague, 2000; 2001; Manmatha et al., 2001; Lillis et al., 2006].

Lee [1997] proposed a modified version of CombMNZ where document ranks are used instead of document scores. Linear combination methods for data-fusion have been investigated [Vogt and Cottrell, 1999; Vogt, 2000; 1999] in which, documents are merged according to their normalized relevance scores. Both training data and document scores are needed for calculating the final score of a document. A logistic regression model is suggested by Calvé and Savoy [2000] to map the document ranks to the probabilities of relevance. Their suggested approach produces slightly better results than linear combination methods.

A probabilistic fusion approach is suggested by Aslam and Montague [2000]. They assume that the the probability of relevance of a document in any given rank is known for each retrieval model. They use those probabilities to calculate the final document scores for merging. Their suggested method—also known as Bayes-fuse [Aslam and Montague, 2001]—can produce comparable results to CombMNZ. Borda-fuse [Aslam and Montague, 2001] is inspired by a voting algorithm. In Borda-fuse, documents are merged according to their ranks and no training data is required. Aslam and Montague [2001] showed that the performance of Borda-fuse is usually slightly lower (but comparable) than that of standard fusion techniques such as CombMNZ. Manmatha et al. [2001] empirically showed that for a given query, the distributions of scores for relevant and nonrelevant documents respectively follow normal and exponential curves. They used curve-fitting techniques to map the document scores to the probabilities of relevance. Then the outputs of search engines can be merged by averaging the probabilities of relevance for documents. They showed that their proposed approach can produce similar results to those of CombMNZ.

In ProbFuse [Lillis et al., 2006], each of the returned rankings is divided into k consecutive segments. Lillis et al. [2006] empirically showed that for the TREC adhoc experiments, where each submitted run (rank list) contains 1000 answers per query, dividing lists into 25 segments produces effective results. That is, for each rank list, the first 40 answers for a query are considered as the first segment. Answers 41–80 generate the second segment and so forth. For each search system, the probability of relevance of documents in each segment is estimated by using training queries as follows:

$$P(rel|d_k^c) = \frac{\sum_{q=1}^{Q} \frac{|R_{k,q}|}{Size_k}}{Q} \qquad where\ Size_k = 25 \tag{3}$$

Here, $P(rel|d_k^c)$ represents the probability of relevance of a document (d), that has been returned in the kth segment of the rank list c. Q is the number of training queries and $|R_{k,q}|$ is the number of relevant documents that are returned in the kth segment of the rank list for the training query q. Once the probabilities of relevance for each system are calculated, ProbFuse uses them to calculate the merging scores of documents. The final document scores are calculated by summing up their probabilities of relevance in all systems (the chorus effect) as:

$$ProbFuse_d = \sum_{c}^{N} \frac{P(rel|d_k^c)}{k} \tag{4}$$

Where N is the total number of search systems used for data-fusion. ProbFuse divides the probability of relevance of a document by its segment number (k). This strategy advantages the documents that appear on the top ranks of each result set (the skimming effect). ProbFuse is one of the most recent and effective data-fusion methods. Thus, we use it as one of the baselines in our experiments.

In the following sections we introduce our data-fusion technique and compare it with two other baselines. We show that our method can produce better merged lists than the competitive approaches.

3 SegFuse

In this section, we describe SegFuse, our novel data-fusion technique. SegFuse is a probabilistic data-fusion method that estimates the probabilities of relevance of documents using a small number of training queries. SegFuse can be summarized in three steps:

1. For each search system, the returned rank list is partitioned into chunks. The size of chunks increases exponentially in the rank lists.
2. For each search system, the probability of relevance for documents in each chunk is estimated using a few training queries.
3. The final score of a document is computed according to its probabilities of relevance in the rank lists. SegFuse merges the results and ranks the documents according to their final scores.

As for ProbFuse [Lillis et al., 2006], the outputs of each ranking function are divided into segments. However, there are two major distinctions between SegFuse and the suggested approach by Lillis et al. [2006].

First, in ProbFuse, the size of segments is always constant (say 40). Equation (4) suggests that a document that is returned in the rank 39 of a result set with the relevance probability of R is advantaged over another document that is returned at the top-rank of another result set with the relevance probability of $R - \epsilon$. This is against the skimming effect which suggests that top-ranked documents should be advantaged. We propose that the size of segments be varied exponentially according to the ranks. Therefore, the rank difference of documents is considered more effectively.

Second, ProbFuse ignores the scores of documents that may be provided by search systems. We propose the use of the normalized document score values.

In SegFuse, the probability of relevance of a document d returned in the kth segment of a result set c is calculated by averaging the number of relevant documents that are returned in that segment for the training queries:

$$P(rel|d_k^c) = \frac{\sum_{q=1}^{Q} \frac{|R_{k,q}|}{Size_k}}{Q} \qquad where\ Size_k = (10 \times 2^{k-1}) - 5 \tag{5}$$

Here, k is the segment number of document d in the result set c. Q is the total number of training queries and $|R_{k,q}|$ is the number of relevant documents that are returned for the training query q in the kth segment of the result set c. $Size_k$ is the number of documents that are available in the kth segment of the rank list. The parameters of the exponential function are chosen empirically according to our preliminary experiments. For a given result set, the first top five documents are included in the first segment. Segment two consists of the documents between ranks 6 to 20. The other segments are defined with an exponential increase in the size. Note that except for the way that $Size_k$ is calculated, this is the same equation that is used by ProbFuse in Eq. (3).

The strategy used by SegFuse for calculating the segment size is inspired by the observations of other researchers regarding that the probabilities of relevance for documents and their importance for users decrease exponentially[Joachims et al., 2005; Manmatha et al., 2001]. In a given rank list, a few top documents are usually more important for users [Joachims et al., 2005] and have the highest probability of relevance [Manmatha et al., 2001]. The difference between documents becomes negligible as we move towards the bottom of a rank list. There might be other exponential functions that produce more effective results. But we leave the investigation of an optimum exponential function as future work. SegFuse calculates the final score of a document as:

$$SegFuse_d = \sum_{c}^{N} P(rel|d_k^c) \times D^c + P(rel|d_k^c) \quad (6)$$

Where N is the total number of rank lists that contain the document d. D^c is the normalized score of document d returned by collection c that is computed by equation (2). Note that, unlike ProbFuse, the scores are not divided by the segment number. That is because the highly ranked documents are already advantaged by Eq. (5). Such documents are picked from smaller populations with higher density of relevant documents compared to those in other segments. In general, the skimming and chorus effects are implicitly considered by equations (5) and (6) respectively.

4 Testbeds

To evaluate our model and compare it with other techniques we created two testbeds from the result sets submitted to the TREC3 and TREC5 conferences. For the TREC3 experiments [Harman, 1994], 40 runs were submitted by different participants to the adhoc track for the TREC topics 151–200. We randomly picked 30 files from the submitted runs. We then assign the selected files into six groups each containing five submissions.

In the TREC5 conference [Voorhees and Harman, 1996], participants were asked to submit their results for the adhoc track for two categories A and B. In *CategoryB* the queries (TREC topics 251–300) were only executed on a subset of data. We use the results submitted for *CategoryA*, where queries are on the

Table 1. Six groups are created from the submitted runs to the TREC3 conference. Each group contains five runs that are chosen randomly from the results submitted to the TREC3 adhoc task for topics 151–200.

	Run1	**Run2**	**Run3**	**Run4**	**Run5**
G1	input.assctv2	input.clartm	input.rutfua2	input.siems2	input.topic4
G2	input.assctv1	input.brkly6	input.citri2	input.eth001	input.virtu1
G3	input.brkly7	input.citya1	input.eth002	input.nyuir1	input.xerox3
G4	input.acqnt1	input.clarta	input.erima1	input.nyuir2	input.padre1
G5	input.citri1	input.lsia0mf	input.padre2	input.topic3	input.vtc2s2
G6	input.citya2	input.inq101	input.pircs1	input.siems1	input.vtc5s2

Table 2. Seven groups are created from the submitted runs to the TREC5 conference. Each group contains five runs that are chosen randomly from the results submitted to the TREC5 adhoc task for topics 251–300.

	Run1	**Run2**	**Run3**	**Run4**	**Run5**
G1	input.Cor5A2cr	input.ibmge2	input.LNmFull2	input.uwgcx1	input.vtwnA1
G2	input.Cor5A1se	input.DCU962	input.ibmge1	input.INQ302	input.pircsAAS
G3	input.anu5man4	input.colm1	input.genrl4	input.ibms96b	input.KUSG2
G4	input.Cor5M2rf	input.DCU964	input.ETHas1	input.gmu96au1	input.ibmgd2
G5	input.Cor5M1le	input.erliA1	input.fsclt3	input.gmu96au2	input.pircsAM1
G6	input.city96a1	input.DCU961	input.genrl3	input.LNaDesc1	input.pircsAAL
G7	input.CLCLUS	input.genrl1	input.gmu96ma2	input.KUSG3	input.LNaDesc2

complete dataset. In total, there are 61 runs submitted by participants from which we picked 35 randomly. We then assign the selected files randomly into seven groups each containing five unique submissions.

Tables 1 and 2 respectively show the names of selected submissions from TREC3 and TREC5 runs that are used in our experiments. For both testbeds, the runs in each group (G) are used together for the data-fusion experiments.

5 Experimental Results

We use CombMNZ and ProbFuse as the baselines of our experiments. The former is a common baseline for data-fusion experiments [Lee, 1997; Aslam and Montague, 2000; 2001; Manmatha et al., 2001; Lillis et al., 2006] while the latter is one of the most recent algorithms proposed in this area.

In both TREC3 and TREC5 testbeds, the rank lists assigned to each group are merged using our SegFuse method and the results are compared with that of CombMNZ and ProbFuse. There are 50 TREC topics available for each testbed. For ProbFuse and SegFuse methods that use training queries, we applied different portions of available query sets for training and testing. For each testbed, we used different training sizes Q where $Q \in \{5, 10, 15, 20\}$. That is, 10%–40% of the available queries in each testbed are used for training and the rest are used for testing. Although CombMNZ does not use training queries, for the sake of fair comparison, it was evaluated solely based on the testing subset of queries.

Data-fusion experiments are usually compared by the mean average precision value and the produced average precision at different recall points [Aslam and Montague, 2001; Manmatha et al., 2001; Lillis et al., 2006; Vogt, 2000].

Figure 1 depicts the 11-point average precision values obtained for varying number of testing queries on both testbeds. The curves in each graph are produced by averaging the precision values of data-fusion runs on the available groups when the specified number of testing queries are used. For example, the precision values for SegFuse on the top-left graph are calculated as follows. First, for each group in Table 1, 10% of the available queries (topics 151–155) are used for calculating the probabilities of relevance. Once the probabilities are

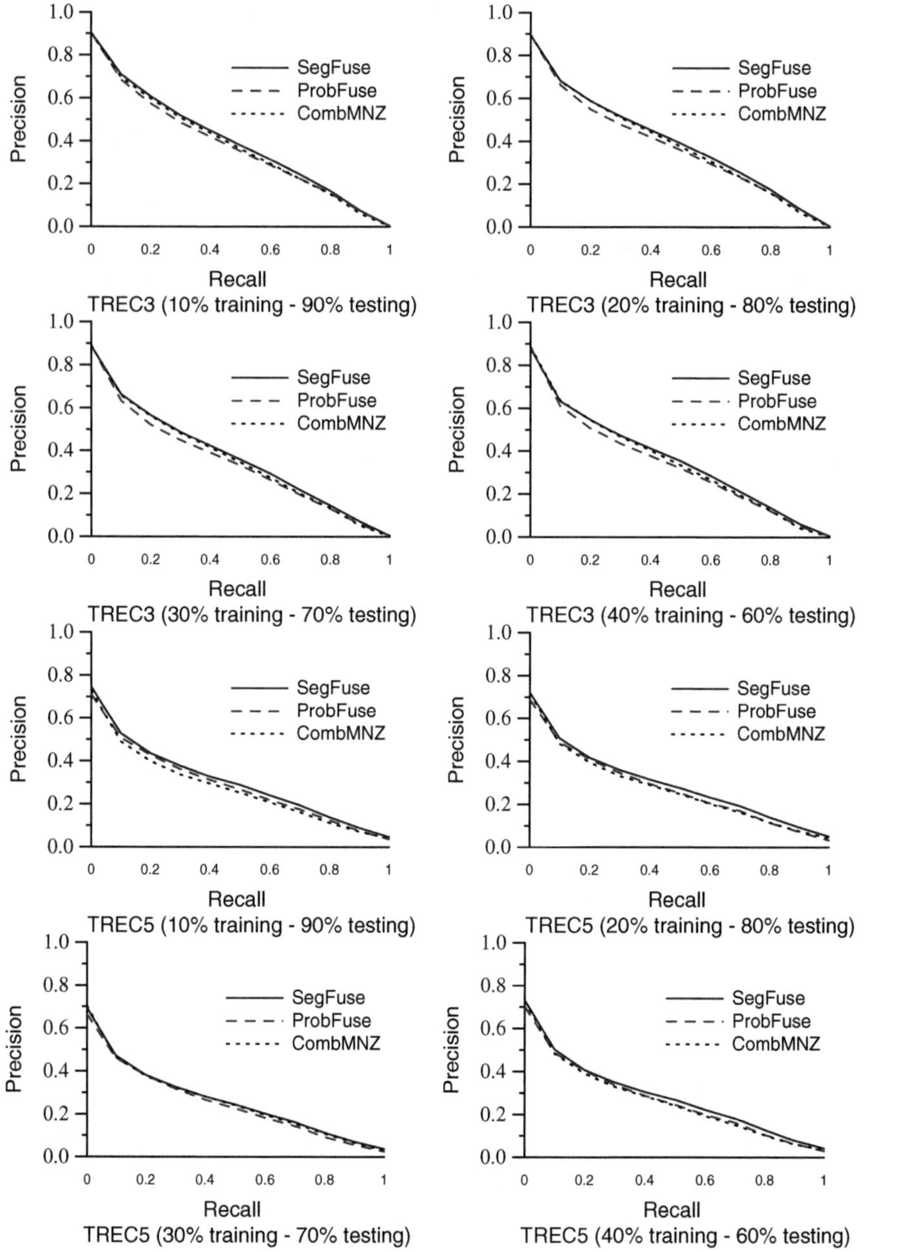

Fig. 1. The precision-recall curves for fusion methods on the TREC3 and TREC5 data

computed, SegFuse merges the submitted runs of each group in a separate list. Finally, the precision values in the six produced merged lists are averaged to draw the final curve. The other curves are produced in the same way.

It can be seen that in all experiments, SegFuse has higher precision values than both baselines. CombMNZ performs slightly better than ProbFuse on the TREC3 data while an opposite observation can be made on the TREC5 data. This is consistent with the experiments reported by Lillis et al. [2006]. Since all graphs in Fig. 1 are produced by calculating the average values among many groups, the gaps between methods are small. The following sections provide a detailed analysis of experiments on individual groups. Due to space limitations, we only include the results for 5 and 20 training queries that are respectively the minimum and maximum training boundaries in our experiments.

5.1 Analysis of Individual Experiments on TREC3 Data

Table 3 shows the precision values obtained by data-fusion methods on different groups of TREC3 data. The TREC topics 151–155 are used for training and computing the probabilities of relevance and topics 156–200 are used for testing the performance of data-fusion techniques.

The methods are compared using different evaluation metrics. MAP represents the mean average precision over 11 recall points and is a common metric for evaluating the data-fusion experiments [Aslam and Montague, 2001; Manmatha et al., 2001; Lillis et al., 2006; Vogt, 2000]. Therefore, we use the t-test to measure the statistical significance of differences betwee the MAP values. The significant differences at the 99% and 99.9% confidene intervals between the MAP values produced by ProbFuse and other methods are represented by † and ‡ respectively. Bpref shows how many times judged nonrelevant documents are returned before the relevant documents [Buckley and Voorhees, 2004]. P@10 is the precision at the top 10 returned documents and MRR is the mean reciprocal rank value showing the position of the first relevant document in the rank list. For each metric, the method with the highest value is specified with an underline. The results show that SegFuse has higher MAP and Bpref values than the baselines for all groups. MRR and P@10 values are comparable for all methods.

In Table 4, 40% percent of available queries (topics 551–570) are used as the training data. The produced MAP and Bpref values by SegFuse are better than the alternatives respectively in all and four of six cases. However, SegFuse is

Table 3. Data-fusion experiments on the TREC3 submitted runs (input files). The TREC topics 151–155 are used for training while topics 156–200 are used for testing.

	CombMNZ				*ProbFuse*				*SegFuse*			
	MAP	Bpref	P@10	MRR	MAP	Bpref	P@10	MRR	MAP	Bpref	P@10	MRR
G1	0.422†	0.410	0.746	0.901	0.402	0.390	0.751	0.879	0.433‡	0.422	0.764	0.915
G2	0.338	0.342	0.673	0.855	0.329	0.331	0.693	0.871	0.342†	0.345	0.695	0.873
G3	0.409	0.395	0.775	0.834	0.417	0.393	0.760	0.892	0.420	0.401	0.771	0.860
G4	0.329‡	0.331	0.673	0.863	0.303	0.300	0.648	0.806	0.339‡	0.337	0.664	0.783
G5	0.318	0.328	0.655	0.758	0.323	0.324	0.688	0.845	0.349‡	0.353	0.686	0.834
G6	0.396	0.389	0.764	0.851	0.389	0.377	0.751	0.818	0.405†	0.393	0.726	0.849

Table 4. Data-fusion experiments on the TREC3 submitted runs (input files). The TREC topics 151–170 are used for training while topics 171–200 are used for testing.

	CombMNZ				*ProbFuse*				*SegFuse*			
	MAP	Bpref	P@10	MRR	MAP	Bpref	P@10	MRR	MAP	Bpref	P@10	MRR
G1	0.386†	0.377	0.710	0.902	0.356	0.349	0.703	0.861	0.394‡	0.387	0.710	0.916
G2	0.320	0.322	0.643	0.870	0.305	0.302	0.653	0.879	0.320†	0.321	0.636	0.863
G3	0.380	0.368	0.723	0.806	0.379	0.356	0.716	0.855	0.388	0.373	0.710	0.815
G4	0.286‡	0.294	0.616	0.819	0.255	0.256	0.583	0.759	0.290‡	0.292	0.606	0.717
G5	0.279	0.287	0.630	0.739	0.273	0.276	0.616	0.840	0.300‡	0.306	0.626	0.811
G6	0.362	0.356	0.726	0.838	0.347	0.337	0.730	0.814	0.366‡	0.356	0.706	0.837

not as successful as other methods for placing the relevant documents at the top ranks. On average, ProbFuse produces the highest MRR values while CombMNZ is the dominant method for producing high precision at the top 10 documents.

5.2 Analysis of Individual Experiments on TREC5 Data

Table 5 shows the results of data-fusion methods on the TREC5 data when 10% of available queries (topics 251–255) are used for training. SegFuse produces the highest MAP and Bpref values in all experiments. It is also the dominant method for the MRR metric. The effectiveness of SegFuse and ProbFuse is comparable in terms of P@10 while CombMNZ clearly performs poorer.

Table 5. Data-fusion experiments on the TREC5 submitted runs (input files). The TREC topics 251–255 are used for training while topics 256-300 are used for testing.

	CombMNZ				*ProbFuse*				*SegFuse*			
	MAP	Bpref	P@10	MRR	MAP	Bpref	P@10	MRR	MAP	Bpref	P@10	MRR
G1	0.316	0.299	0.513	0.793	0.338	0.317	0.557	0.800	0.339	0.327	0.522	0.785
G2	0.198	0.194	0.378	0.578	0.185	0.178	0.333	0.534	0.208	0.199	0.377	0.597
G3	0.290	0.291	0.526	0.799	0.329	0.313	0.520	0.772	0.347	0.334	0.553	0.818
G4	0.212	0.200	0.408	0.631	0.236	0.229	0.413	0.643	0.261	0.255	0.475	0.704
G5	0.248	0.245	0.428	0.692	0.247	0.236	0.453	0.645	0.263†	0.259	0.448	0.680
G6	0.278	0.263	0.467	0.704	0.255	0.230	0.440	0.600	0.297	0.283	0.493	0.720
G7	0.281	0.272	0.510	0.722	0.309	0.293	0.522	0.715	0.320	0.310	0.517	0.693

In Table 6, the TREC topics 251–270 are used for calculating the probabilities of relevance and topics 271–300 are used for testing. As in the previous experiments, SegFuse produces the highest MAP and Bpref values. CombMNZ and SegFuse produce comparable P@10 values and they both outperform ProbFuse in that metric. CombMNZ is the best method for placing the first relevant document on the highest ranks. The MRR values for ProbFuse and SegFuse are comparable although SegFuse is generally better.

Table 6. Data-fusion experiments on the TREC5 submitted runs (input files). The TREC topics 251-270 are used for training while topics 271–300 are used for testing.

	CombMNZ				*ProbFuse*				*SegFuse*			
	MAP	Bpref	P@10	MRR	MAP	Bpref	P@10	MRR	MAP	Bpref	P@10	MRR
G1	0.292	0.274	0.512	0.751	0.293	0.272	0.513	0.721	0.305	0.287	0.500	0.747
G2	0.188	0.183	0.371	0.574	0.183	0.172	0.330	0.561	0.198†	0.188	0.363	0.574
G3	0.275	0.271	0.516	0.750	0.314	0.291	0.523	0.787	0.341†	0.322	0.526	0.819
G4	0.220	0.208	0.425	0.660	0.226	0.229	0.393	0.651	0.239	0.229	0.453	0.694
G5	0.233	0.232	0.438	0.707	0.215	0.207	0.430	0.621	0.236†	0.227	0.416	0.677
G6	0.277	0.261	0.493	0.722	0.246	0.222	0.430	0.606	0.282	0.267	0.500	0.699
G7	0.276	0.262	0.522	0.740	0.290	0.271	0.510	0.711	0.309	0.293	0.506	0.695

One may have noticed that the performance of methods have decreased when using a larger training set. This is due to the distribution of relevant documents for the TREC topics. For both TREC3 and TREC5 topics, the number of relevant documents returned by the TREC runs for the first few queries are higher than the other topics. The fact that the performance of CombMNZ also decreases—although it does not use training—is further evidence of our claim.

To solve this problem, Lillis et al. [2006] suggested that queries can be rearranged in different random orders. Then, data-fusion experiments can be performed for each random order and the final results are calculated by averaging the obtained values for different query distributions. However, our aim is not to prove that more training data can improve the data-fusion but is to show that SegFuse can effectively merge the rank lists using small volume of training data.

Overall, SegFuse produces higher Bpref and MAP values than other methods. In both testbeds, the MAP values produced by SegFuse are significantly higher than ProbFuse. The differences between the MAP values for CombMNZ and ProbFuse are significant on the TREC3 testbed and insignificant on the TREC5 data. For P@10 and MRR metrics none of the methods have a clear advantage over the others and the results are usually comparable.

6 Conclusions

We have proposed a new data-fusion method that estimates the probability of relevance of documents using a small amount of training data. Unlike the major competitive method ProbFuse, which is less sensitive to the top-ranked documents, our method pays special attention to the highly ranked documents. The document scores in SegFuse decrease exponentially, which is consistent with the importance of documents for users and their probabilities of relevance [Joachims et al., 2005; Manmatha et al., 2001]. We have shown that our method always produces higher MAP values than the other two baselines; CombMNZ and ProbFuse, while generating comparable MRR and P@10 values in general. The Bpref values are usually better than ProbFuse and comparable to that of CombMNZ.

As future work, we aim to investigate the impact of using a separate segmentation function for each system on the effectiveness of final merged results.

Acknowledgment

I am grateful to Justin Zobel, for his insightful comments on this work.

References

J. Aslam and M. Montague. Bayes optimal metasearch: a probabilistic model for combining the results of multiple retrieval systems (poster session). In *Proc. ACM SIGIR conf.*, pages 379–381, Athens, Greece, 2000.

J. Aslam and M. Montague. Models for metasearch. In *Proc. ACM SIGIR conf.*, pages 276–284, New Orleans, Louisiana, 2001.

Z. Bar-Yossef and M. Gurevich. Random sampling from a search engine's index. In *Proc. 15th Int. Conf. on the World Wide Web*, Edinburgh, Scotland, 2006.

C. Buckley and E. Voorhees. Retrieval evaluation with incomplete information. In *Proc. ACM SIGIR conf.*, pages 25–32, Sheffield, UK, 2004.

J. Callan. Distributed information retrieval. *Advances in information retrieval, chapter 5*, pages 127–150, 2000.

A. Calvé and J. Savoy. Database merging strategy based on logistic regression. *Information Processing and Management*, 36(3):341–359, 2000.

B. Croft. Combining approaches to information retrieval. *Advances in information retrieval, chapter 1*, pages 1–36, 2000.

D. Dreilinger and A. Howe. Experiences with selecting search engines using metasearch. *ACM Transaction on Information Systems*, 15(3):195–222, 1997.

C. Dwork, R. Kumar, M. Naor, and D. Sivakumar. Rank aggregation methods for the Web. In *Proc. the 10th Int. conf. on World Wide Web*, pages 613–622, Hong Kong, Hong Kong, 2001.

E. Fox and J. Shaw. Combination of multiple searches. In *Proc. the second Text REtrieval Conf.*, pages 243–252, Gaithersburg, Maryland, 1993. NIST Special Publication.

E. Fox and J. Shaw. Combination of multiple searches. In *Proc. the Third Text REtrieval Conf.*, pages 105–108, Gaithersburg, Maryland, 1994. NIST Special Publication.

D. Harman. Overview of the third Text REtrieval Conference (TREC-3). In *Proc. the third Text REtrieval Conf.*, pages 1–19, NIST, 1994.

T. Joachims, L. Granka, B. Pan, H. Hembrooke, and G. Gay. Accurately interpreting clickthrough data as implicit feedback. In *Proc. ACM SIGIR conf.*, pages 154–161, Salvador, Brazil, 2005.

J. Lee. Analyses of multiple evidence combination. In *Proc. the 20th ACM SIGIR conf.*, pages 267–276, Philadelphia, Pennsylvania, 1997.

D. Lillis, F. Toolan, R. Collier, and J. Dunnion. ProbFuse: a probabilistic approach to data fusion. In *Proc. ACM SIGIR conf.*, pages 139–146, Seattle, Washington, 2006.

R. Manmatha, T. Rath, and F. Feng. Modeling score distributions for combining the outputs of search engines. In *Proc. ACM SIGIR conf.*, pages 267–275, New Orleans, Louisiana, 2001.

W. Meng, C. Yu, and K. Liu. Building efficient and effective metasearch engines. *ACM Computing Surveys*, 34(1):48–89, 2002.

Y. Rasolofo, D. Hawking, and J. Savoy. Result merging strategies for a current news metasearcher. *Information Processing and Management*, 39(4):581–609, 2003.

J. Savoy, A. Calvé, and D. Vrajitoru. Information retrieval systems for large document collections. In *Proc. the Fifth Text REtrieval Conf.*, pages 489–502, Gaithersburg, Maryland, 1996.

E. Selberg and O. Etzioni. The MetaCrawler architecture for resource aggregation on the web. *IEEE Expert*, 12(1):8–14, 1997.

L. Si and J. Callan. A semisupervised learning method to merge search engine results. *ACM Transactions on Information Systems*, 21(4):457–491, 2003.

A. Spink, B. Jansen, C. Blakely, and S. Koshman. A study of results overlap and uniqueness among major web search engines. *Information Processing and Management*, 42(5):1379–1391, 2006.

C. Vogt. How much more is better? characterizing the effects of adding more ir systems to a combination. In *Content-Based Multimedia Information Access (RIAO)*, pages 457–475, Paris, France, 2000.

C. Vogt. *Adaptive combination of evidence for information retrieval.* PhD thesis, University of California, San Diego, 1999.

C. Vogt and G. Cottrell. Fusion via a linear combination of scores. *Information Retrieval*, 1(3):151–173, 1999.

E. Voorhees and D. Harman. Overview of the third Text REtrieval Conference (TREC-5). In *Proc. the fifth Text Retrieval Conf.*, pages 1–28, National Institute of Standards and Technology, 1996.

E. Voorhees, K. Gupta, and B. Johnson-Larid. The collection fusion problem. In *Proc. the Third Text REtrieval Conf. (TREC-3)*, pages 95–104, 1994.

E. Voorhees, N. Gupta, and B. Johnson-Laird. Learning collection fusion strategies. In *Proc. ACM SIGIR conf.*, pages 172–179, Seattle, Washington, 1995.

Query Hardness Estimation Using Jensen-Shannon Divergence Among Multiple Scoring Functions

Javed A. Aslam* and Virgil Pavlu

Northeastern University
{jaa,vip}@ccs.neu.edu

Abstract. We consider the issue of query performance, and we propose a novel method for automatically predicting the difficulty of a query. Unlike a number of existing techniques which are based on examining the ranked lists returned in response to perturbed versions of the query with respect to the given collection or perturbed versions of the collection with respect to the given query, our technique is based on examining the ranked lists returned by multiple scoring functions (retrieval engines) with respect to the given query and collection. In essence, we propose that the results returned by multiple retrieval engines will be relatively similar for "easy" queries but more diverse for "difficult" queries. By appropriately employing Jensen-Shannon divergence to measure the "diversity" of the returned results, we demonstrate a methodology for predicting query difficulty whose performance exceeds existing state-of-the-art techniques on TREC collections, often remarkably so.

1 Introduction

The problem of *query hardness estimation* is to accurately and automatically predict the difficulty of a query, i.e., the likely quality of a ranked list of documents returned in response to that query by a retrieval engine, and to perform such predictions in the absence of relevance judgments and without user feedback. Much recent research has been devoted to the problem of query hardness estimation, and its importance has been recognized by the IR community [1,2,3,4,5,6,7]. An accurate procedure for estimating query hardness could potentially be used in many ways, including the following:

- Users, alerted to the likelihood of poor results, could be prompted to reformulate their query.
- Systems, alerted to the difficult query, could automatically employ enhanced or alternate search strategies tailored to such difficult queries.
- Distributed retrieval systems could more accurately combine their input results if alerted to the difficulty of the query for each underlying (system, collection) pair [2].

* We gratefully acknowledge the support provided by NSF grants CCF-0418390 and IIS-0534482.

G. Amati, C. Carpineto, and G. Romano (Eds.): ECIR 2007, LNCS 4425, pp. 198–209, 2007.

In this work, we propose a new method for automatically predicting the difficulty of a given query. Our method is based on the premise that different retrieval engines or scoring functions will retrieve relatively similar ranked lists in response to "easy" queries but more diverse ranked lists in response to "hard" queries. As such, one could automatically predict the difficulty of a given query by simultaneously submitting the query to multiple retrieval engines and appropriately measuring the "diversity" of the ranked list responses obtained. In order to measure the diversity of set of ranked lists of documents, we map these rankings to distributions over the document collection, where documents ranked nearer the top of a returned list are naturally associated with higher distribution weights (befitting their importance in the list) and vice versa. Given a set of distributions thus obtained, we employ the well known Jensen-Shannon divergence [8] to measure the diversity of the distributions corresponding to these ranked lists.

We extensively tested our methodology using the benchmark TREC collections [9,10,11,12,13,14,15]. To simulate the ranked lists returned by multiple retrieval strategies in response to a given (TREC) query, we chose subsets of the retrieval runs submitted in response to that query in a given TREC. We then predicted query difficulty using the methodology described and compared our estimated query difficulty to the difficulty of that query in that TREC, as measured in a number of standard and new ways. Finally, we compared the quality of our query difficulty estimates to state-of-the-art techniques [1,6,7], demonstrating significant, often remarkable, improvements.

The remainder of this paper is organized as follows. We begin by more extensively discussing relevant related work, followed by a presentation of our methodology in detail. We then discuss the results of extensive experiments using the TREC collections. Finally, we conclude with a summary and discussion of future work.

2 Background and Related Work

Existing work on query hardness estimation can be categorized along at least three axes: (1) How is query hardness defined?, (2) How is query hardness predicted?, and (3) How is the quality of the prediction evaluated? In what follows, we describe our work and related work along these dimensions.

2.1 Defining Query Hardness

One can define query hardness in many ways; for example, queries can be inherently difficult (e.g., ambiguous queries), difficult for a particular collection, or difficult for a particular retrieval engine run over a particular collection. Other notions of query difficulty exist as well. In what follows, we discuss two notions of query harness, which we shall refer to as *system query hardness* and *collection query hardness.*

System query hardness captures the difficulty of a query for a given retrieval system run over a given collection. Here the notion of query hardness is system-specific; it is meant to capture the difficulty of the query for a *particular* system, run over a given collection. System query hardness is typically measured by the average precision of the ranked list of documents returned by the retrieval system when run over the collection using the query in question.

Examples of work considering system query hardness include (1) Carmel et al. [3] and Yom-Tov et al. [1] who investigate methods for predicting query hardness, testing against the Juru retrieval system, (2) Cronen-Townsend et al. [6] and Zhou and Croft [7] who investigate methods for predicting query hardness, testing against various language modeling systems, and (3) the Robust track at TREC [13] wherein each system attempted to predict its own performance on each given query.

Collection query hardness captures the difficulty of a query with respect to a given collection. Here the notion of query hardness is meant to be largely *independent* of any specific retrieval system, capturing the inherent difficulty of the query (for the collection) and perhaps applicable to a wide variety of typical systems. Collection query hardness can be measured by some statistic taken over the performance of a wide variety of retrieval systems run over the given collection using the query in question. For example, Carmel et al. [3] consider collection query hardness by comparing the query difficulty predicted by their method to the median average precision taken over all runs submitted in the Terabtye tracks at TREC for a given query.

Our work: We consider both system query hardness and collection query hardness and demonstrate that our proposed methodology is useful in predicting either. In order to test the quality of our methodology for predicting a given system's performance (system query harness), one must fix a retrieval system. In this work, we simply choose the system (retrieval run) whose mean average precision was the median among all those submitted to a particular TREC; thus, we consider a "typical" system, one whose performance was neither extremely high or low. We refer to this measure as the *median system AP* (med-sys AP).

In order to test the quality of our methodology for predicting collection query hardness, one must fix a measure for assessing the hardness of a query for a given collection. In this work, we consider two statistics taken over all runs submitted to a particular TREC with respect to a given query: (1) the *average* of the average precisions for all runs submitted in response to a given query and (2) the *median* of the average precisions for all runs submitted in response to a given query. We refer to the former as *query average AP* (avgAP) and the latter as *query median AP* (medAP).

2.2 Predicting Query Hardness

Cronen-Townsend et al. [6] introduced the *clarity score* which effectively measures the ambiguity of the query with respect to a collection, and they show that clarity scores are correlated with query difficulty. Clarity scores are computed by assessing the information-theoretic distance between a language model associated with the query and a language model associated with the collection. Subsequently, Zhou and Croft [7] introduced *ranking robustness* as a measure of query hardness, where ranking robustness effectively measures the stability in ranked results with respect to perturbations in the collection.

Carmel et al. [3] proposed the use of pairwise information-theoretic distances between distributions associated with the collection, the set of relevant documents, and the query as predictors for query hardness. Yom-Tov et al. [1] proposed a method for predicting query hardness by assessing the stability of ranked results with respect to perturbations in the query; subsequently, they showed how to apply these results to the problem of metasearch [2].

In other related work, Amati et al. [4] studied query hardness and robustness in the context of query expansion, Kwok [5] proposed a strategy for selecting the scoring function based on certain properties of a query, and Macdonald et al. [16] investigated query hardness prediction in an intranet environment.

Our work: While fundamentally different from existing techniques, our work is related to the methodologies described above in a number of ways. A number of existing techniques predict query hardness by measuring the stability of ranked results in the presence of perturbations of the *query* [1] or perturbations of the *collection* [7]. In a similar spirit, our proposed technique is based on measuring the stability of ranked results in the presence of perturbations of the *scoring function*, i.e., the retrieval engine itself. We measure the "stability" of the ranked results by mapping each ranked list of documents returned by a different scoring function to a probability distribution and then measuring the diversity among these distributions using the information-theoretic Jensen-Shannon divergence [8]. In a similar spirit, Cronen-Townsend et al. [6] use the related Kullback-Leibler divergence [17] to compute clarity scores, and Carmel et al. [3] use the Jensen-Shannon divergence to compute their query hardness predictor.

2.3 Evaluating the Quality of Query Hardness Predictions

In order to evaluate the quality of a query hardness prediction methodology, test collections such as the TREC collections are typically used. The system and/or collection hardnesses of a set of queries are measured, and they are compared to predicted values of query hardness. These actual and predicted values are real-valued, and they are typically compared using various parametric and non-parametric statistics. Zhou and Croft [7] and Carmel et al. [3] compute the *linear correlation coefficient* ρ between the actual and predicted hardness values; ρ is a parametric statistic which measures how well the actual and predicted hardness values fit to a straight line. If the queries are *ranked* according to the actual and predicted hardness values, then various non-parametric statistics can be computed with respect to these rankings. Cronen-Townsend et al. [6] and Carmel et al. [3] compute the Spearman rank correlation coefficient. Zhou and Croft [7], Yom-Tov et. al [1], and the TREC Robust track [13] all compute and report the Kendall's τ statistic.

Our work: In the results that follow, we assess the quality of our query hardness predictions using both the linear correlation coefficient ρ and Kendall's τ.

3 Methodology

Our hypothesis is that disparate retrieval engines will return "similar" results with respect to "easy" queries and "dissimilar" results with respect to "hard" queries. As such, for a given query, our methodology essentially consists of three steps: (1) submit the query to multiple scoring functions (retrieval engines), each returning a ranked list of documents, (2) map each ranked list to a distribution over the document collection, where higher weights are naturally associated with top ranked documents and vice versa, and (3) assess the "disparity" (collective distance) among these distributions. We discuss (2) and (3) in the sections that follow, reserving our discussion of (1) for a later section.

3.1 From Ranked Lists to Distributions

Many measures exist for assessing the "distance" between two ranked lists, such as the Kendall's τ and Spearman rank correlation coefficients mentioned earlier. However, these measures do not distinguish between differences in the "top" of the lists from equivalent differences in the "bottom" of the lists; however, in the context of information retrieval, two ranked lists would be considered much more dissimilar if their differences occurred at the "top" rather than the "bottom" of the lists.

To capture this notion, one can focus on the top retrieved documents only. For example, Yom-Tov et al. [1] compute the *overlap* (size of intersection) among the top N documents in each of two lists. Effectively, the overlap statistic places a uniform $1/N$ "importance" to each of the top N documents and a zero importance to all other documents. More natural still, in the context of information retrieval, would be weights which are higher at top ranks and smoothly lower at lesser ranks. Recently, we proposed such weights [19,20] which correspond to the implicit weights which the average precision measure places on each rank, and we use these distribution weights in our present work as well. Over the top c documents of a list, the distribution weight associated with any rank r, $1 \leq r \leq c$, is given below; all other ranks have distribution weight zero.

$$\text{weight}(r) = \frac{1}{2c}\left(1 + \frac{1}{r} + \frac{1}{r+1} + \cdots + \frac{1}{c}\right). \tag{1}$$

3.2 The Jensen-Shannon Divergence Among Distributions

Using the above distribution weight function, one can map ranked lists to distributions over documents. In order to measure the "disparity" among these lists, we measure the disparity or divergence among the distributions associated with these lists. For two distributions $\boldsymbol{a} = (a_1, \ldots, a_n)$ and $\boldsymbol{b} = (b_1, \ldots, b_n)$, a natural and well studied "distance" between these distributions is the Kullback-Leibler divergence [17]:

$$KL(\boldsymbol{p}||\boldsymbol{q}) = \sum_i p_i \log \frac{p_i}{q_i}$$

However, the KL-divergence suffers two drawbacks: (1) it is not symmetric in its arguments and (2) it does not naturally generalize to measuring the divergence among more than two distributions. We instead employ the related Jensen-Shannon divergence [8]. Given a set of distributions $\{\boldsymbol{p_1}, \ldots, \boldsymbol{p_m}\}$, let $\overline{\boldsymbol{p}}$ be the average (centroid) of these distributions. The Jensen-Shannon divergence among these distributions is then defined as the average of the KL-divergences of each distribution to this average distribution:

$$JS(\boldsymbol{p_1}, \ldots, \boldsymbol{p_m}) = \frac{1}{m} \sum_j KL(\boldsymbol{p_j}||\overline{\boldsymbol{p}})$$

An equivalent and somewhat simpler formulation defined in terms of entropies also exists [8]. In this work, we employ the Jensen-Shannon divergence among the distributions associated with the ranked lists of documents returned by multiple retrieval engines in response to a give query as an estimate of query hardness.

4 Experimental Setup and Results

We tested our methodology extensively on multiple TREC datasets: TREC5, TREC6, TREC7, TREC8, Robust04, Terabyte04, and Terabyte05. The performance of our proposed Jensen-Shannon query hardness estimator is measured against three benchmark query hardness statistics: query average AP (avgAP) and query median AP (medAP), both measures of *collection query hardness*, and median-system AP (med-sys AP), a measure of *system query hardness*. When predicting the difficulties of multiple queries in any given TREC, the strength of correlation of our predicted difficulties with actual query difficulties is measured by both Kendall's τ and linear correlation coefficient ρ. We conclude that even when using few input systems, our method consistently outperforms existing approaches [1,6,7], sometimes remarkably so.

The ad hoc tracks in TRECs 5–8 and the Robust track in 2004 each employ a standard 1,000 documents retrieved per system per query on collections of size in the range of hundreds of thousand of documents. For these collections, the weight cutoff was fixed at $c = 20$ in Equation 1; in other words, only the top 20 documents retrieved by each system received a non-zero weight in the distribution corresponding to the retrieved list, as used in the Jensen-Shannon divergence computation. The Terabyte tracks use the GOV2 collection of about 25 million documents, and ranked result lists consist of 10,000 documents each; for this larger collection and these longer lists, the weight cutoff was set at $c = 100$ in Equation 1. This work leaves open the question of how to optimally set the weight cutoff per system, query, and/or collection.

The baseline statistics query avgAP, query medAP, and the fixed system med-sys AP are computed among *all* retrieval runs available. The Jensen-Shannon

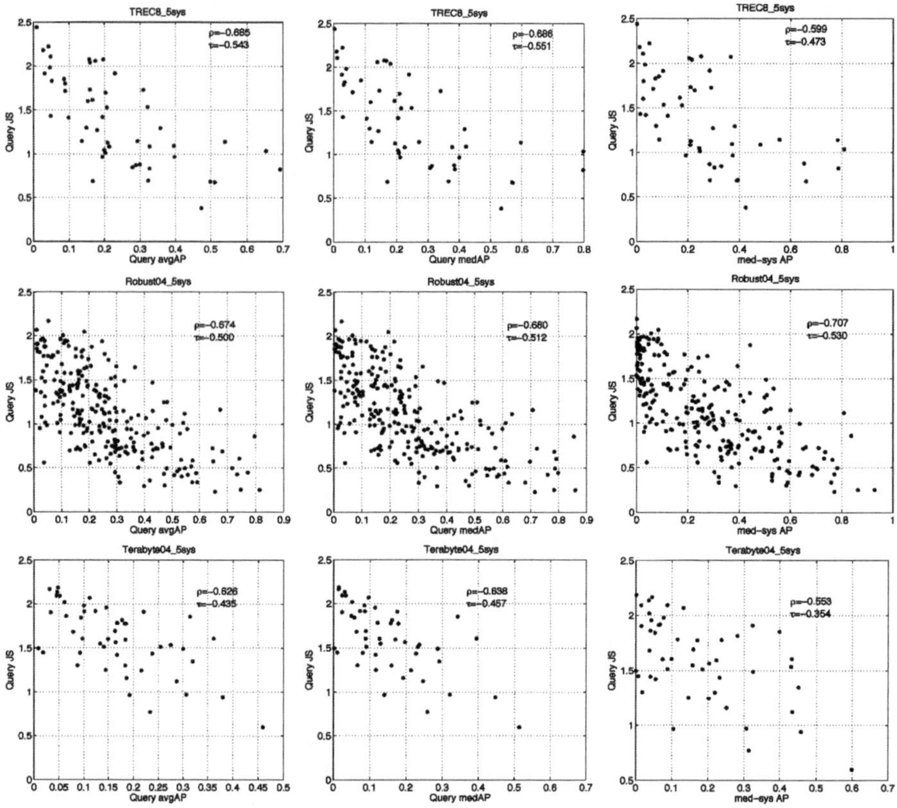

Fig. 1. Query hardness prediction results using five input systems for (top to bottom) TREC8, Robust04 and Terabyte04. Each dot in these scatter plots corresponds to a query. The x-axis is actual query hardness as measured by query average AP (left), query median AP (center), and median-system AP (right). The y-axis is the Jensen-Shannon divergence computed over the ranked results returned by five randomly chosen systems for that query.

divergence is computed among 2, 5, 10, 20, or all retrieval runs available. When less than all of the available runs are used, the actual runs selected are chosen at random, and the entire experiment is repeated 10 times; scatter plots show a typical result among these 10 repetitions, and tables report the average performance over all 10 repetitions. We note that in general, the quality of our query hardness predictions increases rapidly as more system runs are used, with improvements tailing off after the inclusion of approximately 10 systems.

We compare our Jensen-Shannon query hardness predictions with all three baseline statistics, for all queries and all collections; in two isolated cases we excluded queries with zero relevant documents. Figures 1 and 2 show a selection of the results as scatter plots, separately for JS estimation using five system runs and for JS estimation using 10 system runs.

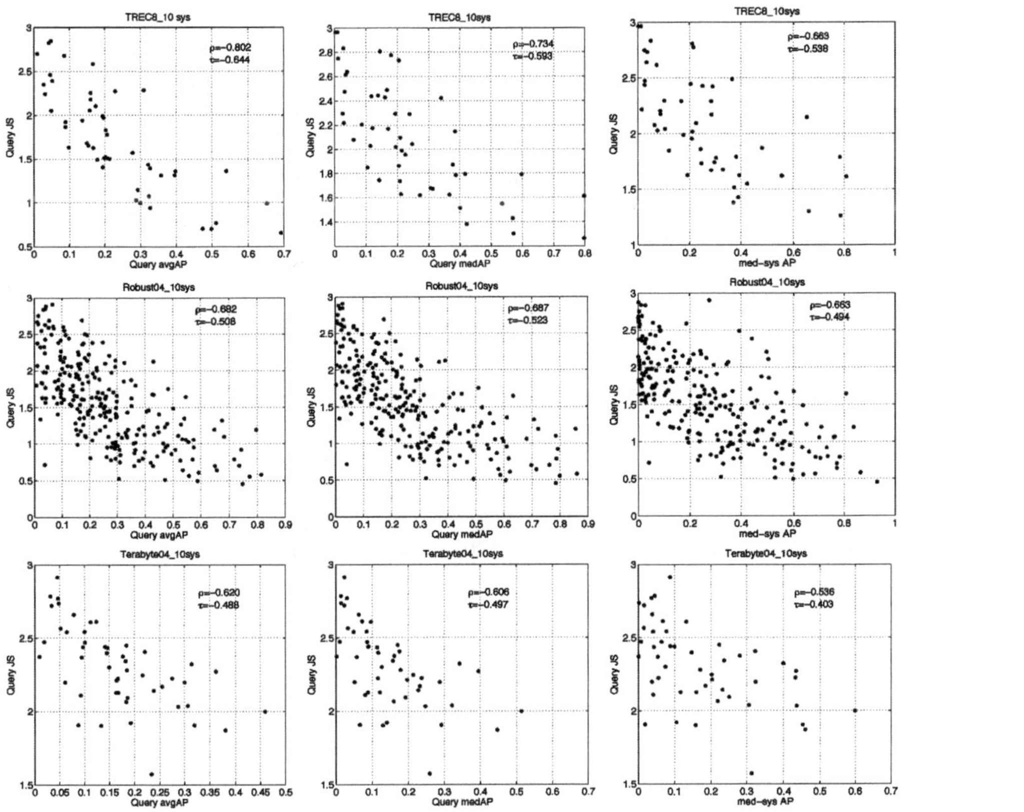

Fig. 2. Query hardness prediction results using 10 input systems for (top to bottom) TREC8, Robust04 and Terabyte04. Each dot in these scatter plots corresponds to a query. The x-axis is actual query hardness as measured by query average AP (left), query median AP (center), and median-system AP (right). The y-axis is the Jensen-Shannon divergence computed over the ranked results returned by 10 randomly chosen systems for that query.

Table 1. Kendall's τ (JS vs. query average AP) for all collections using 2, 5, 10, 20, and all input systems. All but the last row report 10-run average performance.

Prediction Method	TREC5	TREC6	TREC7	TREC8	Robust04	TB04	TB05
JS (2 systems)	0.334	0.353	0.436	0.443	0.393	0.339	0.288
JS (5 systems)	0.420	0.443	0.468	0.551	0.497	0.426	0.376
JS (10 systems)	0.468	0.444	0.544	0.602	0.502	0.482	0.406
JS (20 systems)	0.465	0.479	0.591	0.613	0.518	0.480	0.423
JS (all systems)	0.469	0.491	0.623	0.615	0.530	0.502	0.440

Kendall's τ measures the similarity of two rankings, in our case, the rankings of the queries in terms of a baseline measure (query average AP, query median AP, or median-system AP) and the rankings of the queries in terms of our

Jensen-Shannon estimate. Prediction performance as measured by Kendall's τ is given in Tables 1, 2, and 3, and for visual purposes, we graph Tables 2 and 3 in Figure 3. Note that while our scatter plots seem to indicate negative correlation (high Jensen-Shannon divergence implies low query performance), this indicates positive correlation with the problem as defined (high Jensen-Shannon divergence implies high query difficulty). As such, we report the corresponding "positive" correlations in all tables, and we note the equivalent negative correlations in all scatter plots.

Table 2. Kendall's τ (JS vs. query median AP) for all collections using 2, 5, 10, 20, and all input systems. All but the last row report 10-run average performance.

Prediction Method	TREC5	TREC6	TREC7	TREC8	Robust04	TB04	TB05
JS (2 systems)	0.341	0.385	0.452	0.442	0.401	0.338	0.260
JS (5 systems)	0.448	0.475	0.483	0.547	0.510	0.435	0.340
JS (10 systems)	0.483	0.464	0.556	0.585	0.515	0.485	0.366
JS (20 systems)	0.488	0.503	0.610	0.599	0.533	0.496	0.382
JS (all systems)	0.510	0.530	0.634	0.597	0.544	0.520	0.391

Where we could make a direct comparison with prior results (TREC5, TREC8, Robust04, Terabyte04, and Terabyte05), we indicate the performance reported in prior work along with references. For past results measuring *system query hardness* (i.e., correlations between predicted query hardness and the hardness of the query for a *specific* system), we compare these prior results against our correlations with the median-system AP, as that would be closest to a fair comparison. Using 10 input system runs for the Jensen-Shannon computation yield improvements over best previous results of on average approximately 40% to 50%; the complete Tables 3, 5, and 6 show improvements ranging from 7% to 80%.

Table 3. Kendall's τ (JS vs. median-system AP) for all collections using 2, 5, 10, 20, and all input systems. All but the last row report 10-run average performance.

Prediction Method	TREC5	TREC6	TREC7	TREC8	Robust04	TB04	TB05
[7] clarity	0.311				0.412	0.134	0.171
[7] robust+clarity	0.345				0.460	0.226	0.252
[1] hist. boost. class.				.439			
JS (2 systems)	0.260	0.276	0.384	0.452	0.387	0.298	0.241
JS (5 systems)	0.350	0.370	0.427	0.525	0.472	0.349	0.318
JS (10 systems)	0.355	0.334	0.476	0.552	0.490	0.408	0.359
JS (20 systems)	0.339	0.365	0.516	0.577	0.498	0.403	0.380
JS (all systems)	0.363	0.355	0.509	0.561	0.512	0.427	0.381

The linear correlation coefficient ρ effectively measures how well actual and predicted values fit to a straight line; in our case, these actual and predicted values are the hardness of queries in terms of a baseline measure (query average AP, query

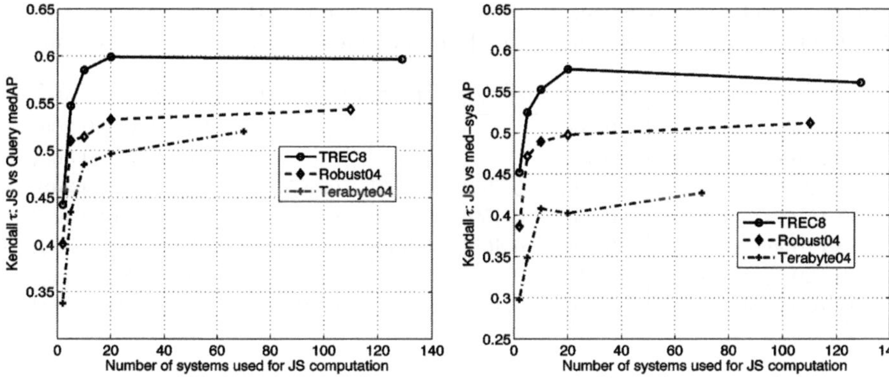

Fig. 3. Kendall's τ for JS vs. query median AP (left) and Kendall's τ for JS vs. median-system AP (right)

Table 4. Correlation coefficient ρ (JS vs. query average AP) for all collections using 2, 5, 10, 20, and all systems. All but the last row report 10-run average performance.

Prediction method	TREC5	TREC6	TREC7	TREC8	Robust04	TB04	TB05
JS (2 systems)	0.498	0.456	0.601	0.627	0.555	0.466	0.388
JS (5 systems)	0.586	0.611	0.637	0.736	0.673	0.576	0.516
JS (10 systems)	0.632	0.651	0.698	0.778	0.672	0.642	0.564
JS (20 systems)	0.645	0.677	0.731	0.784	0.688	0.666	0.577
JS (all systems)	0.623	0.698	0.722	0.770	0.695	0.682	0.581

Table 5. Correlation coefficient ρ (JS vs. query median AP) for all collections using 2, 5, 10, 20, and all systems. All but the last row report 10-run average performance.

Prediction method	TREC5	TREC6	TREC7	TREC8	Robust04	TB04	TB05
[3] Juru(TB04+05)						.476	
JS (2 systems)	0.495	0.467	0.612	0.622	0.557	0.466	0.366
JS (5 systems)	0.592	0.631	0.654	0.727	0.677	0.586	0.477
JS (10 systems)	0.622	0.678	0.715	0.762	0.676	0.646	0.524
JS (20 systems)	0.630	0.703	0.752	0.769	0.694	0.674	0.541
JS (all systems)	0.600	0.727	0.743	0.755	0.701	0.687	0.543

median AP, or median-system AP) and the hardness of these same queries in terms of our Jensen-Shannon estimate. Prediction performance as measured by linear correlation coefficient is presented in Tables 4, 5, and 6. Note the substantial improvements over prior results as shown in Tables 5 and 6.

Table 6. Correlation coefficient ρ (JS vs. median-system AP) for all collections using 2, 5, 10, 20, and all systems. All but the last row report 10-run average performance.

Prediction method	TREC5	TREC6	TREC7	TREC8	Robust04	TB04	TB05
[7] clarity	0.366				0.507	0.305	0.206
[7] robust+clarity	0.469				0.613	0.374	0.362
JS (2 systems)	0.425	0.294	0.553	0.595	0.542	0.435	0.338
JS (5 systems)	0.537	0.459	0.609	0.676	0.645	0.490	0.467
JS (10 systems)	0.556	0.469	0.639	0.707	0.659	0.566	0.524
JS (20 systems)	0.562	0.479	0.679	0.724	0.665	0.585	0.545
JS (all systems)	0.567	0.497	0.657	0.702	0.677	0.603	0.541

5 Conclusion and Future Work

Previous work on query hardness has demonstrated that a measure of the stability of ranked results returned in response to perturbed versions of the query with respect to the given collection or perturbed versions of the collection with respect to the given query are both correlated with query difficulty, both in general and for specific systems. In this work, we further demonstrate that a measure of the stability of ranked results returned in response to perturbed versions of the *scoring function* is also correlated with query hardness, often at a level significantly exceeding that of prior techniques. Zhou and Croft [7] and Carmel et al. [3] demonstrate that *combining* multiple methods for predicting query difficulty yields improvements in the predicted results, and we hypothesize that appropriately combining our proposed method with other query difficulty prediction methods would yield further improvements as well. Finally, this work leaves open the question of how to optimally pick the number and type of scoring functions (retrieval engines) to run in order to most efficiently and effectively predict query hardness.

References

1. Yom-Tov, E., Fine, S., Carmel, D., Darlow, A.: Learning to estimate query difficulty: including applications to missing content detection and distributed information retrieval. In: SIGIR. (2005) 512–519
2. Yom-Tov, E., Fine, S., Carmel, D., Darlow: Metasearch and Federation using Query Difficulty Prediction. In: Predicting Query Difficulty - Methods and Applications. (August 2005)
3. Carmel, D., Yom-Tov, E., Darlow, A., Pelleg, D.: What makes a query difficult? In: SIGIR '06: Proceedings of the 29th annual international ACM SIGIR conference on Research and development in information retrieval, New York, NY, USA, ACM Press (2006) 390–397
4. Amati, G., Carpineto, C., Romano, G.: Query difficulty, robustness and selective application of query expansion. In: Proceedings of the 25th European Conference on Information Retrieval ECIR 2004. (2004)
5. Kwok, K.: An attempt to identify weakest and strongest queries. In: ACM SIGIR'05 Query Prediction Workshop. (2005)

6. Cronen-Townsend, S., Zhou, Y., Croft, W.: Predicting query performance. In: In Proceedings of the ACM Conference on Research in Information Retrieval (SIGIR). (2002)
7. Zhou, Y., Croft, W.B.: Ranking robustness: A novel framework to predict query performance. Technical Report IR-532, Center for Intelligent Information Retrieval, University of Massachusetts, Amherst (2006) URL: `http://maroo.cs.umass.edu/pub/web/674`.
8. Lin, J.: Divergence measures based on the shannon entropy. IEEE Trans. Infor. Theory **37** (1991) 145–151
9. Voorhees, E.M., Harman, D.: Overview of the Fifth Text REtrieval Conference (TREC-5). In: TREC. (1996)
10. Voorhees, E.M., Harman, D.: Overview of the Sixth Text REtrieval Conference (TREC-6). In: TREC. (1997) 1–24
11. Voorhees, E.M., Harman, D.: Overview of the Seventh Text REtrieval Conference (TREC-7). In: Proceedings of the Seventh Text REtrieval Conference (TREC-7). (1999) 1–24
12. Voorhees, E.M., Harman, D.: Overview of the Eighth Text REtrieval Conference (TREC-8). In: Proceedings of the Eighth Text REtrieval Conference (TREC-8). (2000) 1–24
13. Voorhees, E.M.: The TREC robust retrieval track. SIGIR Forum **39**(1) (2005) 11–20
14. Clarke, C., Craswell, N., Soboroff, I.: The TREC terabyte retrieval track. (2004)
15. Clarke, C.L.A., Scholer, F., Soboroff, I.: The TREC 2005 terabyte track. In: Proceedings of the Fourteenth Text REtrieval Conference (TREC 2005). (2005)
16. Macdonald, C., He, B., Ounis, I.: Predicting query performance in intranet search. In: ACM SIGIR'05 Query Prediction Workshop. (2005)
17. Cover, T.M., Thomas, J.A.: Elements of Information Theory. John Wiley & Sons (1991)
18. Aslam, J.A., Pavlu, V., Savell, R.: A unified model for metasearch, pooling, and system evaluation. In Frieder, O., Hammer, J., Quershi, S., Seligman, L., eds.: Proceedings of the Twelfth International Conference on Information and Knowledge Management, ACM Press (November 2003) 484–491
19. Aslam, J.A., Pavlu, V., Yilmaz, E.: A statistical method for system evaluation using incomplete judgments. In Dumais, S., Efthimiadis, E.N., Hawking, D., Jarvelin, K., eds.: Proceedings of the 29th Annual International ACM SIGIR Conference on Research and Development in Information Retrieval, ACM Press (August 2006) 541–548
20. Aslam, J.A., Pavlu, V., Yilmaz, E.: Measure-based metasearch. In Marchionini, G., Moffat, A., Tait, J., Baeza-Yates, R., Ziviani, N., eds.: Proceedings of the 28th Annual International ACM SIGIR Conference on Research and Development in Information Retrieval, ACM Press (August 2005) 571–572

Query Reformulation and Refinement Using NLP-Based Sentence Clustering

Frédéric Roulland, Aaron Kaplan, Stefania Castellani, Claude Roux,
Antonietta Grasso, Karin Pettersson, and Jacki O'Neill

Xerox Research Centre Europe, Grenoble, France

Abstract. We have developed an interactive query refinement tool that helps users search a knowledge base for solutions to problems with electronic equipment. The system is targeted towards non-technical users, who are often unable to formulate precise problem descriptions on their own. Two distinct but interrelated functionalities support the refinement of a vague, non-technical initial query into a more precise problem description: a synonymy mechanism that allows the system to match non-technical words in the query with corresponding technical terms in the knowledge base, and a novel refinement mechanism that helps the user build up successively longer and more precise problem descriptions starting from the seed of the initial query. A natural language parser is used both in the application of context-sensitive synonymy rules and the construction of the refinement tree.

1 Introduction

In order to reduce service costs and downtime, users of complex electronic devices are increasingly being encouraged to solve problems for themselves. In order to do this, users need tools that help them identify appropriate solutions. A number of troubleshooting systems are currently available. Some consist of searchable textual descriptions of problems and associated solutions; searchable documents are relatively easy to create and maintain, but in order to search effectively in such a knowledge base the user must be familiar with the content and the terminology used in it. These systems are therefore more suited to expert troubleshooters than to non-technical users. Other approaches such as expert systems and decision trees provide more guidance, asking the user questions rather than simply responding to a query, but such systems are more expensive to build and maintain.

We have developed a text retrieval system called Pocket Engineer (PE) that is tailored to searching in troubleshooting knowledge bases. When a user's initial query returns too many results, PE automatically proposes a list of expressions that contain either some of the words of the query, or synonyms thereof (as listed in a thesaurus). When the user selects one of the expressions, the system proposes a new list of sentence fragments that are possible extensions of the selected expression. This process can be repeated indefinitely, until the number of results is small enough for the user to read through them conveniently.

G. Amati, C. Carpineto, and G. Romano (Eds.): ECIR 2007, LNCS 4425, pp. 210–221, 2007.

The refinement choices are constructed on the fly, using the output of a natural language parser that has analyzed the knowledge base in a preprocessing step. Since the refinement choices are generated automatically from natural language documents, the cost of maintenance is lower than for knowledge-rich approaches such as expert systems or decision trees; yet by iteratively proposing sets of possible refinements, it provides a kind of step-by-step guidance. Particularly when combined with a thesaurus that maps vague, non-technical terms to more precise terms found in the knowledge base, this guidance can help users with little prior knowledge of the terminology or content of the knowledge base to find the information they need.

In [13], we described an ethnographic study of professional troubleshooters mediating between non-technical users and a collection of problem descriptions and associated solutions, and how the findings from that study informed the design of PE. In the current paper, after recalling the type of interaction that PE provides, we present technical details of the various steps of processing that support this interaction, and explain why we believe this mechanism is particularly appropriate, compared to other query refinement mechanisms that have been described in the literature, for helping non-technical users search in a troubleshooting knowledge base.

2 The Query Refinement Process as Seen by the User

Figure 1 shows the PE screen. Initially only the query formulation section at the top of the screen is visible. The user begins a session by typing a query ("lines across page" in the example) in the search box. Subsets of the query words for which there are hits appear underneath the search box; one of these subsets is preselected, but the user can choose a different one (here the user has chosen "lines").

In the matched problems area, a list of results that match the query is displayed. Each result contains a problem description and possibly a short excerpt from an associated solution description. To the left of this list, in the refinement area, is a navigation tree. Each node of the tree is labeled with an expression from the retrieved documents. Clicking on a node reduces the list of results to those that contain all of the expressions on the path from the root of the tree to the selected node. In the example, clicking on the node "in image area when making copies" causes the matched problem list to be reduced to those results containing both the expressions "white lines" and "in image area when making copies" in the same sentence. Clicking on a node also causes the children of that node to be displayed, to allow further refinement. When the matched problem list has been reduced enough for the user to see a problem description that corresponds to his or her problem, the user can click on the problem description to see a list of solutions.

In its visual aspect, the PE interface resembles those of clustering search engines, which help the user navigate a large set of results by factoring them into groups. Indeed, the term "clustering" is an appropriate description of what the PE refinement tree does to the retrieved documents; but as will be explained in the next section, the hierarchical structure of the PE refinement tree is derived from the structure of individual retrieved sentences, rather than on similarity of vocabulary over entire documents, as is the case in other document clustering methods.

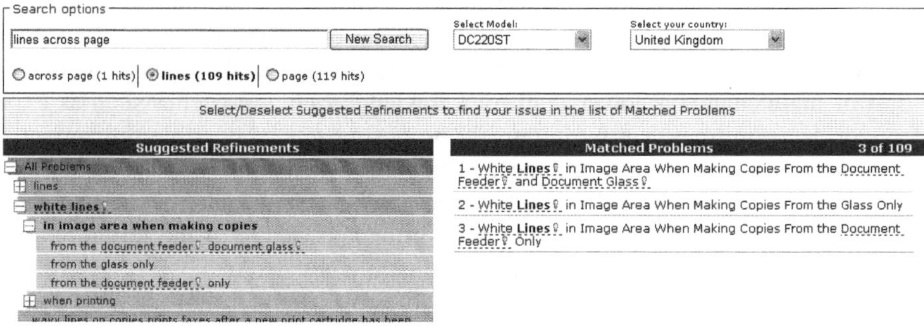

Fig. 1. Pocket Engineer interface

3 Implementation and Rationale

Having given an overview of the main components of the PE refinement process, we will now present some details of the implementation, and explain the motivations for certain decisions in the context of troubleshooting search. The first subsection presents the underlying objects that we build from the content of the knowledge base, and the second describes how we use these objects to implement the refinement process seen by the user.

The knowledge base accessed by PE consists of problem descriptions, each of which is a single sentence, and solutions, which range from a single sentence to roughly a page of text. Each problem description is associated with one or more solutions, and each solution with one or more problem descriptions. Both problems and solutions are searched for query keywords, but the objects displayed in the "matched problems" area of the interface are problem descriptions; if query words are matched in a solution, then all problems associated with that solution are retrieved, and each is considered to be a separate result. Henceforth, we use the terms "result" and "document" interchangeably to mean a single problem, possibly paired with one of its associated solutions.

3.1 Preprocessing

Indexing: Lemmatization and Vocabulary Expansion. Each document in the knowledge base is segmented into sentences, and each sentence into words. Each document is indexed under the lemmata of the words that occur in it, as well as synonyms and related terms, as listed in a thesaurus, in order to improve the recall of initial queries. This expansion is performed at indexing time, rather than at query time, in order to allow synonymy rules that include constraints on a word's context in the knowledge base. A rule can include lexico-syntactic constraints that a sentence must satisfy in order to be indexed under a given synonym. Each sentence of the knowledge base is parsed, and the constraints are verified against the results of this analysis. For example, a rule can state that the synonym "replace" should be added to the index for any sentence in which the word "change" appears with a direct object whose head is "cartridge," but not for sentences where "change" occurs with other direct objects (for example, "change fax settings").

Segmentation of Sentences into Refinement Expressions. As part of the process of indexing a knowledge base for publication via PE, each sentence is segmented into a series of expressions of a granularity appropriate for presentation as refinement choices. This segmentation is performed by rule-based natural language parser [1], using the parser's general-purpose grammar plus a layer of PE-specific rules. Grammars for English, French, Spanish, Portuguese, Italian, and German are in various stages of development, with the English and French grammars being the most mature and well-tested.

In writing the rules for identifying refinement expressions, we used the following rule of thumb: a refinement expression should be a sequence of words that is as short as possible while containing enough detail and context that users will be able to judge whether or not the expression is relevant to the problems they are experiencing.

The extent of the refinement expressions identified by our rules corresponds in large part to the notion of *chunk* as introduced by Abney [2], though the structure of verbal expressions differs from that typically constructed by shallow parsers. Refinement expressions do not have recursive structure–while the general-purpose grammar identifies nested constituents, the rules for identifying refinement expressions perform a flat segmentation based on that structure. When multiple nested expressions in a sentence fulfill the criteria, the maximal such expression is chosen as a refinement expression. The rules for English can be summarized as follows:

- A simple noun phrase consisting of a head noun and any preceding determiners and modifiers, but not including postmodifiers such as prepositional phrases, is a possible refinement expression. For example,
  - *lines*
  - *white lines*
  - *white and black wavy lines*

  are possible refinement expressions, but
  - *white lines on copies*
  - *white lines when printing*

  would each be divided into two refinement expressions.
- A prepositional phrase consisting of a preposition and a simple noun phrase as defined above is a possible refinement expression. For example,
  - *on paper*
  - *from the document feeder*
- An intransitive verb is a possible refinement expression, as is a transitive verb combined with its direct object (more precisely, the simple noun phrase that heads its direct object). Any auxiliary verbs and adverbs related to a main verb are made part of the verb's refinement expression. For example,
  - *will not open*
  - *release toner particles*
- A form of the verb "be" used as a main verb, combined with its subject complement, constitutes a possible refinement expression. For example,
  - *is blank*
- In a passive construction, the subject (the logical object) and the verb are combined to form a refinement expression. For example.
  - *when the document feeder is used*

Rules for other languages differ in the details but follow the same general principles.

Normalization of Refinement Expressions. In order to make the refinement tree more compact and the refinement choices easier to apprehend, PE attempts to group together refinement expressions that are very similar in meaning, and uses a single node in the refinement tree to represent all the members of such a group. The grouping is performed by mapping each expression to a canonical form; expressions with the same canonical form are grouped together. Two types of transformations are applied: morpho-syntactic and lexical. The morpho-syntactic normalization consists of dropping function words such as determiners and auxiliary verbs, transforming passive constructions to active, and lemmatization, which in English consists of replacing plural nouns with their singular forms and conjugated verbs with their infinitive. For example, the normalization process makes the following transformations:

- *white lines* → *white line*
- *when the document feeder is used* → *when use document feeder*

The lexical normalization uses a thesaurus in which equivalence classes of words and multi-word expressions are listed; each class has a distinguished synonym which is used as the canonical form. The system thus uses two distinct thesauri: one lists groups of equivalent terms for use in normalization of refinement expressions, and the other lists looser relationships used to enrich the index in order to improve the recall of initial queries. (The second thesaurus in fact contains the first—equivalences used in normalization are also used for query expansion, but not the reverse.)

3.2 Query-Time Processing

Subquery Selection. If the query contains more than one word, then there may be hits that contain some but not all of the query words. For each retrieved document, we identify the sentences that contain one or more query keywords, and we identify *query matches* within these sentences. A query match is a contiguous sequence of expressions (as defined above) each of which contains at least one word of the query.

Subsets of the query words (henceforth "subqueries") for which there are query matches are listed below the query box. The subqueries are ranked first by the number of words they contain, and secondarily by the average frequencies of the words. The list is truncated to the n best (five in the current prototype) to avoid overwhelming the user with an exponential number of subqueries in the (relatively rare) case of a long query. The top-ranked subquery (which is typically the entire query if there are hits for it) is preselected so that results deemed most likely to be relevant are visible immediately after the initial query, but the user can change this selection.

Typically, when an information retrieval system determines that no documents match the user's entire query, it either returns zero results (in the case of a boolean retrieval system), or displays a ranked list of results that match parts of the query (in the case of systems that use a measure of query-document similarity). Our approach is different: PE gives the user an explicit choice of degraded queries. It is standard practice in information retrieval to give less weight to query words that are very frequent in the document collection, and therefore less discriminative. PE does this, by factoring the number of hits into the ranking of subqueries, but unlike systems that present a single ranked list of all of the results, it gives explicit feedback about which keywords it is considering as most important, and allows the user to modify that decision. We chose this mechanism based on two factors. First, before designing PE, we observed users of its predecessor, called OSA. When no matches for the entire query are found, OSA displays a ranked list of results that match various parts of the query. In such cases, many of the displayed results may be irrelevant, and we observed that users were often confused about why these results were being presented. The PE subquery selection is a way of explaining to users how the results they are viewing were chosen, as well as allowing them to override the frequency-based weighting if they see fit. The second reason for the subquery mechanism is related to our particular clustering method, which will be described next. If we were to build a single refinement tree for all documents that match any part of the query, the tree could end up with an inordinately large number of top-level nodes, which would make it difficult to use.

3.3 The Refinement Tree

The root of the refinement tree is labeled "All problems." This is the node that is selected when the tree is first displayed, and selecting it has the effect of selecting all documents that match the current subquery.

The children of the root are expressions that contain one or more of the words of the subquery, or synonyms thereof. These expressions are extracted from the matching documents. Expression boundaries are determined as previously described in Section 3.1. This level of the tree serves to disambiguate query words that are ambiguous, in two ways.

First, if a query word is ambiguous, and the ambiguity has not already been resolved implicitly by the presence of other query words, then the expressions containing that query word often serve as disambiguation choices. For example, the documents in our knowledge base refer to two different photocopier parts called "document feeder" and "high-capacity feeder." (the former handles originals, the latter handles blank paper). If the user's query is simply "feeder," then the expressions "document feeder" and "high-capacity feeder" will both appear as choices in the first level of the refinement tree. It has often been observed that users of standard search interfaces pick up terms seen in the results of an initial query to formulate a subsequent, more precise query. By presenting multi-word terms that contain one of the user's query words as refinement choices, the PE interface anticipates subsequent queries that the user might want to make and presents them in a concise format, hopefully reducing the time and effort involved in finding the appropriate term and refining the query.

Secondly, the inclusion of expressions containing synonyms of the query words involves the user explicitly in query expansion. The fact that synonyms are offered as choices rather than added automatically to the query allows us to use a more extensive thesaurus: while more aggressive synonym expansion increases recall at the expense of overall precision, grouping the results for a given synonym in a single subtree allows users easily to ignore results for synonyms they consider inappropriate. Ours is certainly not the first system to propose synonyms of query words in a query expansion step (see Section 5), but we believe we have arrived at an elegant integration of interactive query expansion into a more general query refinement mechanism.

Nodes in subsequent levels of the tree are also labeled with expressions extracted from the results, but unlike in the first level these expressions need not contain query words or their synonyms. Instead, if node n of the tree is labeled with expression e, then the children of n are labeled with expressions that appear as continuations of e in the documents associated with n. In the example of Figure 1, "white lines" has the children "in image area when making copies" and "when printing" because these are the continuations of "white lines" found in the results.

Recall that clicking on node n selects those results that contain the expressions of n and all its ancestors (except for the root, whose label is not an expression). Children are generated to cover all of these results; that is, each result associated with n is associated with at least one of n's children. If n's expression e is the last expression in one of the result sentences, *i.e.* e has no continuation in that sentence, then we choose the rightmost expression in that sentence that is to the left of all expressions already chosen. For example, the query "fax" yields results including "black bands on faxes" and "blurred image on faxes." These results will both be associated with a first-level node labeled "on faxes," and each will be associated with a second-level node labeled "black bands on faxes" or "blurred image on faxes."

Each node represents a unique normalized expression, which may be realized by different surface forms in different sentences (see Section 3.1). Since the normalization process sometimes results in expressions that are ungrammatical or unnatural, we use surface forms as the labels displayed in the refinement tree. When there are multiple surface realizations in the knowledge base of a single normalized expression, we choose one of them arbitrarily to serve as the label.

We hope ultimately to use the syntactic relationships between expressions, rather than simply their linear order, to define which expressions will be chosen as refinements for a given node. In principle this could lead to more appropriate choices for results such as "White lines when printing and when copying." Whereas the current strategy presents "when printing" as a refinement of "white lines" and then "and when copying" as a refinement of "white lines when printing," it would be more appropriate to have both "when printing" and "when copying" as refinements of "white lines." The method we currently use for building a refinement tree from linear sequences of refinement expressions does not generalize in a straightforward way to sentences with a branching structure; making this adaptation, particularly in a way that remains comprehensible in the face of the inevitable errors and ambiguities in parsing, is a subject of ongoing work.

Note that while the procedure described here could be used to generate a refinement tree from the results of a query over an arbitrary document collection, the tree generated for a heterogeneous corpus such as the web would be less useful than that generated for troubleshooting documents. In generating the refinement tree, we exploit the recurrent vocabulary and structure that is typical in troubleshooting documents; if the documents in the collection do not exhibit this recurrent structure, the tree will likely contain many nodes that represent only one document each.

4 Experimental Results

We evaluated PE in a comparison test against the system it was designed to replace, called OSA. The testing involved both quantitative measurements and ethnographic observation. We summarize the results here.

OSA provides access to the same troubleshooting knowledge base as PE. When a user submits a query to OSA, the query words are lemmatized, synonyms are applied, and documents that contain one or more of the query words are returned in a ranked list. There is a limited refinement mechanism in the form of a few hard-coded multiple-choice questions (*e.g.* "Select when the image quality problem occurs: copy, fax, print"), but we observed that users rarely avail themselves of this mechanism (we hypothesize that this is related both to the layout of the page and to the choice of questions, which often seems inappropriate relative to a given query). In these tests OSA thus serves chiefly as a representative of "classical" information retrieval systems that accept a keyword query and return a ranked list of results.

Each of fourteen users was asked to play out four predefined troubleshooting scenarios, two judged to be simple and two more complicated. The scenarios were selected after discussion with professional troubleshooters about problems their users frequently have. Only problems for which the knowledge base contains a solution were chosen. (We hypothesize that the PE refinement mechanism would also help users determine more quickly that no solution to their problem is available, since it factors a potentially large number of search results into a smaller number of refinement choices that can be scanned quickly to determine whether any are appropriate. The experiment described here did not test this hypothesis, but some anecdotal evidence supports it.) As much as possible the scenarios were presented pictorially (showing what the copies looked like, what the machine and its interface looked like, etc.) to avoid biasing users' choice of query terms. Each user attempted to solve troubleshooting scenarios with first one system and then the other. The order in which the systems were presented to the user and of which scenario was used with which system was varied, to avoid learning and ordering effects. In addition to recording the number of problems users solved, we administered a usability questionnaire.

PE showed a statistically significant advantage over OSA in solve rate and user preference. 71% of the sessions (20 out of 28) using PE ended with the user finding the right solution, compared to 50% of the sessions (14 out of 28) using OSS[1]. In terms of preference, out of the 14 participants, 10 preferred PE, 3 OSA and 1 was indifferent.[2]

In addition to the quantitative measurements, qualitative observation allowed us to confirm that many users understood the intended purpose of the navigation tree, and to identify areas for improvement.

From these initial tests we conclude that compared to a flat list of results, the PE refinement tree makes it significantly easier for a user to find the document that describes his or her problem. Due to the complexity and expense of performing this type of user testing, the number of scenarios used in the first round of testing was necessarily small; further testing against OSA is planned to confirm the initial results. In addition, comparative testing against a general-purpose clustering information retrieval system is planned, in order to test our hypothesis that the PE refinement mechanism is particularly well-suited to the troubleshooting task.

---

[1] Two-tailed paired samples t-test: p=0,047.

[2] Two-tailed test for difference between proportions: $p = 0,008$; Kolmogorov-Smirnov test for assessing distribution normality: $p = 0,01$.

5 Comparison of PE with Other Troubleshooting and Information Retrieval Tools

Existing approaches to the design of on-line troubleshooting systems can be grouped roughly into two kinds: "knowledge-rich" approaches that make use of domain knowledge *e.g.* in the form of decision trees or models of the machine, and "knowledge-poor" approaches based on information retrieval techniques that do not use extensive domain knowledge. Systems that use knowledge-rich approaches [3,10,14] can prompt the user with relevant questions, rather than simply processing queries as the user formulates them, and thus have the potential to be more helpful for naive users, but the knowledge they require can be very expensive to build and maintain over the lifetime of the system. In Pocket Engineer we have implemented a knowledge-poor approach, but one that provides more guidance than a conventional information retrieval interface.

In IR-based troubleshooting systems, *e.g.* Eureka [7], the standard keyword search paradigm is applied to searching in a collection of solution descriptions. This type of interface is very familiar to most users, and can be effective for someone who is already familiar with the contents of the knowledge base and the way important concepts are expressed therein, but it does not meet the needs of a non-technical user who has a problem but is unable to generate spontaneously a sufficiently precise description of it.

The AI-STARS system [4] addressed the mismatch between user vocabulary and the vocabulary of a troubleshooting knowledge base by automatically expanding queries with synonyms. PE also uses synonyms to bridge the vocabulary gap, but with a refinement mechanism that gives the user more control. We believe that this approach will make the system more robust to contextually inappropriate synonyms. As in any system, there is a precision vs. recall balance to be maintained, but in cases where an inappropriate synonym is added to the query, the inappropriate results will be grouped together in a single branch of the refinement tree, and thus more easily ignored in favor of more appropriate results. AI-STARS included a number of tools for automatically identifying potential synonymy relationships by analyzing the text of the knowledge base and logs of user queries; we are currently developing tools along these lines.

Outside of the troubleshooting domain, there has been a great deal of work on mechanisms for helping users view and explore the results of an initial search in ways that are more useful than a simple ranked list. One early approach was relevance feedback [15], in which the user is asked to indicate whether each of the top-ranked results is relevant or not, and new words chosen automatically from the relevant documents are added to the query. Relevance feedback is designed for retrieval scenarios in which there are many relevant documents, which is typically not the case in troubleshooting. A variation is to ask the user to choose candidate keywords directly [6].

Another class of interaction mechanisms partitions search results into groups and allows the user to narrow down the selection to members of a particular group. Early attempts based on a fixed clustering of the entire document base failed to improve access to relevant documents [8]. The Scatter/Gather technique introduced the possibility of inducing clusters among only the documents retrieved for a given query, with better results [8]. Since the Scatter/Gather paper, a host of alternative algorithms for clustering search results have been proposed, with the goals of improving the speed and/or the quality of the clustering. Whereas Scatter/Gather and many subsequent systems used "polythetic" clustering methods, in which documents are defined to be similar to the degree that their entire vocabularies overlap, a few more recent systems [5,16,12,9] have used "monothetic" clustering methods, in which a cluster is defined by a single term, and all documents in the cluster contain that term. As Sanderson and Croft argue [16], monothetic methods yield clusters in which it is clear what the documents have in common with each other, and finding appropriate names for the clusters is not a problem as it can be in polythetic clustering. In addition, a monothetic method is particuarly well-suited to the troubleshooting scenario, in which the user is looking not for a collection of documents about a general topic, but the unique document that describes a particular problem. By using a monothetic clustering, PE generates a refinement tree that, despite being built by knowledge-poor methods, often helps the user construct a coherent problem description, rather than simply a collection of more-or-less related keywords. For example, in Figure 1, from the initial query "lines," the system proposes the refinement "white lines," and from there "white lines when printing."

Among the systems based on monothetic clustering, that of Edgar et al [9] is perhaps the most similar to ours. The root of their hierarchy is the initial query term, and the children of each phrase are longer phrases that contain the parent phrase; for example, *forest* → *sustainable forest* → *sustainable forest management*. Their notion of phrase is more general than ours, so that what PE treats as a sequence of phrases might be considered a single long phrase in [9]. They define phrase boundaries using some simple heuristics based on stop words and punctuation; we hope that the more sophisticated NLP used to define phrase boundaries in PE leads to more linguistically appropriate phrases (this remains to be tested). During the user testing reported in [9], users complained about the fact that Edgar et al.'s system accepted only a single query term, the term to be used as the root of the refinement tree. In PE, we allow the user to specify any number of query terms. Only one can be used as the root of the tree, but the others are taken into account in determining the result set, and for ranking results.

In summary, PE provides a troubleshooting interface that provides searchers with some guidance, yet doesn't require hand-crafted knowledge resources. PE provides a unique integration of a keyword-based query interface with a refinement tree that groups documents based on both the words and the syntactic structure of individual sentences.

6 Summary and Future Work

We have built a system that enables lay users to find solutions to problems they experience with a machine. This system reuses a knowledge base that was initially designed to support expert users. Since lay users are less familiar with the vocabulary used in the knowledge base, and have less understanding of what sorts of information might be important for troubleshooting, they often need help formulating useful descriptions of their problems.

Knowledge-rich approaches to helping users navigate a space of problems and solutions would have required the creation of a new knowledge base with additional content, *e.g.* a decision tree with sequences of questions that can guide a user to the relevant problem description. This additional content would be expensive to create and maintain; by generating refinement choices automatically from the original problem descriptions, our approach allows a similar type of interaction with lower maintenance costs. The main cost of adapting our system to a new document collection would be the cost of adapting the grammar, and we expect this cost to remain relatively small once a general-purpose grammar for the language in question has been written. In our first trials, the grammar developed for a particular knowledge base proved to be sufficient for processing several other knowledge bases that cover the same subject matter. Adapting the grammar to a different domain would probably require somewhat more work; collection-specific grammar adaptations typically involve additions to the lexicon and rules for specific typographical conventions.

Document clustering approaches are easier to put in place and to maintain than knowledge-rich approaches, but the type of clustering most information retrieval systems perform is not well suited to the troubleshooting domain, where there is typically at most one document in the collection that addresses the user's problem, not an entire class of recognizably similar documents. We have thus developed a method based on finding common expressions among individual sentences, rather than on comparing entire document vocabularies, and makes use of syntactic structure identified by a natural language parser. The parser can apply lexical and syntactic transformations in order to discover concepts that are common to multiple sentences even when they are expressed using different surface forms. This method takes advantage of the particular structure of a troubleshooting knowledge base, namely the fact that when a query in such a knowledge base returns many results, the retrieved sentences often have a significant degree of structural and terminological similarity. Our refinement mechanism helps a user iteratively build up a more and more detailed description of a problem, without requiring prior familiarity with the structure and terminology of the knowledge base.

PE includes infrastructure that supports the expansion of queries with context-sensitive synonymy rules intended to help with the mismatch between user vocabulary and the vocabulary of the knowledge base, but the thesaurus needed for this functionality has not yet been systematically populated. We expect this functionality to be particularly effective in combination with the refinement tree, which is structured to make it easy to ignore results based on inappropriate synonyms. Work is currently ongoing to automate parts of the thesaurus development process.

Initial user testing indicated that users have more success at finding solutions with PE than with a classical search mechanism that simply answers a query with a list of results. Further testing is planned to compare our clustering method with a more traditional clustering based on similarity of document vocabularies, and to evaluate the contribution that individual components of the PE functionality, namely the synonymy mechanism and the novel clustering method, make to the overall performance.

References

1. S. Aït-Mokhtar, J.-P. Chanod and C. Roux: Robustness beyond shallowness: incremental dependency parsing. *NLE Journal*, 2002.
2. S. P. Abney: Parsing by Chunks. In R. C. Berwick, S. P. Abney, and C. Tenny, eds., *Principle-Based Parsing: Computation and Psycholinguistics*, pp. 257–278. Kluwer Academic Publishers, Boston, 1991.

3. D. W. Aha, Tucker Maney, and Leonard A. Breslow: Supporting Dialogue Inferencing in Conversational Case-Based Reasoning. In *Proc. of EWCBR'98*.
4. P. G. Anick: Adapting a full-text information retrieval system to the computer troubleshooting domain. In *Proceedings of the 17th Annual international ACM SIGIR Conference on Research and Development in Information Retrieval*, Dublin, Ireland, July 1994. Springer-Verlag, NY, 349-358.
5. P. G. Anick and S. Tipirneni: The paraphrase search assistant: terminological feedback for iterative information seeking. In *Proceedings of the 22nd annual international ACM SIGIR conference on Research and development in information retrieval*, 1999, 153–159.
6. N. J. Belkin, C. Cool, D. Kelly, S. Lin, S. Y. Park, J. Perez-Carballo, and C. Sikora: Iterative exploration, design and evaluation of support for query reformulation in interactive information retrieval. *Information Processing Management*, Vol. 37, Num 3 (2001), 403-434.
7. D. G. Bobrow and J. Whalen: Community knowledge sharing in practice: the Eureka story. In *Journal of the Society for Organizational Learning*, Vol. 4 Issue 2, 2002.
8. D. R. Cutting, D. R. Karger, J. O. Pedersen, and J. W. Tukey: Scatter/Gather: a cluster-based approach to browsing large document collections. *Proceedings of the 15th Annual international ACM SIGIR Conference on Research and Development in information Retrieval*, Copenhagen, Denmark, June 1992. ACM Press, New York, NY, 318-329.
9. K.D. Edgar, D.M. Nichols, G.W. Paynter, K. Thomson, and I.H. Witten: A user evaluation of hierarchical phrase browsing. In *Proc. European Conference on Digital Libraries*, Trondheim, Norway, 2003.
10. F. V. Jensen, C. Skaanning, and U. Kjaerulff: The SACSO system for Troubleshooting of Printing Systems. In *Proc. of SCAI 2001*, pp. 67-79.
11. B. H. Kang, K. Yoshida, H. Motoda, and P. Compton: Help Desk System with Intelligent Interface. In *Applied Artificial Intelligence*, Vol. 11, Num. 7, 1 Dec. 1997, pp. 611-631(21).
12. D. Lawrie and W. B. Croft and A. Rosenberg: Finding topic words for hierarchical summarization. In *SIGIR '01: Proceedings of the 24th annual international ACM SIGIR conference on Research and development in information retrieval*, 2001, 349–357.
13. J. O'Neill, A. Grasso, S. Castellani, and P. Tolmie: Using real-life troubleshooting interactions to inform self-assistance design. In *Proc. of INTERACT*, Rome, Italy, 12-16 Sep. 2005.
14. B. Peischl and F. Wotowa: Model-based diagnosis or reasoning from first principles. In *IEEE Intelligent Systems* Vol. 18, Num. 3, 2003, pp. 32-37.
15. J. Rocchio: Relevance feedback in information retrieval. In G. Salton (ed.) *The SMART Retrieval System—Experiments in Automatic Document Processing*. Prentice-Hall, 1971.
16. M. Sanderson and B. Croft: Deriving concept hierarchies from text. In *SIGIR '99: Proceedings of the 22nd annual international ACM SIGIR conference on Research and development in information retrieval*, 1999, 206–213.

Automatic Morphological Query Expansion Using Analogy-Based Machine Learning

Fabienne Moreau, Vincent Claveau, and Pascale Sébillot

IRISA, Campus universitaire de Beaulieu, 35042 Rennes cedex, France
{Fabienne.Moreau, Vincent.Claveau, Pascale.Sebillot}@irisa.fr

Abstract. Information retrieval systems (IRSs) usually suffer from a low ability to recognize a same idea that is expressed in different forms. A way of improving these systems is to take into account morphological variants. We propose here a simple yet effective method to recognize these variants that are further used so as to enrich queries. In comparison with already published methods, our system does not need any external resources or *a priori* knowledge and thus supports many languages. This new approach is evaluated against several collections, 6 different languages and is compared to existing tools such as a stemmer and a lemmatizer. Reported results show a significant and systematic improvement of the whole IRS efficiency both in terms of precision and recall for every language.

Keywords: Morphological variation, query expansion, analogy-based machine learning, unsupervised machine learning.

1 Introduction

Information retrieval systems (IRSs) aim at establishing a relation between users' information needs (generally expressed by natural language queries) and the information contained in documents. To this end, a commonly used method consists of making a simple match between the query terms and the document words. A document is said to be relevant if it shares terms with the query. IRSs face two problems with such a mechanism, mainly bound to the inherent complexity of natural language. The first problem is related to polysemy: a single term may have different meanings and represent various concepts (*e.g.* `bug`: `insect` or `computer problem`); because of term ambiguity, IRSs may retrieve non relevant documents. The second and dual issue reflects the fact that a single idea may be expressed in different forms (*e.g.* `bicycle-bike`). Therefore, a relevant document can contain terms semantically close but graphically different. To overcome those two limitations, a rather natural solution is to perform a linguistic analysis of documents and queries. Based on natural language processing (NLP) techniques, it enables to obtain richer and more robust descriptors than simple keywords. These descriptors are able to highlight the fact that a same word can have different meanings or undergo variations of form (`retrieve` ↔ `retrieval`), structure (`information retrieval` ↔ `information that is retrieved`) or meaning (`seek` ↔ `search`). Among the various types of linguistic analysis that

G. Amati, C. Carpineto, and G. Romano (Eds.): ECIR 2007, LNCS 4425, pp. 222–233, 2007.

can be applied (*i.e.* morphological, syntactic or semantic), morphological analysis appears to be one of the most effective ones to improve IRS performances. It leads to recognize that words such as `produce`, `produced`, `producing`, and `producer`, although graphically different, are actually forms of the same word; in other terms they are morphological variants. Enabling the match of these graphically different but semantically close forms can be consequently relevant in information retrieval (IR).

Morphological variation is a well-known problem in IR and has been exhaustively investigated in the literature (for a state-of-the-art, see [1,2,3,4] for instance). Despite those studies, one main issue remains: the non-portability of the methods proposed for detecting morphological variants: a majority of them are developed for one given language and are based on external knowledge (list of endings, recoding rules, lexicon...); consequently, they cannot be re-used out of their framework of creation. Considering the potential impact of morphological information on the performances of IRSs, it is essential to conceive tools that exceed the limits of existing methods and are adapted to IR data specificities.

Therefore, a simple but effective approach using an unsupervised machine learning technique is proposed in order to detect morphological variants. This method has to fulfill the following requirements: it must not require any external knowledge or resources; it must be entirely automatic; it must be directly applicable to various languages. Our acquisition method is used in IR for query expansion. The goal of our approach is to detect, within a collection of texts, words that are in morphological relations with the terms of a query and to add them to it.

The rest of the paper has the following structure: Section 2 presents some of the approaches existing to take into account morphological variation in IR. Section 3 describes the method developed for the detection of morphological variants and its use in an IRS to extend queries. Section 4 details the experiment results obtained on various collections and discusses them. Finally, Section 5 concludes on the relevance of our method to improve IRS performances.

2 Background: Morphological Variation in IR

There are generally two ways for coping with morphological variation in IR: at indexing time (conflation approach) or at retrieval time (query expansion). In the conflation approach, the various forms of a same word (variants) are reduced to a common form (stem, root or lemma). Thus match between documents and query is done on the basis of this canonical form. In the expansion method, documents and queries are indexed with original word forms; and the terms of a user's query are expanded with their morphological variants at retrieval time (see [5] for instance). One usual technique to handle morphological variation is stemming, which is used to reduce variant word forms to common roots (stems) [1, for instance]. Other approaches choose more sophisticated tools based on linguistic methods, like lemmatizers (inflectional morphology) or derivational analyzers [6,7,8].

The principal limit of the existing tools is that they are, in most cases, based on external resources such as affix lists, morphological rules or dictionaries. Consequently, they can only be applied to one very particular language and present a restricted coverage. Many studies yet suggest to use them in IR [1,3,9, *inter alia*]; the experiments tend to show the added-value of taking into account morphological variants to improve both recall and precision of systems. However, obtained results depend on numerous factors, like collection language, query length or document type (general or from specialized fields for example). More generally, among those studies, very few are compatible with the three requirements given in introduction as a framework of our work. Some approaches that meet entirely our constraints rely on statistical techniques, which have the advantage of being independent of the language and may be unsupervised. Thus, several word segmentation tools were developed while being mainly based on frequency criteria [10, for instance] or on a N-grams technique [11]. Generally, those statistical methods, although they answer our requirements, show low reliability for the detection of morphological variants [12] and their contributions to IR has not been really proved.

3 New Automatic Acquisition of Morphological Variants Used to Extend Query in IR

We describe here our method to extract morphological variants from documents. To fulfill the three requirements enumerated in introduction, our approach is based on a rather simple but flexible technique better suited to IR specificities. The principles are the followings: an original technique (*cf.* Section 3.1) is used to detect every morphologically related word pairs (joined up by a link of morphological variation); since we are looking for query extensions, we use it to locate within the document database all the words that are morphologically related to one of the terms of the query. All the detected words are then added to this query for its expansion. The proposed acquisition method is first explained; then its use within IRSs for query expansion is described in details.

3.1 Learning by Analogy

Our approach for morphological variant acquisition of query terms is based on a technique initially developed to be used in the field of terminology [13]. Its principle is simple and based on analogy. Analogy can be formally represented as $A : B \doteq C : D$, which means "A is to B what C is to D"; *i.e.* the couple A-B is in analogy with the couple C-D. The use of analogy in morphology, which is rather obvious, has already been studied [14]. For example, if we have analogies like `connector` : `connect` $\doteq$ `editor` : `edit`, and knowing that `connector` and `connect` share a morpho-semantic link, we can guess a same link between `editor` and `edit`.

The most important feature in learning by analogy is the notion of similarity that is used to determine if two pairs of propositions —in our case, two pairs of words— are analogous. The similarity notion we use, hereafter *Sim*, is quite simple but well fit to many languages in which inflection and derivation are mainly obtained by prefixation and suffixation. Intuitively, *Sim* checks that to go from a word w_3 to a word w_4, the same "path" of deprefixation, prefixation, desuffixation and suffixation is needed as to go from w_1 to w_2. More formally, let us name lcss(X, Y) the longest common substring shared by two strings X and Y (*e.g.* lcss(`republishing`, `unpublished`) = `publish`), X $+_{suf}$ Y (respectively $+_{pre}$) being the concatenation of the suffix (resp. prefix) Y to X, and X $-_{suf}$ Y (respectively $-_{pre}$) being the removal of the suffix (resp. prefix) Y from X. The similarity measure *Sim* can then be defined as follows:

$Sim(w_1\text{-}w_2,\, w_3\text{-}w_4) = 1$ if the four following conditions are simultaneously met:

$$\begin{cases} w_1 = \text{lcss}(w_1, w_2) +_{pre} \text{Pre}_1 +_{suf} \text{Suf}_1, \text{ and} \\ w_2 = \text{lcss}(w_1, w_2) +_{pre} \text{Pre}_2 +_{suf} \text{Suf}_2, \text{ and} \\ w_3 = \text{lcss}(w_3, w_4) +_{pre} \text{Pre}_1 +_{suf} \text{Suf}_1, \text{ and} \\ w_4 = \text{lcss}(w_3, w_4) +_{pre} \text{Pre}_2 +_{suf} \text{Suf}_2 \end{cases}$$

$Sim(w_1\text{-}w_2,\, w_3\text{-}w_4) = 0$ otherwise

Pre_i and Suf_i are any character strings. If $Sim(w_1\text{-}w_2,\, w_3\text{-}w_4) = 1$, the analogy $w_1 : w_2 \doteq w_3 : w_4$ stands, then we can suppose that the morphological relation between w_1 and w_2 is identical to the one between w_3 et w_4.

Our morphological acquisition process checks if an unknown pair is in analogy with one or several given examples. For instance, we can determine that the couple `rediscovering-undiscovered` is in analogy with one example-pair `republishing-unpublished`, since the similarity measure defined as follows:

$$\begin{cases} w_1 = \texttt{publish} +_{pre} \texttt{re} +_{suf} \texttt{ing}, \text{ and} \\ w_2 = \texttt{publish} +_{pre} \texttt{un} +_{suf} \texttt{ed}, \text{ and} \\ w_3 = \texttt{discover} +_{pre} \texttt{re} +_{suf} \texttt{ing}, \text{ and} \\ w_4 = \texttt{discover} +_{pre} \texttt{un} +_{suf} \texttt{ed} \end{cases}$$

worths 1.

For efficiency reasons during analogy search, rather than the word-pair examples, the prefixation and suffixation operations used in the similarity measure are stored. Thus, the example-couple `republishing-unpublished` is not stored as such, but retained according to the following rule:

$$w_2 = w_1 -_{pre} \texttt{re} +_{pre} \texttt{un} -_{suf} \texttt{ing} +_{suf} \texttt{ed}$$

To show the analogy `republishing` : `unpublished` $\doteq$ `rediscovering` : `undiscovered` consists in testing that `rediscovering-undiscovered` verifies the preceding rule.

As already emphasized in [6], prefixation and suffixation operations considered in our approach enable to take into account partly the light variations of roots as long as they are common enough to be present in one of our examples. More complex variations such the one existing in `go-went` are of course not supported. Yet it has been already proved that this simple analogy-based technique is able to detect morphological variants using examples of semantically and morphologically related words with a very good coverage and a high degree of accuracy in a context of computational terminology (*cf.* [13]). It is worth noting that it is moreover possible to identify the semantic link between these variants with excellent rates of success by annotating each rule with a label of semantic relation. Those are not used here: although it was shown that some semantic links are more relevant than others [15], we made the choice to take into account all the kinds of semantic links (synonymy, hyperonymy...) for query expansion.

3.2 Use for Query Expansion

In order to be operational, the previously presented detection method needs examples (*i.e.* morphologically related word couples). Such a supervised property is not well suited to a use within IR and does not correspond to the fully automatic aspect of the system in our requirements. To solve this problem, we substitute this supervision phase by a rustic technique that allows to constitute a set of word pairs that can be used as examples. This example-pair research proceeds in the following way:

1. randomly choose one document in the IRS collection;
2. form all the possible word pairs resulting from this document;
3. add to the example set couples w_1-w_2 such as $lcss(w_1,w_2) > l$;
4. return to step 1.

These steps are repeated until the resulting set of example-couples is large enough; in the experiments described in Section 4, 500 documents were analyzed. Notice that this operation also supposes that derivation and inflection are mainly done by prefixation and suffixation operations.

During this phase, it is necessary to avoid building word pairs that are not valid examples. The correct behavior of our analogy technique relies on it. That is why we have added two constraints. On the one hand, a minimal length of common substring l is fixed at a large enough value (in our experiments, $l = 7$ letters). Thus, the probability to aggregate two words that do not share any link is reduced. On the other hand, like what was already shown [5], variant search within a same document maximizes the probability that the obtained two words belong to the same domain.

At the end of this step, a set of morphologically related word-pair examples is available; analogy rule learning can be conducted (*cf.* Section 3.1). It is then possible to check if unknown word pairs are in derivation or inflection relation. In our case, we precisely want to retrieve query term variants. Each query term is thus confronted with each word of the document collection. If a formed pair is in analogy with one of the example-pairs, then the document word is used to enrich the query. In order to speed up treatments, analogy rules are in fact used in a generative way. Words are produced from the query terms according to prefixation and suffixation operations indicated in the morphological rules and are kept only if they appear in the index of the collection's terms. Rule learning being made off-line, only the morphological variant search for query terms within the index is made on-line. Search complexity is $O(n)$ where n is the number of distinct terms in the collection. In our experiments, it takes some tenths of a second using a Pentium 1.5 GHz (512 MB). For instance, for the original query: `Ineffectiveness of U.S. Embargoes or Sanctions`, the result of the expansion will be: `ineffectiveness ineffective effectiveness effective ineffectively embargoes embargo embargoed embargoing sanctioning sanction sanctioned sanctions sanctionable`.

During expansion, only words directly related to query terms are added; the words themselves related to the extensions are not taken into account. This voluntary absence of transitivity aims at avoiding propagating errors, such as `reduce` → `produce` → `product` → `productions` → `production` → `reproduction`... In our experiments, an average of three variants is added to each query term. No manual filtering is performed; thus, some extensions are not relevant. The quality of the extensions is evaluated by measuring their impact on the IRS performances. An intrinsic evaluation, out of the context of use, turns out to be non relevant to estimate their impact.

4 Experimental Results

This section details the evaluation of our query expansion method. We first present the various document collections that have been used (Section 4.1), and then successively describe different experiments: results obtained from French and English collections (Sections 4.2 and 4.3) are first reviewed; then the impact of the query length (Section 4.4) is analyzed; and finally the portability of our approach on other languages (Section 4.5) is evaluated.

4.1 Document Collections

Three different document collections are used for our experiments. The evaluation of our method is carried out for English on a subset of the TIPSTER collection used in TREC. More precisely the Wall Street Journal subcollection made up of 175,000 documents and a set of 50 queries (from TREC-3) has been chosen. In order to emulate the usual short-query behavior, only the *title* field containing few words has been employed.

The evaluation on French is based on the INIST collection, made up of 30 queries and 163,000 documents, which are paper abstracts from various scientific disciplines. The portability of our method is controlled on the ELRA collection, made up of 30 queries and 3,511 documents that are questions/answers of the European Commission, available in French, English, German, Portuguese, Spanish, and Italian. Short queries (*title* field) are also chosen for these two collections, except in Section 4.4 where the impact of the query length is studied. The IRS used is LEMUR (`http://www.lemurproject.org`), implemented with the well-known Okapi-like (BM-25) weighting scheme.

4.2 French Experiments

The first experiment is performed on the French INIST collection. In order to evaluate the added-value of query expansion with morphological variants detected with our method, results are computed with and without extensions.

Standard IR measures are used for evaluation: precision and recall (computed for several threshold values), interpolated average precision (calculated at 11 recall points (IAP)), R-precision and non-interpolated average precision (MAP). For comparison, we also present the results obtained by applying on the same collection three traditional morphological tools: 2 French stemmers based on a set of fixed rules of desuffixation — one developed by Savoy [16], the other is an CPAN Perl adaptation of the Porter algorithm for French — and a French lemmatizer — part-of-speech tagger TREETAGGER (`http://www.ims.uni-stuttgart.de/projekte/corplex/TreeTagger`).

In contrast with our method, these tools perform by conflation. Results are given in Table 1 . Those considered as being not statistically significant (using paired t-test with the condition p-value < 0.05) are indicated in italic. The average length (number of words, stop-words included) of the queries ($|Q|$) is also indicated.

Table 1. Query expansion performances on the INIST collection

	Without extension	With extension (improvement %)	Stemming (Savoy) (improvement %)	Stemming (Porter) (improvement %)	Lemmatization (improvement %)
|Q|	5.46	16.03	5.2	5.2	5.17
MAP	14.85	18.45 (+24.29%)	*17.31 (+16.63%)*	15.89 (+7.00%)	17.82 (+20.07%)
IAP	16.89	19.93 (+17.97%)	*18.85 (+11.57%)*	17.69 (+5.92%)	19.72 (+16.73%)
R-Prec	17.99	21.63 (+20.24%)	*19.88 (+10.53%)*	18.77 (+4.34%)	*19.71 (+9.56%)*
P(10)	34.33	38.67 (+12.62%)	*36.67 (+6.80%)*	*34.33 (0%)*	39.67 (+15.53%)
P(20)	27.83	31.83 (+14.37%)	*29.00 (+4.19%)*	*26.50 (-4.78%)*	31.6 (+13.77%)
P(50)	18.33	21.27 (+16.00%)	*20.13 (+9.82%)*	*18.33 (0%)*	*20.87 (+13.82)*
P(100)	12.23	14.80 (+20.98%)	15.23 (+24.52%)	13.87 (+13.41%)	14.97 (+22.34%)
P(500)	3.88	4.80 (+23.71%)	4.55 (+17.18%)	4.56 (+17.53%)	4.47 (+15.29%)
P(1 000)	2.21	2.68 (+21.30%)	2.53 (+14.80%)	2.54 (+15.26%)	2.48 (+12.39%)
P(5 000)	0.56	0.67 (+20.38%)	0.63 (+13.47%)	0.62 (+11.81%)	0.64 (+15.14%)
R(10)	8.00	8.99 (+12.36%)	*8.45 (+5.64%)*	*8.19 (+2.38%)*	*9.04 (+13.02%)*
R(20)	12.33	14.50 (+17.59%)	*12.81 (+3.90%)*	*12.00 (-2.75%)*	*13.62 (+10.48%)*
R(50)	19.65	24.07 (+22.47%)	*20.78 (+5.74%)*	*19.71 (+0.31%)*	*21.56 (+9.71%)*
R(100)	26.85	32.87 (+22.41%)	31.32 (+16.64%)	29.28 (+9.05%)	31.58 (+17.59%)
R(500)	43.09	53.83 (+24.92%)	49.31 (+14.43%)	50.16 (+16.42%)	49.35 (+14.54%)
R(1 000)	48.43	59.45 (+22.74%)	55.27 (+14.12%)	56.94 (+17.57%)	55.03 (+13.62%)
R(5 000)	59.32	72.20 (+21.71%)	67.22 (+13.31%)	67.82 (+14.32%)	68.20 (+14.96%)

The reported figures show that, for each measure, our query expansion method obtains very good results that are all statistically significant. For most measures, query expansion appears not only more effective than stemming or lemmatization, but also more stable since several results of the last two techniques have been found not statistically significant. It is also worth noting that improvements are distributed on every precision and recall thresholds (from 10 to 5000 documents). Thus, improvement does not only correspond to a re-ranking of relevant documents at the head of the result list but also to the obtaining of relevant documents that would not have been retrieved without query extensions.

4.3 English Experiments

This experiment proposes to test if the good results obtained for French can also be observed on English. The preceding experiments are reiterated on the English TIPSTER collection. Table 2 shows the results obtained compared with those of a traditional research without extension, and of researches with lemmatization (using TREETAGGER) and stemming (based on Porter's stemmer [17]).

Table 2. Query expansion performances on the TIPSTER collection

	Without extension	With extension (improvement %)	Stemming (improvement %)	Lemmatization (improvement %)
|Q|	6.5	16.6	6.48	6.48
MAP	23.15	27.18 (+17.40%)	28.09 (+21.33%)	*23.85 (+3.02%)*
IAP	25.02	29.44 (+17.65%)	29.71 (+18.73%)	*25.68 (+2.63%)*
R-Prec	27.52	31.96 (+16.15%)	32.66(+18.69%)	*27.68 (+0.59%)*
P(10)	39.60	47.00 (+18.68%)	45.00 (+13.63%)	*41.40 (+4.54%)*
P(20)	36.10	40.90 (+13.29%)	41.00 (+13.57%)	*36.60 (+1.38%)*
P(50)	28.28	32.80 (+15.98%)	31.72 (+12.16%)	*28.68 (+1.41)*
P(100)	21.44	25.50 (+18.93%)	23.76 (+10.82%)	*22.10 (+3.07%)*
P(500)	7.94	9.21 (+15.90%)	*8.72 (+9.81%)*	*8.35 (+5.13%)*
P(1 000)	4.66	5.37 (+15.08%)	*5.09 (+9.12%)*	*4.85 (+3.89%)*
P(5 000)	1.17	1.31 (+12.21%)	1.30 (+12.10%)	*1.22 (+4.09%)*
R(10)	10.20	11.18 (+9.52%)	*12.77 (+25.13%)*	*10.68 (+4.66%)*
R(20)	16.10	16.82 (+4.46%)	19.36 (+20.24%)	*17.79 (+10.45%)*
R(50)	29.68	32.41 (+9.18%)	32.84 (+10.65%)	*30.43 (+2.53%)*
R(100)	39.48	44.59 (+12.95%)	43.86 (+11.09%)	*41.61 (+5.40%)*
R(500)	61.11	67.68 (+10.74%)	67.82 (+10.98%)	*62.46 (+2.20%)*
R(1 000)	68.68	75.50 (+9.92%)	74.81 (+8.92%)	*69.28 (+0.87%)*
R(5 000)	80.59	87.66 (+8.77%)	87.22 (+8.22%)	81.20 (+0.75%)

The results are positive. The contribution of our approach using query expansion on English is important since the observed gain on the IRS performances is ranging from 4 and 18% according to the measures. Although improvements are sometimes slightly lower than those observed for stemming, they are all statistically significant and constant for all measures. These observations highlight the robustness of our method, and its ability of self-adaptation to English. Other experiments are proposed in Section 4.5 in order to precisely evaluate its portability.

4.4 Impact of Query Length

In order to measure impact of the query length on our expansion method, the French experiment is repeated using the other fields of INIST queries so as to cope with increasingly long queries. In this collection, a query is associated with a set of concepts, each one being represented in a distinct field. The fields are added one by one to the original query (*i.e.* the *title* field). Figure 1 shows results according to the query length that is measured in number of words before expansion. The IRS performance is measured by non-interpolated average precision.

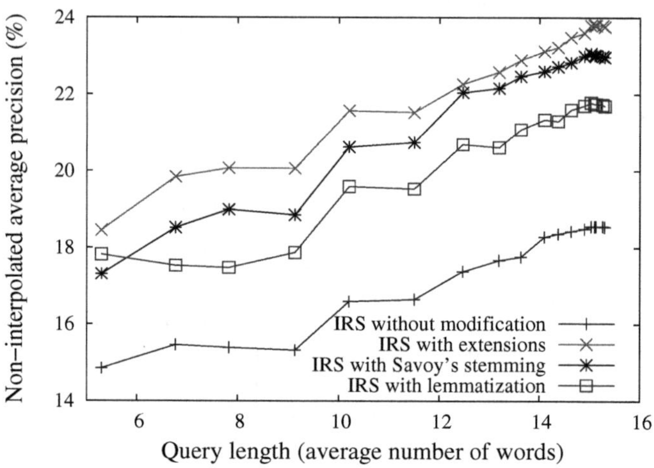

Fig. 1. Precision evolution according to the query length

Broadly speaking, these results prove the interest of taking into account the morphological variants whatever the query length and the morphological process. Among the three evaluated techniques, our approach for query expansion has yet shown better results than stemming and lemmatization.

4.5 Portability

The principal asset of our approach compared with other existing tools is its portability. It is supposed to be directly usable on any language whose morphology is done by prefixation and suffixation. In order to establish the truth of this assertion, Table 3 presents the results obtained on the ELRA collection for German, English, Spanish, French, Italian and Portuguese. For each language, variation (expressed as a percentage) compared to the same search without query extension is indicated.

Results are all very positive since improvements given by query extensions are ranging from 10 to 20% according to languages and measures. As for the other experiments, this gain concerns all precision and recall thresholds. However, for low thresholds (10 to 50 documents), some not statistically significant figures seem to indicate results varying from one query to another. In contradiction with what is usually claimed in some studies, we would like to emphasize here some original remarks. First, query extension with morphological variants has more impact for English, which is generally seen as a morphologically poor language, than for so-called richer languages like Spanish, Italian... It also appears that it is German that

Table 3. Query extension performances on different languages

	Languages					
	German	English	Spanish	French	Italian	Portuguese
MAP	+16.25%	+17.52%	+10.03%	+11.89%	+10.45%	+9.69%
IAP	+15.93%	+16.66%	+8.70%	+10.99%	+9.79%	+9.25%
R-Prec	*+3.03%*	+10.23%	+7.97%	+9.43%	+10.23%	*+6.20%*
P(10)	*+10.68%*	+7.03%	*0%*	*+3.53%*	*+2.54%*	*0%*
P(20)	*+8.33%*	*+3.62%*	+7.41%	+6.85%	+11.15%	*+4.38%*
P(50)	*+6.69%*	+8.23%	+13.40%	+13.85%	+13.48%	+8.31%
P(100)	+9.54%	+14.31%	+16.76%	+16.24%	+18.98%	+14.24%
P(500)	+13.18%	+20.49%	+18.13%	+17.19%	+18.94%	+23.35%
P(1 000)	+12.97%	+21.60%	+20.32%	+18.26%	+22.13%	+24.64%
R(10)	*+6.82%*	+2.90%	*+1.88%*	*+5.43%*	*-0.67%*	*-0.47%*
R(20)	*+5.95%*	*+3.27%*	+7.40%	+7.36%	+7.82%	*+7.55%*
R(50)	*+11.12%*	+8.48%	+7.72%	+10.82%	+7.37%	*+6.21%*
R(100)	+11.87%	+13.23%	+10.14%	+10.11%	+8.93%	+9.39%
R(500)	+16.45%	+21.68%	+14.49%	+12.69%	+14.31%	+17.71%
R(1 000)	+18.15%	+20.93%	+17.38%	+13.20%	+18.35%	+19.23%

benefits the most from the extension technique; this is most probably related to the fact that frequent word agglutinations are better taken into account by our approach (the pair `Menschenrechte-Menschenrechtsorganisation` for instance).

4.6 Discussion

Reported experimental results have shown that our approach for morphological variant detection and its use in query expansion significantly improves IRS performances. Its portability has been demonstrated: good results are observed even for languages that are traditionally found to be morphologically poor. These conclusions are distinct from those in several studies of the same field [18, for instance]. Moreover, contrary to what is sometimes observed in other studies, query length appears to have almost no impact on the results: improvement is constant and comparable for query lengths between 5 and 15 words.

Our method for enriching query terms with their morphological variants is nevertheless not perfect. Some terms actually related to query terms are not detected. For instance, for the English collection, our method did not allow to find the variant `hazard` related to the original term `hazardous` nor the term `paid` related to the conjugated verb `pays` of the initial query. These errors are avoided by the methods based on resources, thus explaining why in some cases results obtained by Porter's stemmer are better. What is on the other hand more prejudicial for query extensions is that non relevant terms can be sometimes added. Concerning this last point, several cases can be distinguished. First, some detected words are not semantically related to the original term; the morphological link is fortuitous or no longer used, like `composition-exposition` for instance. Then, some polysemous terms cause errors that are difficult to avoid.

For example, `production` and `reproduction`, detected as morphologically related, are indeed linked in `result production` and `result reproduction` but not in `fish reproduction`. To limit the impact of these errors, words that are themselves related to extensions are not used to enrich queries. This voluntary absence of transitivity aims at avoiding propagating errors. For this reason, the approach by expansion seems more flexible than the conflation method in which `production` and `reproduction` together with their variants would all be transformed to one single form.

5 Conclusion

In this paper, we have proposed a simple and original technique, relying on an analogy-based learning process, able to automatically detect morphological variants within documents and use them to expand query terms. This morphological expansion approach yields very good results. It rivals and even almost always outperforms results obtained with existing tools such as rule-based stemmer or lemmatizer, and also provides more stable performances. Moreover, contrary to most existing techniques, our method is fully unsupervised and thus can be used for many languages; in this paper, we successfully used it on English, French, German, Italian, Portuguese and Spanish test collections.

From a broader point of view, the conclusions of our experiments confirm those generally claimed in state-of-the-art studies since taking into account morphological variation always improves IRS performances, whatever the language. However, our results go against what is sometimes concluded. Indeed, we have shown that morphology can improve IRS performances whatever the query length or the morphological complexity of language providing that a flexible enough method is used.

This paper opens many future prospects that need further consideration. As further studies, there might be some added-values not to include all variants related to a query term but only retain the most relevant ones. Expansion decision could be thus based on the level of confidence of detected analogy (according to its productivity for instance) and on the importance of the query term directly or indirectly related. It would also be interesting to work on the weighting of the variants that are added to the original query and to integrate it in the ranking function. Reported results on studied languages require to checked and consolidated on other collections and to be extended to other languages. Finally, within a framework of translinguistic IR, a similar approach based on analogy used for translation of specialized terms is being studied.

References

1. Harman, D.: How Effective is Suffixing? Journal of the American Society for Information Science **42** (1991) 7–15
2. Kraaij, W., Pohlmann, R.: Viewing Stemming as Recall Enhancement. In: Proceedings of the 19th ACM International Conference on Research and Development in Information Retrieval (SIGIR), Zürich, Switzerland (1996)

3. Hull, D.: Stemming Algorithms - A Case Study for Detailed Evaluation. Journal of the American Society of Information Science **47** (1996) 70–84
4. Moreau, F., Sébillot, P.: Contributions des techniques du traitement automatique des langues à la recherche d'information. Research report, IRISA, Rennes, France (2005)
5. Xu, J., Croft, W.B.: Corpus-Based Stemming Using Cooccurrence of Word Variants. ACM Transactions on Information Systems **16** (1998) 61–81
6. Gaussier, E.: Unsupervised Learning of Derivational Morphology from Inflectional Corpora. In: Proceedings of Workshop on Unsupervised Methods in Natural Language Learning, 37th Annual Meeting of the Association for Computational Linguistics (ACL), College Park, United-States (1999)
7. Vilares-Ferro, J., Cabrero, D., Alonso, M.A.: Applying Productive Derivational Morphology to Term Indexing of Spanish Texts. In Gelbukh, A., ed.: Computational Linguistics and Intelligent Text Processing. Springer-Verlag (2001) 336–348
8. Moulinier, I., McCulloh, J.A., Lund, E.: West Group at CLEF 2000: Non-English Monolingual Retrieval. In: Proceedings of the Workshop of Cross-Language Evaluation Forum, CLEF 2000, Lisbon, Portugal (2000)
9. Fuller, M., Zobel, J.: Conflation-Based Comparison of Stemming Algorithms. In: Proceedings of the 3rd Australian Document Computing Symposium, Sydney, Australia (1998)
10. Goldsmith, J.A., Higgins, D., Soglasnova, S.: Automatic Language-Specific Stemming in Information Retrieval. In: Proceedings of Workshop of Cross-Language Evaluation Forum (CLEF), Lisbon, Portugal (2001)
11. Frakes, W.B.: Stemming Algorithms. In Frakes, W.B., Baeza-Yates, R., eds.: Information Retrieval: Data Structures and Algorithms. Prentice Hall (1992) 131–160
12. Savoy, J.: Morphologie et recherche d'information. Technical report, Neuchâtel University, Neuchâtel, Switzerland (2002)
13. Claveau, V., L'Homme, M.C.: Structuring Terminology by Analogy Machine Learning. In: Proceedings of the International Conference on Terminology and Knowledge Engineering (TKE), Copenhagen, Denmark (2005)
14. Hathout, N.: Analogies morpho-synonymiques. Une méthode d'acquisition automatique de liens morphologiques à partir d'un dictionnaire de synonymes. In: Proceedings of 8ème conférence annuelle sur le traitement automatique des langues naturelles (TALN), Tours, France (2001)
15. Voorhees, E.M.: Query Expansion Using Lexical-Semantic Relations. In: Proceedings of the 17th ACM International Conference on Research and Development in Information Retrieval (SIGIR), Dublin, Ireland (1994)
16. Savoy, J.: A Stemming Procedure and Stopword List for General French Corpora. Journal of the American Society for Information Science **50** (1999) 944–952
17. Porter, M.F.: An Algorithm for Suffix Stripping. Program **14** (1980) 130–137
18. Arampatzis, A., Weide, T.P.V.D., Koster, C.H.A., Van Bommel, P. In: Linguistically Motivated Information Retrieval. Volume 69. M. Dekker, New York, United-States (2000) 201–222

Advanced Structural Representations for Question Classification and Answer Re-ranking

Silvia Quarteroni[1], Alessandro Moschitti[2], Suresh Manandhar[1], and Roberto Basili[2]

[1] The University of York, York YO10 5DD, United Kingdom
{silvia,suresh}@cs.york.ac.uk
[2] University of Rome "Tor Vergata", Via del Politecnico 1, 00133 Rome, Italy
{moschitti,basili}@info.uniroma2.it

Abstract. In this paper, we study novel structures to represent information in three vital tasks in question answering: question classification, answer classification and answer reranking. We define a new tree structure called PAS to represent predicate-argument relations, as well as a new kernel function to exploit its representative power. Our experiments with Support Vector Machines and several tree kernel functions suggest that syntactic information helps specific task as question classification, whereas, when data sparseness is higher as in answer classification, studying coarse semantic information like PAS is a promising research area.

1 Introduction

Question answering (QA) can be seen as a form of information retrieval where one or more answers are returned to a question in natural language in the form of sentences or phrases. The typical QA system architecture consists of three phases: question processing, document retrieval and answer extraction [1].

In question processing, useful information is gathered from the question and a query is created; this is submitted to an information retrieval engine, which provides a ranked list of relevant documents. From these, the QA system must extract one or more candidate answers, which can then be reranked following various criteria such as their similarity to the query. Question processing is usually centered around question classification (QC), the task that maps a question into one of k expected answer classes. Such task is crucial as it constrains the search space of possible answers and contributes to selecting answer extraction strategies specific to a given answer class. Most accurate QC systems apply supervised machine learning techniques, e.g. Support Vector Machines (SVMs) [2] or the SNoW model [3], where questions are encoded using various lexical, syntactic and semantic features; it has been shown that the question's syntactic structure contributes remarkably to the classification accuracy.

The retrieval and answer extraction phases consist in retrieving relevant documents [4] and selecting candidate answer passages [5,1] from them. A further phase called answer re-ranking is optionally applied. It is especially relevant in the case of non-factoid questions, such as those requiring definitions, where the

G. Amati, C. Carpineto, and G. Romano (Eds.): ECIR 2007, LNCS 4425, pp. 234–245, 2007.

answer can be a whole sentence or paragraph. Here, too, the syntactic structure of a sentence appears to provide more useful information than a bag of words.

An effective way to integrate syntactic structures in machine learning algorithms is the use of tree kernel functions [6]. Successful applications of these have been reported for question classification [2,7] and other tasks, e.g. relation extraction [8,7]. However, such an approach may be insufficient to encode syntactic structures in more complex tasks such as computing the relationships between questions and answers in answer reranking. The information provided by parse trees may prove too sparse: the same concept, expressed in two different sentences, will produce different, unmatching parses. One way to overcome this issue is to try to capture semantic relations by processing shallow representations like predicate argument structures proposed in the PropBank (PB) project [9] (`www.cis.upenn.edu/~ace`). We argue that semantic structures can be used to characterize the relation between a question and a candidate answer.

In this paper, we extensively study new structural representations, namely parse trees, bag-of-words, POS tags and predicate argument structures for question classification and answer re-ranking. We encode such information by combining tree kernels with linear kernels. Moreover, to exploit predicate argument information - which we automatically derive with our state-of-the-art software - we have defined a new tree structure representation and a new kernel function to process its semantics. Additionally, for the purpose of answer classification and re-ranking, we have created a corpus of answers to TREC-QA 2001 description questions obtained using a Web-based QA system.

Our experiments with SVMs and the above kernels show that (a) our approach reaches state-of-the-art accuracy on question classification and (b) PB predicative structures are not effective for question classification but show promising results for answer classification. Overall, our answer classifier increases the ranking accuracy of a basic QA system by about 20 absolute percent points.

This paper is structured as follows: Section 2 introduces advanced models to represent syntactic and semantic information in a QA context; Section 3 explains how such information is exploited in an SVM learning framework by introducing novel tree kernel functions; Section 4 reports our experiments on question classification, answer classification and answer reranking; Section 5 concludes on the utility of the new structure representations and sets the basis for further work.

2 Advanced Models for Sentence Representation

Traditionally, the majority of information retrieval tasks have been solved by means of the so-called bag-of-words approach augmented by language modeling [10]. However, when the task requires the use of more complex semantics the above approach does not appear to be effective, as it is inadequate to perform fine-level textual analysis. To overcome this, QA systems use linguistic processing tools such as syntactic parsers to produce sentence parse trees. In our study we exploited two sources of syntactic information: deep syntactic parsers – the outcome of a well-studied technology [6,11] – and shallow semantic parsers, only recently the object of a consistent body of work.

2.1 Syntactic Structures

The syntactic parse tree of a sentence is a hierarchical representation of the syntactic relationships between its words. In such tree, each node with its children is associated with a grammar production rule, where the symbol at the left-hand side corresponds to the parent and the symbols at the right-hand side are associated with the children. The terminal symbols of the grammar are always associated with the leaves of the tree.

Parse trees have often been applied in natural language processing applications requiring the use of grammatical relations, e.g. extraction of subject/object relations. It has been shown [2,7] that syntactic information outperformed bag-of-words and bag-of-n-grams on question classification in QA. The advantage of computing parse tree-based sentence similarity with respect to purely lexical approaches is that trees provide structural relations hard to compute otherwise.

However, when approaching complex QA tasks, the use of parse trees has some limitations. For instance in definitional QA candidate answers can be expressed by long and articulated sentences or even paragraphs. Since the information encoded in a parse tree is intrinsically sparse, it does not contribute well to computing the similarity between such answers; shallow semantics however, being a more "compact" source of information, could prevent the sparseness of deep structural approaches and the noise of bag-of-word models.

2.2 Semantic Structures

Initiatives such as PropBank (PB) [9] have led to the creation of vast and accurate resources of manually annotated predicate argument structures[1]. Using these, machine learning techniques have proven successful in Semantic Role Labeling (SRL), the task of attaching semantic roles to predicates and their arguments. SRL is a fully exploitable technology: our SVM-based SRL system achieves a 76% accuracy on PB data [12]. Attempting an application of SRL to QA seems natural, as pinpointing the answer to a question relies on a deep understanding of the question and answer's semantics.

Let us consider a typical PB annotation for a sentence, such as: `[`$_{ARG0}$ `Compounded interest] [`$_{pred}$ `computes] [`$_{ARG1}$ `the effective interest rate for an investment] [`$_{ARGM-TMP}$ `during the current year]`.

Such shallow semantic annotation is a useful information source. For instance, the PB annotation of a similar sentence would be: `[`$_{ARGM-TMP}$ `In a year][`$_{ARG1}$ `the bank interest rate] is [`$_{pred}$ `evaluated] by [`$_{ARG0}$ `compounded interest]`. Such annotations can be represented via tree structures as those in Figure 1, which we call PASs. These attempt to capture the semantics of both sentences.

[1] The PB corpus contains 300,000 words annotated with predicative information on top of the Penn Treebank 2 Wall Street Journal texts. For each predicate, the expected arguments are labeled sequentially from $ARG0$ to $ARG5$, $ARGA$ and $ARGM$.

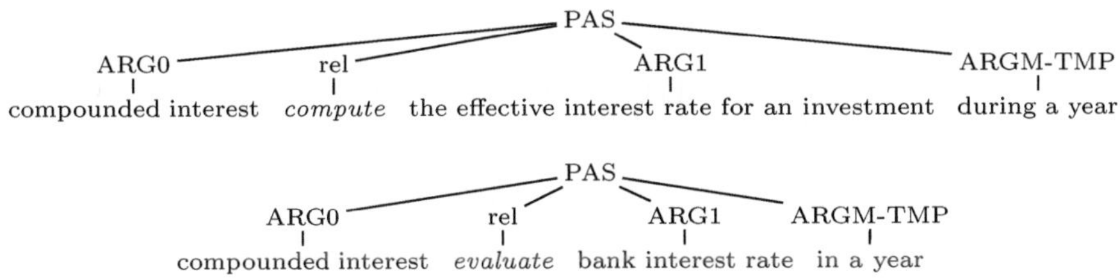

Fig. 1. Predicate argument structures of two sentences expressing similar semantics

We can improve such representation by substituting the arguments with their most important word – often referred to as the semantic head – as in Figure 2. It seems intuitive that data sparseness can be remarkably reduced by using this shallow representation instead of the BOW representation.

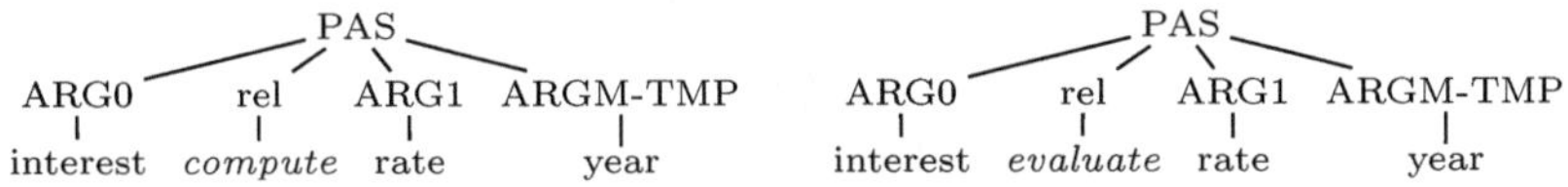

Fig. 2. Improved predicate argument structures of two different sentences

Knowing that parse trees and PASs may improve the simple BOW representation, we face the problem of representing tree structures in learning machines. Section 3 introduces a viable representation approach based on tree kernels.

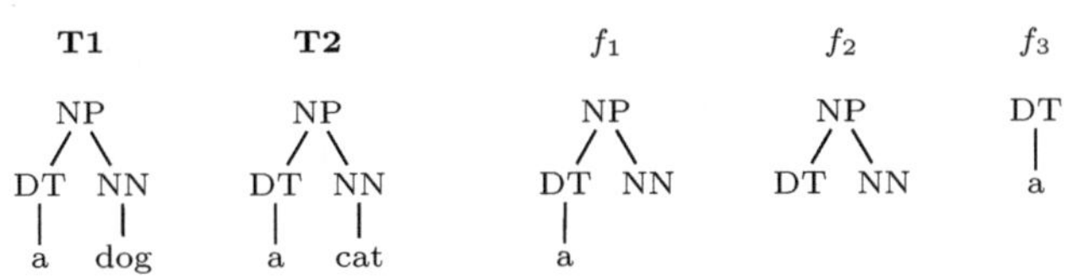

Fig. 3. T1 and T2 with their fragments f_1, f_2 and f_3 derived by the kernel function

3 Syntactic and Semantic Tree Kernels

As mentioned above, encoding syntactic/semantic information represented by means of tree structures in the learning algorithm is problematic. One possible solution is to use as features of a structure all its possible substructures. Given the combinatorial explosion of considering the subparts, the resulting feature space is usually very large. To manage such complexity we can define kernel functions that implicitly evaluate the scalar product between two feature vectors without explicitly computing such vectors.

Below, we report the tree kernel function devised in [6] computing the number of common subtrees between two syntactic parse trees and a new version evaluating the number of semantic structures shared between two PASs.

3.1 Syntactic Tree Kernel

Given two trees T_1 and T_2, let $\{f_1, f_2, ..\} = \mathcal{F}$ be the set of substructures (fragments) and $I_i(n)$ be equal to 1 if f_i is rooted at node n, 0 otherwise. We define

$$K(T_1, T_2) = \sum_{n_1 \in N_{T_1}} \sum_{n_2 \in N_{T_2}} \Delta(n_1, n_2) \quad (1)$$

where N_{T_1} and N_{T_2} are the sets of nodes in T_1 and T_2, respectively and $\Delta(n_1, n_2) = \sum_{i=1}^{|\mathcal{F}|} I_i(n_1) I_i(n_2)$. The latter is equal to the number of common fragments rooted in nodes n_1 and n_2. We can compute Δ as follows:

1. if the productions at n_1 and n_2 are different then $\Delta(n_1, n_2) = 0$;
2. if the productions at n_1 and n_2 are the same, and n_1 and n_2 only have leaf children (i.e. they are pre-terminals symbols) then $\Delta(n_1, n_2) = 1$;
3. if the productions at n_1 and n_2 are the same, and n_1 and n_2 are not pre-terminals then

$$\Delta(n_1, n_2) = \prod_{j=1}^{nc(n_1)} (1 + \Delta(c_{n_1}^j, c_{n_2}^j)) \quad (2)$$

where $nc(n_1)^2$ is the number of children of n_1 and c_n^j is the j-th child of node n. As proved in [6], the above algorithm allows to evaluate Eq. 1 in $O(|N_{T_1}| \times |N_{T_2}|)$. A decay factor λ is usually added by changing the formulae in (2) and (3) to[3]:

2. $\Delta(n_1, n_2) = \lambda$,
3. $\Delta(n_1, n_2) = \lambda \prod_{j=1}^{nc(n_1)} (1 + \Delta(c_{n_1}^j, c_{n_2}^j))$.

For instance, Figure 3 shows two trees and the substructures they have in common. It is worth to note that the fragments of the above Syntactic Tree Kernel (STK) are such that any node contains either all or none of its children. Consequently, `[NP [DT]]` and `[NP [NN]]` are not valid fragments. This limitation makes it unsuitable to derive important substructures from the PAS tree. The next section shows a new tree kernel that takes this into account.

3.2 Semantic Tree Kernel

As mentioned above, the kernel function introduced in Section 2 is not sufficient to derive all the required information from trees such as the PAS in Fig. 2: we would like to have fragments that contain nodes with only part of the children, e.g. to neglect the information constituted by ARGM-TMP. For this, we need to slightly modify the PAS and to define a new kernel function.

First, we change the PAS into the PAS+ structure as shown in Figure 2(a). Each slot node accommodates an argument label in the natural argument order.

[2] Note that, since the productions are the same, $nc(n_1) = nc(n_2)$.

[3] A normalization in the kernel space, i.e. $K'(T_1, T_2) = \frac{K(T_1,T_2)}{\sqrt{K(T_1,T_1) \times K(T_2,T_2)}}$, ensures a similarity score between 0 and 1.

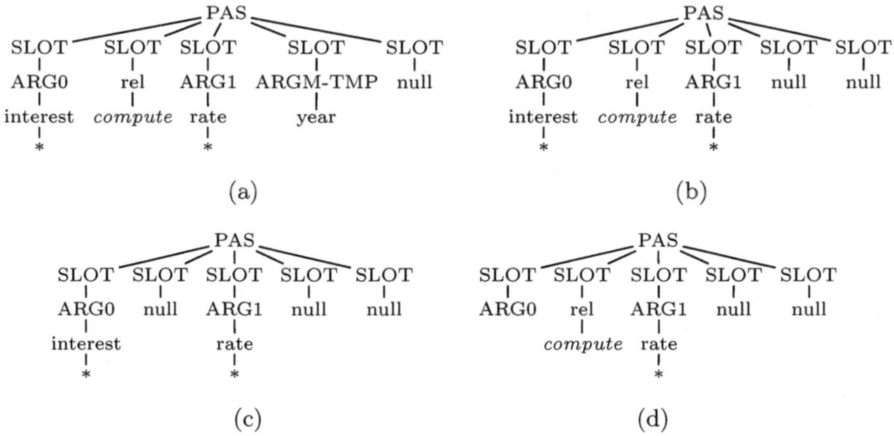

Fig. 4. A PAS+ with some of its fragments

Since diverse predicates or their different use may involve a different number of arguments, we provide additional slots, filled with *null* arguments. The figure shows just one slot to complete a structure of 5 arguments. More slots can be added to manage the maximum number of arguments that a predicate can have. The leaf nodes are filled with a wildcard character, i.e. $*$. They may alternatively accommodate additional information. The slot nodes are used in such a way that the adopted tree kernel function can generate fragments containing one or more children like for example those shown in frames (b), (c) and (d). As previously pointed out, if the arguments were directly attached to the root node, the kernel function would only generate the structure with all children (or the structure with no children, i.e. empty).

Second, we observe that the above approach generates many matches with slots filled with the null label. To solve this problem, we have set a new step 0:

0. if n_1 (or n_2) is a pre-terminal node and its child label is *null*, $\Delta(n_1, n_2) = 0$;

and by subtracting one unit to $\Delta(n_1, n_2)$, in step 3:

3. $\Delta(n_1, n_2) = \prod_{j=1}^{nc(n_1)}(1 + \Delta(c_{n_1}^j, c_{n_2}^j)) - 1$,

The new Δ in Eq. 1 defines a new kernel[4] that we call Shallow Semantic Tree Kernel (SSTK).

4 Experiments

The purpose of our experiments is to study the impact of the new structure representations introduced earlier for QA tasks. In particular, we focus on question classification and answer reranking for Web-based QA systems.

[4] By induction, it can be proven that SSTK applied to PAS+ generates the space of all possible k-ary relations derivable from a set of k arguments.

In the question classification (QC) task, we extend previous studies, e.g. [2,7], by testing a set of previously designed kernels and their combination with our new Shallow Semantic Kernel. SVMs are the learning machines used to build the multi-class classifiers based on the SSK, the kernel combinations being the sum of the individual models. This operation always produces a valid kernel [13].

In the answer reranking task, we approach the problem of detecting description answers (among the most complex in the literature [14,15]). We learn binary answer classifiers based on question-answer pairs constructed by querying our Web QA system, YourQA [16], with the same questions as the test set used in the QC experiment. Our experiments with different kernel combinations on question-answer pairs allow us to select the best performing classifier, which in turn is used to re-rank answers. The resulting ranking is compared with the ranking provided by Google and by YourQA.

4.1 Question Classification

As a first experiment, we focus on question classification (QC), because of its great impact on the quality of a QA system and because it is a widely approached task for which benchmarks and baseline results are available [2,3].

QC is defined as a multi-classification problem which consists in assigning an instance I to one of n classes, which generally belong to two types: factoid, seeking short fact-based answers (e.g. name, date) or non-factoid, seeking e.g. descriptions or definitions (see e.g. the taxonomy in [3]). We design a question multi-classifier by using n binary SVMs combined according to the ONE-vs-ALL scheme, where the final output class is the one associated with the most probable prediction. Question representation is based on the following features/structures: parse tree (PT), bag-of-words (BOW), bag-of-POS tags (POS) and predicate argument structure (PAS). We implemented the proposed kernels in the SVM-light-TK software available at `ai-nlp.info.uniroma2.it/moschitti/` which encodes the tree kernel functions in SVM-light [17][5]. The PAS structures were automatically derived by our SRL system [12].

As benchmark data, we use the question training and test set available at: `l2r.cs.uiuc.edu/~cogcomp/Data/QA/QC/`, where the test set are the TREC 2001 test questions[6] [18]. We refer to this split as UIUC.

The performance of the multi-classifier and the individual binary classifiers is measured with accuracy resp. F1-measure. To collect more statistically significant information, we run 10-fold cross validation on the 6,000 questions.

Question Classification Results. Table 1.(a) shows the accuracy of different question representations on the UIUC split (Column 1) and the average accuracy

[5] We adopted the default regularization parameter (i.e., the average of $1/||\boldsymbol{x}||$) and tried a few cost-factor values to adjust the rate between Precision and Recall on the development set.

[6] The benchmark is manually partitioned according to the coarse-grained question taxonomy defined in [3] – i.e. ABBR, DESC, NUM, HUM, ENTY and LOC – and contains 5,500 training and 500 test instances.

± standard deviation on the cross validation splits (Column 2) whereas Table 1.(b) reports the F1 for the individual classes using the best model, PTBOW. The analysis of the above data suggests that:

Firstly, the STK on PT and the linear kernel on BOW produce a very high result, i.e. about 90.5%. This is higher than the best outcome derived in [2], i.e. 90%, obtained with a kernel combining BOW and PT. When our BOW is combined with STK, it achieves an even higher result, i.e. 91.8%, very close to the 92.5% accuracy reached in [3] by using complex semantic information derived manually from external resources. Our higher results with respect to [2] are explained by a highly performing BOW, the use of parameterization and most importantly the fact that our model is obtained by summing two separate kernel spaces (with separate normalization), as mixing BOW with tree kernels does not allow SVMs to exploit all its representational power.

Secondly, model PTBOW shows that syntactic information can be beneficial in tasks where text classification is vital, such as QA. Here, syntax can give a remarkable contribution in determining the class of a question; moreover, the lexical information (BOW) has a limited impact due to the little number of words forming a question.

Thirdly, the PAS feature does not provide improvement. This is mainly due to the fact that at least half of the training and test questions only contained the predicate "to be", for which a PAS cannot be derived by our PB-based shallow semantic parser. Also, PT probably covers most of the question's semantic information encoded by PAS.

Next, the 10-fold cross-validation experiments confirm the trends observed in the UIUC split. The best model is PTBOW which achieves an average accuracy of 86.1%. This value is lower than the one recorded for the UIUC split. The explanation is that the test set in UIUC is not consistent with the training set (it contains the TREC 2001 questions) and it includes a larger percentage of easily classified question types, e.g. the numeric (22.6%) and description classes (27.6%) while their percentage in training is 16.4% and 16.2%, respectively. This shows the importance of cross-validation results that, given the very low values of the standard deviation, also suggest that the superior accuracy of the PTBOW over the BOW model is statistically significant.

Finally, for individual binary classification, the most accurate is the one carried out for NUM, which generally exhibits easily identified cues such as "how much/many". The more generic ENTY type proves hardest in both the UIUC and cross-validation experiments, while LOC and HUM remain well-classified in both cases also thanks to their regular patterns ("where" and "who" identifiers).

4.2 Answer Classification and Reranking

Question Classification does not allow to fully exploit the predicate argument potential since questions tend to be short and with no predicates. A different scenario is answer classification, i.e. deciding if a passage/sentence correctly answers the question: here, the semantics that the classifier has to generate are

Table 1. Accuracy of the question classifier with different feature combinations and performance of the best classifier by question class

(a)

Features	Acc (UIUC)	Acc (xval.)
PT	90.4	84.8±1.4
BOW	90.6	84.7±1.4
PAS	34.2	43.0±2.2
POS	26.4	32.4±2.5
PTBOW	**91.8**	**86.1±1.3**
PTBOWPOS	91.8	84.7±1.7
PASBOW	90.0	82.1±1.5
PASBOWPOS	88.8	81.0±1.7

(b)

Q. class	P (UIUC)	R (UIUC)	F1 (UIUC)	F1 (xval.)
ABBR	87.5	77.8	82.4	78.5± 7.0
DESC	95.8	99.3	97.5	84.6±2.3
ENTY	73.6	83.0	78.0	75.7±1.3
HUM	89.6	92.3	90.9	86.8±2.0
LOC	86.6	85.2	85.7	88.9±1.5
NUM	99.0	86.7	92.5	94.2±1.4
Multi-Class. Accuracy			91.8	86.1±1.3

not constrained to a small taxonomy and the length of an answer may make the representation based on PT too sparse.

We learn answer classification with a binary SVM which determines if a answer is correct for the target question: consequently, the classification instances are the ⟨question, answer⟩ pairs. Each pair component can be encoded with PT, BOW, POS and PAS representations and processed with the previous kernels.

The output of the binary classifier can be used to rerank the list of candidate answers of a QA system. Starting from the top answer, each instance is classified based on its correctness with respect to the question. If it is classified as correct its rank is unchanged; otherwise it is pushed down, until a lower ranked incorrect answer is found.

As output of the basic QA we use Google rank along with the YourQA [16] system. YourQA uses the Web documents corresponding to the top 20 Google results for the question. Then, each sentence in each document is compared to the question to compute the Jaccard similarity, which, in the answer extraction phase, is used to select the most relevant sentence. A passage of up to 750 bytes is then created around the sentence and returned as an answer.

As test data, we collected the 138 TREC 2001 test questions labeled as "description" and for each, we obtained a list of answer paragraphs extracted from Web documents using YourQA. Each paragraph sentence was manually evaluated according to whether it contained an answer to the corresponding question; moreover, to simplify the classification problem, we isolated for each paragraph the sentence which obtained the maximal judgment (in case more than one sentence in the paragraph had the same judgment, we chose the first one). We collected a corpus containing 1123 sentences, 401 of which – labeled as "+1" – answered the question either concisely or with noise; the rest – labeled as "-1"– were either irrelevant to the question or contained hints relating to the question but could not be judged as valid answers[7].

[7] For instance, given the question "What are invertebrates?", the sentence "At least 99% of all animal species are invertebrates, comprising ..." was labeled "-1" , while "Invertebrates are animals without backbones." was labeled "+1".

Answer Classification and Reranking Results. To gather statistically significant data, we ran 5-fold cross-validation, with the constraint that two pairs $\langle q, a_1 \rangle$ and $\langle q, a_2 \rangle$ associated with the same question q could not be split between training and testing. The answer classification experiment results are in Tab. 2.

We note that: first, the contribution of the POS feature in answer classification is much higher than in question classification and even outperforms the PT feature (see Table (a)). This is because on the one side we work with Web data, a noisy input wich can drastically reduce parser performance; on the other, POS tagging is a more robust operation and yields less errors. Moreover, while question classification is a multi-classification task where the POS feature must be used to determine a semantic category, definition answer classification is a binary classification task – hence statistically simpler.

Second, although the accuracy of PAS as a standalone was inferior to that of PT, when coupled with BOW it yielded higher accuracy[8]; in this case, its ability to generalize the answer information allowed to overcome the erroneous/noisy information provided by the PT on Web data.

Table 2. Answer classifier with different feature combinations, baseline classifiers accuracy and MRR of the best reranker compared to the baseline

(a)

Features	P	R	F1
PT	56.4	70.0	59.6±4.0
BOW	58.5	85.9	69.3±6.6
POS	52.4	84.1	64.0±5.9
PAS	52.4	71.1	58.6±5.6
PTBOW	59.8	79.7	68.1±8.0
PASBOW	**64.1**	**79.2**	**70.7±5.9**
PTBOWPOS	63.8	71.7	67.4±7.6
PASBOWPOS	64.4	75.2	69.2± 6.5

(b)

Baseline	P	R	F1
Gg@1	39.7	9.4	15.2±3.1
QA@1	45.3	10.9	17.6±2.9
Gg@all	35.8	100	52.7±6.2
QA@all	35.8	100	52.7±6.2

	Gg	QA	Reranker
MRR	54.8±6.7	60.1±4.1	79.2±0.9

Third, we compared the answer classifier with two baselines built using the YourQA and Google rankings. For this, we considered the top N ranked results as correct definitions and the remaining ones as incorrect for different values of N. Table 2.(b) shows the results for $N = 1$ and the maximum N (*all*), i.e. all the available answers. Each measure is the average of the Precision, Recall and F1 of the three systems on the cross-validation splits. The F1 of Google (Gg) and YourQA (QA) are greatly outperformed by the classifier, even when all answers are considered ($N = all$) and the low standard deviations ensure the statistical relevance of the results.

[8] Although the standard deviation in this case is high, as the complexity can vary across splits, since the PAS and PASBOW models are similar, the standard deviation of their difference is lower, i.e. 2.03. When we performed the t-test on such value, we confirmed that PASBOW is superior to BOW with a 90% level of confidence.

Finally, we implemented the simple re-ranking algorithm described previously and assessed its performance with the MRR[9] metric adopted in TREC 2001[10].

YourQA's MRR outperforms the Google MRR (Tab. 2.(b), last row) as Google ranks are based on whole documents, not on single passages, so documents where no passage contains all of the question's keywords may be ranked higher than others containing them all. When the answer classifier is applied to improve the QA ranking, MRR reaches .792, rising by nearly 20 points.

Related Work on Definitional QA. Unfortunately, no results are known to the authors concerning a Web-based answer classifier for the same question set and few are available on the performance computed over description questions alone on the NIST corpus; for instance, NTT's system achieved an MRR of .247 on description questions using a heuristic searching for appositions [15].

Interesting related work on definition answer reranking [20] was conducted by comparing the use of an SVM classifier predictions to induce a ranking and of the Ranking SVM algorithm [17]. In [21], ranks were computed based on the probabilities of biterm language models generating candidate answers.

5 Conclusion

In this paper, we introduce new structures to represent textual information in three question answering tasks: question classification, answer classification and answer reranking. We define a new tree structure called PAS to represent predicate-argument relations, which we automatically extract using our SRL system. We also introduce a new kernel function to exploit its representative power.

Our experiments with SVMs and such new functions suggest that syntactic information helps specific tasks such as question classification. On the other hand, the coarse-grained semantic information contained in PAS gives promising results in answer classification, which suffers more from data sparseness. Moreover, our simple answer reranker, based on the answer classifier output, obtains a 20% more accurate ranking than our baseline QA system.

In the future, we will study the utility of PASs for other tasks affected by noisy data and apply a true SVM reranker trained with the proposed information.

References

1. Kwok, C.C.T., Etzioni, O., Weld, D.S.: Scaling question answering to the web. In: WWW. (2001)
2. Zhang, D., Lee, W.S.: Question classification using support vector machines. In: Proceedings of SIGIR, ACM Press (2003)

---

9 The Mean Reciprocal Rank is defined as: $MRR = \frac{1}{n}\sum_{i=1}^{n}\frac{1}{rank_i}$, where n is the number of questions and $rank_i$ is the rank of the first correct answer to question i.

10 Although since the TREC 2003 definition track [19] answers were expected in the form of bags of information "nuggets", we still believe the MRR meaningful for QA.

3. Li, X., Roth, D.: Learning question classifiers: The role of semantic information. Journal of Natural Language Engineering (2005)
4. Collins-Thompson, K., Callan, J., Terra, E., Clarke, C.L.: The effect of document retrieval quality on factoid question answering performance. In: Proceedings of SIGIR, New York, NY, USA, ACM Press (2004)
5. Pasca, M.: Open-Domain Question Answering from Large Text Collections. CSLI Studies in Computational Linguistics (2003)
6. Collins, M., Duffy, N.: New ranking algorithms for parsing and tagging: Kernels over discrete structures, and the voted perceptron. In: ACL. (2002)
7. Moschitti, A.: Efficient convolution kernels for dependency and constituent syntactic trees. In: Proceedings of ECML. (2006)
8. Zelenko, D., Aone, C., Richardella, A.: Kernel methods for relation extraction. JMLR (2003)
9. Kingsbury, P., Palmer, M.: From treebank to propbank. In: Proceedings of LREC. (2002)
10. Allan, J., et al.: Challenges in information retrieval and language modeling. In: Workshop at University of Amherst. (2002)
11. Charniak, E.: A maximum-entropy-inspired parser. In: Proceedings of NAACL. (2000)
12. Moschitti, A., Coppola, B., Giuglea, A.M., Basili, R.: Hierarchical semantic role labeling. In: Proceedings of the CoNLL 2005 shared task. (2005)
13. Shawe-Taylor, J., Cristianini, N.: Kernel Methods for Pattern Analysis. Cambridge University Press (2004)
14. Cui, H., Kan, M.Y., Chua, T.S.: Generic soft pattern models for definitonal question answering. In: Proceedings of SIGIR. (2005)
15. Kazawa, H., Isozaki, H., Maeda, E.: NTT question answering system in TREC 2001. In: Proceedings of TREC. (2001)
16. Quarteroni, S., Manandhar, S.: User modelling for adaptive question answering and Information Retrieval. In: Proceedings of FLAIRS. (2006)
17. Joachims, T.: Making large-scale SVM learning practical. In Schölkopf, B., Burges, C., Smola, A., eds.: Advances in Kernel Methods - Support Vector Learning. (1999)
18. Voorhees, E.M.: Overview of the TREC 2001 QA track. In: TREC. (2001)
19. Voorhees, E.M.: Overview of TREC 2003. In: TREC. (2003)
20. Xu, J., Cao, Y., Li, H., Zhao, M.: Ranking definitions with supervised learning methods. In: Special interest tracks and posters of WWW, New York, NY, USA, ACM Press (2005)
21. Chen, Y., Zhou, M., Wang, S.: Reranking answers from definitional question answering using language models. In: Proceedings of ACL. (2006)

Incorporating Diversity and Density in Active Learning for Relevance Feedback

Zuobing Xu, Ram Akella, and Yi Zhang

University of California, Santa Cruz, CA, USA, 95064

Abstract. Relevance feedback, which uses the terms in relevant documents to enrich the user's initial query, is an effective method for improving retrieval performance. An associated key research problem is the following: Which documents to present to the user so that the user's feedback on the documents can significantly impact relevance feedback performance. This paper views this as an active learning problem and proposes a new algorithm which can efficiently maximize the learning benefits of relevance feedback. This algorithm chooses a set of feedback documents based on relevancy, document diversity and document density. Experimental results show a statistically significant and appreciable improvement in the performance of our new approach over the existing active feedback methods.

1 Introduction

Information retrieval has traditionally been based on retrieving documents which match user's query in content. It is well known that the original query formulation does not always reflect the user's intent. In other words, merely matching words (or "terms") in the original query and the document may not be an effective approach, as the word overlap alone may not capture the semantic intent of a query. In particular, without detailed knowledge of the collection make-up, and of the retrieval environment, most users find it difficult to formulate information queries that are well designed for retrieval purposes. This suggests that the first retrieval operation can be conducted with a tentative initial query, which retrieves a few useful documents for user to evaluate their relevance. Based on the relevance evaluation and the initial query, we construct a new improved query to retrieve more relevant documents in subsequent operations.

The above retrieval process is well known as relevance feedback process [1,2]. There are two major problems while using relevance feedback framework. First, how to select first set of documents to be presented to the user for feedback. Second, how to effectively utilize the relevant feedback information to reformulate the query. Much of the previous research on relevance feedback focuses on the second problem of feedback query updating for a given set of feedback documents by choosing important topic related terms from the relevant documents and expanding the original query based on the chosen terms.

However, how to choose a good set of documents is not well studied in the information retrieval community, although an effective approach has much potential to further enhance retrieval performance. Most of the earlier relevance

G. Amati, C. Carpineto, and G. Romano (Eds.): ECIR 2007, LNCS 4425, pp. 246–257, 2007.

feedback systems usually ignore the first problem and choose top ranked documents for feedback. This ignores many important factors that affect the learning results. Recently, Shen and Zhai [3] presented this problem as an active feedback framework and derived several practical algorithms based on the diversity of the feedback documents. Their algorithms take into account of the document diversity by clustering retrieved documents or choosing documents with a certain ranking gap. In our paper, we proposed a new active feedback approach which comprehensively considers relevance, diversity and density of the feedback documents. We call this new active feedback algorithm **Active-RDD** (denoting Active Learning to achieve **Relevance,Diversity** and **Density**).

Active feedback is essentially an application of active learning in ad hoc information retrieval. Active learning has been extensively studied in supervised learning and other related context. Cohn et al. [4] proposed one of the first statistical analysis of active learning, demonstrating how to construct queries that maximize the error reduction by minimizing learners' variance. They developed their method for two simple regression problems in which this question can be answered in closed form. Both the Query by Committee (QBC) algorithm [5] and Tong's version space method [6] are based on choosing a sample which is close to classification boundary. Both of their methods have been applied to text classification problems. To avoid choosing outliers, McCallum and Nigam [7] modify the QBC method to use the unlabeled pool for explicitly estimating document density. Batch mode active learning, which selects a batch of unlabeled examples simultaneously, is an efficient way to accelerate the learning speed. In [8], Brinker presented a new approach that is especially designed to construct batches by incorporating a diversity measure. Besides the above application area, supervised learning, active learning has also been recently applied to adaptive information filtering [9].

One major drawback of the above methods is their computational complexity, which prevents us from using them directly in the information retrieval task. This paper explores how to overcome this problem by designing an efficient active learning algorithm (Active-RDD) for relevance feedback. Because most of the well motivated active learning approaches choose data samples by implicitly or explicitly considering the uncertainty, density or diversity of data samples, we designed the new algorithm to explicitly capture these important factors by integrating document relevancy, document density measure and document diversity measure. We apply the proposed algorithm to the language modeling retrieval framework and evaluate the effectiveness of the proposed technique on two benchmark data sets. The experimental results demonstrate the statistical validated performance improvement of our algorithm over existing algorithms.

The remainder of this paper is organized as following. In section 2, we first analyze the important elements that influence retrieval performance and derive an efficient active learning algorithm for document selection based on these elements. In section 3, we discuss the experimental setting and the experimental results. In Section 4, we conclude with a description of our current research, and present several future research directions for further work.

2 Active Learning Algorithm

2.1 Algorithm Intuition

The goal of active relevance feedback is to improve retrieval performance by actively selecting feedback documents for user evaluation. Here we will first illustrate the intuition underlying our new approach.

Relevant documents directly reflects a user's search interest, and the current relevance feedback algorithms based on language modeling only rely on the information contained in relevant feedback documents. So choosing relevant documents for evaluation will effectively direct the second round search results to the user's intent. Initially, when a query is input into a retrieval engine, we do not know the true relevance of documents until we get feedback from the user. The only criteria to judge the relevance of a document during an initial pass is the relevance score given by retrieval engine. The relevance score of a document is calculated based on the similarity between the initial query and the document. Considering the above two facts, we will choose documents with high relevance scores. The traditional relevance feedback method Top K selects the top k ranked documents for feedback. Although the Top K algorithm is in line with our hypothesis, which is that relevant documents are good for learning, it is not the best strategy from a learning perspective. For instance, if there are two identical documents among the top ranked documents, the improvement of second round retrieval performance achieved by choosing both documents is equivalent to the improvement achieved by choosing any one of them. In the next section, we will analyze another important factor on choosing feedback documents to avoid this redundancy problem in the previous example.

The Top K approach does not take into account of the redundancy between selected feedback documents: this redundancy results from very similar (and near duplicated) documents. Thus, in our active learning approach, we need to capture diversity of feedback document set in the algorithm. The Gapped Top K algorithm [3] increases the diversity of feedback documents by selecting the top K documents with a ranking gap G in between any two documents. Another heuristic method to increase diversity is the Cluster Centroid algorithm [3], which groups retrieved documents into K clusters and chooses one representative document from each cluster. Our Active-RDD algorithm, which is different from the above two methods, maximizes the diversity of feedback document set by explicitly maximizing the distance between new document and selected documents.

If the selection criterion only takes into account the relevance score and diversity of the batch document set, it loses the benefit of the implicit modeling of the data distribution. For instance, such selection criteria may select documents that lie in unimportant, sparsely populated regions. Labeling documents in high density regions or in low density regions gives the query feedback algorithm different amounts of information. To avoid choosing outliers, we aim to select documents in high density regions. Choosing relevant documents in high probability density regions will retrieve more relevant documents in the subsequent round, which leads to a better retrieval performance.

Finally, in order to combine the above three factors, we build a linear combination of all the measures and proceed in the following way to construct a new feedback document set. To reduce the computation, we select K feedback document from the top L ranked documents. For instance, the reasonable sizes of L and K could be 100 and 6 respectively. Let I denote the set of unlabeled documents that have not yet been selected for evaluation, we incrementally construct a new feedback document set S. The selection scheme can be described as follows:

1 : $S = 0$
2 : **repeat**
3 :

$$d_i = \arg\max_{d_i \in I \notin S} [(\alpha)\text{relevance}(d_i) + (\beta)\text{density}(d_i) + (1-\alpha-\beta)\text{diversity}(d_i, S)] \quad (1)$$

4 : $S = S \cup d_i$
5 : **Until** $size(S) = K$

where relevance(d_i) is the relevance score of document d_i, density(d_i) is the density performance measure around document d_i, and distance(d_i, S) is the distance between document d_i and the existing feedback document set S . $\alpha \in [0, 1]$, $\beta \in [0, 1]$ are weighting parameters. Setting $\alpha = 1$ restores the Top K approach; if $\beta = 1$, the algorithm selects feedback document only based on its density performance measure; whereas if $\alpha = 0$ and $\beta = 0$, the algorithm focuses exclusively on maximizing the diversity of selected document set. In the following sections, we will explain how we calculate the above three factors in detail.

2.2 Relevance Measure

Language modeling approaches to information retrieval have received recognition for being both theoretically well founded, and showing excellent retrieval performance and effective implementation in practice. In this paper, we apply language modeling approach using KL divergence measure for our basic retrieval model. Suppose that a query q is generated by a generative model $p(q|\theta_Q)$ with θ_Q denoting the parameters of the query unigram language model. Similarly, we assume that a document d is generated by a generative model $p(d|\theta_D)$ with θ_D denoting the parameters of the document unigram language model. The query unigram language model and document unigram language model are smoothed multinomial models in language modeling. If $\widehat{\theta}_Q$ and $\widehat{\theta}_D$ are the estimated query language model and document language model respectively, then the relevance score of document d with respect to query q can be calculated by negative KL-divergence[10]. KL-divergence is calculated by the formula below:

$$KL(\widehat{\theta}_Q \| \widehat{\theta}_D) = \sum_w p(w|\widehat{\theta}_Q) \log \frac{p(w|\widehat{\theta}_Q)}{p(w|\widehat{\theta}_D)} \quad (2)$$

Where $p(w|\widehat{\theta}_Q)$ is the probability of generating word w by query language model $\widehat{\theta}_Q$; $p(w|\widehat{\theta}_D)$ is the probability of generating word w by document language model $\widehat{\theta}_D$.

The retrieval engine ranks all the documents according to their negative KL-divergence scores. In the Active-RDD algorithm, we use the negative KL-divergence measure, which is given by first round search, as relevance score.

2.3 Document Density Measure

Document density is one of the important factors in the defined active selection scheme. Owing to the large scale of the document collection, estimating document probability density in the whole collection is computationally unachievable. To reduce the computation, we only measure the density performance of the top L documents in the initial retrieval results.

We approximate the density in a region around a particular document by measuring the average distance from that document to all the other documents. Distance between individual documents is measured by J-Divergence[11]. KL divergence is a non symmetric measure between two probability mass functions, while J-Divergence obtains the symmetry by adding two KL divergences as described in (2). The formula of J-Divergence is as follows:

$$J(d_i||d_j) = KL(d_i||d_j) + KL(d_j||d_i) \tag{3}$$

Consequently, the average J divergence between a document d_i and all other documents measures the degree of overlap between d_i and all other documents. In other words, large average J divergence indicates that the document is in low document density region. Thus we use negative average J divergence (4) to approximate document density performance measure, which reflects the closeness of this document to the other documents. The reason we use this measure is to normalize the value of density performance measure to be on the same scale of the relevance score.

$$\text{density}(d_i) = \frac{-1}{|D|} \sum_{d_h \in D} J(d_i||d_h) \tag{4}$$

2.4 Diversity Measure

The metric we use to measure the distance between a document and a document set is the minimum distance between the document and any document in the set. This method corresponds to the single linkage method in hierarchical clustering literature. The single linkage method has the advantage of efficient time complexity, and it also ensures that the new document is different from all the selected documents.

To normalize all components in the overall metric to be of comparable values, we use J divergence to measure the distance between candidate document and selected documents. To maximize the combined score of relevance score, density performance measure and diversity measure, which is shown in (1), we employ the following incremental strategy: Given a set of unlabeled documents, we start with document d_1 which has the highest combined score of relevance score and density performance measure; then we add a new document d_2 to our set $S = d_1 \cup d_2$, which maximize the combined score of relevance score, density performance measure and diversity measure. We continue by adding new documents until the size of the selected documents reaches the predefined size.

The individual influence of each factor can be adjusted by the weighting parameters α and β. The combined strategy can be implemented very efficiently. Recalculating the distance between an unselected document and every single document already added in the feedback document set to evaluate the maximum distance between the unselected document and the document set results in quadratic computational time depending on the feedback document size. We cache the maximum distance of all the unselected documents from selected document set and update the score only if the distance between the newly added document and the unselected document is larger than the stored maximum. We only need to compute distance once for every unselected document instead of already selected documents number. If we are choosing K documents from top L retrieved documents, the computation complexity in this part is reduced from $O(K^2L)$ to $O(KL)$. The complete pseudo code of an efficient implementation of the algorithm is given in Table 1.

The Maximal Marginal Relevance ranking algorithm [12] (MMR) is a greedy algorithm for ranking documents based on relevance ranking score and at the same time avoiding redundancy. Our Active-RDD algorithm extends the MMR algorithm by adding an extra term, which reflects the document density. In [3], Shen and Zhai proposed the MMR algorithm to solve the active feedback problem, but they have not implemented that algorithm.

2.5 Query Updating Algorithm

Based on user's relevance judgment on feedback document, we use the divergence minimization model [13] to update query. The divergence minimization model minimizes the divergence between the query model and the relevant feedback documents. Let $R = d_1, \ldots, d_n$ be the set of relevant feedback documents. We define the empirical KL-divergence between the feedback query model θ_F and the relevant feedback documents $R = d_1, \ldots, d_n$ as the average divergence between the query model and relevant feedback document model.

$$D_e(\theta_F, R) = \frac{1}{|R|} \sum_{i=1}^{n} D(\theta_F \| \theta_i) \tag{5}$$

We subtract the negative divergence between the query language model and collection model to remove the background information. Considering all the above conditions, we derive the following empirical divergence function of a feedback query model:

$$\theta_F = \arg\min_{\theta_F} \frac{1}{|R|} \left\{ \sum_{i=1}^{n} D(\theta_F \| \theta_i) - \lambda D(\theta_F \| p(.|C)) \right\} \tag{6}$$

Table 1. Active-RDD Algorithm

input:

α	(relevance coefficient)
β	(density coefficient)
K	(size of feedback document set for evaluation)
L	(size of document set from which we choose K documents)
$D = (d_0, \dots d_{L-1})$	(permutation of $0, \dots, L-1$)
$R = (r_0, \dots r_{L-1})$	(relevance score of each document)

output:

$D = (d_0, \cdots d_{L-1})$	(permutation of $0, \dots, L-1$)

relevance = $array[L]$
maxDis = $array[L]$
for $j = 0$ to $L - 1$ **do**
  relevance(j) = $R(j)$
  Calculate document density performance using (4)
  maxDis(j) = 0
end for
for $k = 0$ to $K - 1$ **do**
  maxIndex = k
  maxValue = 0
  for all $j = k$ to L **do**
    value= (α) relevance(j) + (β)density(j) + $(1 - \alpha - \beta)$maxDis(j)
    if value > maxValue **then**
      maxValue = value
      maxIndex = j
    end if
  end for
  swap (d_{maxIndex}, d_k)
  for all $j = k + 1$ to L **do**
    distance = $J(d_j || d_k)$
    if distance > maxDis(j) **then**
      maxDis(j) = distance
    end if
  end for
end for

Here $p(.|C)$ is the collection language model and $\lambda \in [0, 1)$ is the weighting parameter. Taking the first derivative of (6) with respective to $p(w|\theta_F)$, we will get the simple closed form solution.

$$p(w|\theta_F) \propto \exp\left(\frac{1}{1-\lambda}\frac{1}{|R|}\sum_{i=1}^{n} \log p(w|\theta_i) - \frac{\lambda}{1-\lambda}\log p(w|c)\right) \tag{7}$$

To exploit θ_F in our KL-divergence retrieval model, we interpolate it with the original query model θ_Q to obtain updated model θ'_Q ,

$$\theta'_Q = (1 - \mu)\theta_Q + \mu\theta_F \tag{8}$$

and then use the updated query θ'_Q to score document d_i by negative KL-divergence.

3 Experiment Methodology and Experimental Results

To evaluate our Active-RDD algorithm described in previous sections, we use two different TREC data sets. The first one is TREC HARD 2005 Track, which contains the full AQUAINT collection; the second one is TREC HARD 2003 Track, which use part of AQUAINT data plus two additional datasets (Congressional Record (CR) and Federal Register (FR)). We do not have the additional data set in TREC HARD 2003 Track. Our results are comparable to other published TREC HARD 2003 results, although the data is a little different. For both tracks, we use all the 50 topics which have relevance judgments. We use only the titles of the topic description, because they are closer to the actual queries used in real applications.

We employ the Lemur Toolkit[14] as our retrieval system and KL-Divergence language retrieval model as our baseline retrieval model. We compare the Active-RDD algorithm with the existing active feedback algorithms such as Top K, Gapped Top K and Cluster Centroid. For all the algorithms, we select $(K) = 6$ feedback documents from top $(L) = 100$ documents. All the parameters in the query updating model are fixed at the default values in The Lemur Toolkit[14].

To measure the performance of an active relevance feedback algorithm, we use two standard ad hoc retrieval measures: (1) Mean Average Precision (MAP), which is calculated as the average of the precision after each relevant document is retrieved, reflects the overall retrieval accuracy. (2) Precision at 10 documents (Pr@10): this measure does not average well and only gives us the precision for the first 10 documents. It reflects the utility perceived by a user who may only read up to top 10 documents.

In the following sections, we use cross-validation for Active-RDD algorithm and Gapped Top K algorithm, and then statistically compare the Active-RDD algorithm with existing algorithms.

3.1 Cross Validation

Coefficients α and β play an important role on selecting the feedback documents. How to select these coefficients significantly impacts the overall algorithm performance. In order to have a fair comparison, we pursue 5-fold cross-validation on the Active-RDD algorithm and Gapped Top K algorithm, and compare their cross-validation performance (CVP) with Cluster Centroid and Top K algorithm performance,(these algorithms are consequently parameter free in this setting).

We separate 50 queries into 5 parts, where each part contains 10 queries. For the kth set of queries, we train parameters to optimize the retrieval performance for the other 4 sets of queries, and use this set of the parameters to test on kth set of queries to obtain the retrieval performance measure for kth part.

We do this for $k = 1, 2, 3, 4, 5$ and the cross-validation performance is the average performance on the 5 test query sets. The cross-validation experimental results are shown in Table 2.

From Table 2, we conclude that the cross-validation performance of our Active-RDD algorithm is better than the Gapped Top K algorithm. Furthermore, we will compare these cross-validation performances with the Cluster Centroid algorithm and Top K algorithm.

Table 2. Cross-validation comparison of Active-RDD and Gapped Top K approaches. CVP indicates cross-validation performance, which is the average value of the MAP and Pr@10 on test data.

	Active-RDD				Gapped Top K			
HARD 2003	MAP Train	MAP Test	Pr@10 Train	Pr@10 Test	MAP Train	MAP Test	Pr@10 Train	Pr@10 Test
Folder 1	0.3855	0.3566	0.5925	0.6700	0.3676	0.3295	0.5450	0.6400
Folder 2	0.3954	0.3169	0.6325	0.5100	0.3792	0.2831	0.5950	0.4400
Folder 3	0.3966	0.3119	0.6225	0.5300	0.3747	0.3013	0.5925	0.4500
Folder 4	0.3793	0.3812	0.6275	0.5500	0.3594	0.3189	0.5750	0.5100
Folder 5	0.3416	0.5319	0.5650	0.7800	0.3175	0.5299	0.5275	0.7100
CVP		0.3797		0.6080		0.3525		0.55
HARD 2005	MAP Train	MAP Test	Pr@10 Train	Pr@10 Test	MAP Train	MAP Test	Pr@10 Train	Pr@10 Test
Folder 1	0.2675	0.2356	0.5575	0.5400	0.2496	0.2634	0.5450	0.6400
Folder 2	0.2583	0.2722	0.5550	0.5700	0.2309	0.2821	0.5525	0.6100
Folder 3	0.2489	0.3097	0.5325	0.6400	0.2508	0.2584	0.5600	0.5800
Folder 4	0.2673	0.2362	0.5700	0.4900	0.2594	0.2238	0.5875	0.4700
Folder 5	0.2634	0.2519	0.5600	0.5300	0.2569	0.2339	0.5750	0.5200
CVP		0.2611		0.5540		0.2523		0.5640

3.2 Comparison of Different Active Learning Algorithms

To evaluate the effectiveness of different document selecting approaches, we compare the performance of the non-feedback approach baseline with Top K, Gapped Top K , Cluster Centroid and our Active-RDD algorithm, all of which are feedback based algorithms. The performance of the Active-RDD and the Gapped Top K algorithm are the cross-validation performance in the previous section.

From Table 3, we can see that all these feedback algorithms perform better than the baseline non-feedback retrieval. All the results show that the underlying relevance feedback mechanism is very effective. From the results, our active learning algorithm Active-RDD outperforms Top K algorithm significantly, and it also performs better than other active feedback approaches at the statistical significance level 10% in most cases.

Table 3. Average performance of different active learning approaches. The best performance is shown is bold. We compare our Active-RDD algorithm with the Top K algorithm, the Gapped Top K algorithm and the Cluster Centroid algorithm, and percentage improvements over these three existing algorithms are shown in column 7,8,9 respectively. A double star(**) and a single star(*) indicate that the performance of our active learning algorithm is significantly better than the existing method used in the corresponding column (Top K, Gapped Top K or Cluster Centroid) according to Wilcoxon signed rank test at the level of 0.05 and 0.1 respectively.

Method		Baseline	Top K	Gap K	Cluster	RDD	Improv. over Top K	Improv. over Gap K	Improv. over Cluster
HARD 2003	MAP	0.3150	0.3508**	0.3525**	0.3771	**0.3797**	8.07%	7.72%	0.69%
	pr@10	0.5000	0.5380**	0.5500**	0.5760**	**0.6080**	13.01%	10.55%	5.56%
HARD 2005	MAP	0.1919	0.2367**	0.2523	0.2369*	**0.2611**	10.31%	3.49%	10.22%
	pr@10	0.4340	0.4800**	**0.5640**	0.5420**	0.5540	15.42%	−1.77%	2.21%

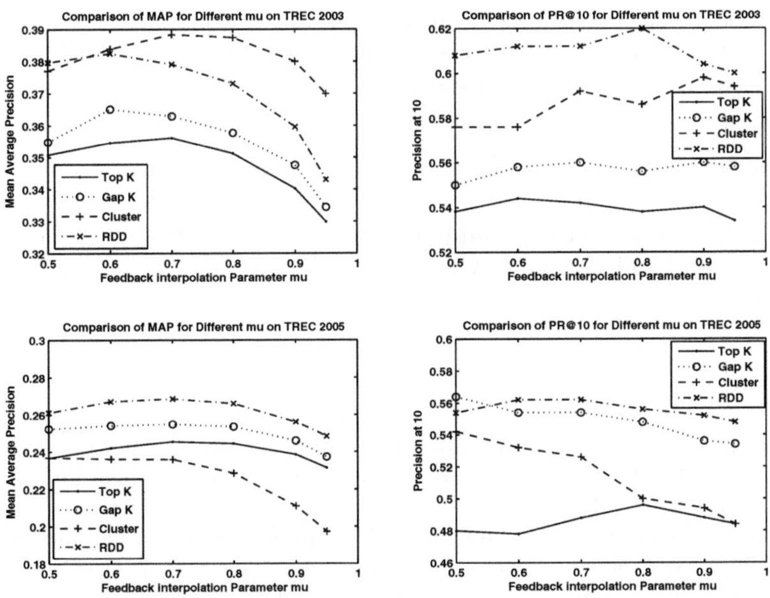

Fig. 1. Sensitivity of average performance of different active learning algorithm on μ

3.3 Performance Sensitivity of Feedback Interpolation Parameter μ

Owing to the nature of explicit feedback, the relevant feedback documents judged by the user are more reliable. This intuition leads to adding more weight to the feedback interpolation parameter μ in (8). In the previous experiments, we set $\mu = 0.5$ as the Lemur Toolkit[14] default setting. We did another set of experiments by increasing μ, and the results are shown in Fig. 1. The results indicate that setting $\mu = 0.7$ gives the Active-RDD algorithm best performance (with performance improvement of $1-2\%$). The curves are fairly flat and indicate relative insensitivity around the optimal value of feedback parameters, which is a desirable pattern.

4 Conclusions

This paper explores the problem of how to select a good set of documents to ask user for relevance feedback. This paper presents a new efficient active learning algorithm, which dynamically selects a set of documents for relevance feedback based on the documents' *relevancy*, *density* and *diversity*. We evaluate the algorithm on TREC2005 HARD dataset and TREC2003 HARD dataset. The experimental results show that our algorithm significantly outperforms the existing Top K, Gapped Top K and Cluster Centroid algorithms.

There are several interesting research directions that may further improve relevance feedback under the active learning framework: first, making full use of users' feedback by learning from non-relevant documents; second, learning different active learning parameters for different queries; and third, combining implicit feedback with active learning.

Acknowledgments

We would like to acknowledge support by Cisco, University of California's MICRO Program, CITRIS, and UARC. We also appreciate discussions with associated colleagues.

References

1. Harman, D.: Relevance feedback revisited. In: Proceedings of the 15th Annual International ACM SIGIR Conference on Research and Development in Information Retrieval. (1992) 1–10
2. Salton, G., Buckley, C.: Improving retrieval performance by relevance feedback. Journal of the American Society for Information Science **41**(4) (1990) 133–168
3. Shen, X., Zhai, C.: Active feedback in ad hoc information retrieval. In: Proceedings of the 28th annual international ACM SIGIR conference on Research and development in information retrieval. (2005) 55–66
4. Cohn, D.A., Ghahramani, Z., Jordan, M.I.: Active learning with statistical models. In: Advances in Neural Information Processing Systems. Volume 7., The MIT Press (1995) 705–712
5. Freund, Y., Seung, H.S., Shamir, E., Tishby, N.: Selective sampling using the query by committee algorithm. Machine Learning **28**(2-3) (1997) 133–168
6. Tong, S., Koller, D.: Support vector machine active learning with applications to text classification. In: Proceedings of 17th International Conference on Machine Learning. (2000) 999–1006

7. McCallum, A., Nigam, K.: Employing EM and pool-based active learning for text classification. In: Proceedings of the Fifteenth International Conference on Machine Learning. (1998) 350 – 358
8. Brinker, K.: Incorporating diversity in active learning with support vector machines. In: Proceedings of the Twentieth International Conference on Machine Learning . (2003) 59–66
9. Zhang, Y., Xu, W., Callan, J.: Exploration and exploitation in adaptive filtering based on bayesian active learning. In: Proceedings of 20th International Conf. on Machine Learning. (2003) 896–903
10. Lafferty, J., Zhai, C.: Document language models, query models, and risk minimization for information retrieval. In: Research and Development in Information Retrieval. (2001) 111–119
11. Lin, J.: Divergence measures based on the shannon entropy. IEEE Transactions on Information Theory (1) (1991) 145–151
12. Carbonell, J.G., Goldstein, J.: The use of MMR, diversity-based reranking for reordering documents and producing summaries. In: Proceedings of the 21st Annual International ACM SIGIR Conference on Research and Development in Information Retrieval. (1998) 335–336
13. Zhai, C., Lafferty, J.: Model-based feedback in the language modeling approach to information retrieval. In: Proceedings of the Tenth ACM International Conference on Information and Knowledge Management. (2001) 403–410
14. (The lemur toolkit) `http://www.lemurproject.org`.

Relevance Feedback Using Weight Propagation Compared with Information-Theoretic Query Expansion

Fadi Yamout, Michael Oakes, and John Tait

School of Computing and Technology,
University of Sunderland, U.K.
{Fadi.Yamout,Michael.Oakes,John.Tait}@sunderland.ac.uk

Abstract. A new Relevance Feedback (RF) technique called Weight Propagation has been developed which provides greater retrieval effectiveness and computational efficiency than previously described techniques. Documents judged relevant by the user propagate positive weights to documents close by in vector similarity space, while documents judged not relevant propagate negative weights to such neighbouring documents. Retrieval effectiveness is improved since the documents are treated as independent vectors rather than being merged into a single vector as is the case with traditional vector model RF techniques, or by determining the documents relevancy based in part on the lengths of all the documents as with traditional probabilistic RF techniques. Improving the computational efficiency of Relevance Feedback by considering only documents in a given neighbourhood means that the Weight Propagation technique can be used with large collections.

Keywords: Relevance Feedback, Rocchio, Ide, Deviation From Randomness.

1 Introduction

In a typical Information Retrieval (IR) system, the terms in a user's query are matched with the terms in each of the documents using a similarity metric, and an initial ranked list of the best matching documents is presented to the user [1]. With Relevance Feedback (RF), the user is able to mark which of the retrieved documents were relevant, and which were not relevant [2]. In some systems, the users are allowed to submit a real-valued relevance score for each document [3], but this paper will consider only binary (relevant / not relevant) feedback.

Using positive RF alone, index terms are taken from documents judged relevant, and added to the initial query. The reformulated query can be resubmitted to the system, and the search process is repeated. Using negative as well as positive RF, index terms found in documents judged not relevant are removed from the query. It has been found that the evidence of negative judgements does not improve the retrieval performance of standard RF techniques [4].

In this paper we describe a new RF technique called Weight Propagation (WP), which is both computationally more efficient and produces better retrieval results. The structure of the remainder of this paper is as follows: in Section 2 we review the RF techniques, and in Section 3 we describe the WP technique in general. Specific variants of WP are described in Section 4, namely WPI and WPR (inspired by Ide and

G. Amati, C. Carpineto, and G. Romano (Eds.): ECIR 2007, LNCS 4425, pp. 258 – 270, 2007.

Rocchio respectively), and WPY (the main focus of this paper, where we consider only the maximum weight propagated to each document). In our experiments, WPY outperformed both WPI and WPR, so WPY alone will be discussed in the rest of this paper. In Section 5 we describe the results of experiments with the TREC WT18G test collection to demonstrate the retrieval effectiveness of the WPY technique.

2 Relevance Feedback Techniques

The earliest work on RF by Rocchio [5] was based on the vector space model. In the vector space model, weights are assigned to the terms to reflect their importance within documents. Salton and McGill [1] showed that the best results were obtained by multiplying the term frequency by the inverse document frequency. Ide [6] produced a modified version. Rocchio and Ide merge relevant and nonrelevant documents into a single query vector rather than assessing each one separately. This might affect precision. In addition, Rocchio inspects all the documents.

Rochio and Ide's techniques compute the mathematical average of the relevant documents term weights and store the results in a single vector, and also compute the mathematical average of the non-relevant documents and store them in a different vector. The 2 vectors and the original query vector are merged into single vectors, which are consequently merged into a single query vector. Clearly, this will probably require more run-time to answer the query than the proposed approach especially for large document collection since the baseline must check all the documents in the collection, whereas WPY needs to check only few documents. We will address both of these problems in the following sections.

In this paper, RF was also tested using other models. Robertson and Sparck-Jones [7] proposed the probabilistic model. They adjusted the individual term weight based on the distribution of these terms in relevant and non-relevant documents retrieved in response to queries; for more explanation of the mathematics involved, see [8]. Documents and queries in the probabilistic model are also viewed as vectors with a weight given by a probability.

Probabilistic models have been extended in different models; Okapi [9], Statistical Language Modeling (SLM) [10], and Deviation from Randomness (DFR) [11][12][13]. The term weighting in the DFR, DFR_{weight}, is derived by measuring the divergence of the actual term distribution from that obtained under a random process. We have chosen Divergence approximation of the binomial (D) normalized with Ratio B of two binomial distributions (B) and the documents length are normalized using normalization 2 (2) [13]. This framework produces different non parametric models forming alternative baselines to the standard tf*idf model.

$$DFR_{weight} = NORM * keyFrequency * (TF*\log (f/p) + 0.5 * \log (2\pi*TF*(termFrequency - TF)/ termFrequency)) \quad (1)$$

where

- $NORM = (termFrequency + 1) / (documentFrequency * (TF + 1))$
- *termFrequency* is the term frequency in the Collection
- *documentFrequency*: The document frequency of the term
- *keyFrequency* is the term frequency in the query

- $TF = tf * log (1 + (c * averageDocumentLength) / docLength)$
- *tf* is the term frequency in the document
- *c* is set to 1.28 (default for the WT18G)
- $f = TF / termFrequency$
- $p = 1 / numberOfDocuments$

For the relevant process, Information-theoretic query expansion [11][12] is used where the topmost documents to be assessed are called "Elite_set T of documents" and the most informative terms are selected by using the information-theoretic Kullback-Leibler divergence function (KL).

$$KL = KL\_f \; * log (KL\_f / KL\_p) \qquad (2)$$

where

- $KL\_f = termFrequencyElite / SumDocumentLengthElite$
- $KL\_p = termFrequency / SumDocumentLength$
- *termFrequencyElite* is the term frequency in the elite_set T
- *SumDocumentLengthElite* is the sum of the document lengths in the elite_set T
- *termFrequency* is the term frequency in the Collection
- *SumDocumentLength* is the sum of the document lengths in the Collection

The values of KL are then combined with the initial term frequency of the term within the query as follows:

$$Tfqexp \; = keyFrequency / Max_{tf\_Q} + \beta * KL / Max_{KL} \qquad (3)$$

where
- *keyFrequency* is the term frequency in a query
- Max_{tf_Q} is the maximum number of occurrences of a term in a query
- β is set to 0.5
- *KL* is the information-theoretic Kullback-Leibler divergence function KL
- Max_{KL} is the highest *KL* value in the elite_set T

and the new term weighting function RF_{weight}, used in the second retrieval, is given by

$$RF_{weight} = Tfqexp * DFR_{weight} \qquad (4)$$

In this paper, Rocchio, Ide, and KL are used as models for our baseline comparisons. Other version of tf*idf might work better than tf*idf or DFR and this will improve the results of the query. However, any improvement in the query will consequently affect the results of the relevance feedback process. Therefore, the baseline and WPY will profit from this improvement since both techniques are tested on the same list of documents retrieved from the initial query.

3 The Weight Propagation Technique

The RF techniques used in this paper use Weight Propagation (WP), where documents judged relevant propagate positive weights to nearby documents in vector similarity space, and documents judged not relevant propagate negative weights to nearby documents [14]. The propagation of relevance in documents is not recursive:

for two relevant documents, the propagation is computed separately and then merged. Although the WP technique has never been used before for query reformulation, similar techniques have been used in other areas of computer science, such as the system presented by Sander et al. [15], used for task allocation in multi-agent systems. The system enables these agents to move toward tasks and away from other agents based on weight propagation founded on a distance between them. In Chakrabarti et al [16] also, a new technique was devised to locate resources relevant to the user information needs from an enormous amount of information found in the World Wide Web. The technique is based on the "HITS algorithm" [17] composed of a sampling component and a weight-propagation component. The pages with the highest weights are considered relevant. Related techniques have been also used by Melnik et al. [18] to assess the similarity of database schemas, and by Cortes in identifying "communities of interest" in statistical fraud detection [19].

The WP technique is illustrated in Figure 1. "doc1" has been judged relevant, and thus propagates positive weights to "doc2", "doc3" and "doc4", while "doc4", which has been judged not relevant, propagates negative weights to each of the other three documents. The weights, both positive and negative, propagated to each document are summed, and the documents with the highest weights are retrieved as a result of this relevance feedback. We call this process a first-order propagation, and the set of documents retrieved with the highest weights, the top_set.

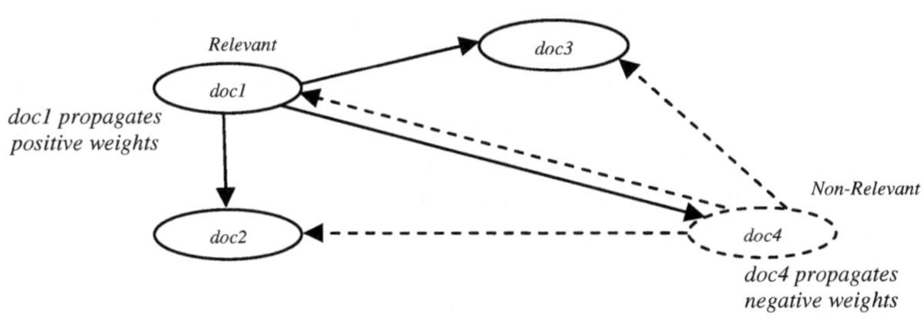

Fig. 1. Documents propagating positive and negative weights

The weights propagated from a document to other documents in the neighbourhood are influenced by how far away those documents are from the given document in vector similarity space. For example, a document that is close to a relevant one is affected more than a distant document. The weight propagated, w_{ij}, from a document *i* to a document *j*, is based on the distance between the two documents as defined by the equation:

$$w_{ij} = 1 / distance\ (document\ i,\ document\ j) \quad (5)$$

where

- w_{ij} is the weight propagated from a document i to a document j
- *distance (document i, document j)* is the conceptual distance between document i and document j.

If the conceptual distance is estimated by the numeric cosine similarity measure, then the propagation is not measured as the inverse distance since documents that have high similarity ought to propagate large weights (equation 6). In this paper, the cosine similarity measure is used.

$$w_{ij} = \textit{similarity (document i, document j)} \quad (6)$$

Since the weight magnitude is based on the distance between documents, therefore, the weight propagated to a distant document is insignificant. To reduce the time complexity, it would be enough for the document to propagate weights only to nearby documents.

The time complexity for performing RF is O (*n*) [20] where *n* is the number of documents in the collection. The time complexity for the WP technique is O (*q n'*), where *q* is the number of documents most highly ranked in the original hit list and marked relevant or non-relevant by the user, and *n'* is the number of documents found in close proximity to each of the retrieved documents and consequently affected by the propagation. The user chooses the value of *q*. Previous experiments [21][22] have shown that the user will mark on average 10 to 20 documents as relevant or non relevant, however the improvement in effectiveness will be quite significant if we can just mark a small number of documents as relevant or non-relevant, and then propagate the weights. The value of *n'* could be used as a threshold to determine the number of documents affected by the propagation, or alternatively one could stipulate the maximum distance allowed for a document to propagate to other documents in the neighbourhood. In our experiments, performance was best when n' was not more than 28. WPY Improves the computational efficiency of Relevance Feedback by considering only documents in a given neighbourhood which means that the Weight Propagation technique is faster. Therefore, WPY is sufficiently efficient to work in a search engine, taking into account the size of the WWW and the speed users expect.

A similarity matrix is constructed to store the distances between the documents. Let G = <N, A> be a connected graph, where *N* is the set of nodes and *A* is the set of edges. Consider *N* to represent the documents in the database and the edges to be the distances between them in a symmetric distance matrix D = [distance (i, j)], used to store the distances between any two documents *i* and *j*. The values stored in the matrix are computed using the cosine similarity. An optimisation, which allows this to be done with acceptable computational efficiency and minimum space, will be described in section 4.2.

4 Experimental Design

The weight propagation technique is inspired by both the Rocchio and the Ide techniques. WP inspired by Rocchio (WPR) takes the following format:

$$w_t = w_i + (1/r)\, \Sigma w_r - (1/n)\, \Sigma w_{n'} \quad (7)$$

where

- w_t is the total weights propagated from neighbourhood documents
- w_i is the weight derived from the initial query
- r is the number of relevant documents

- w_r is the weight propagated from relevant documents
- n is the number of non-relevant documents
- w_n is the weight propagated from non-relevant documents

In a second variant of WP, inspired by Ide (WPI), weight propagation is taken to be

$$w_t = w_i + \Sigma w_r - \Sigma w_n. \tag{8}$$

where the variables $w_t, w_i, w_r,$ and w_n are the same as in (7).

A third variant of WP, WPY, the primary focus of this paper (Figure 2), and counts only the maximum weight propagated to the document. The system, as a result, produces better results than summing all the weights or computing their averages. A third variant of WP, WPY, the focus of this paper, counts only the maximum weight propagated to the document. The system, as a result, produces better results than summing all the weights or computing their averages. The theoretical reason behind this is because the weight propagated from a nearby document, which should evidently be significant, will be affected severely when averaged with a weight propagated from a distant document. This statement is proved experimentally in section 5.

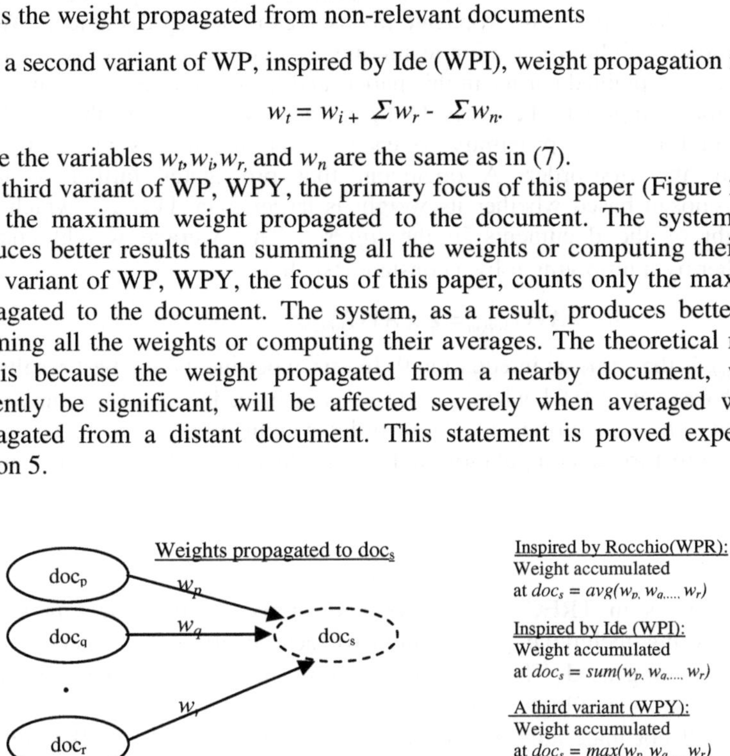

Fig. 2. WP expressed in 3 variant ways

As with the baseline, an additional weight, w_i which is derived from the initial query, affects positively the results. In the Rocchio technique for example, the initial query is added to the results of the reformulated query. To apply a similar concept in the WP technique the average weight w_{avg} propagated from the relevant documents is computed as follow:

$$AVG_{weight} = \Sigma w \,/\, \Sigma top\_set \tag{9}$$

where

Σw is the total weights of the documents in the top_set
Σtop_set is the number of documents in the top_set

Subsequently, AVG_{weight} is added to the *top T* documents proportionally, where the first document at the top receives w_{avg}, the second document receives $(w_{avg} - 1xw_{avg}/T)$, the third receives $(w_{avg} - 2xw_{avg}/T)$ and the Tth document receives w_{avg}/T. The idea is to give more weight to the highly ranked documents originally retrieved by the initial query. The variable T is a threshold that could vary between

1 and n which is the number of documents in the test collection. In our experiments, T gives better results when set to n. The WPY technique can be improved by making the documents, in the top_set, to propagate themselves to nearby documents. We call this a second-order weight propagation technique, as an extension process to the first-order propagation explained earlier in this paper, and have used this technique in this paper. The time complexity becomes O ($q\ n'\ n''$) where n'' if the number of documents found in close proximity to each of the documents affected by the propagation in the first-order. A document that propagates further to nearby documents is chosen based whether its weight is larger than AVG_{weight} which is the average weight of the documents in the *top_set*. The average weight could be modified further by multiplying it by a constant ε such as

$$AVG_{weight} = \varepsilon \times AVG_{weight} \tag{10}$$

where $A\acute{V}G_{weight}$ is the same as in equation 9. In our experiments, the best results were given when ε was set to 0.94 when pseudo-relevance feedback was employed. We could propagate one-step further as a third-order weight propagation technique, but this would require further computation and would degrade further the efficiency of the technique.

4.1 Clustering

The web documents in TREC WT18G were clustered by the bisecting k-means partitional clustering technique [23] to improve search efficiency since large test collections were being used. The k-means algorithm partitions a large collection of documents into s subsets of highly similar documents.

Clustering will not materially affect the retrieval effectiveness of WPY since the weight magnitude is based on the distance between documents and the weight propagated to distant documents is insignificant. Therefore, it is enough for the document to propagate only to the nearest ones. As a result, it is more efficient to cluster the documents into clusters and compute the similarity measure between the documents in each of these clusters rather than computing the similarity measure between all the documents. Consequently, the space required by and the time needed to compute the symmetric distance matrix for all the documents was significantly reduced from n by n to $n\_cluster_i$ by $n\_cluster_i$, where $n\_cluster_i$ is the number of documents found in $cluster_i$, since we will end up having a smaller symmetric distance matrix for each cluster. The space required to store the similarity matrices is further reduced from $n\_cluster_i$ by $n\_cluster_i$ to $n\_cluster_i$ by n' (n' is the maximum number of documents to which weights are propagated for each document, which was 12 in our experiments).

Clustering does not require any additional overhead for the WPY technique since it also takes place in both the baseline techniques (Rocchio, Ide, and KL), where given a query vector q, the closest matching cluster is found by computing and comparing the cosine values (similarity measures) between the query vector and the centroid vectors in order to find the one that exhibits a sufficiently high cosine similarity with the query. Once this is found, only the documents in that cluster are compared to the query. This reduces the number of comparisons needed to retrieve the documents[1].

4.2 Residual Collection

For our experiments, we have employed the residual collection technique [24] where the documents initially judged relevant by the user are removed from the collection before evaluation, and consequently, the precision and recall measures are calculated on the remaining documents in the collection (residual collection).

4.3 Pseudo-relevance Feedback vs. Simulated-Relevance Feedback

The TREC WT18G test collection is provided with query-document relevance judgments. Some techniques do not require users to be physically present and give feedback at every single iteration. The simulated-feedback technique [25], for instance, selects from the top-ranked documents the documents found in the relevance-judgment list, and considers them as relevant before reformulating the query and the remaining documents not selected are considered as negative judgments. Pseudo-relevance feedback [25], or as it was called when it was originally proposed [26], local feedback, is used for the feedback process. In a Pseudo-relevance feedback, the query is reformulated based on the assumption that the top-ranked documents are relevant. That is, as if these documents were shown to a user and the user considers them all relevant. Thus, the reformulation is done without the intervention of the user. In this paper, we have used both pseudo and simulated relevance-feedback.

5 Results and Analysis

The experiments were conducted on the TREC WT18G test collection distributed by CSIRO [27]. Many of the documents in WT18G that are in foreign languages were removed in order to reduce the processing time (35,051 documents). We also have used an enhanced Stemmer algorithm [28] for the indexing process. Rocchio Ide and KL techniques were compared against the WPY method. The assessments were done as follows: For each query, an initial document ranking was obtained. Ten retrieved documents (N=10) were taken from the top of the ranked list and used for one iteration of query reformulation. A comparison between Standard TFIDF and DFR

[1] The similarity matrix must be created as an overhead or it could be avoided if the distances between the documents are calculated online. This involves calculating the distances between the relevant and non-relevant documents chosen by the user and the rest of the documents. This is repeated for each relevance feedback iteration. The time complexity is in O (n) but is further reduced to $O(n\_cluster_i)$ where $n\_cluster_i$ is the number of documents found in $cluster_i$ since the documents are clustered as explained above. Computing the distances online will gradually construct the similarity matrix and the system will avoid recomputing the distance between documents if their results are previously computed and stored in the similarity matrix.

models showed that the precision with DFR was better than the precision with TFIDF at most recall levels.

5.1 Pseudo-relevance Feedback Using Rocchio and Ide Based TFIDF

WPY gave better precision than the baselines when tested on the TREC WT18G collection. When Rocchio and Ide were employed, precision was better at low recall levels. This means that the user will retrieve documents that are more relevant particularly on the first pages of search hits as is preferable in a search engine. It should be noted that WPY gave the best results when ε was equal to 0.80.

An important result is presented in this section. When propagating to few documents, for instance. The baselines performed better at low recall levels. When the number of propagated documents increases, the recall level increases as well (Table 1). The performance of the system kept improving as the number of propagated documents was increased to 11, but no further improvement was noted after that, as shown in Table 1. In Figure 3, WPY performs better than the baselines at all recall levels for 12 propagated documents, which means WPY performs better.

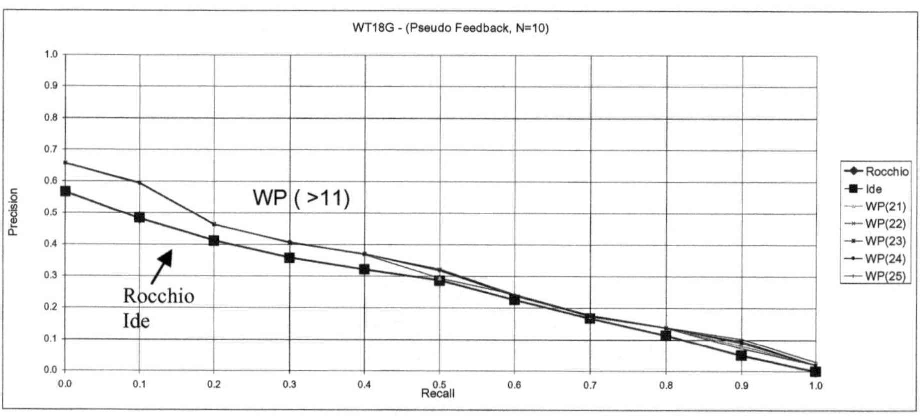

Fig. 3. Propagating over 11 documents ($\varepsilon = 0.80$)

Table 1. Results from propagating to different documents against Rocchio and Ide based TFIDF using Pseudo-relevance Feedback ($\varepsilon = 0.80$)

Recall Level	Rocchio	Ide	WPY(2)	WPY(3)	WPY(4)	WPY(5)	WPY(6)	WPY(7)	WPY(8)	WPY(9)	WPY(10)	WPY(11)	WPY(>11)
Prec@10	0.482	0.484	0.463	0.542	0.575	0.569	0.570	0.575	0.566	0.564	0.560	0.562	0.593
Prec@30	0.358	0.357	0.117	0.232	0.272	0.270	0.287	0.286	0.342	0.361	0.352	0.352	0.407
MAP	0.271	0.271	0.139	0.186	0.217	0.233	0.236	0.243	0.255	0.259	0.261	0.270	0.318

5.2 Pseudo-relevance Feedback with KL Based DFR

When DFR was considered with ε = 0.94, and for fewer than 11 propagated documents, it was not clear whether WPY outperformed KL (see Table 2). However, for 11 propagated documents or more, the precision for WPY performed better than baselines at all recall levels (Figure 4). WP and KL both perform equally well at P@10 when the value of ε was first set to 1.00. However, WP performed better than KL at P@10 when ε = 0.94.

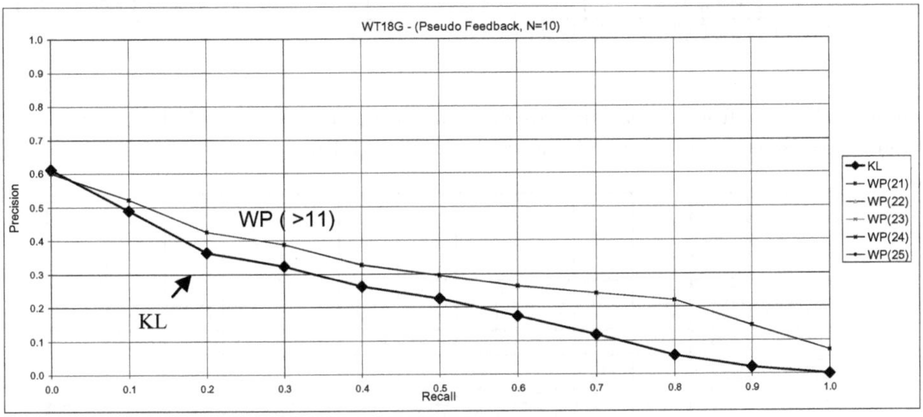

Fig. 4. Propagating over 11 documents (ε = 0.94)

Table 2. Results from propagating to different documents using against KL based DFR using Pseudo-relevance Feedback (ε = 0.94)

Recall Level	KL	WPY(2)	WPY(3)	WPY(4)	WPY(5)	WPY(6)	WPY(7)	WPY(8)	WPY(9)	WPY(10)	WPY(11)	WPY(>11)
Prec@10	0.488	0.493	0.535	0.520	0.512	0.504	0.500	0.500	0.512	0.510	0.510	0.533
Prec@30	0.324	0.167	0.254	0.317	0.387	0.368	0.381	0.391	0.384	0.383	0.379	0.388
MAP	0.240	0.152	0.187	0.217	0.238	0.247	0.259	0.265	0.271	0.279	0.281	0.318

5.3 Relevance Feedback Using Simulated-Relevance Feedback

When the same experiments were repeated with Simulated-relevance feedback (positive and negative), WPY performed better than when pseudo-relevance was employed. With Rocchio and Ide, the performance of the system kept improving as the number of propagated documents was increased to 11, but no further improvement was noted after that (Table 3). However, WPY performed better in simulated-relevance feedback than in pseudo-relevance feedback and outperformed KL at all recall levels (Table 4).

Table 3. Results from propagating to different documents against Rocchio and Ide based TFIDF using Positive and Negative simulated-relevance feedback ($\varepsilon = 0.80$)

	Recall Level	Rocchio	Ide	WPY(2)	WPY(3)	WPY(4)	WPY(5)	WPY(6)	WPY(7)	WPY(8)	WPY(9)	WPY(10)	WPY(11)	WPY(>11)
Pos.	Prec@10	0.507	0.507	0.451	0.538	0.587	0.592	0.580	0.584	0.583	0.581	0.576	0.583	0.580
	Prec@30	0.369	0.372	0.105	0.248	0.294	0.287	0.268	0.288	0.357	0.373	0.365	0.370	0.402
	MAP	0.279	0.278	0.133	0.183	0.219	0.239	0.235	0.241	0.254	0.260	0.260	0.272	0.313
Neg.	Prec@10	0.401	0.429	0.447	0.504	0.549	0.543	0.463	0.481	0.482	0.482	0.467	0.470	0.523
	Prec@30	0.279	0.302	0.097	0.240	0.287	0.282	0.279	0.281	0.329	0.354	0.364	0.372	0.379
	MAP	0.228	0.239	0.133	0.183	0.211	0.227	0.216	0.224	0.226	0.237	0.239	0.251	0.309

Table 4. Results from propagating to different documents against KL based DFR using Positive and Negative simulated-relevance feedback ($\varepsilon = 1.10$)

	Recall Level	KL	WPY(2)	WPY(3)	WPY(4)	WPY(5)	WPY(6)	WPY(7)	WPY(8)	WPY(9)	WPY(10)	WPY(11)	WPY(>11)
Pos.	Prec@10	0.488	0.420	0.514	0.513	0.565	0.550	0.599	0.573	0.605	0.609	0.606	0.605
	Prec@30	0.330	0.058	0.091	0.186	0.253	0.294	0.320	0.309	0.353	0.369	0.403	0.421
	MAP	0.242	0.117	0.148	0.165	0.186	0.212	0.233	0.233	0.251	0.256	0.262	0.318
Neg.	Prec@10	0.445	0.488	0.561	0.581	0.618	0.598	0.612	0.582	0.585	0.588	0.594	0.679
	Prec@30	0.293	0.067	0.126	0.224	0.288	0.322	0.354	0.366	0.410	0.405	0.432	0.529
	MAP	0.218	0.132	0.157	0.181	0.209	0.223	0.245	0.239	0.253	0.267	0.274	0.321

6 Conclusions

In this paper, we have developed a new technique for RF, called WPY, which uses Weight Propagation where positive and negative weights are propagated to documents in a given vicinity. Both the Rocchio and the Ide technique inspire the technique. This new technique improves precision relative to the baselines since the documents are treated as independent vectors rather than being merged into a single vector, as is the case with Ide and Rocchio or by partially determining the documents relevancy based on all documents' lengths as with conventional probabilistic RF techniques. In addition, the WPY approach consumes less computation time since it inspects only nearby documents as opposed to Rocchio, Ide and KL that examine all the documents. A second-order weight propagation technique is employed where the documents that receive the highest weights are to propagate weights themselves to nearby documents. The experiments were conducted on the TREC WT18G test collection using pseudo, positive and negative feedback. In future, the experiments will be repeated using different models in DFR in addition to the Latent Semantic Index model rather than the traditional lexical matching used in this paper. We will also test a third (as opposed to second) order weight propagation to determine whether retrieval performance will improve.

References

1. Salton G & McGill M J (1983). Introduction to modern information retrieval. New York : McGraw Hill
2. Baeza-Yates R & Ribeiro-Neto B (1999). Modern information retrieval (pp. 10). New York: Addison-Wesley.
3. Zhou X S & Hunag T S (2001). Small sample learning during multimedia retrieval using bias map, in Proc. IEEE Conf. Computer Vision and Pattern Recognition, Hawaii, Dec.
4. Dunlop M D (1997). The effect of accessing non-matching documents on relevance feedback. ACM Transactions on Information Systems, 15(2):137-153
5. Rocchio G Jr. (1971). Relevance feedback in information Retrieval. Chapter 14. In G. Salton (Ed.), The SMART Retrieval System: Experiments in Automatic Document Processing. Prentice Hall
6. Ide E (1971). New experiments in relevance feedback. Chapter 16. In G. Salton (Ed.).: The SMART Retrieval System: Experiments in Automatic Document Processing, Prentice Hall
7. Robertson S E & Sparck-Jones K (1976). Relevance weighting of search terms, Journal of the American Society of Information Science, pp.129-146
8. van Rijsbergen C J (1979). Information retrieval, Buttersworth, London
9. Robertson S E, Walker S, & Beaulieu M M (1998). Okapi at TREC-7: automatic ad hoc, filtering, VLC and interactive track. In Proceedings of the 7th Text Retrieval Conference (TREC7), Nist Special Publication 500-242, pages 253-264, Gaithersburg, MD, USA
10. Zhai C and Lafferty J (2001). A study of smoothing methods for language models applied to ad hoc information retrieval. In Proceedings of the 24th Annual International ACM SIGIR Conference on Research and Development in Information Retrieval, pages 334-342, New Orleans, LA, USA
11. Amati G, Carpineto C, & Romano G (1992). Italian monolingual information retrieval with prosit. In Proceedings of CLEF (Cross Language Evaluation Forum 2002, pages 182-191, Rome, Italy
12. Amati G, Carpineto C, & Romano G (2001). Fub at Trec-10 web track: A probabilistic framework for topic relevance term weighting. In Proceedings of the 10th Text REtrieval Conference (TREC-10), NIST Special Publication 500-250, pages 182-191, Gaithersburg, MD, USA
13. Amati G & van Rijsbergen C J (2002). Probabilistic models of information retrieval based on measuring divergence from randomness. ACM Transactions on Information Systems, 20(4):357-389
14. Yamout F, Oakes M & Tait J (2006). Relevance feedback using weight propagation. Proceedings of the 28th Annual International ACM-SIGIR Conference on Research and Development in Information Retrieval
15. Sander P.V., Peleshchuk D., and Grosz B J. (2002) "A Scalable, Distributed Algorithm for Efficient Task Allocation", AAMAS'02, July 15-19, 2002, Bologna, Italy
16. Chakrabarti S., Dom B. E., Gibsony D., Kleinbergz J., Kumar R., Raghavan P., Rajagopalan S., and Tomkins A., (1999) "Mining the Link Structure of the World Wide Web", IEEE Computer, August 1999.
17. Kleinberg J. M. (1999) "Authoritative Sources in a Hyperlinked Environment" J. ACM, 6(5):604–632, September 1999.
18. Melnik S, Garcia M, Hector; Rahm, & Erhard. (2002), Similarity flooding.: A versatile graph matching algorithm and its application to schema matching. In the proceedings of the 18th ICDE Conference

19. Cortes C, Pregibon D & Volinsky C (2001). Communities of interest. Proceedings of IDA 2001 - Intelligent Data Analysis.
20. Yamout F, Moghrabi I & Oakes M (2004). Query and relevance feedback in latent semantic index with reduced time complexity. IASTED International Conference on Database Applications - DBA
21. Harman D (1992). Relevance feedback revisited. In proceedings of the Fifth International SIGIR Conference on Research and Development in IR pp. 1-10
22. Rose D & Stevens C (1996) V-twin: A lightweight engine for interactive use. In NIST Special Publication 500-238: The 5th Text Retrieval Conference (TREC-5), pages 279-290
23. Steinbach M, Karypis G & Kumar V (2000). A comparison of document clustering techniques. In KDD Workshop on Text Mining
24. Ruthven I & Lalmas M (2003). A survey on the use of relevance feedback for information access systems. knowledge engineering review. Vol 18. Issue 2. pp 95-145
25. Efthimiadis N E (1996). Query expansion. In Annual Review of Information Systems and Technology, Vol. 31, pp. 121-187.
26. Croft W B & Harper D J (1979). Using probabilistic models of document retrieval without relevance information. Journal of Documentation, 35(4), 285-295
27. CSIRO TREC Web Tracks homepage. (2001). www.ted.cmis.csiro.au/TRECWeb/
28. Yamout F, Demachkieh R, Hamdan G, Sabra R (2004). Further Enhancement to the Porter's Stemming Algorithm – TIR 2004 - Germany

A Retrieval Evaluation Methodology for Incomplete Relevance Assessments

Mark Baillie[1], Leif Azzopardi[2], and Ian Ruthven[1]

[1] Department of Computing and Information Sciences,
University of Strathclyde, Glasgow, UK
{mb, ir}@cis.strath.ac.uk
[2] Department of Computing Science,
University of Glasgow, Glasgow, UK
leif@dcs.gla.ac.uk

Abstract. In this paper we a propose an extended methodology for laboratory based Information Retrieval evaluation under incomplete relevance assessments. This new protocol aims to identify potential uncertainty during system comparison that may result from incompleteness. We demonstrate how this methodology can lead towards a finer grained analysis of systems. This is advantageous, because the detection of uncertainty during the evaluation process can guide and direct researchers when evaluating new systems over existing and future test collections.

1 Introduction

In this study we revisit the implications on system comparisons that arise from incomplete relevance assessments, and in particular the assumption that unassessed documents are not relevant. Instead of assuming unassessed documents are not relevant [1], or more recently, removing documents that are not judged when estimating performance [2], we propose an alternative approach: to quantify the proportion of unassessed documents in a system's ranked list. This leads to a complementary evaluation methodology, which uses a new set of measures that quantify uncertainty during system comparison. This methodology provides a guide for both researchers who re-use collections, and also for designers of new test collections. Adopting such an approach is important as researchers can detect potential uncertainty during evaluation, and also identify strategies for addressing this uncertainty. In this paper, we illustrate the utility of this evaluation methodology, as well as highlighting the implications of using particular performance metrics, related to the depth of measurement, under incomplete assessments.

Before introducing this new approach we first provide some context by reviewing the running debate on incompleteness, and the subsequent implications for the comparison of systems. We then introduce the proposed methodology which augments the current evaluation protocol (Section 2). Next, we provide an empirical analysis of this approach across a range of existing test collections (Section 3). Finally, we conclude with a discussion of the implications of this study (Section 4).

G. Amati, C. Carpineto, and G. Romano (Eds.): ECIR 2007, LNCS 4425, pp. 271–282, 2007.

1.1 Background

Modern test collections adhere to a well defined model often referred to as the *Cranfield paradigm* [3]. A corpus that follows this model will consist of a collection of documents, statements of information need (named *topics*), and a set of relevance judgements listing the documents that should be returned for each topic. To ensure fair system comparisons, a number of key assumptions are made within this framework such as (i) the topics are independent, (ii) all documents are judged for relevance (completeness), (iii) these judgements are representative of the target user population, and (iv) each document is equally important in satisfying the users information need. These assumptions are made to ensure fair comparison of system performance, although to develop an "ideal" collection where these assumptions hold is unrealistic. Factors such as collection size and available (human) resources dictate to what degree these assumptions do hold.

As a consequence these assumptions are often relaxed while compensating for any potential uncertainty that could be introduced such as system bias. For example, system pooling was proposed to address the intractability of the completeness assumption [4]. Pooling is a focused sampling of the document collection that attempts to discover all potentially relevant documents with respect to a search topic e.g. approximate the actual number of relevant documents for a given topic. To do so, a number of (diverse) retrieval strategies are combined to probe the document collection for relevant documents. Each system will rank the collection for a given topic, then the top λ documents from the subsequent ranked lists are collated [1], removing duplicates, to form a pool of unique documents. All documents in this pool are then judged for relevance by an assessor(s) using specified criteria such as topicality or utility. In the case of TREC [3], the assessor(s) use the topic statement for guidance when determining which documents are topically relevant or not. The remaining unassessed documents are assumed to be not relevant. This assumption follows the argument put forward by Salton, that by using a range of different IR systems the pooled method will discover "the vast majority of relevant items" [1]. This argument is based on the assumption of diminishing returns i.e. because many runs contribute to the system pool it is highly likely that the majority of relevant documents, or those documents representative of relevant documents, are returned through pooling. If this assumption holds then there is little need to assess those documents not pooled.

There has been much debate about the validity of this assumption. Initially this assumption was applied to smaller collections such as Cranfield and CACM. However, as Blair has posited [5], the percentage of documents not assessed for a given topic with respect to a modern collection could be up to 99%, leaving a very large proportion of unassessed relevant documents undiscovered. However, Voorhees and Harman highlight a key point [6], that as pooling is a union of many different ranking approaches and because only relative system performance is measured, if the number of systems contributing to a pool is sufficiently large

[1] The cut off λ is known as the *pooling depth*. The *measurement depth* k refers to the document cut off used when estimating retrieval performance e.g. Precision at k documents.

and these systems are diverse, bias towards one or a set of systems should be minimised, even if all relevant documents were not discovered. Absolute system performance may not be accurately estimated using incomplete relevance assessments, but the relative performances of systems can be fairly compared. This is related to the argument by Salton, where as long as the conditions remain even for all systems, then the relative differences between systems can be compared fairly and with a high degree of certainty.

When using pooling to estimate recall it is difficult to ascertain whether the majority of relevant documents have been discovered. There have been a number of empirical studies that have attempted this across collections such as TREC. Zobel defined a method to extrapolate the potential numbers of unassessed relevant documents, concluding that the assumption of unassessed documents being not relevant was unfounded [7]. He approximated that a large percentage of relevant documents were still to be discovered, especially across topics where a large number of relevant documents were already found through pooling. Therefore, it is not clear what impact the potential proportion of relevant unassessed documents may have on system comparisons.

For this very reason, the effect that pooling has on system comparison has been investigated in the context of relevant document recall, focusing upon several different areas of the completeness assumption and system pooling. These studies have investigated issues such as the effect on system comparison and the subsequent uncertainty when using incomplete relevance judgements [2,3,7], efficient pooling strategies [7,8,9,10], automatically generated relevance assessments [11,12], and the importance of significance testing during system comparison [13]. A running theme throughout these studies is that it is still unclear whether the now standard assessment procedure of pooling, and the resultant evaluation measures adopted, does indeed impact upon the fair and unbiased comparison of retrieval systems and to what extent, if any. For example, a recent investigation of the TREC Robust-HARD 2005 collection identified system bias, which was a result of both a shallow pool depth, and (potentially) similar runs forming the system pool [14]. The outcome was a bias in the collection favouring systems that gave more weight to query terms that appeared in the title of a document over other approaches. Although this does not necessarily indicate a failing of system pooling it motivates the need for a more robust evaluation protocol which considers aspects such as pooling, the measurement depth, and the status of unassessed documents during evaluation.

1.2 Focus of This Study

Based on an analysis of these studies, we posit that uncertainty remains when comparing the relative performance of systems as a result of the status of unassessed documents (being not relevant). One of the cited limitations with laboratory studies is the large amount of subjectivity or uncertainty in such evaluations. The nature of the scientific method demands as much objectivity and certainty as possible. After analysing the history of retrieval evaluation we believe that the status of unassessed documents and the resulting suitability of comparing systems with varying levels of assessed documents is still an open issue. We are especially motivated by the recommendations of Zobel [7], who warned researchers when evaluating new systems across existing test collections for cases where performance could be underestimated. However, a standard protocol for detecting such cases has not been proposed as of yet. We therefore propose a new methodology for quantifying uncertainty during system comparisons that may exist because of incomplete relevance assessments. By doing so, we can determine when it is possible to *fairly* compare two systems using current measures, especially those systems that do not contribute to the pool. Instead of compensating for or ignoring potential uncertainty during system comparisons due to incompleteness, we believe that the proportion of unassessed documents should be captured and reported. Reporting this information can be a useful source of information to help quantify the confidence, accuracy and/or reliability of system performance. By capturing such information, we can determine whether two systems are compared under similar conditions, and flag those cases when one system may have an advantage over another due to a particular condition found in a test collection.

2 Capturing the (un)certainty of System Performance

We hypothesise that the certainty associated with estimating a measurement of a systems performance at a depth k is proportional to the number of documents that have been assessed at that depth. Conversely, the uncertainty is proportional to the number of documents that have not been assessed. The larger proportion of assessed documents contained in a ranked list, the more confident we are in the estimate of a systems performance at the corresponding measurement depth. For example, when comparing the performance of two systems, if all documents have been assessed in the ranked list of both systems then we have the *ideal* situation of completeness i.e. the performance estimates for both systems were made with a high degree of certainty. If the ranked lists of both systems are incomplete, but contain similar proportions of assessed documents, then confidence in the relative comparison of both systems would also be high. However, if one system has a substantially lower proportion of assessed documents than another, then the performance estimate of that system is based on limited information relative to the other. It is these cases that we wish to detect, where the conditions for both systems are not even resulting in a higher degree of uncertainty in the comparison.

2.1 Measure of Assessment

We propose to capture uncertainty by calculating the proportion of assessed documents in a ranked list. Let A be the set of assessed documents in a collection of N documents, and X be the set of retrieved documents by a system for that topic. Then *Assessment Precision* A_p can be defined as the probability of retrieving judged documents:

$$A_p = \frac{|X \cap A|}{|X|}$$

where $|X \cap A|$ is the number of documents in the set defined by the intersection of X and A, and $|X|$ is the number of documents in X. Assessment Precision relates to the confidence we place, or the certainty of a performance estimate, given a ranked result list. Note that uncertainty associated with the estimate is the complement, $1 - A_p$. We now refer to uncertainty and certainty through this measure, where a high Assessment Precision value relates to high certainty and low uncertainty.

This measure is exactly the definition for standard Precision except with respect to assessed as opposed to relevant documents. Consequently, for every Precision and Recall measure there is a corresponding Assessment measure. The Average Assessment can be computed by taking the average over all ranks where an assessed document occurred. The Mean Average Assessment (MAA) then provides a summary of the assessment over all topics placing more emphasis on systems with assessed documents higher in the ranked list. This metric is analogous to Mean Average Precision (MAP), and could be used in situations where it is important to identify whether there is a difference in the proportion of assessed documents at higher ranks between systems when estimating MAP[2].

It should also be noted that the Assessment Precision metrics are functionally related to the corresponding Precision metrics. This relationship is because A is the union of the set of assessed relevant documents and assessed non relevant documents. Therefore a system retrieving more assessed documents is likely to have a higher Precision, because assessed documents are more likely to be relevant. Also, when systems have low levels of A_p there is increased uncertainty in the Precision score, and any subsequent comparison, because of the high proportion of unassessed documents. It is important to consider this context during the evaluation. Assessment Precision provides this context explicitly by capturing the proportion of assessed documents used to estimate the retrieval performance. In this paper, we concentrate on applying A_p to fairly compare systems, and leave these other issues regarding A_p for future research.

2.2 System Evaluation Decision Matrix

We now illustrate how Assessment Precision can be integrated into the current evaluation protocol. We motivate the introduction of the System Evaluation Decision matrix in the form of an example system comparison. We wish to test the performance P(), which denotes the Precision at a given measurement depth k (i.e. P@10, MAP, etc.), of two systems s_1 and s_2 over a test collection with incomplete relevance assessments.

[2] We have focused on Precision based metrics in this paper although quantifying the level of assessment can be extended to included other types of measures. For example, *bpref* has recently been proposed as a measure for incomplete collections [2], which removes all unassessed documents during estimation of performance. Note that, *bpref* itself does not quantify the proportion of assessed documents that are removed but a corresponding A_p measure could be derived to complement such a metric.

Case 1	Case 2
P(s1)==P(s2) A(s1)==A(s2) or P(s1)==P(s2) A(s1)==A(s2)	P(s1)==P(s2) A(s1)<<A(s2) or P(s1)==P(s2) A(s1)>>A(s2)
Accept null hypothesis that s1 is equal to s2, with a low degree of uncertainty	Accept null hypothesis that s1 is equal to s2, with a high degree of uncertainty
Case 3	**Case 4**
P(s1) >>P(s2) A(s1)<<,==A(s2) or P(s1)<<P(s2) A(s1)==,>>A(s2)	P(s1)>>P(s2) A(s1)>>A(s2) or P(s1)<<P(s2) A(s1)<<A(s2)
Reject null hypothesis that s1 is equal to s2, with a low degree of uncertainty	Reject null hypothesis that s1 is equal to s2, with a high degree of uncertainty

Fig. 1. System Evaluation Decision Matrix for system comparison

We have the following research hypothesis:

$$H_0 : P(s_1) = P(s_2)$$
$$H_1 : P(s_1) \neq P(s_2)$$

To determine the level of confidence we can place on this test, we test the supplementary hypothesis using a corresponding Assessment Precision metric A(), which denotes the A_p at a corresponding measurement depth k (i.e. A@10, MAA, etc.):

$$H_0 : A(s_1) = A(s_2)$$
$$H_1 : A(s_1) \neq A(s_2)$$

This forms a contingency table of four possible outcomes of interest displayed in Figure 1. Significance is denoted as either no difference (==) or the significant differences (<<, >>) i.e. s_1 is significantly better (>>) than s_2. We assume that statistical significance is determined using an appropriate test such as Wilcoxon sign rank test, paired T-test or ANOVA [13].

For Case 1, the null hypothesis that $P(s_1) == P(s_2)$ and $A(s_1) == A(s_2)$ cannot be rejected. We define this a "strong" case because the level of assessment for both s_1 and s_2 are equal, that is the proportion of information used to estimate performance was comparable.

For Case 2, the null hypothesis that $P(s_1) == P(s_2)$ cannot be rejected as well, however, the proportion of information used to estimate the performance of both systems was not comparable. In other words, the result list of one system was comprised of a significantly larger proportion of assessed documents than the other, causing a degree of uncertainty in this comparison. It is unknown from this test whether, under comparable conditions, the null hypothesis $P(s_1) == P(s_2)$ would still hold or not. We therefore define this as a "weak" case.

For Case 3, also a "strong" case, the null hypothesis that $P(s_1) == P(s_2)$ is rejected. We can place a high degree of confidence in this outcome as we have either a scenario where both systems share similar proportions of assessed documents, or in special scenarios the system with significantly higher performance has significantly fewer documents assessed than the other system. In other words, even with further information about this system it could not match (or better) the opposing system. Finally, for Case 4, another "weak" case, the null hypothesis that $P(s_1) == P(s_2)$ is rejected, although, we cannot place a high degree of confidence in this outcome, as the system with significantly higher performance also reported a significantly larger proportion of assessed documents. This does not indicate that the system with a smaller proportion of assessed documents would share similar performance under equal conditions, but instead flags a potential problem with this comparison.

Of the weaker outcomes Case 2 is particularly interesting as both systems have similar performance, but this performance is based on different proportions of assessed documents. What is interesting is that the system with significantly less assessed documents could potentially be retrieving a wider diversity of documents, with respect to the pool, and some of these documents may be relevant [7]. A subsequent research question would be to investigate why the systems perform as well as each other. As both systems have equal system performance but unequal levels of assessment, this system may potentially improve performance when compared under even conditions. Further investigation may provide stronger supporting evidence.

At this stage a number of steps could be taken. If the goal of the comparison is precision orientated then system comparison could be made at a shallower measurement depth to ensure the likelihood of parity. By doing so we are assuming that at shallower depths systems will have relatively equal proportions of documents assessed. If both systems have contributed to the pooling process then this assumption would hold up until pooling depth has been reached, however, if a system has not contributed to the pool this may not be the case. The previous step may lead to the creation of test collections with an emphasis on shallow measurement depth [13]. If the goal is to compare a minimal number of systems using shallow measurements, where re-usability of the test collection is not important, such a strategy could also be adopted by research groups. For example, provided with enough resources, further checking of the unassessed documents can be made, adopting strategies such as that outlined by Carterette et al. [8]. Alternatively, comparisons could be made across different test collections where conditions remain even. This step assumes such collection(s) exist, although collections can be evaluated for such properties using the suite of Assessment measures. Finally, this reinforces the need when building test collections to include novel systems in the pooling process [14].

3 Experimental Analysis

To demonstrate the application of the Assessment Precision measure within the evaluation process we conducted an empirical analysis to evaluate both its utility, and to provide further justification for its introduction. Our first objective was to examine the officially submitted runs to TREC over a number of collections, spanning a range of years [3]. By using the official runs we could investigate the level of uncertainty during performance comparisons of runs included in the system pool across these collections. Our second objective was to evaluate the implications of measurement depth with respect to the level of assessment between systems at various cut off values. Using the Assessment Precision metrics, we were investigating what effect using a measurement depth deeper than the pooling depth may have on system comparisons. This is related to the argument that relative system performance can be compared if the conditions remain even for both systems. We then focused our attention on runs from particular collections, such as the Robust-HARD 2005 [4], which has been identified as potentially problematic to use due to title bias [14]. The aim is to better understand the problems cited with this collection, with particular focus on runs that both weight topic titles and runs that do not.

For each collection we first analysed each possible pair-wise system comparison of the officially submitted runs using the decision matrix (see Table 1). To test for significance across all systems we used the ANOVA test. If significant differences in terms of performance and assessment across the systems of a collection were found, we performed a followup Bonferroni multiple comparisons to identify which systems differed significantly both in terms of performance and assessment. We repeated this experiment across numerous Performance and Assessment Precision metrics, spanning a range of measurement depths; including the Performance metrics P@10, P@30, P@100, P@500, P@1000, MAP, and the Assessment Precision metrics A_p@10, A_p@30, A_p@100, A_p@500, A_p@1000, MAA. For each metric, we counted the number of comparisons that fell into each outcome i.e. Cases 1-4 (see Figure 1 for an outline of each case). Table 1 presents the proportion of overall system comparisons that fall into each case. Rows indicate different test collections while columns represent different measures, increasing by measurement depth. Each entry represents the proportion of system comparisons that fall into that case e.g. for the TREC 3 @10 entry, 69% of pair-wise system comparisons fall in Case 1, 7% in Case 2, 17% in Case 3 and 7% in Case 4, where there were 780 pair-wise comparisons performed overall.

The first thing we were interested in was the proportion of significant pair-wise differences in terms of system performance across the various test collections, specifically to test what extent increasing measurement depth had on this proportion. The table provides the proportion of both "strong" and "weak" significant differences between systems for each metric. From the reported results, a noticeable trend was that for the majority of collections where the proportion

[3] See http://trec.nist.gov/results.html

[4] This collection combined runs from Robust 2005 and HARD 2005 to form the pool.

Table 1. The proportion of comparisons for each case across the TREC collections

	@10		@30		@100		@500		@1000		MAP/AA	
TREC 3	**0.69**	0.07	**0.67**	0.14	**0.58**	0.32	**0.44**	0.49	**0.41**	0.57	**0.64**	0.13
780	**0.17**	0.07	**0.11**	0.09	**0.05**	0.05	**0.00**	0.06	**0.00**	0.02	**0.04**	0.19
TREC 4	**0.72**	0.12	**0.65**	0.16	**0.43**	0.39	**0.35**	0.48	**0.33**	0.5	**0.5**	0.25
528	**0.16**	0.00	**0.19**	0.00	**0.05**	0.13	**0.02**	0.15	**0.01**	0.16	**0.02**	0.24
TREC 6	**0.60**	0.16	**0.55**	0.26	**0.45**	0.42	**0.45**	0.44	**0.44**	0.45	**0.6**	0.2
3081	**0.13**	0.11	**0.10**	0.10	**0.033**	0.10	**0.01**	0.10	**0.00**	0.10	**0.02**	0.18
TREC 8	**0.70**	0.04	**0.69**	0.07	**0.52**	0.31	**0.51**	0.36	**0.57**	0.33	**0.57**	0.22
8000	**0.19**	0.06	**0.17**	0.07	**0.01**	0.16	**0.00**	0.13	**0.00**	0.10	**0.04**	0.18
WEB 04	**0.35**	0.00	**0.03**	0.25	**0.02**	0.19	**0.01**	0.19	**0.01**	0.19	**0.02**	0.37
561	**0.65**	0.00	**0.42**	0.30	**0.51**	0.28	**0.53**	0.26	**0.53**	0.27	**0.40**	0.20
ROBUST 03	**0.83**	0.01	**0.74**	0.09	**0.55**	0.2	**0.52**	0.22	**0.57**	0.17	**0.65**	0.14
2145	**0.03**	0.13	**0.01**	0.16	**0.00**	0.25	**0.00**	0.26	**0.00**	0.27	**0.00**	0.21
ROBUST 05	**0.77**	0.12	**0.66**	0.23	**0.53**	0.38	**0.57**	0.35	**0.60**	0.31	**0.68**	0.19
1485	**0.07**	0.03	**0.07**	0.04	**0.00**	0.10	**0.00**	0.08	**0.00**	0.08	**0.00**	0.13
HARD 05	**0.71**	0.20	**0.66**	0.25	**0.57**	0.39	**0.57**	0.42	**0.57**	0.42	**0.72**	0.18
2775	**0.05**	0.05	**0.05**	0.04	**0.00**	0.04	**0.00**	0.02	**0.00**	0.01	**0.00**	0.10

of significant differences decreased as the measurement depth increased, the exception being the Robust 2003 collection. This trend is the converse of the intuition stated in [7], whereby increasing measurement depth also increased discrimination between systems. The intuition is that good systems will continue to retrieve relevant documents beyond the pooling depth, which will have been discovered by other runs. From these results it would appear that discrimination between the set of systems lessens as the measurement depth increases. For some collections such as TREC-3, 6, 8 and the Web 2004 collections this becomes more stated as measurement depth is increased beyond the pooling depth.

We then examined the proportion of significant differences between systems that fall into either the "strong" or "weak" case. A common trend across collections was that, as measurement depth increased, the proportion of "strong" comparisons decreased while the proportion of "weak" cases increased. To illustrate, consider first P@10 for the TREC-3 collection in Table 1. We find a smaller proportion of comparisons falling into the "strong" case in contrast to P@30 (17% to 11%). Conversely there is an increase in "weak" cases from 7% to 8.7%. This trend remains as we continue increasing measurement depth towards P@1000. Using MAP, which is calculated over all 1000 documents, 3.7 % significant differences are "strong" compared to 19% "weak" cases. This trend is common across the other collections, and appears to be an indication of the amount of information that is used to estimate system performance. *Increasing measurement depth results in a higher level of uncertainty in system comparison.*

A similar trend is also followed for system comparisons where the null hypothesis that both systems have equal performance cannot be rejected. As measurement depth increases, the proportion of "strong" cases decreases, resulting in a

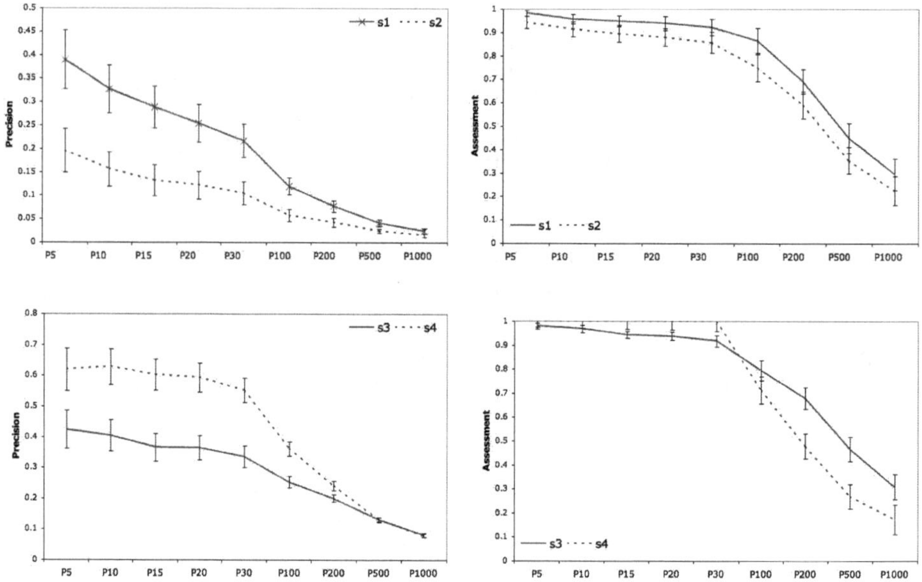

Fig. 2. Comparison of runs from Robust 2003 (top) and Robust 2005 (bottom)

larger proportion of cases where one system has a significantly larger number of assessed documents than another.

We then examined in closer detail what conditions would result in a swap from a "strong" to "weak" comparison and vice versa when increasing measurement depth. As a case study we present a comparison of two sets of systems from the Robust 2003 (Figure 2, top) and Robust 2005 tracks (Figure 2, bottom). In both figures we display both the P@ (left) and A_p@ (right) metrics at various ranks for both systems. Error bars are displayed to show variation across the set of topics and significance between systems.

The first example (Figure 2, top), illustrates a comparison of systems where conditions remain relatively even for both systems across various ranks. System s_1 has significantly higher precision than s_2 up until P@500. If we examine assessment, s_1 also has higher assessment but not significantly so with the exception of A@100. This example illustrates that even with systems that have comparable conditions in terms of assessment, the practice of using a measurement depth larger than the pooling depth can cause uncertainty in comparisons such as at A@100.

We also examined two runs from the Robust 2005 track where it has been identified that there is a bias in the collection towards documents that have query terms in the title [14]. The two runs were from the same research group, with system s_3 placing emphasis on keywords appearing in the title of documents, while s_4 uses an external collection to expand the original query. Initially system s_4 reports significantly higher precision across the 50 topics. As the measurement depth increases this improvement becomes marginal until both runs share similar performance. If we examine the levels of assessment across the same ranks, we find that assessment is equal until we reach A@100, then s_4 decreases in assessment with respect to s_3. For this collection the pool depth is 47.

This result reflects the findings in [14], where the performance of s_4 is underestimated once a larger measurement depth is used. System s_3 begins to return more assessed (and relevant) documents than s_4, which in turn is returning more unique and unassessed documents. From the study of Zobel [7], who investigated the rate of discovering new relevant documents beyond pool depth, it is uncertain if both systems shared similar levels of assessed documents that performance would converge.

4 Discussion and Conclusion

In this paper we argued that uncertainty should be identified during the evaluation process when using incomplete relevance judgements. We therefore proposed a new set of metrics based on the level of assessment which can be used to provide an indication of uncertainty during system comparisons. If the level of assessment between systems is similar, we believe that a fair comparison can be made, otherwise uncertainty has been introduced into the evaluation. By using the System Evaluation Decision matrix we can make stronger claims of significance (or not), and guide subsequent research to decide when further testing is required (e.g. shallower measurement depth, different collections, etc.). The advantage is those comparisons that may not be fair can be detected and investigated accordingly.

During the course of our empirical study, where we employed the extended evaluation methodology, we found evidence to suggest that the use of a measurement depth larger than the pooling depth weakens claims of significance in performance. This was a concern that was previously raised, but not confirmed, by Zobel [7]. Our results indicate that as the measurement depth increases beyond the pooling depth, uncertainty across many system comparisons also increases, and interestingly, the discrimination between systems weakens. Consequently, this supports the conclusions drawn by Sanderson and Zobel [13], who stated that metrics which consider early precision such as P@10 can be used to accurately discriminate between systems.

This decrease in discrimination at deeper measurement depths may result from the higher variation is performance estimates across topics stemming from the lack of assessment at these depths. Also, the majority of systems, what Aslam and Savell refer to as the "popular" systems [11], appear to discover a similar proportion of relevant documents once the measurement depth increases. However, the performance of the best systems may still be underestimated because they return documents than are unique to the documents returned by the "popular" systems. In general the relative rankings of systems tend to remain stable across measurement depths and varying levels of incompleteness. However, it is those cases where a system changes in ranking because of the effects of incompleteness that should to be detected, in particular, comparisons resulting in Case 2 in the decision matrix. This is because it is these systems, which are novel and considerably different to the "popular" systems, that require more consideration during evaluation.

In this paper we have introduced an extended retrieval evaluation methodology which uses Assessment Precision to determine whether comparisons between competing systems are made under similar conditions. The adoption of this methodology leads to a fine grained analysis during the evaluation as the necessary context is provided to draw firmer conclusions. Future work will examine the implications and usage of this methodology in greater detail as well as investigate issues relating to Assessment Precision such as the relationship between Precision and the level of assessment.

Acknowledgements

We would like to thank Fabio Crestani, Kalervo Järvelin, David Harper, and the anonymous reviewers for their valuable comments.

References

1. Salton, G.: The state of retrieval system evaluation. Information Processing Management **28** (1992) 441–449
2. Buckley, C., Voorhees, E.M.: Retrieval evaluation with incomplete information. In: Proceedings of the 27th ACM SIGIR Conference. (2004) 25–32
3. Voorhees, E.M., Harman, D.K., eds.: TREC: Experiment and Evaluation in Information Retrieval. MIT Press, Cambridge, Massachusetts 02142 (2005)
4. Spärck-Jones, K., Van Rijsbergen, C.J.: Report on the need for and provision of an "ideal" information retrieval test collection. Technical report, British Library Research and Development Report 5266, University of Cambridge. (1975)
5. Blair, D.C.: Some thoughts on the reported results of TREC Information Processing Management **38** (2002) 445–451
6. Voorhees, E., Harman, D.: Letters to the editor. Information Processing Management **39** (2003) 153–156
7. Zobel, J.: How reliable are the results of large-scale information retrieval experiments? In: Proceedings of the 21st ACM SIGIR. (1998) 307–314
8. Carterette, B., Allan, J., Sitaraman, R.: Minimal test collections for retrieval evaluation. In: Proceedings of the 29th ACM SIGIR, Seattle, WA (2006) 268–275
9. Cormack, G.V., Palmer, C.R., Clarke, C.L.A.: Efficient construction of large test collections. In: Proceedings of the 21st ACM SIGIR. (1998) 282–289
10. Sanderson, M., Joho, H.: Forming test collections with no system pooling. In: Proceedings of the 27th ACM SIGIR. (2004) 33–40
11. Aslam, J.A., Savell, R.: On the effectiveness of evaluating retrieval systems in the absence of relevance judgments. In: Proceedings of the 26th ACM SIGIR, Toronto, Canada (2003) 361–362
12. Soboroff, I., Nicholas, C., Cahan, P.: Ranking retrieval systems without relevance judgments. In: Proceedings of the 24th ACM SIGIR. (2001) 66–73
13. Sanderson, M., Zobel, J.: Information retrieval system evaluation: effort, sensitivity, and reliability. In: Proceedings of the 28th ACM SIGIR. (2005) 162–169
14. Buckley, C., Dimmick, D., Soboroff, I., Voorhees, E.: Bias and the limits of pooling. In: Proceedings of the 29th ACM SIGIR Conference, Seattle, WA (2006) 619–620

Evaluating Query-Independent Object Features for Relevancy Prediction*

Andres R. Masegosa[1], Hideo Joho[2], and Joemon M. Jose[2]

[1] Department of Computer Science and A.I., University of Granada, Spain
[2] Department of Computing Science, University of Glasgow, UK
andrew@decsai.ugr.es, {hideo,jj}@dcs.gla.ac.uk

Abstract. This paper presents a series of experiments investigating the effectiveness of query-independent features extracted from retrieved objects to predict relevancy. Features were grouped into a set of conceptual categories, and individually evaluated based on click-through data collected in a laboratory-setting user study. The results showed that while textual and visual features were useful for relevancy prediction in a topic-independent condition, a range of features can be effective when topic knowledge was available. We also re-visited the original study from the perspective of significant features identified by our experiments.

1 Introduction

There has been a growing interest in leveraging *contexts* in different aspects of Interactive Information Retrieval (IIR) systems [1,2,3]. While the IR community might not have a consensus regarding what exactly a context is, the progress has been made on the understanding of IR in contexts. For example, Ingwersen and Järvelin [4] propose a model of context stratification which includes a wide range of features in the information seeking and retrieval environment. The model offers structured focus for the work on finding the potentially significant contexts to improve the performance of IIR systems. Some of the proposed strata relevant to this work are: work task features; interaction features; and document features.

One way to identify significant contextual features is to investigate their relationship to the relevancy of retrieved objects. For example, Kelly and Belkin [5] found that the reading time of documents can vary significantly across the topics, thus, it can be difficult to predict the document relevancy. Fox et al. [6] applied a machine learning technique to model the interaction features with respect to the document relevancy. Another way to find significant features is to observe the effect of features in an IR technique such as relevance feedback. For instance, White, et al. [7] investigated the effects of topic complexity, search experience, and search stage in the performance of implicit relevance feedback. Furthermore, the relationship between the context strata is important to understand the significance of features. For example, Freund, et al. [8] suggest that the document genres can be indicative of the type of topics in a workplace environment.

* This work was supported by ALGRA project (TIN2004-06204-C03-02), FPU scholarship (AP2004-4678) and EPSRC (Ref: EP/C004108/1).

G. Amati, C. Carpineto, and G. Romano (Eds.): ECIR 2007, LNCS 4425, pp. 283–294, 2007.

In this paper, we present a series of experiments investigating the effectiveness of query-independent object features to predict document relevancy. Our evaluation was based on experimental data collected in a laboratory-based user study with 24 participants searching four different topics. Participants were asked to bookmark a document when perceived relevant information was found. Object features were extracted from click-through documents and bookmarked documents. The objective of using such data was two-fold: First, we can decrease the uncertainty of users' underlying information needs inferred from interaction data, compared to a search engine's transaction log. Second, we were interested in re-visiting the result of the original study from the perspective of significant features identified by the experiments.

The rest of this paper is structured as follows. Section 2 describes our methodology to extract a set of potential object features, and other operations to improve the performance of the relevancy prediction. Section 3 presents a series of experiments investigating the effectiveness of the object features. Section 4 discusses the implications of our findings. Finally, Section 5 concludes the paper with future work.

2 Methodology

This section describes our methodology used to extract object features and other operations applied to the features to improve the performance of relevancy prediction. An overview of the classifiers used in our experiment and performance measures is also discussed.

2.1 Conceptual Categories of Object Features

The first step to find effective features for relevancy prediction was to identify candidate features that can be extracted from retrieved documents. Based on some informal experimentation and literature survey, we have identified approximately 150 object features. To increase the understanding of candidate features in relevancy prediction, we then grouped them into a set of conceptual feature categories. The structure used for the categorisation is shown in Table 1.

As can be seen, there are four main categories: Document textual features, visual/layout features, structural features, and other selective features. The objective of the main categories was to group candidate features into a set of independent functionality played in a document. Therefore, we do not claim that our categorisation is ideal for all applications. On the contrary, the structure of features should be revised as an investigation progresses. Nevertheless, the structure shown in Table 1 was a good starting point for us to investigate the effectiveness of object features. An overview of the main categories is as follows.

Document textual features: This category consists of features that are related to textual contents of documents. The examples of features include the number of words in a document and anchor texts, number of upper-case words, number of digits, Shannon's entropy value [9] for a document and anchor texts.

Visual/Layout features: This category consists of features that are related to visual or layout aspects of documents. There are three sub categories: *Visual appearance*

Table 1. Conceptual categories for object features

Main Category	Sub category	Code	Feature size
Document textual features		DOC	14
Visual/Layout features	Visual Appearance	V-VS	28
	Visual HTML tags	V-TG	27
	Visual HTML attributes	V-AT	16
Structural features		STR	18
Other selective features	Selective words in anchor texts	O-AC	11
	Selective words in document	O-WD	11
	Selective HTML tags	O-TG	7
	Selective HTML attributes	O-AT	16

includes features such as the number and dimension of (background) images and foreground/background colours; *HTML Tags* includes a set of HTML tags such as `font`, `li`, and `table`; *HTML attributes* includes attributes used across the HTML tags such as `style`, `border`, and `face`.

Structural features: This category consists of features that are related to hyperlink and site structure of documents. The examples include the depth of document in a URL, the number of outlinks, PageRank scores.

Other selective features: This category consists of features that are not necessarily fit into the above categories. There are four sub categories: *Selective words in document* includes the presence of selective words such as address, search, and help; Selective words in anchor texts is the same as above but extracted from only anchor texts; Selective HTML tags includes a set of HTML tags such as `form`, `object`, and `script`; Selective HTML attributes includes `lang`, `onclick`, `src`, etc.

Extraction of the features were carried out by a mixture of tools such as [10] and [11]. The following sections describe a proposed methodology to build classifier, to select significant features, and finally, to validate the results.

2.2 Probabilistic Classification Approach

The classification problem can be seen as an ability of predicting a given feature of an object using another set of features of the same object. In the probabilistic classification paradigm, the classification problem can be described by two types of random variables:

Class Variable: C. This random variable is the variable to be predicted. This variable contains one state for each possible prediction $\Omega_C = \{c_1, ..., c_k\}$. In our case, $\Omega_C = \{Relevance, Non-Relevance\}$.

Predictive/Attribute Variables: $\mathbf{X} = \{X_1, X_2, ..., X_n\}$. Each variable has a set of possible states (discrete variables) or continuous values (continuous variables). For simplicity, we are only going to talk about discrete variables. Then, $\Omega_{X_i} = \{x_{ij_1}, ..., x_{ij_k}\}$ is the set of possible states of the X_i random variable. In our case, $\mathbf{X}$ is the set of variables described in Section 2.1.

In this model, our objective is to learn the following probability distribution:

$$\mathbf{P}(C|X_1, X_2, ..., X_n) = \{\mathbf{P}(c_1|X_1, ..., X_n), ..., \mathbf{P}(c_k|X_1, ..., X_n)\}$$

In other words, the probability of the class variable given the set of attributes variables. The prediction of the class (Relevance or Non-Relevance) is based on the highest *a posteriori* probability. This probability distribution is estimated by a set of data $\mathbf{D} = \{D_1, ..., D_M\}$, where each D_i contains an instantiation of the predictive features and the class for the object number i:

$$D_i = \{x_{1j_1}, x_{2j_2}, ..., x_{nj_n}, c_j\}$$

In our study, a great number of attribute variables were continuous. However, the literature suggests that the performance of the classifiers can be more robust when the variables are discrete data. Therefore, we used the *equal frequency* discretisation method [12] to split the continuous variables into 10 intervals.

Another aspect to consider in the classification was the balance of the class variable distribution (i.e., the portion of relevant and non-relevant documents in a data set). An imbalance data is known to deteriorate the performance of a classifier [13]. We took the following approach to address the issue. When there were a large number of cases, we randomly removed the cases from the larger class until the portion was balanced. When there were a small number of cases, we used a resampling method to balance the data. Although this resampling method was a good technique to correct imbalanced data, it was also possible to over-estimate the performance of the classifier. We used AUC measure [12] to detect the over-estimation.

2.3 Classifiers Used

While a single Bayesian network approach was used by [6], we were interested in using several classifiers and reporting the result of the best performing classifier. This was because a single classifier was unlikely to show the significance of attribute variables in a complex dependency structure. We selected four classifiers that have been proved to be successful in machine learning classification. An overview of the classifiers used in our experiments is as follows.

Naive Bayes [14] **.** This is one of the well known probabilistic classifiers. The main assumption in this model is that all attribute variables depend on the class variable and they are independent of each other.

AODE [15]**.** This classifier can use multiple representations of a problem space to predict the class variable. A disadvantage is that this classifier can not show an explicit relationship between variables.

HNB [16]**.** This classifier creates a hypothetical variable to represent the relationship between the attribute variables. The resulted representation is then used to predict the class variable. HNB inherits the structural simplicity of Naive Bayes and can be trained without mining the dependency structure.

K2-MDL. This classifier is a variant of Bayesian networks classifier [17] where the structure is learnt by the K2 algorithm [18].

2.4 Feature Selection Scheme

The feature selection in the supervised classification paradigm is to find a minimum set of attribute variables that can achieve the best performance. The selection of significant features in the problem space can prevent the classifiers from introducing noisy evidences in the training stage. The feature selection can also reduce the number of variables to be considered in the problem space, thus, it can facilitate our understanding of significant variables.

While several techniques have been proposed for the feature selection [19], we used a wrapper method which can select a set of the best features based on the AODE classifier. The actual selection process was similar to the cross validation method described in the following section. The final set of features was generated by the features that were selected at least N% of the repeated cross validation process. We used 50%, 80%, and 90% as the cutoff levels in the feature selection. We found that the overall performance did not vary significantly over the cutoff levels. Therefore, we only report the results of 90% in the experiment since it consists of the smallest number of significant features.

2.5 Classification Validation Scheme

With the aim of provide a robust estimation of the accuracy of a classifier, the set of data was partitioned in two separated sets. The training data set was used to build the classifier and the test data set was used to estimate the performance. The K-fold-cross validation method was used to partition the data set as follows. The data set **D** was divided in K random subsets with the same size $\{D_1, ..., D_K\}$, thus, the validation process was repeated K times. In other words, in the step i with $i = 1...K$ a training data set was defined $T_i = \mathbf{D} \setminus D_i$ and the subset D_i was used as a test set and the accuracy was measured based on them. The mean of the K accuracy measures was reported as the final estimated performance of the classifier. In our study, a 10 fold-cross validation was repeated 10 times to measure the performance (i.e., based on 100 repeated estimations).

3 Experiments

This section presents a series of experiments which investigated the effectiveness of query-independent contextual features to predict the relevancy of click-through documents. The accuracy of prediction is defined by the portion of correct prediction in the total number of click-through documents. The correct prediction is a sum of the true positive and true negative cases (i.e., predicting a relevant document as relevant, and predicting a non-relevant as non-relevant). For example, when the data consist of 50 relevant and 50 non-relevant documents, and when 30 relevant and 40 non-relevant documents are correctly predicted, then the performance is 70%[1].

Throughout this section, the results are presented in two groups of data set. The first group is based on all click-through data without the distinction of individual topics. The second group is based on the data within individual topics. The former is referred to as

[1] $\frac{30R+40NR}{50R+50NR} = \frac{70}{100} = .7.$

Table 2. Baseline performance of relevancy prediction

	Click-through	Relevant	Non-Rel	Baseline (%)	Balanced (%)
No topic	737	375	362	50.9	
Topic 1	203	123	80	60.6	50.0
Topic 2	173	83	90	52.1	
Topic 3	154	69	85	55.2	
Topic 4	207	100	107	51.7	

the *topic-independent* set, and the latter is referred to as the *topic-dependent* set. This grouping enables us to examine the effect of topic knowledge in relevancy prediction, and how the effectiveness of the features differs in the two conditions.

The section is structured as follows. First, the baseline performance of relevancy prediction is established by looking at the portion of relevant/non-relevant documents in the click-through data set. Second, the effect of contextual features in each category is examined. Then, the effects of several operations on the contextual features are presented: feature selection, feature combination, and use of highly relevant documents.

3.1 Baseline Performance

A total of 1038 click-through documents were extracted from our user study of 24 participants searching four different search topics [20]. Of those, 375 were unique relevant and 362 were unique non-relevant documents. Therefore, the baseline performance of relevancy prediction was set to 50.9% in the topic-independent set (denoted as *No topic* in the tables). The portion of relevant/non-relevant documents varied across the four topics. The baseline performance was taken from whichever the higher portion of relevance, as shown in Table 2.

Note that a relatively large difference was found between the number of relevant and non-relevant documents in Topic 1. To measure an accurate performance of the classifiers, we generated a balanced data set by a random sampling for Topic 1 (shown in the 6th column of Table 2). In the following analysis, the performance based on the balanced set is used for Topic 1. No change was found to be necessarily for the rest of the data set.

3.2 Effect of Contextual Features

The first experiment examined the effect of contextual features in the individual context categories. In this experiment, the classifiers used only the features defined in each category to predict the relevancy, and the same procedure was repeated for all categories. The performance of relevancy prediction was compared to the baseline performance and the relative improvement was shown in Table 3. The bottom row of the table shows the average improvement across the four topics (but not including the topic-independent set). The statistically significant differences are highlighted in bold in the table. We used the t-test ($p \leq .05$) for the statistical tests throughout the study.

As can be seen, the features in the DOC and V-AT categories were found to be useful for improving the relevancy prediction in the topic-independent set. While the V-VS category was found to be effective in Topic 1, the overall effect of individual categories

Table 3. Effect of contextual features

		DOC	V-VS	V-TG	V-AT	STR	O-AC	O-WD	O-TG	O-AT	Mean
No topic	50.9	**+5.7**	+2.5	+2.2	**+3.9**	+0.4	+1.2	-2.1	-0.9	+2.2	+1.7
Topic 1	50.0	+4.1	**+11.3**	-2.7	+4.2	+2.0	+5.7	-4.0	+3.1	+2.7	+2.9
Topic 2	52.1	+0.3	+7.3	+6.1	-2.2	+2.9	-9.2	-4.3	+2.0	+4.8	+0.9
Topic 3	55.2	+0.6	+4.4	-2.6	+5.4	+1.1	+8.1	-0.8	+4.8	+6.3	+3.0
Topic 4	51.7	-5.1	-2.6	+0.6	+2.3	+2.1	+1.4	+3.1	+0.6	-0.7	+0.2
Mean		+1.1	+4.6	+0.7	+2.7	+1.7	+1.4	-1.6	+1.9	+3.1	+1.7

appeared to be weak across the topics. Furthermore, the performance of most categories appeared to be inconsistent across the topics. The exceptions were the STR and O-TG categories, but the differences were not significant. The following sections present the effects of several operations on the features to improve the performance.

3.3 Effect of Feature Selection

In the previous experiment, all features were used to predict the document relevancy in the individual categories. One way to improve the performance is to use only a subset of features that are likely to contribute to the prediction, which is called a *feature selection*. There are several methods of the feature selection. In this study, we used the features that were selected 90% of times in the repeated tests on the training set. The feature selection was carried out in the indivdual categories and the result of relevancy prediction is shown in Table 4.

Table 4. Effect of feature selection with 90% cutoff

		DOC	V-VS	V-TG	V-AT	STR	O-AC	O-WD	O-TG	O-AT	Mean
No topic	50.9	**+4.6**	+1.4	+3.1	**+3.8**	+2.4	+2.9	+2.0	+0.3	+1.1	+2.4
Topic 1	50.0	0.0	+1.4	**+10.8**	+7.4	**+11.5**	+6.1	+5.8	+4.1	+3.8	+5.6
Topic 2	52.1	0.0	-3.8	+5.8	+4.8	+5.2	+3.7	+0.9	+2.6	-9.0	+1.1
Topic 3	55.2	0.0	**+8.4**	0.0	+6.2	+0.7	**+10.1**	-1.4	+6.4	-3.3	+3.0
Topic 4	51.2	+2.3	0.0	+3.5	+1.3	+5.8	+5.4	**+8.6**	-1.5	+5.4	+3.4
Mean		+1.4	+1.5	+4.6	+4.7	+5.1	+5.6	+3.2	+2.4	-0.4	+3.1

From the far right column (Mean) of Table 4, there appears to be an overall positive effect of the feature selection, compared to Table 3. However, in the topic-independent set, the performance of the significant categories (i.e., DOC and V-AT) was degraded by the feature selection. This suggests that a greater number of the features should be considered in the individual categories when no topic knowledge was available for the relevancy prediction.

On the other hand, a significant improvement was found in several categories of the topic-dependent sets when the feature selection was carried out. The results show that, for example, Topic 1 is likely to benefit from the features in the V-TG and STR categories. What is more important is that the significant category is likely differ across the topics.

Table 5. Effect of feature combination

		Best Cat	Combined	High Rel
No topic	50.9	**+5.7**	**+5.6**	**+11.6**
Topic 1	50.0	**+11.5**	**+10.3**	
Topic 2	52.1	+5.8	**+9.2**	
Topic 3	55.2	**+10.1**	**+9.9**	
Topic 4	51.7	**+8.6**	+2.2	
Mean			+7.4	

In fact, no single category contributed to a significant improvement on more than one topic. Topic 2 appeared to be a particularly difficult topic to find effective features. This suggests that the effectiveness of query-independent features is fairly topic-dependent.

3.4 Effect of Feature Combination

So far, we have examined the performance of the individual feature categories. The features selected in the previous experiment appeared to have a varied effectiveness over the topics. We further investigated the performance of the context features by combining the selected features into a single category. The advantage of the feature combination is that the classifiers do not have to find a particular category for the relevancy prediction. The result is shown in Table 5. In the table, the performance of the best category in the previous experiment is shown in the 3rd column (Best Cat), and the performance of the combined feature is shown in the 4th column.

As can be seen, the overall performance of the feature combination appears to be comparable to the best performing category in the previous experiment, except Topic 4. And, the effect appears to be consistent across the two data sets. In particular, Topic 2 was found to benefit from the feature combination significantly. The mean value of four topics (the bottom row of Table 5) suggests that the performance of combined feature is likely to be more consistent than any single feature category. We also tested the different cutoff levels (50%, 80%, and 90%) of the feature selection before the combination, and a similar performance was found over the cutoff levels.

3.5 Effect of Highly Relevant Documents

The last experiment in this paper looked at the effect of highly relevant documents for the relevancy prediction. In the literature, the importance of highly relevant documents has been suggested in the evaluation of IR systems [21]. In this study, the highly relevant documents were determined when the document was judged as relevant by at least two participants in the same topic. While this criterion was not based on a graded relevance judgement, it enabled us to select a reasonable number of relevant documents whose relevancy was shared by participants.

There were a total of 96 documents that were judged by at least two participants. Of those, 69 were relevant and 27 were non-relevant. Similar to Topic 1's data set, we needed to balance the portion of relevant /non-relevant documents for the this analysis.

Table 6. Effect of highly relevant documents (without feature selection)

		DOC	V-VS	V-TG	V-AT	STR	O-AC	O-WD	O-TG	O-AT	Mean
All rel	50.9	**+5.7**	+2.5	+2.2	**+3.9**	+0.4	+1.2	-2.1	-0.9	+2.2	+1.7
High rel	50.0	**+15.2**	**+15.5**	**+16.7**	+8.2	†+12.0	+4.9	+0.3	**+11.6**	+6.2	+9.8

†An over-estimation was detected in this result, thus, not considered.

We used a replacement method which was often used in machine learning (See Section 2.2). The method is known to be robust to measure the performance in a similar situation, and to detect any anomalies in the results.

The effect of highly relevant documents is shown in Table 6. We only report the result of the topic-independent set since the data was too small to measure the individual topic performance. As can be seen, the significant improvements were found in several categories when the highly relevant documents were targeted for the relevancy prediction. The result also shows that the features from a wider range of categories can be considered for the prediction in the topic-independent set. We also measured the effect of feature combination based on the highly relevant documents, and the result is shown in the 5th column of Table 5. As can be seen, a respectable improvement can be achieved without selecting the best performing category. These results show that the use of highly relevant documents can be a more effective way to predict the document relevancy than the other methods examined in this study. This is interesting because the classifiers usually perform worse when the size of the training data decreases.

4 Discussion

This section discusses the implications of our experimental results. We also re-visit the result of the original user study from the perspective of the significant features identified by the experiments.

4.1 Effectiveness of Query-Independent Features

The findings from our experiments have several implications for the use of query-independent object features to predict the document relevancy. First, the set of effective features can be different when the prediction is performed with/without topic knowledge. In the topic-independent set, the textual document features and visual/layout HTML attributes are likely to be significant to predict the document relevancy. In our experimental conditions, the feature selection or feature combination were found to make little improvement on the performance. However, a simple filtering to select highly relevant documents was found to be effective to improve the performance. In the topic-dependent set, on the other hand, many categories can be effective for the relevancy prediction. However, the effectiveness of individual categories can vary across the topics. The results show that the feature selection and feature combination can be effective for improving the performance in this set. Another implication of our results is that an additional classifier for a topic detection should be used supplementary to the relevance prediction. Such a two-stage approach would allow us to use a topic-dependent significant category effectively, thus, can be promising to improve the performance.

Table 7. Minimum set of significant features[†]

	Topic 1	Topic 2	Topic 3	Topic 4
Topic	Dust allergy in workplace	Music piracy on Internet	Petrol price change	Art galleries and museums in Rome
DOC				
V-VS			**imageBDiskSize**	
V-TG	**meta, li**			
V-AT				
STR	**URL-Levels HtmlLink**	PR-Page, link URL-Domain numlinksAnchor		
O-AC	contact email search		**contact help email**	search help tel
O-WD				**search, address**
O-TG				
O-AT				

[†]The significant features with 90% cutoff is highlighted with bold. The rest are based on 80% cutoff in the feature selection.

Our preliminary test to predict the four topics using the relevancy classifier showed an accuracy of between 45 to 60% with the average of 55%. A further investigation is under way for the integrated approach.

The results also suggest that the textual document features such as the entropy are rarely effective within the individual topics. This was contrast to their performance in the topic-independent set. Our speculation is that the entropy and other document level features might have a low discriminating power to separate relevant documents from non-relevant. Other features that occur less frequently in the data set appear to have a higher discriminating power. Therefore, a similar phenomenon that motivates the idea of inverse document frequency [22] might be applicable to indicate the significance of query-independent features. This also supports our approach to use a range of objects' features to predict the document relevancy.

4.2 Re-examination of the Original Study

Our experiments were based on the experimental results of a user study carried out in a laboratory setting. A motivation for using such data was to decrease an uncertainty of users' underlying information needs in the experiments. A distinct objective was to re-examine the result of the original study from the perspective of the significant features identified by the experiments. In this section, we discuss the findings of such analysis. Table 7 shows the minimum set of the query-independent features that contributed to a significant improvement in the individual topics. The minimum set was determined by the multiple cutoff levels (80% and 90%, See Section 2.4) in the feature selection to increase the number of indicative features.

In Topic 1, participants were asked to find the information on the potential solution to dust allergy in a workplace. Some perceived relevant documents contained a list of steps to reduce the dust inside a building. Therefore, the `li` tag in the visual feature category was a significant indicator of the document relevancy. The depth of document in a web site appeared to vary in this topic compared to the other topics. As discussed before, Topic 2 was a difficult topic to find significant features. In this topic, participants were asked to find the information on the damage of music piracy on Internet. The structural features in this topic suggest that participants were able to find relevant information in the top ranked documents from a limited number of URL domains. The most unexpected result was Topic 3 where the disk size of background images (imageB-DiskSize) was found to be a significant indicator of the document relevancy. A close examination showed that the background image information was more helpful for predicting non-relevant documents than relevant documents. The selective words in the anchor texts appeared to be useful for this topic. The result of Topic 4 was also interesting. We initially expected that the visual features were likely to be significant in this topic, but this was not the case. Instead, the selective words in a document were found to be a significant indicator of the document relevancy. We speculate that since most click-through documents contained a variety of images in this topic, their discriminating power was lower than we had expected. However, since participants were asked to find the information on a particular location, the words such as *address* was found to be significant.

As can be seen, the result of the re-examination of the original work was a mixture of re-assurance and puzzlement. More importantly, however, the significant features appeared to offer us a pointer for the further examination of the original study. In this sense, the re-examination of the original work based on the significant features can supplementarily used in the evaluation of user studies.

5 Conclusion and Future Work

This paper presented a series of experiments which investigated the effectiveness of the query-independent features to predict the document relevancy. The experimental results from a user study were used to extract the various features of retrieved objects. Our results show that the document-level textual features and visual features can be indicative of the document relevancy in an topic-independent situation. The use of highly relevant documents can improve the performance significantly. When the type of topics was known, a wider range of features can be effective for the relevancy prediction. However, the effectiveness of the features is likely to vary across the topics. Overall, these findings highlight the importance of investigating the significance of objects' features from the perspective of the topics and aggregated relevance assessments.

In this study, we investigated the features from retrieved objects. We are conducting a similar experiment based on the interaction features in the other parts of the user logs, searchers features gained from the participants" background information and finally, subjective perceptions on the topic characteristics established by the questionnaires. We anticipate that the features from the additional context strata can facilitate the understanding of the original user study. We also plan to evaluate the features extracted from another user study to investigate the robustness of the significant features identified by this study.

References

1. Ingwersen, P., Belkin, N.: Information retrieval in context - IRiX: workshop at SIGIR 2004. SIGIR Forum **38**(2) (2004) 50–52
2. Ingwersen, P., Järvelin, K.: Information retrieval in context: IRiX. SIGIR Forum **39**(2) (2005) 31–39
3. Ruthven, I., Borlund, P., Ingwersen, P., Belkin, N., Tombros, A., Vakkari, P., eds.: Proceedings of the 1st IIiX Symposium, Copenhagen, Denmark (2006)
4. Ingwersen, P., Järvelin, K.: The Turn: Integration of Information Seeking and Retrieval in Context. Springer (2006)
5. Kelly, D., Belkin, N.J.: Display time as implicit feedback: understanding task effects. In: Proceedings of the 27th SIGIR Conference, Sheffield, United Kingdom, ACM Press (2004) 377–384 1009057.
6. Fox, S., Karnawat, K., Mydland, M., Dumais, S., White, T.: Evaluating implicit measures to improve web search. ACM Transactions on Information Systems **23**(2) (2005) 147–168 1059982.
7. White, R.W., Ruthven, I., Jose, J.M.: A study of factors affecting the utility of implicit relevance feedback. In: Proceedings of the 28th SIGIR Conference, Salvador, Brazil, ACM (2005) 35–42
8. Freund, L., Toms, E.G., Clarke, C.L.A.: Modeling task-genre relationships for ir in the workspace. In: Proceedings of the 28th SIGIR Conference, Salvador, Brazil, ACM (2005) 441–448
9. Shannon, C., Weaver, W.: The Mathematical Theory of Communication. University of Illinois Press, Urbana, Illinois (1949)
10. (Html parser. "http://htmlparser.sourceforge.net/")
11. (Firefox add-ons. "https://addons.mozilla.org/")
12. Duda, R.O., Hart, P.E.: Pattern Classification. Wiley Interscience (2000)
13. Japkowicz, N., Stephen, S.: The class imbalance problem: A systematic study. Intelligent Data Analysis **6**(5) (2002) 429–449
14. Duda, R.O., Hart, P.E.: Pattern Classification and Scene Analysis. John Wiley Sons, New York (1973)
15. Webb, G.I., Boughton, J.R., Wang, Z.: Not so naive bayes: aggregating one-dependence estimators. Mach. Learn. **58**(1) (2005) 5–24
16. H. Zhang, L.J., Su, J.: Hidden naive bayes. In: Proceedings of the Twentieth National Conference on Artificial Intelligence (AAAI-05), (AAAI Press(2005).)
17. Pearl, J.: Probabilistic Reasoning with Intelligent Systems. Morgan & Kaufman, San Mateo (1988)
18. Cooper, G., Herskovits, E.: A bayesian method for the induction of probabilistic networks from data. Machine Learning **9** (1992) 309–347
19. Kohavi, R., John, G.H.: Wrappers for feature subset selection. Artificial Intelligence **97**(1-2) (1997) 273–324
20. Joho, H., Jose, J.M.: Slicing and dicing the information space using local contexts. In: Proceedings of the First Symposium on Information Interaction in Context (IIiX), Copenhagen, Denmark (2006) 111–126
21. Järvelin, K., Kekäläinen, J.: Ir evaluation methods for retrieving highly relevant documents. In: Proceedings of the 23rd SIGIR Conference., Athens, Greece, ACM (2000) 41–48
22. Sparck Jones, K.: A statistical interpretation of term specificity and its application in retrieval. Journal of Documentation **28**(1) (1972) 11–21

The Utility of Information Extraction in the Classification of Books

Tom Betts, Maria Milosavljevic, and Jon Oberlander

School of Informatics, University of Edinburgh,
2 Buccleuch Place, Edinburgh EH8 9LW, UK
tom@4angle.com, {mmilosav,jon}@inf.ed.ac.uk
http://www.inf.ed.ac.uk

Abstract. We describe work on automatically assigning classification labels to books using the Library of Congress Classification scheme. This task is non-trivial due to the volume and variety of books that exist. We explore the utility of Information Extraction (IE) techniques within this text categorisation (TC) task, automatically extracting structured information from the full text of books. Experimental evaluation of performance involves a corpus of books from Project Gutenberg. Results indicate that a classifier which combines methods and tools from IE and TC significantly improves over a state-of-the-art text classifier, achieving a classification performance of $F_{\beta=1} = 0.8099$.

Keyword: Information Extraction, Named Entity Recognition, Book Categorisation, Project Gutenberg, Ontologies, Digital Libraries.

1 Introduction

Books have long been classified in ontologies to help users locate relevant titles. The emergence of digital libraries and on-line resellers has released the classification of books from the constraints of a physical space, allowing the use of multi-labelling, and even folksonomies.[1] This is very useful, since books are often inconsistently classified in different systems, or not satisfactorily classified at all.

Text categorisation (TC) is the task of classifying texts, based on their content, into one or more predefined categories. Popular applications include filtering e-mail messages for spam, or indexing newswire articles. Information Extraction (IE) is a process by which we can identify structure within unstructured natural language text. The IE task consists of a series of subtasks, one of which is Named Entity Recognition (NER), which is concerned with finding entities within a text such as people, locations and organisations. Intuitively, we expect that reducing unstructured text into a structured format should provide leverage in TC. Although there has been considerable research in both TC and IE, only a limited amount of work has combined them. We have found no other work which uses content-based categorisation on full-text books or large texts. Our

[1] See, for example www.librarything.com

G. Amati, C. Carpineto, and G. Romano (Eds.): ECIR 2007, LNCS 4425, pp. 295–306, 2007.

approach is closest to [12] in terms of representation, combining phrases (where we use named entities) with bag-of-words; cf. the models discussed in Section 3.

Here, we set out to show that by using IE, information derived automatically from the full text of books can lead to good TC performance. We first determine the type of information that can easily be extracted from books using NER. Second, we explore how this information can be incorporated into a TC task and develop a framework for its inclusion. Finally, we show that the use of this information leads to a statistically significant improvement in performance when compared to a state-of-the-art text classifier.

2 Motivation and Context

Major players in the text industry have a considerable interest in the digitisation of books. For example, Amazon.com are scanning books for their 'Search Inside the Book' programme, the Million Book Project at Carnegie Mellon has scanned over 600,000 texts[2] and Google is expected to digitise 10.5 million books for their Google Books project [4]. For the purpose of this work, we use books from Project Gutenberg,[3] which contains over 19,000 copyright-free electronic texts.

Categorising books is non-trivial due to subjectivity and the sheer volume and variety of titles available. Manual classification is time consuming and, therefore, impractical for large collections such as digital libraries or online retailers, who stock millions of different titles. Hence, content-based classification may prove a useful alternative. However, the challenges of heterogeneity and scalability distinguish this problem from traditional automated text categorisation. For example, the length of full-text books may be orders of magnitude greater than newswire stories or e-mail messages classified in past work.

In this work, we use the Library of Congress Classification[4] (LoCC), an ontology for categorising books according to their subject, used by most academic and research libraries. Although LoCC was created to satisfy the constraints of a physical library, it can also be used to categorise digital collections. Because a library cannot contain many copies of the same book, books are traditionally placed into only one or at most two LoCC categories. But digital collections have no such constraints, and in fact, assigning multiple categories is highly desirable in order to help diverse users locate relevant books. Nürnberg et al. [13] provide further detail on the differences between digital and physical libraries.

2.1 Approaches to Book Categorisation

Manually categorising books has long been the concern of librarians. They assign books to predefined categories from hierarchical ontologies such as LoCC or the Dewey Decimal Classification. Early work on the automated categorisation of library texts (not necessarily books) focussed on the content of abstracts [2].

[2] `http://www.library.cmu.edu/Libraries/MBP_FAQ.html`

[3] `http://www.gutenberg.org`

[4] As outlined at `http://www.loc.gov/catdir/cpso/lcco/lcco.html`

A more recent approach to categorising books has been to use structured metadata, such as titles or subject headings—a less hierarchically structured form of subject annotation for books [8,9,6]. Frank and Paynter's [6] work is closest to ours, taking a machine learning approach to categorisation. Using Library of Congress Subject Headings, they report an accuracy of $\sim 55\%$ when categorising 50,000 books into 4214 categories. However, their classification is based on curated metadata, which may be susceptible to annotator inconsistency; by contrast, our approach is content-based. Mooney and Roy [11] use content-based text categorisation in a recommender system for books. They use both structured metadata and unstructured text for categorisation, but limit the use of text to product descriptions, editorial and user reviews.

3 Methods and Tools

Our central hypothesis is that structured information that can be automatically extracted from unstructured book texts is useful for the purpose of categorising them. In testing this hypothesis, we use Named Entity Recognition (NER) in addition to a standard bag-of-words approach. NER can be achieved with good performance, suffers acceptable domain coupling, and should be computationally tractable [5]. Furthermore, named entities can provide important cues to the topic of a text. Thus, we pursue NER as a technique for the extraction of structured information, and we use pre-existing tools to extract entities from each book text. We explore the use of names of people, organisations, locations and dates, based on the assumption that these entities should help to discriminate between the categories in our corpus.

3.1 Corpora

We use full-text books from Project Gutenberg, using their published LoCC as the 'gold standard'. Gutenberg texts have been assigned a LoCC including schedule, denoted by the first letter (e.g. D: History: General and Eastern Hemisphere), and subclass, denoted by subsequent letters (e.g. DA: History: General and Eastern Hemisphere: Great Britain, Ireland, Central Europe). In order to experiment with the extraction of structured information from the available texts, and effectively assess its usefulness in categorisation, we created two corpora containing all English texts from a subset of Gutenberg categories.

Our **development corpus** is used to experiment with and develop our classifiers. It includes 810 texts from 28 categories, including subclasses of schedules B (Philosophy, Psychology, Religion) and D (History: General and Eastern Hemisphere). These categories are selected based on their availability, the intuition that named entities may discriminate classes, and their content-based similarity.

Our **extended corpus** (comprising 1029 texts in 52 categories) combines the development corpus with texts from schedules H (Social sciences) and Q (Science). This provides further texts which may exhibit similarity with the development corpus; however, we do not manually alter our models specifically to discriminate H or Q categories. There are no features that exploit domain knowledge specific to these categories. Schedule Q, containing physical science subcategories, poses challenges by adding subject diversity, disparate linguistic style and texts which may not be easily discriminated by named entities.

3.2 Models for Representing Texts

In order to assess the utility of named entities in our categorisation task, we created three models that can be used to represent a text:

BOW Model: Bag-of-Words The first model uses a multinomial bag-of-words to represent the text. This model is considered state-of-the-art for TC, and is the baseline against which we evaluate other techniques.

NER Model: Bag-of-Named-Entities There have been several efforts to find more sophisticated representations of texts that capture additional 'knowledge' or 'meaning' present in a text. Techniques include the use of n-gram phrases [10,15,7,12,3] and word senses [15,12] instead of words. We create a representation of each book text as a variant of a bag-of-phrases, using n-gram entities extracted with NER. The classification of entities is exploited so that entities with the same string value, but different types (e.g. Chelsea-LOC and Chelsea-PER), are treated as different features.

GAZ Model: Generalising Named Entities One apparent difficulty with the use of phrases in TC is that they suffer from inferior statistical qualities when compared to word-based models [12,14]. A named entity model will have similar problems, since it is not clear that the same named entity will be observed across texts. An attempt is made to reduce sparse data difficulties by generalising the observed named entities into a smaller number of synthetic features. We incorporate domain knowledge to create features that exploit ontological structure. For example, we observe that the subclasses of schedule D (History) are organised by geographical region. Features are created using geographic gazetteers, defined according to the region of each category. For subclasses of schedule B (Philosophy, Psychology, Religion), we can use gazetteers of names of relevant people and organisations.

Although we evaluate the baseline BOW, NER and GAZ models individually, previous work with phrases for TC [12,10] has combined a bag-of-words feature space with phrases to overcome their inferior statistical qualities [14]. We use a similar approach, creating a continuous feature space by combining the feature spaces of individual models. The models are:

NER-GAZ combines the feature spaces of NER and GAZ, exploiting GAZ as smoothing, as discussed above;

BOW-NER combines NER features with the baseline BOW feature space, helping to reduce the negative effects of phrase-based representations, such as lack of partial matching and misclassified or missing named entities; and,

BOW-NER-GAZ involves the combination of all three individual models.

3.3 Classifier Configurations

This task is a multiclass problem because Gutenberg texts are only assigned one of the available LoCC. We evaluate two approaches to multiclass classification.

Multiclass. This requires training a single classifier to distinguish between all categories, where the objective function we optimise is the accuracy of classification when assigning one of many categories. This classifier exploits the mutual exclusivity of class labels during training, whereas binary classifiers combined in one-vs-all classification (described below) treat category assignment decisions independently (it is only during the final assignment that mutual exclusivity is exploited). The classifier used in these experiments is a multiclass generalisation of SVM, SVM^{multiclass}[17].

One-vs-All. We evaluate one-vs-all classification because it can assign multiple labels to a book (motivated in Section 2). Unfortunately, training and evaluating classifiers combined using one-vs-all may be computationally expensive, due to the need to use one classifier per category. However, it follows that the complexity of individual classifiers should be relatively low (compared to the direct approach described above). We use multiple instances of SVM^{light}[16].

We make classifications by selecting the classification that we are most confident about[5] and pursue two approaches to category assignment. A **relaxed** classification is made by assigning the category of the classifier outputting the largest positive value. A **forced** classification is made by assigning the 'least bad' classification—selecting the largest value, regardless of whether positive or negative. The relaxed classifier, which comes closest to generalising to a multi-label classification task, guarantees that at most one category is assigned, but does not guarantee to assign any. The intuitive result of forcing a classification would be an improvement in recall, but at the expense of precision.

3.4 Term Reduction and Feature Selection

Initial experiments were conducted using the entire feature space, and even with our relatively small corpus, execution was prohibitively slow. Two processes can reduce the size of models—the removal of infrequent terms and feature selection.

Term reduction involves removing terms that occur with high or low frequency. Due to the scale of texts being represented, we opt for maximum reduction, removing terms that occur fewer than five times in the corpus [14]. This provides an initial reduction in model size but is not sufficiently aggressive. However, it is a vital process because the technique that we use for feature selection, χ^2 [19], cannot provide accurate estimates for low frequency terms.

We use χ^2 to estimate the lack of independence between class labels and features. In particular, we use Yang and Pedersen's [19] definition of χ^2_{max} to obtain a single value for each feature. Features are sorted by χ^2_{max}, and the n (a variable in our experiments) largest are selected for inclusion in a model.

[5] Direct comparisons are made between binary classifiers even though we would not expect these confidences to be normalised, as discussed in Section 5.

3.5 Evaluation

There is considerable choice of metrics for performance evaluation, including micro-averages, a per-document metric, and macro-averages, performing averaging over categories. Given that a motivation for this work is a reduction in human effort required to categorise texts, it follows that a goal is accurate classification of as many *texts* as possible, rather than accurate classification of as many *categories*, since the latter could result in the accurate categorisation of very few texts. Evaluation and optimisation, therefore, focus on micro-average performance. $F_{\beta=1}$ is used as the primary metric for binary decisions and we occasionally discuss this in terms of precision and recall to provide further insight.

4 Results

Due to the quantity of metrics used for analysing experimental output, it is not feasible to comprehensively document all our results in this paper. For clarity, situation-specific visualisations and statistics are included to facilitate this discussion, and an overview of micro-$F_{\beta=1}$ for each classifier, displaying results for every model, is provided in Fig. 1. A comprehensive account of the results can be found in [1]. Unless otherwise stated, metrics quoted are micro-averages. The best result obtained using our baseline model was $F_{\beta=1} = 0.7914$, and this was improved to $F_{\beta=1} = 0.8099$ when incorporating named entities.

4.1 Baseline: Bag-of-Words

We begin this evaluation with our baseline, noting that micro-$F_{\beta=1}$ is indicated in Fig. 1. Both classifiers that guarantee to make exactly one classification, one-vs-all forced (Fig. 1A) and direct multiclass (Fig. 1C) classifiers, exhibit relatively straightforward optimisation of $F_{\beta=1}$, which increases as features are added. However, for the forced one-vs-all classifier, performance begins to fall after 5000 features, labelled (a). In the relaxed one-vs-all classifier (Fig. 1B), $F_{\beta=1}$ performance begins a continuous decline after only 500 features, indicated (b). Between points (b) and (c), $F_{\beta=1}$ falls from 0.7264 to 0.6564, which corresponds to a fall in the number of correct classifications from 507 to 405. This can be explained by a fall in certainty of classifications as the number of features increases. It follows that the classifier makes fewer category assignments, and precision increases, because the most 'uncertain' texts were those most likely to be incorrectly classified. Increasing the number of model features eventually reduces the ability of classifiers to generalise because the trained classifiers will tend to overfit, given our limited training data. This is also true for forced one-vs-all classifiers. Continuing to force classifications leads to a decrease in $F_{\beta=1}$, which we observe at point (a). From Fig. 1C we can see that multiclass classification appears to avoid overfitting (in the limited dimensions that we explore), and may be more robust because class assignment decisions are dependent on one another, whereas local decisions in one-vs-all classifiers are independent.

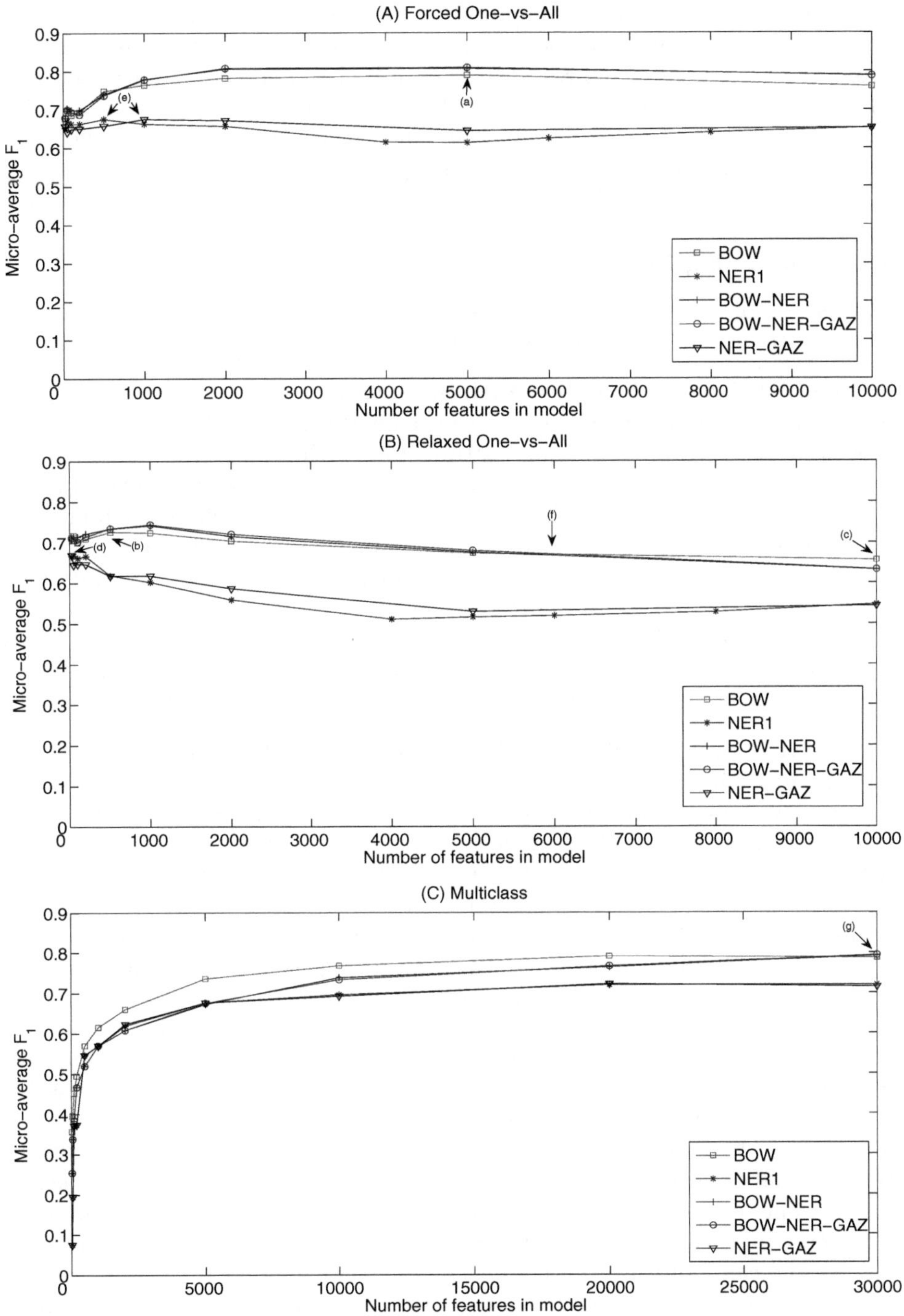

Fig. 1. Micro-average $F_{\beta=1}$ as number of features is varied for each classifier configuration (one per plot) and document model (all models per plot)

Comparing the classifier configurations, we observe that one-vs-all classifiers achieve respectable performance with very few dimensions, whereas multiclass classifiers require larger representations. Multiclass and one-vs-all forced classifiers exhibit the best $F_{\beta=1}$, which is broadly similar between them (~ 0.79), suggesting that the requirement for assigning exactly one class can be effectively exploited. Although the relaxed classifier performs worst overall, it achieves a precision of 0.955, the highest in the evaluation, but at the expense of recall.

4.2 Combining Models

In this section we evaluate the hybrid models, discussed in Section 3.2, that facilitate the combination of features from individual models.

NER-GAZ: Named Entities With Smoothing. Motivating the GAZ model in Section 3.2, we discussed the need for smoothing of NER model features to prevent overfitting. By combining features from NER and GAZ, we hope to improve recall in the NER model. In Fig. 1A and 1B we observe a general rise in $F_{\beta=1}$ when comparing the NER-GAZ and NER models, which we attribute to to an increase in average recall. Although generally NER-GAZ offers improvements over the NER model, optimal $F_{\beta=1}$, indicated (d) and (e), is only marginally improved (relaxed), or not at all (forced). However, the generally increased $F_{\beta=1}$ offered by NER-GAZ is desirable in order to aid generalisation without exhaustive searches of parameter space. As seen in Fig. 1C, multiclass classification appears largely unaffected by the addition of GAZ features to the NER model.

BOW-NER and BOW-NER-GAZ: Combining Word and Named Entity Based Models. By including named entities in addition to word-based features, we hope to improve precision without reducing recall. When evaluating performance of optimised models (summarised in Table 1), both appear to improve performance over any of their constituents. Furthermore, for one-vs-all classifiers this improvement appears substantial. We can see from the model performance in Fig. 1A and 1B that these models tend to perform similarly to their BOW counterparts. Given that the BOW feature space is denser, this is perhaps unsurprising, and indicates that BOW features have greater influence over the performance of these hybrid models than NER or GAZ features.

In Fig. 1A we see that forced one-vs-all classifiers exhibit consistent improvement for BOW-NER and BOW-NER-GAZ models, compared to the baseline BOW model. They also improve relaxed classifiers when the number of features included in the model is < 6000, labelled (f); however, with more features a BOW classifier outperforms the combined models. As previously discussed for the baseline, overall performance of relaxed classifiers suffers from poor recall. Compounding this with the precision-improving characteristics of NER features may be the cause of this eventual decline in combined model performance. The opposite appears true of multiclass classification, in which $F_{\beta=1}$ grows more slowly for the combined models than the baseline. In fact, multiclass does not benefit from the combined models until the final dimension that we sample, labelled (g), where the combined models offer marginal improvement.

Table 1. Development corpus: optimal micro-$F_{\beta=1}$

Model	One-vs-All (Forced)		One-vs-All (Relaxed)		Multiclass	
	Features	$F_{\beta=1}$	Features	$F_{\beta=1}$	Features	$F_{\beta=1}$
BOW	5000	0.7901	500	0.7264	20000	0.7914
NER	500	0.6765	50	0.6676	20000	0.7198
GAZ	28	0.6407	28	0.6499	28	0.5802
NER-GAZ	1000	0.6765	25	0.6689	20000	0.7222
BOW-NER	5000	0.8074	1000	0.7424	30000	0.7926
BOW-NER-GAZ	5000	0.8099	1000	0.7447	30000	0.7938

4.3 Development Corpus: Discussion of Results

Table 1 contains the optimal parameters (number of features in the model) for every model and classifier configuration, and the respective development corpus $F_{\beta=1}$. For comparison, a naïve classifier could obtain $F_{\beta=1} = 0.1843$ on the development corpus by classifying every document as the most numerous category. We note that the BOW-NER-GAZ model consistently produces the best $F_{\beta=1}$. Although the improvements for the combined feature spaces do not always appear substantial over their discrete constituents, it is satisfying that the combined models at least exhibit the improvements intended by design. NER-GAZ improves recall and categorisation performance on texts from unpopular categories (over the NER model), and the BOW-NER and BOW-NER-GAZ models improve the baseline model by enhancing precision, which in many cases leads to an increase in $F_{\beta=1}$.

Statistical Significance. In order to determine whether the performance increases reported between models and our baseline are likely to have occurred by chance, we perform statistical significance testing. We use the micro sign test (or s-test) [18], which compares two systems (A and B) based on the binary decisions that they make. This test evaluates models at a micro level, which aligns with our aim to optimise micro-$F_{\beta=1}$. The s-test produces a (one-sided) P-value for the hypothesis that system A performs better than system B. A subset of the results of this testing are given in Table 2. A smaller P-value indicates a more significant result, and we assume that a P-value > 0.1 indicates that the improvement reported for system A over B is not statistically significant. Using the s-test, we find that two results are significant, and these are both forced one-vs-all classifications. The BOW-NER-GAZ model offers the most significant improvement over the baseline, indicated by the smallest P-value.

4.4 Extended Corpus Evaluation

We analyse classification performance on our extended corpus, as introduced in Section 3.1. Models were trained and evaluated using the parameters established on the development corpus. The results are given in Table 3.

Table 2. Statistical significance test results for development corpus using s-test

System A	System B	Classifier	s-test[a]
BOW-NER	BOW	Forced One-vs-All	$>$
BOW-NER-GAZ	BOW	Forced One-vs-All	$\gg$
BOW-NER	BOW	Relaxed One-vs-All	$\sim$
BOW-NER-GAZ	BOW	Relaxed One-vs-All	$\sim$
BOW-NER	BOW	Multiclass	$\sim$
BOW-NER-GAZ	BOW	Multiclass	$\sim$

[a] where "$\gg$" indicates P-value ≤ 0.05; "$>$" indicates $0.05 <$ P-value ≤ 0.10; and "$\sim$" indicates P-value > 0.10

Table 3. Extended corpus: micro-$F_{\beta=1}$ based on development corpus parameters

Model	One-vs-All (Forced)		One-vs-All (Relaxed)		Multiclass	
	Features	$F_{\beta=1}$	Features	$F_{\beta=1}$	Features	$F_{\beta=1}$
BOW	5000	0.6673	500	0.6359	20000	0.6683
NER	500	0.2917	50	0.4761	20000	0.6341
GAZ	28	0.3522	28	0.3506	28	0.3844
NER-GAZ	1000	0.5141	25	0.4907	20000	0.55
BOW-NER	5000	0.7092	1000	0.6360	30000	0.6761
BOW-NER-GAZ	5000	0.7092	1000	0.6360	30000	0.6761

It is encouraging to see similar patterns between these results and those for the development corpus given in Table 1. In particular, we see that both BOW-NER and BOW-NER-GAZ models achieved equal or better performance than the baseline, and in the case of forced one-vs-all and multiclass classifiers, this improvement was larger than that reported for the development corpus. It is interesting that the BOW-NER and BOW-NER-GAZ models performed identically for all classifiers. Further inspection reveals that no GAZ features were selected for use in the BOW-NER-GAZ model, presumably because these features are noisy, given that they are not tailored to the new categories.

Although performance is universally worse than for the development corpus, this is to be expected, and we cannot isolate the cause of this penalty. The extended corpus is a harder corpus to categorise: it has more categories; the new categories are poorly populated with texts; it contains subcategories that are less likely to be easily discriminated using named entities; and we do not tailor our approach to this corpus as we did during development.

5 Conclusions and Further Work

Our classification system makes progress towards the classification of books using their full text. The extraction of structured data from full-text books, using

Information Extraction, is explored and evaluated for the purpose of assigning a single LoCC to each text. Furthermore, the techniques developed can be generalised to assign multiple category labels to a single text, although this is not evaluated here and is left for further work. In addition to assigning a categorisation, the extracted metadata (such as dates, and names of people and locations) may be useful for creating domain specific interfaces, such as maps for browsing the distribution of geographic entities, or timelines for browsing dates.Examples can be seen in the Perseus Digital Library and Gutenkarte.[6]

We have seen that there are implications for precision and recall when selecting classifiers, and that we can make very precise classifications using a relaxed one-vs-all classifier, or gain higher recall by selecting a forced architecture. Furthermore, we have also seen that one-vs-all classifiers require far fewer features than multiclass classifiers to perform optimally. This is also true of NER models, which required fewer dimensions to achieve optimal performance. This observation may be critical for the scalability of a solution to larger corpora.

We also found that in one-vs-all classifiers, adding features tended to be precision improving, but at the expense of recall, leading to a fall in $F_{\beta=1}$. This overfitting is most likely an artifact of limited training data. Of the architectures, direct multiclass classification appears to be the most robust and in the feature space sampled, we found little evidence of overfitting. However, given that the data is partitioned differently for each binary classifier used in one-vs-all, the respective hyperplanes and distances from hyperplanes to data points (and hence confidences) will not be normalised by default. Although it is not expected to affect the consistency of our results, it may be suboptimal to make decisions by direct comparison of these output confidences.

We found that combining our models based on named entities with bag-of-words representations resulted in an increase in performance over the baseline system, and furthermore, that this improvement has some statistical significance. We expect that these techniques could be widely applied to text categorisation and that the results described may not be specific to books.

Following the work in [6], we intend to develop mechanisms for evaluating the degree of incorrectness in misclassifications of books. In particular, we note that many misclassifications are in fact too general or too specific rather than strictly incorrect. In order to assess the utility of a less strict classification metric, we aim to define a 'lenient' metric as partial success, whereby, for example, a text from category DA may have been classified as D, or vice versa. We will then use this in order to assess the extent to which a classifier appears to make at least partially correct decisions.

Acknowledgments

The authors would like to thank Robert Japp and Sharon Givon for their input in this work.

[6] `http://www.perseus.tufts.edu` and `http://www.gutenkarte.org`

References

1. Betts, T.: Using Text Mining to Place Books into an Ontology. Masters thesis, University of Edinburgh, Edinburgh, UK. (2006)
2. Borko, H.: Measuring the reliability of subject classification by men and machines. American Documentation. **15** (1964) 268–273
3. Caropreso, M.F., Matwin, S., Sebastiani, F.: A learner-independent evaluation of the usefulness of statistical phrases for automated text categorization. In: Text Databases and Document Management: Theory and Practice. Idea Group Publishing. (2001) 78–102
4. Crane, G.: What Do You Do with a Million Books? D-Lib Magazine. **12**:3 (2006)
5. Curran J.R., Clark, S.: Language independent NER using a maximum entropy tagger. In: Proceedings of CoNLL-03, the Seventh Conference on Natural Language Learning. Edmonton, Canada. (2003) 164–167.
6. Frank, E., Paynter, G.W.: Predicting library of congress classifications from library of congress subject headings. J. of the American Society for Information Science and Technology. **55**:3 (2004) 214–227
7. Fürnkranz, J.: A study using n-gram features for text categorization. Technical Report OEFAI-TR-9830, Austrian Institute for Artificial Intelligence. (1998)
8. Hamill, K.A., Zamora, A.: The Use of Titles for Automatic Document Classification. J. of the American Society for Information Science. **31**:6 (1980) 396–402
9. Larson, R.R.: Experiments in automatic Library of Congress Classification. J. of the American Society for Information Science. **43**:2 (1992) 130–148
10. Mladenić, D., Globelnik, M.: Word sequences as features in text learning. In: Proceedings of ERK-98, the Seventh Electrotechnical and Computer Science Conference. Ljubljana, Slovenia. (1998) 145–148
11. Mooney, R.J., Roy, L.: Content-based book recommending using learning for text categorization. In: Proceedings of DL-00, 5th ACM Conference on Digital Libraries. San Antonio, US. (2000) 195–204
12. Moschitti, A., Basili, R.: Complex linguistic features for text classification: a comprehensive study. In: Proceedings of ECIR'04, Sunderland, UK. (2004)
13. Nürnberg, P.J., Furuta, R., Leggett, J.J., Marshall, C.C., Shipmann, F.M.: Digital Libraries: Issues and Architectures. In: Proceedings of the 1995 ACM Digital Libraries Conference. Austin, TX. (1995) 147–153
14. Sebastiani, F.: Machine learning in automated text categorization. ACM Computing Surveys. **34**:1 (2002) 1–47
15. Scott, S., Matwin, S.: Feature engineering for text classification. In: Proceedings of ICML-99. Bled, Slovenia. Morgan Kaufmann Publishers. (1999) 379–388
16. Joachims, T.: SVM^{light} 6.01, `http://svmlight.joachims.org` (2004)
17. Joachims, T.: SVM^{multiclass} 1.01, `http://svmlight.joachims.org/svm_multiclass.html` (2004)
18. Yang, Y., Liu, X.: A re-examination of text categorization methods. Proceedings of the 22th ACM SIGIR. Berkley, US. (1999) 42–49
19. Yang, Y., Pedersen, J.O.: A comparative study on feature selection in text categorization. In: Proceedings of ICML-97. Nashville, US. (1997) 412–420

Combined Syntactic and Semantic Kernels for Text Classification

Stephan Bloehdorn[1] and Alessandro Moschitti[2]

[1] Institute AIFB, University of Karlsruhe, Germany
bloehdorn@aifb.uni-karlsruhe.de
[2] University of Rome 'Tor Vergata', Italy
moschitti@info.uniroma2.it

Abstract. The exploitation of syntactic structures and semantic background knowledge has always been an appealing subject in the context of text retrieval and information management. The usefulness of this kind of information has been shown most prominently in highly specialized tasks, such as classification in Question Answering (QA) scenarios. So far, however, additional syntactic or semantic information has been used only individually. In this paper, we propose a principled approach for jointly exploiting both types of information. We propose a new type of kernel, the Semantic Syntactic Tree Kernel (SSTK), which incorporates linguistic structures, e.g. syntactic dependencies, and semantic background knowledge, e.g. term similarity based on WordNet, to automatically learn question categories in QA. We show the power of this approach in a series of experiments with a well known Question Classification dataset.

1 Introduction

Text Classification (TC) systems [1], which aim at automatically classifying textual input into categories according to some criterion of interest are a major application domain of modern machine learning techniques. Pioneered by [2], Support Vector Machines (SVMs) have been heavily used for text classification tasks, typically showing good results. In the simplest case, such systems use the standard *bag-of-words* feature representation which encodes the input as vectors whose dimensions correspond to the terms in the overall training corpus. The inner product (or the cosine) between two such vectors is used as kernel hence making the similarity of two documents dependant only on the amount of terms they share. This approach has an appealing simplicity and has produced good results in cases where sufficient training data is available. However, several modifications to this rather flat representation have been shown to improve the overall performance in selected scenarios. In particular, there has been interest in incorporating information about (i) the syntactic structure of the input texts and (ii) the semantic dependencies within the used terminology.

Kernel-based learning algorithms like Support Vector Machines have become a prominent framework for using such a-priori knowledge about the problem domain by means of a specific choice of the employed kernel function. On the one

G. Amati, C. Carpineto, and G. Romano (Eds.): ECIR 2007, LNCS 4425, pp. 307–318, 2007.

hand, *Tree Kernels* [3,4] have been used as a powerful way to encode the syntactic structure of the textual input in the form of parse trees and have shown good results in many natural language applications. On the other hand, *Semantic Smoothing Kernels* [5,6,7] exploit background information from semantic reference structures such as WordNet to make different, though semantically similar, terms contribute to the overall similarity of the input instances. Intuitively, the power of this kind of kernels is most apparent when too little training data is available to build stable models by counting occurrences of identical terminology only. While both approaches seem intuitive and powerful, language draws from both syntax and lexical semantics, therefore, finding principled techniques for combining kernels on linguistic structures, e.g. Tree Kernels, with Semantic Kernels appears to be a promising research line.

In this paper, we propose a new type of kernel, the *Semantic Syntactic Tree Kernel (SSTK)*, which exploits linguistic structure and background knowledge about the semantic dependencies of terms at the same time. More technically, the proposed kernel uses semantic smoothing to improve the matching of tree fragments containing terminal nodes.

We show the power of this approach in a series of experiments from the Question Classification (QC) domain. Question Classification [8] aims at detecting the type of a question, e.g. whether it asks for a person or for an organization which is critical to locate and extract the right answers in question answering systems. A major challenge of Question Classification compared to standard Text Classification settings is that questions typically contain only extremely few words which makes this setting a typical victim of data sparseness. Previous work has shown that Tree Kernels as well as Semantic Smoothing Kernels were individually capable of improving effectiveness in QC tasks. Our evaluation studies confirm these findings and indicate a consistent further improvement of the results when the proposed combined kernel is used. Our new Syntactic Semantic Tree Kernel improves the state-of-the-art in Question Classification, which makes it a prototype of a possible future full-fledged natural language kernel.

The remainder of this paper is structured as follows. Section 2 introduces kernel methods and some related work. Sections 3, 4 and 5 describe the design of the Semantic Smoothing Kernels, Tree Kernels and the combined Semantic Syntactic Tree Kernels, respectively. Section 6 gives an account on the performance of these in a series of evaluation experiments from the QC domain. We conclude in section 7 with a summary of the contributions, final remarks and a discussion of envisioned future work.

2 Kernel Methods and Related Work

Support Vector Machines [9] are state-of-the-art learning methods based on the older idea of linear classification. The distinguishing feature of SVMs is the theoretically well motivated and efficient training strategy for determining the separating hyperplane based on the margin maximization principle. In our context, however, the interesting property of SVMs is their capability of naturally incorporating data-specific notions of item similarity by means of a corresponding kernel function. Formally, any function κ that $\mathcal{X}$ satisfies $\kappa(x, z) = \langle \phi(x), \phi(z) \rangle$, is a valid kernel, whereby $\mathcal{X}$ is the input domain under consideration and ϕ is a suitable mapping from $\mathcal{X}$ to a feature (Hilbert-) space $\mathcal{F}$. Kernels can be designed by either choosing an explicit mapping function ϕ and incorporating it into an inner product or by directly defining the kernel function κ while making sure that it complies with the requirement of being a positive semi-definite function. Several closure properties aid the construction of valid kernels from known valid kernels. In particular, kernels are closed under sum, product, multiplication by a positive scalar and combination with well-known kernel modifiers. In particular, a given kernel κ can be normalized using the *cosine normalization modifier* given by $\kappa'(x, y) = (\kappa(x, y)) \,/\, (\sqrt{\kappa(x, x)}\sqrt{\kappa(y, y)})$ to produce kernel evaluations (i.e. similarity measures) normalized to absolute values between 0 and 1. The reader is referred to the rich literature for further information on SVMs and kernel methods, e.g. [10] for a comprehensive introduction.

Lexical semantic kernels were initially introduced in [5] using inverted path length as a similarity measure and subsequently revisited in [11,12], each time based on different design principles. Semantic kernels based on superconcept representations were investigated in [6] and [7]. As an alternative, [11] have put Semantic Kernels into the context of Latent Semantic Indexing.

Tree Kernels were firstly introduced in [3] and experimented with the Voted Perceptron for the parse-tree re-ranking task. The combination with the original PCFG model improved the syntactic parsing. In [13], two kernels over syntactic shallow parser structures were devised for the extraction of linguistic relations, e.g. *person-affiliation*. To measure the similarity between two nodes, the Contiguous String Kernel and the Sparse String Kernel were used. In [14] such kernels were slightly generalized by providing a matching function for the node pairs. The time complexity for their computation limited the experiments on a data set of just 200 news items. In [15], a feature description language was used to extract structural features from the syntactic shallow parse trees associated with named entities. The experiments on named entity categorization showed that too many irrelevant tree fragments may cause overfitting. In [16] Tree Kernels were firstly proposed for semantic role classification. The combination between such kernel and a polynomial kernel of standard features improved the state-of-the-art.

To our knowledge, no other work has so far combined the syntactic and semantic properties of natural language in a principled way as proposed in our approach.

3 Semantic Similarity Kernels

In this section we describe the first component of our new kernel, the Semantic Smoothing Kernel, which combines semantic similarity of terms with the standard bag-of-words representation.

Table 1. Measures of semantic similarity

Inverted Path Length:

$$sim_{IPL}(c_1, c_2) = \frac{1}{(1 + d(c_1, c_2))^{\alpha}}$$

Wu & Palmer:

$$sim_{WUP}(c_1, c_2) = \frac{2\,dep(lso(c_1, c_2))}{d(c_1, lso(c_1, c_2)) + d(c_2, lso(c_1, c_2)) + 2\,dep(lso(c_1, c_2))}$$

Resnik:

$$sim_{RES}(c_1, c_2) = -\log\, P(lso(c_1, c_2))$$

Lin:

$$sim_{LIN}(c_1, c_2) = \frac{2\,\log\, P(lso(c_1, c_2))}{\log\, P(c_1) + \log P\,(c_2)}$$

3.1 Semantic Networks and Similarity

The formal description of semantic kernels requires the introduction of some definitions. We denote terms as $t_1, t_2, \ldots \in \mathcal{T}$ and concepts as $c_1, c_2, \ldots \in \mathcal{C}$; we also sometimes use the somewhat informal disambiguation operator $c(\cdot)$ to map terms to concept representations. To compute useful notions of semantic similarity among the input terms, we employ semantic reference structures which we call, for simplicity, *Semantic Networks.* These can be seen as directed graphs semantically linking concepts by means of taxonomic relations (e.g. *[cat] is-a [mammal]*). Research in Computational Linguistics has led to a variety of well-known measures of semantic similarity in semantic networks.

The measures relevant in the context of this paper are summarized in table 1. These measures make use of several notions. (i) The *distance* (d) of two concepts c_1 and c_2, is the number of superconcept edges between c_1 and c_2. (ii) The *depth* (dep) of a concept refers to the distance of the concept to the unique root node[1]. (iii) The *lowest super ordinate (lso)* of two concepts refers to the concept with maximal depth that subsumes them both. (iv) The probability $P(c)$ of encountering a concept c which can be estimated from corpus statistics. When probabilities are used, the measures follow the trail of information theory in quantifying the *information concept (IC)* of an observation as the negative log likelihood. We point the interested reader to [17] for a detailed and recent survey of the field.

3.2 Semantic Similarity Kernels Based on Superconcepts

In this section, we introduce a class of kernel functions defined on terms that can be embedded in other kernels that make (in whatever way) use of term matching[2].

[1] If the structure is not a perfect tree structure, we use the minimal depth.

[2] For simplicity, we will restrict our attention on defining kernel functions for *concepts* and leave the details of the consideration of lexical ambiguity to the next section.

Table 2. Weighting Schemes for the Superconcept Kernel κ_S

full	No weighting, i.e. $SC(\bar{c})_j = 1$ for all superconcepts c_j of $\bar{c}$ and $SC(\bar{c})_j = 0$ otherwise.
full-ic	Weighting using information content of $SC(\bar{c})_j$, i.e. $SC(\bar{c})_j = sim_{RES}(\bar{c}, c_j)$.
path-1	Weighting based on inverted path length, i.e. $SC(\bar{c})_j = sim_{IPL}(\bar{c}, c_j)$ for all superconcepts c_j of $\bar{c}$ and $SC(\bar{c})_j = 0$ otherwise using the parameter $\alpha = 1$.
path-2	The same but using the parameter $\alpha = 2$.
lin	Weighting using the Lin similarity measure, i.e. $SC(\bar{c})_j = sim_{LIN}(\bar{c}, c_j)$.
wup	Weighting using the Wu&Palmer similarity measure, i.e. $SC(\bar{c})_j = sim_{WUP}(\bar{c}, c_j)$.

Definition 1 (Superconcept Kernel). *The Superconcept Kernel κ_S for two concepts $c_i, c_j \in \mathcal{C}$ is given by $\kappa_S(c_i, c_j) = \langle SC(c_i), SC(c_j) \rangle$, whereby $SC(\cdot)$ is a function $\mathcal{C} \to \mathrm{R}^{|\mathcal{C}|}$ that maps each concept to a real vector whose dimensions correspond to (super-)concepts present in the employed semantic network and the respective entries are determined by a particular weighting scheme.*

This idea, recently investigated in [7], is based on the observation that the more two concepts are similar the more common superconcepts they share. A similar approach has been proposed in [6], however focusing on the simple case of giving the superconcepts in the mapping full and equal weight while varying the number of superconcepts that are considered.

Obviously, κ_S is a valid kernel as it is defined explicitly in terms of a dot product computation. So far, however, we have left the details of the function $SC(\cdot)$ that maps concepts to its superconcepts unspecified. In the same way as in earlier work [7], we have investigated the use of different weighting schemes for the representation of the superconcepts motivated by the following considerations:

1. The weight a superconcept $SC(\bar{c})_j$ receives in the vectorial description of concept $\bar{c}$ should be influenced by its distance from $\bar{c}$.
2. The weight a superconcept $SC(\bar{c})_j$ receives in the vectorial description of concept $\bar{c}$ should be influenced by its overall depth in the semantic network.

We have used the measures of semantic similarity introduced in table 1 as weighting schemes, summarized in table 2. The different weighting schemes behave differently wrt the above motivations. While full does not implement any of them, full-ic considers rationale 2 while path-1 and path-2 consider rationale 1. The schemes lin and wup reflect combinations of both rationales. The superconcept kernel κ_S can normalized to $[0, 1]$ in the usual way using the cosine normalization modifier.

The concept kernel $\kappa_S(\cdot, \cdot)$ can be used directly in conjunction with the standard linear kernel by means of a simple *Semantic Smoothing Kernel.*

Definition 2 (Semantic Smoothing Kernel). *The Semantic Smoothing Kernel κ_S for two term vectors (input texts) $x, z \in X$ is given by $\kappa_{\hat{S}}(x, z) = x'C'K_S C z$*

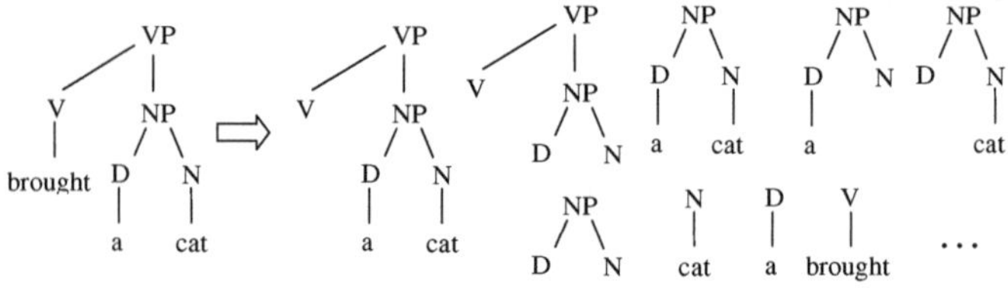

Fig. 1. A tree with some of its fragments

where $K_{\mathcal{S}}$ is a square symmetric matrix whose entries represent the kernel evaluations between the concepts and C denotes the matrix that encodes the evaluations of the disambiguation function C that maps concept dimensions to term dimensions that constitute the input space X.

4 Tree Kernels for Syntactic Structures

The main rationale behind Tree Kernels is to represent trees in terms of their substructures (fragments). The kernel function then counts the number of tree subparts common to both argument trees. We define a tree as a connected directed graph with no cycles. *Trees* are denoted as $T_1, T_2, \ldots$; *tree nodes* are denoted as $n_1, n_2, \ldots$; and the set of nodes in tree T_i are denoted as N_{T_i}. We denote the set of all *substructures (fragments)* that occur in a given set of trees as $\{f_1, f_2, \ldots\} = \mathcal{F}$. As the structures we will work with are parse trees, each node with its children is associated with the execution of a grammar production rule. The labels of the leaf nodes of the parse trees correspond to terms, i.e. *terminal symbols*, whereas the *preterminal symbols* are the parents of leaves. As an example Figure 1 in section 4 shows a parse tree of the sentence fragment ``‘‘brought a cat’’`` with some of its substructures.

Definition 3 (Tree Kernel (Collins & Dufy, 2001)). *Given two trees T_1 and T_2 we define the (Subset-) Tree Kernel as:*

$$\kappa_T(T_1, T_2) = \sum_{n_1 \in N_{T_1}} \sum_{n_2 \in N_{T_2}} \Delta(n_1, n_2)$$

where $\Delta(n_1, n_2) = \sum_{i=1}^{|\mathcal{F}|} I_i(n_1) I_i(n_2)$, and where $I_i(n)$ is an indicator function which determines whether fragment f_i is rooted in node n.

Δ is equal to the number of common fragments rooted at nodes n_1 and n_2. We can compute it more efficiently as follows:

1. if the productions at n_1 and n_2 are different then $\Delta(n_1, n_2) = 0$;
2. if the productions at n_1 and n_2 are the same, and n_1 and n_2 only have leaf children (i.e. the argument nodes are pre-terminals symbols) then $\Delta(n_1, n_2) = 1$;

3. if the productions at n_1 and n_2 are the same, and n_1 and n_2 are not pre-terminals then

$$\Delta(n_1, n_2) = \prod_{j=1}^{nc(n_1)} (1 + \Delta(ch_{n_1}^j, ch_{n_2}^j)).$$

where $nc(n_1)$ is the number of children of n_1 and ch_n^j is the j-th child of node n. Note that, since the productions are the same, $nc(n_1) = nc(n_2)$. Of course, the kernel can again be normalized using the cosine normalization modifier. Additionally, a decay factor λ can be added by modifying steps (2) and (3) as follows:

2. $\Delta(n_1, n_2) = \lambda$,
3. $\Delta(n_1, n_2) = \lambda \prod_{j=1}^{nc(n_1)} (1 + \Delta(ch_{n_1}^j, ch_{n_2}^j))$.

As an example, Figure 1 shows a parse tree of the sentence (fragment) ``bought a cat'' with some of the substructures that the tree kernel uses to represent it[3].

5 Designing Semantic Syntactic Tree Kernels

The Tree Kernel introduced in the previous section relies on the intuition of counting all common substructures of two trees. However, if two trees have similar structures but employ different though related terminology at the leaves, they will not be matched. From a semantic point of view, this is an evident drawback as ``brought a cat'' should be more related to ``brought a tomcat'' than to ``brought a note''.

In analogy with the semantic smoothing kernels for the bag-of-words kernel as described in section 3.2, we are now interested in also counting *partial matches* between tree fragments. A partial match occurs when two fragments differ only by their terminal symbols, e.g. [N [cat]] and [N [tomcat]]. In this case the match should give a contribution smaller than 1, depending on the semantic similarity of the respective terminal nodes. For this purpose, we first define the similarity of two such tree fragments.

Definition 4 (Tree Fragment Similarity Kernel). *For two tree fragments $f_1, f_2 \in \mathcal{F}$, we define the Tree Fragment Similarity Kernel as*[4]*:*

$$\kappa_{\mathcal{F}}(f_1, f_2) = comp(f_1, f_2) \prod_{t=1}^{nt(f_1)} \kappa_S(f_1(t), f_2(t))$$

where $comp(f_1, f_2)$ (compatible) is 1 if f_1 differs from f_2 only in the terminal nodes and is 0 otherwise, $nt(f_i)$ is the number of terminal nodes and $f_i(t)$ is the t-th terminal symbol of f_i (numbered from left to right).

[3] The number of such fragments can be obtained by evaluating the kernel function between the tree with itself.

[4] Note that, as the tree fragments need to be compatible, they have the same number of terminal symbols at compatible positions.

Conceptually, this means that the similarity of two tree fragments is above zero only if the tree fragments have an identical structure. The fragment similarity is evaluated as the product of all semantic similarities of corresponding terminal nodes (i.e. sitting at identical positions). It is maximal if all pairs have a similarity score of 1. We now define the overall tree kernel as the sum over the evaluations of $\kappa_{\mathcal{F}}$ over all pairs of tree fragments in the argument trees. Technically, this means changing the summation in the second formula of definition 3 as in the following definition.

Definition 5 (Semantic Syntactic Tree Kernel). *Given two trees T_1 and T_2 we define the Semantic Syntactic Tree Kernel as:*

$$\kappa_T(T_1, T_2) = \sum_{n_1 \in N_{T_1}} \sum_{n_2 \in N_{T_2}} \Delta(n_1, n_2)$$

where $\Delta(n_1, n_2) = \sum_{i=1}^{|\mathcal{F}|} \sum_{j=1}^{|\mathcal{F}|} I_i(n_1) I_j(n_2) \kappa_{\mathcal{F}}(f_i, f_j)$.

Obviously, the naive evaluation of this kernel would require even more computation and memory than for the naive computation of the standard kernel as also all *compatible* pairs of tree fragments would need to be considered in the summation. Luckily, this enhanced kernel can be evaluated in the same way as the standard tree kernel by adding the following step

0. if n_1 and n_2 are pre-terminals and $label(n_1) = label(n_2)$ then $\Delta(n_1, n_2) = \lambda \kappa_{\mathcal{S}}(ch^1_{n_1}, ch^1_{n_2})$,

as the first condition of the Δ function definition (Section 4), where $label(n_i)$ is the label of node n_i and $\kappa_{\mathcal{S}}$ is a term similarity kernel, e.g. based on the superconcept kernel defined in section 3.2. Note that: (a) since n_1 and n_2 are pre-terminals of a parse tree they can have only one child (i.e. $ch^1_{n_1}$ and $ch^1_{n_2}$) and such children are words and (b) Step 2 is no longer necessary.

Beside the novelty of taking into account tree fragments that are not identical it should be noted that the lexical semantic similarity is constrained in syntactic structures, which limit errors/noise due to incorrect (or, as in our case, not provided) word sense disambiguation.

6 Experimental Evaluation

In a series of experiments we aimed at showing that our approach is effective for IR and text mining applications. For this purpose, we experimented with the TREC question classification corpus for advanced retrieval based on the Question Answering paradigm.

6.1 Experimental Setup

The long tradition of QA in TREC has produced a large question set used by several researchers which can be exploited for experiments on Question Classification. According to [8], we can define question classification *"to be the task that, given a question, maps it to one of k classes, which provide a semantic constraint on the sought-after answer"*. Such questions are categorized according to different taxonomies of different *grains*. We consider the same dataset and classification problem as described in [18,8]. The dataset consists of free text questions and is freely available[5]. It is divided into 5,500 questions[6] for training and the 500 TREC 10 questions for testing. Each of these questions is labeled with exactly one class of the *coarse grained* classification scheme (see [18]) consisting of the following 6 classes: Abbreviations, Descriptions (e.g. *definition* and *manner*), Entity (e.g. *animal*, *body* and *color*), Human (e.g. *group* and *individual*), Location (e.g. *city* and *country*) and Numeric (e.g. *code* and *date*).

We have implemented the kernels introduced in sections 3–5 within the SVM-light-TK software available at `ai-nlp.info.uniroma2.it/moschitti/` which encodes tree kernel functions in SVM-light [19]. In all experiments, we used the noun hierarchy of WordNet[7] as the underlying semantic network. For word sense disambiguation, we used a simplifying assumption in mapping each term to its most frequent noun sense (if it exists). Note that this approach implies an inherent word sense disambiguation side effect, likely to have a negative impact on the results. The results can also be seen as a pessimistic estimate. Kernel similarities that were undefined because of a missing mapping to a noun synset were implicitly assumed to take the default values (i.e. zero for distinct and one identical terms respectively).

6.2 Evaluation of Superconcept Smoothing Kernels

In a first experiment we investigated the effect of simply smoothing term features based on the idea of the Semantic Smoothing Kernel. Questions were preprocessed using the usual steps, namely tokenization, lemmatization and stoopword-removal leading to a total number of 8,075 distinct TFIDF-weighted features. We performed *binary classification experiments* on each of the 6 question types for different settings of the 'soft margin' parameter c. Table 3 summarizes the absolute macro F_1 as well as the micro F_1 values obtained in the question classification setting. The best values per setting of c are highlighted.

The results indicate that the smoothing of the term features considerably affects the performance compared to the simple linear kernel baseline. According to the results, the lin scheme seems to achieve the best overall performance with a relative improvement of 9.32% for the macro F_1 value in the case of $c = 3$ (i.e. the setting for which the linear kernel achieves its maximum). For a more detailed description of these experiments, refer to [7]. For QC, however, the traditional bag-of-words (bow) appears to be somewhat insufficient as it ignores structure and function words that are likely to influence the results.

[5] `http://l2r.cs.uiuc.edu/~cogcomp/Data/QA/QC/`

[6] These are selected from the 4500 English questions published by USC (Hovy et al., 2001), 500 questions annotated for rare classes and the 894 questions from TREC 8 and TREC 9.

[7] `http://wordnet.princeton.edu/`

Table 3. Absolute macro and micro F1 results for QC, for different values of c and different semantic smoothing kernels. The best results per setting of c are highlighted.

macro-averaging						
	soft margin parameter c					
kernel	**0.1**	**0.2**	**0.3**	**1.0**	**2.0**	**3.0**
linear	0.21	0.38	0.47	0.62	0.63	0.64
full	0.38	0.49	0.55	0.61	0.61	0.68
full-ic	**0.53**	**0.53**	0.53	0.62	0.55	0.55
path-1	0.25	0.42	0.51	**0.64**	0.64	0.64
path-2	0.22	0.39	0.47	0.63	**0.65**	0.64
lin	0.36	0.49	**0.56**	**0.64**	0.62	**0.70**
wup	0.34	0.49	0.54	0.62	0.61	0.69

micro-averaging						
	soft margin parameter c					
kernel	**0.1**	**0.2**	**0.3**	**1.0**	**2.0**	**3.0**
linear	0.09	0.25	0.34	0.55	0.57	0.58
full	0.27	0.38	0.45	0.55	0.56	0.68
full-ic	**0.47**	**0.46**	**0.47**	**0.60**	0.49	0.48
path-1	0.14	0.32	0.40	0.57	0.58	0.59
path-2	0.08	0.28	0.37	0.57	**0.59**	0.58
lin	0.27	0.37	**0.47**	0.57	0.57	**0.69**
wup	0.23	0.37	0.45	0.56	0.56	0.68

Table 4. Accuracy of SSTK on QC, for different values of λ parameter and different semantic smoothing kernels. The best λ settings are highlighted.

	Accuracy				
λ parameter	**0.4**	**0.05**	**0.01**	**0.005**	**0.001**
linear (bow)	0.905				
string matching	0.890	0.910	**0.914**	**0.914**	0.912
full	0.904	**0.924**	0.918	0.922	0.920
full-ic	0.908	**0.922**	0.916	0.918	0.918
path-1	0.906	**0.918**	0.912	**0.918**	0.916
path-2	0.896	0.914	0.914	**0.916**	**0.916**
lin	0.908	**0.924**	0.918	0.922	0.922
wup	0.908	**0.926**	0.918	0.922	0.922

6.3 Evaluation of Syntactic Semantic Tree Kernels

In these experiments, we used the same experimental setup as used in [18] as it contains the most comprehensive comparison of experiments on the QC corpus introduced above. As literature results are given in terms of the accuracy of the *multi-classification* of the TREC questions, to compare with them, we designed a multiclassifier based on the scores of the SVMs. For each questions, we selected the class associated with the maximum score.

In this experiment, we compared the linear kernel based on bag-of-words (this time including function words), the original STK and the new SSTK as introduced in Section 5 with different term similarities[8]. The question parse trees were obtained by running the Charniak's parser. Table 4 reports the results of the multiclassifier based on SSTK. Column 1 shows the type of similarity used in the semantic syntactic tree kernel function, where *string matching* means

[8] We again used the superconcept kernels as term similarities, however, in contrast to the previous section we used normalized versions.

that the original tree kernel is used. Columns from 2 to 6 report the accuracy of the multiclassifier according to several values of the λ parameter[9] (see Section 5).

We note that (a) our basic Syntactic Tree Kernel improves the state-of-the-art accuracy, i.e. 91.4% vs. 90% of [18], (b) this is further improved when we use one of the semantic smoothing kernel and (c) the Wu-Palmer similarity achieves the highest accuracy, i.e. 92.6%.

7 Conclusion

In this paper, we have investigated how the syntactic structures of natural language texts can be exploited simultaneously with semantic background knowledge on term similarity. For this purpose, we have proposed a new family of kernels called Semantic Syntactic Tree Kernels (SSTK) that is based on Tree and Semantic Smoothing Kernels. We have motivated this class of kernels by counting all compatible tree fragments of two parse trees weighted by their joint terminology. To our knowledge, no other work has so far combined the syntactic and semantic properties of natural language in such a principled way. We conducted a series of experiments on the TREC question classification data. Our results indicate that the newly proposed Semantic Syntactic Tree Kernels outperform the conventional linear/semantic kernels as well as tree kernels improving the state of the art in Question Classification. In the future, it would be interesting to study different models of lexical semantics, e.g. latent semantic kernels, with kernels based on different syntactic/semantic structures.

References

1. Sebastiani, F.: Machine learning in automated text categorization. ACM Computing Surveys **34**(1) (2002) 1–47
2. Joachims, T.: Text categorization with Support Vector Machines: learning with many relevant features. In: Proceedings of ECML, Chemnitz, DE (1998)
3. Collins, M., Duffy, N.: Convolution kernels for natural language. In: NIPS, MIT Press (2001)
4. Moschitti, A.: Efficient convolution kernels for dependency and constituent syntactic trees. In: Proceedings of ECML, Berlin, Germany. (2006)
5. Siolas, G., d'Alche Buc, F.: Support Vector Machines based on a semantic kernel for text categorization. In: IJCNN. Volume 5. (2000)
6. Mavroeidis, D., Tsatsaronis, G., Vazirgiannis, M., Theobald, M., Weikum, G.: Word sense disambiguation for exploiting hierarchical thesauri in text classification. In: PKDD. (2005)

[9] We preliminary selected the best cost-factor (parameter j) on the validation set and then experimented with different λ values. We also noted that the parameter c is not critical when tree kernels are used. This means that the accuracy does not vary too much and the highest value seems to be achieved with the default SVM-light setting.

7. Bloehdorn, S., Basili, R., Cammisa, M., Moschitti, A.: Semantic kernels for text classification based on topological measures of feature similarity. In: Proceedings of ICDM. (2006)
8. Li, X., Roth, D.: Learning question classifiers. In: Proceedings of the 19th International Conference on Computational Linguistics (COLING). (2002)
9. Vapnik, V., Golowich, S.E., Smola, A.J.: Support vector method for function approximation, regression estimation and signal processing. In: NIPS. (1996)
10. Shawe-Taylor, J., Cristianini, N.: Kernel Methods for Pattern Analysis. Cambridge University Press (2004)
11. Cristianini, N., Shawe-Taylor, J., Lodhi, H.: Latent Semantic Kernels. Journal of Intelligent Information Systems **18**(2-3) (2002) 127–152
12. Basili, R., Cammisa, M., Moschitti, A.: A semantic kernel to exploit linguistic knowledge. In: AI*IA: Advances in Artificial Intelligence. (2005)
13. Zelenko, D., Aone, C., Richardella, A.: Kernel methods for relation extraction. Journal of Machine Learning Research (2003)
14. Culotta, A., Sorensen, J.: Dependency tree kernels for relation extraction. In: Proceedings of ACL. (2004)
15. Cumby, C., Roth, D.: Kernel methods for relational learning. In: Proceedings of the Twentieth International Conference (ICML 2003). (2003)
16. Moschitti, A.: A study on convolution kernels for shallow semantic parsing. In: proceedings of ACL. (2004)
17. Budanitsky, A., Hirst, G.: Evaluating wordnet-based measures of lexical semantic relatedness. Computational Linguistics **32**(1) (2006) 13–47
18. Zhang, D., Lee, W.S.: Question classification using Support Vector Machines. In: Proceedings of SIGIR. (2003)
19. Joachims, T.: Making large-scale SVM learning practical. In: Advances in Kernel Methods. (1999)

Fast Large-Scale Spectral Clustering by Sequential Shrinkage Optimization

Tie-Yan Liu[1], Huai-Yuan Yang[2,⋆], Xin Zheng[3,⋆], Tao Qin[3,⋆], and Wei-Ying Ma[1]

[1] Microsoft Research Asia, 4F, Sigma Center, No. 49, Zhichun Road, Haidian District, Beijing, 100080, P.R. China
[2] Peking University, Beijing, 100871, P.R. China
[3] Tsinghua University, Beijing, 100084, P.R. China

Abstract. In many applications, we need to cluster large-scale data objects. However, some recently proposed clustering algorithms such as spectral clustering can hardly handle large-scale applications due to the complexity issue, although their effectiveness has been demonstrated in previous work. In this paper, we propose a fast solver for spectral clustering. In contrast to traditional spectral clustering algorithms that first solve an eigenvalue decomposition problem, and then employ a clustering heuristic to obtain labels for the data points, our new approach sequentially decides the labels of relatively well-separated data points. Because the scale of the problem shrinks quickly during this process, it can be much faster than the traditional methods. Experiments on both synthetic data and a large collection of product records show that our algorithm can achieve significant improvement in speed as compared to traditional spectral clustering algorithms.

1 Introduction

Clustering, which refers to the unsupervised identification of groups of similar data objects, has been a long-studied research topic in the machine learning communities. And many different clustering algorithms have been proposed. Among these algorithms, the recently-proposed spectral clustering algorithm is one of the most promising methods, especially for those datasets with latent structure [2][8].

However, in many applications nowadays, we need to cluster large-scale data objects. In this scenario, traditional spectral clustering algorithms will suffer from high computational complexity, which restricts their wide applications. To our knowledge, most of the existing work on spectral clustering only focused on the theoretical part and leave the computational issues to the numerical solvers. As a result, there is no efficient implementation of spectral clustering.

In order to tackle this problem, we need to understand why spectral clustering is time consuming. Traditional spectral clustering algorithms consist of two steps: 1) relax the original formulation of spectral clustering that is a discrete

⋆ This work was performed at Microsoft Research Asia.

G. Amati, C. Carpineto, and G. Romano (Eds.): ECIR 2007, LNCS 4425, pp. 319–330, 2007.

optimization problem, to continuous optimization, and solve it to get the continuous embedding of data objects by eigenvalue decomposition (EVD); 2) employ a clustering heuristic, such as k-means, to obtain the discrete label from the continuous embedding. The computational complexities of both steps are not negligible. Step 1 (EVD) is an expensive process even if employing the state-of-the-art EVD solvers. For example, the computational complexity of Lanczos [4] and Preconditioned Conjugate Gradient methods [3][5][6] is $O(kn^2m)$, where k is the number of eigenvectors to be computed, n is the number of data points and m is the number of iterations. It is clear that this is a heavy computational load. And in Step 2, if k-means [13] is employed, the computational complexity is $O(knr)$, where r is the number of iterations. This is also a considerable part of the overall complexity.

Based on the above analysis, if we want to accelerate spectral clustering, we need to accelerate EVD solver, and/or save the computations for the clustering heuristics such as k-means. For this purpose, we monitored the iterative process of spectral clustering. As a result, we found that some data points converge to their true embeddings quickly, while it take a relatively longer time for others to converge. Furthermore, it is those data points which are easy to separate (i.e. they have very strong correlations with some points and very weak correlations with others) that converge fast. Based on this observation, we propose that it is not necessary to treat all the data points equally while getting the solution of spectral clustering. The iterations for those well-separated points can stop early while others stop later. In addition, considering that the original clustering task is actually defined in the discrete space (i.e. what we want is the label of the data objects rather than the embeddings), we even argue that it is not necessary to spend so much time on getting precise solution for the continuous relaxation. Instead, once the embeddings are good enough for the clustering purpose, we can stop without any influence on the final results.

Based on the above discussions, we propose a very fast spectral clustering algorithm in this paper. Technically, instead of solving the relaxed continuous optimization and clustering the embeddings, we recover the discrete solution of the original problem directly by an iterative method. That is, we start the optimization process with an EVD solver. Note that, instead of using it to get the precise eigenvectors, we will stop early after only a few iterations. Then we identify some already-well-converged data points, and fix their embedding values to discrete labels. Note that their values will not be changed in the future iterations of the optimization, so the scale of the subsequent optimization problem is reduced. By use of this shrinkage strategy sequentially, we can get the discrete labels of all the data points eventually. In such a way, we not only avoid over-computing of precise EVD solution, but also avoid additional computations of the clustering heuristics. Experimental results show that our algorithm can achieve significant improvement in speed as compared with traditional spectral clustering algorithms.

The rest of paper is organized as follows: in Section 2, some concepts of spectral clustering are reviewed; in Section 3, our algorithm, named Sequential

Shrinkage Optimization (SSO), is introduced. Experimental results are reported in Section 4, which show that our algorithm can improve the efficiency of spectral clustering by much without significant accuracy reduction. Conclusions and future work are given in Section 5.

2 Review of Spectral Clustering

In spectral clustering, the dataset to be clustered is usually represented by a weighted graph $G = (V, E, W)$, where V is the set of data objects; E is the set of edges between data objects and e_{ij} is used to denote the weight of an edge in E; and W denotes a diagonal matrix which diagonal elements are the weights of the data objects. We further define the adjacency matrix of the graph G as M:

$$M_{ij} = \begin{cases} e_{ij} & \text{if} < i, j > \in E \\ 0 & \text{otherwise} \end{cases} \tag{1}$$

Then the clustering problem can be regarded as a graph partitioning problem, which objective is to find clusters V_1, V_2, V_3, ... such that the sum of the edge weights between different clusters are minimized.

2.1 Mathematical Formulation of Spectral Clustering

The simplest case of spectral clustering is two-way spectral clustering, which minimizes the objective function defined as below:

$$obj(V_1, V_2) = \frac{cut(V_1,V_2)}{weight(V_2)} + \frac{cut(V_2,V_1)}{weight(V_1)} \tag{2}$$

where

$$cut(V_1, V_2) = \textstyle\sum_{i\in V_1, j\in V_2, <i,j>\in E} e_{ij} \tag{3}$$

$$weight(V_i) = \textstyle\sum_{j\in V_i} W_j \tag{4}$$

Different definitions of W will yield different spectral clustering algorithms [2]. In particular, if the weight of a data object is defined by its weighted degree, the corresponding algorithm is called Normalized Cut; while if the weight is defined as the unweighted degree (number of edges), the corresponding algorithm is named Ratio Cut, etc. For simplicity, we will only discuss Normalized Cut in this paper, but all the results apply to other versions of spectral clustering.

By defining indicators of data objects as follows,

$$q^{(i)} = \begin{cases} +\sqrt{\frac{\eta_2}{\eta_1}} & \text{if } i \in V_1 \\ -\sqrt{\frac{\eta_1}{\eta_2}} & \text{if } i \in V_2 \end{cases} \tag{5}$$

where $\eta_1 = weight(V_1)$, $\eta_2 = weight(V_2)$, We can rewrite the objective (2) in a formulation of Rayleigh quotient,

$$\min \frac{q^T L q}{q^T W q}, s.t. \quad q^T W e = 0 \tag{6}$$

where $L = W - M$. If we relax the above discrete optimization problem to a continuous version, the solution of q is equal to the eigenvector associated with the second smallest eigenvalue of the generalized eigenvalue problem $Lv = \lambda Wv$. Similarly, one can prove that k-way spectral clustering corresponds to solving the k smallest eigenvalues and their corresponding eigenvectors.

2.2 Existing Solvers for Spectral Clustering

As mentioned in the introduction, most spectral clustering algorithms leave EVD problem to existing numerical EVD solvers. Although EVD has been extensively investigated for decades, it is still the most time-consuming component of spectral clustering. State-of-the-art EVD solvers can be categorized into three classes: local EVD solvers, global EVD solvers and extreme EVD solvers. Power method is the simplest local EVD solver, which only computes the maximum eigenvalue and its associated eigenvector [1]. QR-decomposition method is a typical global EVD solver [4], which computes all the eigenvalues and eigenvectors. Extreme EVD solvers only compute several extreme (smallest or largest) eigenvalues and the corresponding eigenvectors [12]. Typical examples include Lanczos [10] and Preconditioned Conjugate Gradient (PCG) methods [5][6][12]. Considering the process of spectral clustering, it is clear that extreme EVD solvers are most suitable for spectral clustering. To our knowledge, the computation complexity of Lanczos and PCG is $O(kn^2m)$, where k is the number of eigenvalues to compute, n is the number of data objects and m is the number of iterations. Besides, the obtained eigenvectors should be further clustered, which yield additional computational load. For example, if we employ k-means to fulfill this task, another $O(knr)$ computation complexity is added, where r denotes the number of iterations. In order to make spectral clustering algorithms scalable enough for real-world applications, we need to accelerate both of the aforementioned computations. This is the motivation of our proposed sequential shrinkage optimization method.

3 Sequential Shrinkage Optimization (SSO)

In this section, we describe our fast implementation for spectral clustering, i.e. Sequential Shrinkage Optimization (SSO). First of all, we will recall Conjugate Gradient (CG) algorithm and then propose an SSO algorithm based on non-linear CG. After that, we provide an even faster SSO algorithm based on linear CG.

3.1 Linear and Non-linear Conjugate Gradient (CG)

CG[4] has been one of the most important algorithms in numerical algebra and continuous optimization. Original (linear) CG algorithms solve the following quadratic optimization problem:

$$\min f(q) = \tfrac{1}{2}q^T Aq - b^T q + c \tag{7}$$

After applying some preconditioning techniques, Preconditioned CG (PCG) is recognized as one of the state-of-the-art methods for quadratic optimization. In the rest of this paper, we use CG and PCG interchangeably, because all the preconditioning techniques can be apply to any version of CG. Although (linear) CG is originally designed for quadratic problem (7), many CG-based algorithms have also been developed to solve general continuous optimization problems. These algorithms are called non-linear CG (NLCG) [4][7]. As a special case, generalized EVD problem can also be solved by NLCG, because it is equivalent to a continuous optimization problem as follows

$$\min q^T Aq, s.t. \quad q^T Wq = 1 \tag{8}$$

To our knowledge, many variations of NLCG-based EVD solvers are among the state-of-the-art EVD solvers, including PCG [5], LOBPCG [5][6], etc. Also Lanczos, the most famous extreme EVD solver, has also been shown to be a kind of NLCG method [4]. To summarize, both CG optimizer and NLCG-based EVD solver are the state of the art in their fields. Therefore, we will base our further discussions upon these two algorithms.

3.2 Non-linear SSO

In this subsection, we propose a fast spectral clustering algorithm, which is named as non-linear sequential shrinkage optimization (NLSSO). When using NLCG to solve eigenvectors, some interesting and informative phenomena imply that it is possible to accelerate the clustering procedure. Actually, we monitored the process how the variable q of problem (6) gradually approximate the precise eigenvector of $Lv = \lambda Wv$. As a result, we observed that the variables corresponding to those well-separated data objects will converge relatively faster than other variables. And their absolute values are usually larger than other variables. This shows that these variables can be easily identified and then the optimization of them can be stopped earlier without influencing the final clustering results by much. In this way, the scale of the subsequent optimization can be reduced and the efficiency can be improved. Actually this is not only an implementation trade-off. It is also reasonable in the theoretical view, which is to be elaborated on in the next subsection. Based on the above observations, our NLSSO method works as follows. In the beginning of our algorithm, we start with NLCG to optimize the objective (6). After a few iterations (e.g. $< 1\%$ of the problem scale n), we obtain an initial guess of the final solution to q. According to this initial guess, we fix those largest positive variables to a certain positive value (positive label) and the smallest negative variables to a certain negative value (negative label). In other words, the well-separated points will be fixed early and will not be further optimized. In particular, we denote q_1 and q_2 the fixed and unfixed part of q, containing p and $n-p$ elements respectively. Without loss of generality, we suppose that

$$q = [q_1, q_2]^T \tag{9}$$

Since the solution q should be conjugate orthogonal to e (See (6)), when we fix the values of the variables in q_1, we need to guarantee that at the end of each iteration step, q_1 is conjugate orthogonal to e_1 [9], i.e.

$$q_1^T W e_1 = 0 \tag{10}$$

For this purpose, for each fixed variable i, we set its value as follows.

$$q_1^{(i)} = \begin{cases} +\sqrt{\frac{\eta_2}{\eta_1}} & q_1^{(i)} > 0 \\ -\sqrt{\frac{\eta_1}{\eta_2}} & q_1^{(i)} < 0 \end{cases} \tag{11}$$

where $\eta_1 = \sum_{q_1^{(i)}>0, i \in V^F} W_i$, $\eta_2 = \sum_{q_1^{(i)}<0, i \in V^F} W_i$

After that, according to (9), we divide the matrix L and W into blocks as below

$$L = \begin{bmatrix} L_1 & L_{12} \\ L_{21} & L_2 \end{bmatrix}, W = \begin{bmatrix} W_1 & \\ & W_2 \end{bmatrix} \tag{12}$$

where L_1, L_{12}, L_{21}, L_2 are p-by-p, p-by-$(n-p)$, $(n-p)$-by-p, and $(n-p)$-by-$(n-p)$ matrices; and rewrite the objective function (6) as a function of q_2.

$$\begin{aligned} \min & \frac{\begin{bmatrix} q_1 \\ q_2 \end{bmatrix}^T \begin{bmatrix} L_1 & L_{12} \\ L_{21} & L_2 \end{bmatrix} \begin{bmatrix} q_1 \\ q_2 \end{bmatrix}}{\begin{bmatrix} q_1 \\ q_2 \end{bmatrix}^T \begin{bmatrix} W_1 & \\ & W_2 \end{bmatrix} \begin{bmatrix} q_1 \\ q_2 \end{bmatrix}}, \\ s.t. & \begin{bmatrix} q_1 \\ q_2 \end{bmatrix}^T \begin{bmatrix} W_1 & \\ & W_2 \end{bmatrix} \begin{bmatrix} e_1 \\ e_2 \end{bmatrix} = 0 \end{aligned} \tag{13}$$

Or,

$$\min T(q_2) = \frac{q_2^T L_2 q_2 + 2 q_1^T L_{12} q_2 + q_1^T L_1 q_1}{q_2^T W_2 q_2 + q_1^T W_1 q_1} \tag{14}$$

$$s.t. \quad q_2^T W_2 e_2 + q_1^T W_1 e_1 = 0 \tag{15}$$

As mentioned above, equation (10) holds because of the settings in (11). Therefore, constraint (15) can be gradually satisfied when more and more variables are fixed. Approximately speaking, we leave constraint (15) apart. Then problem (14) itself becomes a continuous optimization problem. It is clear that this problem can be solved using NLCG algorithm. If we further solve (14) with NLCG, after another several iterations, we can fix more variables according to the same criterion as mentioned above. In this way, the scale of the optimization problem is reduced throughout the whole optimization procedure. Moreover, the labels of data objects are naturally obtained without further clustering (see (11)).

3.3 Linear SSO

Although NLSSO can speed up spectral clustering, the nonlinear nature of NLCG still makes NLSSO inefficient. Actually, the reason why we use NLCG to solve problem (14) is that this problem has been beyond the capability of linear CG. In order to tackle this challenge, we propose adopting some linearization strategies so that NLCG in each step of NLSSO can be replaced with linear CG, which is much faster. In particular, we remove the denominator of (14) and only preserve its numerator which has the similar format to (7)

$$H(q_2) = q_2^T L_2 q_2 + 2q_1^T L_{12} q_2 + q_1^T L_1 q_1 \tag{16}$$

We can prove that in certain conditions, problem (14) and problem (16) are almost equivalent. More precisely, the optima of problem (16), q_2^*, and the optima of problem (14), q_2^{**}, satisfie $q_2^{**} = \lambda q_2^*$. It is not difficult to understand that this kind of scaling transformation will not affect the clustering results. Therefore, in such a way, we actually get a linear optimization algorithm which can lead to an almost-the-same solution with the original non-linear problem. In the following paragraphs, we will discuss in which condition (14) and (16) can be equivalent. For this purpose, we introduce proposition 1.

Proposition 1. *Denote q_2^{**} the optima of (14), and q_2^* the optima of (16). Then we have $q_2^{**} \approx \lambda q_2^*$, if $W_2 L_2^{-1} \approx I$.*

Proof. Omit due to space restriction.

It is clear that if L_2 is strongly dominant diagonal, we will have $W_2 L_2^{-1} \approx I$. However, in most cases, this is not natually guaranteed. To solve the problem, we propose using a preprocessing step to mandatorily make the condition satisfied. And we prove that such a preprocessing will not change the final embeddings of spectral clustering owing to preposition 2.

Proposition 2. *If q and λ are a pair of eigenvector and eigenvalue of the generalized eigenvalue problem:*

$$Lq = \lambda W q \tag{17}$$

Then q is also an eigenvector of the following generalized eigenvalue problem:

$$(L + tW)q = \tfrac{\lambda+t}{t+1}(W + tW)q \tag{18}$$

Proof. Omitted due to space restriction.

With proposition 2, we can add tW to both W and L with a sufficiently large t, without affecting the resultant eigenvectors. This makes L dominant diagonal and thus $W_2 L_2^{-1} \approx I$, i.e. the condition of preposition 1 is satisfied. As a result, except the first NLCG initialization, all other NLCG can be replaced with CG. In such a way, the speed of convergence can hence be improved a lot since (14) is a quadratic problem and CG performs very well on it.

4 Experimental Results

In this section, we report our experimental results to show the efficiency and effectiveness of our proposed LSSO algorithm. For comparison, CG-based and Lanczos-based Normalized Cut methods were also tested. Note that since we mainly discussed the two-way clustering problem in this work, all the algorithms are compared based on their two-way versions. LSSO and CG-based EVD solver were implemented using Fortran 90 without any advanced code optimization. Lanczos-based EVD solver was implemented by the function eigs() in MATLAB 6.5. In order to get as high speed as possible, we generate a C++ project using Visual C++ 6.0 which calls the function eigs() from MATLAB Library. Note that MATLAB Library has been commercially optimized. In our experiments, two criteria are used for evaluation: error rate and time cost. The error rate is defined as the ratio of miss-classified data points with respect to the groundtruth labels, and time cost is measured with "seconds". In the first experiment, we show the working procedure of our algorithm with a toy problem, and then compare the precision and speed of our LSSO algorithm with the reference algorithms on a larger and more complicated synthetic dataset. In the second experiment, we used a real-world product database from a commercial e-shopping website for further evaluation.

4.1 Toy Problems

In this subsection, we first constructed a two-circle dataset with 500 data points (see *Fig.1*) to show how our approach works. Note that the former 300 points belong to the outer circle, while the latter 200 belong to the inner circle.

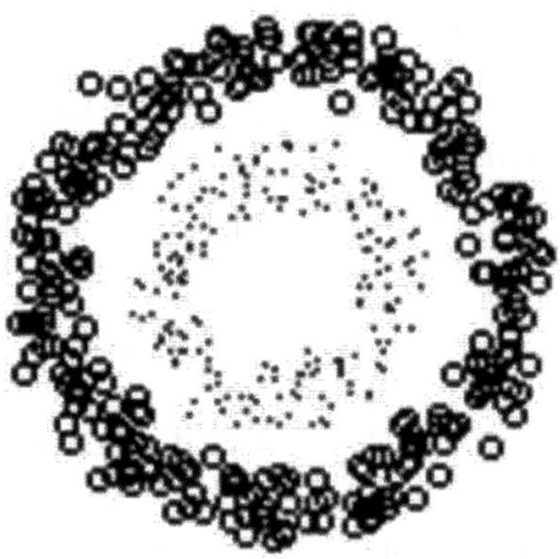

Fig. 1. A two-circle dataset

For this dataset, the second smallest eigenvector of $Lv = \lambda Wv$ is plotted in dashed line, while the clustering label is plotted in solid line in *Fig.2(a)*. Then, we plot the working procedure of LSSO in *Fig.2(b)-(d)*. In these subfigures, x-axis represents the number of data points, and the y-axis represents their embedding

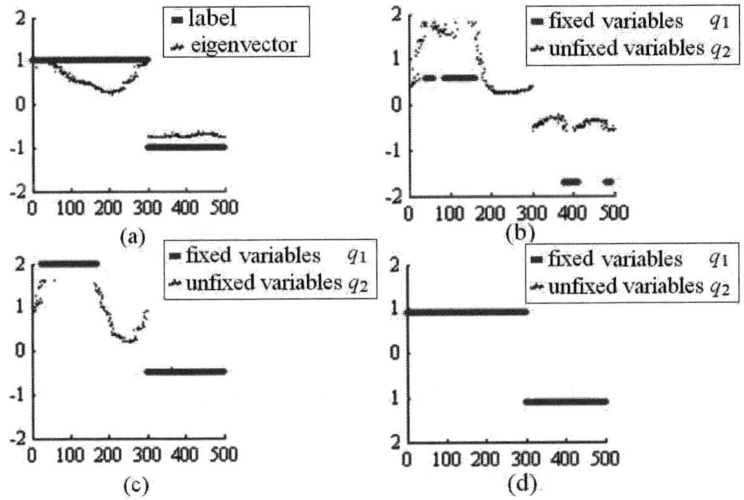

Fig. 2. Working procedure of EVD-based spectral clustering (a) and LSSO at the end of iterations 3 (b), 6 (c) and 16 (d)

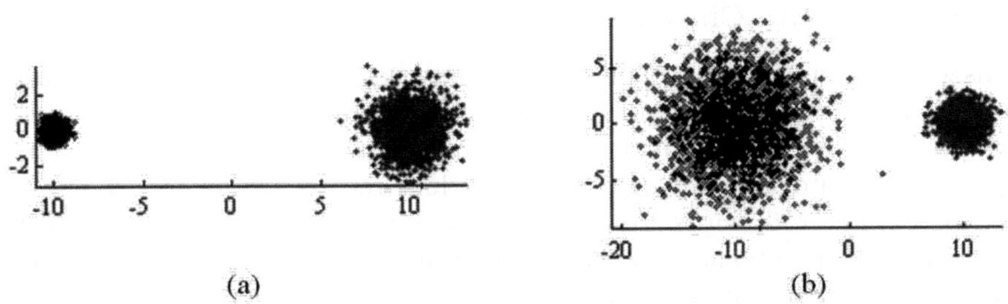

Fig. 3. Two extreme cases in the Density Datasets: (a) the No.1 Density Dataset and (b) the No.20 Density Dataset

values. The unfixed variables are plotted in dashed line, while the fixed are in solid line. We can see that at the end of iteration 3, 6, and 16, the unfixed variables sequentially shrink and the fixed variables converge to the groundtruth labels.

Secondly, we compare LSSO with CG-based and Lanczos-based spectral clustering methods. For this purpose, we constructed a group of datasets, which covers different kinds of data distributions. Each dataset contains two Gaussian balls in the 2-D Euclidean space. For ease of reference, we refer these datasets as "Density Datasets". In "Density Datasets", each ball contains 2000 data points. The distance between the centers of the two balls is 20. The variance of the right ball is fixed to 1, while the variance of the left ball changes from 1/10 to 10 so as to generate 20 datasets. The two extreme datasets (the No.1 dataset and the No.20 dataset) are shown in *Fig.3*, and the comparison of error rate / time cost is plotted in *Fig.4*, in which the x-axis represents the number of the data set in this group (i.e. from No.1 to No.20).

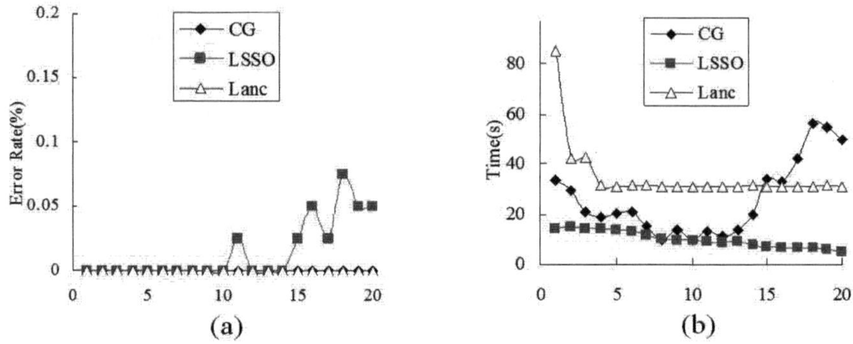

Fig. 4. (a) Error rate and (b) time cost comparison for the Density Datasets

From the above experimental results we can see that on the one hand, LSSO has similar or even better accuracy (in terms of error rate) as compared to Lanczos and CG based spectral clustering methods. On the other hand, in most cases, LSSO is much faster than Lanczos and CG based methods.

4.2 Product Data Clustering

In this subsection, we report the experimental results on a real-world product database, which is the backend database of a commercial e-shopping website. There are over 5 million product records in this database, which have been manually classified into 20 categories. Each product record has a short description of no more than 256 words. As for the descriptions of these records, we first lowered the upper-case characters and then removed stop words such as articles, prepositions, pronouns and conjunctions. Then we used term frequency to build the feature vectors so as to use the inner product in the vector space to calculate the weights in the adjacency matrix. In order to evaluate our algorithm, we sampled the product database and constructed a subset with 5000 records for each category. For ease of reference, this subset is named by "Dataset 5000". Note that the records in this dataset may have multiple labels. We conducted experiments on every pair of categories, and got ten different sub experiments. The error rate and time cost for each sub experiment are listed in *Fig.5*, where the x-axis corresponds to the ID of the sub experiments (see *Table.1*). Note that for the implementation of our LSSO method, we fix 50% of the remaining variables when performing sequential shrinkage.

From *Fig.5*, we can see that our algorithm can lead to a significant improvement in speed over the CG and Lanczos-based spectral clustering methods. At the same time, the error rate of our approach is also lower than CG and Lanczos-based algorithms for some sub experiments. This is very interesting, since we have expected that there will be more or less accuracy loss for LSSO because of the sequential shrinkage. The reason for this result may be that, when the well-separated data points are fixed, they will serve as prior knowledge, and the successive clustering may become semi-supervised and thus has better clustering results.

Table 1. ID of experiments

ID	Pair of Categories	ID	Pair of Categories
1	$C_1 - C_2$	6	$C_2 - C_4$
2	$C_1 - C_3$	7	$C_2 - C_5$
3	$C_1 - C_4$	8	$C_3 - C_4$
4	$C_1 - C_5$	9	$C_3 - C_5$
5	$C_2 - C_3$	10	$C_4 - C_5$

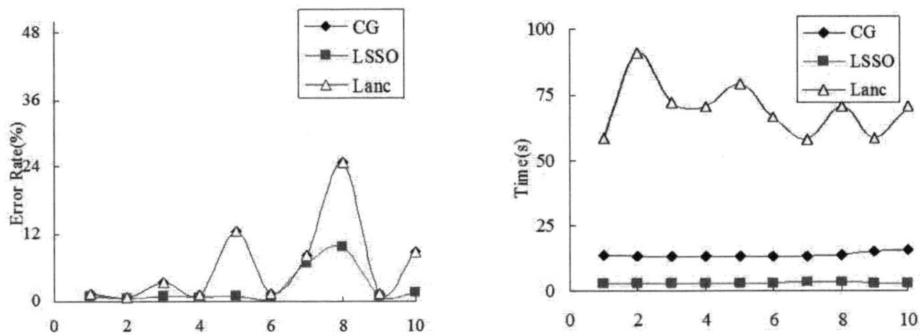

(a) (b)

Fig. 5. (a) Error rate and (b) time cost comparison for Dataset 5000

To sum up, through the above experiments, we find that LSSO is better than the reference algorithms in most cases, especially for large scale datasets, in terms of both faster speed and higher performance. We also assert that our approach can scale up for much larger datasets if the sparseness of data is employed. Due to parallelizability of matrix-vector multiplication, our approach can also be easily applied to distributed system.

5 Conclusions

Spectral clustering has met great challenges when handling large datasets since both its space and computational complexities have been bottlenecks. In this paper, we proposed our solution to speeding up spectral clustering, and demonstrated the effectiveness of our approach in many two-way clustering examples. Although we have not covered how our approach applies to k-way spectral clustering due to space restriction, it is actually not very difficult to generalize the idea of sequential shrinkage optimization to the k-way case. Besides, regarding that we only discuss how to accelerate spectral clustering in this paper, we will also work on reducing the space complexity of spectral clustering in the future. Our ultimate goal is to make spectral clustering not only a theoretically beautiful algorithm, but also a practically workable technique.

References

1. Bunk B.: Conjugate Gradient Algorithm to Compute the Low-lying Eigenvalues of the Dirac Operator in Lattice QCD. Computer Physics Communication (1994)
2. Ding C.: A Tutorial on Spectral Clustering. ICML **21** (2004)
3. Feng, Y. T., Owen, D. R. J.: Conjugate Gradient Methods for Solving the Smallest Eigenpair of Large Symmetric Eigenvalue Problems. International Journal for Numerical Methods in Engineering (1996)
4. Golub, G. H., Loan, C. F. V.: Matrix Computations. John Hopkins University Press (1996)
5. Knyazev, A. V.: Toward the Optimal Preconditioned Eigensolver: Locally Optimal Block Preconditioned Conjugate Gradient Method. SIAM Journal of Scientific Computing (2001)
6. Knyazev, A. V.: Preconditioned eigensolvers: practical algorithms. Technical Report: UCD-CCM 143, University of Colorado at Denver (1999)
7. Nocedal, J., Wright, S. J.: Numerical Optimization. Springer Series in Operations Research (2000)
8. Ng, A. Y., Jordan, M. I., Weiss, Y.: On Spectral Clustering: Analysis and an Algorithm. NIPS **14** (2001)
9. Shi J., Malik J.: Normalized Cuts and Image Segmentation. IEEE Transactions on Pattern Analysis and Machine Intelligence, **22** (2000): 888-905
10. Sorensen, D. C.: Implicitly Restarted Arnoldi/Lanczos Methods for Large Scale Eigen-value Calculations. Parallel Numerical Algorithms (1995)
11. Stanimire, T., Langou, J., Canning, A., et al.: Conjugate-Gradient Eigenvalue Solvers in Computing Electronic Properties of Nanostructure Architectures. Technical Report, UT-CS-05-559, The University of Tennessee Knoxville (2005)
12. Yang, X. P., Sarkar, T. K., Arvas, E.: A Survey of Conjugate Gradient Algorithms for Solution of Extreme Eigen-problems of a Symmetric Matrix. IEEE Transactions on Acoustics, Speech, and Signal Procesing, **37** (1989): 1550-1556
13. Yu, S. X., Shi, J. B.: Multi-class Spectral Clustering. ICCV **9** (2003)

A Probabilistic Model for Clustering Text Documents with Multiple Fields

Shanfeng Zhu[1], Ichigaku Takigawa[1], Shuqin Zhang[2], and Hiroshi Mamitsuka[1]

[1] Bioinformatics Center, Institute for Chemical Research, Kyoto University, Japan
{zhusf,takigawa,mami}@kuicr.kyoto-u.ac.jp
[2] Department of Mathematics, The University of Hong Kong, Hong Kong
sqzhang@hkusua.hku.hk

Abstract. We address the problem of clustering documents with multiple fields, such as scientific literature with the distinct fields: title, abstract, keywords, main text and references. By taking into consideration of the distinct word distributions of each field, we propose a new probabilistic model, Field Independent Clustering Model (FICM), for clustering documents with multiple fields. The benefits of FICM come not only from integrating the discrimination abilities of each field but also from the power of selecting the most suitable component probabilistic model for each field. We examined the performance of FICM on the problem of clustering biomedical documents with three fields (title, abstract and MeSH). From the genomics track data of TREC 2004 and TREC 2005, we randomly generated 60 datasets where the number of classes in each dataset ranged from 3 to 12. By applying the appropriate configuration of generative models for each field, FICM outperformed a classical multinomial model in 59 out of the total 60 datasets, of which 47 were statistically significant at the 95% level, and FICM also outperformed a multivariate Bernoulli model in 52 out of the total 60 datasets, of which 36 were statistically significant at the 95% level.

1 Introduction

Document clustering has become an increasingly important text mining and information retrieval technique due to the abundance of text documents, such as emails, news, web pages, and scientific articles [1]. As an unsupervised learning method, it can explore text collections without any prior knowledge by grouping documents into different topics, which assists in the navigation and location of the documents of interest. Clustering approaches can be divided into *partitional (flat)* clustering and *hierarchical* clustering [6]. In a partitional clustering approach, based on some optimization criterion, documents are binned into K disjoint groups (often given as an *a priori* parameter). On the contrary, in a hierarchical clustering method, the obtained clusters are organized into a tree structure. In terms of the algorithms used for clustering, we can also divide the partitional clustering methods into two categories, *discriminative (similarity-based)* methods, and *generative (model-based)* methods [14].

G. Amati, C. Carpineto, and G. Romano (Eds.): ECIR 2007, LNCS 4425, pp. 331–342, 2007.

In contrast to similarity-based methods, (probabilistic) model-based methods usually have less computational complexity, and they provide an intuitive explanation for each cluster through a corresponding model [14]. Various model-based clustering algorithms have already been proposed to tackle the problem of clustering high dimensional and very sparse text documents, such as multivariate Bernoulli [8], multinomial [8], and von Mises-Fisher models [2].

Here we focus on probabilistic model-based partitional clustering algorithms. Existing model-based algorithms consider each document as an integrated object, whereas in many cases a document actually consists of several distinct fields. One typical example is scientific documents, which are composed of multiple fields, such as title, abstract, keywords, main text, and references. Since each field has a distinguished role in presenting the idea of a document, these fields have different distributions over different vocabularies. In fact, the title, which consists of around 10 words, usually summarizes the topic of a document, while the abstract, which uses between 100 to 200 words, often briefly describes the motivation, a proposed solution, and experiment results. Assuming that each field has a distinct word distribution, we propose a new probabilistic model, which we call the Field Independent Clustering Model (FICM), where each field of a document is independently generated given the specific cluster to which the document belongs. In addition, each field can be modelled by the most suitable probabilistic model, instead of using the same model for all fields. Thus FICM takes advantage of both the discrimination ability of each field and the power of selecting the best model for each field. Rigouste et al. proposed the multinomial mixture model for document clustering, where each document is generated by a mixture of multinomial models [11]. In contrast, we focus on documents with multiple fields, where different fields can be generated by different models.

We examined our model by applying it to the problem of clustering biomedical scientific documents, which has been gaining increasing interest in bioinformatics. The largest public biomedical literature database is MEDLINE [12], which provides many informative fields, including PubMED ID, Title, Authors, Institution, Source, Journal Subset, MeSH (Medical Subject Headings) and Abstract. Out of these fields, we selected the three most important ones: Title, Abstract and MeSH. MeSH is a controlled vocabulary thesaurus defined by the National Library of Medicine [10] and includes a set of description terms organized in a hierarchical structure which are used to index articles in the MEDLINE database. Considering the wide usage of classic multivariate Bernoulli and multinomial models, we compared the performance of FICM with these models in clustering 60 randomly generated datasets from TREC2004 and TREC2005 genomics track data, where the number of classes in each datasets ranges from 3 to 12. Our experimental results showed that with suitable configurations of generative models for each field, FICM outperformed the classical multivariate Bernoulli model and the multinomial model significantly in the majority of cases.

2 FICM for Clustering Documents with Multiple Fields

In this section after introducing the classical multivariate Bernoulli model and multinomial model for clustering text documents, we propose FICM for clustering text documents with multiple fields.

2.1 Notations

We first give the notations used in this paper. Let D be a set of documents, and d be a random variable for a document in D. Let Z be the set of classes (topics) of D, and z be a random class variable in Z. C is the set of fields, and c is a variable representing a specific field in C. In this work, this could be a title, abstract or MeSH. Let D_c be a set of documents only considering field c, and d_c be field c (e.g. title) of a document d. We denote W as the set of words appearing in D, W_c as the set of words appearing in D_c, and w as a word. Let $N_{w,d}$ be the frequency of w appearing in d, and N_{w,d_c} be the frequency of w appearing in d_c. Let $B_{w,d}$ be 1 if w appears in d, otherwise 0. Let B_{w,d_c} be 1 if w appears in d_c, otherwise 0.

2.2 Multivariate Bernoulli Model

The multivariate Bernoulli model only considers whether a word occurs or not in a document. That is, each document is represented by a binary vector. For a document d in class z, the probability of w appearing in d is $p(w|z)$, while the probability of the absence of w in d is $1 - p(w|z)$. Each word in a document is assumed to be independent generated given the cluster the document belongs to, and each document in the data set is also assumed to be independent generated. We then try to maximize the log-likelihood of generating the whole set of documents D. The maximum likelihood estimators of this model can be obtained by the following EM (Expectation and Maximization) algorithm, which repeats the E- and M-steps alternately until some stopping condition is satisfied. In the M-step, we employ a Laplacian prior to avoid zero probabilities, which also belongs to maximum a-posteriori (MAP) parameter estimation. In the later models, similar techniques will also be employed.

Probabilistic Structure:

$$L(D) = \sum_{d \in D} \log p(d) = \sum_{d \in D} \log\Big(\sum_{z \in Z} p(z)p(d|z)\Big)$$
$$= \sum_{d \in D} \log\Big(\sum_{z \in Z}\Big(p(z)\prod_{w \in d} p(w|z)^{B_{w,d}}(1 - p(w|z))^{1-B_{w,d}}\Big)\Big)$$

E-step:

$$p(z|d) \propto p(z)\, p(d|z) = p(z)\prod_{w \in d}\Big(p(w|z)^{B_{w,d}}(1 - p(w|z))^{1-B_{w,d}}\Big)$$

M-step: (with Laplacian smoothing)

$$p(z) \propto \sum_{d \in D} p(z|d), \qquad p(w|z) = \frac{1 + \sum_{d \in D} p(z|d) \cdot B_{w,d}}{2 + \sum_{d \in D} p(z|d)}$$

2.3 Multinomial Model

In contrast to the multivariate Bernoulli model, the multinomial model considers the number of occurrences of each word w in document d and assumes that given a cluster z, it can generate each word in a document independently with constraint $\sum_{w \in W} p(w|z) = 1$. Please note that both multivariate Bernoulli model and multinomial model treat all occurrences of a word in the same way, without considering what field they occur in. The probabilistic structure of multinomial model and its corresponding E- and M-steps are shown below.

Probabilistic Structure:

$$L(D) = \sum_{d \in D} \log p(d) = \sum_{d \in D} \log\Big(\sum_{z \in Z} p(z) p(d|z)\Big)$$
$$= \sum_{d \in D} \log\Big(\sum_{z \in Z}\Big(p(z) \prod_{w \in d} p(w|z)^{N_{w,d}}\Big)\Big)$$

E-step: $$p(z|d) \propto p(z)\, p(d|z) = p(z) \prod_{w \in d} p(w|z)^{N_{w,d}}$$

M-step: (with Laplacian smoothing)

$$p(z) \propto \sum_{d \in D} p(z|d), \quad p(w|z) = \frac{1 + \sum_{d \in D} p(z|d) \cdot N_{w,d}}{|W| + \sum_{w' \in W} \sum_{d \in D} p(z|d) \cdot N_{w',d}}$$

2.4 FICM

The basic idea of FICM is that each component field of a document is independently generated by an underlying probabilistic model given the cluster to which the document belongs. Although in practice a document may not fully obey this rule, this kind of independence assumption has been widely used and has been found to be very successful in machine learning algorithms [9]. In fact, this kind of assumption is also utilized in classical multivariate Bernoulli and multinomial models, which assume that each word in a document is independently generated by some underlying probabilistic model. Under this assumption, the probability of generating a document d is given as follows:

$$p(d) = \sum_{z \in Z} p(z) p(d|z) = \sum_{z \in Z} p(z) \prod_{c \in C} p(d_c|z)$$

Here $p(d_c|z)$ is the probability of generating the set of words that appear in field c given the underlying cluster z. The strength of FICM comes from the integration of the clustering ability of each field, which can be improved by choosing a good probabilistic model for each field. FICM is more like a framework, since the specific implementation of FICM depends on the probabilistic model used for each field. Let C_b be the set of fields modelled by the multivariate Bernoulli model, and C_m be the set of fields modelled by the multinomial model. We can derive the probabilistic structure of FICM as below, and can show the E- and M-steps of the EM algorithm to estimate the parameters of this model.

Probabilistic Structure:

$$\begin{aligned} L(D) &= \sum_{d\in D} \log p(d) = \sum_{d\in D} \log\Big(\sum_{z\in Z} p(z)p(d|z) \Big) \\ &= \sum_{d\in D} \log\Big(\sum_{z\in Z} \Big(p(z) \prod_{c\in C_b} p(d_c|z) \times \prod_{c'\in C_m} p(d'_c|z) \Big) \Big) \\ &= \sum_{d\in D} \log\Big(\sum_{z\in Z} \Big(p(z) \prod_{c\in C_b} \prod_{w\in d_c} p(w|z,c)^{B_{w,d_c}} (1-p(w|z,c))^{1-B_{w,d_c}} \\ &\times \prod_{c'\in C_m} \prod_{w\in d_{c'}} (p(w|z,c')^{N_{w,d_{c'}}}) \Big) \Big) \end{aligned}$$

E-step:

$$\begin{aligned} p(z|d) \propto p(z)\, p(d|z) &= p(z) \prod_{c\in C_b} p(d_c|z) \times \prod_{c'\in C_m} p(d'_c|z) \\ &= p(z) \prod_{c\in C_b} \prod_{w\in d_c} (p(w|z,c)^{B_{w,d_c}} \times (1-p(w|z,c))^{1-B_{w,d_c}}) \\ & \prod_{c'\in C_m} \prod_{w\in d_{c'}} (p(w|z,c')^{N_{w,d_{c'}}}) \end{aligned}$$

M-step: (with Laplacian smoothing)

$$\begin{aligned} p(z) &\propto \sum_{d\in D} p(z|d), \\ p(w|z,c) &= \frac{1+\sum_{d\in D} p(z|d)\cdot B_{w,d_c}}{2+\sum_{d\in D} p(z|d)} \qquad \text{if} \quad c\in C_b, \\ p(w|z,c') &= \frac{1+\sum_{d\in D} p(z|d)\cdot N_{w,d_{c'}}}{|W_{c'}|+\sum_{w'\in W}\sum_{d\in D} p(z|d)\cdot N_{w',d_{c'}}} \qquad \text{if} \quad c'\in C_m \end{aligned}$$

In the simplest case, we can use the same probabilistic model for all fields, such as $C_b = C$ or $C_m = C$, which we call a Fielding Bernoulli Model or Fielding Multinomial Model, respecitively.

3 Experimental Results

3.1 Evaluation Criteria

We used an external measure to assess the quality of the clustering algorithms by assuming that we already know the correct labeling of each document. That is, we evaluated the performance of the clustering algorithms by comparing the predicted clustering result with the correct (true) class labeling. Several external measures have been proposed to evaluate the quality of clustering, such as purity, average entropy, precision, recall, F-measure and mutual information [3]. Since mutual information has been deemed as a superior measure over other measures, such as purity and entropy which have a bias of favoring a large number of clusters, we also used it as the measure in our work [14, 15]. The definition of normalized mutual information is given by the following formula,

$$NMI = \frac{I(X;Y)}{\sqrt{H(X) \cdot H(Y)}},$$

where X and Y are the predicted clusters and the correct class labels, respectively, $I(X;Y)$ is the mutual information between X and Y, and $H(X)$ and $H(Y)$ are the entropy of X and Y, respectively. Zhong et al. [14, 15] proposed a sample estimate to calculate *NMI*,

$$NMI = \frac{\sum_{h,l} n_{h,l} \log(\frac{n \cdot n_{h,l}}{n_h n_l})}{\sqrt{(\sum_h n_h \log \frac{n_h}{n})(\sum_l n_l \log \frac{n_l}{n})}}$$

where n is the total number of documents in the whole collection, n_h is the number of documents in class h (standard), n_l is the number of documents in cluster l (predicted), and $n_{h,l}$ is the number of documents in both class h and cluster l. The NMI value ranges from zero to one, where an NMI value of zero means that the result is equal to the almost random partitioning, and an NMI value of one means that the result is almost identical to the true class labeling.

3.2 Dataset

We used the datasets obtained from the TREC[1] genomics track 2004 and 2005 [4, 5]. Both of these include an ad hoc retrieval task, which requires an information retrieval system to return documents relevant to specific information needs. The documents for both the 2004 and 2005 ad hoc retrieval tasks are a 10-year (from 1994 to 2003) MEDLINE subset, which has a total of 4,591,008 records. In the genomics tracks of each year, 50 topics were formulated by real biologists as information needs. For each topic, the top records returned by each retrieval system were pooled and further assessed by biologists to determine the relevance of the result.

After eliminating the documents that are related to more than one topic or have any empty fields (title, abstract or MeSH), we obtained 39 (total 4400

[1] http://trec.nist.gov/

documents) and 24 (total 2317 documents) topics that have 10 or more relevant documents in TREC 2004 and 2005, respectively. We then randomly generated 30 different datasets for each year by choosing three or more topics[2] for each generated dataset. For each specific class number, three different datasets have been generated. Similar methods of randomly generating test datasets from a base dataset have also been adopted by other researchers [13]. We emphasize that our datasets must have high quality, since the topic of each document was examined by the biologists. The name of each dataset is encoded by the combination of the year of the TREC data from which it was generated, the number of topics in the dataset and the order of the dataset in all datasets with the same number of topics from the same year, such as "200503a", "200408b" and "200512c". For example, "200503a" means that this dataset, which is generated from TREC 2005, is the first one of those having three classes.

From the TREC data, we extracted not only the whole text of each record, but also three distinct fields (title, abstract and MeSH) of each record. In the pre-processing step, we removed stop words and used the Porter's stemming algorithm to tokenize documents. Similar to the work by Zhong et al. [15], we also removed those words that appear in less than three documents. For each field, we also removed the words that appear in less than three documents. Due to space limitations, we only show the statistical information of six example datasets, and the minimum, maximum and mean of all the 60 datasets in Table 1. We can see that these 60 datasets vary greatly in some important characteristics: the number of documents varies from 92 to 1608, the number of classes from 3 to 12, the balance from 0.021 to 0.789 and the number of distinct words in each dataset from 710 to 3993, which makes it a very good testbed for comparing different clustering algorithms.

3.3 Experimental Settings

We used NMI to compare the performance of FICM with the others. When we use the multivariate Bernoulli model and documents with the title field only, the abstract field only, the MeSH field only and all fields, we call the models (and the results) *bert, bera, berm,* and *berw*, respectively. We call the Fielding Bernoulli model *fber*. Similarly, when we use the multinomial model, the corresponding names are *mnst, mnsa, mnsm, mnsw* and *fmns*. For fair comparisons, we fixed the number of iterations in the EM algorithms at 30 for all cases. In addition, K, the number of classes, was given as *a priori* parameter for all models. Furthermore, to reduce the possible bias caused by a random initial partition, we ran each experiment 100 times, and the averages, standard deviations and paired t-test were computed for comparison.

FICM allows us to assign a different probabilistic model to each field. In this experiment, we focused on the following two settings: one is to assign the multinomial model to the title and abstract fields, and the multivariate Bernoulli model to the MeSH field, which is denoted as *fmmb*; and the other is to assign

[2] The maximum number is 12.

Table 1. Summary of some example datasets and the minimum, maximum and mean of all 60 datasets randomly generated from the TREC genomics track 2004 and 2005. N_d is the number of documents, W is the number of distinct words (tokens), K is the number of classes, N_l is the average number of words in each document, Balance is the size ratio of the smallest class to the largest class, N_t (N_a, N_m) is the average number of words in the title(abstract, MeSH) field, and $W_t(W_a, W_m)$ is the number of distinct words in the title(abstract, MeSH) field.

Data	N_d	W	K	N_l	Balance	N_t	N_a	N_m	W_t	W_a	W_m
200403a	102	711	3	126.1	0.7895	5.7	81.6	33.7	88	616	206
200503a	570	2109	3	142.3	0.2203	7.7	97.5	34.6	336	1920	532
200408a	710	2593	8	156.1	0.1217	7.8	108.7	37.2	480	2401	685
200508a	453	2013	8	143.5	0.0798	7.6	101.6	31.3	328	1825	551
200412a	1295	3625	12	166.2	0.0216	9	112.3	43	776	3349	1013
200512a	1357	3187	12	145.7	0.0211	8.7	101.6	34	658	2937	852
min of all 60	92	710	3	126.1	0.0211	5.7	81.6	30.3	76	604	205
max of all 60	1608	3993	12	177.8	0.7895	9.8	120.3	49	902	3710	1171
mean of all 60	776.5	2511.9	7.5	154.3	0.0889	8.2	105.5	38.1	470.3	2304.7	697.1

the multinomial model to the title field, and the multivariate Bernoulli model to the abstract and MeSH fields, which is denoted as *fmbb*. That is, we fixed the multinomial and multivariate Bernoulli models to the title and MeSH fields, respectively. The reason for this is as follows: MeSH terms are originally organized in a hierarchical structure, but this hierarchical information is lost in the multinomial model and the multivariate Bernoulli model, which treat every word in a flat way. Some general words, such as "human" and "genetics", would appear in the MeSH field very frequently, meaning that these words might not be useful for clustering. We think that the multinomial model, which favors frequent terms in clustering, might get stuck in this situation, while the multivariate Bernoulli model, which considers binary occurrences only, might be more robust against this situation. On the other hand, words in the title field are all important to capture the document topic. In this situation, the multinomial model, in which the probability for generating each distinct word sums to 1, would be more suitable than the multivariate Bernoulli model, which has to deal with each word independently.

3.4 Results

Comparison of bert, bera, berm, berw, fber and fmbb

We first examined the performance of the clustering models based on the multivariate Bernoulli model, and *fmbb* which replaces the Bernoulli model for the MeSH field in *fber* with the multinomial model. Table 2 shows the result from six example datasets as well as the average over all 60 datasets. In this table, the model achieving the highest NMI is bolded. The *fmbb* model achieved the highest average NMI of 0.736 for all 60 datasets, which was followed by *fber* achieving 0.733, and *berw* achieving 0.714. Not surprisingly, *berw*, which uses the Bernoulli

Table 2. Performance in terms of NMI (*mean±standard deviation*) for six example datasets and the average of all 60 datasets

Data	*bert*	*bera*	*berm*	*berw*	*fber*	*fmbb*
200403a	.667±.12	.818±.12	.886±.13	.846±.12	.876±.12	**.888 ±.09**
200503a	.667±.11	**.784±.17**	.578±.11	.733±.18	.752±.18	.760±.19
200408a	.671±.07	.765±.06	.756±.05	.808 ±.04	.820±.05	**.834±.04**
200508a	.705±.05	.842±.06	.817±.04	.860±.05	.849±.05	**.864±.04**
200412a	.557±.06	.708±.05	.668±.04	.741±.05	.739±.05	**.763±.05**
200512a	.534±.06	.604±.06	.520±.04	.595±.05	.635±.05	**.643±.05**
Mean of all 60	.638±.06	.700±.07	.670±.05	.714±.06	.733±.06	**.736±.06**

Table 3. Paired t-test in performance comparison over all 60 sets at the 95% level

berw> *bert*	*berw* > *bera*	*berw*> *berm*	*fber*> *berw*	*fmbb*> *berw*	*fmbb*> *fber*
$(+\frac{47}{49}, -\frac{9}{11})$	$(+\frac{29}{42}, -\frac{5}{18})$	$(+\frac{44}{51}, -\frac{4}{9})$	$(+\frac{30}{52}, -\frac{0}{8})$	$(+\frac{36}{52}, -\frac{1}{8})$	$(+\frac{7}{40}, -\frac{3}{20})$

model on the whole text was better than any of *bert, bera,* or *berm*, which uses only one field. Furthermore, Table 3 shows the paired t-test indicating the statistical significance of improvements. The threshold of the p-value for the paired t-test is set to 0.05. For each examined hypothesis, e.g. that model x outperforms model y ($x > y$), we compared the performance of x with y in each dataset by paired t-test, and then retrieve a pair of two numbers $(+\frac{a}{b}, -\frac{c}{d})$, which means that x outperforms y in b datasets, of which a cases are statistically significant, and y outperforms x in d datasets, of which c cases are statistically significant. For our experiment, $b + d = 60$, and $x > y$ is true only when $a > c$ and $b > d$. Hereafter we will call this pair SPair. Table 3 indicates that *berw* outperformed *bert, bera and berm* significantly. We are especially interested in the comparison among *berw, fber* and *fmbb*, which utilize all three fields for clustering. We found that both *fber* and *fmbb* outperformed *berw*, being statistically significant in the majority of cases, while *fmbb* was slightly better than *fber*, which was proved by SPair $(+\frac{7}{40}, -\frac{3}{20})$. The results can be summarized as the performance ordering[3] of the compared models: $fmbb > fber > berw > (bert, bera, berm)$.

Comparison of mnst, mnsa, mnsm, mnsw, fmns and fmmb

Second, we examined the performance of the clustering models based on the multinomial model, and *fmmb* which replaces the multinomial model for the MeSH field in *fmns* with the multivariate Bernoulli model. Similarly, the experimental results are shown in Tables 4 and 5. Based on the average result of all 60 datasets in Table 4, *fmmb* achieved the highest NMI of 0.735, which was followed by *fmns* achieving 0.718 and *mnsw* achieving 0.712. We further investigated the significance of these improvements in Table 5. The *mnsw* method

[3] We put *bert, bera* and *berm* in the same bracket, since we have no interest in comparing the performance of Bernoulli models over different fields.

Table 4. Performance in terms of NMI ($mean \pm standard\ deviation$) for some example datasets and average of all 60 datasets

Data	*mnst*	*mnsa*	*mnsm*	*mnsw*	*fmns*	*fmmb*
200403a	.850±.09	.865±.13	.858±.07	**.879±.10**	.878±.12	.872±.14
200503a	.667±.13	.700±.16	.507±.06	.716±.18	.715±.18	**.756±.18**
200408a	.769±.04	.798±.04	.737±.04	.809±.05	.818±.04	**.834±.05**
200508a	.821±.04	.825±.05	.771±.04	.822±.04	.834±.04	**.861 ±.04**
200412a	.746±.04	.765±.04	.678±.03	.774±.03	.784±.04	**.800±.04**
200512a	.645±.03	.670±.04	.458±.04	.670 ±.04	.674±.04	**.692±.04**
mean of all 60	.711±.05	.704±.06	.623±.04	.712±.05	.718 ±.06	**.735 ±.06**

Table 5. Paired t-tests over all 60 datasets at the 95% level

mnsw>*mnst*	*mnsw*>*mnsa*	*mnsw*>*mnsm*	*fmns*>*mnsw*	*fmmb*>*mnsw*	*fmmb*>*fmns*
$(+\frac{25}{32}, -\frac{21}{28})$	$(+\frac{19}{41}, -\frac{7}{19})$	$(+\frac{55}{58}, -\frac{1}{2})$	$(+\frac{11}{50}, -\frac{1}{10})$	$(+\frac{47}{59}, -\frac{0}{1})$	$(+\frac{38}{56}, -\frac{0}{4})$

outperformed *mnsa* and *mnsm* clearly, but surpassed *mnst* slightly, suggesting that only a few distinguished words in the title can be very effective for document clustering. On the other hand, *fmmb* outperformed both *mnsw* and *fmns* significantly by SPairs $(+\frac{47}{59}, -\frac{0}{1})$ and $(+\frac{38}{56}, -\frac{0}{4})$, while *fmns* was slightly better than *mnsw* by SPair $(+\frac{11}{50}, -\frac{1}{10})$. In this round of comparison, we obtained the following: $fmmb > fmns > mnsw > (mnst, mnsa, mnsm)$.

Summarization

In the above two rounds of experiments, compared with the classical multivariate Bernoulli model *berw* and multinomial model *mnsw*, the direct extension of them via FICM (*fber* and *fmns*, respectively) could improve the clustering performance in the majority of the datasets. This was especially significant in the case of *fber*, which achieved a higher NMI value than *berw* in 52 out of all 60 datasets, of which 30 are statistically significantly at the 95% level. Furthermore, by choosing the most suitable probabilistic model for each field other than the same model for all fields, we could obtain even better clustering results. This was proved by the performance improvement of *fmmb* over *fmns*, and the performance improvement of *fmbb* over *fber*. Overall, *fmmb* outperformed *mnsw* in 59 out of all 60 datasets, of which 47 are statistically significant at the 95% level, and *fmbb* outperformed *berw* in 52 out of 60 datasets, of which 36 are statistically significant. This indicates that FICM is methodologically significantly better against the existing models, which is attributed to the integration of discrimination ability of each field that consists of distinct word distribution, and the power of choosing the suitable generative model for each field.

Discussion

To maximize the performance of FICM, the choice of a suitable generative model for each field becomes a crucial problem. We have shown that choosing the multinomial model and the multivariate Bernoulli model for the title and MeSH fields, respectively, achieves best clustering results. This choice was also justified by the signicant improvement of *mnst* over *bert* in terms of SPair ($+\frac{47}{51}$, $-\frac{6}{9}$) and that of *berm* over *mnsm* in terms of SPair ($+\frac{51}{55}$, $-\frac{2}{5}$). Regarding the abstract field, the choice of the model did not affect the significant difference in performance. We summarize these results in the first three columns in Table 6. We then compared the performance of *fmbb* with *fmmb* with respect to the number of classes (topics) in the datasets. More precisely, we split the 60 datasets into two, by the number of topics. If the number of topics was less than 8, *fmbb* outperformed *fmmb* in 27 out of all 30 cases, of which 19 were statistically significant at the 95% level, and on the contrary, when the number of topics was 8 or more, *fmmb* outperformed *fmbb* in 24 out of 30 cases, of which 20 were statistically significant at the 95% level. This result is summarized in Table 6 as well. Using the same split datasets, we found a similar tendency of the performance between *mnsa* and *bera*. That is, *bera* outperformed *mnsa*, having SPair ($+\frac{23}{25}$, $-\frac{3}{5}$) for datasets with less than eight topics, while *mnsa* outperfomed *bera*, having SPair ($+\frac{19}{22}$, $-\frac{2}{8}$) for the other datasets. This interesting correspondence further verified that the strength of FICM comes from the power of the discrimination ability of each field. As shown in Table 6, we can choose *fmbb* for clustering datasets that are relatively small with few topics, and *fmmb* for clustering datasets that are relatively large with more topics. This is also consistent with McCallum and Nigam's findings that the multivariate Bernoulli model performs very well on datasets of small vocabulary size, but that the multinomial model usually performs better on datasets of large vocabulary size [8].

Table 6. Paired *t*-test results in the comparison of *bert* with *mnst*, *bera* with *mnsa*, *berm* with *mnsm*, and *fmbb* with *fmmb* over all 60 sets

mnst>*bert*	*mnsa*>*bera*	*berm*>*mnsm*	*fmbb*>*fmmb*	*fmbb*>*fmmb* # topics < 8	*fmmb*>*fmbb* # topics >= 8
($+\frac{47}{51}$, $-\frac{6}{9}$)	($+\frac{25}{33}$,$-\frac{22}{27}$)	($+\frac{51}{55}$,$-\frac{2}{5}$)	($+\frac{21}{33}$,$-\frac{22}{27}$)	($+\frac{19}{27}$,$-\frac{1}{3}$)	($+\frac{20}{24}$,$-\frac{2}{6}$)

4 Conclusion and Future Work

The strength of FICM comes from the integration of the discrimination ability of each field and the power of choosing the most suitable generative model for each field. Our experimental results show that a direct extension of the classical multivariate Bernoulli and multinomial models using all fields could already achieve a better performance, and by configuring each field with a suitable model, we could obtain further better clustering results. Here the number of documents in each dataset ranges from 92 to 1680. Although it may be the most frequent case in practice, we would like to examine the performance of FICM on some larger datasets in the future. It would also be very interesting to apply FICM on other types of text documents, which requires us to use some prior knowledge or preliminary experiments to choose a suitable model for each field.

Acknowledgements

This work is supported in part by JSPS (Japan Society for the Promotion of Science) Postdoctoral Fellowship. We also would like to thank anonymous reviewers for very helpful suggestions.

References

1. Baeza-Yates, R., Ribeiro-Neto, B.: Modern information retrieval. Addison Wesley, New York. (1999)
2. Banerjee, A., Dhillon, I., Ghosh, J., Sra, S.: Generative Model-based Clustering of Directional Data. the proceedings of the SIGKDD'03, (2003) 19-28
3. Ghosh, J.: Scalable clustering methods for data mining, Handbook of data mining, editor: Ye, N., Erlbaum, L. (2003)
4. Hersh, WR., Bhupatiraju, RT., Ross, L., Johnson, P., Cohen, AM., Kraemer, DF.: TREC 2004 Genomics Track Overview. the proceedings of the Thirteenth Text REtrieval Conference (TREC 2004). editor: Voorhees, EM. and Buckland, LP. (2004)
5. Hersh, WR., Cohen, A., Bhupatiraju, RT., Johnson, P., Hearst, M.: TREC 2005 Genomics Track Overview. the proceedings of the Fourteenth Text REtrieval Conference (TREC 2005), editor: Voorhees, EM. and Buckland, LP. (2005)
6. Jain, AK., Murty, MN., Flynn, PJ.: Data Clustering: A Review. ACM Computing Surveys. **31**(3) (1999) 264–323
7. Meila, M., Heckerman, D.: An Experimental Comparison of Model-Based Clustering Methods. Machine Learning. **42**(1/2) (2001) 9–29
8. McCallum, A., Nigam, K.: A comparsion of event models for naive Bayes text classification. AAAI Workshop on Learning for Text Categorization. (1998) 41–48
9. Lewis, DD.: Naive (Bayes) at forty: the independence assumption in information retrieval. Proceedings of ECML-98.
10. Nelson, SJ., Schopen, M., Savage, AG., Schulman, J., Arluk, N.: The MeSH Translation Maintenance System: Structure, Interface Design, and Implementation. Proceedings of the 11th World Congress on Medical Informatics. editor: Fieschi, M. et al. (2004) 67–69
11. Rigouste, L., Cappé, O., Yvon, F.: Evaluation of a Probabilistic Method for Unsupervised Text Clustering. International Symposium on Applied Stochastic Models and Data Analysis,Brest, France. (2005)
12. Wheeler, D. et al.: Database resources of the National Center for Biotechnology Information Nucl. Acids Res. **33** (2005) D39–D45
13. Yoo, I., Hu, X.: A comprehensive comparison study of document clustering for a biomedical digital library MEDLINE. ACM/IEEE Joint Conference on Digital Libraries, JCDL 2006, editor: Marchionini, G. et al. (2006) 220–229.
14. Zhong, S., Ghosh, J.: A unified framework for model-based clustering. Journal of Machine Learning Research. **4** (2003) 1001–1037
15. Zhong, S., Ghosh, J.: Generative model-based document clustering: a comparative study. Knowledge and Information Systems. **8**(3) (2005) 374–384

Personalized Communities in a Distributed Recommender System

Sylvain Castagnos and Anne Boyer

LORIA - Université Nancy 2
Campus Scientifique - B.P.239
54506 Vandoeuvre-lès-Nancy Cedex, France
{sylvain.castagnos, anne.boyer}@loria.fr

Abstract. The amount of data exponentially increases in information systems and it becomes more and more difficult to extract the most relevant information within a very short time. Among others, collaborative filtering processes help users to find interesting items by modeling their preferences and by comparing them with users having the same tastes. Nevertheless, there are a lot of aspects to consider when implementing such a recommender system. The number of potential users and the confidential nature of some data are taken into account. This paper introduces a new distributed recommender system based on a user-based filtering algorithm. Our model has been transposed for Peer-to-Peer architectures. It has been especially designed to deal with problems of scalability and privacy. Moreover, it adapts its prediction computations to the density of the user neighborhood.

1 Introduction

With the development of information and communication technologies, the size of information systems all over the world has exponentially increased. Consequently, it becomes harder and harder for users to identify relevant items in a reasonable time, even when using a powerful search engine. Collaborative filtering techniques [1] are a good way to cope with this difficulty. It amounts to identifying the active user to a set of persons having the same tastes, based on his/her preferences and his/her past actions. This system starts from the principle that users who liked the same items have the same topics of interest. Thus, it is possible to predict the relevancy of data for the active user by taking advantage of experiences of a similar population.

There are several fundamental problems when implementing a collaborative filtering algorithm. In this paper, we particularly pay attention to the following significant limitations for industrial use:

- scalability and system reactivity: there are potentially several thousand users and items to manage in real time;
- intrusions into privacy: we have to be careful to be as unintrusive as possible and at least to guarantee the anonymity of users;

G. Amati, C. Carpineto, and G. Romano (Eds.): ECIR 2007, LNCS 4425, pp. 343–355, 2007.

- novelty in predictions: according to the context, users want to have more or less new recommendations. Sometimes their main concern is to retrieve the items that they have high-rated, even if it means having less new recommendations. This is why we introduce an adaptive minimum-correlation threshold of neighborhood which evolves in accordance with user expectations.

We propose an algorithm which is based on an analysis of usage. It relies on a distributed user-based collaborative filtering technique. Our model has been integrated in a document sharing system called "SofoS".[1]

Our algorithm is implemented on a Peer-to-Peer architecture because of the document platform context. In a lot of companies, documents are referenced using a common codification that may require a central server[2] but are stored on users' devices. The distribution of computations and contents matches the constraints of scalability and reactivity.

In this paper, we will first present the related work on collaborative filtering approaches. We will then introduce our Peer-to-Peer user-centered model which offers the advantage of being fully distributed. We called this model "Adaptive User-centered Recommender Algorithm" (AURA). It provides a service which builds a virtual community of interests centered on the active user by selecting his/her nearest neighbors. As the model is ego-centered, the active user can define the expected prediction quality by specifying the minimum-correlation threshold. AURA is an anytime algorithm which furthermore requires very few computation time and memory space. As we want to constantly improve our model and the document sharing platform, we are incrementally and modularly developing them on a JXTA platform[3].

2 Related Work

In centralized collaborative filtering approaches, finding the closest neighbors among several thousands of candidates in real time without offline computations may be unrealistic [2]. By contrast, decentralization of data is practical to comply with privacy rules, as long as anonymity is fulfilled [3]. This is the reason why more and more researchers investigate various means of distributing collaborative filtering algorithms. This also presents the advantage of giving the property of profiles to users, so that they can be re-used in several applications.[4] We can mention research on P2P architectures, multi-agents systems and decentralized models (client/server, shared databases).

There are several ways to classify collaborative filtering algorithms. In [4], authors have identified, among existing techniques, two major classes of algorithms: memory-based and model-based algorithms. Memory-based techniques

1 SofoS is the acronym for "Sharing Our Files On the System".

2 This allows to have document IDs and to identify them easily.

3 http://www.jxta.org/

4 As the owner of the profile, the user can apply it to different pieces of software. In centralized approaches, there must be as many profiles as services for one user.

offer the advantage of being very reactive, by immediately integrating modifications of users profiles into the system. They also guarantee the quality of recommendations. However, Breese *et al.* [4] are unanimous in thinking that their scalability is problematic: even if these methods work well with small-sized examples, it is difficult to change to situations characterized by a great number of documents or users. Indeed, time and space complexities of algorithms are serious considerations for big databases. According to Pennock *et al.* [5], model-based algorithms constitute an alternative to the problem of combinatorial complexity. Furthermore, they perceive in these models an added value beyond the function of prediction: they highlight some correlations in data, thus proposing an intuitive reason for recommendations or simply making the hypotheses more explicit. However, these methods are not dynamic enough and they react badly to insertion of new contents into the database. Moreover, they require a penalizing learning phase for the user.

Another way to classify collaborative filtering techniques is to consider user-based methods in opposition to item-based algorithms. For example, we have explored a distributed user-based approach within a client/server context in [6]. In this model, implicit criteria are used to generate explicit ratings. These votes are anonymously sent to the server. An offline clustering algorithm is then applied and group profiles are sent to clients. The identification phase is done on the client side in order to cope with privacy. This model also deals with sparsity and scalability. We highlight the added value of a user-based approach in the situation where users are relatively stable, whereas the set of items may often vary considerably. On the contrary, Miller *et al.*[7] show the great potential of distributed item-based algorithms. They propose a P2P version of the item-item algorithm. In this way, they address the problems of portability (even on mobile devices), privacy and security with a high quality of recommendations. Their model can adapt to different P2P configurations.

Beyond the different possible implementations, we can see there are a lot of open questions raised by industrial use of collaborative filtering. Canny [3] concentrates on ways to provide powerful privacy protection by computing a "public" aggregate for each community without disclosing individual users' data. Furthermore, his approach is based on homomorphic encryption to protect personal data and on a probabilistic factor analysis model which handles missing data without requiring default values for them. Privacy protection is provided by a P2P protocol. Berkovsky *et al.* [8] also deal with privacy concern in P2P recommender systems. They address the problem by electing super-peers whose role is to compute an average profile of a sub-population. Standard peers have to contact all these super-peers and to exploit these average profiles to compute predictions. In this way, they never access the public profile of a particular user. We can also cite the work of Han *et al.*[9], which addresses the problem of privacy protection and scalability in a distributed collaborative filtering algorithm called PipeCF. Both user database management and prediction computation are split between several devices. This approach has been implemented on Peer-to-Peer overlay networks through a distributed hash table method.

In this paper, we introduce a new hybrid method called AURA. It combines the reactivity of memory-based techniques with the data correlation of model-based approaches by using an iterative clustering algorithm. Moreover, AURA is a user-based model which is completely distributed on the user scale. It has been integrated in the SofoS document platform and relies on a P2P architecture in order to distribute either prediction computations, content or profiles. We design our model to tackle, among others, the problems of scalability, and privacy.

3 SofoS

SofoS is a document platform, using a recommender system to provide users with content. Once it is installed, users can share and/or search documents, as they do on P2P applications like Napster. We conceive it in such a way that it is as open as possible to different existing kinds of data: hypertext files, documents, music, videos, etc. The goal of SofoS is also to assist users to find the most relevant sources of information efficiently. This is why we add the AURA recommender module to the system. We assume that users can get pieces of information either by using our system or by going surfing on the web. SofoS consequently enables to take visited websites into account in the prediction computations.

We are implementing SofoS in a generic environment for Peer-to-Peer services, called JXTA. This choice is motivated by the fact it is greatly used in our research community.

In [7], the authors highlight the fact that there are several types of possible architectures for P2P systems. We can cite those with a central server (such as Napster), random discovery ones[5] (such as Gnutella or KaZaA), transitive traversal architectures, content addressable structures and secure blackboards.

We conceived our model with the idea that it could be adapted to different types of architectures. However, in this paper, we will illustrate our claims by basing our examples on the random approach even if others may have an added value. The following subsection aims at presenting the AURA Algorithm.

3.1 AURA Algorithm

We presume that each peer in SofoS corresponds to a single user on a given device.[6] For this reason, we have conceived the platform in such a way that users have to open a session with a login and a password before using the application. In this way, several persons can use the same computer (for example, the different members of a family) without disrupting their respective profiles. That is why each user on a given peer of the system has his/her own profile and a single ID. The session data remain on the local machine in order to enhance privacy. There is no central server required since sessions are only used to distinguish users on a given peer.

[5] Some of these architectures are totally distributed. Others mixed centralized and distributed approaches but elect super-peers whose role is to partially manage subgroups of peers in the system.

[6] We can easily distinguish devices since SofoS has to be installed on users' computers.

For each user, we use a hash function requiring the IP address and the login in order to generate his/her ID on his/her computer. This use of a hash function H is suitable, since it has the following features:

- non-reversible: knowing "y", it is hard to find "x" such as $H(x) = y$;
- no collision: it is hard to find "x" and "y" such as $H(x) = H(y)$;
- knowing "x" and "H", it is easy to compute $H(x)$;
- H(x) has a fixed size.

In this way, an ID does not allow identification of the name or IP address of the corresponding user. The communication module uses a IP multicast address to broadcast the packets containing addressees' IDs. In order to reduce the information flow, we can optionally elect a super-peer which keeps a list of IDs whose session is active: before sending a message, a peer can ask if the addressee is connected. If the super-peer has no signal from a peer for a while, it removes the corresponding ID from the list.

Users can both share items on the platform and integrate a feedback about websites they consult. Each item has a profile on the platform. In addition to the available documents, each peer owns 7 pieces of information: a personal profile, a public profile, a group profile and 4 lists of IDs (list "A" for IDs of peers belonging to its group, list "B" for those which exceed the minimum-correlation threshold as explained below, list "C" for the black-listed IDs and list "O" for IDs of peers which have added the active user profile to their group profile). An example of the system run is shown on figure 1.

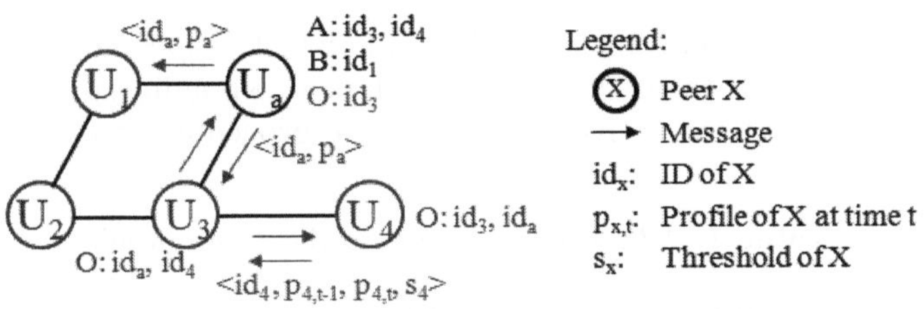

Fig. 1. Run of AURA

In order to build the personal profile of the active user u_a, we use both explicit and implicit criteria. The active user can always check the list of items that he/she shares or has consulted. He/She can explicitly rate each of these items on a scale of values from 1 to 5. The active user can also initialize his/her personal profile with a set of criteria[7] proposed in the interface in order to partially face the cold start problem. This offers the advantage of completing the profile with more consistency and of finding similarities with other users more quickly, since everyone can fill the same criteria rating form.

[7] Ideally, the set of items in the criteria set should cover all the implicit categories that users can find on the platform.

We assume that, despite the explicit voluntary completion of profiles, there are a lot of missing data. We consequently add to AURA a user modeling function, as we did in [6]. The explicit ratings and the estimated numerical votes constitute the active user's personal profile. The public profile is the part of the personal profile that the active user accepts to share with others.

The algorithm also has to build a group profile. It represents the preferences of a virtual community of interests, and has been especially designed to be as close as possible to the active user's expectations. In order to do that, the peer of the active user asks for the public profiles of all the peers it can reach through the platform. Then, for each of these profiles, it computes a similarity measure with the personal profile of the active user. The active user can indirectly define a minimum-correlation threshold which corresponds to the radius of his/her trust circle (cf. infra, figure 2).

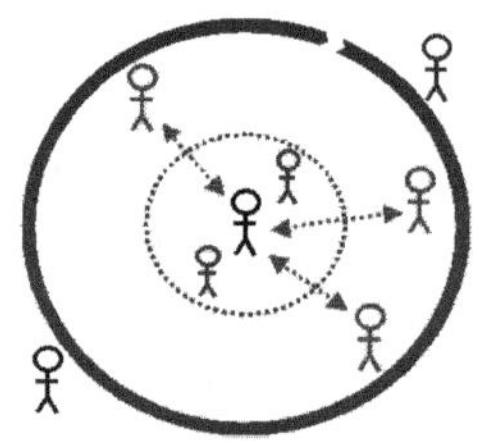

Fig. 2. Virtual community centered on u_a

If the result is lower than this fixed threshold which is specific to each user, the ID of the peer is added to the list "A" and the corresponding profile is included in the group profile of the active user, using the procedure of table 1.

Table 1. Add a public profile to the group profile

Procedure AddToGroupProfile(public profile of u_n)
  $W = W + |w(u_a, u_n)|$
  for each item i **do**
    $(u_{l,i}) = (u_{l,i}) * (W - |w(u_a, u_n)|)$
    $(u_{l,i}) = ((u_{l,i}) + w(u_a, u_n) * (u_{n,i}))/W$
  end for

With: $(u_{l,i})$ the rating for item i in the group profile;
$(u_{n,i})$ the rating of user n for item i;
W the sum of $|w(u_a, u_i)|$, which is stored;
$w(u_a, u_n)$ the correlation coefficient between the active user u_a and u_n.

We used the Pearson correlation coefficient to compute similarity, since the literature shows it works well [10]. Of course, if this similarity measure is higher than the threshold, we add the ID of the peer to the list "B". The list "C" is used to systematically ignore some peers. It enables to improve trust – that is to say the confidence that users have in the recommendations – by identifying malicious users. The trust increasing process will not be considered in this paper.

When his/her personal profile changes, the active user has the possibility to update his/her public profile p_a. In this case, the active peer has to contact every peer[8] whose ID is in the list "O". Each of these peers re-computes the similarity measure. If it exceeds the threshold, the profile p_a has to be removed from the group profile, using the procedure of table 2. Otherwise, p_a has to be updated in the group profile, that is to say the peer must remove the old profile and add the new one.

Table 2. Remove a public profile from the group profile

Procedure RemoveToGroupProfile(old profile of u_n)
  $W = W - |w(u_a, u_n)|$
  for each item i **do**
    $(u_{l,i}) = (u_{l,i}) * (W + |w(u_a, u_n)|)$
    $(u_{l,i}) = ((u_{l,i}) - w(u_a, u_n) * (u_{n,i}))/W$
  end for

By convention, we use the notation $< id, p >$ for the peer-addition packet, that is to say new arrivals. $< id, p, s >$ corresponds to the packet of a peer which is already connected and sends data to a new arrival. "s" is the threshold value. There is no need to specify the threshold value in the peer-addition packet, since there is a default value ($|correlation| >= 0$). At last, $< id, p_{t-1}, p_t, s >$ is the notation for the update packet. In each of these packets, the first parameter corresponds to the ID of the source of the message. In order to simplify the notation, we do not include the addressees' ID in figure 3.

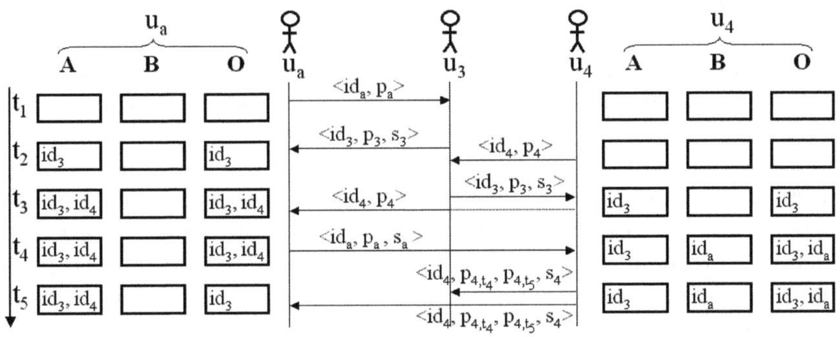

Fig. 3. Example of user interactions

Figure 3 illustrates how the system works. In this example, we consider 3 of the 5 users from figure 1. We show the registers of the active user u_a and

[8] A packet is broadcasted with an heading containing peers' IDs, the old profile and the new public profile.

the user u_4. At time t_1, the active user u_a tries to contact, for the first time, other peers by sending his/her public profile and his/her ID to neighbors. This is the packet $< id_a, p_a >$. u_3 receives the packet and answers at t_2. u_a computes the distance between the public profiles p_3 and p_a. As the Pearson coefficient is inevitably within the default threshold limit, u_a adds id_3 to his/her list "A". If the computed correlation coefficient is higher than "s_3" which is the threshold of u_3, u_a adds id_3 to his/her list "O". Meanwhile, some of the reached peers will add p_a to their list "A" if the correlation is higher than their threshold (this is the case for u_3). At time t_2, u_4 arrives on the platform and sends a packet to u_3. At time t_3, u_3 replies to u_4 and sends the packet of u_4 to peers that he/she already knows. u_a receives it and adds id_4 to his/her list "A". He/She also adds id_4 to the list "O", since u_4 is a new arrival and has a default threshold. At time t_4, u_a consequently gives his/her public profile to u_4. At the same time, u_4 has changed his/her threshold and considers that u_a is too far in the user/item representation space, that is to say the correlation coefficient between u_a and u_4 exceeds the limit. Thus, u_4 adds id_a in the list "B". In the packet $< id_a, p_a, s_a >$, "s_a" allows u_4 to know that he/she must complete the list "O" with id_a. At last, u_4 updates his/her public profile. Afterwards, he/she notifies the change to the IDs in the list "O". This is the packet $< id_a, p_{4,t_4}, p_{4,t_5}, s_4 >$. p_{4,t_4} and p_{4,t_5} are respectively the old and new public profiles of u_4. When u_a receives this packet, he/she updates the list "O" by removing id_4 since s_4 is too high for him/her.

3.2 Adaptive Minimum-Correlation Threshold

As shown in the previous subsection, the active user can indirectly define the minimum-correlation threshold that other people must reach in order to be a member of his/her community (radius of the circle on figure 2). Concretely, a high correlation threshold means that users taken into account in prediction computations are very close to the active user. Recommendations will be consequently extremely similar to his/her own preferences. On the contrary, a low correlation threshold sets forth the will of the active user to stay aware of generalist information by integrating distant users' preferences. In this way, the user avoids freezed suggestions by accepting novelty. In the SofoS interface, a slide bar allows the active user to ask for personalized or generalist recommendations. This allows AURA to know the degree to which it can modify the threshold[9]. The default threshold value is 0, which means that we take all the peers into account. The default step of threshold is 0.1, but it can be adapted to the density of population.

As shown in figure 4, we split the interval of the Pearson coefficient's possible values $[-1; +1]$ into subsets. For each subset, we keep the count of peers which have got in touch with the active user and whose correlation coefficient is

[9] By "threshold", we mean the minimum absolute value of Pearson coefficients to consider in the group profile computation. For example, if the system sets the threshold to 0.1, it means that only peers u_i whose correlation coefficient $|w(u_a, u_i)|$ is higher than 0.1 will be included in the group profile of the active user.

contained in the interval corresponding to the subset. Thus, when a user sends a packet to u_a, the Pearson coefficient is computed in order to know if the active user's group profile has to be updated according to the threshold value. At the same time, we update the corresponding values in the population distribution histogram. For example, if u_a receives an update packet and the Pearson coefficient changes from 0.71 to 0.89, we decrement the register of the interval [0.7; 0.8) and we increment the register of the interval [0.8; 0.9). In this way, we constantly have the population density for each interval.

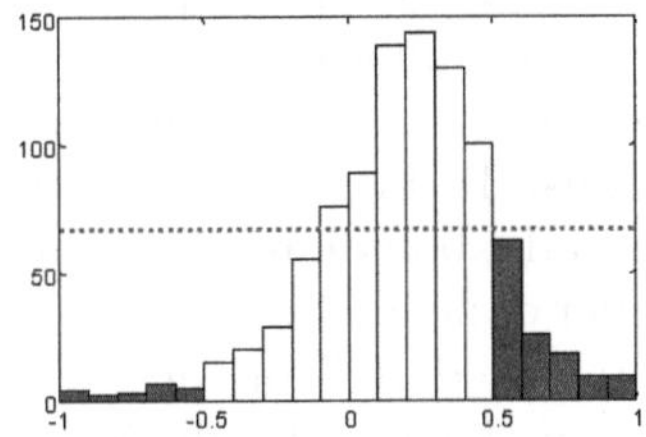

Fig. 4. Adaptive threshold based on density

When the total number of users whose Pearson coefficient is higher than ($threshold + 0.1$) exceeds a given limit (dashed line on figure 4), we increase the threshold. If there are too many users in the next subset, the threshold increase is lower. For the moment, the maximum threshold value is 0.2 for users who want a high degree of novelty and 0.9 for those who expect recommendations close to their preferences.[10] These values have been arbitrarily chosen. We plan to do statistical tests to automatically determine the ideal thresholds according to the context.

4 Discussion

In order to define the degree of privacy of our recommender system, we refer to 4 axes of personalization [11]. Cranor assumes that an ideal system should be based on an explicit data collection method, transient profiles, user initiated involvment and non-invasive predictions. In our system, the users have complete access to their preferences. They have an effect on what and when to share with others. Only numerical votes are exchanged and the logs of user actions are transient. Even when the active user did not want to share his/her preferences, it is possible to do predictions since public profiles of other peers are temporarily available on the active user device. Each user has a single ID, but the anonymity is ensured by the fact that there is no table linking IDs and identities. This privacy-enhanced process requires more network traffic than in [8], but it allows the system to perform user-centered rather than community-centered predictions.

[10] That is to say they want to retrieve items that they have high-rated

As regards scalability, our model no longer suffers from limitations since the algorithms used to compute group profiles and predictions are in $o(b)$, where b is the number of commonly valuated items between two users, since computations are made incrementally in a stochastic context. In return, AURA requires quite a lot of network traffic. This is particularly true if we use a random discovery architecture. Other P2P structures can improve communications [7].

Furthermore, we assume that quality of predictions in real situation should be better – providing that we found enough neighbors – since the virtual community of interests on each peer is centered on the active user (cf. infra, figure 2). We can influence the degree of personalization by adjusting the threshold according to the density of the active user's neighborhood. The system just has to increase the threshold in order to ensure users to retrieve the items that they have high-rated among their recommendations. To highlight this phenomenon, we generated a rating matrix of 1,000 users and 1,000 items. The votes follow a gaussian law and we can see the average number of neighbors as regards Pearson coefficient scaling on figure 5. We randomly removed 20% of these votes and applied the AURA algorithm. Then, we compute the Recall which measures how often a list of recommendations contains an item that the user have already rated in his/her top 10. When increasing the threshold in the system, this measure becomes higher.

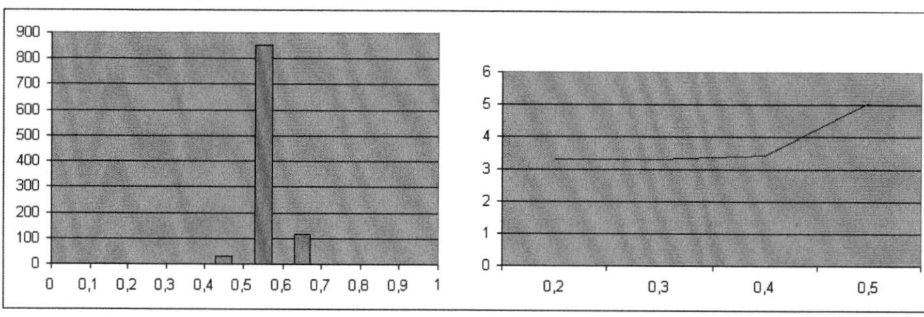

Fig. 5. On the left, average distribution of users as regards Pearson coefficient. On the right, recall as threshold grows.

We have also evaluated our model in terms of prediction relevancy. We used the *Mean Absolute Error* (MAE). MAE is a widely used metric which shows the deviation between predictions and real user-specified values. Consequently, we computed the average error between the predictions and 100,000 ratings of the GroupLens test set[11].

We simulate arrivals of peers by progressively adding new profiles. As shown on figure 6, we get predictions as good as using the PocketLens algorithm [7]. PocketLens relies on a distributed item-based approach. This comparison con-

[11] http://www.grouplens.org/

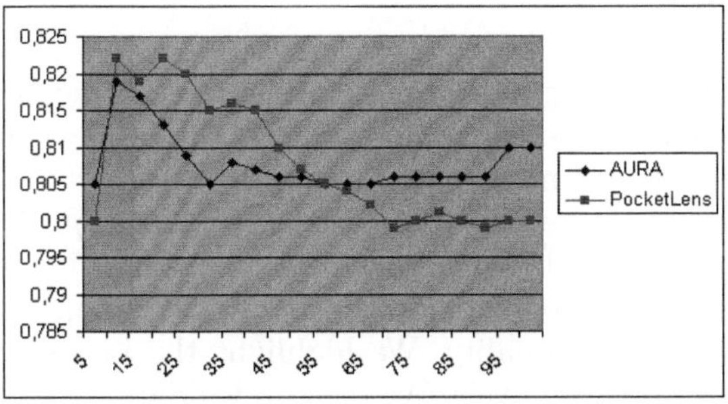

Fig. 6. MAE as neighborhood size grows

sequently demonstrates that AURA provides as relevant results as a performant item-based collaborative filtering.

At last, we compared our recommender system with two centralized algorithms (Item-Item [2] and the Correlation-based Collaborative Filter CorrCF [12]) to illustrate the added value of the distributed approach. In order to determine the computation times of these algorithms, we have generated random public profiles with different numbers of items. In this simulation, the votes of each user follow a Gaussian distribution centered on the middle of the representation space. Moreover, only 1% of data in the generated profiles is missing.[12] Since the Item-Item and CorrCF are centralized, we first aggregate the profiles in a vote matrix.

The results of the tests in term of computation time are shown in the table 3. The announced times for the AURA algorithm do not include the duration required to scan the network in search of public profiles. Of course, the difference between AURA and the two others is mainly due to the fact that we use as many peers as users for computations. However, these results illustrate the considerable gain in comparison with centralized techniques. AURA allows to do real-time predictions. There is no need to do offline computations since we can take into account 10,000 profiles and 150 items in less than an half-second. Moreover, the system does not have to wait until all similarity measures end. As the algorithm is incremental, we can stop considering other peers at any moment.

Table 3. Computation times of three collaborative filtering algorithms

Items	100			150			1000		
Users	AURA	CorrCF	It-It	AURA	CorrCF	It-It	AURA	CorrCF	It-It
200	0”01	2”60	2”14	0”01	3”17	2”71	0”07	11”09	52”74
1,000	0”03	30”22	8”56	0”05	40”68	12”84	0”30	3’06”	3’25”
10,000	0”31	7:30’	1’22”	0”48	-	2’05”	1”90	-	49’28”
100,000	3”04	-	-	-	-	-	-	-	-

[12] Only 1% of missing data is not realistic but can potentially increase the computation time what is interesting in this case.

5 Conclusion

SofoS is a document sharing platform including a recommender system. To cope with numerous problems specific to information retrieval, we proposed a Peer-to-Peer collaborative filtering model which is totally distributed. It allows real-time personalization and manages the degree of personalization that users want. We implement it on a JXTA platform which has been used by researchers all over the world. We show in this paper that we can deal with important problems such as scalability, privacy and quality. We highlight the benefits of our system by doing offline performance analysis. We plan on validating these points by testing our model with real users in real conditions.

Our algorithm is anytime and incremental. Contrary to PocketLens, our model is user-based because we consider that the set of items can change. Even if an item is deleted, we can continue to exploit its ratings in the prediction computations. Moreover, the stochastic context of our model allows the system to update the modified profiles instead of resetting all the knowledge about neighbors. At last, our model is very few memory-consuming because it does not need to store any neighbors' ratings, similarity matrix, dot product matrix and so on. It only requires the sum of pearson coefficients and four lists of user IDs.

Currently, we are developing our protocols further to cope with other limitations, such as trust and security aspects by using specific communication protocols as in [13].

References

1. Goldberg, D., Nichols, D., Oki, B., Terry, D.: Using collaborative filtering to weave an information tapestry. In: Communications of the ACM, Special Issue on Information Filtering. Volume 35(12)., ACM Press (1992) 61–70
2. Sarwar, B.M., Karypis, G., Konstan, J.A., Reidl, J.: Item-based collaborative filtering recommendation algorithms. In: World Wide Web. (2001) 285–295
3. Canny, J.: Collaborative filtering with privacy. In: IEEE Symposium on Security and Privacy, Oakland, CA (May 2002) 45–57
4. Breese, J.S., Heckerman, D., Kadie, C.: Empirical analysis of predictive algorithms for collaborative filtering. In: Proceedings of the fourteenth Annual Conference on Uncertainty in Artificial Intelligence (UAI-98), San Francisco, CA (July 1998)
5. Pennock, D.M., Horvitz, E., Lawrence, S., Giles, C.L.: Collaborative filtering by personality diagnosis: a hybrid memory- and model-based approach. In: Proceedings of the sixteenth Conference on Uncertainty in Artificial Intelligence (UAI-2000), San Francisco, USA, Morgan Kaufmann Publishers (2000)
6. Castagnos, S., Boyer, A.: A client/server user-based collaborative filtering algorithm: Model and implementation. In: Proceedings of the 17th European Conference on Artificial Intelligence (ECAI2006), Riva del Garda, Italy (August 2006)
7. Miller, B.N., Konstan, J.A., Riedl, J.: Pocketlens: Toward a personal recommender system. In: ACM Transactions on Information Systems. Volume 22. (July 2004)
8. Berkovsky, S., Eytani, Y., Kuflik, T., Ricci, F.: Hierarchical neighborhood topology for privacy enhanced collaborative filtering. In: in CHI 2006 Workshop on Privacy-Enhanced Personalization (PEP2006), Montreal, Canada (April 2006)

9. Han, P., Xie, B., Yang, F., Wang, J., Shen, R.: A novel distributed collaborative filtering algorithm and its implementation on p2p overlay network. In: Proc. of the Eighth Pacific-Asia Conference on Knowledge Discovery and Data Mining (PAKDD04), Sydney, Australia (May 2004)
10. Shardanand, U., Maes, P.: Social information filtering: Algorithms for automating "word of mouth". In: Proceedings of ACM CHI'95 Conference on Human Factors in Computing Systems. Volume 1. (1995) 210–217
11. Cranor, L.F.: Hey, that's personal! In: the International User Modeling Conference (UM05). (2005)
12. Resnick, P., Iacovou, N., Suchak, M., Bergstorm, P., Riedl, J.: Grouplens: An open architecture for collaborative filtering of netnews. In: Proceedings of ACM 1994 Conference on Computer Supported Cooperative Work, Chapel Hill, North Carolina, ACM (1994) 175–186
13. Polat, H., Du, W.: Svd-based collaborative filtering with privacy. In: Proc. of ACM Symposium on Applied Computing, Cyprus (2004)

Information Recovery and Discovery in Collaborative Web Search*

Maurice Coyle and Barry Smyth

Adaptive Information Cluster, School of Computer Science & Informatics,
University College Dublin, Belfield, Dublin 4, Ireland
{maurice.coyle,barry.smyth}@ucd.ie

Abstract. When we search for information we are usually either trying to *recover* something that we have found in the past or trying to *discover* some new information. In this paper we will evaluate how the collaborative Web search technique, which personalizes search results for communities of like-minded users, can help in recovery- and discovery-type search tasks in a corporate search scenario.

1 Introduction

Finding new ways to improve the quality of Web search engines has the potential to have a significant impact when it comes to helping people to find the right information at the right time. Modern web search engines face a number of challenges, such as the scale and growth rate of the Web and the needs and expertise of the average Web user. For example, Web search engines experience difficulties in disambiguating the vague queries that are now commonplace in Web search; the typical search query might only contain 2 or 3 keywords, which rarely help to clearly identify the searcher's needs [11].

The result has been an intensive research effort to develop new techniques for improving Web search quality. Early successes are now well known with, for example, Page & Brin [2] famously incorporating page connectivity information into result ranking in what came to be a unique feature of Google. Many researchers have sought recently to include context information as a way to improve the selection and ranking of search results with some *explicitly* requesting such information from searchers [1,8], while others infer context *implicitly* by observing a user's actions or activities [3,6]. Indeed to many, the next generation of search technology will focus on the provision of *personalized* search, by learning about the long-term preferences of users and inferring their short-term needs, in order to promote results that are likely to be more relevant to individuals.

When we think about search it is often useful to distinguish between two types of search activity. Many search sessions are focused on information *recovery* in the sense that the searcher is looking for information that they have seen before or that they have previously searched for and found. Alternatively, there

* This material is based on works supported by Science Foundation Ireland under Grant No. 03/IN.3/I361.

G. Amati, C. Carpineto, and G. Romano (Eds.): ECIR 2007, LNCS 4425, pp. 356–367, 2007.

are those sessions that are focused on information *discovery* in the sense that the searcher is looking for new information. O'Day and Jeffries [10] highlight how searcher behaviour often combines elements of recovery and discovery as searchers often engage in series' of interconnected searches over time and explore topics in an undirected fashion. [7] presents the results of a limited laboratory study which shows that searchers' ability to refind information is based on their familiarity with the search task and the frequency with which they perform the same or similar tasks. [14] presents the results of an analysis of query logs which show that users frequently engage in *re-finding* behaviour. While the analyses in [7,10,14] focus on the individual searcher, in this paper we consider similar behaviours within a community of related searchers and show how the search patterns of such communities can be harnessed to provide useful assistance during individual search sessions.

The field of *exploratory search* (see [16]) seeks to aid users in acquiring and aggregating new information. In that context, Marchionini [9] distinguishes between three related types of search activity: *lookup* is concerned with fact retrieval or known item search and is analogous to recovery-oriented search, while *learning* and *investigating* activities are more exploratory in nature and can be viewed as being discovery-oriented. In [15] the concept of *interaction histories* is introduced to demonstrate that users' information discovery through browsing can be enhanced by leveraging past users' actions. As stated in that work, "we all benefit from experience; preferably someone else's".

In this paper we utilise an approach to personalized Web search known as *collaborative Web search* (CWS), described in detail in [12,13], which is specifically designed to aid in the recovery and discovery of information by harnessing the search patterns of a community of like-minded searchers. CWS is designed to work in concert with some underlying search engine(s), post-processing result-lists to better match the inferred interests of the community. In short, results that have been selected by community members in the past are likely to be promoted for similar queries in the future. Since, using CWS, searchers benefit from the searching histories of their community members, the learning- and discovery-oriented searching tasks that have been performed by community members in the past can be leveraged so that when future users perform similar tasks, they are more recovery-oriented.

Other work using the core CWS technique [12,13] has focused on a particular architecture and promotion mechanism. In this paper we will make a number of contributions. First, to tackle some problems experienced by CWS implementations in the past such as reduced usage due to user mistrust and poor performance, we will describe a new proxy-based architecture that facilitates improved integration opportunities with underlying search engines. Second, we will describe recent interface work that goes beyond the simple promotion of results to explore how such promotions should be communicated or explained to searchers and the type of information that can make up such explanations. Finally, we will describe the results from a recent live-user trial of CWS in a corporate search context that, for the first time, looks specifically at how CWS promotions aid in the recovery and discovery of information and shows how leveraging the search histories of other users for these promotions is preferable to using only individuals' searching histories.

2 Collaborative Web Search

Collaborative Web search (CWS) complements the results of an underlying search engine by promoting results that are likely to be relevant to the community's interests and preferences. While the basic CWS approach has been described in other work [13], we will review the core technique in this section and describe how we have extended it to accommodate a more flexible deployment architecture and a novel explanation-oriented interface.

2.1 Community-Based Promotion

For a target search query, CWS combines a default result-list, R_S, from a standard search engine with a set of recommended (*promoted*) results, R_P, drawn from the community's past search history. To do this the search histories of a given community, C, of users ($C = \{U_1, ..., U_n\}$) are stored anonymously in a *hit-matrix*, H^C, such that each row corresponds to some query q_i and each column to some selected result page p_j. The value stored in H^C_{ij} refers to the number of times that page p_j has been selected for query q_i by members of C.

Each hit-matrix acts as a repository of community search experiences: the results that the community members have found to be relevant for their queries. When responding to a new target query, q_T, H_C is used to identify and rank results that have been regularly selected in the past. These results can then be promoted ahead of any results returned by the underlying search engine according to their estimated relevance with respect to the community's interests.

The relevance of a result p_j in relation to a query q_i can be estimated by the relative frequency that p_j has been selected for q_i in the past, as shown in Equation 1. More generally, we can pool the results that have been selected for queries that are similar to q_T (see Equation 2) and rank each result according to the weighted model of relevance (see Equation 3, where $WRel(p_j, q_T, q_1, ..., q_n)$ represents the weighted relevance score of page p_j for query q_T and $Exists(p_j, q_i)$ is a function whose value is 1 if H_{ij} is non-zero and 0 otherwise), which weights each individual result's relevance by the similarity of the associated query to q_T.

$$Relevance(p_j, q_i) = \frac{H_{ij}}{\sum_{\forall j} H_{ij}} \tag{1}$$

$$Sim(q, q') = \frac{|q \cap q'|}{|q \cup q'|} \tag{2}$$

$$WRel(p_j, q_T, q_1, ..., q_n) = \frac{\sum_{i=1...n} Relevance(p_j, q_i) \bullet Sim(q_T, q_i)}{\sum_{i=1...n} Exists(p_j, q_i) \bullet Sim(q_T, q_i)} \tag{3}$$

Although CWS depends on the underlying search service to provide relevant results *somewhere* in the result list for a given query, it enhances any search service in a number of ways. Firstly, since CWS can return results that have been selected for similar queries, relevant results which may not have appeared at all for the target query can be included. Also, if a relevant result is not returned in the first page of results, the chances of a searcher finding it diminish greatly. With CWS, there only needs to be a small minority of determined searchers willing to forage for these relevant results and select them so that in subsequent searches, they can be promoted to the first result page.

2.2 A Proxy-Based Architecture for Collaborative Web Search

In this paper we detail an architecture for implementing CWS which provides for a more seamless and flexible integration with existing search engines. CWS has previously been implemented as a form of meta-search, post-processing the results of some underlying search engine(s) and presenting these re-ranked results through a dedicated search interface. The problem with this approach is the need to convince users to try an alternative search service and learn a new search interface. The architecture presented here facilitates a very different form of integration with an underlying search engine, by using a proxy-based approach (see Figure 1) to intercept queries on their way to the underlying search engine and manipulate the results that are returned from this engine back to the searcher; basically users get to use their favourite search engine in the normal way, but CWS promotions are incorporated into the result-lists directly via the proxy.

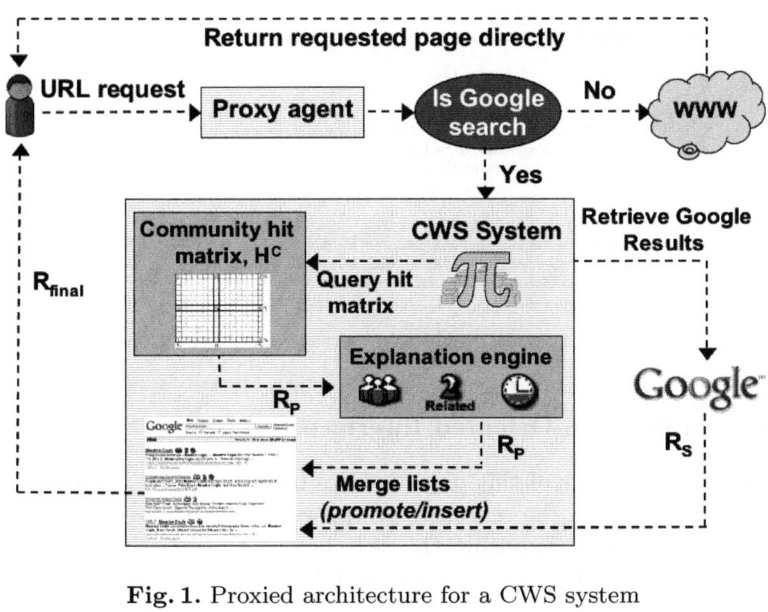

Fig. 1. Proxied architecture for a CWS system

Briefly, the *proxy agent* (currently a standard Squid[1] proxy) operates to transparently filter URL requests from all configured browsers. In this case, the proxy redirects Google requests (e.g. query submissions, result selections, etc.) to the CWS server. For example, consider a user U_i submitting query q_T to Google. This request is redirected to the CWS whereupon two things happen. First, the query is passed on to Google and the result-list R_S is returned in the normal way. Second, in parallel the query is also used to generate a ranked set of promotion candidates, R_P from $U_i's$ community hit matrix, as outlined above. These promotion candidates are annotated by the *explanation engine* (see Section 2.3) to present the searcher with a graphical representation of their community history. Result lists R_P and R_S are merged and the resulting list R_{final} is returned to the user; typically this merge involves promoting the k (e.g., $k = 3$) most relevant promotions to the head of the result-list.

Figures 2 and 3 present example screen shots for the result-list returned by Google for the query *'Michael Jordan'*. In the case of Figure 2 we see the default Google result-list, with results for the basketball star dominating. In Figure 3, however, we see a result-list that has been modified by our proxy-based version of CWS, trained by a community of artificial intelligence and machine learning researchers. The results appear to be standard Google results but now we see that the top 3 results are promotions for the well-known Berkeley professor.

Fig. 2. The result list returned by Google in response to the query 'michael jordan'

2.3 An Explanation-Oriented Interface

One of the contributions of this paper relates to the manner in which promoted results are presented to the searcher. We propose that there are potential benefits to be gained from exposing the searcher to the search history of a promoted result, to help them better understand the reasons behind its promotion.

[1] http://www.squid-cache.org

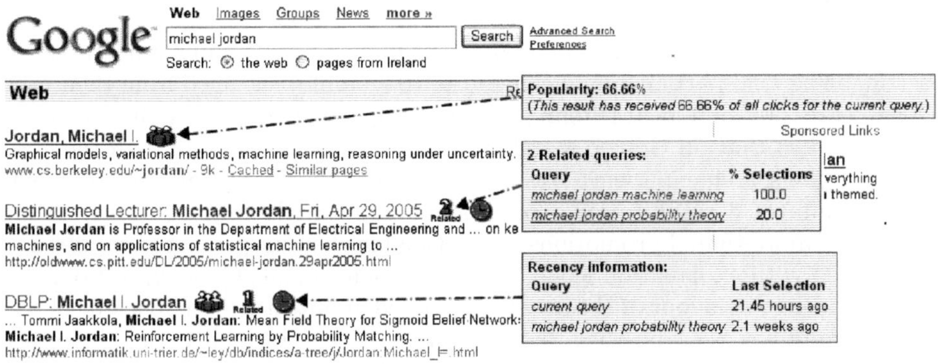

Fig. 3. The result list returned by CWS in response to the query 'michael jordan' issued within a community with a shared interest in computer science. The extra explanation information available by mousing over each promoted result icon type is also shown.

Thus, we introduce an *explanation engine* to the CWS architecture, which annotates promoted results with one or more explanation icons designed to capture different aspects of the result's community history. Specifically, three basic types of explanation icon are used: *popularity* icons convey the overall selection popularity of a result among the target community; *recency* icons indicate how recently the result has been selected by a community member and can help the searcher to judge how current a particular promotion might be; finally, *related queries* icons indicate whether the result in question has been previously selected by the community for any queries that are related (similar) to the current target query.

In Figure 3 we can see how each of the top 3 promotions are annotated with different subsets of the above icons, depending on the selection history of the result in question. Moreover, 'mousing-over' an explanation icon will reveal additional explanatory detail to the searcher; see Figure 3. For example, mousing-over the related queries icon will present a list of the most common related queries that have led to selections for the result in question. These related queries can themselves be selected by the searcher to initiate a new search.

3 Evaluation

The current proxy-based architecture and explanation-oriented interface has been used as the basis of a long-term trial of the CWS approach in a corporate search scenario. In this section we will describe recent results drawn from this trial, paying particular attention to promoted results and the origin of these promotions as indicators of recovery and discovery behaviour on the part of searchers. Specifically, we are interested in how CWS can help to aid recovery and discovery tasks among the community of searchers. For an evaluation of the utility of the explanation icon enhancements to the CWS interface, see [4].

3.1 Setup

The objective of this trial was to explore the impact of CWS in a realistic search setting. The trial participants included the 70+ employees, aged from their early 20s to early 50s, of a local Dublin software company. The CWS architecture was configured so that all Google search requests were redirected through the CWS system. The search experience was based on the standard Google interface, with a maximum of 3 results promoted (and annotated with explanations) in any session; if more than 3 promotions were available then un-promoted results were annotated with explanation icons but left in their default Google position.

As evidence in favour of the new proxy-based architecture, feedback from trial participants has been positive, both in terms of the availability of the familiar Google interface and search response times. This contrasts with previous trials, which involved a custom interface, as participants were often reluctant to learn the new search interface or to depart from their trusted Google. The results presented here are drawn from just over 10 weeks of usage (from 27-07-2006 to 09-10-2006) and cover a total of 12,621 individual search sessions.

One of the challenges in evaluating new search technologies in a natural setting is how to evaluate the quality of individual search sessions. Ideally we would like to capture direct feedback from users as they search when it comes to the relevance of search results (promotions versus standard results in this case). For example, it would be relatively straightforward to ask users to provide feedback during each session or as they selected specific results. Unfortunately, this was not possible in the current trial because participants were eager to ensure that their search experience did not deviate from the norm, and were unwilling to accept pop-ups, form-filling or any other type of direct feedback.

As an alternative, we will use an indirect measure of search session relevance, based on the concept of a *successful session* [12,13]. A successful session is defined as one where at least one search result was selected, indicating that the searcher found at least one (apparently) relevant result. While this is a relatively crude measure of overall search performance it allows us to compare search sessions in a systematic way. Finally, in investigating recovery and discovery behaviours, we will refer to the *clickthru probability* of *promotion types* (see Section 3.3 for a description of the different types of promotions that were considered). This represents the chance that a promotion of a given type will be selected, and is calculated as the number of promotions of that type that were selected as a percentage of the total number of promotions of that type that were displayed.

3.2 Success Rates: Discovery vs Recovery Tasks

To begin with we compare search sessions with promotions (*promoted sessions*) to search sessions without promotions (*standard sessions*) in terms of their success rates. During the course of the 10 week trial, approximately 22% of sessions were promoted sessions and the success rates are presented in Figure 4. These results indicate that, on average, sessions with promotions are more likely to be successful (62%) than standard sessions (48%) containing only Google results, a relative benefit of almost 30% due to the community-based promotion of results.

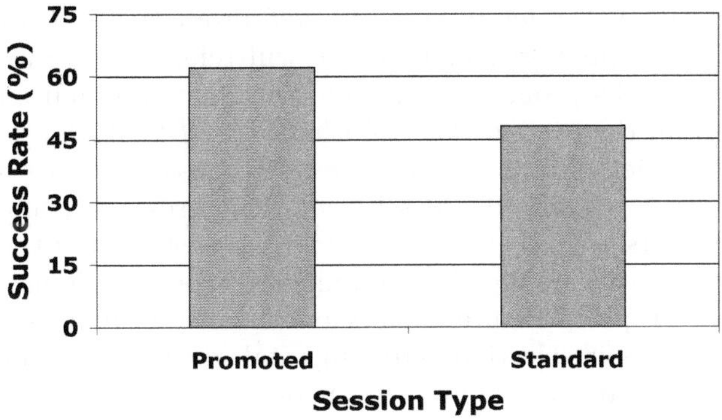

Fig. 4. The success rates for sessions containing promotions compared to those without promotions

This difference in success rates also speaks to some basic differences when it comes to recovery and discovery tasks. Standard sessions can be considered to be examples of users engaging in discovery tasks because neither they nor their colleagues have searched for similar information previously, otherwise promotions would have been made based on these past sessions. Of course this view is somewhat simplified because potentially there are standard sessions which are recovery tasks but whose related queries occurred before the start of the trial.

Intuitively, discovery is likely to represent a more challenging search task compared to recovery. In the case of the latter the searcher is likely to remember something about queries that they may have used in the past to locate a particular result. The same is obviously not true in discovery, since there are no such past queries. Thus, we would expect a higher success rate for recovery type sessions than discovery type sessions. This intuition is borne out by the higher success rates of promoted sessions. Promoted sessions are likely to include many recovery type sessions because the promotions will come from the past sessions of the current searcher. However some promoted sessions are also likely to contain promotions from the past sessions of other users. From the current searcher's perspective these are discovery sessions, because they themselves have not previously searched for this information before. However, from the perspective of the community as a whole, they are more like recovery sessions, because others have successfully searched for this information before. We will see in a later section how CWS can effectively transform a challenging discovery task (for a given searcher) into a more manageable recovery task, by promoting the selections of others, and in so doing improve their success rates.

3.3 Promotion Frequency and Clickthru Probability

Ultimately in this paper we are interested in better understanding the source of promotion success as it relates to recovery and discovery tasks. For any given search session, promotions may originate from the target searcher's own past search sessions (we will refer to these as *self* promotions) or from the search sessions of other community members (we will refer to these as *other* promotions). We also consider promotions that have originated both from the searcher's own past and from the past sessions of others; a *self & other* promotion.

As mentioned briefly in the section above, sessions containing self promotions are related to recovery tasks for the searcher, whilst sessions containing only the promotions of others are discovery tasks from the viewpoint of the searcher. The key issue is the relative value of these different types of promotions. To begin with, in the bar graph in Figure 5, we show the overall frequency of each type of promotion and their selection rates. In total, during the course of the trial nearly 5000 individual promotions were made and Figure 5 tells us that about 75% of these were self promotions with just over 20% coming from the search sessions of other users. We also see that over all of the promotions that were selected, just under 60% of these were self promotions whereas nearly 40% were from other users. This is interesting since the frequency of self promotions tell us that searchers are often engaged in recovery-type search activity. However, despite this apparent focus on recovery we find users frequently selecting results that have been previously selected by other users (other promotions).

In this way CWS promotions are helping users to discover new information even during recovery tasks; information that, presumably, they had previously missed. This is clarified in the line graph in Figure 5, where we present the overall probability that a given promotion type will be selected. We see that self promotions are only selected about 12% of the time, whereas the promotions from other community members are selected more than twice as often (25%).

We are finding that in a very practical sense, the past search behaviour of other users is proving to be twice as useful - when it is available - as the past search behaviour of the current searcher. In a corporate setting this is important, given the time that the average knowledge-worker spends (or wastes) searching

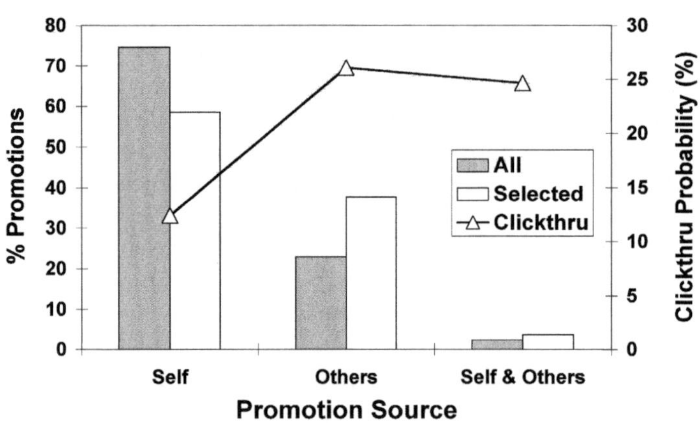

Fig. 5. Percentage of all promotions and selected promotions that had each source type (bars); probability that a result from each source will be selected (line)

for information. According to a recent IDC report [5], the average knowledge-worker spends 25% of their time searching for information and an enterprise employing 1,000 knowledge workers will waste nearly $2.5 million per year (at an opportunity-cost of $15 million) due to an inability to locate and retrieve information. Different employees have different levels of search expertise and these results suggest that CWS has an important role to play when it comes to sharing this expertise, because everyone can benefit from the searches of others.

3.4 A Session-Based Analysis

The previous results present averages over all promotions and as such do not tell us about the promotion success at the level of individual search sessions, which is arguably more important when it comes to understanding granular search tasks. To begin with, we measure the relative success rates of sessions containing the 3 types of promotions (see the bar graph in Figure 6). For sessions that contain self promotions only (approximately 75% of promoted sessions), we find that they are successful just under 60% of the time. In contrast, about 23% of sessions contain promotions that only come from other users, and more than 65% of these sessions are successful. Finally, for the 2% of sessions that contain both types of promotions (self & other) we find success rates in excess of 71%.

These results reconfirm the value of promotions (of other users in particular). Remember from Figure 4 that the baseline average success rate of standard sessions (in which discovery tasks are likely to dominate) is only 48%. Using (self) promotions to aid recovery increases success rates from 48% to almost 60%, a 25% relative increase. Using the promotions of others to effectively convert a discovery task into a recovery task (from the standpoint of the searcher) increases the success rate even further, to over 65%, a relative increase of more than 35%.

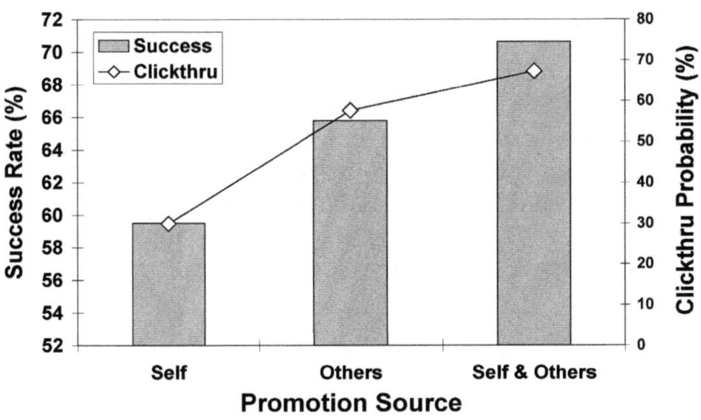

Fig. 6. The success rates for sessions containing promotions for each promotion source type (bars) and the probability that a session containing promotions from a given source will result in one of the promotions being selected (line)

Finally, it is worth considering the probability that *at least one* promotion will be selected from these promoted session types (see the line graph in Figure 6), as an analogue to the clickthru probabilities presented in the previous section. For example, we see that on average, in sessions that only contain self promotions, there is a 30% probability that any one of these promotions will be selected. In contrast, for sessions containing the promotions of other searchers, this selection probability almost doubles to nearly 60%. And in sessions with both self and other promotions, the selection probability grows to nearly 70%.

4 Conclusions

Collaborative Web search seeks to improve the quality of an underlying search engine by promoting results that have been previously selected by a community of searchers. In this paper we have introduced a new architecture for collaborative Web search and an explanation-based interface for the presentation of promoted results. In addition, we have presented the results of a live-user trial and considered their implications from the perspective of how they relate to recovery and discovery activity among searchers in a corporate search setting.

The strength of CWS is twofold: (1) it helps to highlight previously selected results (from similar past queries) in recovery tasks; (2) it helps to convert some discovery tasks into recovery tasks by highlighting the promotions of other community members. In relation to (1) the success rate increases by 25% (from 48% to 60%) and in relation to (2) the success rate increases by more than 35% (from 48% to above 65%). Thus we can conclude that this type of community-based result promotion is providing useful assistance to searchers and, in particular, that the past sessions of other users are playing a particularly powerful role when it comes to making result recommendations; searchers are more than twice as likely to select the promotions of others - when they are made - than their own.

These benefits can have a significant impact on many search scenarios. Our results indicate that CWS can assist searchers in at least 25% of their sessions, improving their success rates by an average of 30%, which is an improvement of nearly 8% across all sessions. Translating these improvements into saved search time can have a significant impact on enterprise productivity. For example, using the data in [5] suggests an annual saving for a 1,000-employee company of up to $200,000 and a recovered opportunity-cost of up to $1.2 million.

References

1. Krishna Bharat. SearchPad: Explicit Capture of Search Context to Support Web Search. In *Proceedings of the Ninth International World-Wide Web Conference (WWW '00)*, pages 493–501. North-Holland Publishing Co., 2000.
2. Sergey Brin and Larry Page. The anatomy of a large-scale hypertextual web search engine. *Computer Networks and ISDN Systems*, 30(1–7):107–117, 1998.

3. J. Budzik and K. Hammond. User Interactions with Everyday Applications as Context for Just-In-Time Information Access. In *Proceedings of the 3rd International Conference on Intelligent User Interfaces (IUI '00)*, pages 44–51. ACM Press, 2000.
4. M. Coyle and B. Smyth. On the Community-Based Explanation of Search Results. In *Proceedings of the 10th International Conference on Intelligent User Interfaces (IUI '07)*, page (in press). ACM Press, 2007. Hawaii, U.S.A.
5. Susan Feldman and Chris Sherman. The High Cost of Not Finding Information. In *(IDC White Paper)*. IDC Group, 2000.
6. L. Finkelstein, E. Gabrilovich, Y. Matias, E. Rivlin, Z. Solan, G. Wolfman, and E. Ruppin. Placing search in context: the concept revisited. In *Proceedings of the 10th International Conference on the World Wide Web (WWW '01)*, pages 406–414. ACM Press, 2001.
7. Robert G. Capra III and Manuel A. Pérez-Quiñones. Using web search engines to find and refind information. *Computer*, 38(10):36–42, 2005.
8. E. Glover, S. Lawrence, M. D. Gordon, W. P. Birmingham, and C. Lee Giles. Web Search - Your Way. *Communications of the ACM*, 44(12):97–102, 2000.
9. Gary Marchionini. Exploratory search: from finding to understanding. *Communications of the ACM*, 49(4):41–46, 2006.
10. Vicki L. O'Day and Robin Jeffries. Orienteering in an information landscape: how information seekers get from here to there. In *Proceedings of the SIGCHI conference on Human factors in computing systems (CHI '93)*, pages 438–445, New York, NY, USA, 1993. ACM Press.
11. Craig Silverstein, Monika Henzinger, Hannes Marais, and Michael Moricz. Analysis of a Very Large AltaVista Query Log. Technical Report 1998-014, Digital SRC, 1998.
12. Barry Smyth, Evelyn Balfe, Oisín Boydell, Keith Bradley, Peter Briggs, Maurice Coyle, and Jill Freyne. A Live-user Evaluation of Collaborative Web Search. In *Proceedings of the 19th International Joint Conference on Artificial Intelligence (IJCAI '05)*, pages 1419–1424. Morgan Kaufmann, 2005. Edinburgh, Scotland.
13. Barry Smyth, Evelyn Balfe, Jill Freyne, Peter Briggs, Maurice Coyle, and Oisín Boydell. Exploiting query repetition and regularity in an adaptive community-based web search engine. *User Modeling and User-Adapted Interaction: The Journal of Personalization Research*, 14(5):383–423, 2004.
14. Jaime Teevan, Eytan Adar, Rosie Jones, and Michael Potts. History repeats itself: repeat queries in yahoo's logs. In *Proceedings of the 29th annual international ACM SIGIR conference on Research and development in information retrieval (SIGIR '06)*, pages 703–704, New York, NY, USA, 2006. ACM Press.
15. A. Wexelblat and P. Maes. Footprints: History-Rich Web Browsing. In *Proceedings of the Third International Conference on Computer-Assisted Information Retrieval (RIAO '97)*, pages 75–84, 1997. Montreal, Quebec, Canada.
16. Ryen W. White, Bill Kules, Steven M. Drucker, and Monica M.C. Schraefel. Supporting exploratory search. *Communications of the ACM Special Issue.*, 49(4), 2006.

Collaborative Filtering Based on Transitive Correlations Between Items

Alexandros Nanopoulos

Department of Informatics, Aristotle University of Thessaloniki, Greece

Abstract. With existing collaborative filtering algorithms, a user has to rate a sufficient number of items, before receiving reliable recommendations. To overcome this limitation, we provide the insight that correlations between items can form a network, in which we examine transitive correlations between items. The emergence of power laws in such networks signifies the existence of items with substantially more transitive correlations. The proposed algorithm finds highly correlative items and provides effective recommendations by adapting to user preferences. We also develop pruning criteria that reduce computation time. Detailed experimental results illustrate the superiority of the proposed method.

1 Introduction

Collaborative Filtering (CF) is a method that provides personalized recommendations, based on suggestions of users with similar preferences. CF helps users to identify interesting information, especially when the number of items renders their individual examination prohibitive.

CF was initially developed for information-filtering systems, which allowed users to find relevant documents based on previous comments by other users, e.g., the Information Tapestry project [5]. During the recent years, following its success, the use of CF has been extended to several new domains. In the World-Wide-Web, non-guided search usually returns numerous irrelevant results. *Active Filtering* (AF) [10] is a type of CF that tries to solve this problem by allowing users to form lists of commonly used links and recommend them to others. CF has also attracted significant attention in the area of retail and especially in e-commerce [9,14]. In these areas, CF matches preferences of the target user with those of similar users, based on purchase or browsing patterns, e.g., Amazon, TiVo. Finally, CF has been proposed for Web personalization, by customizing the content and structure of a web site to meet the specific needs of users [11].

CF algorithms often are categorized either as *memory-based* or as *model-based*. Memory-based algorithms operate directly on user-preference data and make predictions based on the most similar past users. In the following, data pertaining to user preferences are called *ratings*. Model-based algorithms first develop a model from the training data, which is then used to make predictions. Research results and practical experience have reported that memory-based algorithms attain good performance in terms of accuracy [2,7]. Moreover, they are simple, since in contrast to model-based algorithms, they do not require

G. Amati, C. Carpineto, and G. Romano (Eds.): ECIR 2007, LNCS 4425, pp. 368–380, 2007.

complex tuning of parameters. These characteristics have yield to the wide acceptance of memory-based CF algorithms in many real-world applications, most notably Amazon [9]. Recent research has attempted to combine these two classes, e.g., [17,19].

Motivation. With a memory-based CF algorithm, a new user has to rate a sufficient number of items, before the algorithm is able to provide reliable and accurate recommendations. However, in most real-world applications of CF, e.g., e-commerce, web-personalization, it is crucial for users to receive good recommendations at the early steps of their interaction. Without this requirement, factors like customer attraction, retention, or satisfaction are significantly affected. Therefore, the aforementioned problem consists one of the main limitations of memory-based algorithms. The reason is that these algorithms use only direct correlations between items or users, which are imprecise when the new user has not rated many items. The problem is further intensified by the effect called *sparsity*, which occurs because past users tend to rate only a very small fraction of items.

Following a different direction, we identify that direct correlations between items can be viewed as a kind of network. In this framework, the way items are connected through paths, can be considered as transitive correlations between them. Moreover, we recognize that some items are significantly more connected with others via paths. This property is a valuable source of information when existing ratings do not suffice, and designates the suitability of such items for recommendation purposes. These issues have not been examined so far in a similar framework. What is, therefore, required is a new method that will take them into account in order to significantly improve the quality of recommendations.

Contribution. In this paper, we propose a novel algorithm, which considers the issues discussed previously. Our contributions are summarized as follows: (a) We provide the insight that, when considering the direct correlations between items as a network, some items clearly emerge to be substantially more accessible from others through paths. (b) We define the notion of *accessibility matrix*, which helps to find separately for each item, the items that are more reachable from it. This differs from existing network analysis methods, which only find highly accessible items at global level. The reason is that, given a new user, we want to recommend items that are most reachable from (correlated with) the items he has already rated, thus to adapt to his preferences. (c) To clearly understand the characteristics of the proposed approach, we analyze the distribution of values in accessibility matrixes and detect that it tends to form power laws. This fact verifies our assumption about the existence of items that are much highly accessible than others. (d) To address issues pertaining to the time required to form the accessibility matrix, we develop pruning criteria that reduce the search space. (e) We present detailed experimental results that compare the performance of the proposed method against existing memory-based CF algorithms and existing network analysis methods, using real and synthetic data sets.

The rest of this paper is organized as follows. Section 2 summarizes the related work. The proposed approach is described in Section 3. In Section 4 we describe the qualitative characteristics of our method, whereas experimental results are given in Section 5. Finally, Section 6 concludes this paper.

2 Related Work

Memory-based CF includes user-based methods and item-based methods. User-based methods [2,6,13] use ratings from similar users to make predictions for a new user. In contrast, item-based methods [4,15] rely on similarity between items instead of users. Previous research [4,9,15] has shown the clear advantage of item-based versus user-based methods. Thus, for comparison purposes, we consider the method proposed in [15], which is henceforth denoted as IB.

Preference about items can be expressed either explicitly on a rating scale, e.g., 1–5, or implicitly, by recording user actions, like "page visited" or "item purchased". The latter approach results to binary-valued log-archives, which can be conceptualized as ratings on the 0/1 scale. In this scale, 1 represents 'relevance' or 'likeness' similarly to the notion of 'relevance' in text retrieval [18]. Since users are unlikely to provide explicit ratings (they may consider it as annoying or time demanding) [3], the implicit, so-called "log-based", acquisition of user preferences makes CF more favorable in practice. For this reason, we focus on this category and assume that ratings are binary. However, it is possible to extend the proposed approach for multi-scale ratings as well.

Recent research has considered the sparsity problem. Schein *et al.* [16] propose to combine content and collaborative data under a single probabilistic framework. For binary ratings, Huang *et al.* [8] use transitive associations between user and items in the bipartite user-item graph. Transitive associations are formed by spreading activation algorithms. Experiments in [8] show that this method outperforms IB (and other classic CF algorithms) when data are very sparse, but otherwise its performance degrades due to the *over-activation effect*. Differently from [8], our approach considers transitive correlations between items, not between users and items. Thus, it does not rely on complex spreading activation schemes, but naturally extends the widely accepted item-based algorithms with the notion of transitive correlations. Also, our focus is not on data sparsity in general, but on how to provide good recommendations when the new user has not rated a sufficient number of items (a case of the cold-start problem [16]). Finally, to address sparsity, Papagelis *et al.* [12] propose a kind of social network, and they find transitive associations between users in the form of *trust inferences*. Their approach extends user-based methods, whereas we focus on item-based methods, because of their advantage over user-based methods.

3 Proposed Methodology

3.1 Network Formation

In this section we describe how to form a network based on direct correlations between items. Assume that $\mathcal{I}$ is the domain of all items. As described, we assume boolean transactions. A set, Y, of items rated with 1 by a user, is denoted as *transaction*, thus $Y \subseteq \mathcal{I}$. D henceforth denotes a collection of user transactions. To measure correlation between items, we need some simple definitions.

Definition 1 (Occurrence probability). *The occurrence probability for a set of items $X \subseteq \mathcal{I}$, denoted as $P(X)$, is defined as the relative number of D's transactions that contain X, i.e.: $P(X) = \frac{|\{Y|Y \in D, X \subseteq Y\}|}{|D|}$.*

When $X = \{i\}$, $i \in \mathcal{I}$ (X is a singleton), then $P(X) = P(\{i\})$ denotes the occurrence probability of item i.

Definition 2 (Conditional occurrence probability). *Let i, j be two items of $\mathcal{I}$. Given the occurrence of i, the conditional occurrence probability of j is defined as $C(j|i) = \frac{P(\{i,j\})}{P(\{i\})}$.*

$C(j|i)$ measures the correlation between each pair of items i, j. A network G can be formed as a directed weighted graph, where for each item $i \in I$ we have a corresponding vertex in G. Between two items i, j we have an arrow in G if $C(j|i) > 0$. To an arrow between i and j we assign a weight equal to $C(j|i)$.

Example 1. Assume a collection of 7 past transactions, depicted in Figure 1a (with IDs $a - g$). The corresponding network is depicted in Figure 1b. For instance, there is an arrow from vertex 1 to vertex 3 with weight 0.33, because $P(\{1\}) = 3/7$, $P(\{3\}) = 4/7$, and $P(\{1, 3\}) = 1/7$. Therefore, the weight is equal to $C(3|1) = P(\{1, 3\})/P(\{1\}) = 1/3$. In the inverse direction, the arrow is from vertex 3 to vertex 1 with weight 0.25. There is no arrow from vertex 1 to vertex 2, because $C(2|1) = 0$.

(a)	**5 1 6**
(b)	**1 5**
(c)	**3 7 2**
(d)	**4 7 3**
(e)	**2 3**
(f)	**7 4 2**
(g)	**1 3 7 6**
new	**6 (2 3)**

(a)

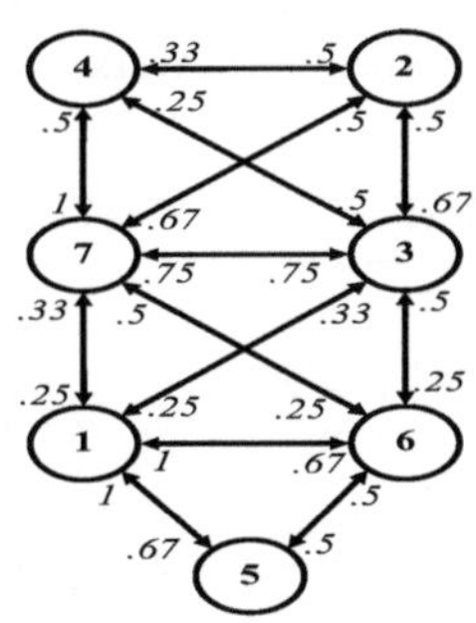

(b)

Fig. 1. Example of network formation

3.2 Computing the Accessibility

In a network G, a path $p = \langle i_1, \ldots, i_n \rangle$ of length n, is a sequence of successive arrows of the form $i_j \rightarrow i_{j+1}$ between n items of G. For a path p, $s(p)$ denotes the starting item of p and $e(p)$ the ending item. A path p implies a transitive correlation between items $s(p)$ and $e(p)$. To measure the amount of correlation, we assume that the transition from item $s(p)$ to item $e(p)$ through path p, represents a Markov chain, thus we define the corresponding transitive probability.

Definition 3 (Transitive probability). *For two items i_1 and i_n for which there exist a path $p = \langle i_1, \dots, i_n \rangle$, we define the transitive probability through p as $T(i_1, i_n|p) = \prod_{j=1}^{n-1} C(i_{j+1}|i_j)$, $i_j \in p, 1 \le j \le n$.*

Example 2. In the network of Figure 1b, for the path $p = \langle 1, 3, 2 \rangle$, we have that $T(1, 2|p) = C(3|1) \cdot C(2|3) = 1/3 \cdot 1/2 = 0.167$.

To measure how accessible is an item from another item, we define the accessibility matrix based on the notion of *acyclic* path. A path $p = \langle i_1, \dots, i_n \rangle$ is acyclic when no vertex appears more than once within it, i.e., $i_j \ne i_k$ for $j \ne k$.

Definition 4 (Accessibility matrix). *For two items $i, j \in \mathcal{I}$, let $\mathsf{P}(i, j)$ be the set of all acyclic paths for which $s(p) = i$ and $e(p) = j$. We form the accessibility matrix A, whose elements are given as $A(i, j) = \sum_{\forall p \in \mathsf{P}(i,j)} T(i, j|p)$.*

$A(i, j)$ measures the overall transitive correlation between items i and j. Notice that A is not symmetric. Since we are interested in transitive correlations, the diagonal of A matrix is of no use and henceforth is omitted.

Example 3. For the network of Figure 1b, to compute $A(1, 2)$, we measure the transitive probability of all 34 acyclic paths that connect item 1 to item 2 (their lengths are between 3 and 7). This is depicted in Figure 2a. Next, we take the summation, which results to $A(1, 2) = 2.27$. In the same way, we can compute the entire accessibility matrix A, which is given in Figure 1b.

Path	T(1,2|Path)
1 → 3 → 2	0.167
1 → 7 → 2	0.167
1 → 3 → 4 → 2	0.042
1 → 3 → 7 → 2	0.125
1 → 6 → 3 → 2	0.167
⋮	⋮
1 → 5 → 6 → 7 → 4 → 3 → 2	+ 0.021
	$A(1, 2) = 2.27$

(a)

	1	2	3	4	5	6	7
1	-	2.27	2.19	1.83	1.22	1.43	2.19
2	2.27	-	2.21	1.44	1.98	1.81	2.17
3	1.65	1.66	-	1.39	1.43	1.31	2.04
4	2.75	2.16	2.78	-	2.39	2.19	2.51
5	1.82	2.96	2.86	2.39	-	1.60	2.86
6	2.15	2.71	2.63	2.19	1.60	-	2.62
7	1.65	1.63	2.04	1.25	1.43	1.31	-

(b)

Fig. 2. (a) Example of how to compute $A(1, 2)$. (b) The resulting A matrix.

To form the accessibility matrix, for the moment we simply assume that we take into account all acyclic paths between each pair of vertexes. These paths are derived by performing a traversal over the network. In Section 3.4 we describe the use of pruning criteria that constrain the number of examined paths.

3.3 Generating Recommendations

To find which items to recommended to a new user, we propose the PB algorithm:

> *Algorithm*: PB
> *Input*: the domain of items $\mathcal{I}$, the current transaction X of the new user, the accessibility matrix A, the required number of recommendations M.
> *Output*: M recommended items.
> 1. For each $j \in \mathcal{I}$ we define its $R(j)$ value as follows:
>
> $$R(j) = \begin{cases} \sum_{\forall i \in X} A(i,j), & j \notin X \\ 0, & \text{otherwise.} \end{cases}$$
>
> 2. Recommend the M items with the highest $R(j)$ value.

We call this algorithm *Path-based* (PB), because it is based on the consideration of accessibility through paths. Intuitively, PB recommends the items with the highest overall accessibility from the items of the current transaction. In case that only $M' < M$ items have a positive $R(j)$ value, we resolve the tie by selecting the remaining $M - M'$ items j ($j \notin X$) that have the highest overall occurrence probability $\sum_{\forall i \in X} P(\{i, j\})$.

Example 4. In the example of Figure 1a, assume that the transaction of the new user is $X = \{6\}$. Also assume that we want to recommend $M = 2$ items. From Figure 2b we find that $A(6, 2)$ and $A(6, 3)$ are the two highest values in the 6-th row of matrix A. Therefore, the two recommended items are 2 and 3.

3.4 Pruning Criteria

The number of all acyclic paths between the vertexes of a directed graph (network) grows exponentially with the size of the graph. To manage the computational cost, we develop three pruning criteria that drastically reduce the number of considered paths without affecting the detection of the most accessible items. Recall that paths are formed with a traversal of the network. All criteria are applied to the current path that is extended at each step of this procedure. The description of the criteria follows.

Maximum path length (L). We stop the extension of the current path when its length exceeds a parameter L. In a long path, the transitive probability between its two ends decreases rapidly. By omitting a long paths, we exclude small transitive probability, thus the values in the accessibility matrix are not much affected.

Maximum conditional probability (C). We stop the extension of the current path p when the value $C(s(p), e(p)|p)$ is less than a parameter C ($0 \leq C \leq 1$). This criterion acts complementary to the previous one and filters out short paths which do not have a significant transitive probability between their ending vertexes.

Maximum fraction of followed neighbors (N). When extending the current path p, we consider only a fraction N ($0 \leq N \leq 1$) of neighboring vertexes, i.e., those for which exists an arc from $e(p)$ to them. This criterion reduces the search space of the graph-traversal procedure. We have examined several heuristics about how to select neighbors, e.g., select the N neighbors with the highest conditional probability w.r.t. the current node. However, we did not find significant differences among them. For this reason we use the most simple one, that is, we select at random among them.

4 Understanding the Characteristics of PB

4.1 Distribution of Values in the Accessibility Matrix

To understand the characteristics of PB, and the motivation behind it, we examine the distribution of values inside the accessibility matrix A. We use several real data sets that are related with various application areas of CF (retail, web personalization, etc). Due to space constaints, we present results for the following: (a) The BMS-POS data set (POS): contains several years worth of point-of-sale data from a large electronics retailer.[1] (b)The Clarknet log file (Clarknet): two week's worth of HTTP requests to the site of ClarkNet access provider (log was processed to create transactions).[2] (c) The Entree Chicago Recommendation Data (Entree): a record of user interactions with the Entree Chicago restaurant recommendation system.[3]

We start by examining the distribution of occurrence probabilities $P(i)$ for all items i. The results are illustrated in Figure 3. In all plots, the horizontal axis contains the values of occurrence probabilities, whereas the vertical axis contains the number of items (denoted as count) that have a particular occurrence probability. Both axes are plotted in logarithmic scale. We additionally include the corresponding fitting curves (mean square error, linear or quadratic, depending on the case).

For the POS data set, the distribution follows a power law, because in the log-log scale it shows itself to be linear. This means that we can easily distinguish a set of few items that appear substantially more frequently than the others. For the Entree data set, although the distribution is not uniform, it does not follow a power law. The distribution for Clarknet is between the previous two cases: it does not follow a power law, but it is more skewed than the case of Entree. Therefore, the three examined data sets represent three different degrees of skewness in terms of the distribution of occurrence probabilities.

Next, we examine the distribution of values inside the accessibility matrix. The results are depicted in Figure 4 (to ease the presentation, the zero values are omitted; in all measurements L was set to 6). In the horizontal axis, each point represents a value in the accessibility matrix (denoted as accessibility),

[1] Available at http://fimi.cs.helsinki.fi/data/
[2] Available at http://ita.ee.lbl.gov/html/contrib/ClarkNet-HTTP.html
[3] Available at: http://kdd.ics.uci.edu/databases/entree/entree.html

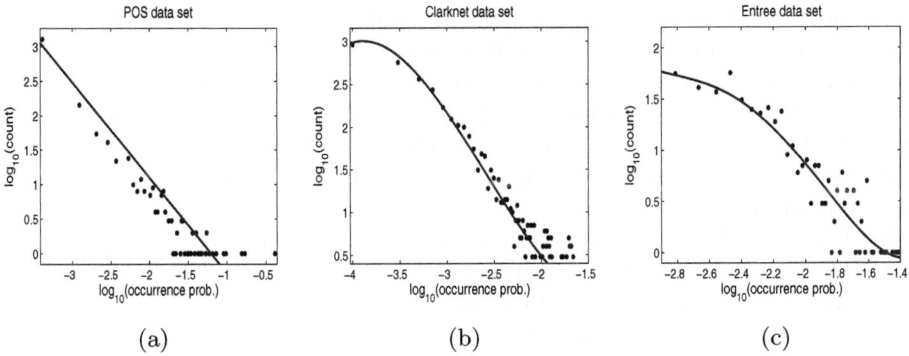

Fig. 3. Distribution of occurrence probabilities for (a) POS, (b) Clarknet, (c) Entree

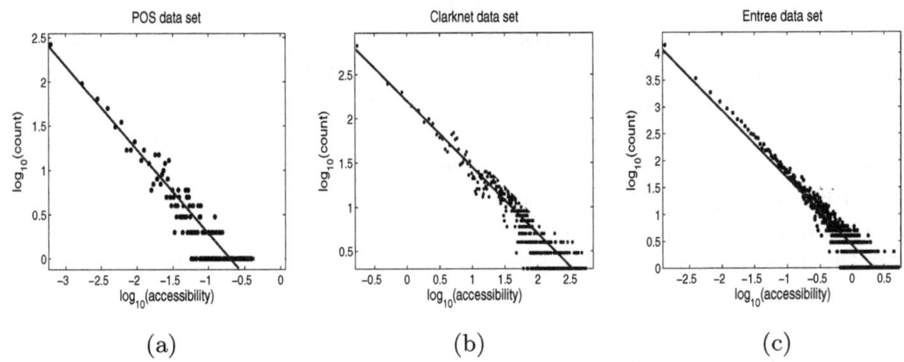

Fig. 4. Distribution of accessibility values for (a) POS, (b) Clarknet, (c) Entree

whereas the vertical axis contains the number (denoted as count) of items that have a particular accessibility. As previously, both axes are in logarithmic scale and the plots include the corresponding fitting curves. For all three data sets, the distribution follows a power law, as it presents a linear trend in the log-log scale. It means that correlations between items are formed in way that some few items become clearly distinguishable from the others by having a significantly larger number of transitive correlations. Therefore, it is possible for PB to detect such items.

4.2 Qualitative Comparison

In this section we present a qualitative comparison of PB, through which we understand better the characteristics of PB and its advantage against other possible solutions.

A simple heuristic is to recommend the items with the highest occurrence probability values. Henceforth, this heuristic is called MF. For instance, in the example of Figure 1, MF recommends items 3 and 7. MF opts that the most "popular" items will be preferred by the new user as well. We previously observed that $P(i)$ values do not always follow a power law. Hence, we may not be able to find a small set of distinguished items based on their occurrence probability. Moreover, MF provides recommendations independently from the items that the new user has already rated. In contrast, PB takes into account the preferences of the new user, because it recommends the items that are more correlated with those already rated.

IB [4,15] considers only direct correlations between items. For example, for the case in Figure 1, IB recommends items 1 and 5, because they have the smallest cosine distance with the already rated item 6. However, as described, these direct correlations may be imprecise when the user has not rated many items. Since IB looks only for direct correlations, differently from PB, it cannot reach items like 2.

Finally, we can develop a heuristic that is based on existing methods for network analysis, and the most prominent case is Page Rank. Therefore, we can form a matrix with probabilities of direct transition between each pair of items[4] and recommend the items with the highest Page Rank. Henceforth, this heuristic is called PR. In the example of Figure 1, the two items with the highest Page Rank are 1 and 3. Although PR considers transitive correlations, similarly to MF, it does not take into account the items that the new user has already rated. Thus, PR ignores the fact that item 2 is the more accessible from item 6 (the already rated) but is not the most globally accessible item.

In the following section we present experimental results that validate the presented qualitative comparison.

5 Experimental Results

In our experiments we compared all examined algorithms (PB, MF, IB, PR) using the real data sets that were described in the previous section. Since we consider binary ratings (see Section 2), our objective is to predict the relevant items. Thus, the examined evaluation metric is not the mean square error (MSE). Instead, we employ precision, recall, and F measure (denoted as F_1), which are widely accepted in information retrieval. Assume that an algorithm recommends to a user a set of items S_1, whereas the user, after the recommendation, actually rates a set of items S_2. Precision is the ratio $(S_1 \cap S_2)/|S_1|$, whereas recall is $(S_1 \cap S_2)/|S_2|$. F_1 is an even combination of them: $F_1 = 2 \cdot \text{precision} \cdot \text{recall}/(\text{precision} + \text{recall})$. We separately measured the sensitivity of PB against the pruning criteria and their impact on its execution time.

For PB, the default value for L is 4, for C is 0.3, and for N is 0.6. The rest parameters for all algorithms were tuned to the best values, e.g., the number of nearest neighbors required by IB, convergence tollerance for PR, etc. Each measurement is the result of 5-fold cross validation. Every transaction in the evaluation set is divided into two parts. The first part, denoted as *past*, contains the items that are treated as known and are used to generate recommendations.

[4] For each item, the sum of probabilities has to be normalized to the range [0, 1].

Their number is denoted as *past size* and is given as percentage relatively to the transaction size. The second part contains the items that are treated as unknown, i.e., those we want to predict.

In our first measurement we examine the precision versus recall that results by varying the number of recommended items. For the POS data set the results are illustrated in Figure 5a (past size is 10%). As mentioned, for POS the distribution of $P(i)$ values follows a power law. This is the best case for MF, because it can effectively distinguish the appropriate items to recommend. In contrast, because of the small *past size* and the skewed distribution of $P(i)$, IB cannot detect precise direct correlations and its resulting precision and recall are very low. PB is able to overcome this problem, because it considers transitive correlations. Thus, its performance is comparable to MF's. With increasing *past size* for the POS data set (Figure 5b), IB improves its performance but is still clearly outperformed by the others. For larger *past size*, PB attains a small improvement over MF, as it can find more and better transitive correlations.

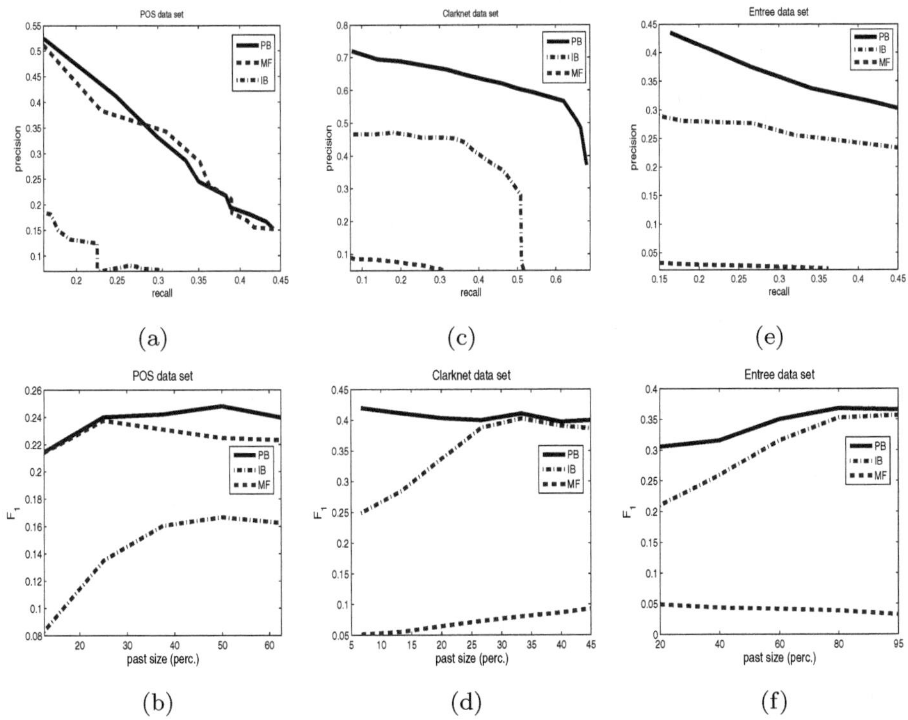

Fig. 5. Results for precision vs. recall and F_1

Next we examine the Clarknet data set. The results for precision versus recall are illustrated in Figure 5c (past size is 10%). Since the distribution of $P(i)$ values does not follow a power law, the precision/recall values of MF substantially degrade, compared to the case of POS data set. In contrast, for IB we notice a significant improvement. Nevertheless, PB compares favorably to the other algorithms. The reason is that, as we saw, both in POS and Clarknet, there exist items with much higher accessibility than others, thus PB can detect and provide them as good recommendations, in both cases. Focusing on the impact of *past size* (Figure 5d), we see that for the problem we are interested, i.e., quality of recommendation for small *past size*, PB is superior. When *past size* gets significantly large, as expected, PB and IB converge to the same performance, because IB becomes able to detect precise direct correlations. Analogous conclusions are derived for the Entree data set (Figures 5e–f).

To further understand the impact of data set's characteristics, we examined synthetic data generated with the system of [1] and in particular the T10I4D100K data set (see [1] for the notation). We modified the generator and by assuming a mapping of items into consecutive integer numbers, we impose that, for item i, the occurrence probability is $P(i) = 1/\sum_{\forall j \in \mathcal{I}}(1/j^{-\theta}) \cdot 1/i^{-\theta}$. The first factor is the normalization constant, whereas θ $(0 \leq \theta \leq 1)$ controls the amount of skewness ($\theta = 0$ results to uniform distribution, $\theta = 1$ to pure Zipf distribution). The resulting F_1 values versus θ are illustrated in Figure 6a. As expected, with increasing θ, the performance of MF improves substantially, because the distribution of $P(i)$ becomes more skewed. In contrast, F_1 for IB and PB slightly reduces, as the number of direct and transitive correlations reduces when some items appear significantly more frequently than the others. Nevertheless, PB present the best F_1 in all cases.

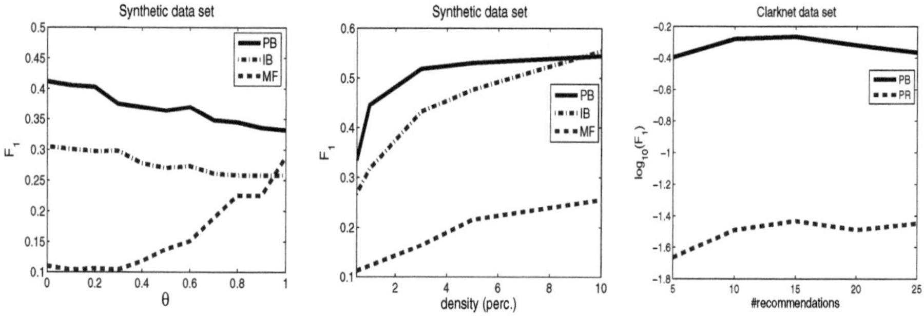

(a) (b) (c)

Fig. 6. (a) F_1 vs. θ (amount of skewness), (b) F_1 vs. density, (c) PB vs. PR

We also examined the impact of density, i.e., the percentage of rated items, by varying the average number of items in the transactions. The resulting F_1 versus density (given as percentage of $|\mathcal{I}|$) are illustrated in Figure 6b. In this measurement, the *past size* increases proportionally to the average transaction size. Therefore, for higher density, F_1 values for IB and PB converge to the same point. However, in most real-world applications, density values are quite low (sparsity).

The comparison of PB against PR is presented separately, because PR presented the worst performance among all algorithms. For the Clarknet data set, we measured F_1 against the number of provided recommendations. Figure 6c depicts the results, where F_1 values are given in logarithmic scale. This results shows the inadequacy of PR, which, differently from PB, does not take into account the user preferences.

Finally, we measured the sensitivity of PB against the pruning criteria. Due to space limitations, we present results only for the Clarknet data set, whereas analogous results were obtained for the other data sets. For Clarknet data set, the resulting F_1 versus threshold L, C, and N are given in Figures 7a, b, and c, respectively. In the same figures we also plot (using the right vertical axis) the relative execution time for the formation of the accessibility matrix. For L, very small values do not allow PB to capture transitive correlations. However, after an adequate value, F_1 stabilizes. In contrast, execution time increases rapidly with increasing L. Therefore, we need to select an appropriately small value. Analogous reasoning holds for N. Regarding C, higher values reduce execution time. However, very high values should be avoided, because they prevent the formation of transitive relations. In general, we notice that for a wide range of all threshold values, F_1 for PB does not change abruptly, a fact that indicate its robustness. To support the feasibility of PB, we mention that for the Clarknet data set, when using the default values for the pruning thresholds, the absolute execution time for the formation of accessibility matrix was less than 1 min.

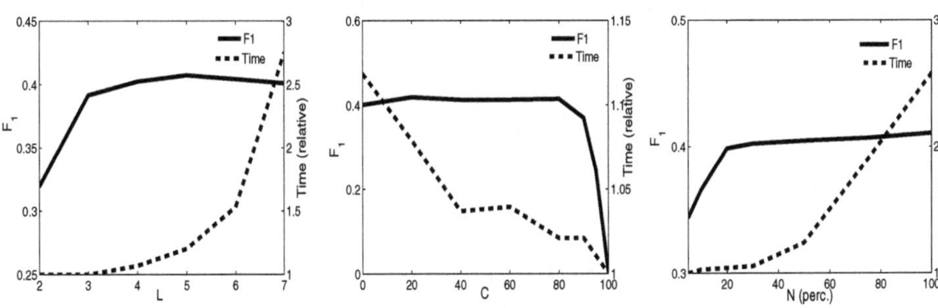

(a) (b) (c)

Fig. 7. F_1 vs.: (a) L, (b) C, (c) N

6 Conclusions

We have addressed the problem of how to provide qualitative recommendations when the new user has not rated a sufficient number of items. We have proposed a novel CF algorithm that finds transitive correlations in a network formed by the direct correlations between the items. With measurements on real data we have examined the characteristics of the proposed method and justified our approach. Detailed experimental results have illustrate its superiority over existing CF methods. In our future work we will examine how to enrich the proposed network of correlations with information about the content, and the combination of the proposed method with a transitive closure algorithm.

References

1. R. Agrawal and R. Srikant. Fast algorithms mining association rules in large databases. Technical Report RJ 9839, IBM Almaden Research Center, San Jose California, 1994.
2. J. Breese, D. Heckerman, and C. Kadie. Empirical analysis of predictive algorithms for collaborative filtering. In *Proc. Conf. on Uncertainty in Artificial Intelligence*, pages 43–52, 1998.
3. M. Claypool, M. W. P. Le, and D. C. Brown. Implicit interest indicators. In *Proc. IUI Conf.*, 2001.
4. M. Deshpande and G. Karypis. Item-based top-n recommendation algorithms. *ACM Trans. on Information Systems*, 22(1):143–177, 2004.
5. D. Goldberg, D. Nichols, M. Brian, and D. Terry. Using collaborative filtering to weave an information tapestry. *ACM Communications*, 35(12):61–70, 1992.
6. J. Herlocker, J. Konstan, A. Borchers, and J. Riedl. An algorithmic framework for performing collaborative filtering. In *Proc. ACM SIGIR Conf.*, pages 230–237, 1999.
7. J. Herlocker, J. Konstan, L. Terveen, and J. Riedl. Evaluating collaborative filtering recommender systems. *ACM Trans. on Information Systems*, 22(1):5–53, 2004.
8. Z. Huang, H. Chen, and D. Zeng. Applying associative retrieval techniques to alleviate the sparsity problem in collaborative filtering. *ACM Trans. on Information Systems*, 22(1):116–142, 2004.
9. G. Linden, B. Smith, and J. York. Amazon.com recommendations: Item-to-item collaborative filtering. *IEEE Internet Computing*, 7(1):76–80, 2003.
10. D. Maltz and K. Ehrlich. Pointing the way: Active collaborative filtering. In *Proc. CHI Conf.*, pages 202–209, 1995.
11. B. Mobasher, H. Dai, T. Luo, and M. Nakagawa. Improving the effectiveness of collaborative filtering on anonymous web usage data. In *Proc. Workshop Intelligent Techniques for Web Personalization*, pages 53–60, 2001.
12. M. Papagelis, D. Plexousakis, and T. Kutsuras. Alleviating the sparsity problem of collaborative filtering using trust inferences. In *Proc. iTrust Conf.*, pages 224–239, 2005.
13. P. Resnick, N. Iacovou, M. Suchak, P. Bergstrom, and J. Riedl. Grouplens: An open architecture for collaborative filtering on netnews. In *Proc. Conf. Computer Supported Collaborative Work*, pages 175–186, 1994.
14. B. Sarwar, G. Karypis, J. Konstan, and R. J. Analysis of recommendation algorithms for e-commerce. In *Proc. ACM Electronic Commerce Conf.*, pages 158–167, 2000.
15. B. Sarwar, G. Karypis, J. Konstan, and J. Riedl. Item-based collaborative filtering recommendation algorithms. In *Proc. WWW Conf.*, pages 285–295, 2001.
16. A. I. Schein, A. Popescul, and L. H. Ungar. Methods and metrics for cold-start recommendations. In *Proc. ACM SIGIR Conf.*, 2002.
17. J. Wang, A. P. de Vries, and M. J. Reinders. Unifying user-based and item-based collaborative filtering approaches by similarity fusion. In *Proc. ACM SIGIR Conf.*, 2006.
18. J. Wang, A. P. de Vries, and R. M. J. T. A user-item relevance model for log-based collaborative filtering. In *Proc. ECIR Conf.*, page 3748, 2006.
19. G. Xue, C. Lin, and Q. e. Yang. Scalable collaborative filtering using cluster-based smoothing. In *Proc. ACM SIGIR Conf.*, pages 114 – 121, 2005.

Entropy-Based Authorship Search in Large Document Collections

Ying Zhao and Justin Zobel

School of Computer Science and Information Technology, RMIT University
GPO Box 2476V, Melbourne, Australia
{yizhao,jz}@cs.rmit.edu.au

Abstract. The purpose of authorship search is to identify documents written by a particular author in large document collections. Standard search engines match documents to queries based on topic, and are not applicable to authorship search. In this paper we propose an approach to authorship search based on information theory. We propose relative entropy of style markers for ranking, inspired by the language models used in information retrieval. Our experiments on collections of newswire texts show that, with simple style markers and sufficient training data, documents by a particular author can be accurately found from within large collections. Although effectiveness does degrade as collection size is increased, with even 500,000 documents nearly half of the top-ranked documents are correct matches. We have also found that the authorship search approach can be used for authorship attribution, and is much more scalable than state-of-art approaches in terms of the collection size and the number of candidate authors.

1 Introduction

The purpose of authorship search (AS) is to find within a large collection the documents that appear to have been written by a given author. That is, given documents of known authorship, the task is to find other documents by the same author. AS has not previously been widely investigated, but is related to authorship attribution (AA), the task of identifying the authorship of unknown documents given a corpus of known authorship. AA and AS are valuable in applications such as plagiarism detection, literary analysis, and forensics. However, none of the AA approaches has been scaled to large document collections. For example, for the Federalist Papers, the authorship of 65 documents is explored [17]. Holmes et al. used 17 journal articles [13], while Koppel et al. used 21 English books [19]. Diederich et al. used a collection with seven authors and around 100 texts for each author [8]. Hoover's collection had 50 documents [14]. We previously used 4900 documents, so far the largest AA collection [28,30].

Our method for AS is motivated by information retrieval (IR) techniques, where matching is determined by computing the similarity between queries and documents, but there are significant differences. A key difference is choice of index terms; IR techniques make use of content-bearing words, while in AS it is necessary to identify style markers. We explore use of function words and part-of-speech (POS) tags. Another

G. Amati, C. Carpineto, and G. Romano (Eds.): ECIR 2007, LNCS 4425, pp. 381–392, 2007.

potential difference is choice of similarity measure. We propose relative entropy as the similarity measure for AS, inspired by the language models used in IR.

For data, we use collections of 10,700 to 500,700 newswire articles from the TREC collections. Our results show that, with sufficiently large queries, matching documents can be found from within large collections. While results are mixed, with even the largest document collection precision in the top 10 is 44.2%. The best results were achieved with stop words; use of style markers was less effective, while standard similarity measures were, compared to relative entropy, nearly useless. We also investigated the applicability of the proposed AS approach to AA, finding that, with a large volume of text by an unattributed author, authorship could be identified with reasonable reliability, greatly improving on previous methods. With less training text and on larger collections, the accuracy of attribution was not as good. Overall, however, we have demonstrated for the first time the feasibility of AA and AS on large document collections. As we have since shown for a small collection, the method is also highly effective for literature [29].

2 Document Search

IR systems are used to search for documents that satisfy users' information needs [2]. Current IR systems usually deal with large and heterogeneous collections and typically take as input queries of a few words, returning as a response a list of documents deemed most likely to be relevant. Search involves two stages, index term extraction and similarity computation. For documents in English, indexing involves separating the text into words, case-folding, stopping, and stemming [32].

Various models have been proposed as bases for measurement of similarity between documents and queries. One is the vector-space model [2], where items are represented as vectors. The assumption is that similar documents should be separated by a relatively small angle. Plausibly a similar assumption would apply to AS: given appropriate style markers, the distribution of style markers in the documents by an author should be similar. An alternative are models used to derive estimates for the probability that a document is relevant to a query. The BM25 model is one of the most successful probabilistic models in IR [24]. Whether such a model is suitable for AS is, intuitively, not clear, but given the success of BM25 in IR it is reasonable to consider its use for AS.

Language models, also based on probability theory, were originally motivated by tasks such as speech recognition. These models are used to estimate the probability distributions of words or word sequences. In IR, language models are used to estimate the likelihood that a given document and query could have been generated by the same model [7]. Given a document d and a model Θ_d inferred from d, language models estimate the probability that model Θ_d could have generated the query q. Smoothing techniques are applied to assign probabilities for missing terms [6,12,27].

Although language models have elements that are counter-intuitive (suggesting, for example, that queries comprised of common words are more likely than queries comprised of words that are specific to a given document), they are a high effective approach to IR. In this paper a key contribution is exploration of whether language models are suitable for AS.

3 Authorship Attribution

The purpose of AA is to identify documents that are written by a particular author. A range of AA methods have been proposed in recent research. Despite the differences amongst all these approaches, the framework of an AA technique involves two stages in general: extraction of document representations and making attribution decisions.

Document representations are comprised of style makers. Both lexical and grammatical style markers have been proposed for AA. A simple approach is to use lexical markers, that is, function words and punctuation symbols [8,13,15]. More sophisticated syntactic and grammatical components can be extracted by natural language processing (NLP) techniques, such as part-of-speech (POS) tags [16,25]. However, these more advanced style markers do not necessarily produce better performance for AA [30]. A particular issue is that the idiosyncratic grammatical patterns that are particular to an author may not be identified due to the lack of observations of such patterns in the process of training the parser. That is, NLP is not only error prone, but is likely to make errors on the most significant elements of the data.

In the attribution stage a variety of classification methods have been investigated. Principle component analysis (PCA) has been used in several approaches to AA [5,13]. Hoover [14] examined the scalability of PCA to large corpora or multi-class AA, in which the number of author candidates is greater than 2, finding only 25% accuracy given 50 samples by a total of 27 authors, suggesting that PCA would not scale to large numbers of authors.

Machine learning approaches such as support vector machines (SVMs) [8,18] are considered to be competitive alternatives for AA. SVMs are effective when provided with sufficient samples and features. However SVMs are not always superior to other methods when given small number of samples for training, which is often the case in AA. Computational cost is another issue of SVMs. Bayesian networks are less effective and more computationally expensive than SVMs [11,23].

Language models have also been proposed for AA. Benedetto et al. used a compression-based language model [4], based on a standard compression suite; however, Goodman [9] was unable to reproduce the result, and the method is not plausible. A Markov model has also been applied to AA by Khmelev et al. [17], in which the features are individual characters. Good accuracy was achieved on data collected from the Gutenberg project, but the accuracy may be overestimated, due to the duplicate texts provided by Gutenberg. For example the number of distinct texts by Burroughs is only 9, but Khmelev et al. included 25 of his works in their experiments.

Most of these AA methods are not directly applicable to search tasks. In a search system a query is evaluated by ranking the similarities measured between the query and each document individually in the collection. The result is a list of top-ranked documents. In contrast to search, there is no ranking required for AA; instead, an explicit decision is made for each unknown document individually. Documents are required for training to learn a model for a particular author in AA. There is no document-by-document calculation involved. AA techniques have not been applied to search problems. In this paper we propose what we believe is the first AS mechanism.

4 Relative Entropy for Authorship Search

In AA and AS, the underlying assumption is that there are patterns or characteristics of an author's writing that can be automatically extracted and then used to distinguish their work from that of others. Given appropriate style markers, the distribution with which they are observed should be similar in all of the author's documents, regardless of topic. Distributions can be compared via their entropy, and we therefore propose use of the Kullback-Leibler divergence (KLD, or relative entropy) as a similarity measure for AA. The distributions need to be estimated, and we propose use of the language models that have been successfully applied in IR [6,20,27]. We used a similar approach for AA [30], but it was not clear that such an approach could be used for AS.

Entropy measures the uncertainty of a random variable X, where, in this application, each $x \in X$ could be a token such as a word or other lexical feature, and $p(x)$ is the probability mass function of X. The KLD quantifies the dissimilarity between two distributions. In the context of AS, we can build entropy models for the queries and the documents. The differences between query models and document models can be measured by relative entropy as:

$$KLD(d\|q) = \sum_{x \in X} p_d(x) \log_2 \frac{p_d(x)}{p_q(x)} \tag{1}$$

The divergence is calculated between the query and every document in the collection. The documents whose entropy has the lowest divergence from the query are the most likely to share authorship and thus should be the highest ranked. However, if $p(x)$ is zero for some symbol the divergence is undefined. To address this issue, we use Dirichlet smoothing to estimate probabilities [27]:

$$\hat{p}_d(x) = \frac{f_{x,d}}{\mu + |d|} + \frac{\mu}{\mu + |d|} p_B(x) \tag{2}$$

Here x are the style markers used for document representations and $f_{x,d}$ is the frequency of token x in document d. The notation $|d| = \sum_{x \in d} f_{x,d}$ represents the number of token occurrences in d, and $p_B(x)$ is the probability of the token x in the *background model*, which provides statistics on the tokens. The parameter μ controls the mixture of the document model and the background model. The background probabilities dominate for short documents, in which the evidence for the in-document probabilities is weak; when the document is longer then the influence of the background model is reduced. In principle the background model could be any source of typical statistics for token occurrences.

Additionally, a set of style markers is required. Some researchers have found that function words are effective [1,8,13,15,18], while use of part-of-speech tags has also been considered [30]. We make use of both kinds of marker, but, in agreement with our earlier work [30], find that function words are superior.

5 Experiments

As data, we use collections of documents extracted from the TREC corpus [10]. There are large numbers of documents included in TREC that can be used to evaluate the proposed search technique, as the author is identified. We believe that this data presents a difficult challenge for AA or AS, as, compared to novelists or poets, journalists do not necessarily have a strong authorial style, and the work may have been edited to make it consistent with a publication standard.

We develop three collections of documents to evaluate the proposed AS system, which consist of 10,700, 100,700, and 500,700 documents respectively. We call these the 10k-collection, 100k-collection, and 500k-collection. The documents in the 10k-collection and 100k-collection are from the AP subcollection of TREC; the 500k-collection consists of documents from AP, WSJ, and SJM. Metadata, including author names is discarded. As authors, we select the seven[1] that we earlier used for AA [28,30]. These authors are regular contributors to AP and each of them has over 800 documents in the TREC corpus. We randomly select 100 documents of each author and include them as part of each of the collections, giving in total the extra 700 documents in each case. All queries and documents are pre-processed to obtain the style markers that are applied to the system; query construction is discussed below. The background models of different types of style markers used in all experiments are derived from the AP collection of over 250,000 documents; an alternative would have been to use the collection as the background model in each case, but we decided to hold the background model constant across all experiments.

We evaluate our proposed authorship search system from several perspectives. Scalability is examined by considering effectiveness on collections of different sizes. We run the experiments with different kinds of style marker. The differences between KLD-based search and other retrieval techniques are tested. Finally we explore use of the AS approach as an AA method.

Feasibility and scale in size. In the first experiment we examine whether AS is feasible for small and large collections. The first seven queries used in this experiment are generated by concatenating 500 randomly selected documents written by each of the seven authors. These documents are distinct from the 100 documents that are included as part of each collection. We call these the 500-document queries. The style markers are function words. The next seven queries are formed by concatenating the 100 documents that are included in the collection; we call these the 100-included queries.

The numbers of correct matches in the top-100 ranked documents are in Table 1, for the 10k-collection. Amongst the 500-document queries, those based on the documents of Currier and Dishneau are the most effective, while the query based on the documents of Beamish is much less effective. The 100-included queries are slightly better than 500-document queries in most cases, despite being based on less text, and are highly consistent with the 500-document queries, suggesting that the style of some authors is easier to identify than that of others.

Overall precision and recall is plotted in Figure 1, based on the 500-document queries on all three collections. We achieve average $p@10$ (precision at 10 documents retrieved) of 84.2% on the 10k-collection, 74.2% on the 100k-collection, and 30.0% on the 500k-collection. Thus, while the density of correct matches falls from 1% to 0.02%,

[1] The authors are Barry Schweid, Chet Currier, Dave Skidmore, David Dishneau, Don Kendall, Martin Crutsinger, and Rita Beamish.

Table 1. The number of correct matches in the top 100 documents in response to each query, on the 10k-collection

	Number of correct answers in top 100						
	Schweid	Currier	Skidmore	Dishneau	Kendall	Crutsinger	Beamish
500-document	48	61	35	61	44	52	30
100-included	59	58	49	61	46	56	37

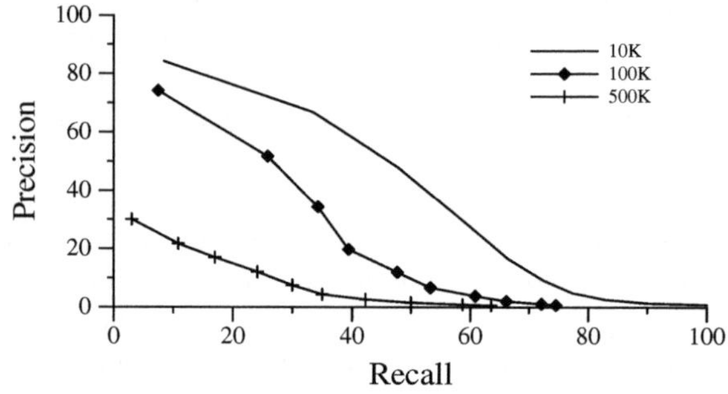

Fig. 1. Precision versus recall for 500-document queries on each of the three collections: 10k, 100k, and 500k

effectiveness drops more slowly. Achievement of high recall is much more difficult with the largest collection, but the results show that, with a large query, AS is indeed feasible on even half a million documents.

Another dimension of scale is the volume of training data available. In the experiments above we had a large volume of text per author. With less text, effectiveness may decline. For each author we constructed 5 100-document queries and 25 20-document queries; average results are shown in Figure 2. It can be seen that reducing the amount of training data does indeed reduce effectiveness. For low levels of recall, queries of 100 documents (whether 100-included or queries comprised of another 100 documents) lead to reasonable effectiveness; indeed, whether or not the documents are included has a surprisingly low effect on the results, demonstrating that style as measured by function words must be moderately consistent within the work of an author. However, queries of 20 documents are much less effective. While reasonable numbers of correct documents are still found in the top 10 to 50 answers, subsequent results are poor.

Style markers. In text categorization, documents are usually indexed or represented by topic words occurred in the documents [3,21,22,26]. However, in AA whether topic words are appropriate style markers is controversial; some researchers have used them, but most have not. In this experiment we contrasted use of function words and topic words for AA, using the 10k-collection. Results are shown in Figure 3. In this figure, the uppermost curve uses the 500-document queries and is the same as in Figure 1; the dashed line is the comparable results for queries of topic-words; and the solid line is based on topic words and the 100-included queries. As can be seen, AS with topic

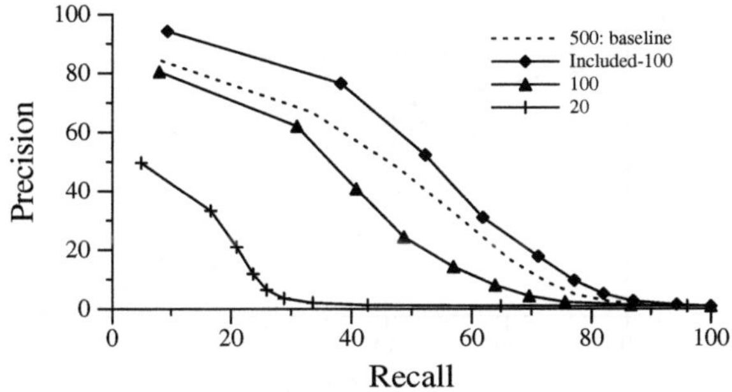

Fig. 2. Effectiveness for queries composed of 20–500 documents, on the 10k-collection

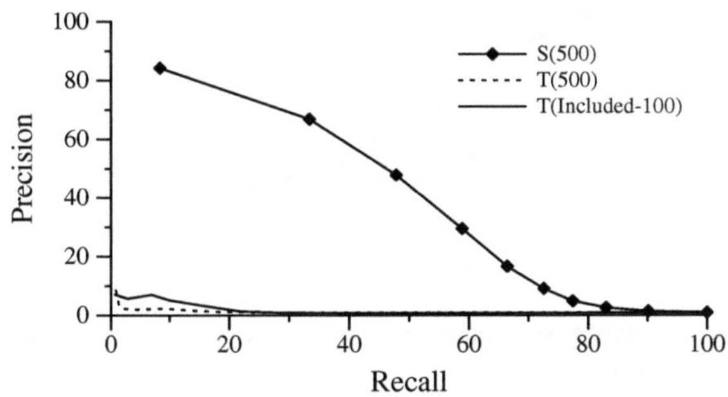

Fig. 3. Comparison of using different indexing methods: function words versus topic words on 10k-collection

words completely failed for authorship search; results are little better than random. The results show that the topic words are misleading in characterizing authors' writing style.

Other kinds of style marker are more plausible. For the next experiment, we used NLTK (a natural language toolKit)[2] has been applied to extract part-of-speech (POS) tags from documents. That is, in this approach, each document is represented by a stream of POS tags. We used a list of 183 POS tags, then indexed documents with function words, POS tags, and both combined. Results are shown in Figure 4.

Function words consistently lead to greater effectiveness than POS tags, which is consistent with our previous work in AA [30]. The combination of function words and POS tags leads to even greater effectiveness. With the smallest 10k-collection, function words are almost as good as the combined features, and both of them achieve the same $p@10$ of 84.2%. However, with larger collections the advantage of combination increases. On the 500k collection, function words achieve 30.0% $p@10$; addition of POS tags increases this to 44.2%. These results show that, even though POS tags by themselves do not yield good effectiveness, they are helpful additional evidence of style.

[2] Available from `http://nltk.sourceforge.net/index.html`

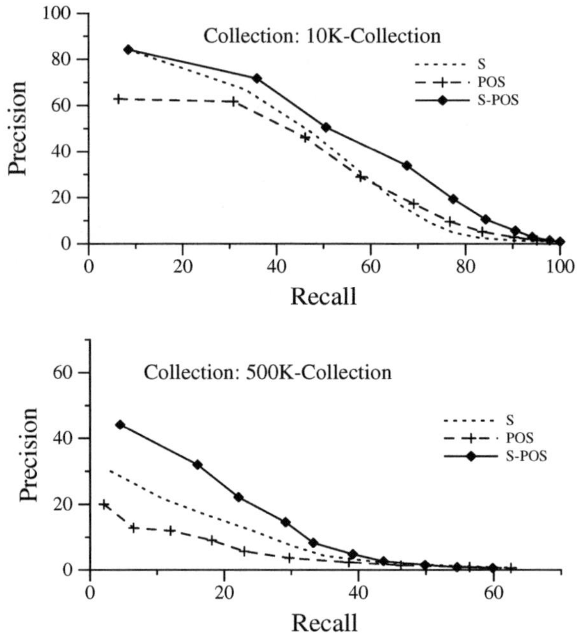

Fig. 4. Effectiveness of different style markers on the 10k (upper) and 500k (lower) collections, using the 500-included queries

KLD ranking versus other measures. In this experiment we compare similarity measures. In addition to KLD we used three measures that have been successfully used in IR, including BM25 and the vector-space measures BB-BCI-BCA and BB-ACB-BCA [31,32].

Results are in Figure 5. The IR similarity measures are surprisingly poor — none has proved suitable for AS. The BM25 measure is slightly better than the other two vector space models but none is usable. The reason why these measures are ineffective for AS is unclear and needs further investigation.

Applicability to authorship attribution. In this experiment we examine whether our AS approach can be used for AA. Instead of returning a list of documents that are judged likely to have the same authorship as to the query, an explicit authorship is returned corresponding to the query.

The proposed AA approach is as follows. We have a query for which authorship is unknown. Using search, a list of l top-ranked documents is returned. These are of known authorship, with k distinct authors and for each author a a count f_a of the number of documents by a in the list; thus $l = \sum_a f_a$.

A simple way to attribute authorship is to select a with the largest f_a. More strictly, a threshold t where $0 \leq t \leq 1$ can be selected so that the query can be assigned to a particular author a if $a = argmax_a(f_a)$ and $f_a/l > t$. Increasing t should reduce the likelihood of incorrect attribution.

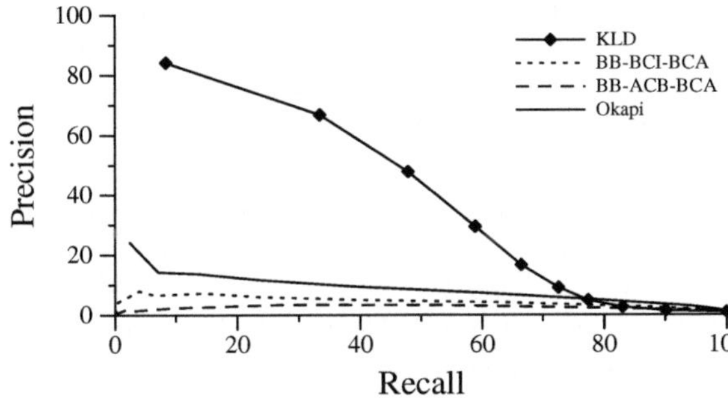

Fig. 5. Effectiveness of different similarity measures on 10k-collection, using the 500-document queries

To test these methods we built two collections from the AP data. The 10k-vote collection includes 10,000 documents from 342 authors, and the 100k-vote collection consists of 100,000 documents by 2229 authors. In both collections, 100 documents of each of the seven test authors are included. Overall the number of texts per author varies from 1 to 835. In both collections more than 10% of the distinct authors have written over 100 documents each. All documents in 10k-vote have identified authorship, while in the 100k-vote collection more than 90% of the texts have identified authorship. As style markers we use the combination of function words and POS tags.

Results from previous experiments show that it is feasible to search for documents written by the same author as that of the query, given a group of documents of known authorship as the query. In this experiment the authorship of the query is unknown and is to be identified. In this experiment, 500-document queries are unreasonably large. We experimented with queries that are formed from individual documents and from 10-document sets; none of the query documents are in the collections.

Results are shown in the Table 2, using the threshold $t = 0$ so that attribution is made to the authorship of the biggest f_a. Evaluation is based on the top l ranked documents, for l from 10 to 100. As can be seen, queries can be effectively attributed using the 10k-vote collection using only the top 10 documents retrieved; with both 1-document and 10-document queries, increasing l is not helpful.

With 1-document queries, the overall correctness of attribution is 51.0%. Previous methods achieve this accuracy only on small collections. Greater attribution effectiveness is achieved with 10-document queries, giving overall 74.3% correct attribution.

Table 2. Voting results for authorship attribution, showing the number of queries (1-document and 10-document queries) correctly attributed, on the 10k-vote collection, in the top 10, 20, 40, 60, 80, and 100 answers retrieved. There were 700 1-document queries and 70 10-document queries.

		Number of answers retrieved					
Queries	N_q	10	20	40	60	80	100
1-doc	700	357	343	334	346	335	337
10-doc	70	52	55	58	56	55	56

Table 3. Voting-based AA results for each author; for each author there are 100 1-document queries and 10 10-document queries on 10k-vote and 20 1-document queries and 5 10-document queries on 100k-vote. On the 100k-vote collection, for some authors only negligible numbers of correct documents were found; these are shown as *negl.*

		Number correctly attributed / Average correct in top 10						
Collection		Schweid	Currier	Skidmore	Dishneau	Kendall	Crutsinger	Beamish
10k-vote	Q_{1-doc}/100	39/3.2	69/9.2	36/4.4	76/9.8	58/4.8	54/5.5	25/2.7
	Q_{10-doc}/10	8/3.6	10/8.0	1/2.0	10/10.0	10/7.4	10/6.3	3/3.0
100k-vote	Q_{1-doc}/20	*negl.*	14/4.8	*negl.*	15/5.2	8/2.9	*negl.*	*negl.*
	Q_{10-doc}/5	*negl.*	3/7.0	*negl.*	5/7.4	5/4.4	*negl.*	*negl.*

There has been no previous attempt at multi-class AA with more than a few authors. Both the number of authors and the size of the collection are much more substantial than in all previous AA work.

We have observed strong inconsistencies amongst queries based on the work of different authors. Results extracted from top-10 lists are shown in Table 3. As can be observed, queries using documents by Currier and Dishneau are more effective than other queries, not only in accuracy of AA but also in confidence. This observation is consistent with results from previous search experiments.

The confidence is indicated by the average number of correct documents in the top-k ranked list. For instance, on the 10k-vote collection, the 100 1-document queries of Dishneau can be correctly attributed at 76% accuracy, providing around 98% confidence. Note that, unsurprisingly, the effectiveness of attribution for the 10-document queries is generally better than for the 1-document queries.

We also tested the proposed method on the 100k-vote collection, which has over 2000 known authors. This experiment is much less successful, with near-zero accuracy in four of the seven cases. Interestingly, these failures correspond to the results of lower confidence on the 10k-vote collection. For queries based on documents by Currier and Dishneau, the attribution accuracies are respectively 70% and 75%, suggesting 48% and 52% confidence. Again, use of 10-document queries leads to greater effectiveness. However, it can be seen that AA on large collections with large numbers of authors remains a challenge.

6 Conclusion

We have explored the novel task of authorship search. Our proposal is that simple entropy-based statistics and characterization of documents by distributions of style markers can be used to find documents by an author, given some training documents by that author.

Our experiments show that such a method can be highly successful for collections of moderate size. The proposed similarity measure, the Kullback-Leibler divergence, which is used to compute relative entropy, is far more effective than standard measures drawn from information retrieval. As style markers, both function words and part-of-speech tags are effective; for large collections, combined use of both kinds of marker led to even better results. Reasonable effectiveness can be achieved on collections of even half a million documents.

To our knowledge our approach is the first that is able to search a large collection for documents written by a particular author. The success of the method is highlighted by the fact that we have used experimental data, newswire articles, that we regard as challenging for this task: in contrast to material drawn from sources such as literature, we would not expect human readers to be aware of strong stylistic differences between the authors.

The proposed search approach can also be applied to author attribution. Previous methods struggle to correctly attribute authorship when given more than a few hundred documents or more than a few authors. Our method has reasonable accuracy with 10,000 documents and several hundred authors. While it did not successfully scale further in our experiments, this approach is nonetheless much more effective than previous methods and is a clear demonstration that authorship attribution can be applied on realistic collections.

References

1. H. Baayen, H. V. Halteren, A. Neijt, and F. Tweedie. An experiment in authorship attribution. *6th JADT*, 2002.
2. R. Baeza-Yates and B. Ribeiro-Neto. *Modern Information Retrieval.* Addison-Wesley Longman, May 1999.
3. R. Bekkerman, R. El-Yaniv, N. Tishby, and Y. Winter. Distributional word clusters vs. words for text categorization. *J. Mach. Learn. Res.*, 3:1183–1208, 2003.
4. D. Benedetto, E. Caglioti, and V. Loreto. Language trees and zipping. *The American Physical Society*, 88(4), 2002.
5. J. N. G. Binongo. Who wrote the 15th book of Oz? an application of multivariate statistics to authorship attribution. *Computational Linguistics*, 16(2):9–17, 2003.
6. S. F. Chen and J. Goodman. An empirical study of smoothing techniques for language modeling. In A. Joshi and M. Palmer, editors, *Proc. 34th Annual Meeting of the Association for Computational Linguistics*, pages 310–318. Morgan Kaufmanns, 1996.
7. W. B. Croft and J. Lafferty. *Language Modeling for Information Retrieval.* Kluwer Academic Publishers, Norwell, MA, USA, 2003.
8. J. Diederich, J. Kindermann, E. Leopold, and G. Paass. Authorship attribution with support vector machines. *Applied Intelligence*, 19(1-2):109–123, 2003.
9. J. Goodman. Extended comment on language trees and zipping, 2002.
10. D. Harman. Overview of the second text retrieval conf. (TREC-2). *Information Processing & Management*, 31(3):271–289, 1995.
11. D. Heckerman, D. Geiger, and D. Chickering. Learning bayesian networks: the combination of knowledge and statistical data. *Machine Learning*, 20:197–243, 1995.
12. D. Hiemstra. Term-specific smoothing for the language modeling approach to information retrieval: the importance of a query term. In *Proc. 25th ACM SIGIR Conf. on Research and Development in Information Retrieval*, pages 35–41. ACM Press, 2002.
13. D. I. Holmes, M. Robertson, and R. Paez. Stephen Crane and the New York Tribune: A case study in traditional and non-traditional authorship attribution. *Computers and the Humanities*, 35(3):315–331, 2001.
14. D. L. Hoover. Statistical stylistics and authorship attribution: an empirical investigation. *Literary and Linguistic Computing*, 16:421–444, 2001.

15. P. Juola and H. Baayen. A controlled-corpus experiment in authorship identification by cross-entropy. *Literary and Linguistic Computing*, 2003.
16. A. Kaster, S. Siersdorfer, and G. Weikum. Combining text and linguistic doument representations for authorship attribution. In *SIGIR workshop: Stylistic Analysis of Text For Information Access*, August 2005.
17. D. V. Khmelev and F. Tweedie. Using markov chains for identification of writers. *Literary and Linguistic Computing*, 16(4):229–307, 2002.
18. M. Koppel and J. Schler. Exploiting stylistic idiosyncrasies for authorship attribution. In *Exploiting Stylistic Idiosyncrasies for Authorship Attribution. In IJCAI'03 Workshop on Computational Approaches to Style Analysis and Synthesis*, 2003.
19. M. Koppel and J. Schler. Authorship verification as a one-class classification problem. In *Proc. 21st Int. Conf. on Machine Learning*. ACM Press, 2004.
20. O. Kurland and L. Lee. Corpus structure, language models, and ad hoc information retrieval. In *Proc. 27th ACM SIGIR Conf. on Research and Development in Information Retrieval*, pages 194–201. ACM Press, 2004.
21. Y. S. Lai and C. H. Wu. Meaningful term extraction and discriminative term selection in text categorization via unknown-word methodology. *ACM Transactions on Asian Language Information Processing*, 1(1):34–64, 2002.
22. D. D. Lewis, Y. M. Yang, T. G. Rose, and F. Li. Rcv1: A new benchmark collection for text categorization research. *J. Mach. Learn. Res.*, 5:361–397, 2004.
23. B. Scholkopf and A. J. Smola. *Learning with Kernels: Support Vector Machines, Regularization, Optimization and Beyond.* MIT Press, 2002.
24. K. Spark Jones, S. Walker, and S. E. Robertson. A probabilistic model of information retrieval: development and comparative experiments. *Inf. Process. Manage*, 36(6):779–840, 2000.
25. E. Stamatatos, N. Fakotakis, and G. Kokkinakis. Computer-based authorship attribution without lexical measures. *Computers and the Humanities*, 35(2):193–214, 2001.
26. Y. M. Yang. A study on thresholding strategies for text categorization. In *Proc. 24th ACM SIGIR Conf. on Research and Development in Information Retrieval*, pages 137–145. ACM Press, 2001.
27. C. X. Zhai and J. Lafferty. A study of smoothing methods for language models applied to information retrieval. *ACM Transaction on Information System*, 22(2):179–214, 2004.
28. Y. Zhao and J. Zobel. Effective authorship attribution using function word. In *Proc. 2nd AIRS Asian Information Retrieval Symposium*, pages 174–190. Springer, 2005.
29. Y. Zhao and J. Zobel. Search with style: authorship attribution in classic literature. In *Proc. 30th ACSC Thirtieth Australasian Computer Science Conference*, page to appear. ACM Press, 2007.
30. Y. Zhao, J. Zobel, and P. Vines. Using relative entropy for authorship attribution. In *Proc. 3rd AIRS Asian Information Retrieval Symposium*, pages 92–105. Springer, 2006.
31. J. Zobel and A. Moffat. Exploring the similarity space. *ACM SIGIR Forum*, 32(1):18–34, Spring 1998.
32. J. Zobel and A. Moffat. Inverted files for text search engines. *ACM Computing Surveys*, 38:1–56, 2006.

Use of Topicality and Information Measures to Improve Document Representation for Story Link Detection

Chirag Shah* and Koji Eguchi**

National Institute of Informatics (NII)
Tokyo 101-8430, Japan
{chirag,eguchi}@nii.ac.jp

Abstract. Several information organization, access, and filtering systems can benefit from different kind of document representations than those used in traditional Information Retrieval (IR). Topic Detection and Tracking (TDT) is an example of such a domain. In this paper we demonstrate that traditional methods for term weighing does not capture topical information and this leads to inadequate representation of documents for TDT applications. We present various hypotheses regarding the factors that can help in improving the document representation for Story Link Detection (SLD) - a core task of TDT. These hypotheses are tested using various TDT corpora. From our experiments and analysis we found that in order to obtain a *faithful* representation of documents in TDT domain, we not only need to capture a term's importance in traditional IR sense, but also evaluate its *topical* behavior. Along with defining this behavior, we propose a novel measure that captures a term's importance at the corpus level as well as its discriminating power for topics. This new measure leads to a much better document representation as reflected by the significant improvements in the results.

1 Introduction

Document representation is one of the most common and crucial stages of an information organization and access system. Several methods and models of document representation have been proposed based on the target application. Examples include vector space representations [1], probabilistic language models [2], graph-based [3,4], etc. Some of them are general enough to be applicable to almost any IR-based application. However, some tasks demand a different approach to document representation. Topic Detection and Tracking (TDT) [5] is one such domain. In this paper we analyze the peculiarities of TDT and propose a novel approach for document representation. In particular, we focus on term weighing and use Story Link Detection (SLD), a core task of TDT, as the target application.

We can identify three major components in an IR implementation on the system side:

1. The units of representation
2. The method of weighing these units
3. Matching criteria between a query and documents or document and document

* Now with University of North Carolina at Chapel Hill, USA.

** Now also with Kobe University, Japan.

G. Amati, C. Carpineto, and G. Romano (Eds.): ECIR 2007, LNCS 4425, pp. 393–404, 2007.

There have been a few studies focusing on the first of these components that demonstrated that using named entities as units of representation is a good idea for TDT applications [6,7,8]. Some studies suggested using noun phrases along with named entities [9]. There are also many popular methods for weighing the terms such as TF [10], TFIDF [1], etc. A considerable amount of work has also been done on comparing two streams of text that includes cosine similarity [1], information-theoretic distance [11,12], etc. In this paper our focus is on the second aspect - the ways to weigh the given units. Therefore, throughout our experiments we will use the same units (all the words), and the same matching criteria (cosine similarity of two texts).

We started this work by using some traditional IR methods for document representation for SLD task. As we moved from one method of term weighing to another, we realized their shortcomings. Our experiments and analysis showed that none of the traditional methods captured the *topical* information to weigh the terms, which we found was essential for TDT domain. This led us into investigating a better way to capture and incorporate such information. The main contributions of our work reported here are this realization and the methods that emerged from it.

The rest of the paper is organized as the following. In the next section we provide some background of TDT domain and more details of SLD task. This section highlights the uniqueness of TDT and motivates for considering non-traditional measures for term weighing. In section 3 we present our hypotheses regarding term weighing and describe the systems that we implemented to test these hypotheses. The results obtained from TDT corpora are shown and a unique approach that combines a term's *usual* importance as well as its *topical* significance is proposed. Additional results and analysis are given in section 4. We finally conclude the paper with some pointers to the future work in section 5.

2 Background

In this section we review the TDT research and provide more specifics of SLD task. We also show the uniqueness of TDT tasks and contrast with traditional IR systems. With this explanation, we try to motivate the use of topical information to represent the news stories.

2.1 TDT

The Topic Detection and Tracking (TDT) research has provided a standard platform for addressing event-based organization of broadcast news and evaluating such systems [5]. The governing motivation behind such research was to provide a core technology for a system that would monitor broadcast news and alert an analyst to new and interesting events happening around the world. The research program of TDT focuses on five tasks: story segmentation, first story detection, cluster detection, tracking, and story link detection. Each is viewed as a component technology whose solution will help address the broader problem of event-based news organization. The details of each of these tasks are beyond the scope of this paper. We, instead, want to focus on the general characteristics of TDT and specific details of the story link detection task.

To appreciate the unique nature of TDT, it is important to understand the notion of a topic. In TDT, a topic is defined to be a set of news stories that are strongly related by some seminal real-world event. For instance, when hurricane Katrina hit the coast, it became the seminal event that triggered a new topic. The stories that discussed the origin of this hurricane, the damage it did to the places, the rescue efforts, etc. were all parts of the original topic. Stories on another hurricane (occurring in the same region or time) could make up another topic. This shows an important contrast with typical IR. Along with hurricane Katrina, a query "hurricane" will bring up the documents about other hurricane related events. On the other hand, for the query "hurricane Katrina", some of the stories that followed the original event of hurricane Katrina may not be considered as "about" hurricane Katrina by the traditional IR measures and would be ranked very low in the retrieval set.

This contrast indicates that the notion of an event-based topic is narrower than a subject-based topic; it is built upon its triggering event. Hereafter, we focus on dealing with an event-based topic rather than a subject-based. A typical *topic* would have a start time and it would fade off from the news at some point of time. Since it is a specific event, it happened not only at some particular time, but in a specific location, and usually with an identifiable set of participants. In other words, a topic is well defined in scope and possibly in time. In this paper we would ignore the temporal nature of a topic and just focus on the scope.

2.2 SLD

The Story Link Detection task evaluates a TDT system that detects if two stories are "linked" by the same event. For TDT, two stories are linked if they discuss the same event. Unlike other TDT tasks, link detection was not motivated by a hypothetical application, but rather the task of detecting when stories are linked is a "kernel" function from which the other TDT tasks can be built.

The Uniqueness of SLD. For this task, a set of pairs of documents are given to compare with each other and we need to declare if they are on the same *topic*, i.e., event-based topic as defined in section 2.1. It is important to note that although this might seem similar to finding the document similarity, there is a fine difference here. It is possible that two documents do not share many common terms but belong to the same topic. It is also possible that two documents may have several terms that match with one another, but they talk about different topics. We illustrate this with an example. Figure 1 shows excerpts from two stories of different topics. The matching terms between them are highlighted. If we run any regular term-based representation for these documents, we are likely to get a match while comparing them. However, we can see that they are actually talking about different events and therefore, not on the same topic.

Evaluation. The performance of any TDT system is characterized in terms of the probabilities of missed detection (P_{Miss}) and false alarm errors (P_{Fa}) [13]. These error probabilities are linearly combined into a single detection cost, C_{Det}, by assigning costs to missed detection and false alarm errors and specifying an *a priori* probability of a target. The resulting formula is

$$C_{Det} = (C_{Miss} * P_{Miss} * P_{Target} + C_{Fa} * P_{Fa} * (1 - P_{Target})) \tag{1}$$

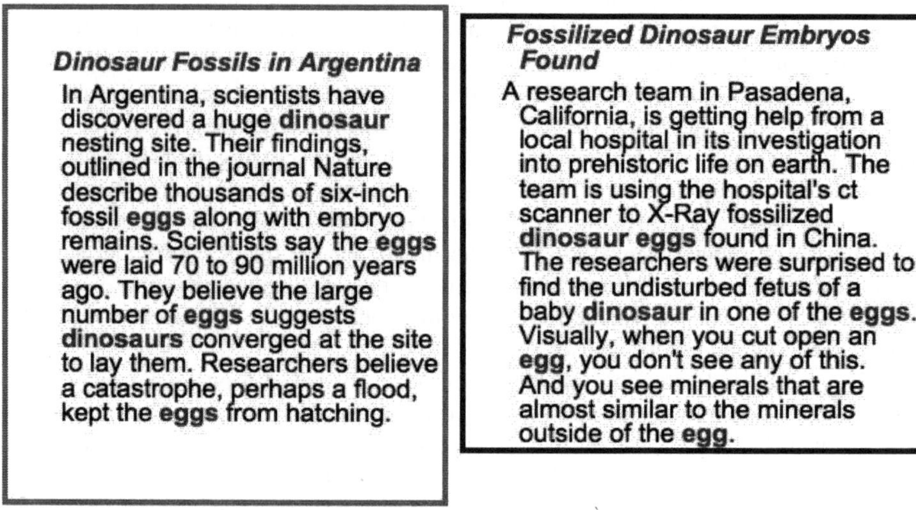

Dinosaur Fossils in Argentina

In Argentina, scientists have discovered a huge dinosaur nesting site. Their findings, outlined in the journal Nature describe thousands of six-inch fossil eggs along with embryo remains. Scientists say the eggs were laid 70 to 90 million years ago. They believe the large number of eggs suggests dinosaurs converged at the site to lay them. Researchers believe a catastrophe, perhaps a flood, kept the eggs from hatching.

Fossilized Dinosaur Embryos Found

A research team in Pasadena, California, is getting help from a local hospital in its investigation into prehistoric life on earth. The team is using the hospital's ct scanner to X-Ray fossilized dinosaur eggs found in China. The researchers were surprised to find the undisturbed fetus of a baby dinosaur in one of the eggs. Visually, when you cut open an egg, you don't see any of this. And you see minerals that are almost similar to the minerals outside of the egg.

Fig. 1. Example of two stories on different topics getting a match with a word-based technique

where P_{Miss} = No. of missed detection/No. of targets, P_{Fa} = No. of false alarms/No. of non-targets, C_{miss} and C_{Fa} are the costs of a missed detection and a false alarm, respectively, and are pre-specified, P_{Target} is the *a priori* probability of finding a target.

This detection cost is then normalized as given below.

$$(C_{Det})_{Norm} = \frac{C_{Det}}{MIN(C_{Miss} * P_{Target}, C_{Fa} * (1 - P_{Target}))} \quad (2)$$

Recent Work on SLD. Although SLD is considered to be the core technology for almost all of the tasks in TDT, not much has been done specifically for SLD. The efforts have rather been focused on more practical applications such as new event detection and tracking. The following list enumerates some of the latest trends as demonstrated by various sites for SLD task in TDT 2004 [10].[1]

- *Vector Space Model.* This was used by CMU1, NEU1, UIowa4, and UMass2 systems. NEU used TF weighting unlike more popular TFIDF weighting. CMU's system was same as their 2002 submission with re-tuned decision thresholds. CMU also submitted two variations of their baseline: HIGHP submission, which manually adjusts the decision thresholds of the baseline system towards higher precision (i.e. lower false alarm rate), and HIGHR submission, which manually adjusts the decision thresholds of the baseline system towards higher recall (i.e. lower miss rate). Another variation was UIowa3 system, which created the vectors using noun phrases.

[1] The systems are identified with run-tags here as per their official NIST submission. For details, see [10].

- *Relevance Models.* UMass1 and UMass3 systems used this technique. Relevance model [11] uses the new story as the query and all the past stories as the corpus [14]. It then performs retrieval and expands the new story with top words. Finally, the similarity between the models of the new story and the training stories (from past stories) is computed using Kullback-Leibler (KL) divergence.
- *Voting Scheme.* CMU's MULTI submission used a majority voting scheme among five separate similarity measures. Three of these measures were similarity ratio, Mountford coefficient, and the Lance-Williams measure. The other two similarity measures were windowed versions of the Mountford coefficient and a BLEU-like symmetric n-gram overlap measure (giving equal weight to unigrams and bigrams).
- *Graph Comparison.* UIowa1 system used Cook-Holder graph similarity on noun phrases for finding the similarity between stories.
- *Edit Distance Based Similarity.* UIowa2 system used normalized Levenshtein edit distance-based similarity for comparison.

3 Hypotheses and Proposed Methods

In this section we present various hypotheses about the factors that can affect the document representation. To be specific, these hypotheses are relating to the term weighing schemes.

3.1 Hypothesis-1: Capturing a Term's Importance at the Document and/or Collection Level Provides a Faithful Representation

Method-1. TFIDF on all the words - Baseline TFIDF based representation of documents is widely used in document similarity [1], document classification [15], and document clustering [16] literature. We adopt this approach as our baseline, which is a typical bag-of-words approach. TF values were found using the following equation.

$$TF(t,d) = \frac{TF\_raw(t,d)}{TF\_raw(t,d) + 0.5 + \frac{1.5*DocLen(d)}{Avg\_DocLen}} \quad (3)$$

where $TF\_raw(t,d)$ is the raw frequency of term t in a given document d, $DocLen(d)$ is the length of the given document, and Avg_DocLen is the average length of the documents. IDF values were found using the following formulation.

$$IDF(t) = log\left(\frac{N+1}{N_t + 0.5}\right) \quad (4)$$

where N is the total number of documents in the corpus and N_t is the number of documents in which term t occurs.

We construct vectors for each document using $TF(t,d) \times IDF(t)$ and then find cosine between two vectors. This score becomes the similarity measurement for the given documents. If this score is above the threshold, then the given pair of stories are said to be on the same topic. Later we shall see how to derive this threshold from the training data.

Method-2. Information Content (IC) Information content (IC) of a term t is found using the following equation.

$$IC(t) = -log_2(P(t|C)) \tag{5}$$

where $P(t|C)$ is the probability of term t occurring in corpus C which is defined as

$$P(t|C) = \frac{freq(t)}{\text{Total number of terms in } C} \tag{6}$$

where $freq(t)$ is the frequency of term t in the given corpus.

Method-3. Pointwise KL (PKL) divergence scores KL divergence or its variations are widely used in language modeling framework [11,12]. In general, it is useful for finding how different two probability distributions are. We conjectured that the further a term's *behavior* is different from the *behavior* of the world, the more useful it is for the representation. In other words, if a term is unique, it is more important than those terms that are not, such as stopwords. This intuition can easily be converted in a language modeling framework that uses pointwise KL divergence [17,18]. The traditional KL divergence, modified for pointwise computation, results in the following formulation.

$$PKL(t,d) = P(t|d)log\left(\frac{P(t|d)}{P(t|C)}\right) \tag{7}$$

where $PKL(t,d)$ is the pointwise KL divergence between document d and collection C with respect to term t, $P(t|d)$ is the probability of term t occurring in a given document d, and $P(t|C)$ is the probability of t occurring in C, which is calculated as given in equation (6). $P(t|d)$ is computed as the following.

$$P(t|d) = \frac{TF(t,d)}{\text{Total number of terms in } d} \tag{8}$$

We implemented these three weighing schemes using TDT2 corpus as training and TDT3 corpus for testing[2]. This means that for method 1, IDF values for all the terms were found using TDT2 (training corpus) and combined with TF values found using TDT3 (testing) to compute the TFIDF scores. For method 2, we found the information content of each term using TDT2 and used it to represent the terms of TDT3. For method 3, $P(t|C)$ values were calculated using TDT2 corpus and $P(t|d)$ using TDT3. The results of these systems are displayed in Figure 2. It is important to note here that the lower the cost is, the better the system is.

3.2 Hypothesis-2: Capturing the Topical Nature of a Term Provides a Better Representation for TDT-Like Applications

We realized that in none of the methods tried before, the information about topics is explicitly captured. It seemed interesting and useful to us to see what would happen if

[2] This is based on the assumption that we cannot know how the whole corpus is, in advance, in the context of TDT.

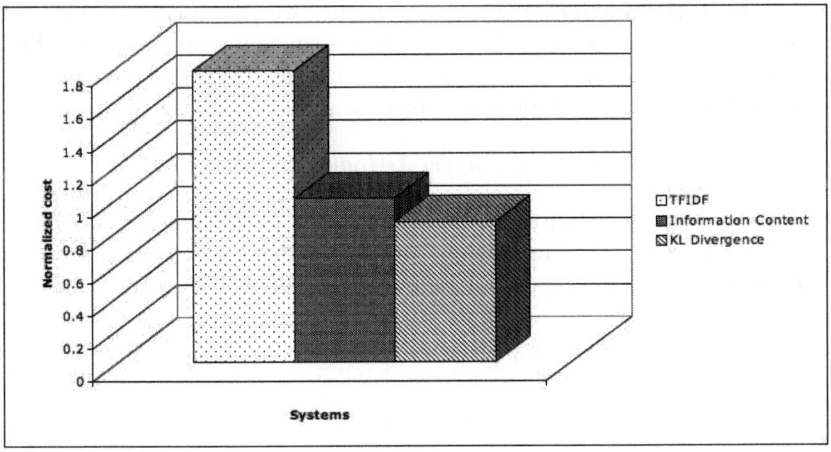

Fig. 2. Comparison of normalized detection cost for various systems and corpora

we incorporate such information while weighing the terms. However, it was not clear to us how exactly we could go about defining and capturing *topicality*. Following are two possible methods of doing so. They are based on some intuition, but by no means the best ways to capture *topicality* of terms.

Method-4. Topicality scores We started with a very simple approach of defining and capturing *topicality*. Our formulation was basically derived from the following intuition.

- In the documents on the same topic, a term that occurs frequently is more useful than the terms that are not very frequent.
- A term that is more frequent in a topic and less frequent in other topics is more useful.

It can be seen that the desired characteristics of a topical terms are very similar to the formulation of TFIDF - the first point is similar to finding TF and the second point is similar to finding IDF. However, there are some basic differences here, which are mainly in point two. In case of normal IDF, we are interested in calculating how frequently the term occurs in the entire corpus, whereas in the formulation that we gave here, we are interested in finding how frequent the term is in the documents of different topics. This is translated in the following formula.

$$\text{Topicality}(t) = (\text{Probability of } t \text{ occurring in a topic}) \tag{9}$$

$$\times(\text{Probability of } t \text{ not occurring in any other topic})$$

$$= max_C \left(P(t|T_i) \cdot \overline{P_{j \neq i}(t|T_j)} \right) \tag{10}$$

$$= max_C \left(P(t|T_i) \cdot (1 - P_{j \neq i}(t|T_j)) \right) \tag{11}$$

Here, T_i is i-th topic. We can compute $P(t|T_i)$ as the following.

$$P(t|T_i) = \frac{\text{Frequency of } t \text{ in the documents of topic } T_i}{\text{Total number of terms in topic } T_i} \tag{12}$$

Method-5. Topical Information Content. Earlier we proposed to use information content of a term as a mean to find its weight in document representation. This information content was measured with respect to the corpus. We now change it slightly to evaluate it with respect to the topics. This new measure is defined below.

$$TIC(t) = max_C\,(-log(P(t|T_i))) \qquad (13)$$

where $P(t|T_i)$ is the probability of term t given topic T_i. Once again, we used TDT2 corpus for training and TDT3 for testing. This means that for method 4, the *Topicality* was computed using TDT2 corpus and used for representing TDT3 terms. Similarly, for method 5, the information content with respect to the topics was found on TDT2 and used for TDT3. The results along with the previous three systems are shown in Figure 3.

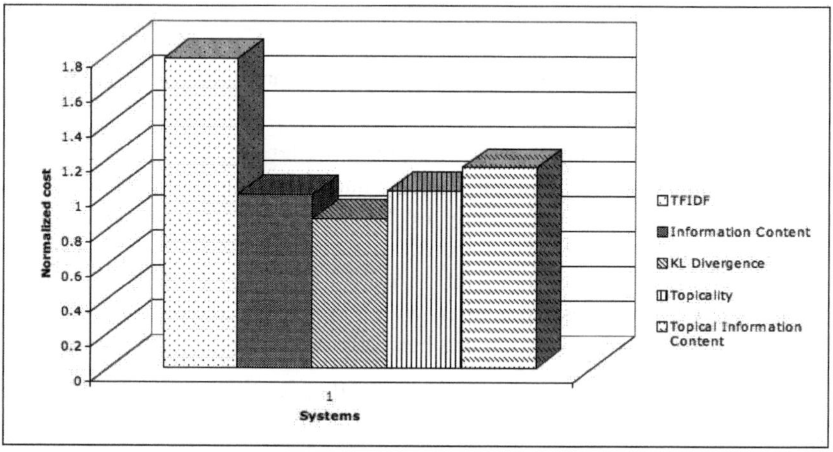

Fig. 3. Comparison of normalized detection cost for various systems and corpora

As we can see from these results that using only *topical* information resulted in better performance over the baseline, but did not give any advantage over the other methods of term weighing that did not use *topical* information. However, if the *importance* of a term in the collection and that in a topic are two relatively orthogonal factors, then we might be able to combine them in a way that would outperform a method that uses only one of them. Thus, we came up with the following hypothesis.

3.3 Hypothesis-3: Combining a Term's Importance at Collection Level and Its Topicality Provides a Better Representation Than That of Either of Them Used Separately

Method-6. Topical KL (TKL) divergence scores. We now approach our pointwise KL (PKL) divergence calculation with topicality in the formulation. Earlier we defined the PKL divergence with respect to the document and the corpus. Now we will measure it with respect to the topic and the corpus. This new formulation is given below.

$$TKL(t) = max_C \left[P(t|T_i) log \left(\frac{P(t|T_i)}{P(t|C)} \right) \right] \tag{14}$$

where T_i is a topic and C is the corpus. Once again, we find the scores for each topic for a given term and take the maximum. While representing a document, each term is weighted by its corresponding topical KL divergence score.

We found both $P(t|T)$ and $P(t|C)$ using our training corpus TDT2 and used these values to compute a term's weight on TDT3 corpus. The result obtained by this system along with all the previous systems is displayed in Figure 4. As we can see, our proposed approach obtains the least cost. The actual values of these costs are given in Table 1.

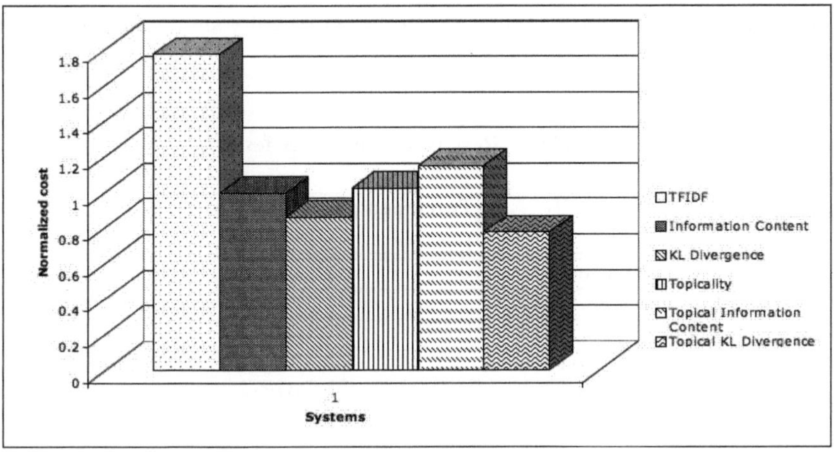

Fig. 4. Comparison of normalized detection cost for various systems and corpora

To measure how significant these changes were, we used standard two-tailed paired t-test (parametric) as well as McNemar's statistical significance test [19] (non-parametric). McNemar's test uses only the number of discordant pairs, that is, the pairs for which we have different decisions in given two techniques. Let us define the pairs that transferred from YES to NO to be R and the pairs that transferred from NO to YES to be S. We then calculated the Chi-Square value using the following equation.

$$\chi^2 = \frac{(|R - S| - 1)^2}{R + S} \tag{15}$$

Since McNemar's test also calculates the p value, we can put these two things together and claim that if there were really no association between the given techniques, there is probability p of chance that the observed odds ratio would be so far from 1.0 (no association) as given by the Chi-Square above.

Table 1. Cost for various systems. Training on TDT2, testing on TDT3

System	Normalized detection cost	Improvement over the baseline
TFIDF scores (baseline)	1.7758	–
Information content	0.9971	43.85%
PKL divergence	0.8570	51.74%
Topicality scores	1.0197	42.58%
Topical information content	1.1526	35.09%
Topical KL scores	0.7749	56.36%

Table 2 shows results of significance tests using these two methods on the basis of the baseline results using TFIDF scores for term weighing. The results indicate that the improvements of our methods can be said to be statistically significant than that of the baseline system.

Table 2. Results of significance tests

System	Two tailed paired t-test p-value	McNemar's test p-value
Information content	0.0000	0.0000
PKL divergence	0.0000	0.0000
Topicality scores	0.0000	0.0000
Topical information content	1.13e-103	7.47e-307
Topical KL scores	0.0000	0.0000

4 Additional Experiments and Analysis

In order to support our hypothesis-3 with additional empirical results, we carried out experiments using TDT4 corpus, too. This corpus is significantly different than TDT2 and TDT3 corpora. The results of our baseline, plain topicality, and topical KL systems with TDT4 corpus are given in Tables 3 and 4. As we can see, using merely topicality scores does worse than the baseline, but our proposed system of topical KL scores still outperforms the baseline. Since the testing corpus differs quite a bit from the training corpora in these experiments, the improvements achieved by our proposed method are not as dramatic as the results reported in Table 1, but they are still significant. Table 5 shows how the improvements given by our topical KL scores are significant comparing with the baseline method using TFIDF scores. The results indicate that the improvements of our method can still be said to be statistically significant than that of the baseline method.

We found these results interesting and tried to do further analysis. It turns out that although the vocabulary of TDT4 is not very different than those of TDT2 and TDT3, the nature of the topics is quite different. Thus, using only topical information from TDT2

Table 3. Training on TDT2, testing on TDT4

System	Normalized Cost	Improvement over the baseline
TFIDF scores (baseline)	1.1104	–
Topicality scores	1.1392	-2.59%
Topical KL scores	1.0383	6.49%

Table 4. Training on TDT3, testing on TDT4

System	Normalized Cost	Improvement over the baseline
TFIDF scores (baseline)	0.9254	–
Topicality scores	0.9915	-7.14%
Topical KL scores	0.8983	2.93%

Table 5. Results of significance tests

Training	Testing	Two tailed paired t-test p-value	McNemar's test p-value
TDT2	TDT4	0.0000	0.0000
TDT3	TDT4	0.0000	2.46e-05

or TDT3 on TDT4 corpus hurts the performance. On the other hand, our proposed system that uniquely combines the topical as well as the overall corpus information still does the best. This indicates the robustness of our proposed approach.

5 Conclusion

In this paper we presented a novel approach for term weighing that incorporated a term's importance at the collection level at the same time capturing its topicality. With our analysis and experiments, we showed that traditional IR techniques of term weighing do not consider the topical information, which is essential for TDT tasks. We selected SLD as our target application, which is a core task of TDT. Through a set of hypotheses testing, experiments, and analysis we realized that while traditional IR methods of term weighing do not capture topical information explicitly, the information that they provide is still very useful. We then proposed a unique way of combining topical information with the information that a traditional IR term weighing scheme provides. This method consistently outperformed the baseline across all the TDT corpora. Another advantage of this model is that since it is based on well-studied information theoretic and probabilistic frameworks, it becomes easier and more effective to analyze and understand it.

The experiments reported here are done for SLD. However, since SLD is at the core of TDT, we conjecture that our proposed methods, which gave significant improvements, should help in other TDT tasks as well.

References

1. Salton, G., ed.: Automatic Text Processing: The Transformation, Analysis and Retrieval of Information by Computer. Addison-Wesley (1989)
2. Ponte, J.M., Croft, W.B.: A language modeling approach to information retrieval. In: Research and Development in Information Retrieval. (1998) 275–281
3. Tomita, J., Hayashi, Y.: Improving Effectiveness and Efficiency of Web Search by Graph-based Text Representation. In: Proceedings of The Ninth World Wide Web Conference. (2000)
4. Collins-Thompson, K., Callan, J.: Query expansion using random walk models. In: CIKM. (2005)
5. Allan, J., ed.: Topic Detection and Tracking. Kluwer Academic Publishers (2002)
6. Allan, J., Jin, H., Rajman, M., Wayne, C., Gildea, D., Lvrenko, V., Hoberman, R., Caputo, D.: Topic-based novelty detection. Technical report, Center for Language and Speech Processing, John Hopkins University (1999)
7. Kumaran, G., Allan, J.: Text classification and named entities for New Event Detection. In: Proceedings of ACM SIGIR. (2004) 297–304
8. Shah, C., Croft, W.B., Jensen, D.: Representing Documents with Named Entities for Story Link Detection (SLD). In: CIKM. (2006)
9. Eichmann, D.: Experiments with tracking / detection / etc. using entities and noun phrases. Technical report, University of Iowa (2001)
10. Fiscus, J., Wheatley, B.: Overview of the tdt 2004 evaluation and results. Technical report, NIST (2004)
11. Lavrenko, V., Croft, W.B.: Relevance-based language models. In: Proceedings of the 24th annual international ACM SIGIR conference. (2001) 120–127
12. Zhai, C., Lafferty, J.: Model-based feedback in the language modeling approach to information retrieval. In: Proceedings of the 10th International Conference on Information and Knowledge Management. (2001) 403–410
13. Fiscus, J.G., Doddington, G.R.: Topic detection and tracking evaluation overview. In Allan, J., ed.: Topic Detection and Tracking. Kluwer Academic Publishers (2002) 17–32
14. Lavrenko, V., Allan, DeGuzman, J.E., LaFlamme, D., Pollard, V., Thomas, S.: Relevance models for topic detection and tracking. In: Proceedings of Human Language Technologies Conference. (2002) 104–110
15. Lewis, D.D.: Evaluating and optmizing autonomous text classification systems. In Fox, E.A., Ingwersen, P., Fidel, R., eds.: Proceedings of SIGIR-95, 18th ACM International Conference on Research and Development in Information Retrieval, Seattle, US, ACM Press, New York, US (1995) 246–254
16. Kanungo, T., Mount, D.M., Netanyahu, N., Piatko, C., Silverman, R., Wu, A.Y.: An efficient k-means clustering algorithm: Analysis and implementation. IEEE Trans. Pattern Analysis and Machine Intelligence **24** (2002) 881–892
17. Tomokiyo, T., Hurst, M.: A Language Model Approach to Keyphrase Extraction. In: Proceedings of the ACL 2003 Workshop on Multiword Expressions: Analysis, Acquisition and Treatment. (2003) 33–40
18. Kelly, D., Diaz, F., Belkin, N.J., Allan, J.: A User-Centered Approach to Evaluating Topic Models. In: Proceedings of the 26th European Conference on Information Retrieval (ECIR 2004). (2004) 27–41
19. Gillick, L., Cox, S.: Some Statistical Issues in the Comparison of Speech Recognition Algorithms. In: Proceedings of IEEE's ICASSP 1989. (1989) 532–535

Ad Hoc Retrieval of Documents with Topical Opinion

Jason Skomorowski and Olga Vechtomova

University of Waterloo, 200 University Avenue West, Waterloo, ON, Canada, N2L 3G1
{jcskomor, ovechtom}@uwaterloo.ca

Abstract. With a growing amount of subjective content distributed across the Web, there is a need for a domain-independent information retrieval system that would support ad hoc retrieval of documents expressing opinions on a specific topic of the user's query. In this paper we present a lightweight method for ad hoc retrieval of documents which contain subjective content on the topic of the query. Documents are ranked by the likelihood each document expresses an opinion on a query term, approximated as the likelihood any occurrence of the query term is modified by a subjective adjective. Domain-independent user-based evaluation of the proposed method was conducted, and shows statistically significant gains over the baseline system.

1 Introduction

Users searching for information on the Web may have more complex information needs than simply finding any documents on a certain subject matter. For instance they may want to find documents containing other people's opinions on a certain topic, e.g. product reviews, as opposed to documents with objective content, such as technical specifications. In this work we address the problem of ad hoc retrieval of documents that express opinion on a specific topic. There exist a large number of documents with opinionated content on the Web, however they are scattered across multiple locations, such as individual websites, Usenet groups and web logs ("blogs"). If a person wants to find opinions on a certain subject they have to go to specific websites which might contain such content, for instance, IMDb for film reviews or Amazon for the reviews of books and CDs. Alternatively, they may add words with subjective connotation, such as "review" and "opinion", to their queries. However, it is obvious that only a small fraction of documents expressing opinion on a topic would actually contain words such as "review" or "opinion". There is a clear need for a domain-independent search engine that would support ad hoc retrieval of documents containing opinion about the topic expressed in the user's query. This paper sets to fill this need by proposing a domain-independent method for ad hoc retrieval of documents containing opinion about a query topic.

We propose a lightweight method for ad hoc retrieval of documents which express subjective content about the topic of the query. Documents are ranked by the likelihood each document expresses an opinion on a query term, approximated as the likelihood the query term occurrences in a document are modified by subjective adjectives. For our experiments we use a manually constructed list of subjective adjectives, proposed in [1]. Our method calculates the probability of a noun at a certain distance from an adjective being the target of that adjective. Probabilities at different

G. Amati, C. Carpineto, and G. Romano (Eds.): ECIR 2007, LNCS 4425, pp. 405–417, 2007.

distances are precomputed using a parsed training corpus. As part of our approach we have also developed a method of locating a noun modified by an adjective (i.e. resolving an adjective target), which demonstrated high accuracy in our evaluation.

While many elements of language can be used to express subjective content, adjectives are one of the major means of expressing value judgement in English. Our approach of using adjectives as markers of subjective content targeted at the concept expressed in the query relies on the assumption that users frequently want to find opinions about a single entity, such as a product, person, company, travel destination, activity, etc. Such an entity is typically expressed as a noun, a noun phrase or a gerund (a verb with -ing suffix which can act as a noun), and consequently queries of this type consist of either a single query term or a phrase. While it is true that users may be interested in opinions about more complex subjects, such as "The effect of global warming on the environment", opinions on such subjects are likely to be expressed by a greater diversity of more complex language structures (clauses, entire sentences or even paragraphs), and therefore require more sophisticated discourse processing tools. These types of queries are outside the scope of the current work.

In this work we also propose a method of topical opinion ranking by the likelihood a document expresses opinions on the *collocates* of the query terms, i.e. words strongly associated with them in the corpus. The rationale is that an author may express opinion about an entity indirectly, by referring to its related concepts, such as parts or attributes of the car as opposed to the car itself.

The proposed approach is well suited to real-time document retrieval: the computationally expensive task of resolving adjective targets in the training corpus and calculating probabilities of subjective adjectives modifying nouns at various distances is done once at pre-search time, whereas at search time the system only needs to find instances of query terms and subjective adjectives, as well as distances between them.

The rest of the paper is organised as follows: in section 2 we review related work, in section 3 we describe our methodology, including adjective target resolution algorithms and document ranking methods. Section 4 presents evaluation, section 5 discusses the evaluation results, and section 6 concludes the paper and outlines future research directions.

2 Related Work

Although sentiment and subjective language represent a growing research area, work on identifying language that is both subjective and on topic is limited. Hurst and Nigam [2] propose a method of identifying sentences that are relevant to some topic and express opinion on it. First, to determine if a document is relevant to a topic, they use a machine learning approach, trained on hand-labeled documents, and if the classifier predicts the whole document as topically relevant, they apply the same classifier to predict topical relevance of each sentence. For the sentences predicted topically relevant, they apply sentiment analyser, which relies on a set of heuristic rules and a hand-crafted domain-specific lexicon of subjective words, marked with polarity (positive or negative). Yi et al. [3] propose to extract positive and negative opinions about specific features of a topic. By feature terms they mean terms that have either a part-of or attribute-of relationships with the given topic or with a known feature of the topic. Their method first determines candidate feature terms based on structural heuristics then narrows the selection using either the mixture language model, or the log-likelihood ratio. A pattern-dependent comparison is then made to a sentiment lexicon gathered from a variety of linguistic resources.

There exists a larger body of research directed towards document classification by sentiment polarity [4, 5, 6, 7]. The focus of these works is on classifying reviews as either positive or negative. A review can be viewed as an example of topical subjectivity with the writer's opinion being a subjective expression on the topic of the item being reviewed. Pang et al. [6] evaluate several machine learning algorithms to classify film reviews as either containing positive or negative opinions. Dave et al. [4] propose and evaluate a number of algorithms for selecting features for document classification by positive and negative sentiment using machine learning approaches. Turney [7] proposes an unsupervised algorithm for classifying reviews as positive or negative. He proposes to identify whether a phrase in a review has a positive or negative connotation by measuring its mutual information with words "excellent" and "poor". A review's polarity is predicted from the average semantic orientation (positive or negative) of the phrases it contains. The method, evaluated on 410 reviews from Epinions in four different domains, showed accuracy between 66% and 84% depending on the domain. Hu and Liu [5] developed a method of identifying frequent features of a specific review item, and finding opinion words from reviews by extracting adjectives most proximate to the terms representing frequent features. This paper is most closely related to our approach because of its use of adjective proximity.

3 Methodology

In order to determine whether an opinion is given on a topic, we need not only to identify subjectivity in the document, but determine if that subjectivity is being directed at the topic in question. Adjectives have often been defined in terms of their use as a direct noun modifier, and while Baker favours a more general definition for his crosslinguistic study, he agrees that this generalisation holds across "a great many languages" [8]. Not only do adjectives tend to have clear targets, they also are one of the primary means of expressing opinions. While the role played by the adjective can vary widely between languages, value judgement is among the four core semantic types associated with this part of speech [9]. Support for this is found in a study by Bruce and Wiebe which shows the presence of adjectives correlates with subjectivity [10].

The general approach of our work is to rank documents by the likelihood that a document expresses an opinion on a query term, approximating it as the likelihood that the query term occurrences in a document are modified by subjective adjectives. Instead of applying syntactic parsing at search time in order to determine whether a query term instance is the target of a subjective adjective in a document, which is computationally expensive, we instead chose to use a training corpus with marked adjective targets to calculate probabilities that each position outstanding from a subjective adjective contains its target noun. At search time we only have to determine the distance between an instance of the query term and the nearest subjective adjective, and look up the probability that the adjective modifies a noun at this distance. The document score is then calculated as the sum of such probabilities. For this approach we need: a list of subjective adjectives; positional information of index terms in documents; the probability that an adjective modifies a noun at a given distance from it; a corpus where adjectives and their targets are marked for calculating such probabilities.

A list of subjective adjectives can be created manually or automatically, for example, using machine learning techniques. In our work we used a list of 1336 subjective adjectives manually composed by Hatzivassiloglou and McKeown [1]. There also exist many automatic methods of learning subjective language [e.g., 11, 12], which can be used instead. Positional information of index terms in a document is recorded in a typical IR system's index, and therefore is easily obtainable. To calculate the probability that an adjective modifies a noun at a certain distance we need a corpus with marked adjectives and their targets. Such corpus, however, is not available. The method of resolving adjective targets also does not exist. Therefore we developed our own method of resolving adjective targets, which is presented in Section 3.1.

3.1 Resolving Adjective Targets in English

English adjectives are characteristically used either attributively or predicatively [13]. Attributive usage is where a noun is modified directly, typically premodified (e.g., the blue sky). Predicative usage links the adjective to the subject with a copular verb such as "be" (e.g., the sky is blue). Other, less frequent constructions include objective complements of verbs, such as "make" and "prove" (e.g., made the sky blue), resultative secondary predicates [8] (e.g., dyed the sky blue), and degree phrases [14] (e.g., blue as the sky; more blue than the sky).

Since we do not require maximum precision for our application, we will focus our target resolution on only the most frequent usages, attributive and predicative. For identifying resultative secondary predicates we need to have a list of verbs that can be used in such constructs, which is unavailable. Determining the specifics of other usages of adjectives is complicated by the numerous syntactic applications of "as", "than", "make" and other words involved in these constructs.

In order to identify what part of a sentence is being modified by a given adjective, syntactic information is needed. For our approach, we need to know the part of speech (POS) of words and the boundaries of noun phrases, therefore we require a POS tagger and a parser. After evaluating a variety of tools, the SNoW Shallow Parser [15] was found to have a good balance of precision and speed.

3.1.1 Resolving Attributive Use of Adjectives

In the attributive case, a noun phrase to which the adjective refers is the one containing it. In order to determine noun phrase boundaries we use the parser. Manually examining a random sample of 200 subjective adjectives used attributively, we found that the parser fails to find appropriate phrase boundaries in 6.5% of these instances. Most errors involve the parser ending the noun phrase because it has mistagged a noun usage as verb, or erroneously saw an adjective where none exists. A notable limitation of this approach is that it does not account for other noun phrases potentially modified by the adjective via coordinate conjunctions, prepositional phrases, and other constructs. However, it is difficult to identify the correct target in such constructs without the knowledge of their meaning, as demonstrated by the following examples:

- Sudbury is famous for its colourful culture and people. (the people are colourful);
- The artist uses colourful pastels and charcoal. (the charcoal is not colourful);
- A delicious bowl of ice cream. (the ice cream is delicious);
- A ridiculous amount of pasta. (the pasta is not ridiculous).

3.1.2 Resolving Predicative Use of Adjectives

If an adjective occurs outside of a noun phrases, it is likely to be used predicatively. In this case we then read back from the adjective to see if there is a copular verb[1] present before it and, if so, assume the preceding noun phrase to be the subject of that verb and thus predicatively modified by the adjective in question. We employ a variety of measures to improve the accuracy of this approach:

- Only cases where the parser tags the copular verb as actually being used as a verb are considered.
- Clauses delimited from the verb by commas are bypassed when searching for the subject (e.g. The ice-cream, prepared fresh this afternoon, is delicious).
- Situations where there is an intervening noun between the adjective and copular verb are not counted as a predicative construct, because it is most likely that the adjective is used as an objective complement of a different verb (e.g., The ice-cream is made with strawberries and is quite delicious).
- Noun phrases preceded with prepositions are skipped when looking for a potential target as these form a prepositional phrase and are not the subject of the link verb (e.g., The ice-cream in the fridge is old.).

The evaluation of the predicative adjective target resolution algorithm was conducted on the random sample of 200 subjective adjectives used predicatively in the AQUAINT corpus. The target noun phrase was identified correctly in the 86% of cases. The 11% of errors were due to the parser error. One frequent cause of the parser error was that contractions of "not" such as "wasn't" and "didn't" were erroneously tagged as nouns. Only 3% of the errors were caused by our method. While some potential heuristics present themselves, further refinement will be left to later work as additional precision is not necessary to explore search applications and is made irrelevant by parser error.

3.2 Statistics on Adjective Usage

Using the above method and a corpus of text, we can calculate the probability of a noun being the target of an adjective at a certain distance from it. A noun is considered to be the target of an adjective when it is the head of the noun phrase that the adjective modifies as determined by the method described in Section 3.1. We consider the last noun in the noun phrase as the head.

The probability P_i that a noun is the target of (i.e. modified by) an adjective at distance i is calculated according to Eq. 1:

$$P_i = \frac{T_i}{K_i} \quad (1)$$

---

[1] We used a list of copular verbs from [16].

Where: T_i – the total number of nouns which are targets of any subjective adjective separated by distance i; K_i - total number of nouns separated by distance i from a subjective adjective.

For example, in calculating the probability P_1 that a noun is the target of an adjective which immediately precedes it in text (i.e. noun is located in position 1 relative to the adjective), "deep sea" would count towards T_1 because the adjective "deep" modifies the noun "sea". On the other hand, "exciting sea adventure" would not count towards T_1, because the adjective "exciting" does not modify "sea", but "adventure", which is the head of the noun phrase "sea adventure". Both examples would count towards K_1, because in both of them we have a noun immediately preceded by an adjective.

Table 1. Probabilities of a noun being modified by an adjective at different distances

Distance (i) of noun from adjective	All adjectives			Subjective adjectives		
	Proper nouns	Common nouns	All nouns	Proper nouns	Common nouns	All nouns
-10	0.0026	0.0011	0.0012	0.007	0.0024	0.0026
-9	0.003	0.0016	0.0017	0.0084	0.0033	0.0036
-8	0.0037	0.0021	0.0022	0.0098	0.0048	0.0051
-7	0.0052	0.0031	0.0032	0.0141	0.0068	0.0072
-6	0.0073	0.0045	0.0047	0.0194	0.01	0.0105
-5	0.0112	0.0065	0.0069	0.031	0.0147	0.0156
-4	0.0206	0.0105	0.0112	0.061	0.025	0.027
-3	0.0414	0.0218	0.0232	0.1265	0.0545	0.0585
-2	0.0568	0.0294	0.0313	0.1657	0.0712	0.0765
-1	0.0077	0.0029	0.0033	0.0068	0.0014	0.0017
1	0.331	0.6689	0.6451	0.1971	0.5886	0.5666
2	0.1775	0.1741	0.1743	0.1283	0.1517	0.1504
3	0.1761	0.0489	0.0579	0.1133	0.04	0.0441
4	0.0911	0.0143	0.0197	0.0441	0.0123	0.0141
5	0.0326	0.0041	0.0061	0.017	0.0034	0.0042
6	0.0109	0.0014	0.0021	0.0073	0.0011	0.0014
7	0.0041	0.0005	0.0008	0.0028	0.0004	0.0005
8	0.0022	0.0002	0.0004	0.0021	0.0002	0.0003
9	0.0012	0.0001	0.0002	0.0013	0.0001	0.0001
10	0.0004	0.0001	0.0001	0.0002	0	0

Table 1 contains the probabilities of nouns which, at some position relative to an adjective at position 0, are the target of that adjective. We only calculated probabilities for positions of +/-10 words away from an adjective, based on the average sentence size of 21 words. The probabilities were calculated from the AQUAINT corpus.

As can be seen from Table 1, the position immediately following a subjective adjective (position 1) has the highest probability (0.5666) of containing the target of the adjective (see the last column of Table 1). Position with the next highest probability of containing the target is one word away following the adjective (position 2), which is due to the cases where the target is the head of a longer noun phrase with an intervening modifier. Position -2 has the next highest probability of containing the target noun of a subjective adjective, which represents predicative use of adjectives.

Out of all adjectives, 77% are used attributively and 9% predicatively. When restricted to subjective adjectives, the count becomes 65% attributive and 20% predicative. One explanation for the larger proportion of subjective adjectives used predicatively compared to all adjectives may be that subjectivity is more often directed at proper nouns. Proper nouns do not usually take prenominal adjectives [17], so this attributive usage would need to be written predicatively instead (e.g., one is more likely to say "tall person" or "Jane is tall", but less likely "tall Jane").

3.3 Document Ranking

The goal of our method is to rank documents by the likelihood that they express opinions on the query concept. Our method, therefore, attempts to rank documents by topical subjectivity, i.e. expression of opinion about the query topic. Document ranking is performed by locating all instances of subjective adjectives[2] in the document and computing the aggregate probability that they refer to occurrences of the query term based on the precomputed probabilities described in the previous section.

In more detail a document score is calculated as follows: first the user's query term (or phrase) is used to find a set of top N ranked documents using a best-match IR system. In each document all instances of the query term and subjective adjectives are identified. For each occurrence of the query term, we determine if there are any subjective adjectives within 10 words either side, and note the distance separating them. The probability of a subjective adjective referring to a query term instance occurring i words away is referenced from precomputed statistics[3] (Table 1). We use the sum of these probabilities as the probability that the document contains an opinion on the query topic. The sum of probabilities is calculated according the inclusion-exclusion formula [18] for n non-mutually exclusive events (Eq. 2):

$$P(\bigcup_{i=1}^{n} A_i) = \sum_{i} P(A_i) - \sum_{i<j} P(A_i A_j) + \sum_{i<j<k} P(A_i A_j A_k) - ... + (-1)^{n+1} P(A_1 ... A_n) \quad (2)$$

Where, A_i – co-occurrence of a query term with a subjective adjective at distance i in the document; $P(A_i)$ – precomputed probability (from Table 1) that at distance i a subjective adjective modifies a noun.

The instance of the inclusion-exclusion formula for three events (i.e. three *query term – subjective adjective* co-occurrence pairs) is presented in Eq. 3:

$$P(A_i \cup A_j \cup A_k) = P(A_i) + P(A_j) + P(A_k) - P(A_i A_j) - P(A_i A_k) - P(A_j A_k) + P(A_i A_j A_k) \quad (3)$$

3.4 Collocates of Query Terms as Opinion Targets

A document can express opinion not directly about the concept represented by the query term, but about related concepts, which can be more general or specific. For example an author may talk subjectively about a film by expressing opinions on the actors' performance or a particular scene or work of the director in general. Another example would be

---

2 We used a list of manually tagged subjective adjectives from [1].

3 In the evaluation we used proper noun statistics as most of the user queries were proper nouns.

someone giving a review of an automobile model by talking about its specific features or components, such as fuel efficiency, comfort, engine or accumulator.

In this work we propose a method of using collocates, words significantly co-occurring in the contexts of query terms in the corpus, as representatives of concepts related to the query topic. Specifically, our approach consists of first finding collocates of a query term, and then calculating a document score which is an aggregate probability that subjective adjectives modify the original query term instances plus instances of their collocates. The next section describes the method used for collocate selection.

3.4.1 Collocate Selection Method

A large number of statistical methods have been used to find and rank collocates, such as Mutual Information [19], Z-score [20], Log-Likelihood ratio and chi-square test [21]. We can view the problem of finding related terms for opinion scoring as similar to query expansion. The difference is that we do not explicitly add additional terms to the query, but use their probabilities of being the target of a subjective adjective as additional evidence that the document expresses opinion on the query topic.

It is outside of the scope of the present work to evaluate different term association measures, therefore we chose to use one term association measure, Z-score, which showed good performance in query expansion experiments [20]. Systematic comparison of different term selection measures is left for future work. Z-score is a statistic for hypothesis testing, i.e. for assessing whether a certain event is due to chance or not. When used for collocation selection, Z-score tests whether the co-occurrence of two words is due to other factors than chance. It is similar to a t-score measure as proposed by Church et al. [19].

We used the method for extracting collocates and calculating Z-score as proposed in [20]. The procedure and parameters we used for selecting collocates are as follows: in the 50 top ranked documents retrieved in response to the user's query term, all terms surrounding instances of the query term within the windows of 20 words (10 words either side of the query term instance) are extracted. In cases where windows surrounding query term instances overlap, terms are extracted only once. All extracted terms are then ranked according to the modified Z-score in Eq. 4 [20], and up to 12 top-ranked terms are selected for the use in our method. All collocates with Z-score less than the significance threshold of 1.6 were rejected.

$$Z = \frac{f_r(x,y) - \frac{f_c(y)f_r(x)v_x(R)}{N}}{\sqrt{\frac{f_c(y)f_r(x)v_x(R)}{N}}} \tag{4}$$

Where: R – the set of top retrieved documents; $f_r(x,y)$ – joint frequency of x and y in R; $f_c(y)$ – frequency of y in the corpus; $f_r(x)$ – frequency of x in R, $v_x(R)$ – average window size around x in the relevant documents; N – corpus size.

More information about the modified Z-score and its derivation can be found in [20]. The values chosen for the parameters in our study (the window size and the number of Z-ranked collocates selected) are those that showed best results in the query expansion experiments by [20]. It is left for future work to systematically evaluate which parameters perform best in our task.

Table 2 shows a list of collocates selected for the sample of queries submitted by users in our evaluation experiment, which will be described in the next section.

Table 2. Collocates selected for a sample of queries submitted by users in the evaluation experiment

Bill Gates	wealth, dynamite, Microsoft, rich, interview, Napoleon, dollars, he, man, person, short, say
Egypt	ancient, guardian, pyramids, tour, arab, egyptian, travel, Nile, Cairo, modern, country, history
J.K. Rowling	Bloomsbury, Potter, Harry, author, interview, book, books, site
JDeveloper	Oracle, 10g, soa, oc4j, Webgalileo, Oracle9i, ide, bpel, clover, jsf, adf, java
iPod	nano, Apple, iTunes, grayscale, 30gb, generation, mini, dock, gb, shuffle, applecare, playback

3.4.2 Document Ranking Using Collocates

After the set of collocates of the query term is selected, a score is calculated for each of the top *N* documents retrieved in response to the user's query as follows: in each document all instances of the query term, collocates and subjective adjectives are identified. For each occurrence of the query term and collocates, determine if there are any subjective adjectives within 10 words and note the distance separating them. For each subjective adjective get the probability from precomputed statistics (section 3.2) that it refers to a query term or a collocate instance occurring *i* words away. Aggregate probabilities are calculated according to Eq. 2 (section 3.3).

4 Evaluation

We conducted a user-based evaluation of the proposed approach. Altogether 33 users, solicited from the University of Waterloo graduate student mailing list, voluntarily participated in the evaluation. The form requesting users to submit their queries contained the following instructions:

"Please enter a word or phrase identifying some person/item/entity, about which you are curious to see opinions. This should complete the sentence: "I'd like to know what the Web thinks of ____________".

The form also contained a text field where users were asked to enter a more detailed description of their information need for future analysis.

We used Google to retrieve the initial set of documents in response to the users' queries. The retrieved documents consist of all results obtainable via the Google API up to one thousand. Because of the limit on the number of documents that Google can return per day via its API, it was not possible to simulate the search process in real time. Users were therefore asked to come back in a few days after query submission in order to do the relevance judgements. Each user submitted one query, and in total for 33 queries 1192 documents were judged.

We evaluated two methods of ranking documents by topical opinion: "Opinion" method using only original query terms (section 3.3); "Collocation opinion" method using original query terms plus their collocates (section 3.4.2).

The baseline against which the above two methods are evaluated is the original Google ranking. The reason for selecting Google as the baseline is that it is one of the most widely used state-of-the-art Web search engines, which users may use to satisfy their opinion information needs. Also, to our knowledge there is no publicly available Web search engine which specifically retrieves documents containing opinions on the subject of the user's query.

For each topic, up to the top 1000 documents retrieved by Google are re-ranked using "Opinion" and "Collocation Opinion" methods. Top 15 ranked documents from each of the three ranked document sets, the "Opinion", "Collocation Opinion" and Google, are extracted, and presented in the random order to the user. By randomizing the order in which documents in the three sets are presented to the user, we ensure that the user is unable to infer which method was used to retrieve each document. It also removes the possibility of user bias due to ranking order. The decision of including only the top 15 documents from each of the three retrieved sets in the results list was made so that the relevance judgement task was not too time-consuming for users. Users were asked to judge the full text of each document in the list as one of the following:

1. Containing an opinion about the query topic ("query relevance");
2. Containing an opinion about something closely related to the query topic ("relevance to a related topic");
3. Containing no opinion about the query or related topics.

5 Results

The performance of two methods "Opinion" and "Collocation Opinion" was evaluated by means of Precision at 10 retrieved documents (P@10) and Precision at 15 retrieved documents (P@15) using "query relevance" and "relevance to a related topic" judgements. Results are presented in Table 3.

Table 3. Evaluation results (* indicates that a run has a statistically significant difference from the baseline, paired t-test, $P<0.05$)

Method	Query relevance		Relevance to a related topic	
	P@10	P@15	P@10	P@15
Google (baseline)	0.3758	0.3717	0.5424	0.5455
Opinion	0.5182*	0.4990*	0.6636*	0.6626*
Collocation opinion	0.4727*	0.4747*	0.6363*	0.6404*

As can be seen from Table 3 all runs significantly (paired t-test, $P<0.05$) improved performance of topical opinion retrieval over the baseline. The use of only query terms in estimating the likelihood of the document expressing opinion on the query topic performs better than the use of collocates. Using query relevance judgements, out of 33 queries, "opinion" ranking method improves P@15 of 24, deteriorates 6, and does not affect 3 queries, while "collocation opinion" method improves P@15 of

21 queries, deteriorates 6, and does not affect 6 (Fig. 1). In P@10 "opinion" method improves performance in 21 cases, deteriorates 6, and does not affect 6, while "collocation opinion" improves the performance of 20 queries, deteriorates 10, and does not affect 3. Using related topic judgements, "opinion" improves P@15 of 24, deteriorates 6, and does not affect 3 queries, while "collocation opinion" method improves P@15 of 23 queries, deteriorates 7, and does not affect 3 queries. In P@10 "opinion" method improves performance in 22 queries, deteriorates 6, and does not affect 5. "Collocation opinion" also improves P@10 of 22 queries, but deteriorates 11.

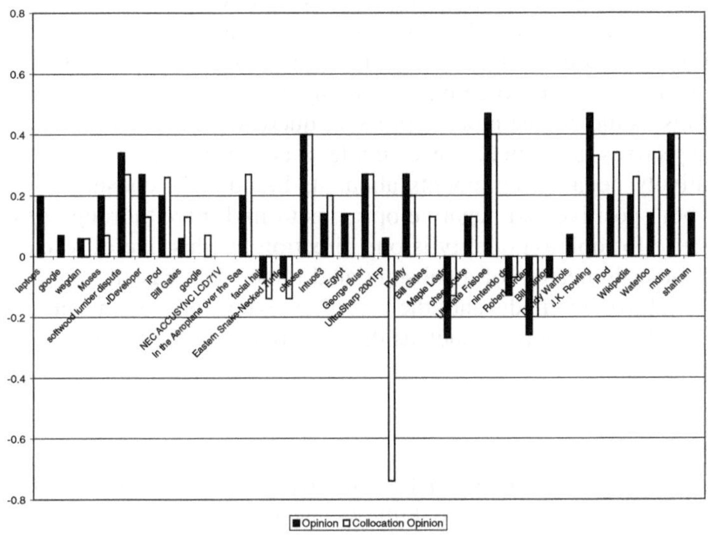

Fig. 1. Precision at 15 (P@15) results for Opinion and Collocation Opinion ("Query Relevance" judgements) relative to the baseline

Our analysis also shows that "collocation opinion" method has better P@10 than "opinion" in 8 queries in the "query relevance" judgements, however it has better P@10 in 10 queries in the "related topic relevance" judgements. This suggests that the "collocation opinion" method may be more suitable for situations where the user's information need is broad, and she may be interested in documents expressing opinions on subjects related to the query topic.

As can be seen from Fig. 1, the topics that benefit most from the "Collocation Opinion" method are those where the author is likely to express opinion about related concepts, such as features or characteristics of the topic entity. For instance, the topic "iPod" performed better with the use of collocates. Some of the collocates selected for this topic represent features of iPod, namely "playback" and "shuffle", and it is likely that authors express their opinion about iPod through such features. Examples of other topics that performed better with "Collocation Opinion" are "Google", "Wikipedia", "Intuos3" and "Bill Gates".

6 Conclusions and Future Work

In this paper we proposed a computationally lightweight algorithm for topical opinion retrieval. As an element of our technique, we developed a method for adjective target resolution in English, which demonstrated high accuracy. We conducted a thorough user-based evaluation of the developed method in an unrestricted domain using Web as the corpus. Comparison of the developed method to a state-of-the-art search engine (Google) shows statistically significant gains, demonstrating that the system is useful in resolving genuine topical opinion needs of real users. Since at present no domain-independent topical opinion search engine exists, our experiment demonstrates potential uses of such a system, and the types of queries that people may ask.

In addition to the subjectivity of adjectives, the system could incorporate additional metainformation on each adjective, including polarity (positive and negative) and intensity. This would enable more expressive queries to be formulated, limiting which subset of adjectives is applied. For example, a company might be most interested in the superlatively negative comments about its brand, or a consumer might prefer a balance of both positive and negative opinions to find more thorough product evaluations. A metric for opinion quality is one direction for this line of research and could incorporate other indicators of a substantiated rather than casual opinion. We plan to evaluate the above methods and their further extensions by means of user-based evaluations and test collections such as the one created in the Blog track of TREC.

References

1. V. Hatzivassiloglou and K. R. McKeown, Predicting the semantic orientation of adjectives. In Proceedings of the Thirty-Fifth Annual Meeting of the Association for Computational Linguistics, pp. 174-181, 1997.
2. M. Hurst and K. Nigam, Retrieving topical sentiments from online document collections. In Proceedings of the 11th conference on document recognition and retrieval, 2004.
3. J. Yi, T. Nasukawa, R. Bunescu, and W. Niblack, Sentiment analyzer: extracting sentiments about a given topic using natural language processing techniques. In Proceedings of the 3rd IEEE International Conference on Data Mining, 2003.
4. K. Dave, S. Lawrence, and D. M. Pennock, Mining the Peanut Gallery: Opinion Extraction and Semantic Classification of Product Reviews. In Proceedings of the 12th World Wide Web Conference, 2003.
5. M. Hu and B. Liu, Mining opinion features in customer reviews. In Proceedings of the ACM SIGKDD International Conference on Knowledge Discovery & Data Mining, 2004.
6. B. Pang, L. Lee, and S. Vaithyanathan, Thumbs up? Sentiment classification using machine learning techniques. In Proceedings of the 2002 conference on empirical methods in natural language processing, 2002.
7. P. D. Turney, Thumbs up or thumbs down? Semantic orientation applied to unsupervised classification of reviews. In Proceedings of the 40th Annual Meeting of the Association for Computational Linguistics (ACL'02), pp. 417-424, 2002.
8. M. C. Baker, Lexical categories: verbs, nouns and adjectives. Cambridge University Press, 2003.
9. R. M. W. Dixon and A. Y. Aikhenvald, Adjective classes. A crosslinguistic typology. Oxford University Press, 2004.

10. R. F. Bruce and J. M. Wiebe, Recognizing subjectivity: a case study in manual tagging. Natural Language Engineering, vol. 5, no. 2, pp. 187-205, 1999.
11. J. Wiebe, Learning subjective adjectives from corpora. In Proceedings of the 17th National Conference on Artificial Intelligence, pp. 735-740, 2000.
12. P. D. Turney and M. L. Littman, Unsupervised learning of semantic orientation from a hundred-billion-word corpus. Tech. Rep. EGB-1094, National Research Council Canada, 2002.
13. S. Greenbaum, The Oxford English grammar. Oxford University Press, 1996.
14. P. Rijkhoek, On Degree Phrases and Result Clauses. PhD thesis, University of Groningen, Groningen, 1998.
15. X. Li and D. Roth, Exploring evidence for shallow parsing. In Proceedings of the Annual Conference on Computational Natural Language Learning, 2001.
16. J. Sinclair (Ed.) Collins Cobuild English Grammar. Harper Collins, 1990.
17. Z. Vendler, Adjectives and nominalizations, p. 86. Mouton & Co. N.V., The Hague, 1968.
18. J. Pitman, Probability, p. 559. Springer-Verlag, New York, 1993.
19. K. Church, W. Gale, P. Hanks, and D. Hindle, Lexical substitutability. In: Atkins B.T.S. and Zampoli A. (eds.) Computational Approaches to the Lexicon. Oxford University Press, 1994. pp. 153-177.
20. O. Vechtomova, S. E. Robertson, and S. Jones, Query expansion with long-span collocates. Information Retrieval, vol. 6, pp. 251-273, 2003.
21. D. Manning and H. Schütze, Foundations of Statistical Natural Language Processing, The MIT Press, Cambridge, Massachusetts, 1999.

Probabilistic Models for Expert Finding

Hui Fang and ChengXiang Zhai

University of Illinois at Urbana-Champaign, IL, USA
{hfang,czhai}@cs.uiuc.edu

Abstract. A common task in many applications is to find persons who are knowledgeable about a given topic (i.e., *expert finding*). In this paper, we propose and develop a general probabilistic framework for studying expert finding problem and derive two families of generative models (candidate generation models and topic generation models) from the framework. These models subsume most existing language models proposed for expert finding. We further propose several techniques to improve the estimation of the proposed models, including incorporating topic expansion, using a mixture model to model candidate mentions in the supporting documents, and defining an email count-based prior in the topic generation model. Our experiments show that the proposed estimation strategies are all effective to improve retrieval accuracy.

1 Introduction

The problem of *expert finding* is concerned with finding the experts on a specified topic. This problem has many real-world applications. For example, organizers of a conference need to assign submissions to the PC members based on their research interests and expertise. Customer service of a company needs to decide which staff should be assigned to solve a given complaint. Currently, people have to manually identify the experts, which is obviously labor-intensive and time-consuming. Thus, it would be very interesting to study how to automatically identify experts for a specified expertise area.

As a retrieval task, expert finding has recently attracted much attention mostly due to the launching of the Enterprise track [4,11] of TREC [12]. The task setup in the Enterprise track includes the following three components: (1) a supporting document collection; (2) a list of expert candidates, which are specified by names and emails; (3) a set of topics (i.e., descriptions of expertise). The task of expert finding is to rank the expert candidates for a given topic query based on the information from the data collection. Expert finding is similar to the traditional ad hoc retrieval task in the sense that both tasks are to find relevant information items for a given topic. The key challenge in expert finding is to infer the association between a person (i.e., candidate expert) and an expertise area (i.e., topic) from the supporting document collection.

Participants in the Enterprise track have tried various methods. The methods mainly fall into two categories: profile-based methods and document-based methods. In profile-based methods [5,1,7,2], researchers would first build a term-based

G. Amati, C. Carpineto, and G. Romano (Eds.): ECIR 2007, LNCS 4425, pp. 418–430, 2007.

expertise profile (called document reorganization in [5]) for each candidate, and rank the candidate experts based on the relevance scores of their profiles for a given topic by using traditional ad hoc retrieval models. In document-based methods [3,2,8], instead of creating such term-based expertise profiles, the researchers use the supporting documents as a "bridge" and rank the candidates based on the co-occurrences of topic and candidate mentions in the supporting documents. However, the existing methods are not general and usually rely on heuristics, such as rule-based methods to detect the candidate mentions in the supporting documents, to achieve reasonable retrieval accuracy.

In this paper, we develop a general probabilistic framework for studying expert finding. We derive two families of generative models based on the framework - candidate generation models and topic generation models. The derived models are analogous to the probabilistic models derived in [6] for traditional ad hoc retrieval. These models cover most existing probabilistic models for expert finding, including probabilistic versions of both profile-based and document-based methods [3,2,1,8]. We further develop several techniques to improve the estimation of the proposed models, including incorporating topic expansion, using a mixture model to put different weights on the matching of different representations of an expert candidate, and defining a candidate prior for topic generation models based on the counts of email matches in the supporting documents. Evaluation on two standard TREC test collections shows that both families of models perform well empirically. In addition, we discover that how the judgements are made can affect the relative performance of different models. Experiment results also show that putting different weights on the matching of different representations of candidates with a simple mixture model is beneficial. Furthermore, it is also beneficial to compute the candidate prior for topic generation models based on the email counts in the supporting documents. However, topic expansion only improves the performance slightly when optimized.

2 A Probabilistic Framework for Expert Finding

Recognizing the similarity between expert finding and the traditional ad hoc retrieval task, we can apply the probabilistic ranking principle [10] to develop a general probabilistic framework for expert finding. Specifically, we will rank the candidates according to the probability that a candidate is "relevant" to the topic (i.e., expertise) specified in a query, and the key challenge is to compute this probability.

Formally, suppose $S = \{d_1, ..., d_{|S|}\}$ is a collection of supporting documents. Let $t = t_1, t_2, ..., t_n$ be the description of a topic, where t_i is a term in the description. Let c be an expert candidate whose email and name are denoted as $e(c)$ and $n(c)$, respectively. Let R be a binary random variable to denote relevance (1 for relevant and 0 for non-relevant). Given a query t and an expert candidate c, we are interested in estimating the conditional probability $p(R = 1|c, t)$, i.e., the probability that candidate c is relevant to topic t. After using odds ratio to rank the candidates and applying the Bayes' Theorem we have

$$p(R=1|c,t) \stackrel{\text{rank}}{=} \frac{p(R=1|c,t)}{p(R=0|c,t)} \stackrel{\text{rank}}{=} \frac{p(c,t|R=1)}{p(c,t|R=0)} \tag{1}$$

where $\stackrel{\text{rank}}{=}$ means "equivalence for ranking the candidates".

We now discuss two different ways to factor the conditional probabilities $p(c,t|R=1)$ and $p(c,t|R=0)$. They correspond to two different families of probabilistic models, which are referred to as *candidate generation models* and *topic generation models*, respectively. The high-level derivation is in spirit the same as in [6] if we take a topic (i.e., t) as Q and a candidate (i.e., c) as D.

2.1 Candidate Generation Models

One way to factor the conditional probabilities in Equation 1 is as follows:

$$\frac{p(c,t|R=1)}{p(c,t|R=0)} = \frac{p(c|t,R=1)p(t|R=1)}{p(c|t,R=0)p(t|R=0)}$$

Here we assume that an expert candidate c is "generated" by a probabilistic model based on a query t. Thus, this family of probabilistic models is referred to as *candidate generation models.*

Since $p(t|R=1)$ and $p(t|R=0)$ are independent of the candidates, they can be ignored for the purpose of ranking candidates. Thus, the general retrieval function of the *candidate generation model* is:

$$p(R=1|c,t) \stackrel{\text{rank}}{=} \frac{p(c|t,R=1)}{p(c|t,R=0)},$$

where $p(c|t,R=1)$ is the probability of candidate c given the "expert generative model" topic t, while $p(c|t,R=0)$ is the probability of c given the "non-expert generative model" of t. Thus the main question is how to estimate $p(c|t,R=1)$ and $p(c|t,R=0)$.

For $p(c|t,R=1)$, we may use the supporting documents to connect t and c in the following way:

$$p(c|t,R=1) = \sum_{d \in S} p(c|d,t,R=1) \times p(d|t,R=1) \approx \sum_{d \in S} p(c|d,R=1) \times p(d|t,R=1) \tag{2}$$

Here we assume that t and c are independent given the document d and the event $R=1$. Intuitively, this formula is quite reasonable: $p(d|t,R=1)$ allows us to model the probability that a document d matches a topic t while $p(c|d,R=1)$ allows us to model the probability that a supporting document mentions a candidate c. A document d with higher values for both $p(c|d,R=1)$ and $p(d|t,R=1)$ would contribute more to the estimation of $p(c|t,R=1)$, which intuitively makes sense. Indeed, this is essentially to exploit the co-occurrences of the topic terms and candidate mentions in the supporting documents, an idea already used in the existing work [2,3],

Unfortunately, it is not immediately clear how we should estimate $p(c|t,R=0)$, the "non-expert" model for topic t. The difficulty comes from the fact that we do not have evidence for a candidate not to be an expert for t. Thus as a possibly inaccurate simplification, we simply assume that $p(c|t, R = 0)$ is uniformly distributed, leaving more accurate estimation as our future work. This assumption allows us to drop $p(c|t, R = 0)$ and rank candidates solely based on $p(c|t, R = 1)$, which we estimate using Equation 2.

To compute $p(d|t, R = 1)$ efficiently, we apply Bayes' Theorem and rewrite the equation in the following way:

$$p(c|t, R = 1) = \sum_{d \in S} p(c|d, R = 1) \times \frac{p(t|d, R = 1)p(d|R = 1)}{\sum_{d' \in S} p(t|d', R = 1)p(d'|R = 1)}$$

Since $\sum_{d' \in S} p(t|d', R = 1)p(d'|R = 1)$ is the same for all the candidates, it can be dropped for ranking. $p(d|R = 1)$ can be regarded as a prior on d that can be exploited to favor a certain type of documents (e.g., email messages) in S. For simplicity, we assume that $p(d|R = 1)$ is uniform, which leads to

$$p(R = 1|c, t) \stackrel{\text{rank}}{=} \sum_{d \in S} p(c|d, R = 1) \times p(t|d, R = 1), \tag{3}$$

where $p(t|d, R = 1)$ is the probability that the topic t is relevant to the document d and $p(c|d, R = 1)$ is the probability that the candidate c is mentioned in the document d. Both of them can be computed efficiently by using an existing probabilistic retrieval model[14].

The candidate generation model shown in Equation 2 covers the two-stage model proposed in [3] as a special case and can be regarded as a probabilistic version of the document-based approaches to expert finding.

2.2 Topic Generation Models

The other way to factor the conditional probabilities in Equation 1 is as follows:

$$\frac{p(c, t|R = 1)}{p(c, t|R = 0)} = \frac{p(t|c, R = 1)p(c|R = 1)}{p(t|c, R = 0)p(c|R = 0)}$$

Here a topic t is assumed to be "generated" by a probabilistic model based on an expert candidate c, thus we call this family of probabilistic models *topic generation models*. The general retrieval function of *topic generation models* is:

$$p(R = 1|c, t) \stackrel{\text{rank}}{=} \frac{p(c|R = 1)}{p(c|R = 0)} \times \frac{p(t|c, R = 1)}{p(t|c, R = 0)},$$

where $p(t|c, R = 1)$ is the probability of topic t according to the "expertise topic model" of candidate c, $p(t|c, R = 0)$ is the probability of topic t according to "non-expertise topic model" of candidate c, $p(c|R = 1)$ is the prior probability that c is an expert, and $p(c|R = 0)$ is the prior probability that c is not an expert. The expert finding problem is thus reduced to the problem of estimating these probabilities.

As in the case of candidate generation models, we make a possibly inaccurate simplification assumption that $p(t|c, R = 0)$ is uniform due to the lack of appropriate data for estimating it. Thus the major task is to estimate $p(t|c, R = 1)$, the probability that t describes the expertise of candidate c. We discuss two possible ways to estimate it: *profile-based estimation* and *document-based estimation*, which correspond to the two categories of the existing methods for expert finding.

Profile-based estimation: The idea of profile-based estimation is to first estimate an expertise profile language model θ_c for every expert candidate c, and then compute the likelihood of t given the profile language model θ_c, i.e.:

$$p(t|c, R = 1) \approx p(t|\theta_c, R = 1) \tag{4}$$

Naturally, the key challenge is to estimate expertise profile θ_c for a candidate.

One possible estimation method proposed in [2] is as follows:

$$p(t|\theta_c, R = 1) = \prod_{t_i \in t} p(t_i|\theta_c, R = 1)^{count(t_i,t)} \tag{5}$$

$$= \prod_{t_i \in t} (\sum_{d \in S} p(t_i|d, R = 1)p(d|c, R = 1))^{count(t_i,t)} \tag{6}$$

where $count(t_i, t)$ is the count of term t_i in the query t. This model (i.e., Model 1 in [2]) was shown to perform consistently worse than the other model in [2], but no clear explanation was given. When viewing the method in our probabilistic framework, we see that a main reason why this method does not work well is because the estimation of $p(t_i|\theta_c, R = 1)$ is not accurate. Specifically, the problem lies in that a supporting document matching candidate c well (i.e., with a high value of $p(d|c, R = 1)$) may not necessarily support that the candidate is an expert on a topic (i.e., $p(t|d)$ may be low). Based on Equation 6, as long as document d contains one query term t_i and mentions the candidate c, the document would be regarded as a useful document to support that c is an expert on topic t. This is clearly inaccurate because the document might not match the whole query concept even though it matches one query term very well.

Based on this analysis, it would be reasonable to hypothesize that when we selectively use the documents that truly reflect the expertise of candidate c, instead of using every document in the collection as shown in Equation 7, such profile-based estimation will perform better. We have not further explored this direction; instead, we will use the document-based estimation method (to be described below) in our experiments.

Document-based estimation: Instead of creating an expertise profile language model, we could use supporting documents to connect a candidate and a topic as in the candidate generation models. Specifically, making similar assumptions as we have made in estimating candidate generation models, we have

$$p(t|c, R = 1) = \sum_{d \in S} p(t|d, c, R = 1) \times p(d|c, R = 1) \approx \sum_{d \in S} p(t|d, R = 1) \times p(d|c, R = 1).$$

Thus, the expert candidates can be ranked according to

$$p(R = 1|c, t) \stackrel{\text{rank}}{=} \frac{p(c|R = 1)}{p(c|R = 0)} \times \sum_{d \in S} (p(t|d, R = 1) \times p(d|c, R = 1)) \tag{7}$$

Similarly, we rewrite $p(d|c, R = 1)$ in terms of $p(c|d, R = 1)$, which can be efficiently computed by treating representations of the candidate c as a query and using a standard probabilistic retrieval method to compute $p(c|d, R = 1)$. In addition, as in the candidate generation models, we assume $p(d|R = 1)$ is uniform, which leads to

$$p(R = 1|c, t) \stackrel{\text{rank}}{=} \frac{p(c|R = 1)}{p(c|R = 0)} \times \sum_{d \in S} (p(t|d, R = 1) \times \frac{p(c|d, R = 1)}{\sum_{d' \in S} p(c|d', R = 1)}), \quad (8)$$

where $p(c|R = 1)$ is the prior probability that a candidate c is an expert, $p(c|R = 0)$ is the prior probability that a candidate is not an expert, $p(t|d, R = 1)$ is the probability that topic t is discussed in document d, and $p(c|d, R = 1)$ is the probability that an expert candidate c is mentioned in the document d.

Comparing the two models in Equations (3) and (8), a main difference is that the topic generation model contains a candidate normalizer, $N(c) = \sum_{d \in S} p(c|d, R = 1)$, while the candidate generation model does not. Since a larger $N(c)$ indicates a more popular c, this normalizer would penalize a popular candidate. As we will show in Section 3, this normalization factor could provide an explanation for the performance difference between the two models.

The Model 2 proposed in [2] and the model used in [8] are both special cases of the model we presented in Equation 7 when $p(c|R = 1)$ and $p(c|R = 0)$ are assumed to be uniform. As we will show later, it is possible to specify a non-uniform prior to improve retrieval accuracy.

2.3 Estimation of Component Models

So far, we have derived two families of generative models, as shown in Equation 3 and Equation 8. They contain the following three component models: (1) $p(c|d, R = 1)$, the probability that candidate c is mentioned in document d; (2) $p(t|d, R = 1)$, the probability that topic t is discussed in document d; (3) $p(c|R = 1)$ and $p(c|R = 0)$, the prior probabilities of a candidate. We now discuss how we estimate each of them.

Modeling candidate mentions $p(c|d, R = 1)$: In general, $p(c|d, R = 1)$ can be computed by treating the description of candidate c (e.g., name and/or email) as a query and using a standard retrieval method to score document d.

The simplest method is to concatenate the email $e(c)$ and the name $n(c)$ to form a query to represent candidate c. $p(c|d, R = 1)$ could thus be computed as

$$p(c|d, R = 1) = p(\text{“}e(c), n(c)\text{''}|d, R = 1) \quad (9)$$

However, intuitively, a name and an email have different characteristics. For example, using email alone to identify an expert could generate high-precision results, while using name alone to identify an expert may lead to high-recall results due to partial matching of names. Thus intuitively, to achieve the best results, it would be reasonable to combine them with appropriate weights. To capture this intuition, we propose to model $p(c|d, R = 1)$ using a mixture model involving both $p(e(c)|d, R = 1)$ and $p(n(c)|d, R = 1)$. That is, $p(c|d, R = 1)$ can be computed as

$$p(c|d, R = 1) = \lambda_e \cdot p(e(c)|d, R = 1) + (1 - \lambda_e) \cdot p(n(c)|d, R = 1), \quad (10)$$

where λ_e is the weight of the email model. Since $e(c)$ and $n(c)$ are both text, $p(e(c)|d, R = 1)$ and $p(n(c)|d, R = 1)$ can be computed using query generation retrieval model with Dirichlet prior smoothing as described in [14].

Modeling topic-document relationship $p(t|d, R = 1)$: Since t is a piece of text, $p(t|d, R = 1)$ can be computed using the query generation retrieval method with Dirichlet prior smoothing [14]. However, the original topic description t tends to be quite short, so it may not be informative. We thus propose to use some pseudo feedback method (e.g., the model-based feedback method proposed in [13]) to estimate an enriched topic query model θ_t, and incorporate this query model into our candidate finding model through generalizing the topic likelihood $p(t|d, R = 1)$ as the cross entropy of the query model θ_t and the document model θ_d estimated based on d using Dirichlet prior smoothing. That is,

$$p(t|d, R = 1) \propto exp(\sum_w p(w|\theta_t) log(p(w|\theta_d)). \quad (11)$$

Clearly, if we set θ_t to the empirical word distribution in t, this would be equivalent to the original topic likelihood.

Setting the candidate prior $\frac{p(c|R=1)}{p(c|R=0)}$: $p(c|R = 1)$ and $p(c|R = 0)$ are the prior probabilities of a candidate. In most existing work [3,2], they are assumed to be uniform. However, as shown in Section 3, reasonable prior can help improve the retrieval accuracy.

Based on Bayes' Theorem, we can rewrite the candidate prior as

$$\frac{p(c|R = 1)}{p(c|R = 0)} \overset{rank}{=} \frac{p(R = 1|c)}{p(R = 0|c)} = \frac{p(R = 1|c)}{1 - p(R = 1|c)},$$

where $p(R = 1|c)$ is the probability that a candidate is an expert. To estimate it, we may reasonably assume that a candidate whose email has been mentioned many times has a high prior probability of being an expert. We adapt a formula that is similar to BM25 term frequency normalization formula [9], i.e., $p(R = 1|c) \propto \frac{count(e(c),S)}{2 \times count(e(c),S)+\beta}$. Thus, the candidate prior is

$$\frac{p(c|R = 1)}{p(c|R = 0)} \propto \frac{count(e(c), S)}{count(e(c), S) + \beta}, \quad (12)$$

where $count(e(c), S)$ is the count of mentions of the email of candidate c in the collection S, and β is the parameter to control the skewness of the prior. A larger β would reduce the skewness of the prior (i.e., leading to a weaker prior), thus it can be interpreted as being inversely proportional to our confidence in this prior.

3 Experiments

We evaluate the proposed models on two TREC enterprise collections [4] - (1) **Ent05:** W3C corpus with topics EX01-EX50. The topics are the names of working groups. This is the collection used in the Enterprise track of 2005. (2) **Ent06:** W3C corpus with topics EX51-EX105. The topics are contributed by the participants of enterprise track in 2006, and represent the real world information need. This is the collection used in the Enterprise track of 2006.

We use the entire corpus, and only the titles of topic descriptions. We have done minimal preprocessing, where we apply stemming with a Porter stemmer and no stop word is removed. We evaluate the methods with mean average precision (MAP), which is the official evaluation measure of expert finding task in enterprise track. We conduct six sets of experiments. First, we evaluate the proposed models, and compare them with the baseline models. Second, we examine the effectiveness of using a mixture model (i.e., Equation 10) to model the candidate mentions. Third, we study the effectiveness of topic expansion described in Equation 11. Fourth, we demonstrate the effectiveness of candidate prior proposed in Equation 12. Fifth, we examine the parameter sensitivity. Finally, we compare our results to the official TREC runs.

3.1 Comparison of Proposed Models

In Table 1, we compare the optimal performance of the proposed two models with a state-of-the-art baseline model. "Cand-gen (mixture)" is the candidate generation model (Equation (3)) estimated using the mixture model in Equation (10). "Topic-gen (mixture+prior) " is the topic generation model (Equation (8)) estimated using the mixture model in Equation (10) and the prior in Equation (12). In both models, $p(t|d, R = 1)$ is computed using query generation model with Dirichlet prior smoothing as described in [14]. "Baseline"is the topic generation model in Equation (8) estimated using Equation 9 and a uniform prior; this model is essentially similar to the model 2 proposed in [2].

Table 1. Performance comparison of different models

Models	Ent05	Ent06
Baseline	0.151	0.191
Cand-gen (mixture)	0.186	**0.449**
Topic-gen (mixture+prior)	**0.196**	0.334

Table 1 shows that our proposed models perform much better than the baseline model. In addition, the optimal performance of topic generation model on Ent05 is 0.196, which is better than the best reported performance (i.e., 0.1894) of a similar model using rule-based name matching (i.e., Model 2 in [2]).

Compared with topic generation model, the candidate generation model performs slightly worse on Ent05 but much better on Ent06. When looking into it,

Table 2. Average popularity of candidates

	Judgements	Results of Topic-gen	Results of Cand-gen
Ent05	0.91	1.18	2.27
Ent06	2.37	1.21	2.02

we found that such performance difference may be caused by the different ways of creating judgements in these two years' Enterprise track. In Ent05, the topics are the names of working groups, and the judgements are independent of the document collection. But in Ent06, the topics are contributed by the participants, and the judgements are created based on the information from the document collection. Such different ways of creating judgements directly affect the characteristics of the relevant experts, especially their occurrences in the collection (i.e., the popularity of the experts). Using the normalizer $N(c) = \sum_{d \in S} p(c|d, R = 1)$ as a measure of candidate popularity, we show some popularity statistics in Table 2. We see that, as expected, the relevant experts in Ent06 are more popular in the collection compared with those identified in Ent05, which means that we should expect penalizing popular experts to help more (or harm less) for Ent05 than for Ent06. Indeed, the fact that the topic-generation model performs better than the candidate-generation model on Ent05 but worse on Ent06 suggests that such penalization is beneficial for Ent05 but harmful for Ent06; this is because the two models differ mainly in the penalization of popular candidates as discussed in Section 2. From Table 2, we can also see that the average popularity of the candidates returned by the topic-generation model is indeed much lower than that returned by the candidate-generation model.

It is surprising that the different ways of how Ent05 and Ent06 are created can affect the relative performance of the two models, making it interesting to further explore how to create appropriate test collections for expert finding.

3.2 Effectiveness of Mixture Model for Candidate Mentions

We discussed two ways to model candidate mentions as shown in Equation 9 (denoted as "merge") and in Equation 10 (denoted as "mixture"). We compare these two estimations and report the results in Table 3. The results show that it is more effective to use a mixture model to model the candidate mentions.

Table 3. Effectiveness of mixture model for candidate mentions

		Ent05	Ent06
Cand-gen	merge	0.130	0.280
	mixture	0.186 (43%)	0.449 (60%)
Topic-gen (prior)	merge	0.155	0.193
	mixture	0.196 (26%)	0.334 (73%)

3.3 Effectiveness of Topic Expansion

We proposed two possible ways to model the topic-document relationship: (1) "query likelihood", which is the query generation model described in [14]; (2) "feedback", which is the model-based feedback method described in [13]. We compare these two strategies and report the optimal performance in Table 4. The results show that topic expansion consistently improves the performance when optimized, but the performance improvement is smaller compared with the performance improvement in traditional ad hoc retrieval problem [13].

Table 4. Effectiveness of topic expansion

	Ent05		Ent06	
	query likelihood	expansion	query likelihood	expansion
Cand-gen(mixture)	0.186	0.196 (5%)	0.449	0.465 (4%)
Topic-gen(mixture+prior)	0.196	0.204 (4%)	0.334	0.359 (7%)

3.4 Effectiveness of Candidate Prior in Topic Generation Models

In the topic generation models, we proposed to compute the candidate prior based on the counts of email using Equation 12, which is denoted to as "email-prior". In contrast, most existing work assumes that the candidate prior is uniform, which is denoted as "uniformprior". Table 5 shows the performance difference between these two strategies. As shown in Table 5, incorporating email-based prior could improve the performance. It is interesting that when we estimate the candidate mentions with the mixture model as in Equation 10, the email-based prior improves the performance more.

Table 5. Performance comparison for candidate prior

Topic-gen		Ent05	Ent06
merge	uniformprior	0.151	0.191
	emailprior	0.155 (3%)	0.193 (1%)
mixture	uniformprior	0.172	0.204
	emailprior	0.196 (14%)	0.334 (64%)

3.5 Parameter Sensitivity

There are four parameters in both candidate generation models and topic generation models: μ_t, μ_e, μ_n and λ_e. Topic generation models have one extra parameter for the candidate prior, i.e., β.

μ_t, μ_e and μ_n are the Dirichlet prior smoothing parameters used to compute $p(t|d, R = 1)$, $p(e(c)|d, R = 1)$ and $p(n(c)|d, R = 1)$, respectively. We examine the parameter sensitivity for these three parameters in Figure 1. In every case, we

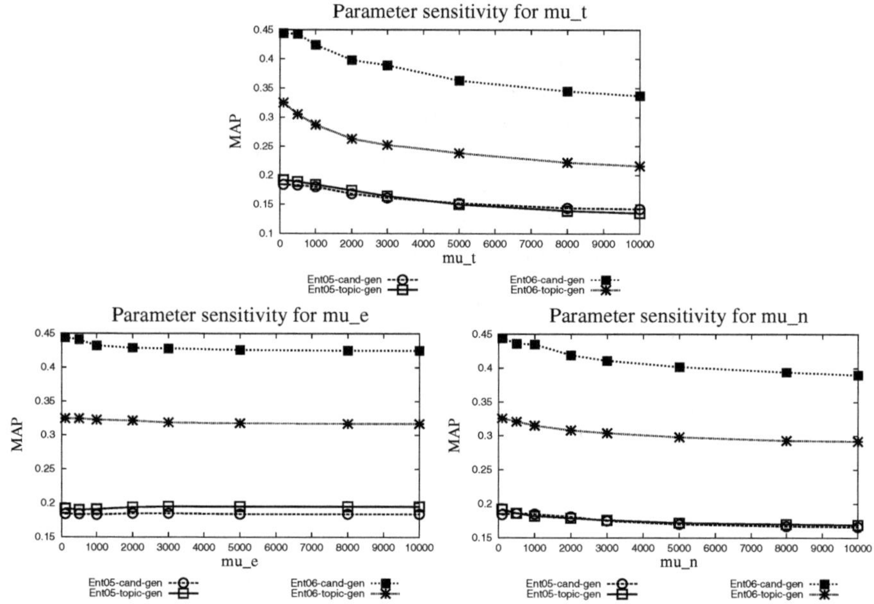

Fig. 1. Performance Sensitivity of μ_t(upper), μ_e (lower left) and μ_n (lower right)

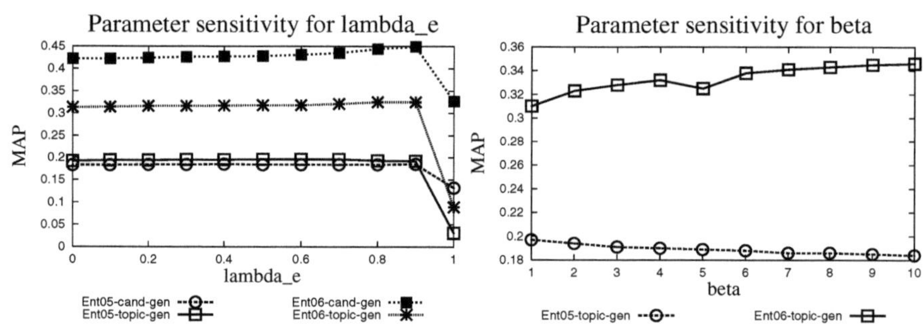

Fig. 2. Performance Sensitivity of λ_e(left) and β (right)

change the value of one parameter while fixing the value of the other parameters. The plots show that the performance is relatively more stable to the change of μ_e compared with the change of μ_t and μ_n. It is interesting that the optimal values of μ's are generally around 100, which is much smaller than 2000, the recommended value in traditional ad hoc retrieval [14].

λ_e controls the relative weight on email matching to name matching (as in Equation 10). We now examine how it affects the performance. The left plot in Figure 2 shows that the performance is relatively stable when $\lambda_e < 0.9$.

β is a prior confidence parameter defined in Equation 12. The right plot in Figure 2 shows the parameter sensitivity curve of this parameter. We observe that the optimal values of β are different for the two data sets. Larger β leads to better performance in Ent06, while it leads to worse performance in Ent05. Such observation implies that we could be more confident in the prior on Ent05 than on Ent06, which might be related to the different characteristics of two sets of queries. We leave the further analysis as our future work.

3.6 Comparison with TREC Results

Our best results are 0.204 for Ent05 and 0.465 for Ent06 as shown in Table 4. Compared with the official results of the TREC Enterprise track [4,11], our best results would be in the top 5 for Ent05 and top 10 for Ent06. The top performing systems tend to use various kinds of heuristics, which we did not use. Since our models are general and orthogonal to many of these heuristics, we can expect our models to perform even better when we add these heuristics.

4 Conclusions and Future Work

In this paper, we present a general probabilistic framework to solve the problem of expert finding. We derive two families of generative models, i.e., candidate generation models and topic generation models. These models cover most existing probabilistic models for expert finding. To improve the estimation of the proposed models, we further propose the following three techniques: (1) a mixture model for modeling the candidate mentions, which allows us to put different weights on different representations of an expert candidate; (2) topic expansion for modeling the topic document relationship, which expands the original queries with more informative terms; (3) email-based candidate prior, which provides a better estimation of prior probability that a candidate is an expert. Empirical results have demonstrated the effectiveness of the proposed models and estimation strategies.

There are many interesting future research directions. First, we need to study how to automatically set all the parameters through statistical estimation. Second, we could explore alternative ways to estimate the components in the models. Finally, it would be interesting to study how to estimate a reasonable expertise profile for every expert candidate in a principled way since such profiles can potentially be used for many other tasks in addition to expert finding.

Acknowledgments

This material is based in part upon work supported by the National Science Foundation under award numbers IIS-0347933 and IIS-0428472. We thank Lixin Zhou and three anonymous reviewers for their useful comments.

References

1. L. Azzopardi, K. Balog, and M. de Rijke. Language modeling approaches for enterprise tasks. In *Proceedings of TREC-05*, 2006.
2. K. Balog, L. Azzopardi, and M. de Rijke. Formal models for expert finding in enterprise corpora. In *Proceedings of SIGIR-06*, 2006.

3. Y. Cao, J. Liu, S. Bao, and H. Li. Research on expert search at enterprise track of trec2005. In *Proceedings of TREC-05*, 2006.
4. N. Craswell, A. P. de Vries, and I. Soboroff. Overview of the trec-2005 enterprise track. In *Proceedings of TREC-05*, 2006.
5. Y. Fu, W. Yu, Y. Li, Y. Liu, M. Zhang, and S. Ma. Thuir at trec 2005: Enterprise track. In *Proceedings of TREC-05*, 2006.
6. J. Lafferty and C. Zhai. Probabilistic relevance models based on document and query generation. *Language Modeling and Information Retrieval, Kluwer International Series on Information Retrieval*, 13, 2003.
7. C. Macdonald, B. He, V. Plachouras, and I. Ounis. University of glasgow at trec 2005: Experiments in terabyte and enterprise tracks with terrier. In *Proceedings of TREC-05*, 2006.
8. D. Petkova and W. B. Croft. Umass notebook 2006: Enterprise track. In *Proceedings of TREC-06*, 2007.
9. S. Robertson and S. Walker. Some simple effective approximations to the 2-poisson model for probabilistic weighted retrieval. In *Proceedings of SIGIR'94*, 1994.
10. S. E. Robertson. The probability ranking principle in ir. *Journal of Documentation*, 33(4):294–304, Dec. 1977.
11. I. Soboroff, A. P. de Vries, and N. Craswell. Overview of the trec 2006 enterprise track. In *Proceedings of TREC-06*, 2007.
12. E. Voorhees and D. Harman, editors. *Proceedings of Text REtrieval Conference (TREC1-9)*. NIST Special Publications, 2001. http://trec.nist.gov/pubs.html.
13. C. Zhai and J. Lafferty. Model-based feedback in the language modeling approach to information retrieval. In *Proceedings of CIKM-01*, 2001.
14. C. Zhai and J. Lafferty. A study of smoothing methods for language models applied to ad hoc information retrieval. In *Proceedings of SIGIR-01*, 2001.

Using Relevance Feedback in Expert Search

Craig Macdonald and Iadh Ounis

Department of Computing Science,
University of Glasgow, G12 8QQ, UK
{craigm,ounis}@dcs.gla.ac.uk

Abstract. In Enterprise settings, expert search is considered an important task. In this search task, the user has a need for expertise - for instance, they require assistance from someone about a topic of interest. An expert search system assists users with their "expertise need" by suggesting people with relevant expertise to the topic of interest. In this work, we apply an expert search approach that does not explicitly rank candidates in response to a query, but instead implicitly ranks candidates by taking into account a ranking of document with respect to the query topic. Pseudo-relevance feedback, aka query expansion, has been shown to improve retrieval performance in adhoc search tasks. In this work, we investigate to which extent query expansion can be applied in an expert search task to improve the accuracy of the generated ranking of candidates. We define two approaches for query expansion, one based on the initial of ranking of documents for the query topic. The second approach is based on the final ranking of candidates. The aims of this paper are two-fold. Firstly, to determine if query expansion can be successfully applied in the expert search task, and secondly, to ascertain if either of the two forms of query expansion can provide robust, improved retrieval performance. We perform a thorough evaluation contrasting the two query expansion approaches in the context of the TREC 2005 and 2006 Enterprise tracks.

1 Introduction

In large Enterprise settings with vast amounts of digitised information, it is often important that a user is not only be able to identify documents that are relevant to a topic of interest, but also to find people that have relevant expertise to the topic. People are a critical source of information because they can explain and provide arguments about why specific decisions were made [5]. Hence, in addition to classical document Information Retrieval (IR) systems, there is a growing interest in the research community to build accurate expert search systems. An *expert search* system aids a user in their "expertise need" by identifying people with relevant expertise to the topic of interest.

The retrieval performance of an expert search system is very important. If an expert search system suggests incorrect experts, then this could lead the user to contacting these people inappropriately. Similar to document IR systems, the accuracy of an expert search system can be measured using the traditional

G. Amati, C. Carpineto, and G. Romano (Eds.): ECIR 2007, LNCS 4425, pp. 431–443, 2007.

IR evaluation measures: *precision*, the accuracy of suggested candidate experts; and *recall*, the number of candidate experts with relevant expertise retrieved. Expert search was a retrieval task in the Enterprise tracks of the Text REtrieval Conferences (TREC) since 2005 [4], aiming to evaluate expert search approaches.

Pseudo-relevance feedback (PRF) [10] has been used in adhoc search tasks to improve the performance of document IR systems. PRF describes the process of automatically examining top-ranked documents in an IR system ranking, and using information from these documents to improve the ranking of documents. This is done by assuming that the top-ranked documents are relevant, and using information from this 'pseudo-relevant set' to improve the accuracy of the ranking by expanding on the initial query and re-weighting the query terms[1].

In this paper, we explore how query expansion can be applied in an expert search task. To this end, we experiment with an expert search system that is based on the voting model for expert search [8]. In this model, documents are firstly associated with candidates to represent the candidates expertise. Then the voting model considers the ranking of documents with respect to the query, in order to generate an accurate ranking of candidates. The voting model for expert search is interesting for these experiments, as we can apply query expansion using the underlying ranking of documents as the pseudo-relevant set. Moreover, we investigate how query expansion can be applied if the ranking of candidates is used as the pseudo-relevant set. We call these approaches document-centric, and candidate-centric query expansion respectively.

In this work, our objectives are two fold: firstly, to determine if query expansion can be successfully applied in expert search; and secondly, to analyse both forms of query expansion, allowing conclusions to be drawn concerning the applicability and effectiveness of both approaches. In order to fully understand the applicability of query expansion, we experiment using two statistically different models from the Divergence from Randomness (DFR) framework for extracting informative terms from the pseudo-relevant set - one model based on the Bose-Einstein statistics and is similar to Rocchio [10], and one based on Kullback-Leibler divergence [1]. Furthermore, we experiment using two different voting techniques for ranking candidates, using the topics and relevance assessments for the W3C collection from the TREC 2005 and 2006 Expert Search tasks. Conclusions are drawn across the two voting techniques applied.

Section 2 provides further detail on the model for expert search that we employ in this work, and demonstrates the baselines achieved using this approach. Section 3 defines how query expansion can be applied to expert search. We experimentally investigate both approaches for query expansion in Section 4. Section 5 investigates the effect of varying the parameters of query expansion, to assess the maximum potential and stability of each approach. Section 6 provides concluding remarks and suggestions for future work.

---

[1] In this work, we use the terms pseudo-relevance feedback and query expansion interchangeably.

2 Expert Search

Modern expert search systems for Enterprise settings work by using documents to form the profile of textual evidence for each candidate. The candidate's profile represent the expertise of the candidate expert to the expert search system. This documentary evidence can take many forms, such as intranet documents, documents or emails authored by the candidates, or even emails sent by the candidate or web pages visited by the candidate (see [8] for an overview). In this work, the profile of a candidate is considered to be the set of document associated with the candidate. These candidate profiles can then be used to rank candidates automatically in response to a query.

This work uses the voting approach for expert search proposed by Macdonald & Ounis in [8], which considers the problem of expert search as a voting process. Instead of directly ranking candidates, it considers the *ranking of documents*, with respect to the query Q, denoted by $R(Q)$. The ranking of candidates can then be modelled as a voting process, from the retrieved documents in $R(Q)$ to the profiles of candidates: every time a document is retrieved and is associated with a candidate, then this is a vote for that candidate to have relevant expertise to Q. Each document retrieved by the IR system, that is associated with the profile of a candidate, can be seen as a vote for that candidate to have relevant expertise to the query topic. The ranking of the candidate profiles can then be determined by aggregating the votes of the documents. Eleven voting techniques for ranking experts were defined in [8], which each employ various sources of evidence that can be derived from the ranking of documents with respect to the query topic. In this work, we only use the CombSUM and expCombMNZ voting techniques, because they provide robust results on the W3C collection. The CombSUM technique ranks candidates by considering the sum of the relevance scores of the documents associated with each candidate's profile. Hence the relevance score of a candidate expert C with respect to a query Q, $score\_cand(C,Q)$, is:

$$score\_cand(C,Q) = \sum_{d \in R(Q) \cap profile(C)} score(d,Q) \tag{1}$$

where $profile(C)$ is the set of documents associated with candidate C, and $score(d,Q)$ is the relevance score of the document in the document ranking $R(Q)$. For expCombMNZ, the relevance score of a candidate C's expertise to a query Q is given by:

$$score\_cand_{expCombMNZ}(C,Q) = \|R(Q) \cap profile(C)\| \cdot \sum_{d \in R(Q) \cap profile(C)} exp(score(d,Q)) \tag{2}$$

where $\|R(Q) \cap profile(C)\|$ is the number of documents from the profile of candidate C that are in the ranking $R(Q)$, and $exp()$ is the exponential function. expCombMNZ is similar to CombSUM, but also includes a second component which takes into account the number of documents in $R(Q)$ associated to each candidate, hence explicitly modelling the number of votes made by the documents for each candidate. The exponential function boosts candidates that are associated to highly scored documents (strong votes).

In the remainder of this section, we define the strong baselines that we deploy for our experiments. Secondly, we provide details on two statistically different query expansion (QE) techniques based on the Divergence from Randomness (DFR) framework. We employ two QE techniques in our experiments to ensure our drawn conclusions are general.

2.1 Baselines

In this section, we define our experimental setup, and the baselines we use in this work. Our experiments are carried out in the setting of the Expert Search tasks of the TREC Enterprise track, 2005 and 2006. The TREC W3C collection is indexed using Terrier [9], removing standard stopwords and applying the first two steps of Porters stemming algorithm. Initial experimental results have shown that applying only this weaker form of stemming results in increased high precision without degradation in mean average precision (MAP) for this task.

Next, we generate the profiles of documentary evidence of expertise for the candidates: for each candidate, documents which contain an exact match of the candidates full name are used as the profile of the candidate.

From the two TREC expert search tasks, we have a total of 99 topics with relevance assessments. Documents are ranked using the DLH13 document weighting model from the DFR framework. The DLH13 document weighting model is a generalisation of the parameter-free hypergeometric DFR model in a binomial case [2,7]. The hypergeometric model assumes that the document is a sample, and the population is from the collection. For the DLH13 document weighting model, the relevance score of a document d for a query Q is given by:

$$score(d,Q) = \sum_{t \in Q} \frac{qtw}{tf + 0.5} \cdot \Big(\log_2(\frac{tf \cdot avg\_l}{l} \cdot \frac{N}{F}) + 0.5 \log_2 \big(2\pi tf(1 - \frac{tf}{l})\big) \Big) \quad (3)$$

where tf is the term frequency of the term t in document d, F is the frequency of the query term in the collection and N is the number of documents in the collection. l is the length of the document d in tokens, and avg_l is the average document length in the whole collection. The query term weight qtw is given by qtf/qtf_{max}. qtf is the query term frequency. qtf_{max} is the maximum query term frequency among the query terms.

We chose to experiment using DLH13 because it has no term frequency normalisation parameter that requires tuning, as this is assumed to be inherent to the model. Moreover, DLH13 performs robustly on many collections and tasks without any need for parameter tuning [7]. By applying DLH13, we remove the presence of any term frequency normalisation parameter in our experiments.

In this work, we could also experiment with other weighting models. However, it was shown that the relative performance rankings of the voting techniques were concordant across a selection of weighting models, on the same W3C collection [8]. This infers that conclusions drawn using one document weighting model should be applicable to any other state-of-the art model.

Table 1 shows the retrieval performances achieved by the baseline expert search approach we employ in this paper. We report MAP and P@10 evaluation measures. The retrieval performance is reported on the TREC 2005 and 2006 topics. In addition, we also report the median run of MAP for each year (the median P@10 runs are not available). As apparent from Table 1, the voting

Table 1. Baseline performances of CombSUM and expCombMNZ, using the DLH weighting model, on the 2005 and 2006 TREC Enterprise track, expert search tasks. For TREC 2005, topics only had one fields, while for TREC 2006, we use title-only (short) queries. Mean average precision (MAP) and Precision at 10 (P@10) measure are reported. The MAP median runs of all participants from the respective year of TREC are given. Moreover, the best result for each measure are emphasised.

	TREC 2005		TREC 2006	
	MAP	P@10	MAP	P@10
Median	0.1402	-	0.3412	-
CombSUM	**0.2037**	**0.3240**	0.5188	0.6388
expCombMNZ	**0.2037**	0.3100	**0.5502**	**0.6837**

techniques are clearly performing well above the median run for both years. Moreover, these results for the TREC 2005 Enterprise task are similar to those of the 3rd top group participating that year. For TREC 2006, the ranking results are not yet publicly available, but with such a large margin over the median run, these results appear strong. The voting approach is robust and general, as it is not dependent on heuristics based on the used enterprise collection.

In [3], Balog et al. defined a language modelling approach for expert search. However, in contrast to the voting approach by Macdonald & Ounis, their approach can only be applied in a language modelling setting. The voting model approach is more flexible, because any approach (including language modelling) can be used to generate the document ranking $R(Q)$. However, there are some similarities between the two approaches. In particular, for the voting approach, if Hiemstra's language modelling approach [6] was used to generate $R(Q)$, and CombSUM applied to combine the scores for candidates, then this would be identical to the candidate ranking formula of Equation (4) in [3]. For this reason, we do not experiment using the Balog et al. language modelling approach, as its characteristics are encapsulated in the CombSUM voting technique[2].

[2] In fact, experimental evaluation of Balog et al's approach on the same profile set, provides similar results to Hiemstra's language model combined with CombSUM voting technique.

Because the voting approach allows any IR technique to be used to generate the ranking of documents $R(Q)$, we wish to determine the extent to which the performance of the approach can be improved by increasing the quality of the document ranking. An obvious way to apply QE is to use the top-ranked documents in $R(Q)$ as the pseudo-relevant set. However, we also propose an alternative approach for applying query expansion, namely using the top-ranked candidates as the pseudo-relevant set.

2.2 Query Expansion Models

In our investigation into query expansion (QE) in expert search, we need to determine if the QE model employed has any effect on the conclusions concerning our two approaches for query expansion. Hence, we employ two statistically different QE models from the DFR framework [1], known as term weighting models, for extracting informative terms from the pseudo-relevant set of top-ranked documents. DFR term weighting models measure the informativeness of a term by considering the divergence of the term occurrence in the pseudo-relevant set from a random distribution.

Terrier provides various DFR-based term weighting models for query expansion. We experiment with two term weighting models to understand the importance of the choice of model. One term weighting model, known as Bo1, is based on Bose-Einstein statistics and is similar to Rocchio [1]. The other is based on the Kullback Leibler (KL) divergence between the pseudo-relevant set sample and the collection. In Bo1, the informativeness $w(t)$ of a term t is given by:

$$w(t) = tf_x \cdot log_2 \frac{1 + P_n}{P_n} + log_2(1 + P_n) \tag{4}$$

where tf_x is the frequency of the term in the pseudo-relevant set, and P_n is given by $\frac{F}{N}$. F is the term frequency of the query term in the whole collection and N is the number of documents in the collection.

Alternatively, $w(t)$ can be calculated using the Kullback Leibler divergence term weighting model [1]:

$$w(t) = P_x \cdot log_2 \frac{P_x}{P_c} \tag{5}$$

where $P_x = \frac{tf_x}{l_x}$ and $P_c = \frac{F}{token_c}$. l_x is the size in tokens of the pseudo-relevant set, and $token_c$ is the total number of tokens in the collection. Note that unlike Bo1, KL uses the size of the pseudo-relevant set while measuring divergence.

Using either Bo1 or KL to define $w(t)$, the top *exp_term* informative terms are identified from the top *exp_doc* ranked documents, and these are added to the query ($exp\_term \geq 1$, $exp\_doc \geq 2$). The default setting for these parameters is $exp\_doc = 3$ and $exp\_term = 10$, suggested by Amati in [1] after extensive experiments. Finally, for both the Bo1 and KL term weighting models, the query term frequency qtw of an expanded query term is given by [1]:

$$qtw = qtw + \frac{w(t)}{w_{max}(t)} \tag{6}$$

where $w_{max}(t)$ is the maximum $w(t)$ of the expanded query terms. qtw is initially 0 if the query term was not in the original query.

3 Applying QE in Expert Search Task

Our work concerns the applicability of QE to expert search. The application of QE in adhoc search tasks is known to improve retrieval performance. Using the voting model described in Section 2, it can be seen that the quality of the generated ranking of candidates is dependent on how well $R(Q)$ ranks documents associated with relevant candidates. Then any improvement in the quality of the document ranking should improve the accuracy of the retrieved candidate ranking, because the document ranking votes will be more accurate, and hence the aggregated ranking of candidates will be more accurate.

We call *document-centric query expansion*, the approach that considers the top-ranked documents of the document ranking $R(Q)$ as the pseudo-relevant set. We hypothesise that the candidate ranking generated by applying a voting technique to the refined document ranking will have increased retrieval performance, when compared to applying the voting technique to the initial $R(Q)$.

Moreover, we propose a second approach called *candidate-centric query expansion* where the pseudo-relevant set is taken from the final ranking of candidates generated by a query. If the top-ranked candidates are defined to be the pseudo-relevance set, then we can extract informative terms from the corresponding candidate profiles, and use these to generate a refined ranking of documents. In using this expanded query, we hypothesise that the document ranking will become nearer to the expertise area of the initially top-ranked candidates, and hence the generated candidate ranking will likely include more candidates with relevant expertise.

In the following section, we assess the usefulness of both forms of QE compared to the baseline approaches defined in Section 2. It is of note that typically, each candidate profiles will many associated documents. Hence, applying candidate-centric QE will consider far more tokens of text in the top-ranked candidates, than applying document-centric QE. In particular, Table 2 details the statistics of the documents of the W3C collection, and the document candidate associations we use in this work. Of particular note is the size in tokens of profiles compared to documents - the average profile size is 76 times larger than the average document, while the largest candidate profile is a massive 444 times larger than the largest document in the collection. Due to the large difference between candidate profiles and documents, it is possible that the default settings of $exp\_doc = 3$ and $exp\_term = 10$ may not be suitable for candidate-centric query expansion. In the remainder of the paper, we assess whether the default settings are in fact suitable for document-centric and, in particular, candidate-centric query expansion.

Table 2. Statistics of the W3C collection, and of the candidate-document associations used in this work

W3C Collection	
Number of Documents	331,037
Size of Collection (tokens)	310,720,411
Average size of a Documents (tokens)	9,385
Largest Document (tokens)	50,001
Number of Candidates	1,092
Size of all Candidate Profiles (tokens)	779,840,190
Average size of a Candidate Profile (documents)	913
Average size of a Candidate Profile (tokens)	714,139
Largest Candidate Profile (documents)	88,080
Largest Candidate Profile (tokens)	22,182,816

4 Experimental Results

Table 3 shows the results of document-centric and candidate-centric forms of QE, using both the Bo1 and KL term weighting models. For both Bo1 and KL, the default setting of extracting the top $exp\_term = 10$ most informative terms from the top $exp\_doc = 3$ ranked documents or candidates [1] is applied. Statistically significant improvements from the baselines are shown using the Wilcoxon signed rank test. At first inspection, it appears that query expansion can be successfully applied in an Expert Search task to increase retrieval performance.

Table 3. Results for query expansion using the Bo1 and KL term weighting models. Results are shown for the baseline runs, with document-centric query expansion (DocQE) and candidate-centric query expansion (CandQE). The best results for each measure, term weigthing model and voting technique combination are emphasised. Statistically significant improvements ($p \leq 0.05$) over the corresponding baseline are marked by *, while significant improvements ($p \leq 0.01$) are denoted **.

		TREC 2005		TREC 2006	
		MAP	P@10	MAP	P@10
Baselines	CombSUM	**0.2037**	**0.3240**	0.5188	0.6388
	expCombMNZ	0.2037	0.3100	0.5502	0.6837
Bo1					
DocQE	CombSUM	0.1742	0.2860	**0.5216**	**0.6510**
	expCombMNZ	**0.2185**	**0.3340***	**0.5606**	**0.6959**
CandQE	CombSUM	0.1473	0.2240	0.4203	0.5388
	expCombMNZ	0.1760	0.2500	0.4554	0.5939
KL					
DocQE	CombSUM	0.1805	0.2880	**0.5296**	**0.6490**
	expCombMNZ	**0.2231***	**0.3400****	**0.5689***	**0.7020**
CandQE	CombSUM	0.1627	0.2560	0.5195	0.6265
	expCombMNZ	0.2031	0.3100	0.5600	0.6592

Moreover, the document-centric QE outperforms the candidate-centric QE on both MAP and P@10, in all settings. It is possible that the default setting of *exp_doc* and *exp_term* used is not suitable for candidate-centric query expansion, because of the size of the candidate profiles being considered in the pseudo-relevant set. In particular, it can be seen that applying document-centric QE results in an increase over the baseline for the TREC 2006 topics, and when using expCombMNZ for the TREC 2005 topics - some of these improvements are statistically significant ($p <= 0.05$). Compared to the respective baselines, applying candidate-centric QE results in a degradation in performance for most settings using the TREC 2005 topics. Document-centric QE provides an increase in MAP and P@10 over the baselines, except when using CombSUM for the TREC 2005 topics.

Overall, the KL term weighting model performs better in terms of MAP and P@10 when compared to the baselines, than Bo1 achieves. This is interesting as previous thorough experiments on various test collections shows that Bo1 performs consistently better than KL on adhoc search tasks [1]. Note also, that applying document-centric QE to expCombMNZ will always results in an increase in performance if it increased the performance of the CombSUM baseline. This can be explained by the fact that the generated refined document ranking by applying QE is identical. It appears then that expCombMNZ is better than CombSUM at converting the refined document ranking into a ranking of candidates, in line with the same results for unrefined document rankings. Moreover, this follows the persistent high performance of expCombMNZ observed by Macdonald & Ounis in [8]. QE using documents has been well tested in classical IR systems, so it is no surprise that it performs well here in increasing the quality of the document ranking. However, as discussed in Section 3, candidate profiles are many times larger than standard documents, so it is possible that the default setting of $exp\_term = 10, exp\_doc = 3$ is not as suitable for candidate-centric QE. In the next section, we assess the extent to which the setting of the QE parameters can affect the retrieval performance of either forms of QE.

5 Effect of Query Expansion Parameters

In this section, we investigate the extent to which the parameters for QE have an effect on the retrieval performance of the expert search task. The parameters of query expansion are *exp_doc*, the number of top-ranked documents or candidates to be considered as the pseudo-relevance set, and *exp_term*, the number of informative terms to be added to the query. To fully investigate their effect, we perform a large-scale evaluation of many parameter combinations. We aim to conclude if one of document-centric or candidate-centric QE is more stable with respect to various parameter settings, and to have a better comparison of the two forms of QE, as well as the term weighting model employed.

For these experiments, we use the expCombMNZ voting technique, using only the TREC 2006 topics, as this is the best performing setting (see Section 4). To assess the stability of the approaches with respect to *exp_term* and *exp_doc*, we

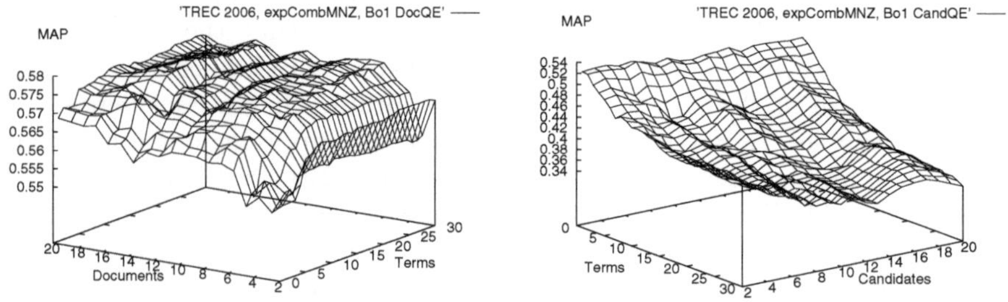

(a) Document-Centric Query Expansion (b) Candidate-Centric Query Expansion

Fig. 1. Surface plots of MAP for expCombMNZ, using the Bo1 term weighting model, when the *exp_doc* documents or candidates and *exp_term* terms query expansion parameters are varied

vary them and record the MAP of the generated run. In particular, we vary $2 \leq exp\_doc \leq 20$ and $1 \leq exp\_term \leq 30$. This generates a matrix of 570 points per setting. Figures 1 & 2 present surface plots of the Bo1 and KL QE settings, using the expCombMNZ voting techniques. In each figure, (a) uses document-centric query expansion, and (b) uses candidate-centric query expansion[3]. From the Figure 1 (a), shows that the number of document used as the pseudo-relevant set in document-centric QE has some effect on the retrieval performance of the generated ranking of candidates. In particular, it appears that using the 3 top-ranked documents is not a good setting, as can be seen from the crevice running across the surface plot on $exp\_doc = 3$; $exp\_doc = 2$ and $exp\_doc \geq 4$ are better settings. With respect to terms considered in the document-centric QE, using less than 10 terms means a drop-off in MAP, while for $exp\_term \geq 10$, the retrieval performance is stable. Indeed, the best performance achieved in Figure 1 (a) is MAP 0.5799, for $exp\_term = 16$ and $exp\_doc = 15$, compared to 0.5606 from Table 3, with default setting ($exp\_term = 10$, $exp\_doc = 3$).

Figure 1 (b) shows that as more terms are considered in candidate-centric QE, the MAP degrades. In particular, expanding the query by only 1 term ($m = 1$), still does not achieve the baseline MAP of 0.5502 from Table 3. In this case, varying the number of candidate profiles considered by the QE mechanism has little affect for a low number of terms. As the number of terms increases to 30, considering less profiles is favoured. The best setting on this figure is $exp\_term = 2$ and $exp\_doc = 6$, which gives a markedly better MAP of 0.5306, compared to the default setting of 0.4554.

For Figure 2, the patterns are similar for the Bo1 term weighting model as in Figure 1 (a). Again, the crevice for $exp\_term = 3$ is apparent. In addition, there is also a slight crevice in MAP at $exp\_term = 11$ for $exp\_doc > 10$. For candidate-centric query expansion (Figure 2 (b)), as the number of terms considered increases, there is again a decrease in MAP, but not as noticeable as in Figure 1 (b). Moreover, MAP is not as stable as exp_term increases.

[3] Note that some figures have different orientation to allow easier viewing.

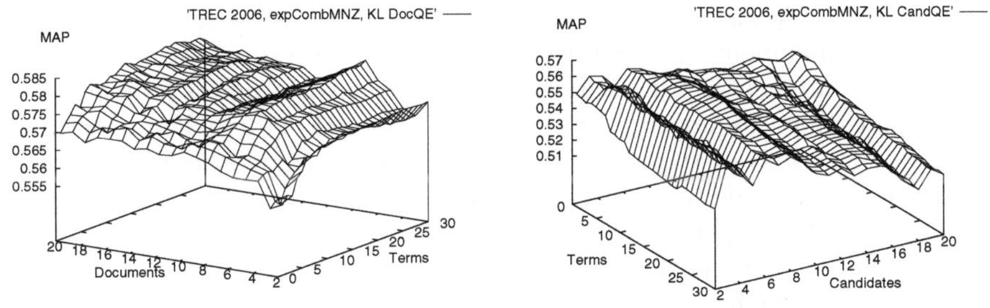

(a) Document-Centric Query Expansion (b) Candidate-Centric Query Expansion

Fig. 2. Surface plots of MAP for expCombMNZ, using the KL term weighting model, when the *exp_doc* documents or candidates and *exp_term* terms query expansion parameters are varied

Best settings for Figure 2 are (a) $exp\_term = 24$, $exp\_doc = 20$ (MAP 0.5827), and (b) $exp\_term = 6$, $exp\_doc = 3$ (MAP 0.5627), compared to the default settings of 0.5689 and 0.5600 respectively.

Overall, our large-scale experiments has allowed us to draw some conclusions concerning the applicability and stability of both forms of query expansion. Document-centric QE performs robustly, although *exp_doc* and *exp_term* should not be too small - in particular a fairly flat MAP surface is exhibited for $exp\_term \geq 6$ and $exp\_doc \geq 10$. For candidate-centric query expansion, more profound influencing of MAP is apparent as *exp_doc* and *exp_term* are varied. From our experiments, $3 \geq exp\_doc \geq 8$ and $exp_t erm \leq 5$, exhibit the most stable MAP surfaces for this form of query expansion. In particular, the quality of terms decreases rapidly, which is possibly due to the large and varied size of candidate profiles. In summary, overall it appears that document-centric QE is the more stable and effective of the two approaches.

6 Conclusions

We have investigated pseudo-relevance feedback QE in an Enterprise expert search setting. It was shown how query expansion can be applied in two different manners in the context of the voting approach for expert search, namely document-centric and candidate-centric QE. Experiments were carried out using two different voting techniques, and two different query expansion term weighting models. Topics from the TREC 2005 and 2006 Enterprise track Expert Search tasks were used. The results showed that firstly, QE can be successfully applied in expert search and secondly, using the default setting for query expansion, document-centric QE outperforms candidate-centric QE.

By performing a large-scale evaluation of the effect of the QE parameter settings, we observed that document-centric QE is stable with $exp\_term \geq 6$ and $exp\_doc \geq 10$. In contrast, candidate-centric QE was observed to be stable with respect to the number of candidate profiles considered (*exp_doc*), but increasing the number of expansion terms caused degradations in retrieval performance. Overall, the document-centric QE was more stable and consistently outperformed candidate-centric QE. The major difference when performing candidate-centric QE is that candidate profiles can be extremely large when compared to the documents considered in document-centric QE. We hypothesise that modern QE techniques struggle to identify informative terms when presented with such a large sample. In particular, the more terms identified by candidate-centric QE, the worse the retrieval performance. This also explains the better performance of the KL term weighting model for candidate-centric QE, as KL accounts for the size of the pseudo-relevant set when measuring the informativeness of terms.

Another possible explanation for the less stable performance of candidate-centric QE is due to 'topic drift'. A candidate profile contains many documents that represent the various interests of a candidate. When candidate-centric QE is performed, the expanded query terms may describe other common, not relevant interests of the candidates in the pseudo-relevant set, causing more candidates with these incorrect interests to be retrieved erroneously. Topic drift is less likely to occur with document-centric QE as documents are smaller and more likely to be about a single topic.

In the future, we would like to develop advanced forms of QE suitable for use for candidates. This would combine the best properties of document-centric and candidate-centric QE by only considering the top-ranked documents from the top-ranked candidates profiles as the pseudo-relevant set. An alternative possible approach for extracting informative terms from top-ranked candidate profiles might involve clustering the profile documents in each profile, to identify important interest areas of the candidates.

References

1. G. Amati. *Probabilistic Models for Information Retrieval based on Divergence from Randomness.* PhD thesis, Department of Computing Science, University of Glasgow, 2003.
2. G. Amati. Frequentist and bayesian approach to information retrieval. In *Advances in Information Retrieval, 28th European Conference on IR Research, ECIR 2006*, pages 13–24. Springer, April 2006.
3. K. Balog, L. Azzopardi, and M. de Rijke. Formal models for expert finding in enterprise corpora. In *Proceedings of the 29th ACM SIGIR 2006*, pages 43–50, Seattle, WA. August 2006.
4. N. Craswell, A. P. de Vries, and I. Soboroff. Overview of the TREC-2005 Enterprise Track. In *Proceedings of the 14th Text REtrieval Conference (TREC-2005)*, 2005.
5. M. Hertzum and A. M. Pejtersen. The information-seeking practices of engineers: searching for documents as well as for people. *Inf. Process. Manage.*, 36(5):761–778, 2000.
6. D. Hiemstra. *Using language models for information retrieval.* PhD thesis, Centre for Telematics and Information Technology, University of Twente, 2001.
7. C. Macdonald, B. He, V. Plachouras, and I. Ounis. University of Glasgow at TREC 2005: Experiments in Terabyte and Enterprise tracks with Terrier. In *Proceedings of 14th Text REtrieval Conference (TREC 2005)*, 2005.

8. C. Macdonald and I. Ounis. Voting for candidates: Adapting data fusion techniques for an expert search task. In *Proceedings of the 15th ACM CIKM 2006*. Arlington, VA. November 2006.
9. I. Ounis, G. Amati, V. Plachouras, B. He, C. Macdonald, and C. Lioma. Terrier: A high performance and scalable information retrieval platform. In *Proceedings of the OSIR Workshop 2006*, pages 18–25, August 2006.
10. J. Rocchio. *Relevance feedback in information retrieval.* Prentice-Hall, Englewood Cliff. NJ.

Using Topic Shifts for Focussed Access to XML Repositories

Elham Ashoori and Mounia Lalmas

Queen Mary, University of London
London, E1 4NS, UK
{elham,mounia}@dcs.qmul.ac.uk

Abstract. In focussed XML retrieval, a retrieval unit is an XML element that not only contains information relevant to a user query, but also is specific to the query. INEX defines a relevant element to be at the right level of granularity if it is exhaustive and specific to the user's request – i.e., it discusses fully the topic requested in the user's query and no other topics. The exhaustivity and specificity dimensions are both expressed in terms of the "quantity" of topics discussed within each element. We therefore propose to use the number of topic shifts in an XML element, to express the "quantity" of topics discussed in an element as a mean to capture specificity. We experimented with a number of element-specific smoothing methods within the language modelling framework. These methods enable us to adjust the amount of smoothing required for each XML element depending on its number of topic shifts, to capture specificity. Using the number of topic shifts combined with element length improves retrieval effectiveness, thus indicating that the number of topic shifts is a useful evidence in focussed XML retrieval.

1 Introduction

Content-oriented XML[1] retrieval systems aim at supporting more precise access to XML repositories by retrieving XML document components (the so-called XML elements) instead of whole documents in response to users' queries. Therefore, in principle, XML elements of any granularity (for example a paragraph or the section enclosing it) are potential answers to a query, as long as they are relevant. However, the child element (paragraph) may be more focussed on the topic than its parent element (the section), which may contain additional irrelevant content. In this case, the child element is a better element to retrieve than its parent element, because not only it is relevant to the query, but it is also specific to the query. Thus the aim of an XML retrieval system is to provide a *focussed access* to XML repositories by returning the most appropriate units of retrieval for a given query.

To identify what constitutes a most appropriate unit of retrieval, INEX, the initiative evaluation for XML retrieval (e.g., [6]) defined relevance in terms of

[1] XML stands for eXtensible Markup Language – see http://www.w3.org/

G. Amati, C. Carpineto, and G. Romano (Eds.): ECIR 2007, LNCS 4425, pp. 444–455, 2007.

two dimensions, *exhaustivity* and *specificity*, each defined on a graded scale. These two dimensions are respectively defined as "how exhaustively an element discusses the topic of request" and "how focussed an element is on the topic of request (i.e., discusses no other irrelevant topics)" [5]. The combination of these two relevance dimensions is used to identify relevant elements that are both exhaustive and specific to the user's query.

In IR, the main factors that affect the importance of a term in a text – in terms of how good it is at representing the content of the text – are the term frequency, the inverse document frequency, and the document length. To incorporate the "specificity" dimension in XML retrieval, we propose to exploit the "number of topic shifts" as another factor affecting the importance of a term in a text. We incorporate the "number of topic shifts" in the smoothing process within the language modelling framework. In the language modelling approach, smoothing refers to adjusting the maximum likelihood estimator for the element language model so as to correct inaccuracy arising from data sparseness. In the smoothing process, the probability of terms seen in an element are discounted mainly by combining the element language model with the collection language model, thus assigning a non-zero probability to the unseen terms.

We use "Dirichlet smoothing" approach [14] as the smoothing framework. Dirichlet smoothing is one of the popular document-dependent smoothing methods. Using this approach in XML retrieval enables us to adjust the amount of smoothing dynamically by the features of elements (e.g. length, topic shifts). In this work, we explore various ways to incorporate topic shifts in the smoothing process, either individually or combined with the length of XML elements. We investigate whether using topic shifts individually, or combined with length provides a better smoothing approach for the focussed access to XML documents.

The paper is organised as follows. Section 2 discusses related work. In section 3, we define the notion of topic shifts and how we calculate it. In section 4, we describe the language modelling formalism used to perform focussed retrieval. Section 5 describes the methodology used to compare the proposed topic shifts-based smoothing process, including the INEX test collection used to carry out this investigation. The experiments and results are discussed in Section 6. Section 7 concludes the paper, with some thoughts for future work.

2 Related Work

The language modeling approach to IR is a sound and flexible framework, not only in content-oriented XML retrieval (e.g., [9,13,12]), but also for IR research in general [3]. The basic idea of this approach is to estimate a language model for each document, and rank documents with respect to the likelihood that the query can be generated from the estimated language models. Retrieval performance of language modelling approaches have been shown to be sensitive to the smoothing parameters both in ad hoc IR [14] and in XML retrieval [9]. Although smoothing is essential in language modelling due to data sparseness, Zhai et al. [14] introduced another role for it. They showed that the effects of smoothing is very sensitive to the type of queries (long, short) which results in a new role for smoothing, query modelling, to "explain the common and non-informative words in a query". Following the query modelling role of smoothing, Hiemstra [8] introduced the term-specific smoothing and used feedback documents for estimating term-specific smoothing parameters.

In the context of ad hoc XML element retrieval, Kamps et al. [9] used a multinomial language model, with Jelinek-Mercer smoothing which is a linear interpolation of the element language model and the collection model. In this approach the smoothing parameter is fixed for all elements. They showed that the smoothing parameter indirectly introduces a length bias by increasing the importance of the presence of query terms in the retrieved elements. In this approach, the optimal amount of smoothing depends on the relevance assessments. If during assessment, the assessors favor the longer elements in the collection, little smoothing is required. More precisely, it was shown that a high amount of smoothing leads to the retrieval of shorter elements. This work is different from ours as we propose an element-dependent smoothing approach, where the amount of smoothing can be dynamically adjusted according to features of elements (in this paper, length, topic shifts). This allows us to investigate the effect of topic shifts (and element length) by making them "parameters" of the language modeling framework[2].

3 Topic Shifts

In this section, we describe how we measure the number of topic shifts of the elements forming a XML document. We use the number of topic shifts in an XML element, to express the "quantity" of topics discussed in an element. For this purpose, both the logical structure and a semantic decomposition of the XML document are needed. Whereas the logical structure of XML documents is readily available through their XML markup, their semantic decomposition needs to be extracted. To achieve that, we apply a topic segmentation algorithm based on lexical cohesion, TextTiling[3] [7], which has been successfully used in several IR applications (e.g., [2]). The underlying assumption of topic segmentation algorithms based on lexical cohesion is that a change in vocabulary signifies that a topic shift occurs. This results in topic shifts being detected by examining the lexical similarity of adjacent text segments.

TextTiling is a linear segmentation algorithm that considers the discourse unit to correspond to a paragraph, and therefore subdivides the text into multi-paragraph segments. TextTiling is performed in three steps. In the first step, after performing tokenization, the text is divided into pseudo-sentences of size W, called token-sequences. Next, these token-sequences are grouped together into blocks of size K. The gap between two adjacent blocks constitutes a potential

[2] This work is also different from ours, as the former is concerned with the task of estimating the relevance of XML elements, and not the focussed access to XML elements (see Section 5).

[3] http://elib.cs.berkeley.edu/src/texttiles/

boundary for a semantic segment. To identify the actual boundaries, a depth score is computed for each potential boundary, by using the similarity scores assigned to the neighbouring gaps between blocks, and by applying a smoothing process. The algorithm determines the number of segments, referred to as tiles, by considering segment boundaries to correspond to gaps with depth scores above a certain threshold. The detected boundaries are then adjusted to correspond to the actual discourse unit breaks, i.e., the paragraph breaks.

The semantic decomposition of an XML document is used as a basis to calculate the number of topic shifts in each XML element forming that document. We consider that a topic shift occurs (i) when one segment ends and another segment starts, or (ii) when the starting (ending) point of an XML element coincides with the starting (ending) point of a semantic segment.

The *number of topic shifts* in an XML element e in document d is defined as:

$$actual\_topic\_shifts(e, d) + 1 \tag{1}$$

where $actual\_topic\_shifts(e, d)$ are the actual occurrences of topic shifts in element e of document d. We are adding 1 to avoid zero values. With the above definition, the larger the number of topic shifts, the more topics are discussed in the element, which would indicate that the content of element is less focussed with respect to the overall topic discussed in the element. By considering the number of topic shifts occurring in an element instead of the number of topics discussed (in our case modelled as the number of tiles), we are able to distinguish the cases where the topic shift occurs not within the actual content of an element, but at its boundaries.

4 Element-Specific Smoothing Using Topic Shifts

Since XML elements of any granularity are potential answers to a query, we estimate a language model for each XML element in the collection. The element language model is smoothed using a Dirichlet prior [14] with the collection language model as the reference model.

If we estimate a language model for each element, then the relevance of an element e to a given query q is computed as how likely the query can be generated from the language model for that element. We rank elements based on the likelihood for a query $q = (t_1, t_2, ..., t_n)$ to be generated from an element e as:

$$P(t_1, .., t_n|e) = \prod_{i=1}^{n} (\frac{c(t_i, e) + \mu P(t_i|C)}{\mu + |e|}) \tag{2}$$

$$= \prod_{i=1}^{n} ((1 - \frac{\mu}{\mu + |e|}) \frac{c(t_i, e)}{|e|} + \frac{\mu}{\mu + |e|} P(t_i|C)) \tag{3}$$

$$= \prod_{i=1}^{n} ((1 - \alpha_e) P_{ml}(t_i|e) + \alpha_e P(t_i|C)) \tag{4}$$

where

- t_i is a query term in q,
- $c(t_i, e)$ is the number of occurrences of the query term t_i in element e,
- μ is a constant,
- $|e|$ is the number of terms in element e,
- $P_{ml}(t_i|e) = \frac{c(t_i,e)}{|e|}$ is the probability of observing term t_i in element e, estimated using the maximum likelihood estimation,
- $P(t_i|C) = \frac{ef(t_i)}{\sum_t ef(t)}$ is the probability of observing query term t_i in the collection where $ef(t)$ is the number of XML elements in which the term t occurs.
- $\alpha_e = \frac{\mu}{\mu+|e|}$ is an element-dependent constant which is related to how much probability mass will be allocated to unseen query terms, i.e., the amount of smoothing.

Since the maximum likelihood estimator will generally underestimate the probability of any term unseen in the element, the main purpose of smoothing is to improve the accuracy of the term probability estimation. If we are concerned with the exhaustivity dimension of relevance, then we may expect that most of the query terms to appear in an element for the element to be retrieved. In this case, one would expect that the term probability estimates are more reliable for long elements as they contain more terms compared to the short elements. Therefore, a shorter element needs to be more smoothed with the collection model compared to a longer element. This shows that a higher value of α_e is needed to capture exhaustivity in small elements. Smoothing with Dirichlet prior (Equation 4) satisfies this requirement as the value of α_e depends on the length of the elements.

The above smoothing process is reasonable if we are not concerned with the specificity dimension. In INEX, specificity is automatically measured by calculating the ratio of the relevant content in an XML element (see Section 5). This implies that unseen terms are less of an issue for small elements compared to the above case. Therefore a smaller amount of smoothing (a lower value of α_e) is needed to capture specificity in small elements than the amount of smoothing required to capture exhaustivity. Due to this contradictory behaviour in the required amount of smoothing – if we want to capture both exhaustivity and specificity – Equation 4 in its current version cannot be used to capture both relevance dimensions if only length is taken into account.

To accommodate for the specificity dimension, we propose to set α_e, the amount of smoothing, to be proportional to the number of topic shifts in the element. The idea of incorporating topic shifts in this manner originates from the fact that if the number of topic shifts in an element is low and an element is relevant, then it is likely to contain less non-relevant information compared to the case where a high number of topic shifts exists.

It might be argued that in general, when the length of an element increases, it is highly likely that it will discuss more topics. However, this is not always the case, as it was shown in [1], where the number of topic shifts of parent elements was compared to that of their children. Even though the length from children to their parents increases, the number of topic shifts in the majority of cases stays the same, i.e. it does not vary when the length increases. As topic shifts and length are two distinct evidences [1], we explore several ways to compute the element-dependent constant α_e (amount of smoothing) in Equation 4 as a function of its length, its number of topic shifts and a combination of both. We replace length by the number of topic shifts in the original formula to compare how these two retrieval settings are useful to capture exhaustivity. Next, we replace the length with the inverse of length and inverse of topic shifts to capture specificity. Finally we combine length and topic shifts in a retrieval setting to capture both exhaustivity and specificity. We, thus, experiment with five different retrieval approaches:

1. $\alpha_e = \frac{\mu}{\mu+|e|}$ implies that longer elements need less smoothing. This approach is the original Bayesian smoothing with Dirichlet priors. We refer to this approach as our baseline approach (L).
2. $\alpha_e = \frac{\mu}{\mu+1/|e|}$ implies that shorter elements need less smoothing. This means that the presence of a query term in an element is rewarded if the number of terms in the element is small. We refer to this approach as (1/L).
3. $\alpha_e = \frac{\mu}{\mu+|T|}$ implies that elements with a high number of topic shifts need less smoothing. We refer to this approach as (T).
4. $\alpha_e = \frac{\mu}{\mu+1/|T|}$ implies that elements with a lower number of topic shifts need less smoothing. This means that the presence of a query term in an element is rewarded if the number of topic shifts in the element is low. We refer to this approach as (1/T).
5. $\alpha_e = \frac{\mu}{\mu+\frac{|e|}{|T|}}$ implies that elements should be smoothed based on the average number of terms per topic shifts of element. This is an approximation of the average number of terms per topic in an XML element. In this way, we return back to the normal interpretation of smoothing in Equation 4 but we consider a more refined version of length. In this case we differentiate between two elements with equal length and different numbers of topic shifts so that the presence of a query term in element with a lower number of topic shifts is rewarded. We refer to this approach as (L/T).

5 Methodology

In our work, we use the INEX-2005 test collection. The INEX collection, *Version 1.8*, contains 16,819 scientific articles from 24 IEEE Computer Society journals, marked up in XML, consisting of over 10 million elements of varying length.

We use the title field of the 29 content-only (CO) topics of *Version 2005-003* of the INEX 2005 data set[4]. CO topics are requests that ignore the document structure and contain only content related conditions, and at this stage of our work, are sufficient for our investigation. We evaluate our approaches against

[4] In INEX 2005, these topics are referred to as CO+S.

the relevance assessments *Version adhoc2005-assessments-v7*. For INEX 2005, exhaustivity is measured on a *3 + 1*-point scale: highly exhaustive (e=2), somewhat exhaustive (e=1), not exhaustive (e=0) and too small (e=?). In this work, we ignore too small elements. The specificity dimension is automatically measured on a continuous scale [0,1], by calculating the ratio of the relevant content of an XML element after the assessors highlighted text fragments containing only relevant information.

The official evaluation metrics employed in INEX 2005 are the eXtended Cumulated Gain (XCG) metrics [10], which include the user-oriented measures of extended cumulated gain ($nxCG[i]$) and the system-oriented effort-precision/gain-recall measures ($MAep$). For a given rank i, $nxCG[i]$ reflects the relative gain the user accumulated up to that rank, compared to the gain he/she could have attained if the system would have produced the optimum best ranking.

The effort-precision *ep* at a given gain-recall value *gr* is defined as the number of visited ranks required to reach a given level of gain relative to the total gain that can be obtained. The non-interpolated mean average effort-precision, $MAep$, is calculated by averaging the effort-precision values measured at natural recall-point, i.e., whenever a relevant XML element is found in the ranking.

INEX employs quantization functions to combine the two graded relevance dimensions, by providing a relative ordering of the various combinations of e-s values and a mapping of these to a single relevance scale in [0, 1], as required by the XCG metrics. In INEX 2005, two quantization functions were used:

$$f_{strict}(e,s) := \begin{cases} 1 \text{ if } e = 2 \text{ and } s = 1, \\ 0 \text{ otherwise.} \end{cases} \tag{5}$$

$$f_{generalised}(e,s) := \begin{cases} e * s \text{ if } e \in \{1,2\}, \\ 0 \quad\quad \text{otherwise.} \end{cases} \tag{6}$$

The strict quantization function f_{strict} is used to evaluate retrieval methods with respect to their capability of retrieving highly exhaustive and highly specific elements (e=2, s=1). The generalised quantization function $f_{generalised}$ credits elements according to their degree of relevance, hence allowing modelling varying levels of user satisfaction gained from not fully specific and highly exhaustive elements, i.e., less relevant elements.

The retrieval task addressed in this work is focussed XML retrieval. INEX has defined various XML retrieval scenarios, each corresponding to a specific task. In the *focussed* task, the aim is for XML retrieval systems to return to the user a non-overlapping ranked list of the most exhaustive and specific elements on a relevant path[5]. The five approaches described in the previous section will rank elements, for example, considering the number of topic shifts. They will, however, not produce an overlap-free ranking. There are sophisticated ways to remove overlapping elements (e.g., [11]). In this work we restrict ourselves to a

[5] A relevant path is a path within the XML tree of a given XML document, whose root node is the root element and whose leaf node is a relevant element that has no or only irrelevant descendants.

post-filtering on the retrieved ranked list by selecting the highest scored element from each of the paths, as our main interest here is to investigate how the proposed smoothing approaches can help retrieval effectiveness.

To calculate the number of topic shifts in each XML element, our first step is to decompose the INEX XML documents into semantic segments through the application of TextTiling. We consider the discourse units in TextTiling to correspond to *paragraph* XML elements. We considered paragraph elements to be the lowest possible level of granularity of a retrieval unit. Although this can be viewed as collection-dependent and might change from one collection to the next, it is likely that for many XML content-oriented collections, meaningful content will occur mainly at paragraph level and above.

We set the TextTiling parameters to $W = 10$ and $K = 6$, which is based on a heuristic setting $W * K$ to be equal to the average paragraph length (in terms of the number of terms) [7]. After the application of TextTiling, we compute the number of topic shifts in elements.

6 Experiments and Results

In this section, we report on the experiments, and their results, that were carried out in order to investigate the effects of topic shifts in the smoothing process. We experiment with a wide range for μ between [0, 20000] to study the behaviour of each individual retrieval approach. To compare the five smoothing approaches (*L, 1/L, T, 1/T, L/T*), we select a best run (in terms of MAep) for each approach and then compare the behaviour of these best runs based on nxCG.

For each of the approaches, the top 1500 ranked elements are returned as answers for each of the CO topics. For the user-oriented evaluation, we report nxCG at three different early cut-off points (10, 25, 50). In addition, the nxCG graphs for both the full rank and the early rank levels are given. For the system-oriented evaluation, MAep is reported. For both evaluations, both strict and generalised quantization functions are used.

To determine whether the differences in performance between two approaches are statistically significant, we use the bootstrapping significance testing [4]. Improvements at confidence levels 95% and 99% over the baseline are respectively marked with + and ++. Similarly, decreases in performance at confidence level of 95% and 99% are marked with – and – –.

Table 1 shows a summary of the results. This table presents, for each quantization function, the results for both the user- and the system-oriented evaluation of the five retrieval approaches, with the L approach acting as the baseline.

We first consider the results in terms of mean average effort precision (MAep), shown in Figure 1 and the last column of Table 1.

Under the *generalised quantization function*, the MAep ranks L/T approach followed by L above the other approaches reflecting that on average the user needs to spend less effort when scanning the output of L/T to achieve the same level of gain. However the difference is not significant. To obtain a better understanding we look at the performance at different values for parameter μ.

Table 1. Focussed retrieval task: MAep and nxCG at different cut-off points considering L as baseline

Approach	μ	nxCG@10	nxCG@25	nxCG@50	MAep
			General		
L	$\mu = 256$	0.2634	0.2387	0.2258	0.0938
$1/L$	$\mu = 0.01$	0.2333(-11%)	0.2358(-1.2%)	0.2198(-2.5%)	0.0911(-2.9%)
T	$\mu = 7$	**0.2638**(0.1%)	0.2391(1.6%)	0.2187(-3.1%)	0.0899(%-4.2)
$1/T$	$\mu = 0.15$	0.2388(-9.3%)	0.2297(-3.7%)	0.2138(-5.3%)	0.0894(-4.7%)
L/T	$\mu = 448$	0.2603(-1.1%)	**0.2506(4.9%)**	**0.2424(7.4%)**	**0.0992**(5.8%)
			Strict		
L	$\mu = 256$	0.069	0.1295	0.1351	0.029
$1/L$	$\mu = 0.01$	**0.0863**(25%)	0.15(15.8%)	0.1513(12%)	0.0305(5.2%)
T	$\mu = 8$	0.069(0%)	0.1333(2.9%)	0.1411(4.4%)	0.0251(-13.4%)
$1/T$	$\mu = 0.7$	**0.0863(25%)**	**0.1515(17%)**	0.1688(25%)	0.0304(4.8%)
L/T	$\mu = 1280$	0.0687(-0.4%)	0.1446(11.7%)	**0.1843(36.4%)(++)**	**0.0308(6.2%)**

Figure 1(b) shows the mean average effort precision for μ between [0, 20000]. We observe that L/T approach shows better performance than L regardless of the values of μ, which indicates that elements with equal length and smaller number of topic shifts require less smoothing. This is due to the fact that in the L/T approach, the presence of a query term in an element with a lower number of topic shifts is rewarded.

Under the *strict quantization function*, results show that L/T, $1/L$, and $1/T$ are the most effective approaches with almost the same MAep, whereas T did not perform particularly well. The $1/T$ approach considers less smoothing for elements with fewer number of topic shifts represented by the lower value for α_e. The $1/L$ approach is in the opposite direction of the standard Dirichlet smoothing and considers more smoothing for the larger elements. These results support the argument in Section 4 in which we suggested that the Dirichlet smoothing in its standard formulation is not sufficient to satisfy the "specificity" dimension of relevance. These results show that for retrieving highly specific and highly exhaustive elements, in the strict case, less smoothing is required for elements that are either small or contain fewer number of topic shifts than those that are longer or contain a higher number of topic shifts. Similar to the observed behaviour for the generalised quantization function, L/T shows better performance than L in most of the values of μ.

Overall, the L/T approach, where the number of topic shifts combined with length affects the amount of smoothing, performs better than other retrieval approaches when evaluated using the system-oriented measures.

Next, we focus on the *user-oriented evaluation* and discuss the results obtained using the nxCG measure, shown in Figure 2 and Table 1. Under the *generalised quantization function*, the baseline approach, L, shows better performance at the very early ranks (1% ranks) (see Figure 2(d)). For the other rank positions, the combination of length and topic shifts (L/T) is the most effective approach (see Figure 2(c)). These approaches are useful to satisfy a user who also gains from less relevant elements, i.e., not fully specific and highly exhaustive elements. However, for retrieving highly specific and highly exhaustive elements, the strict case, both $1/L$ and $1/T$ approaches improve the results at the early cut-

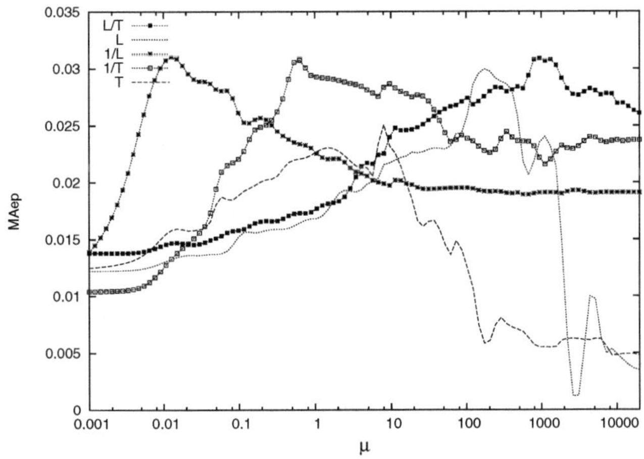

(a) strict

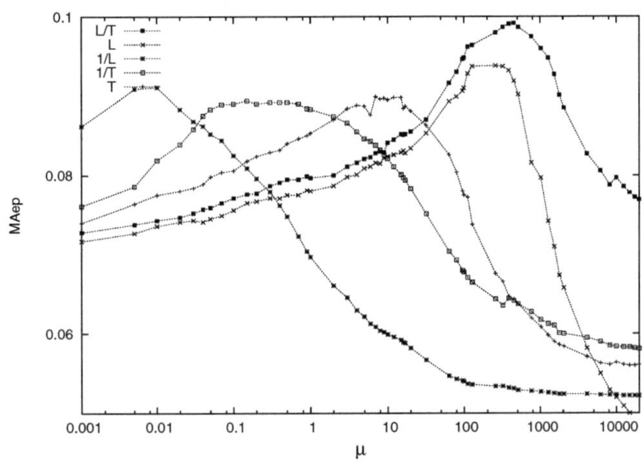

(b) general

Fig. 1. Mean Average effort precision (MAep) for different element-specific smoothing approaches

off points 10 and 25, where we get (+25%,17.8%) and (25%,15%) improvements, respectively (see Table 1). Similar to the system-oriented evaluation, these results again support our argument in Section 4, i.e. the smoothing process should be treated differently depending on the relevance dimensions.

To conclude, the use of the number of topic shifts led to improvement of performance particularly when combined with element length, thus indicating that the number of topic shifts is a useful evidence in focussed XML retrieval.

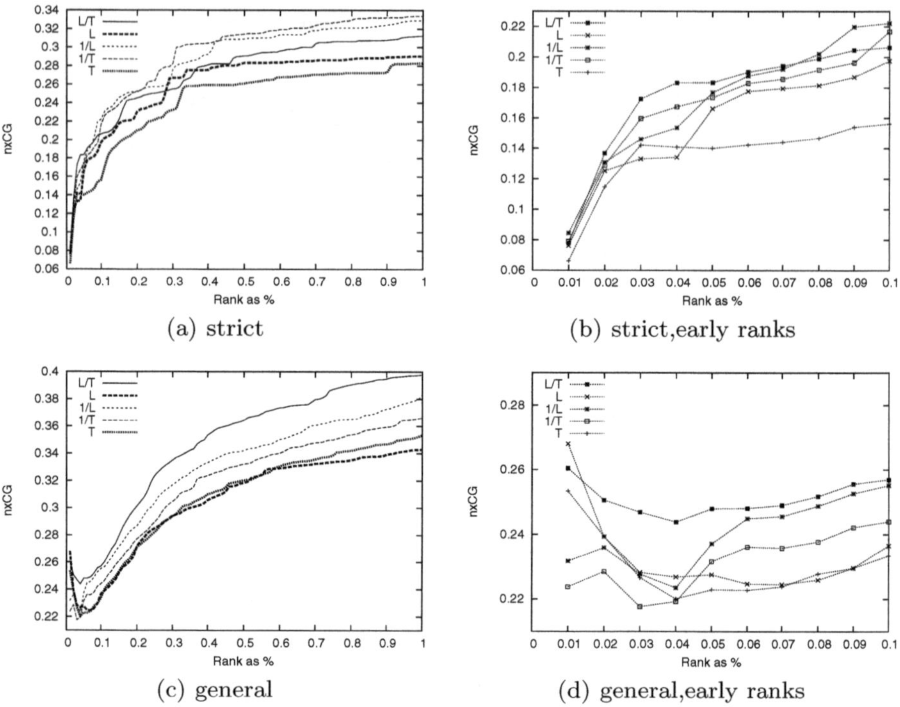

Fig. 2. Evaluation based on normalised eXtended Cumulated Gain (nxCG)

7 Conclusion

INEX defines a relevant element to be at the right level of granularity if it is exhaustive to the user request – i.e., it discusses fully the topic requested in the user's query – *and* it is specific to that user request – i.e., it does not discuss other topics. The exhaustivity and specificity dimensions are both expressed in terms of the "quantity" of topics discussed within each element. We therefore use the number of topic shifts in an XML element, to express the "quantity" of topics discussed in an element. We experimented with a number of element-specific smoothing methods within the language modelling framework. These methods enable us to adjust the amount of smoothing required for each XML element with respect to the specificity and exhaustivity dimensions of relevance. Using the number of topic shifts combined with element length improved retrieval effectiveness, thus indicating that the number of topic shifts is a useful evidence in focussed XML retrieval. Our other finding was that the smoothing process should be treated differently if we are concerned with the specificity dimension.

For our future work, our first aim is to go beyond the element level for smoothing and provide a term-specific smoothing based on the number of topic shifts and the distribution of the terms inside XML elements. Secondly, we will investigate the effects of applying the proposed smoothing approach on the Wikipedia XML collection, which is the collection used in INEX 2006[6]. On another direction, we will investigate whether other segmentation algorithms are better suited for XML documents, and whether we can eventually obtain other, more effective means to calculate the number of topic shifts of XML elements.

Acknowledgments

This work was carried in the context of INEX, an activity of the DELOS Network of Excellence.

References

1. E. Ashoori, M. Lalmas and Theodora. Tsikrika. Examining Topic Shifts in Content-Oriented XML Retrieval, submitted, 2006.
2. C. Caracciolo and M. de Rijke. Generating and retrieving text segments for focused access to scientific documents. In *Proceedings ECIR 2006*, pages 350–361, 2006.
3. W. B. Croft and J. Lafferty. *Language Modeling for Information Retrieval.* Kluwer Academic Publishers, 2003.
4. B. Efron and R. J. Tibshirani. *An Introduction to the Bootstrap.* Chapman & Hall, 1993.
5. N. Fuhr and M. Lalmas. Report on the INEX 2003 Workshop, Schloss Dagstuhl, 15-17 December 2003. *SIGIR Forum*, 38(1):42–47, June 2004.
6. N. Fuhr, M. Lalmas, S. Malik, and G. Kazai, editors. *Advances in XML Information Retrieval and Evaluation: Fourth Workshop of the INitiative for the Evaluation of XML Retrieval (INEX 2005)*, 2006.
7. M. A. Hearst. Multi-paragraph segmentation of expository text. In *Proceedings of the 32nd Association for Computational Linguistics*, pages 9–16, 1994.
8. D. Hiemstra. Term-specific smoothing for the language modeling approach to information retrieval: the importance of a query term. In *Proceedings of the 25th ACM SIGIR Conference*, pages 35–41, 2002.
9. J. Kamps, M. de Rijke, and B. Sigurbjörnsson. The importance of length normalization for XML retrieval. *Information Retrieval*, 8(4):631–654, 2005.
10. G. Kazai and M. Lalmas. INEX 2005 Evaluation Metrics. In Fuhr et al. [6].
11. V. Mihajlovic, G. Ramírez, T. Westerveld, D. Hiemstra, H. E. Blok, and A. P. de Vries. TIJAH Scratches INEX 2005: Vague Element Selection, Image Search, Overlap, and Relevance Feedback. In Fuhr et al. [6].
12. P. Ogilvie and J. Callan. Hierarchical language models for XML component retrieval. In *Proceedings of the INEX 2004 Workshop*, pages 224–237, 2005.
13. G. Ramirez, T. Westerveld, and A. P. de Vries. Using structural relationships for focused XML retrieval. In *Proceedings FQAS 2006*, 2006.
14. C. Zhai and J. Lafferty. A study of smoothing methods for language models applied to ad hoc information retrieval. In *Proceedings of the 24th ACM SIGIR conference*, pages 334–342, 2001.

[6] http://inex.is.informatik.uni-duisburg.de/2006/index.html

Feature- and Query-Based Table of Contents Generation for XML Documents

Zoltán Szlávik, Anastasios Tombros, and Mounia Lalmas

Department of Computer Science,
Queen Mary University of London

Abstract. The availability of a document's logical structure in XML retrieval allows retrieval systems to return document portions (elements) instead of whole documents. This helps searchers focusing their attention to the relevant content within a document. However, other, e.g. sibling or parent, elements of retrieved elements may also be important as they provide context to the retrieved elements. The use of table of contents (TOC) offers an overview of a document and shows the most important elements and their relations to each other. In this paper, we investigate what searchers think is important in automatic TOC generation. We ask searchers to indicate their preferences for element features (depth, length, relevance) in order to generate TOCs that help them complete information seeking tasks. We investigate what these preferences are, and what are the characteristics of the TOCs generated by searchers' settings. The results have implications for the design of intelligent TOC generation approaches for XML retrieval.

1 Introduction

As the *eXtensible Markup Language (XML)* is becoming increasingly used, retrieval engines that allow search within collections of XML documents are being developed. In addition to textual information, XML documents provide a markup that allows the representation of the logical structure of XML documents in content-oriented retrieval. The logical units, called elements, are encoded in a tree-like structure by XML tags. The logical structure allows retrieval systems to return document portions that may provide better searcher satisfaction, as the retrieved information will be more focussed.

XML retrieval has received large interest over the last few years, mainly through the INEX initiative [5,4]. The interactive aspect of XML IR has recently been investigated through the interactive track at INEX (iTrack) [16,10]. One of the findings of the iTrack was the importance of the hierarchically structured presentation of documents and elements in the result list, and the table of contents of the documents [9,8]. The table of contents (TOC), in particular, provided context to the retrieved elements. The use of TOCs offered an overview of the document structure and showed the relevant elements and their relations to each other. This was appreciated by many searchers [11].

G. Amati, C. Carpineto, and G. Romano (Eds.): ECIR 2007, LNCS 4425, pp. 456–467, 2007.

In the systems used in the iTrack, TOCs had two main limitations: i) they were static, i.e. the same TOCs for a given document were displayed for all searcher queries, and ii) they were manually defined, i.e. before the documents were used in the systems, they had to be analysed and several (types of) elements selected to be included in TOCs, and this selection was not automatic.

These limitations were verified in [13,14], where it was also investigated what a useful TOC should be like according to searchers of an interactive IR system. It was found that a TOC should reflect which elements are possibly relevant to the searchers' queries, i.e. TOC items of relevant elements are more important to be shown in the TOC than those of non relevant ones. It was also found that the display of a TOC should depend on the length of the elements, e.g. longer sections are more important to include in the TOC. Another finding was that it is important to consider how deeply an element is in the document structure, i.e. for most of the XML documents currently being used in XML retrieval, a one or two-level deep TOC is probably too shallow, while four or five level deep TOCs may sometimes be too deep to help searchers with their information seeking task.

Based on the above, in this work we follow an approach to automatically generate "dynamic" TOCs by considering the characteristics of XML documents and the searcher's query. The TOC generation algorithm is inspired by early text summarisation (sentence extraction) systems (e.g. [3,17]). We consider features of elements such as their length, depth and relevance and combine these in order to determine whether an element should have a reference in the TOC. Our approach allows us to create a TOC for any XML document, and the TOC will be biased towards the searcher's query. In this paper, we investigate how this approach can be used in a user-oriented system, and what features are more important in TOC generation than others according to searchers. We also examine what the size of a TOC should be so that searchers can find the relevant information effectively.

To answer these questions, we created a system. Searchers were asked to consider information seeking tasks and to find relevant information within XML documents. They were allowed to adjust the importance of element features (i.e. length, depth, relevance). By adjusting these, searchers were able to alter the characteristics of the current TOC and the aim was to generate an appropriate TOC for documents in the context of the current query. We recorded these searcher preferences along with questionnaires and analysed them.

The system and the methodology followed are described in Section 2, followed by the detailed description of the TOC generation algorithm (Section 3). Section 4 presents and analyses the results, and we close with discussion and conclusions in Section 5.

2 Experimental Setup

We followed a methodology that is based on simulated work task situations [1]. Searchers were given work task descriptions so they could search for relevant information. During their search, they were asked to identify the best TOC of the current document with respect to the current work task by adjusting the importance of element features.

Searchers were asked to read the work task descriptions, proceed to the document view of as many documents as they wish, and adjust their preferences for three element features (length, relevance, depth) and a threshold by moving sliders on the interface. By adjusting the sliders, searchers were able to alter the characteristics of the current TOC. When they felt that the displayed TOC was helpful enough to assist them in finding relevant information, they could move on to the next document or topic.

Participants were asked to fill in questionnaires before and after the experiment, they were given detailed introduction to what their task was and no time restrictions to finish the experiment were imposed. After filling in the entry questionnaire and having read the introduction, searchers were presented with the first topic description and links to the corresponding relevant documents (Figure 1). The order of the displayed topics was randomised to avoid any effect caused by one particular order.

Let us suppose that you work as teaching assistant for operating systems. This inspired your interest in microkernel operating systems, and you are looking for documents that talk about operating systems with microkernel. Documents or document segments talking only about operating systems or microkernel are not of your interest. (A Microkernel is a highly modular collection of powerful OS-neutral abstractions, upon which can be built operating system servers.)

The following documents contain possibly relevant information. Use the sliders to get the best table of contents and explore the document through clicking on the items in the left window.
CLICK, CLICK or CLICK.

Fig. 1. A topic description and links to its documents

After choosing a document, the document view was shown (Figure 2). This consisted of four main parts:

- on the left, sliders associated with element features were shown. These needed to be adjusted by searchers to generate TOCs;
- on the bottom left, the generated TOC was shown;
- on the right hand side, the contents of the document or element was presented. These changed when searchers clicked on an item in the TOC;
- on the top left corner, links to the topic description, next topic and final page were displayed.

By clicking on the 'finish' link, the exit questionnaire was shown, where we recorded information about the searchers' perception of the system and TOC generation, e.g. the strategies searchers used when adjusting the sliders on the main screen.

Documents from two XML document collections - IEEE and Wikipedia [2] - were used[1]. Ten topics that were available for these collections were selected by random and converted into work task descriptions. This gave us five work

[1] These are the test collections used in INEX.

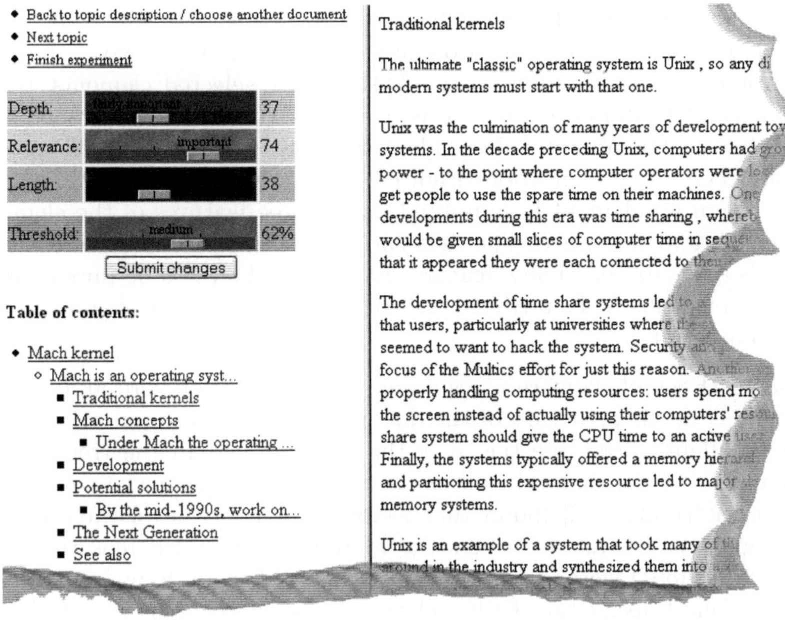

Fig. 2. Screen shot of the main screen with sliders, TOC and element display

tasks for each collection. For each topic, three to five relevant documents were selected. Relevant documents were obtained by formulating queries from the topic descriptions, running these queries in the TopX system[2] [15] and selecting the most relevant documents from the result list. The retrieval status values of elements were saved to be used in the TOC generation (Section 3). Documents of various sizes were selected, the shortest document contained 334 bytes of text while the longest 49KB. We made an effort to select documents from both collections for a particular topic; there were only two topics where relevant documents were found in both collections.

As a result of the topic and document selection, 33 documents and 10 topics were selected from the two collections. These provide an appropriate level of diversity thus ensuring that our results are not biased by topic and document selection.

3 TOC Generation

In this section, the algorithm to generate TOCs is described. The algorithm aims to identify among a set of XML elements, those that will form the TOC. It makes use of an element score that is calculated for every element in consideration. If the score of an element is higher than a certain threshold value (described below), the element is considered as a TOC element. Ancestors of such elements, i.e.

[2] TopX is the system used in INEX for the collection exploration phase.

elements higher in the XML hierarchy, are also used to place the TOC elements into context. For example, a section reference in a TOC without the chapter it is in would be just 'floating' in the TOC. These selected elements' titles are displayed as TOC items. If no title is available, the first 25 characters of the text are shown. The ancestor-descendant relation of elements is reflected, as in a standard TOC, by indentation (Figure 2).

The score of an element is computed using three features of the element: its depth, length and relevance. The first two are element-based features, whereas the third is query-based. These features have been shown to be important characteristics in various XML retrieval tasks [4], although other features can be also taken into account.

Depth score. Each element receives a depth score between zero and one, based on where it is in the structure of the document. In our case, an *article* element is always at depth level one (i.e. it is the root element in the tree structure). Descendants of a depth level one element are at depth level two (e.g. /article[1]/section[4]), etc. [13] found that elements at depth level three of a TOC were the most important to access the relevant content whereas the adjacent levels (two and four) were found less important, and so on. Sigurbjörnsson ([12], Ch. 8) also found, using the IEEE collection, that searchers mostly visited level 2-3 elements while looking for relevant information. Hammer-Aebi et al.[6] confirmed that searchers found the highest number of relevant elements at levels two to four. Since the latter work used a different XML collection (Lonely Planet [18]), the importance of these levels seems to be general for XML collections. To reflect these findings, the following scoring function was used to calculate an element's depth score (Equation 1):

$$S_{depth}(e) = \begin{cases} 1 & \text{if } depth(e) = 3, \\ 0.66 & \text{if } depth(e) \in \{2,4\}, \\ 0.33 & \text{if } depth(e) \in \{1,5\}, \\ 0 & \text{otherwise} \end{cases} \tag{1}$$

where $S_{depth}(e)$ denotes the depth score of element e.

Length score. Each element receives a length score, which is also normalised to one. The normalisation is done on a logarithmic scale [7], where the longest element of the document, i.e. the root element, receives the maximum score of one (Equation 2):

$$S_{length}(e) = \frac{\log(TextLength(e))}{\log(TextLength(root))} \tag{2}$$

where $S_{length}(e)$ is the length score of element e, $root$ is the root element of the document structure and $TextLength$ denotes the number of characters of the element.

Relevance score. A score between zero and one is used to reflect how relevant an element is to the current search topic. The scores were those given by the search engine used in INEX for the collection exploration [15] (i.e. a normalised retrieval status value). The RSVs are obtained in the document selection phase (described in Section 2).

Feature weighting. The scores of the above three features are combined so that we can emphasize the importance of a feature over another. This is done by using a weighted linear combination of the feature scores (Equation 3). Searchers are allowed to set the weights themselves. This allows us to investigate what searchers find important in TOC generation, and also, to determine what weights should be used to generate TOCs based on such features.

$$S(e) = \sum_{f \in F} W(f) \cdot S_f(e) \tag{3}$$

where $S(e)$ denotes the overall score of element e, F is the set of the three features, $W(f)$ is the weight of feature f and $S_f(e)$ denotes the score that is given to element e based on its feature f.

Threshold. To determine the lowest score an element must achieve in order to be included in the TOC, we use a threshold value. As well as the feature weights described above, this value is set by the searchers of the system. This allows us to determine what the desirable size of a TOC should be. In our algorithm, if the threshold is set to 100% only elements with the maximum depth, relevance and length scores will be included in the TOC (i.e. $\sum S_f(e) = 1$). If the threshold is set to zero, every element with greater than zero score will be in the TOC. We use a default value of 50%.

4 Results and Analysis

In this section we describe and analyse the results of our experiment. We start with results regarding participation (Section 4.1) followed by a detailed analysis of the collected data. We investigate what slider values were set and what were the main characteristics of the generated TOCs when searchers were finished with a document (Section 4.2), whether there were differences in terms of preferences among searchers (Section 4.3) and whether differences could be found among documents of the two collections and among the topics used (Section 4.4).

4.1 Participation and Questionnaires

50 searchers, mainly with a computer science background, responded to our call for participation. To record information only from searchers who spent a significant amount of time participating in the experiment (thus providing usable data), we filtered the log data so that only those involving at least three different documents were kept and analysed. As a result, 31 searcher sessions were analysed where participants used an average of 7.74 (out of the the maximum ten) topics and 15.58 (out of the maximum 33) documents. This gave us 483

different settings. The slider values after a searcher finished with a document were considered as a single setting.

In the post experiment questionnaires, searchers indicated how easy they had found to learn and understand the usage of the system. On a seven point scale, they indicated an average of 4.85 and 4.64 respectively, where 1 meant 'was not easy at all' and 7 meant 'it was extremely easy'. This shows that the use of the system was not an issue in our study. In the same scale, searchers indicated an average value of 3.71 for the question 'How easy was it to set up the sliders to get a good table of contents?' which shows that getting the desired TOC took some effort. In their comments, searchers indicated various understandings of the sliders, but most of them had a strategy how to set them up and, according to the logs, all of the searchers used the sliders extensively.

4.2 Sliders and TOC Characteristics

On average, the depth, length and threshold slider values were around the default value (50), while the relevance slider was in a higher average position (75.79). The standard deviation of the depth, relevance and length sliders was the same (25), and the threshold slider had a lower (15) standard deviation. This indicates that searchers found the relevance feature to select TOC elements more important than the other two. Indeed, the importance of relevance to access the relevant content is the most intuitive of the three features, but results indicate that other features also needed to be considered if one wanted to place the relevant elements into context, i.e. to show related contents.

The average threshold slider value of 50% with standard deviation of 15 shows that the agreement among searchers regarding the threshold was higher than those of the three features. In addition, we did not find a tendency of different slider values for documents of various sizes, thus the average values seem to be appropriate for XML documents of any size in our setting.

An average TOC consisted of 19 items which is, on average, 8.16% of all the elements in the original XML document. We examined how the size of the TOCs is related to the size of the documents. Since there was high correlation between the length of the documents and the number of elements they contained, we use these two measures as 'size' interchangeably. We found that the more elements a document contained, the smaller the proportion of the number of TOC elements to the number of document elements was (Figure 3). This means that a long document does not necessarily need a long TOC. This is because a too long TOC does not help searchers gain an overview of the contents of the document because they need to gain an overview of the contents of the TOC first. This clearly indicates that a TOC generation algorithm has to perform particularly well for longer documents, as the TOC algorithm is much more selective in element items in longer documents.

To also examine the distribution of the length of elements in the TOCs and in the documents, we created five size categories based on the length of text in elements (Figure 4.a). We found that the distribution of element lengths of documents (light gray in Figure 4.a) follows a bell curve where most of the elements

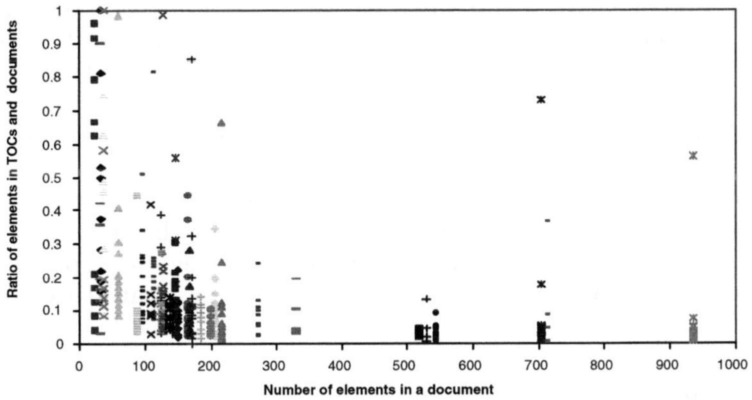

Fig. 3. Number of elements in the documents and the TOC elements' ratio to it

cover 10-100 characters of text, i.e. there are many elements with the size of a short sentence. Elements longer than 10KB are very rare in our documents, such elements can be e.g., the root (i.e. *article*) element of a long document.

The distribution of the length of elements included in the TOCs (dark gray in Figure 4.a) is also a bell curve having its peak at the category of 100-1000 character long elements. Slightly shorter and longer elements are less frequent in the TOCs while very long and very short elements do not occur frequently in the TOCs. This shows that the length distribution of the elements in the documents and TOCs is of the same nature, only the parameters are different. The implication of this is that a TOC has to be constructed to reflect the original structure of the document.

We also examined the distribution of depth levels in the TOCs and document elements. (Figure 4.b). We considered eight depth levels because deeper than the eighth level there were very few, if any, elements in a document, and none of the displayed TOCs were deeper than seven levels. The distribution of element depths of documents follows, similar to the length distribution, a bell curve: most of the elements are between the fourth and sixth level, there is only one

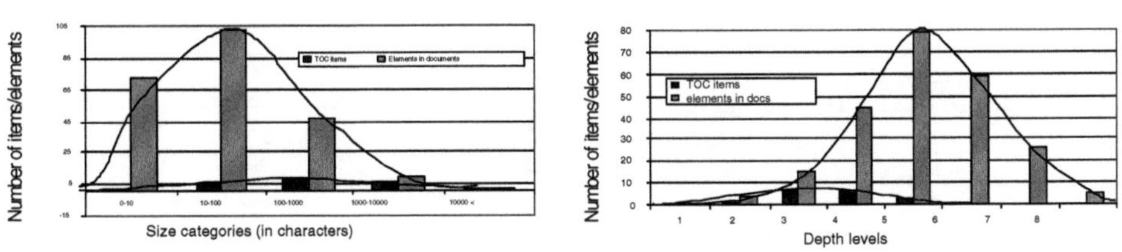

(a) (b)

Fig. 4. TOC items and document elements at various size categories (a) and depth levels (b)

element at level one (which is the root element of a document's tree structure) and very few elements are below the seventh level. The depth distribution of the TOC elements is still a bell curve with the peak at level three and four, which is consistent with a finding in [14]. In other words, the depth distribution found in the TOCs and documents are the same, only the parameters are different.

Considering both the length and depth distributions, the TOCs reflect the main characteristics of the documents. The TOCs can therefore be viewed as extracts of the document structure, so not only is the algorithm based on summarisation but the output of it is also a summary (i.e. that of the document structure).

4.3 Searchers

Searchers used different slider setting strategies to generate the best TOC. Nonetheless, the majority of them did set the relevance slider high. This high view of relevance was confirmed in the post experiment questionnaires. The questionnaires show that the relevance slider was usually set first, which also shows its importance. Some of the searchers set the length, depth and relevance sliders first and changed the threshold slider slightly document after document. In this process, setting the most appropriate threshold was found difficult for most searchers, especially because they had an 'ideal' TOC in their minds that was to be reached by adjusting the sliders. According to the questionnaires, an ideal TOC contained those, and only those, elements that had been found useful when searchers had been experimenting with the settings for the TOC of the current document. Although not all of them followed these strategies, the TOCs generated did not differ very much for different searchers: apart from a few searchers, the TOCs were not longer than twenty items, and searchers also seemed to agree in terms of length categories and depth levels (Figure 5). Based on the above, it seems that the size of a TOC does not significantly depend on individuals. Indeed, a TOC should rarely contain more element references than

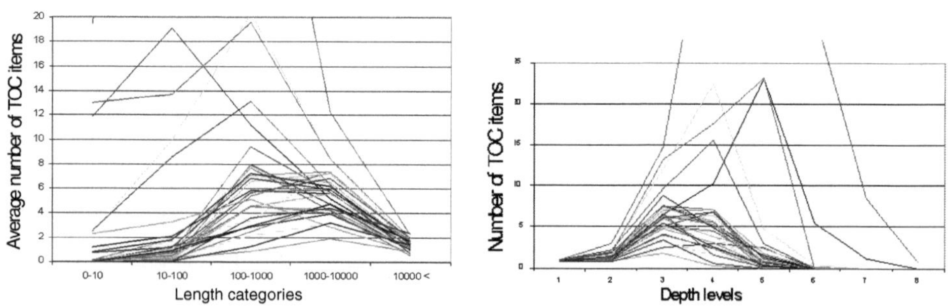

Fig. 5. Number of TOC items at (a) various length categories and (b) depth levels for searchers

some fixedvalue, in our case twenty. Our study shows that it is important to select the best (maximum) twenty elements, appropriately.

4.4 Collections and Topics

We investigated the possible differences between TOCs generated for documents of the two collections. There were slightly more documents used from the Wikipedia collection (20) than from the IEEE (13), and therefore 158 slider values were examined for the IEEE collection and 325 for Wikipedia documents. We did not find differences in the average slider values with respect to the two collections; settings for one collection also seemed satisfactory for the other.

The generated TOCs' characteristics were not significantly different either. Although the structure of the two collections' documents are similar to each other, we do not expect that these results will be different for other XML documents. This is because documents of XML collections must have exactly one root element, which has several child elements etc., so we expect that the algorithm described in this paper can be used for any XML documents.

We also investigated whether there were differences in settings and TOCs among the ten topics. The distribution of the slider values did not reveal great differences: the relevance values were always higher than that of other sliders and the depth-length-threshold triplet's order was slightly different for some topics but these were always closely around the default value of 50. This shows that for the ten topics we used, searchers did not need to use very different settings to obtain an acceptable TOC. However, the number of topics does not guarantee that we covered a wide enough range of topics and task types. To compare the results for different task types (e.g. finding background information vs. answering a question), more topics are needed from different task types.

The number of TOC elements were between 14 and 27 for 8 of the 10 topics, which is around 9% of the number of elements in the documents. There was one topic (*w2*) for which the average number of TOC elements was as low as 7, and for another topic (*w5*), this number was 32. These two extreme values are closely related to the number of elements the topics' documents had, i.e. documents for (*w2*) were very short while documents for (*w5*) contained the longest document. This shows that longer documents may require longer TOCs, but since the ratio of the number of elements in TOCs and documents is different for shorter and longer documents (see Figure 3), size differences in the TOCs should not be linearly proportional to document sizes.

5 Discussion and Conclusions

We have studied searchers' preferences of element features in automatic TOC generation for XML retrieval. The features and searchers' preferences (as weights) can be used to select those elements that will form the TOC. We have considered three features: depth, length of the elements and their relevance to the current query. We have conducted a user study to investigate which of these features are important in TOC generation, and what are the characteristics of TOCs that were generated by searchers' feature preferences.

Our algorithm offers a mapping from the set of elements in the document to that of the TOC, where the most important elements of the documents are selected as TOC elements and the distribution of length and depth of elements remains very similar. The result of the mapping is an extract of the document structure, which, by organising it into a table of contents, can be used to help searchers find relevant content inside an XML document.

In this study, we found that a TOC generation algorithm like the one introduced in this paper has to consider the relevance of an element, i.e. it should be highly query-biased. The other two element features used, i.e. length and depth of an element, should also be considered and the weight of these two should be lower than that of the relevance feature. It is also understood that TOCs should not be large in size, i.e. longer documents should still have a relatively small TOC. This also shows that automatic TOC generation has to be more carefully designed when dealing with longer documents. To ensure better results, the TOC algorithm can be extended to include other element features such as e.g. tag names, titles of elements.

Our data also suggest that the size of a TOC does not significantly depend on individual searchers. The selection of the most important elements is much more important, and here may be worth considering searchers' individual preferences. We suggest that if a TOC algorithm selects more than a certain number (e.g. 20) of TOC elements (e.g. 30), the top scored elements (i.e. top 20) should be kept regardless of what threshold value the algorithm uses. If the number of TOC elements is lower than this number (e.g. 10), these elements should be used to construct the TOC.

During the analysis of the collected data we learned that searchers had several strategies to get the best TOCs. Although not every one of the searchers understood the concepts of the features completely, they actively used the sliders and created TOCs that, according to questionnaires, were suitably good to access the relevant content more easily.

To conclude, we have developed an algorithm to automatically generate tables of contents for XML documents. The algorithm uses features of elements to select those that will form the TOC. Different TOCs are generated for different queries which we think may help searchers access the relevant content more quickly. In our experiment, we investigated which features are important in TOC generation, and what are the characteristics of TOCs that are generated by searchers' feature preferences.

The work presented here is part of a wider work that aims at developing and evaluating methods that summarise the content and the structure of documents for structured document retrieval. We believe that the effective combination of the two types of summarisation can help searchers focus on only the useful contents of the documents, decrease the time searchers spend on finding relevant elements, and thus, increase user satisfaction.

Acknowledgments

The authors wish to acknowledge the participants of the study. This work was partly funded by the Nuffield Foundation (grant NAL/01081/G) and the DELOS, Network of Excellence in Digital Libraries.

References

1. P. Borlund. The IIR evaluation model: a framework for evaluation of interactive information retrieval systems. *Information Research*, 8(3), 2003.
2. L. Denoyer and P. Gallinari. The Wikipedia XML Corpus. *SIGIR Forum*, 40(1):64–69, 2006.
3. H. P. Edmundson. New methods in automatic extracting. *J. ACM*, 16(2):264–285, 1969.
4. N. Fuhr, M. Lalmas, S. Malik, and G. Kazai, editors. *Proceedings of INEX 2005*, volume 3977, 2006.
5. N. Fuhr, M. Lalmas, S. Malik, and Z. Szlávik, editors. *Proceedings of INEX 2004*, volume 3493, 2005.
6. B. Hammer-Aebi, K. W. Christensen, H. Lund, and B. Larsen. Users, structured documents and overlap: interactive searching of elements and the influence of context on search behaviour. In *Proceedings of IIiX*, pages 46–55, 2006.
7. J. Kamps, M. de Rijke, and B. Sigurbjörnsson. Length normalization in XML retrieval. In *Proceedings of ACM SIGIR*, pages 80–87, 2004.
8. J. Kamps and B. Sigurbjörnsson. What do users think of an XML element retrieval system? In Fuhr et al. [4], pages 411–421.
9. H. Kim and H. Son. Users interaction with the hierarchically structured presentation in XML document retrieval. In Fuhr et al. [4], pages 422–431.
10. B. Larsen, S. Malik, and A. Tombros. The interactive track at INEX 2005. In Fuhr et al. [4], pages 398–410.
11. S., C.-P. Klas, N. Fuhr, B. Larsen, and A. Tombros. Designing a user interface for interactive retrieval of structured documents - lessons learned from the INEX interactive track. In *Proceedings of ECDL 2006*, pages 291–302, 2006.
12. B. Sigurbjörnsson. *Focused Information Access using XML Element Retrieval.* PhD thesis, Faculty of Science, University of Amsterdam, 2006.
13. Z. Szlávik, A. Tombros, and M. Lalmas. Investigating the use of summarisation for interactive XML retrieval. In F. Crestani and G. Pasi, editors, *Proceedings of ACM SAC-IARS'06*, pages 1068–1072, 2006.
14. Z. Szlávik, A. Tombros, and M. Lalmas. The use of summaries in XML retrieval. In *Proceedings of ECDL 2006*, pages 75–86, 2006.
15. M. Theobald, R. Schenkel, and G. Weikum. An efficient and versatile query engine for TopX search. In *Proceedings of VLDB*, pages 625–636, 2005.
16. A. Tombros, B. Larsen, and S. Malik. The interactive track at INEX 2004. In Fuhr et al. [5], pages 422–435.
17. A. Tombros and M. Sanderson. Advantages of query biased summaries in information retrieval. In *Proceedings of ACM SIGIR*, pages 2–10, 1998.
18. R. van Zwol, G. Kazai, and M. Lalmas. INEX 2005 multimedia track. In Fuhr et al. [4], pages 497–510.

Setting Per-field Normalisation Hyper-parameters for the Named-Page Finding Search Task

Ben He and Iadh Ounis

Department of Computing Science
University of Glasgow
United Kingdom
{ben,ounis}@dcs.gla.ac.uk

Abstract. Per-field normalisation has been shown to be effective for Web search tasks, e.g. named-page finding. However, per-field normalisation also suffers from having hyper-parameters to tune on a per-field basis. In this paper, we argue that the purpose of per-field normalisation is to adjust the linear relationship between field length and term frequency. We experiment with standard Web test collections, using three document fields, namely the body of the document, its title, and the anchor text of its incoming links. From our experiments, we find that across different collections, the linear correlation values, given by the optimised hyper-parameter settings, are proportional to the maximum negative linear correlation. Based on this observation, we devise an automatic method for setting the per-field normalisation hyper-parameter values without the use of relevance assessment for tuning. According to the evaluation results, this method is shown to be effective for the body and title fields. In addition, the difficulty in setting the per-field normalisation hyper-parameter for the anchor text field is explained.

1 Introduction

In Information Retrieval (IR), it is a crucial issue to rank retrieved documents in decreasing order of relevance. A recent survey on the query logs from real Web search engine users concluded that the users rarely look beyond the top returned documents [9]. Therefore, it is important to rank the highly relevant documents at the top of the retrieved list. Usually, the document ranking is based on a *weighting model*. In particular, most weighting models apply a *term frequency (tf) normalisation* method to normalise term frequency, the number of occurrences of the query term in the document.

Various *tf* normalisation methods have been proposed in the literature, e.g. the pivoted normalisation [16] in the vector space model [15], the normalisation method of the BM25 weighting model [13], normalisation 2 [1] and normalisation 3 [1,8] in the Divergence from Randomness (DFR) framework [1]. All the above mentioned normalisation methods normalise term frequency according to document length, i.e. the number of tokens in the document. Each of the above

G. Amati, C. Carpineto, and G. Romano (Eds.): ECIR 2007, LNCS 4425, pp. 468–480, 2007.

mentioned normalisation methods involve the use of a hyper-parameter. The setting of these hyper-parameter values usually has an important impact on the retrieval performance of an IR system [2,7,8].

Recently, Robertson et al. and Zaragoza et al. proposed the per-field normalisation technique, which normalises term frequency on a per-field basis [14,18], by extending BM25's normalisation method [13]. The resulting field-based weighting model is called BM25F. Using BM25F, the retrieval process is performed on indices of different document fields, such as body, title, and anchor text of incoming links. Following [14,18], Macdonald et al. extended the PL2 DFR weighting model, by employing the per-field normalisation 2F [10]. Compared with *tf* normalisation on a single field, on one hand, per-field normalisation can significantly boost the retrieval performance, particularly for Web search [12,18]. On the other hand, per-field normalisation has a hyper-parameter for each document field used. Therefore, per-field normalisation has more hyper-parameters than *tf* normalisation on a single index of the whole collection, which requires a heavier training process to set the hyper-parameter values. Similarly to *tf* normalisation on a single field, the setting of the hyper-parameter values of per-field normalisation can significantly affect the retrieval performance. In particular, the optimal hyper-parameter setting of a document field, which provides the best retrieval performance, varies across different collections [12]. As a consequence, training is required on each given new collection to guarantee an effective retrieval performance.

In this paper, we study how the per-field normalisation hyper-parameter setting is related to the resulting retrieval performance. Our study follows Harter and Amati's idea that there is a linear relationship between term frequency and document length [1,6]. This linear relationship is indicated by the linear correlation between these two variables [1,6]. The study of this paper is focused on a typical Web search task, namely the named-page finding search task [17]. The main contributions of this paper are two-fold. First, we provide a better understanding in per-field normalisation. we suggest that the purpose of per-field normalisation is to adjust the linear relationship between term frequency and field length, i.e. the number of tokens in the field. This is our main argument. Experiments are conducted to study how per-field normalisation adjusts this linear relationship. Second, we devise and evaluate an automatic hyper-parameter setting method for the per-field normalisation hyper-parameters. The proposed method does not need relevance assessment for tuning, making it particularly practical in an operational and realistic setting.

The rest of this paper is organised as follows. Section 2 introduces the current main per-field normalisation techniques. Section 3 describes our main argument in this paper. Sections 4 and 5 describe the experimental methodology for investigating the linear relationship between field length and term frequency, and analyse the experimental results, respectively. Section 6 devises an automatic method for setting the per-field normalisation hyper-parameter values, which is evaluated is Section 7. Finally, Section 8 concludes the paper and suggests possible future work.

2 Per-field Normalisation

In the context of field-based retrieval, Robertson et al. proposed the idea of normalising term frequency on a per-field basis [14]. The extended BM25 field-based weighting model, called *BM25F*, assigns the relevance score of a document d for a query q as follows:

$$score(d, q) = \sum_{t \in q} w^{(1)} \frac{(k_1 + 1)tfn}{k_1 + tfn} \frac{(k_3 + 1)qtf}{k_3 + qtf} \tag{1}$$

where qtf is the query term frequency. k_1 and k_3 are parameters. The default setting is $k_1 = 1.2$ and $k_3 = 1000$ [13]. $w^{(1)}$ is the *idf* factor, which is given by:

$$w^{(1)} = \log_2 \frac{N - N_t + 0.5}{N_t + 0.5}$$

where N is the number of documents in the whole collection. N_t is the document frequency of term t.

The above BM25F's weighting function is the same as the one of BM25 [13]. Instead of normalising term frequency on a single index, BM25F applies a per-field normalisation method to assign the normalised term frequency *tfn* [18]:

$$tfn = \sum_f w_f \cdot tfn_f = \sum_f w_f \cdot \frac{tf_f}{(1 - b_f) + b_f \cdot \frac{l_f}{avg\_l_f}} \tag{2}$$

where w_f is the weight of a field f, which reflects the relative contribution of a field to the document ranking. tfn_f is the normalised term frequency on field f. tf_f is the frequency of the query term in the field f of the document. b_f is the term frequency normalisation hyper-parameter of field f. l_f is field length, namely the number of tokens in field f of the document. avg_l_f is the average length of field f in the collection.

Moreover, following [18], Macdonald et al. extended the PL2 weighting model to cope with different document fields, within the Divergence from Randomness (DFR) probabilistic framework [1]. The idea of the DFR framework is to infer the informativeness of a query term in a document by measuring the divergence of the term's distribution in the document from a random distribution. The larger the divergence is, the more informative the query term is in the document. The *PL2F* field-based weighting model has the following weighting function:

$$score(d, q) = \sum_{t \in q} qtw \cdot \frac{1}{tfn + 1} \Big(tfn \cdot \log_2 \frac{tfn}{\lambda} + (\lambda - tfn) \cdot \log_2 e + 0.5 \cdot \log_2(2\pi \cdot tfn)\Big) \tag{3}$$

where λ is the mean and variance of a Poisson distribution. It is given by $\lambda = tf_c/N$. tf_c is the frequency of the query term in the collection, and N is the number of documents in the collection. In PL2F, the normalised term frequency *tfn* is given by the so-called *Normalisation 2F* as follows:

$$tfn = \sum_f w_f \cdot tfn_f = \sum_f (w_f \cdot tf_f \cdot \log_2(1 + c_f \cdot \frac{avg\_l_f}{l_f})), (c_f > 0) \quad (4)$$

where c_f is the hyper-parameter of field *f*. tfn_f is the normalised term frequency on field f. l_f is the length of field f in the document, and avg_l_f is the average field length in the collection. w_f is the weight of field f. In the above Normalisation 2F, the term frequency is normalised in each field, and each field f has a hyper-parameter c_f with a field weight w_f. The above normalisation 2F is based on the assumption that term density is decreasing with document length [1].

In addition, the PL3 weighting model [1,8], which applies the Dirichlet priors for *tf* normalisation, can be extended in a similar way to deal with field-based retrieval. The resulting field-based PL3 weighting model has the same weighting function as PL2F in Equation (3). The normalised term frequency *tfn* is given by *Normalisation 3F* as follows:

$$tfn = \sum_f w_f \cdot tfn_f = \sum_f (w_f \cdot \frac{tf_f + \mu_f \cdot \frac{tf_{cf}}{l_{cf}}}{l_f + \mu_f} \cdot \mu_f) \quad (5)$$

where w_f and tf_f are the weight and term frequency of field f in the document, respectively. tfn_f is the normalised term frequency on field f. μ_f is the hyper-parameter of field f. l_{cf} is the number of tokens in field f in the whole collection. l_f is the field length in the document. tf_{cf} is the frequency of the query term in field f in the whole collection.

As shown by previous experiments, compared with applying *tf* normalisation on a single index, per-field normalisation is particularly effective for Web search, such as named-page finding [12,18]. However, per-field normalisation has an associated hyper-parameter for each document field. The setting of these hyper-parameters is a crucial issue, which has an important impact of the retrieval performance. In this paper, this issue is studied by following the idea of measuring the linear relationship between field length and term frequency [6,1]. Based on this idea, we further understand the purpose of per-field normalisation, as described in the next section.

3 The Purpose of Per-field Normalisation

In the context of *tf* normalisation on a single index, Harter [6] and Amati [1] suggested that document length and term frequency have a linear relationship. Such a linear relationship can be indicated by the linear correlation between these two variables. Following their idea, in the context of per-field normalisation, we suggest that field length and term frequency also have a linear relationship, which can be indicated by the linear correlation between them. In this paper, following the definition of correlation in [4], the linear correlation between field length and normalised term frequency is given by:

$$\rho(tfn_f, l_f) = \frac{COV(tfn_f, l_f)}{\sigma(tfn_f)\sigma(l_f)} \quad (6)$$

where tfn_f is the normalised term frequency on field f, and l_f is the field length. COV stands for covariance and σ stands for the standard deviation. Note that the use of a *tf* normalisation method changes term frequency. Therefore, the normalised term frequency, instead of term frequency, is considered in our study.

We suggest that the aim of *tf* normalisation on a document field is to adjust the linear relationship between field length and term frequency. Applying different hyper-parameter settings results in different correlation values, which indicate different degree of linear dependence between field length and term frequency. In our study, we investigate how per-field normalisation affects the correlation $\rho(tfn_f, l_f)$. In particular, from our experiments, we expect to find a pattern that may help in proposing an automatic method in setting the hyper-parameter values, without using relevance assessment for tuning.

4 Experimental Setting and Methodology

Our experiments are conducted using Terrier [11]. Two TREC Web test collections, namely the .GOV and the .GOV2 collections, are used in our experiments. These two collections are the only currently available ones for the named-page finding task. The .GOV collection is a 1.25 million pages crawl of the .gov domain. The .GOV2 collection is a later crawl of the .gov domain, which contains 25,205,179 Web documents and 426 Gigabytes of uncompressed data[1]. This collection has been employed in the TREC Terabyte track since 2004. In addition, a named-page finding task has been run in the Terabyte track since 2005. .GOV2 is currently the largest TREC test collection. The indices of these two collections are created with the body, anchor text and title fields, respectively. Porter's stemmer and standard stopword removal are applied.

The test queries used are the 525 topics used in the TREC 2002-2004 Web track named-page finding tasks [17] on the .GOV collection, and the 252 topics used in the TREC 2005 Terabyte track named-page finding task [3] on the .GOV2 collection. The evaluation measure used is mean reciprocal rank (MRR), which is the official measure in TREC for the named-page finding task [17].

In our experiments, we investigate the linear relationship between field length and normalised term frequency. This linear relationship is indicated by the linear correlation between these two variables. Three field-based weighting models in the literature, namely PL2F, BM25F and PL3F, are used in this study.

The first step of the experiments is to optimise the three weighting models used, which provides a basis for our study. For each of the field-based weighting models, we need to optimise six parameters, namely the hyper-parameters

[1] Information of the collections can found at http://ir.dcs.gla.ac.uk/test_collections/

Table 1. The optimal hyper-parameter settings and weights of the body, anchor text, and title fields, using three field-based weighting models, on the two collections used

Coll.	Body	Anchor	Title	Body	Anchor	Title	Body	Anchor	Title
	PL2F (c_f)			BM25F (b_f)			PL3F (μ_f)		
.GOV	0.8	15.0	5.5	0.85	0.10	0.45	300	50000	10
.GOV2	1.2	2.6	2.6	0.85	0.90	0.50	300	40	20
	PL2F (w_f)			BM25F (w_f)			PL3F (w_f)		
.GOV	1	1.1	4.6	1	8.1	12.4	1	0.4	18.2
.GOV2	1	3.4	2.6	1	6.0	8.6	1	1.1	12.0

and the weights of the three indexed document fields. Our optimisation process follows the one for BM25F applied in [18]. However, we apply manual data sweeping, instead of automatic optimisation, as applied in [18]. This is because in our previous experiments, we found that the manual data sweeping with a small enough granularity can usually lead to a better optimised retrieval performance than automatic optimisation. Following [18], we set the field weight of body to 1 to reduce the cost of optimisation. For the remaining five parameters, the optimisation process is described as follows:

1. On each field, we optimise the hyper-parameter of the field, while disabling the other two fields. The optimised hyper-parameter setting are obtained by multiple-step data sweeping with from-large-to-small granularities. Data sweeping is performed within a reasonable range of values. This range is [1, 32] for PL2F, (0, 1] for BM25F and (0, 100000] for PL3F. The minimal granularity is 0.1 for PL2F, 0.05 for BM25F and 10 for PL3F.
2. We optimise the field weights of body and title by a two-step two-dimensional data sweeping within [0, 20], while setting the hyper-parameter values to the ones optimised in the first step. The granularities in the two data sweeping steps are 1 and 0.1, respectively.

We only briefly describe the optimisation process, for lack space. The obtained optimised parameter values are provided in Table 1.

The second step of the experiments is to investigate the linear relationship between field length and normalised term frequency. This linear relationship is indicated by $\rho(tfn_f, l_f)$, the linear correlation between these two variables. In particular, we study how the optimised hyper-parameter values are related to $\rho(tfn_f, l_f)$. For the three different document fields used, we plot the hyper-parameter values against $\rho(tfn_f, l_f)$, in order to study how the linear relationship between tfn_f and l_f varies on different document fields. We also look at the proportion of the optimal $\rho(tfn_f, l_f)$ value to the maximum $\rho(tfn_f, l_f)$ value with respect to all possible hyper-parameter values. By doing this, we expect to find a pattern that may indicate the optimal hyper-parameter setting, which can lead to a practical approach for setting the hyper-parameter values. The analysis of the related experimental results are provided in the next section.

5 The Linear Relationship Between Field Length and Normalised Term Frequency

Table 1 contains the optimised hyper-parameter values and field weights after the data sweeping process. From Table 1, we observe that, across the two collections used, on one hand, for the body and title fields, the optimised hyper-parameter settings are relatively similar. On the other hand, for the anchor text field, the hyper-parameter settings are largely different across the two collections used (e.g. 40 vs. 50000 using PL3F). A possible explanation is that the weighting models used assume a random distribution of a query term in the collection. This assumption is usually true for written text, such as body and title. Differently from these two fields, the anchor text of a Web page is extracted from its incoming links. Eiron & McCurley concluded that the anchor text of a Web page usually has only one or two repeatedly occurring unique terms [5]. Consequently, in the anchor text field, the curve of a query term's distribution looks like a Beta(0.5, 2) distribution[2] [4], because a query term usually has a large number of occurrences in the anchor text of some Web pages, and does not appear at all in the anchor text of other Web pages. Therefore, the optimised hyper-parameter setting for the anchor text field is unpredictable and can be largely different across different collections.

Next, we study the linear relationship between field length and normalised term frequency. Figure 1 (see page 475) plots the linear correlation between these two variables. From Figure 1, we find that the linear correlation $\rho(tfn_l, l_f)$ varies with the use of different hyper-parameter values. In particular, in all the cases, the plotted curve has a lowest point, which corresponds to the maximum negative $\rho(tfn_l, l_f)$ value. To further analysis the linear relationship between field length and normalised term frequency, Table 2 provides the resulting $\rho(tfn_f, l_f)$ values of the optimised hyper-parameter settings. The values in parenthesis are the $ratio_f$ that is given by:

$$ratio_f = \frac{\rho_{opt}(tfn_f, l_f)}{\rho_{max}(tfn_f, l_f)} \tag{7}$$

where $\rho_{opt}(tfn_f, l_f)$ is the $\rho(tfn_f, l_f)$ value given by the optimised hyper-parameter setting. $\rho_{max}(tfn_f, l_f)$ is the maximum negative $\rho(tfn_f, l_f)$ value that corresponds to the lowest points in the curves in Figure 1.

From Table 2, for the body and title fields, we find that the $\rho_{opt}(tfn_f, l_f)$ value seems to be proportional to the maximum negative $\rho(tfn_f, l_f)$ value. The $ratio_f$ for the body and title fields are similar to each other. For the anchor text field, we do not have the same observation, probably because of the repeatedly occurring tokens of the query terms in this field. Based on the observation from Table 2, in the next section, we devise an automatic method for estimating the hyper-parameter values of the body and title fields. For the anchor text field, we simply apply the optimised hyper-parameter setting after data sweeping.

[2] Beta(0.5, 2) distribution refers to a Beta distribution with $\alpha = 0.5$ and $\beta = 2$.

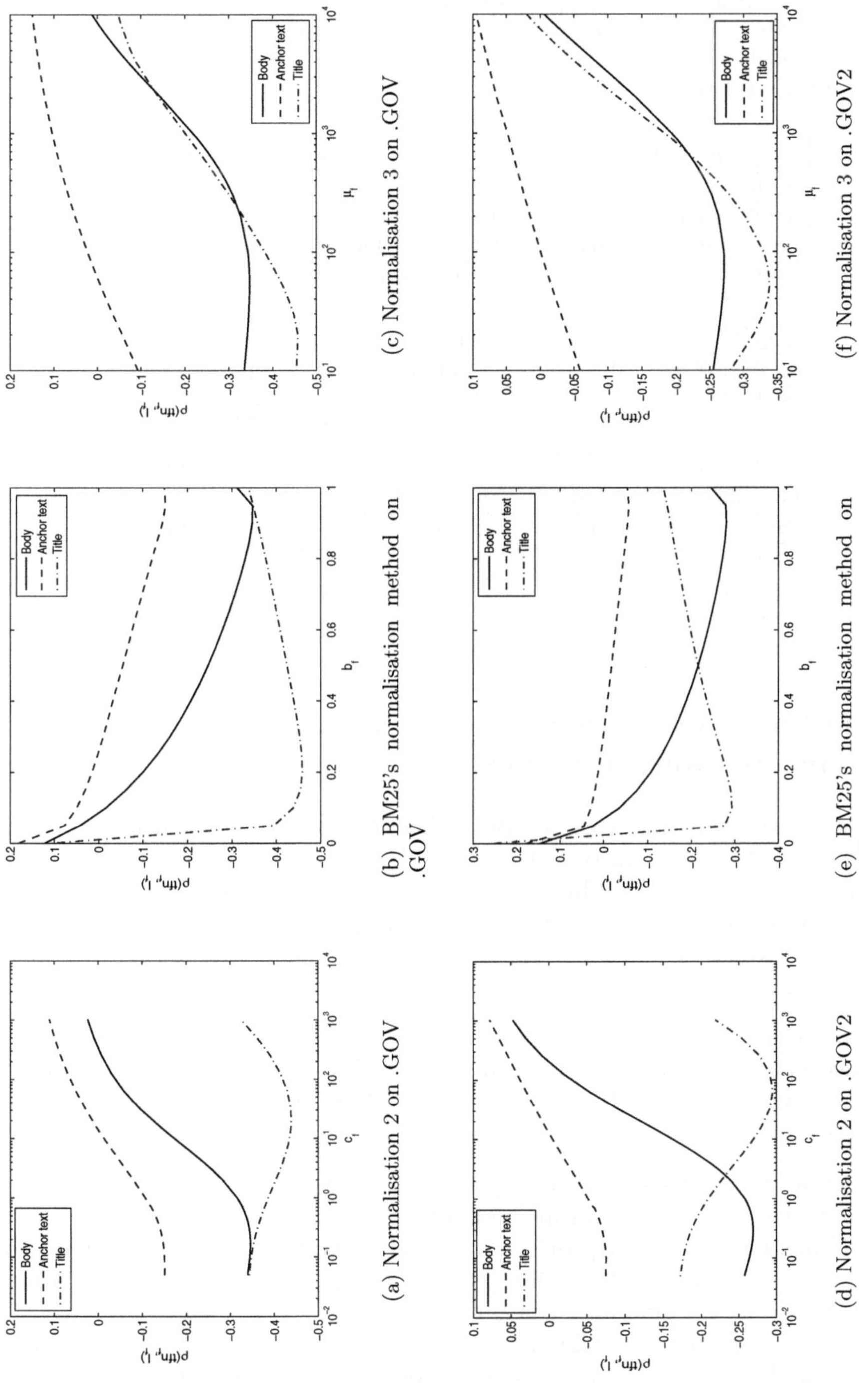

(a) Normalisation 2 on .GOV

(b) BM25's normalisation method on .GOV

(c) Normalisation 3 on .GOV

(d) Normalisation 2 on .GOV2

(e) BM25's normalisation method on .GOV

(f) Normalisation 3 on .GOV2

Fig. 1. The linear correlation $\rho(tfn_f, l_f)$ against the *tf* normalisation hyper-parameters on the two collections used

Table 2. The optimal $\rho(tfn_f, l_f)$ values and the corresponding $ratio_f$ for the three document fields. Mean ratio refers to the mean of the $ratio_f$ values over the three weighting models used. Side specifies on which side of the curves (in Figure 1) the optimal $\rho(tfn_f, l_f)$ value locates.

	ρ_{body} $(ratio_{body})$	ρ_{anchor} $(ratio_{anchor})$	ρ_{title} $(ratio_{title})$
Coll.		PL2F	
.GOV	-0.3232 (0.9364)	0.002456 (-0.01629)	-0.4265 (0.9699)
.GOV2	-0.2543 (0.9447)	-0.03282 (0.4361)	-0.2332 (0.9688)
Side	Increasing	Increasing	Decreasing
Coll.		BM25F	
.GOV	-0.3389 (0.9763)	0.04932 (-0.3263)	-0.4325 (0.9414)
.GOV2	-0.2697 (0.9947)	-0.07142 (0.9393)	-0.2489 (0.7385)
Side	Decreasing	Decreasing	Increasing
Coll.		PL3F	
.GOV	-0.3097 (0.8918)	0.1658 (-1.0967)	-0.4546 (0.9886)
.GOV2	-0.2527 (0.9313)	-0.02301 (0.3898)	-0.3157(0.9312)
Side	Increasing	Increasing	Decreasing
Coll.		Mean ratio	
.GOV	0.9348	-0.4798	0.9666
.GOV2	0.9568	0.5884	0.8795

6 Method *prop* for Setting the Per-field Normalisation Hyper-parameter Values

In the previous section, across the two collections used, we found that on both the body and title fields, the optimal $\rho(tfn_f, l_f)$ value is proportional to the maximum negative $\rho(tfn_f, l_f)$ value. Therefore, on these two fields, we can estimate the hyper-parameter value, which gives a $\rho(tfn_f, l_f)$ value that is proportional to the $ratio_f$ value (see Table 2). Using the above suggested solution, we make the following hypothesis:

> Hypothesis H(*prop*): For the given body or title field, across different collections, the optimal hyper-parameter values provide a constant ratio of the optimal $\rho(tfn_f, l_f)$ value divided by the maximum negative $\rho(tfn_f, l_f)$ value.

The above Hypothesis H(*prop*) implies that, for a given body or title field, the optimal $\rho(tfn_f, l_f)$ value is proportional to the maximum negative $\rho(tfn_f, l_f)$ value. Using Hypothesis H(*prop*), for a given collection, we can estimate the hyper-parameter value that satisfies $\rho(tfn_f, l_f) = \rho_{max}(tfn_f, l_f) \cdot ratio_f$, where $\rho_{max}(tfn_f, l_f)$ is the maximum negative $\rho(tfn_f, l_f)$ value. $ratio_f$ is given by Equation (7). On the two collections used, the $ratio_f$ values of body and title are listed in Table 2. We denote the above described approach by method *prop*.

To apply method *prop*, we need to create a bidirectional mapping between a hyper-parameter and $\rho(tfn_f, l_f)$. Each $\rho(tfn_f, l_f)$ value should correspond to a unique hyper-parameter value, and vice versa. In fact, from Figure 1, we can see that a $\rho(tfn_f, l_f)$ value corresponds to two different hyper-parameter values: One is on the increasing side of the curve, and the other is on the decreasing side of the curve. Therefore, by looking at the curves in Figure 1, we identify at which side of the curve the optimal $\rho(tfn_f, l_f)$ value is located (see Table 2).

For the application of method *prop*, we need a training collection to obtain the assumed constant $ratio_f$. The training process for computing this constant $ratio_f$ needs to be done only once. For a given new collection, we do not need any associated relevance judgement. Finally, we apply the hyper-parameter setting that results in the $ratio_f$ value on the given query set, instead of optimising the hyper-parameter by maximising the retrieval performance using relevance judgement. For a given collection, the tuning process takes place before the retrieval process. There is no additional overhead during retrieval. We evaluate method *prop* in the next section.

7 Evaluation of Method *prop*

For evaluating method *prop*, we conduct a two-fold holdout evaluation on the two collections, namely .GOV and .GOV2. In each fold of the holdout evaluation, we use one collection for training, in order to compute the assumed constant $ratio_f$ value. The other collection is used for testing. The assumed constant $ratio_f$ value is the mean of the $ratio_f$ values of the three weighting models used, obtained on the training collection (see the mean $ratio_f$ values in Table 2). In addition to the test queries used in Section 5, we also experiment with the 181 latest TREC topics used in the TREC 2006 named-page finding task. Note that method *prop* is only applied for the body and title fields. For the anchor text field, we apply the optimised hyper-parameter setting obtained by data sweeping.

We compare the retrieval performance obtained using the hyper-parameter setting, estimated by method *prop*, with the optimised retrieval performance using data sweeping. We suggest that a large (>5%) and statistically significant difference between the obtained MRRs indicates a failure of method *prop* in estimating the hyper-parameter setting. Otherwise, we conclude that method *prop* is effective for the named-page finding retrieval task, when different document fields are used. The statistical test used is the sign test[3].

The evaluation results are listed in Table 3. In two cases, we observe that the estimated hyper-parameters result in MRRs that are higher than the optimised ones by data sweeping (see the MRR values in *italic* in Table 3). We suggest that this is because the optimisation procedure optimises the hyper-parameter of each document field separately. However, the optimised hyper-parameter setting of each individual field may not necessarily lead to the optimised retrieval performance, when different fields are summed up together. From Table 3, we observe that in all the nine cases, method *prop* provides a retrieval performance that is as good as the one optimised by data sweeping. We find no large (5%) difference

[3] For MRR, the sign test is more appropriate than the Wilcoxon test.

Table 3. Evaluation results for method *prop*. Columns body and title provides the hyper-parameter settings for the body and title fields, estimated by method *prop*. MRR(opt) and MRR(prop) are the MRRs obtained by data sweeping and by method *prop*, respectively. *diff.* is the difference between the two MRR values in percentage. p-value is given by the sign test.

Topics	body	title	MRR(opt)	MRR(prop)	*diff* (%)	p-value
			PL2F			
TREC 2002-2004 on .GOV	0.62	1.03	0.7294	0.7018	-3.78	3.03e-05
TREC 2005 on .GOV2	1.34	21.65	*0.4341*	*0.4522*	+4.17	0.0422
TREC 2006 on .GOV2	1.06	20.54	0.4736	0.4733	≈ 0	0.630
			BM25F			
TREC 2002-2004 on .GOV	0.81	0.63	0.7142	0.7145	≈ 0	0.497
TREC 2005 on .GOV2	0.69	0.17	0.4738	0.4522	-4.56	0.583
TREC 2006 on .GOV2	0.74	0.19	0.4405	0.4392	≈ 0	0.798
			PL3F			
TREC 2002-2004 on .GOV	170	2.40	0.6390	0.6140	-3.91	0.0115
TREC 2005 on .GOV2	291	27.23	*0.3721*	*0.3751*	≈ 0	0.00259
TREC 2006 on .GOV2	234	23.59	0.4470	0.4259	-4.72	1.03e-07

between MRR(opt) and MRR(prop). Therefore, we conclude that method *prop*, based on Hypothesis H(prop), is effective on the two collections used.

To summarise, in a practical setting, the assumed constant $ratio_f$ value is obtained on a training collection for once. For a given new collection, we recommend applying method *prop* for the body and title fields without the use of relevance assessment. For the anchor text field, we recommend applying an empirical hyper-parameter setting.

8 Conclusions and Future Work

In this paper, we have provided a better understanding of per-field normalisation, based on Harter and Amati's idea that there is a linear relationship between term frequency and document length. We argue that the purpose of per-field normalisation is to adjust the linear relationship between term frequency and field length. Based on this argument, we have conducted a study of setting the per-field normalisation hyper-parameters, based on experimentation on two TREC Web collections for named-page finding. From the experiments, we have the following important finding: For the body and title fields, using three different field-based weighting models, the optimal $\rho(tfn_f, l_f)$ value, given by the optimised hyper-parameter value, is proportional to the maximum negative $\rho(tfn_f, l_f)$ value across the two collections used. Another important finding is that the optimised hyper-parameter setting for the anchor text field are largely different across the two collections used. We suggest that this is because of the repeatedly occurring tokens of the query terms in the anchor text field. Based on the above findings, we proposed an automatic setting method for setting the per-field normalisation hyper-parameters, called method *prop*, for the body and title fields. The proposed method does not require relevance assessment for tuning. According to the evaluation results, method *prop* was shown to be effective on the two test collections used, with 958 associated test queries. For the application of method *prop* in practise, we recommend applying method *prop* for the document fields of written text, such as the body and title fields. For the anchor text field, we recommend applying an empirical hyper-parameter setting, obtained by training using relevance assessment.

The reported experiments in this paper were conducted for the named-page finding retrieval task, on two different TREC collections, including the large-scale .GOV2 collection. We have also conducted experiments for ad-hoc retrieval on various TREC test collections, from which we had similar observations with those in this paper. For lack of space, we only focus on the named-page finding retrieval task in this paper. In the future, We will further study if method *prop* is general enough to cope with other Web search tasks, in the context of field-based retrieval. Moreover, because of the repeatedly occurring terms in anchor text, it is difficult to estimate the hyper-parameter setting for this field. A possible solution is to apply an absolute discount on the term frequency in this field, before per-field normalisation is applied. We will also investigate this issue in future work.

References

1. G. Amati. *Probabilistic Models for Information Retrieval based on Divergence from Randomness.* PhD thesis, University of Glasgow, 2003.
2. A. Chowdhury, M. C. McCabe, D. Grossman, and O. Frieder. Document normalization revisited. In *Proceedings of ACM SIGIR 2002.*
3. C. Clarke, F. Scholer, and I. Soboroff. Overview of the TREC-2005 Terabyte Track. In *Proceedings of TREC 2005.*
4. M. DeGroot. *Probability and Statistics.* Addison Wesley, 2nd edition, 1989.
5. N. Eiron and K. McCurley. Analysis of anchor text for web search. In *Proceedings ACM SIGIR 2003*, 2003. URL: http://mccurley.org/papers/anchor.pdf.
6. S. Harter. *A probabilistic approach to automatic keyword indexing.* PhD thesis, The University of Chicago, 1974.
7. B. He and I. Ounis. Term frequency normalisation tuning for BM25 and DFR model. In *Proceedings of ECIR 2005*, 2005.
8. B. He and I. Ounis. A study of the Dirichlet Priors for term frequency normalisation. In *Proceedings of ACM SIGIR 2005*, 2005.
9. B. Jansen and A. Spink. How are we searching the World Wide Web? a comparison of nine search engine transaction logs. *Information Processing & Management*, 42(1), 2006.
10. C. Macdonald, B. He, V. Plachouras, and I. Ounis. University of Glasgow at TREC 2005: experiments in Terabyte and Enterprise tracks with Terrier. In *Proceedings of TREC 2005.*
11. I. Ounis, G. Amati, V. Plachouras, B. He, C. Macdonald, and C. Lioma. Terrier: A high performance and scalable Information Retrieval platform. In *Proceedings of ACM SIGIR OSIR Workshop 2006.*

12. V. Plachouras. *Selective Web Information Retrieval.* PhD thesis, University of Glasgow, 2006.
13. S.E. Robertson, S. Walker, and M. Beaulieu. Okapi at TREC-7: automatic ad hoc, filtering, VLC and interactive. In *Proceedings of TREC 7*, 1998.
14. S.E. Robertson, H. Zaragoza, and M. Taylor. Simple BM25 extension to multiple weighted fields. In *Proceedings ACM CIKM 2004.*
15. G. Salton. *The SMART Retrieval System.* Prentice Hall, New Jersey, 1971.
16. A. Singhal, C. Buckley, and M. Mitra. Pivoted document length normalization. In *Proceedings of ACM SIGIR 1996.*
17. E. Voorhees. *TREC: Experiment and Evaluation in Information Retrieval.* The MIT Press, 2005.
18. H. Zaragoza, N. Craswell, M. Taylor, S. Saria, and S.E. Robertson. Microsoft Cambridge at TREC 13: Web and Hard Tracks. In *Proceedings of TREC 2004.*

Combining Evidence for Relevance Criteria: A Framework and Experiments in Web Retrieval

Theodora Tsikrika and Mounia Lalmas

Department of Computer Science, Queen Mary, University of London, UK
{theodora, mounia}@dcs.qmul.ac.uk
http://qmir.dcs.qmul.ac.uk

Abstract. We present a framework that assesses relevance with respect to several relevance criteria, by combining the query-dependent and query-independent evidence indicating these criteria. This combination of evidence is modelled in a uniform way, irrespective of whether the evidence is associated with a single document or related documents. The framework is formally expressed within Dempster-Shafer theory. It is evaluated for web retrieval in the context of TREC's Topic Distillation task. Our results indicate that aggregating content-based evidence from the linked pages of a page is beneficial, and that the additional incorporation of their homepage evidence further improves the effectiveness.

Keywords: Dempster-Shafer theory, topic distillation, best entry point.

1 Motivation, Background, and Aim

In ad hoc Information Retrieval (IR), multiple *criteria* are applied when assessing the relevance of documents. The relevance criterion at the heart of IR, and the one usually employed by IR systems, is the *topical* relevance (or *topicality*) of documents [1]. From a user's perspective, though, empirical studies have reached a consensus that users are influenced by beyond topical factors when assessing retrieved documents [1]. Therefore, IR systems need to consider beyond topical relevance criteria. For instance, on the Web, due to its size and unregulated nature, users desire authoritative information, without explicitly stating so.

An IR system assesses relevance by using *evidence of relevance* in its retrieval function. In essence, each source of evidence indicates relevance with respect to a specific criterion. For instance, content-based evidence is used for capturing a document's *topicality*. In web environments, link-based query-independent evidence, such as a page's PageRank [2], indicates a page's *authority*. Algorithms such as HITS [11], on the other hand, express a query-dependent view of a page's authority, i.e. its *topical authority*. In addition, URL-based query-independent evidence (e.g. URL length [14] or URL types [12]) is used for assessing a page's "*homepageness*" [5] (i.e. how likely it is for a page to be a site's homepage).

To assess relevance that reflects various criteria, IR systems combine evidence indicating the criteria of interest. The predominant *combination of evidence* approaches that incorporate, in a principled manner, evidence indicating

G. Amati, C. Carpineto, and G. Romano (Eds.): ECIR 2007, LNCS 4425, pp. 481–493, 2007.

beyond topical criteria are probabilistic frameworks. These estimate the belief in relevance given query-dependent and query-independent features. For instance, in language modelling frameworks (e.g. [12]), prior probabilities of relevance, estimated using query-independent features, are embedded in the framework, and combined with the (content-based) language modelling probability. Other frameworks (e.g. [7]) transform each feature's value into a feature-based relevance score, and subsequently linearly combine all available relevance scores.

Our aim is similar: to estimate, in a principled manner, the belief in relevance, when various criteria are of interest, by combining the (query-independent and query-dependent) features indicating these (or any combination of these) criteria. However, unlike others, our aim also is to estimate the belief in each of the relevance criteria of interest by decomposing relevance into the criteria involved. For instance, for web retrieval, by decomposing relevance into topicality, authority and homepageness, the *combination of evidence for relevance criteria* allows us to estimate the belief, for each page, in being each of the following: on the topic, a topical authority, a topical homepage, a topical authoritative homepage, and any other possible combination of these criteria.

In addition, since on the Web, and other hyperlinked environments, users browse, they assess a web page in terms of the information it contains, and the information it provides access to [4,11], i.e. as an entry point to the Web's structure. In particular, when many interlinked pages from the same site are retrieved, users would rather not be presented with all of them, but with only a *Best Entry Point* (BEP) [4] to the site, i.e. a page at a suitable level in the site's hierarchy providing access, by browsing, to the relevant information in the site.

For instance, BEPs could correspond to homepages, viewed as good entry points for users to follow the flow of information in the site, or to be presented with an overview of its content [9]. To identify BEPs as *topical homepages*, web IR systems could combine content-based and homepage evidence. Alternatively, they could assess each page in terms of its own features and those of the pages it provides access to. Such approaches aggregate the features of each page with those of its linked pages, by propagating them through the site's structure [4].

The aggregation can be performed by propagating: (i) *term weights* [10,15] or (ii) *relevance scores* [13,15]. The former identifies BEPs with respect to topicality, whereas the latter is able to capture multiple relevance criteria, depending on the relevance scores incorporated in the aggregation. For instance, by aggregating relevance scores indicating the topicality and authority of pages, we can model BEPs that provide access to pages containing authoritative information on a topic. This flexibility has not been fully exploited in the context of the Web, where relevance scores reflecting only a single criterion, e.g. content-based relevance scores reflecting topicality [13,15,5,6], are usually aggregated.

Therefore, our aim is twofold: (i) to assess relevance with respect to relevance criteria, by combining the evidence indicating these criteria, and (ii) to model this combination of evidence in a uniform way, irrespective of whether the evidence is associated with a single information item (e.g. a single web page) or with related information items (e.g. linked web pages). Although our aim is to provide a framework applicable to various environments, we focus on web retrieval, where we assess each web page as an entry point with respect to any relevance criterion, given either its own features, or also those of its linked pages.

To estimate the belief in relevance by combining the available evidence, various formalisms for reasoning with uncertainty can be employed. We explore the possibility of modelling our framework using Dempster-Shafer theory of evidence [17], an alternative formalism to probability theory. We consider this theory to be useful at the conceptual design level, and for providing guidance in expressing and performing the combination of evidence. We apply our framework to a web retrieval task, TREC's Topic Distillation [5,6], an informational task concerned with retrieving for a broad topic key resources, interpreted as BEPs (that correspond to *homepages*) of sites providing *credible* information on the *topic.*

Section 2 introduces Dempster-Shafer theory. The framework, expressed within this theory, is described in Section 3. It is evaluated for TREC's Topic Distillation task. Section 4 describes the experimental setting, and presents and discusses the results of the experiments. Section 5 provides some concluding remarks.

2 Dempster-Shafer Theory of Evidence

Dempster-Shafer (DS) theory of evidence is a formalism for representing, manipulating and revising *degrees of belief* rendered by multiple sources of evidence to a common set of propositions. It concerns the same concepts as those considered by Bayesian probability theory. It does not rely, though, on the probabilistic quantification of degrees of belief, but on a more general system based on *belief functions.* This theory was developed by Shafer [17], based on Dempster's earlier work [8]. We summarise the necessary background of the theory, by adopting Shafer's [17] initial terminology, notation and interpretation of the formalism.

Frame of discernment. Suppose we are concerned with the value of some quantity θ and the set of its possible values is Θ. In DS theory, this set Θ of exhaustive and mutually exclusive events is called *frame of discernment.* There is an one-to-one correspondence between subsets of Θ and propositions. The propositions of interest could be: "the value of θ is in A", $A \subseteq \Theta$. If $A = \{a\}$, $a \in \Theta$, the proposition is expressed as "the value of θ is a" and constitutes an *elementary proposition. Non-elementary propositions* are disjunctions of elementary ones.

Basic probability assignment. The belief committed to a proposition given some evidence is quantified by a function $m : 2^{\Theta} \rightarrow [0,1]$ called a *basic probability assignment* (bpa). Bpas can assign belief to any proposition in the frame and not only to the elementary ones. No belief can ever be assigned to the false proposition ($m(\emptyset) = 0$) and the sum of all bpas must equate 1: $\sum_{A \subseteq \Theta} m(A) = 1$. The quantity $m(A)$ represents the belief committed *exactly* to A, which due to lack of evidence (ignorance) cannot be committed to any proper subset of A.

Belief assignments are carried out only for propositions for which there is evidence. Consequently, committing belief to a proposition A does not necessarily imply that the remaining belief is committed to its negation $\neg A$. Therefore, if $m(A)=0.6$, and there is no further evidence for or against A or any other proposition in Θ, then, the remaining $1-0.6=0.4$ is assigned to the frame: $m(\Theta)$=0.4. This represents a state of ignorance and implies that this remaining belief could be assigned to any proposition in Θ, when new evidence becomes available. Complete ignorance with respect to the frame Θ is represented by the *vacuous* bpa: $m(\Theta)$=1 and $m(A)$=0, $\forall A\subset\Theta$. In any case, if $m(A)>0$, A is called a *focal element*. The focal elements and associated bpa define a *body of evidence*.

We can also obtain a *δ-discounted bpa* m^δ $(0\leq\delta\leq 1)$ from the original bpa m as follows: $m^\delta(A)=\delta * m(A)$, $\forall A\subset\Theta$ and $m^\delta(\Theta)=\delta * m(\Theta)+1-\delta$. The discounting factor δ represents a form of knowledge on the reliability of the body of evidence.

Belief function. Given a body of evidence with bpa m, one can compute the *total* belief committed to a proposition $A\subseteq\Theta$. This is done with a *belief function* $Bel : 2^\Theta \mapsto [0,1]$ defined upon m, so that it considers the belief assigned to the more specific propositions (i.e. to the subsets) of A: $Bel(A)=\sum_{B\subseteq A} m(B)$.

Dempster's combination rule. This rule aggregates two distinct bodies of evidence, with bpas m_1 and m_2, defined within the same frame Θ, into one body of evidence defined by a bpa m on the same frame: $m(A)=m_1\oplus m_2(A)=\frac{\sum_{B\cap C=A} m_1(B)m_2(C)}{1-\sum_{B\cap C=\emptyset} m_1(B)m_2(C)}$. The rule is commutative and associative. It computes a measure of agreement between two bodies of evidence concerning propositions discerned from a common frame. It focuses only on propositions that both bodies of evidence support. The numerator is the sum over all conjunctions that support a proposition. The denominator is a normalisation factor ensuring m is a bpa. Combining bpa m_1 with a vacuous bpa m_v, has no effect on m_1: $m_1\oplus m_v=m_1$.

3 Combining Evidence for Relevance Criteria

This section presents our framework, expressed within DS theory, for modelling the combination (and aggregation) of evidence (features) for relevance criteria. Our presentation focuses on web retrieval. In Section 3.1, we assess the relevance of each page, given either its own features, or also those of its linked pages, without considering what the underlying criteria are. In Section 3.2, we extend our framework and explicitly consider the criteria of interest. In both cases, we consider that the features' values have been transformed to relevance scores.

3.1 The Basic Framework: Combining Evidence for Relevance

We define the frame of discernment Θ in terms of the relevance criteria of interest. When we only consider the relevance of web pages without explicitly specifying the underlying criteria, the elements of Θ are defined as the mutually exclusive propositions $\theta_0=\{\neg R\}$ and $\theta_1=\{R\}$. Proposition $\{R\}$ reflects "a good point to enter for accessing R information", R being relevant. Each page x, referred to as *object* o_x, is represented by a body of evidence defined in Θ. Its associated bpa $m_x(A)$ quantifies the belief in $A\subseteq\Theta$, given all available evidence for x.

Representation. When a single source of evidence of relevance is available, $m_x(\{R\})$ (denoted $m_x(R)$ for simplicity) quantifies the degree to which this evidence indicates that this is a good point to enter to access relevant information. Suppose page x has a relevance score 0.6 given evidence e, then $m_x(R)=0.6$. The remaining belief is assigned to Θ, $m_x(\Theta)=0.4$, representing that, at this stage, we have no further evidence for any other proposition in Θ. The total belief is $Bel_x(R)=m_x(R)$. Suppose page y has a zero relevance score given evidence e. A first approach is to associate o_y with a *vacuous bpa* $m_y(\Theta)=1$. This expresses complete ignorance with respect to Θ, i.e. we consider that our evidence does not allow us to assign any belief in $\{R\}$ or $\{\neg R\}$. However, we do know that, given evidence e, page y was assessed as non relevant. This can be used to express our belief in $\{\neg R\}$. Therefore, a second approach is to set $0<m_y(\neg R)<1$.

When more than one source of evidence is available for each page, a separate bpa is defined in terms of each source of evidence taken into account. Suppose that page z is assigned relevance score 0.6 given evidence e_1, and relevance score 0.7 given evidence e_2. Then, page z is represented by 2 separate bpas: $m_{z:e_1}(R)=0.6$ ($m_{z:e_1}(\Theta)=0.4$) and $m_{z:e_2}(R)=0.7$ ($m_{z:e_2}(\Theta)=0.3$).

Combination. To combine the available evidence associated with a page, we combine the bodies of evidence using Dempster's combination rule. For instance, given page z as above, the combination yields $m_z=m_{z:e_1}\oplus m_{z:e_2}$, with $m_z(R)=(m_{z:e_1}(R)*m_{z:e_2}(R)+m_{z:e_1}(R)*m_{z:e_2}(\Theta)+m_{z:e_1}(\Theta)*m_{z:e_2}(R))/1=0.6*0.7+0.6*0.3+0.4*0.7=0.88$ and $m_z(\Theta)=(m_{z:e_1}(\Theta)*m_{z:e_2}(\Theta))/1=0.4*0.3=0.12$.

Aggregation. To assess each page in terms of its own features and those of the pages it provides access to, we aggregate the bodies of evidence of linked pages belonging to the same site using Dempster's combination rule.

Consider the web sites in Figure 1. Each page i, referred to as object o_i, is represented by a body of evidence in Θ and its associated bpa is m_i. Given the evidence from page p (site A) and its linked children pages c_k, $k=1,...,5$, the aggregation is expressed as: $m_{p,c_{1-5}}=m_p\oplus m_{c_1}\oplus...\oplus m_{c_5}$.

As the user enters site A at page p, the actual information accessed is the one contained in p. The information contained in its children should be considered as "potential" [13], since the user needs to traverse the links in order to fully access it. Hence, the contribution of evidence from the children as a whole should be weighed appropriately, to reflect the uncertainty associated with their propagation to the parent page. This is expressed with a *propagation* (or *fading* [13]) factor, and is modelled by a *discounted bpa.* For instance, the bpa associated with the aggregate $o_{c_{1-5}}$, formed from the children of page p, is $m_{c_{1-5}}^{prop}$, where *prop* is the propagation factor. This is expressed as: $m_{p,c_{1-5}}=m_p\oplus m_{c_{1-5}}^{prop}$.

We can also express the contribution of each child o_{c_k} in forming the aggregate. The extent of this contribution, referred to as *accessibility* (*acc*) [16], is modelled by a discounted bpa $m_{c_k}^{acc_k}$. The bpa for c_{1-5} is: $m_{c_{1-5}}=m_{c_1}^{acc_1}\oplus...\oplus m_{c_5}^{acc_5}$. Thus, the belief in $\{R\}$ is (see also the definition of discounted bpas):

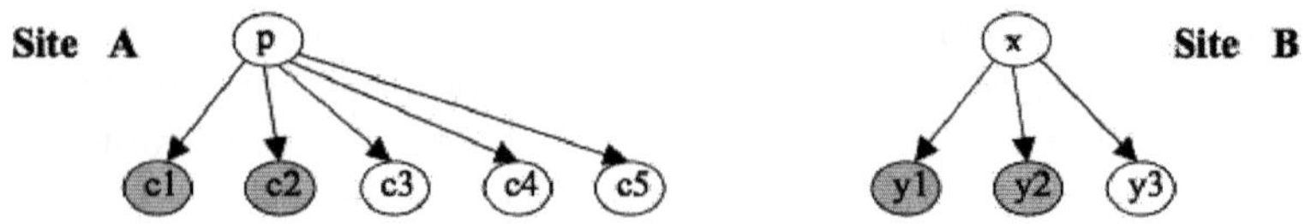

Fig. 1. Examples of linked pages in web sites

Table 1. Examples of aggregation methods applied to site A

Site A	Aggregation method **acc1**				Aggregation method **accn**				Aggregation method **notR**			
	$m_i(\cdot)$			$Bel_i(\cdot)$	$m_i(\cdot)$			$Bel_i(\cdot)$	$m_i(\cdot)$			$Bel_i(\cdot)$
o_i	R	$\neg R$	Θ	R	R	$\neg R$	Θ	R	R	$\neg R$	Θ	R
o_{c_1}	0.8	0	0.2	0.8	0.16	0	0.84	0.16	0.8	0	0.2	0.8
o_{c_2}	0.6	0	0.4	0.6	0.12	0	0.88	0.12	0.6	0	0.4	0.6
$o_{c_{1-2}}$	0.92	0	0.08	0.92	0.261	0	0.739	0.261	0.92	0	0.08	0.92
o_{c_3}	0	0	1	0	0	0	1	0	0	0.1	0.9	0
$o_{c_{1-3}}$	0.92	0	0.08	0.92	0.261	0	0.739	0.261	0.91	0.01	0.08	0.91
o_{c_4}	0	0	1	0	0	0	1	0	0	0.1	0.9	0
$o_{c_{1-4}}$	0.92	0	0.08	0.92	0.261	0	0.739	0.261	0.90	0.02	0.08	0.90
o_{c_5}	0	0	1	0	0	0	1	0	0	0.1	0.9	0
$o_{c_{1-5}}$	0.92	0	0.08	0.92	0.261	0	0.739	0.261	0.89	0.03	0.08	0.89
o_p	0	0	1	0	0	0	1	0	0	0.1	0.9	0
$o_{p,c_{1-5}}$	0.92	0	0.08	0.92	0.261	0	0.739	0.261	0.88	0.04	0.08	0.88

$$m_{c_{1-5}}(R) = (acc_1 * m_{c_1}(R)) \oplus ... \oplus (acc_5 * m_{c_5}(R)) \quad (1)$$

$$m_{p,c_{1-5}}(R) = m_p(R) \oplus (prop * m_{c_{1-5}}(R)) \quad (2)$$

To determine the BEP in a site with respect to relevance, we rank the pages in the site by their total belief in $\{R\}$: $Bel(R) = m(R)$.

Aggregation methods. By appropriately setting the accessibility and propagation factors, we can express various aggregation methods.

Suppose that only pages c_1, c_2 (site A) and pages y_1, y_2 (site B) are assigned non-zero relevance scores given all available evidence e, i.e. are retrieved given evidence e. Suppose also that the bpas for these retrieved pages (given evidence e) are: $m_{c_1}(R)=m_{y_1}(R)=0.8$ and $m_{c_2}(R)=m_{y_2}(R)=0.6$. and that we associate the non-retrieved pages with vacuous bpas: $m_j(\Theta)=1$, $j=\{p,c_3,c_4,c_5,x,y_3\}$.

Method **acc1** sets the accessibility of each child, i.e. its individual contribution to the aggregation, equal to 1. The aggregation of objects o_{c_1}, o_{c_2} (site A) yields object $o_{c_{1-2}}$ (Table 1). The belief of the aggregate object in $\{R\}$, $m_{c_{1-2}}(R)=0.92$, is greater than that of either of its component objects. Since, the non-retrieved children, o_{c_3}, o_{c_4} and o_{c_5}, are associated with vacuous bpas, their aggregations with $o_{c_{1-2}}$, for forming $o_{c_{1-5}}$, leave $m_{c_{1-2}}$ unaffected, i.e. $m_{c_{1-5}}(R)=m_{c_{1-2}}(R)=0.92$. Similarly for site B, $m_{y_{1-3}}(R)=m_{y_{1-2}}(R)=0.92$. If the propagation factor is uniformly set across sites (e.g. $prop=1$), the belief in pages p and x is the same, despite having different numbers of non-retrieved children. Method **acc1** considers only the contribution of the retrieved children.

To model page x as a better BEP than page p, since it provides access to less non-relevant information, we need to consider the non-retrieved pages. One way is to set the propagation factor $prop = \frac{1}{n}$ (n is the number of children). Another is to set the accessibility $acc = \frac{1}{n}$ (method **accn** in Table 1). With a propagation factor uniformly set across sites (e.g. $prop = 1$), page x is now considered a better BEP than page p ($m_{x,y_{1-3}}(R) = 0.416 > 0.261 = m_{p,c_{1-5}}(R)$). Method **accn** greatly discounts the contribution of the children in the aggregation.

Method **notR** explicitly takes into account the non-retrieved pages, by modelling them not with a vacuous bpa, but with a bpa that assigns belief to $\{\neg R\}$. Suppose $m_j(\neg R) = 0.1$, $j = \{p, c_3, c_4, c_5\}$. We set $acc = 1$ and form object $o_{c_{1-2}}$ as before. Objects $o_{c_{1-2}}$, o_{c_3} support conflicting propositions. Their aggregation erodes the beliefs in them, and $m_{c_{1-2}}(R) = 0.92$ becomes $m_{c_{1-3}}(R) = (m_{c_{1-2}}(R) * m_{c_3}(R) + m_{c_{1-2}}(R) * m_{c_3}(\Theta) + m_{c_{1-2}}(\Theta) * m_{c_3}(R))/(1 - m_{c_{1-2}}(R) * m_{c_3}(\neg R) - m_{c_{1-2}}(\neg R) * m_{c_3}(R)) = (0.92*0 + 0.92*0.9 + 0.08*0)/(1 - 0.92*0.1 - 0*0) = 0.91$ (Table 1). Greater values of $m_i(\neg R)$ for non-retrieved pages lead to even greater erosion. Also, the more non-retrieved children are included in the aggregation, the more the belief in $\{R\}$ is reduced. By setting $prop = 1$, $m_{p,c_{1-5}}(R) = 0.88 < 0.90 = m_{x,y_{1-3}}(R)$. The values of $m(\neg R)$ can be determined experimentally or by evidence reflecting, for instance, the system's reliability or the query's difficulty.

This aggregation of linked pages belonging to the same site can be performed in an ascending manner (**bottom-up** propagation), starting from the pages deepest in the site's hierarchy. To remove the cycles from the site's structure, we construct a sitemap tree (similarly to [15]), using only *Down* type links (i.e. those linking pages with those below in the site's directory path [9]). Alternatively, we can perform an **1step** propagation, by considering for each page only its immediate neighbours (not necessarily just those connected with *Down* links).

3.2 The Extended Framework: Combining Evidence for Relevance Criteria

In this section, we explicitly consider the criteria underlying relevance.

Frame of discernment. The frame Θ is constructed based on the set of criteria of interest: $\mathbb{E} = \{e_1, \cdots, e_E\}$. The mutually exclusive elementary propositions of Θ are all the possible Boolean conjunctions of all the elements $e_i \in \mathbb{E}$, containing either e_i or $\neg e_i$. There are 2^E elements in Θ, each denoted as $\theta_{b_1 b_2 \cdots b_n}$, with $b_1 b_2 \cdots b_n$ an n-bit binary number, such that $\theta_{b_1 b_2 \cdots b_n}$ corresponds to the proposition "$x_1 \wedge x_2 \wedge \cdots \wedge x_n$", where $x_i = e_i$ if $b_i = 1$, and $x_i = \neg e_i$ if $b_i = 0$.

Suppose the criteria of interest are topicality (T), authority (A), and homepageness (HP): $\mathbb{E} = \{T, HP, A\}$. Then, the propositions forming the frame Θ are listed in Table 2. For instance, θ_{111} corresponds to $\{T \wedge HP \wedge A\}$, reflecting that a page is "a good point to enter to access *homepages* containing *authoritative* information *on the topic*". Analogously, $\{T \wedge A\}$ reflects that a page is "a good point to enter to access *topical* and *authoritative* information". Therefore, θ_{111} provides a more refined representation of the notion of topical relevance compared to $\{T \wedge A\}$, $\{T \wedge HP\}$ or $\{T\}$. In this work, we focus on $\mathbb{E} = \{T, HP\}$. (Due to space limitations, we do not report on criterion $\{A\}$, that we also considered.)

Representation. Consider we have two sources of evidence for each page: one capturing topicality $\{T\}$, and the other homepageness $\{HP\}$. Then, each page is

Table 2. Propositions forming the frame of discernment Θ in the extended framework

θ_{000}	$\neg T \wedge \neg HP \wedge \neg A$	θ_{010}	$\neg T \wedge HP \wedge \neg A$	θ_{100}	$T \wedge \neg HP \wedge \neg A$	θ_{110}	$T \wedge HP \wedge \neg A$
θ_{001}	$\neg T \wedge \neg HP \wedge A$	θ_{011}	$\neg T \wedge HP \wedge A$	θ_{101}	$T \wedge \neg HP \wedge A$	θ_{111}	$T \wedge HP \wedge A$

Table 3. Combination in the extended frame

Site A	$m_i(\cdot)$				$Bel_i(\cdot)$		
o_i	T	HP	$T \wedge HP$	Θ	T	HP	$T \wedge HP$
$o_{c_1:C}$	0.8	0	0	0.2	0.8	0	0
$o_{c_1:U}$	0	0.6	0	0.4	0	0.6	0
o_{c_1}	0.32	0.12	0.48	0.08	0.8	0.6	0.48
$o_{c_2:C}$	0.6	0	0	0.4	0.6	0	0
$o_{c_2:U}$	0	0.7	0	0.3	0	0.7	0
o_{c_2}	0.18	0.28	0.42	0.12	0.6	0.7	0.42

Table 4. Aggregation in the extended frame

Site A	$m_i(\cdot)$				$Bel_i(\cdot)$		
o_i	T	HP	$T \wedge HP$	Θ	T	HP	$T \wedge HP$
o_{c_1}	0.32	0.12	0.48	0.08	0.8	0.6	0.48
o_{c_2}	0.18	0.28	0.42	0.12	0.6	0.7	0.42
$o_{c_{1-2}}$	0.11	0.07	0.81	0.01	0.92	0.88	0.81
o_{c_3}	0	0	0	1	0	0	0
...							
$o_{p,c_{1-5}}$	0.11	0.07	0.81	0.01	0.92	0.88	0.81

represented by two separate bpas. Suppose the content-based (**C**) score, for page c_1 (site A), reflecting its topicality, is 0.8 and its URL-based (**U**) one, reflecting its homepageness, is 0.6. Then, we have $m_{c_1:C}$ and $m_{c_1:U}$ (Table 3).

Combination. The combination $m_{c_1} = m_{c_1:C} \oplus m_{c_1:U}$ (Table 3) assigns belief to propositions $\{T\}$, $\{HP\}$, and their conjunction $\{T \wedge HP\}$. Given the initial belief $m_{c_1:C}(T)$=0.8, we were unable to draw any finer distinction about the type of topicality supported, i.e $\{T \wedge HP\}$ or $\{T \wedge \neg HP\}$. Once evidence for $\{HP\}$ became available, some of this initial belief was assigned to subset $\{T \wedge HP\}$, but the total belief in $\{T\}$, $Bel_{c_1}(T) = m_{c_1}(T) + m_{c_1}(T \wedge HP) = 0.8$, remained the same. We also combine the evidence for o_{c_2}: $m_{c_2} = m_{c_2:C} \oplus m_{c_2:U}$ (Table 3).

Aggregation. Suppose we use aggregation method acc1. The aggregation $m_{c_{1-2}} = m_{c_1} \oplus m_{c_2}$ (Table 4) further redistributes the belief among non-disjoint propositions. The aggregation in terms of *Bel* is not affected by this distribution of belief, since it is only concerned with the total belief assigned to propositions. For instance, $Bel_{c_{1-2}}(T) = Bel_{c_1}(T) \oplus Bel_{c_2}(T) = 0.8 \oplus 0.6 = 0.92$. Furthermore, this, in essence, corresponds to $Bel_{c_1:C}(T) \oplus Bel_{c_2:C}(T) = 0.8 \oplus 0.6 = 0.92$, i.e. the aggregation in terms of *Bel*, irrespective of the additional evidence incorporated, produces the same results as if a single source is considered. Aggregating with o_{c_3}, o_{c_4}, o_{c_5}, and o_p, while setting *prop*=1, leads to $m_{p,c_{1-5}}$ (Table 4).

When multiple evidence are aggregated, we can produce many rankings using the belief *Bel* in different propositions, and determine BEPs with respect to different criteria. For instance, $Bel(T)$ identifies BEPs for accessing topically relevant pages, while $Bel(T \wedge HP)$ BEPs for accessing topical homepages.

Advantages of using DS theory include assigning belief to criteria for which there is evidence, rather than, of necessity, to every criterion. Also, we can assign belief to a set of propositions, without having to distribute belief among its individual propositions. Finally, the relaxation of the law of additivity ($m(A) + m(\neg A) \leq 1$) allows us to flexibly represent web pages given the available evidence.

4 Experiments

We perform evaluation experiments using the .GOV corpus and the topics and relevance assessments from TREC's Topic Distillation (TD) task (50 topics from TD2003 [5] and 75 topics from TD2004 [6]). We index the pages in the collection by combining their content and incoming anchor text. We apply stopword removal and stemming and use the weighting scheme and retrieval component employed in InQuery [3]. This content-based retrieval approach (**C**) is our baseline. To select the BEP from each site, we group, by their domain name, the top 500 pages retrieved by C, and apply aggregation approaches to each group.

We perform the aggregation in our extended DS framework with criteria of interest *topicality* (T) and *homepageness* (HP), i.e. we form Θ based on $\mathbb{E} = \{T, HP\}$. In Section 4.1, we focus on topicality and rank the pages by their $Bel(T)$. In Section 4.2, we also consider their homepageness and rank them by their $Bel(T \wedge HP)$. The belief $m\{T\}$ is quantified by the *content-based relevance score* (C), whereas $m\{HP\}$ by a query-independent *URL-based relevance score* (**U**), computed using each page's URL path length: $\frac{1}{log_2(urlpathlen+1)}$ [14].

We apply the following *aggregation methods*: **DS acc1**(*prop*), **DS accn**(*prop*), and **DS notR**(*prop*, not$T$), where *prop* is the propagation factor, and not$T$ the belief experimentally assigned to proposition $\{\neg T\}$ for pages not in the top 500 retrieved by C: $m(\neg T)$=not$T$. We compare these DS aggregations to linear combination (**LC**) aggregations, which can be considered to derive from equations (1) and (2) (Section 3.1), by replacing DS combination ($\oplus$) with addition (+). These aggregation methods are **LC acc1**(*prop*) and **LC accn**(*prop*).

For each of these aggregation methods (DS acc1,LC acc1,DS accn,LC accn,DS notR), we apply the *propagation strategies*: *bottom-up* and *1step Down*. These two strategies consider only the *Down* type links and our results indicate that they perform similarly. Therefore, we only present the more efficient *1step Down* propagation. We also apply *1step* propagation by aggregating linked pages connected with all, not only *Down*, types of intra-site links (*1step All*).

To tune parameters *prop* and not$T$, we use TD2003 as our training set, with TD2004 becoming our test set. Tuning *prop* involved an exploration from 0.1 to 1 at step 0.1, and tuning not$T$ an exploration from 0.1 to 0.9 at step 0.1. These tunings aimed at maximising P@10. We select P@10 because, in TD2003, mean average precision (MAP) and R-precision (precision at R, R = number of relevant documents for a query) are more sensitive than P@10 [18]. We also set *prop*=$\frac{1}{n}$ (n=number of children) which led to poor results and is not reported.

In all the presented tables, the effectiveness values improving over the baseline are depicted in **bold**. Statistically significant results, indicated by a $*$, are determined by applying a Wilcoxon matched-pairs signed ranks test (α=0.05).

4.1 Experiments in Aggregating Evidence for Topicality

First, we select the BEP from each site with respect to topical relevance criteria, i.e. we select pages that provide access to topically relevant information. To this end, we aggregate, in a DS or linear manner, the content-based relevance scores of

Table 5. Aggregating content-based evidence (top 500 pages retrieved by C)

TD2004		MAP	P@5	P@10	R-Prec.
C	(baseline)	0.1237	0.2187	0.1893	0.1622
BEP DS acc1	1step Down *prop* = 0.1	0.1064	**0.2480***	**0.2013**	**0.1628**
	1step All *prop* = 0.1	**0.1347**	**0.2827***	**0.2187***	**0.1974***
BEP LC acc1	1step Down *prop* = 0.1	0.0998	**0.2213**	**0.1947**	0.1537
	1step All *prop* = 0.1	0.1136	**0.2453***	**0.2120***	**0.1873***
BEP DS accn	1step Down *prop* = 0.9	0.0977	**0.2213**	0.1880	0.1541
	1step All *prop* = 0.1	0.1121	**0.2373***	**0.2027**	**0.1638**
BEP LC accn	1step Down *prop* = 0.8	0.0981	**0.2213**	0.1880	0.1541
	1step All *prop* = 0.1	0.1121	**0.2453***	**0.2013**	0.1579
BEP DS notR	1step Down *prop* = 0.2 *notT* = 0.1	0.1069	**0.2373**	**0.2000**	0.1566
	1step All *prop* = 0.1 *notT* = 0.1	0.1116	**0.2533***	**0.2067***	**0.1669**

linked pages. Previous research has already indicated that the within-site linear aggregation of content-based relevance scores is effective for Topic Distillation [15]. Our objectives are: (i) to examine the effectiveness of this aggregation when modelled within our DS framework (and also compare it to a linear aggregation) and (ii) to gain an insight into the workings of the aggregation, by studying the effect of the various aggregation methods and propagation strategies.

In the training set, for most propagation strategies (except for DS accn and LC accn for *1step Down*), the lower the contribution of the children as a whole (determined by the propagation factor), the better the results. The best results were achieved for *prop*=0.1. Also, the more links were considered (*1step All* vs. *1step Down*), the more the effectiveness improved, suggesting that evidence from pages connected with all types of links is beneficial. We applied each propagation strategy, with its most effective parameter(s) for each aggregation method, to our test set (Table 5). Our training set observation, that considering low contributing evidence from all children is beneficial, is confirmed by our test set results.

While the contribution of the children pages as a whole is determined by *prop*, the contribution of each individual child is determined by the aggregation method. Method acc1 considers only the children retrieved by C. Method accn greatly discounts the contribution of each child and thus is *indirectly* affected by the non-retrieved children. Method DS notR is *directly* affected by non-retrieved children, with their contribution expressed through not$T$.

In the training set, acc1 was the most effective method, followed by notR, whereas accn did not perform particularly well. These observations are confirmed by the test set results, indicating that although the contribution of the retrieved children should be low (expressed through low *prop* values), it should not be too greatly discounted (as achieved by accn). This is further supported by DS notR being most effective for low not$T$ values, i.e. not$T$=0.1, which discount the contribution of retrieved pages more gradually than accn (see Table 1). The most effective methods, DS acc1, LC acc1, and DS notR, for *1step All*, improve P@10 significantly over the baseline, with DS acc1 *1step All* also improving MAP.

Overall, our results confirm previous findings that aggregating content-based evidence from the retrieved children of a web page is beneficial for Topic Distillation [15,5,6]. Our framework allowed us to study these aggregations further, indicating that the contribution of the retrieved children should be low, but not too greatly discounted. In addition, considering only the immediate neighbours is sufficient, with the most effective and robust strategy (*1step All*) taking into account all linked (immediate) children. These findings apply for both DS and linear aggregations, with the DS aggregation performing comparatively better. Our DS framework also provides the expressiveness and flexibility to incorporate evidence for additional relevance criteria, e.g. homepageness, discussed next.

4.2 Experiments in Combining and Aggregating Evidence for Topicality and Homepageness

These experiments aim at assessing relevance with respect to topicality (T) and homepageness (HP) relevance criteria, by considering the available evidence capturing each criterion, i.e. the content-based (C) and URL-based (U) scores of web pages. First, we combine, for each page, its two scores, producing, in essence, a reranking of the C baseline. Next, we express within our DS framework the aggregation of these two scores of linked pages belonging to the same site. In that way, we identify each site's BEP as the page that provides access to homepages containing topically relevant information. We denote this aggregation as T⊕HP.

The combination of the C and U scores is performed in our DS framework and compared to a linear combination. The DS combination is expressed as $m(T) \oplus m(HP)$ resulting in belief also assigned to $m(T \wedge HP)$ (see Table 3). We rerank the top 500 pages retrieved by C in terms of $Bel(T \wedge HP) = m(T \wedge HP)$. Since this corresponds, in essence, to a multiplication of the C and U scores, we denote it as CU. The linear combination is expressed as $C + w * U$. We tune w in TD2003 for values 0.1 to 1 at step 0.1, and achieve the best results for $w = 0.2$.

Both combinations improve significantly over C in TD2004 (Table 6), confirming the usefulness of homepage evidence for this task [5,6]. They are also more effective than the aggregations of content-based evidence (see Table 5). Next, we perform the T⊕HP aggregation, using CU and C+0.2U as baselines.

Our training set results for the T⊕HP aggregation indicate that, when also considering homepage evidence, the contribution of the retrieved children is still beneficial, but should be greatly discounted. In fact, accn was the most effective followed by notR and then acc1, with all achieving their best results for $prop = 0.1$. Also considering only few of the children might be sufficient, since *1step Down* performed comparably to *1step All*. We apply the most effective approaches for the T⊕HP aggregation to the test set (Table 6). Our training set observations are confirmed by our test set results. The most effective method is accn, with *1step Down* improving P@10 significantly over all baselines. Method notR is slightly less effective, but still improves, though not significantly, over the baselines, whereas acc1 only improves over the content-based baseline (C).

Previous research has examined either the aggregation of content-based evidence from linked pages, or the combination of content-based and homepage evidence for a single page. We integrate these approaches, and indicate that by incorporating beyond content-based evidence when aggregating linked pages, as modelled by our DS framework, we can further improve the effectiveness.

Table 6. Combining/Aggregating content- and URL-based evidence (top 500 pages retrieved by C)

TD2004		MAP	P@5	P@10	R-Prec.
C	(baseline for CU and C+0.2U)	0.1237	0.2187	0.1893	0.1622
CU	(baseline for BEP T⊕HP approaches)	**0.1478***	**0.2827***	**0.2227***	**0.1881***
C+0.2U	(baseline for BEP T⊕HP approaches)	**0.1504***	**0.2827***	**0.2213***	**0.1752**
BEP T⊕HP DS acc1	1step Down $prop = 0.1$	0.0971	0.1947	0.1800	0.1623
	1step All $prop = 0.1$	0.1014	0.2267	0.1920	0.1663
BEP T⊕HP DS accn	1step Down $prop = 0.1$	0.1312	0.2720	**0.2413***	**0.1985**
	1step All $prop = 0.1$	0.1287	0.2773	**0.2373**	**0.2027**
BEP T⊕HP DS notR	1step Down $prop = 0.1$ $notT = 0.9$	0.1238	0.2560	**0.2387**	**0.1934**
	1step All $prop = 0.1$ $notT = 0.1$	0.1245	0.2773	0.2200	0.1876

5 Conclusions

We proposed a framework that assesses relevance with respect to any of the relevance criteria of interest, by combining the evidence indicating these criteria, derived both from a web page and its linked web pages. We estimate the belief in relevance and perform this combination using Dempster-Shafer (DS) theory of evidence. The expressiveness and flexibility of the framework is demonstrated by the ease with which the combination with respect to any relevance criterion is expressed, the aggregation of evidence from linked pages is incorporated, and negated evidence can be considered. We evaluated the framework in the context of TREC's Topic Distillation task, and in terms of the topicality and homepageness relevance criteria. Our experiments indicated the effectiveness of aggregating content-based evidence on their own, or together with homepage evidence, and allowed us to study the workings of aggregation methods.

References

1. C. L. Barry. User-defined relevance criteria: An exploratory study. *JASIS*, 45(3):149–159, 1994.
2. S. Brin and L. Page. The anatomy of a large-scale hypertextual Web search engine. *Computer Networks and ISDN Systems*, 30(1–7):107–117, 1998.
3. J. P. Callan, W. B. Croft, and S. M. Harding. The INQUERY retrieval system. In *DEXA'92, pp. 78-83.*
4. Y. Chiaramella. Information retrieval and structured documents. In *European Summer School in IR*, volume 1980 of *LNCS*, pages 286–309, 2001.
5. N. Craswell and D. Hawking. Overview of the trec-2003 web track. In *TREC-2003.*
6. N. Craswell and D. Hawking. Overview of the trec-2004 web track. In *TREC-2004.*
7. N. Craswell, S. Robertson, H. Zaragoza, and M. Taylor. Relevance weighting for query independent evidence. In *SIGIR'05*, pages 416–423, 2005.
8. A. Dempster. A generalization of bayesian inference. *Journal of Royal Statistical Society*, 30:205–247, 1968.
9. N. Eiron and K. S. McCurley. Untangling compound documents on the web. In *ACM Hypertext and Hypermedia conference*, pages 85–94, 2003.
10. N. Fuhr, M. Lalmas, S. Malik, and Z. Szlavik, editors. *Proceedings of INEX 2004.*
11. J. Kleinberg. Authoritative sources in a hyperlinked environment. *Journal of the ACM*, 46(5):604–632, 1999.

12. W. Kraaij, T. Westerveld, and D. Hiemstra. The importance of prior probabilities for entry page search. In *SIGIR'02*, pages 27–34, 2002.
13. M. Marchiori. The quest for correct information on the web: hyper search engines. In *Proceedings of the 6th WWW conference*, pages 1225–1235, 1997.
14. V. Plachouras and I. Ounis. Usefulness of hyperlink structure for query-biased topic distillation. In *SIGIR'04*, pages 448–455, 2004.
15. T. Qin, T.-Y. Liu, Z. X.-D., Z. Chen, and W.-Y. Ma. A study of relevance propagation for web search. In *SIGIR'05*, pages 408–415, 2005.
16. T. Roelleke, M. Lalmas, G. Kazai, I. Ruthven, and S. Quicker. The accessibility dimension for structured document retrieval. In *ECIR'02*, pages 382–402, 2002.
17. G. Shafer. *A mathematical theory of evidence.* Princeton University Press, 1976.
18. I. Soboroff. On evaluating web search with very few relevant documents. In *SIGIR '04*, pages 530–531, 2004.

Classifier Fusion for SVM-Based Multimedia Semantic Indexing

Stéphane Ayache, Georges Quénot, and Jérôme Gensel

Laboratoire d'Informatique de Grenoble (LIG)
385 rue de la Bibliothéque - BP 53
38041 Grenoble - Cedex 9 France

Abstract. Concept indexing in multimedia libraries is very useful for users searching and browsing but it is a very challenging research problem as well. Combining several modalities, features or concepts is one of the key issues for bridging the gap between signal and semantics. In this paper, we present three fusion schemes inspired from the classical early and late fusion schemes. First, we present a kernel-based fusion scheme which takes advantage of the kernel basis of classifiers such as SVMs. Second, we integrate a new normalization process into the early fusion scheme. Third, we present a contextual late fusion scheme to merge classification scores of several concepts. We conducted experiments in the framework of the official TRECVID'06 evaluation campaign and we obtained significant improvements with the proposed fusion schemes relatively to usual fusion schemes.

1 Introduction

In order to retrieve multimedia documents from huge digital libraries, the needs for concept-based indexing are rapidly growing. Finding concepts in multimedia documents, such as video sequences, is one of the main objectives of the content-based semantic indexing community. Hence, new issues are arising on the combination (fusion) of several features, modalities and/or intermediate concepts to obtain a better accuracy of concept detection. For instance, an efficient fusion scheme must enhance concept indexing in multimedia documents by merging visual and textual modalities, color and texture modalities, or global and local features. Using a generic framework, usual approaches propose either to merge data on a concatenated vector before achieving classification [1, 2], or to perform several classification and then to merge confidence scores using a higher level classifier [6, 11] by the means of a stacking technique [16]. Called "early" and "late" fusion [13], those approaches are easy to implement and provide state of the art performance. However, such fusion schemes are not always able to outperform unimodal classifiers, especially when one of the modalities provide much better accuracy than the others or when one has to handle imbalanced input features. Such situations are particularly frequent in the field of multimedia indexing due to the diversity of concepts with regard to the extracted features.

G. Amati, C. Carpineto, and G. Romano (Eds.): ECIR 2007, LNCS 4425, pp. 494–504, 2007.

Using kernel-based classifier, for instance a Support Vector Machine, recent approaches have been proposed to take advantage of some useful kernel properties. They aim to merge features at the kernel level before performing the concept classification. Kernel-based data fusion has been successfully applied in biology to the problem of predicting the function of yeast proteins [8]. [8, 15] propose efficient algorithms to learn simultaneously the parameters of each unimodal kernel and the parameters of the combining function.

In this paper, we study and compare three fusion schemes in the scope of semantic video indexing. The first one takes advantage of some useful kernel properties, we present a simple algorithm which merges unimodal kernels before performing the concept classification using a SVM classifier. In such a way, features are combined at the earliest possible step using a kernel-based classifier. The second fusion scheme is derived from the early fusion scheme. We normalized each individual feature vectors so that their average norm becomes equal in order to reduce the problem of imbalanced input features. The third fusion scheme is a late-like fusion scheme; it performs fusion at the concept level taking into account the classification scores of 39 concepts from visual and textual modalities.

In Section 2, we briefly present Support Vectors Machines and some required knowledge about kernels. In section 3, we describe the proposed fusion schemes and give some background information for formally comparing them with other fusion schemes. In section 4, we describe the experiments conducted using the TRECVID'06 [7] corpus and metrics.

2 Kernel-Based Classifier

Kernel-based methods have provided successful tools for solving many recognition problems, such as KFD, KPCA or SVM [12]. One of the reasons of this success is the use of kernels which overcome the problem of non-linearly separable data sets by mapping the initial problem into a higher dimensional space. The main idea behind kernel-based classifiers is that the similarity between examples in a data set gives much information about the patterns that may be present in these data.

2.1 Support Vector Machines

Support Vector Machines (SVM) have shown their capacities in pattern recognition and have been widely used for classification in CBIR. SVM is formalized as an optimization problem which finds the best hyperplane separating relevant and irrelevant vectors by maximizing the size of the margin between both sets. The use of a kernel allows the algorithm to find the maximum-margin hyperplane in a transformed feature space. The transformation may be non-linear and the transformed space may be of higher dimensionality than the original one. Thus, though the classifier separator is a hyperplane in the high-dimensional feature space it may be non-linear in the original input space. Furthermore, if the kernel used is a Gaussian radial basis function, the corresponding feature space is a Hilbert space of infinite dimension. Maximum margin classifiers are well regularized and the infinite dimension does not spoil the results. In a two-class case, the decision function for a test sample $\mathbf{x}$ has the following form:

$$g(\mathbf{x}) = \sum_i \alpha_i y_i K(\mathbf{x}_i, \mathbf{x}) - b$$

where $K(\mathbf{x}_i, \mathbf{x})$ is the value of a kernel function for the training sample $\mathbf{x}_i$ and the test sample $\mathbf{x}$, y_i the class label of $\mathbf{x}_i$ (+1 or −1), α_i the learned weight of the training sample $\mathbf{x}_i$, and b is a learned threshold parameter. The training samples with weight $\alpha_i > 0$ are usually called "support vectors".

2.2 Kernel Matrices

A kernel matrix is a similarity matrix where each entry represents a measure of similarity between two sample vectors, and must be positive definite (i.e. satisfy the Mercer's condition) to ensure that the optimization problem is convex. Therefore, a clever choice of the kernel (or similarity) function is essential to obtain a positive definite kernel matrix. Nevertheless, some unproved positive definite kernels such as EMD-based kernels or Log kernels have been successfully used in image recognition [18, 3].

Kernel matrices satisfying the Mercer's condition have interesting properties which provide modularity and derivation of kernels from other kernels. If K and K' are kernels, then the following are also kernels (not exhaustive):

- $aK + bK'$, for $a > 0$ and $b \geq 0$
- $K \times K'$, where $\times$ is the entrywise product

The most commonly used kernel function with SVM classifier in multimedia indexing is the RBF kernel defined as follow:

$$K(\mathbf{x}, \mathbf{y}) = e^{-\frac{\|\mathbf{x}-\mathbf{y}\|^2}{2\sigma^2}}$$

where $\| \,.\, \|$ denotes the L_2 norm, $\mathbf{x}$ and $\mathbf{y}$ are two sample vectors, and σ the width of the Gaussian kernel, generally determined using cross-validation. RBF kernels have exhibited good generalization properties in many classification problems. However, the use of a simple Euclidian distance implies small variations on the kernel value in high dimensional feature spaces.

3 Fusion Schemes

We present in this section three fusion schemes inspired from the usual early and late fusion schemes. Those schemes use a classifier to learn the relations between modality components at different abstraction levels.

Figure 1 describes the process of early and late fusion schemes. The feature extraction (FE) process extracts and creates a vector for each modality of the video item. We show the SVM process as two main steps: first, the construction of the Kernel, then the Learning or Classification (L / C) processes aims to assign a classification score to the video item.

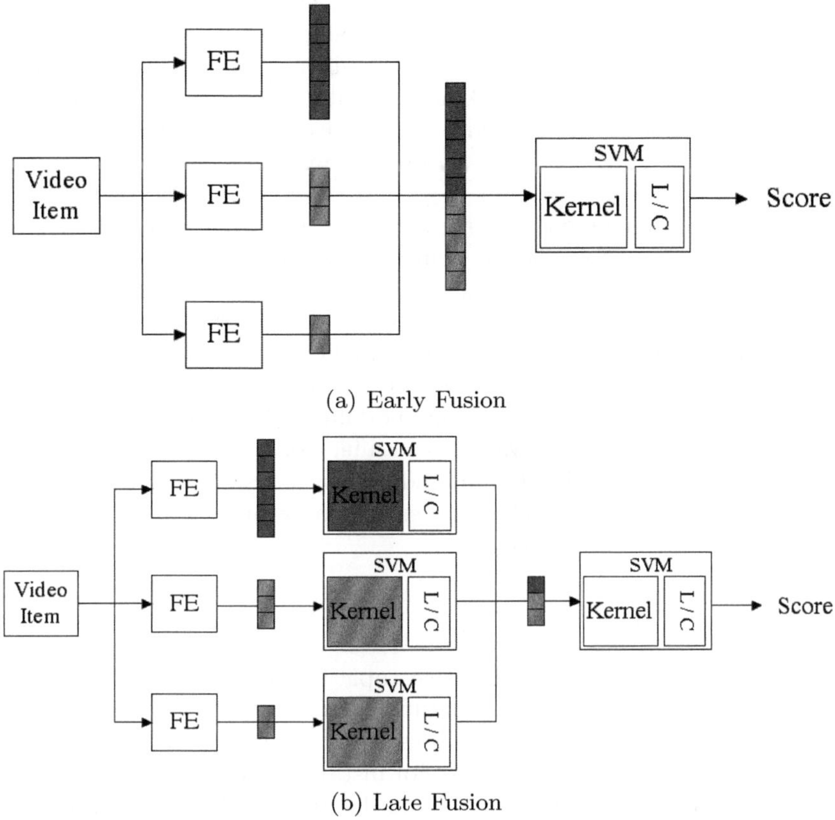

Fig. 1. Classical Early and Late Fusion schemes

Merging all the descriptors into a single flat classifier leads to a fully integrated fusion strategy since the fusion classifier obtains all the information from all sources. The advantage of such a scheme is its capacity to learn the regularities formed by the components independently from the modalities. Also, it is easy to use as it just consists in concatenating the various data in a single vector. The main disadvantage is the use of a unique approach (classifier and/or kernel) to merge different types of information. Assuming a RBF kernel and two sample vectors $\mathbf{x}$ and $\mathbf{y}$ from sets of features 1 and 2, the classical early fusion scheme leads to the following kernel:

$$K(\mathbf{x},\mathbf{y}) = e^{-\frac{\|\mathbf{x}-\mathbf{y}\|^2}{2\sigma^2}} = e^{-\frac{\|\mathbf{x}_1-\mathbf{y}_1\|^2+\|\mathbf{x}_2-\mathbf{y}_2\|^2}{2\sigma^2}}$$
$$= e^{-\frac{\|\mathbf{x}_1-\mathbf{y}_1\|^2}{2\sigma^2}} e^{-\frac{\|\mathbf{x}_2-\mathbf{y}_2\|^2}{2\sigma^2}}$$

This formulation shows that using a SVM classifier with RBF kernels, an early fusion scheme is equivalent to multiply unimodal kernels which share the same σ parameter. The σ parameter is often fixed by cross validation, it is then optimal for the concatenated vectors, but not necessary for each modality

A late Fusion is performed on top of several classifiers. It has been presented using different formalisms, such as meta-classification which aims to re-classify the classification results made by other classifiers [9]. The closest theory to illustrate a late Fusion is the Stacking Ensemble learning [16] which is part of the ensemble methods [5]. The idea behind Ensemble learning methods (e.g. bagging, boosting, stacking) is to improve the generalization by training more than one model on each problem (e.g. train 10 SVM instead of just one) and then to combine their predictions by averaging, by voting or by other methods. Using staking, the combination is achieved by a final classifier which provides the final result. Hence, in the context of multimedia indexing, the late fusion scheme consists in performing a first classification separately on each modality and then in merging the outputs using a higher level classifier. In such a way, in contrast with the early fusion, one can use different classifier algorithms and different training sets according to the modalities. Furthermore, the late fusion scheme also allows to combine various classifiers for the same modality. However, the significant dimensional reduction induced by the stacked classifiers might be a disadvantage as the fusion classifier cannot fully benefit from the correlation among the sources of information.

3.1 Kernel Fusion

Kernel combination is a current active topic in the field of machine learning. It takes benefit of Kernel-based classifier algorithms. Advantages of merging modalities at kernel level are numerous. First, it allows to choose the kernel functions according to the modalities. For instance, histograms of colors can take advantage of specific histogram matching distances. Likewise, textual modality can be categorized using appropriate kernels such as String Kernels [10] or Word-Sequence kernels [4].

Kernel fusion also allows to model the data with more appropriate parameters. Merging modalities using an early fusion scheme leads to model the data using a single kernel function. Consequently, when using a RBF kernel, a single σ parameter is expected to "fit" properly the sample vectors relations, whereas it makes much more sense to train a combined RBF kernel using one σ per modality. Combination of unimodal kernels leads to keep as much information as possible from each modality. A combined RBF kernel has the following form:

$$K_c(\mathbf{x}, \mathbf{y}) = F(K_m(\mathbf{x}_m, \mathbf{y}_m)_{(1 \leq m \leq M)})$$

where $K_c(\mathbf{x}, \mathbf{y})$ is the combined kernel value for samples $\mathbf{x}$ and $\mathbf{y}$, $(K_m)_{1 \leq m \leq M}$ are the considered unimodal RBF kernels, F is the combining function over the M modalities, $\mathbf{x}_m$ and $\mathbf{y}_m$ are the sample vectors for modality m. Figure 2 shows the kernel fusion process, the unimodal kernels are merged using a fusion function in order to create the multimodal kernel. Then, learning and classification steps aim to assign a classification score to the video item.

One of the main issues in the current kernel research is the learning of such combined kernels. Called *Multiple Kernels Learning*, it aims to learn at the

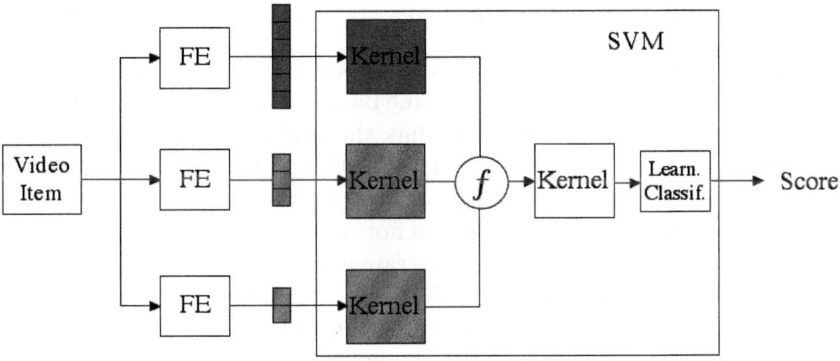

Fig. 2. Kernel Fusion scheme

same time the parameters of all the unimodal kernels and the parameters of the combining function [15]. In our experiments, we used a very simple strategy to create combined kernels. The following algorithm describes the steps to simply create combined kernels:

1. Construct each unimodal kernels K_m,
2. Perform cross-validation on each unimodal kernels to fix their parameters,
3. Construct the combined kernel using the F combining function,
4. Perform cross-validation to optimize the parameters of F.

This algorithm assumes that the best parameters of unimodal kernels are suitable enough to allow efficient generalization of the combined kernel.

Combining individual kernels using a product operator is highly comparable to the classic early scheme where feature vectors are just concatenated. The difference is that by performing kernel fusion, each modality m is associated to its own kernel parameters (ie: σ_m. Furthermore, due to the product operator, this combination might lead to sparse kernels and provide poor generalization. We used the sum operator instead of the product operator to try to avoid too sparse kernel representations. Summing unimodal kernels should be more suitable for concept detection when extracted features from a single modality are noisy and lead to incorrect detection.

We actually combine unimodal kernels by linear combination (weighted sum). Using RBF unimodal kernels, combined kernels are defined by the following formula:

$$K_c(\mathbf{x}, \mathbf{y}) = \sum_m w_m \, e^{-\frac{\|\mathbf{x}_m - \mathbf{y}_m\|^2}{2\sigma_m^2}}$$

where σ_m is the RBF parameter of kernel m and w_m is the weight of the associated modality. The w_m's can be fixed *a priori* or by cross-validation. In the conducted experiments, we optimized the w_m's on the training set.

3.2 Normalized Early Fusion

The number of extracted features depends upon the modalities and the type of the features. Hence, an early fusion scheme based on simple vector concatenation is much affected by the vector which has the highest number of inputs. Such fusion should have an impact on the classification, especially with a RBF kernel which is based on Euclidian distance between each training sample.

In traditional SVM implementation, a normalization process is integrated and aims to transform each input in the same range (e.g. [0..1], [−1..1]) in order to unbias the Euclidian distance. But, for the scope of merging features, this normalization doesn't take into account the number of input from the source features. The goal of normalized early fusion scheme is to avoid the problem of imbalanced features inputs by reprocessing each feature vectors before concatenation. We normalized each entry of the concatenated vector so that the average norm of each source vector is about the same. The normalization formula becomes:

$$x_i\prime = \frac{x_i - min_i}{(max_i - min_i) \times \sqrt{Card(x_i)}}$$

where x_i is an input of the feature vector x, min_i and max_i are respectively the minimum and maximum value of the ith input among the training samples and $Card(x_i)$ is the number of dimensions of the source vector of x_i.

3.3 Contextual-Late Fusion

Usual late fusion scheme first classify each concept using individual modalities and then merge the scores in a second layer of classifier. Here, we generalize this scheme by considering more than a single concept. Contextual information has been widely exploited in multimedia indexing [14, 11]. Here, the second layer (stacked) classifier is able to exploit the contextual relation between the different concepts. This proposed scheme merges each unimodal classification score from a set of several concepts, in order to exploit both multimodal and conceptual contexts.

Assume that we have M modalities (e.g. visual, audio and text) and C concepts (e.g. Car, Face, Outdoor, Bus, etc). The stacked classifier merges M scores to classify the C concepts in the classic late fusion scheme. The late context fusion scheme merges $M \times C$ classification scores to classify the C concepts.

4 Experiments

We have evaluated and compared the presented fusion schemes in the framework of the TRECVID'06 evaluation campaign. The objective of the "high level feature extraction task" is to find video shots containing a visual appearance of 39 predefined concepts (high level features). For each concept, an ordered list of 2000 relevant shots should be returned by the competing systems. The Inferred Average Precision (IAP) [17] on the returned lists computed using the `trec_eval` tool is used as the evaluation metric. We compare the three proposed fusion schemes with the commonly used early and late fusion schemes, as well as with unimodal approaches. We have extracted features from visual and textual modalities; we present them in the following section.

4.1 Visual and Text Features

The features used in this evaluation are mid-level semantic features. Visual features are based on the concatenation of several intermediate concept classification scores detected at a patch level. Those visual "local concepts" are automatically extracted in each key frame, which are split into 260 (20×13) overlapping patches of 32×32 pixels. Local descriptors (low-level features) include: color (9 color momentum in RGB space), texture (8 orientation $\times$ 3 scales Gabor filters) and motion vector (extracted by optical flow). An SVM classifier is trained in order to detect a set of 15 visual concepts (eg: vegetation, sky, skin, etc.) selected from the LSCOM ontology. Those intermediate concepts have been selected as they can be extracted at patch level. For each of the 39 concepts, we manually associated a subset of 6 intermediate visual concepts. Thus, visual feature vectors contain 1560 dimensions (6×260).

Text features are derived from speech transcription result. We used 100 categories of the TREC Reuters collection to classify each speech segment. The advantages of extracting such concepts from the Reuters collection are that they cover a large panel of news topics like the TRECVID collection and they are obviously human understandable. Thus, they can be used for video search tasks. Examples of such topics are: economics, disasters, sports and weather. The Reuters collection contains about 800000 text news items in the years 1996 and 1997.

We constructed a vector representation for each speech segment by applying stop-list and stemming. Also, in order to avoid noisy classification, we reduced the number of input terms. While the whole collection contains more than 250000 terms, we have experimentally found that considering the top 2500 frequently occurring terms gives the better classification results on Reuters collection. We built a prototype vector of each topic category on Reuters corpora and apply a Rocchio classification on each speech segment. Such granularity is expected to provide robustness in terms of covered concepts as each speaker turn should be related to a single topic. Our assumption is that the statistical distributions of the Reuters corpus and of TRECVID transcriptions are similar enough to obtain relevant results. Finally, the vector of text features has 100 dimensions. More explanation about those features can be found in [2].

4.2 Comparison of Fusion Schemes

The goal of the experiment is to study how imbalanced input features and large difference on the performance of unimodal classifiers are managed by the various fusion schemes. We show the results for unimodal runs and we compare all proposed fusion schemes and the usual early and late schemes. The 20 following concepts have been assessed for the TRECVID'06 evaluation campaign: SPORTS, WEATHER, OFFICE, MEETING, DESERT, MOUNTAIN, WATERSCAPE, CORPORATE LEADER, POLICE / SECURITY, MILITARY, ANIMAL, COMPUTER / TV SCREEN, US FLAG, AIRPLANE, CAR, TRUCK, PEOPLE MARCHING, EXPLOSION / FIRE, MAPS, CHART.

The results presented in this paper are based on those 20 concepts. We do not study here each individual concepts result due to lack of space. The table 1 shows the Mean Inferred Average Precision (MIAP) obtained from the 20 assessed concepts. The two first entries refer to the unimodal runs, the two following correspond to the state of the art fusion schemes. The three fusion schemes described in this paper are shown in bold. We also show the median MIAP obtained from all of the TRECVID'06 participants.

Table 1. Mean IAP of the 20 TRECVID'06 concepts

Visual	0.0634
Text	0.0080
Classical Early Fusion	0.0735
Classical Late Fusion	0.0597
Normalized Early Fusion	**0.0884**
Kernel Fusion	**0.0805**
Contextual Late Fusion	**0.0753**
Median	0.0680

Unimodal Runs

We observe that the two unimodal runs are very different in terms of accuracy; the visual based classification is almost 7 times higher than text based concept detection. This is probably due to the nature of the assessed concepts, which seems to be hard to detect using text modality. The difficulty to detect concepts from the text modality is also probably due to the poor quality of automatic speech transcription (and translation) in some videos of the collection. This point is actually interesting for the evaluation of the ability of the various fusion schemes to handle such heterogeneous data. The features we want to merge lead to different accuracies and are also imbalanced regarding the number of input features.

Classic Early and Late Fusion Schemes

The two classical fusion schemes do not merge unimodal features similarly. While early fusion is able to outperform both unimodal runs, the late fusion scheme achieves poorer accuracy than the visual run. It might be due to the low number of dimensions handled by the stacked classifier. The early fusion scheme exploits context provided by all of the local visual features and the textual features. The gain obtained by such fusion means that those two modalities provide distinct kind of information. The merged features are, somehow, complementary.

Early Based Fusion Schemes

The gain obtained by the normalized fusion schemes is the most important compared to other fusion schemes. Processing the unimodal features by re-equilibrating them according to the number of dimensions is determining factor in order to significantly outperform unimodal runs. In such a way, despite the different number of dimensions, both the visual and textual modalities have the same impact on concept classification. This normalization process leads to a gain of almost 17% (in MIAP) comparing to the classic early fusion scheme, which simply normalizes input in a common range, and 28% comparing to the best unimodal run.

The gain obtained by the kernel fusion scheme is less significant than the gain obtained by the normalized fusion run. However, when comparing to the classic early fusion, it seems that a combination using sum operator leads to better accuracy than multiplying kernels (which is somehow what the classic early fusion do). Furthermore, it is important to notice that the σ parameters are selected first by cross-validation on unimodal kernels and that we optimize the linear combination separately. We can expect that an integrated framework which learns simultaneously σ_m and w_m parameters should lead to better results.

Contextual-Late Fusion Scheme
Contextual-Late fusion is directly comparable with the classical late fusion scheme. This fusion scheme take into account the context from the score of other concepts detected in the same shot. By doing so, the context from other concepts leads to a gain of 26%. Furthermore, we observe that the MIAP obtained using the late contextual fusion scheme is almost the same as the one obtained for the classical early fusion scheme. In order to go further in this study, it could be interesting to evaluate the impact of the number and/or accuracy rate of concepts used in the context.

We notice that both of unimodal runs lead to poorer accuracy than the median of TRECVID'06 participants. This may be due to the basic and not so optimized features used in our experiments. However, the gain induced by the three fusion schemes presented in this paper lead to better accuracy than the median. We think that an optimization in the choice of descriptors for each modality could enhance the accuracy rate of both unimodal and multimodal runs.

5 Conclusion

We investigated three fusion schemes derived from the classical early and late fusion schemes when using SVM classifier. We have shown that all of the presented strategies perform in average better than the best unimodal run on the concept detection task of TRECVID'06. Furthermore, those fusion schemes outperform the median of TRECVID'06 participants over all of their runs. Kernel fusion schemes make it possible to take advantage of individual modalities, with a set of suitable parameters. Normalized early fusion is a good way to re-equilibrate the influence of individual modalities. Finally, the Contextual-Late fusion allows integration of context information from unimodal classification score of other concepts.

We studied influences of those fusion schemes on a set of 20 concepts, and did not analyzed individual concepts variations. As argued in [14], it is possible that one strategy performs differently than other depending the nature of the concepts. It could be interesting to go further in this direction. Also, the nature of the combined feature differs depending of the fusion schemes: early fusion is based on low- or intermediate-level features, where late fusion merges unimodal classification scores of high-level features. It could be interesting to merge those two heterogeneous kind of features in an integrated fusion scheme.

Bibliography

[1] S. Ayache, G. Quénot, J.Gensel, and S. Satoh. CLIPS-LSR-NII experiments at TRECVID 2005. *In proceedings of TRECVID Workshop*, 2005.

[2] S. Ayache, G. Quénot, J.Gensel, and S. Satoh. Using topic concepts for semantic video shots classification. *In proceedings of CIVR*, 2006.

[3] S. Boughorbel, J. Tarel, and N. Boujemaa. Conditionally positive definite kernels for SVM based image recognition. *In proceedings of ICME*, 2005.

[4] N. Cancedda, E. Gaussier, C. Goutte, and J.-M. Renders. Word-sequence kernels. *Journal of Machine Learning Research*, 2003.

[5] T. G. Dietterich. Ensemble methods in machine learning. In *Lecture Notes in Computer Science*, 2000.

[6] G. Iyengar and H. Nock. Discriminative model fusion for semantic concept detection and annotation in video. *In proceedings of ACM Multimedia*, 2003.

[7] W. Kraaij, P. Over, T. Ianeva, and A. F. Smeaton. *TRECVID 2006 – An Introduction*, 2006. http://www-nlpir.nist.gov/projects/tvpubs/tv6.papers/tv6intro.pdf.

[8] G. R. Lanckriet, M. Deng, N. Cristianini, M. I. Jordan, and W. S. Noble. Kernel-based data fusion and its application to protein function prediction in yeast. In *Proceedings of the Pacific Symposium on Biocomputing*, pages 300–311, 2004.

[9] W. Lin, R. Jin, and A. Hauptmann. Meta-classification of multimedia classifiers. In *Proceedings of First International Workshop on Knowledge Discovery*, 2002.

[10] H. Lodhi, J. Shawe-Taylor, N. Cristianini, and C. J. C. H. Watkins. Text classification using string kernels. *NIPS*, 2000.

[11] M. Naphade. On supervision and statistical learning for semantic multimedia analysis. *Journal of Visual Communication and Image Representation*, 2004.

[12] J. Shawe-Taylor and N. Cristianini. *Kernel Methods for Pattern Analysis*. Cambridge University Press, 2004.

[13] C. Snoek, M. Worring, and A. Smeulders. Early versus late fusion in semantic video analysis. *In proceedings of ACM Multimedia*, 2005.

[14] C. G. Snoek, M. Worring, J.-M. Geusebroek, D. C. Koelma, F. J. Seinstra, and A. W. Smeulders. The semantic pathfinder for generic news video indexing. In *Proceedings of ICME*, 2006.

[15] S. Sonnenburg, G. Ratsch, and C. Schafer. A general and efficient multiple kernel learning algorithm. *In proceedings of NIPS*, 2005.

[16] D. H. Wolpert. Stacked generalization. *Journal of Neural Networks*, 1990.

[17] E. Yilmaz and J. A. Aslam. Estimating average precision with incomplete and imperfect judgments. In *Proceedings of CIKM*, 2006.

[18] J. Zhang, M. Marszalek, S. Lazebnik, and C. Schmid. Local features and kernels for classification of texture and object categories: A comprehensive study. *Beyond Patches workshop, In proceedings of conjunction with CVPR*, 2006.

Search of Spoken Documents Retrieves Well Recognized Transcripts

Mark Sanderson and Xiao Mang Shou

Department of Information Studies, University of Sheffield, Western Bank, Sheffield, S10 2TN, UK
{m.sanderson, x.m.shou}@shef.ac.uk

Abstract. This paper presents a series of analyses and experiments on spoken document retrieval systems: search engines that retrieve transcripts produced by speech recognizers. Results show that transcripts that match queries well tend to be recognized more accurately than transcripts that match a query less well. This result was described in past literature, however, no study or explanation of the effect has been provided until now. This paper provides such an analysis showing a relationship between word error rate and query length. The paper expands on past research by increasing the number of recognitions systems that are tested as well as showing the effect in an operational speech retrieval system. Potential future lines of enquiry are also described.

1 Introduction

The Spoken Document Retrieval (SDR) track was part of TREC from 1997 (TREC 6) to 2000 (TREC 9). During this period, substantial research and experimentation was conducted in speech retrieval. The work focused on retrieval of radio and TV broadcast news: high quality recordings of generally clearly spoken scripted speech. The overall result of the track (as reported in the summary paper by Garofolo et al, 2000) was that retrieval of transcripts generated by a speech recognition system was almost as effective as retrieval of transcripts generated by hand with proper expansion techniques. Garofolo et al also presented results showing that there appeared to be a relationship between WER and retrieval effectiveness. They showed that for topics where retrieval was effective, WER of retrieved items tended to be low. The authors speculated that hard to recognize documents may also be hard to retrieve.

A more detailed analysis of the reasons for the success of spoken document retrieval was described by Allan in his review of SDR research (2002). Allan pointed out that documents that were most relevant to a query were ones that had query words repeated many times (i.e. the words had a high term frequency - *tf* - within the document). The repetition of query words within a document provided to the recognition system multiple opportunities to spot the query words correctly. Documents that contained query words only once may not have had such word occurrences spotted by a recognizer and therefore were less likely to be retrieved, however, such documents were also less likely to be relevant to the query; failing to retrieve them was not particularly important. Actually the failure to recognize single occurrences of terms in non relevant documents may offer an advantage in SDR over text retrieval as the speech document will not be retrieved. Allan reported that

G. Amati, C. Carpineto, and G. Romano (Eds.): ECIR 2007, LNCS 4425, pp. 505 – 516, 2007.

retrieval from spoken document collections with a high Word Error Rate (WER) resulted in poorer effectiveness than that resulting from retrieval over a collection with a low WER. Allan also reported that this inverse relationship between WER and retrieval effectiveness was linear.

Following on from those the two review papers, additional analysis of the SDR track data was conducted by Shou, Sanderson and Tuffs (2003) who reported work describing the variation of the word error rates of retrieved documents across ranking. In the paper, it was shown that across the groups who submitted runs to the TREC SDR track, top ranked documents in each run had a lower WER than documents that were further down the ranking. Some speculations on the reasons for this effect were provided, but little evidence of a reason was reported. This paper provides such evidence.

The paper starts with an overview of past work, followed by a series of experiments that expand on the work reported in the 2003 paper.

2 Past Work

Beyond Garofolo et al's observation of a relationship between effective topics and WER, little past work on the relationship between effectiveness, document rank and word error in recognized transcripts has been reported. However, some related research has been published, which is now described.

In the internal working of a speech recognition system, an audio segment of speech is recognized into a lattice of possible text strings, each string a hypothesis of what was spoken. The hypotheses are compared to the acoustic and language models stored in the speech recognizer. Based on both models, a confidence score is assigned to each word in each hypothesis, signifying the probability that the word was spoken. The sequence of words with the highest scores is chosen as the text string the recognizer will output. It can be expected that the higher score assigned to a word, the more confident one can be that the recognizer's selection was correct. Zechner and Waibel investigated summarization of spoken documents (2000) and use of confidence scores to improve summarization quality. Their summarizer ranked passages of a spoken document by their similarity to the overall document. Summary quality was computed by counting the number of relevant words (manually identified in human transcription) found within the summary. It was found that if the ranking formula was adjusted to prefer passages holding words with high confidence scores, the quality of the summaries increased by up to 15%. With Zechner and Waibel an approximation of word error rate (i.e. the confidence scores) was used to influence a ranking algorithm to improve the quality of the top ranked passages. Given such success, one might assume that similar use of confidence scores in information retrieval ranking algorithms would also be beneficial. However, attempts to improve retrieval effectiveness through use of the scores have at best been marginally successful (see Siegler et al, 1998, Johnson et al, 1999).

Sanderson and Crestani conducted preliminary investigations of retrieval from a collection composed of both hand transcribed (containing only human errors) and speech recognized documents (with a level of word error within them) (1998). Two versions of each spoken document were placed into a collection, one hand transcribed and one speech recognized. By having pairs of identical documents in the collection, the only difference in the two sub-sets of the mixed collection was the errors in the speech recognized set. If one was to retrieve on such a collection, any difference in rank positions of documents from the two subsets would be due to the error in the second set. Sanderson and Crestani reported that retrieval from such a collection resulted in the hand transcribed documents being retrieved at higher rank positions than the speech recognized documents. By experimenting with two retrieval ranking algorithms, Sanderson and Crestani were able to show the predominant reason for the hand transcribed documents being ranked higher than the recognized was due to word errors reducing the *tf* weight assigned to words in the recognized documents, therefore making such a document receive a lower score than that assigned to hand transcribed when ranking documents relative to a query. Sanderson and Crestani assumed that documents in the recognized collection had a uniform word error rate and did not explore the effect of different word error rates across such a collection. Neither was the investigation run across a range of retrieval systems or outputs from other speech recognition systems. Further research in retrieval from similar forms of collection was conducted by Jones and Lam-Adesina (2002).

3 Experiments on the Extent of the Effect of WER and Rank Position

In their paper, Shou, Sanderson and Tuffs (2003) presented evidence of variation of WER across rankings. That work is expanded on here. In the past paper, the speech recognized transcripts of the one hundred hours of audio data making up the TREC-7 SDR collection were collected from six of the groups participating in the speech track. In addition, the runs submitted by each group were also gathered: these hold the ranked list of documents retrieved for each topic by each group's retrieval system. The collection had an accompanying accurate manually generated text transcript, which allowed WERs to be computed for each document at each rank position for each topic within each collected transcript. A scatter plot of the WER of retrieved documents against their rank position was produced for each of the six transcripts. In addition, one of the six transcripts, from AT&T, had two forms of retrieval system search over it, which resulted in seven plots. The seven data sets are now described.

1. derasru-s1, UK Defence Evaluation and Research Agency (DERA, Nowell, 1998). Here a large vocabulary continuous speech recognizer (50,000 word vocabulary plus 500 bigrams) developed by DERA was used to generate the transcript. Its average word error rate was 66.4%. Retrieval was based on the Okapi system. The topics of the TREC track were syntactically tagged. Certain syntactic patterns were used to identify keywords of the topic text. Selected topic keywords were expanded with synonyms and sometimes with hypernyms taken from the WordNet thesaurus. When keywords were ambiguous, the commonest synset was chosen to provide expansion terms.

2. derasru-s2: using the same retrieval set up as derasru-s1, the speech recognizer had an additional processing step, which reduced the error rate to 61.5%. Here the audio data was segmented into different streams depending on the quality of audio recording found within parts the TREC spoken document collection. Audio recordings identified as being speech over telephones for example were recognized differently from segments judged to be recorded to a higher quality.
3. att-s1, AT&T. Recognition was performed using an in-house speech recognition system that produced transcripts with a 32.4% WER. The vocabulary size of the system was not stated in the paper describing the AT&T submission to TREC (Singhal et al, 1998). Retrieval was based on the SMART retrieval system with a phrase identification process operating on TREC topic text and pseudo-relevance feedback used to expand topics with additional terms. The form of feedback used was a method referred to as collection enrichment: here the first search of the pseudo-relevance feedback stage was conducted on a large collection of news articles and not the relatively small SDR collection.
4. att-s2. For the second AT&T submitted run, the same recognition system was used, but retrieval was altered to include a document expansion step. Here in the same manner that topic text was expanded using pseudo relevance feedback, each recognized transcript was expanded, by searching a large collection of newspaper texts with the transcript text as a query. The transcript was expanded with terms found to commonly co-occur in top retrieved newspaper articles. This run produced better retrieval results than att-s1.
5. dragon-s1, Dragon systems and the University of Massachusetts. This was a combined submission using a speech recognizer from Dragon and retrieval using the UMass Inquery retrieval system (Allan et al, 1998). The recognizer used a 57,000 word vocabulary. It produced transcripts with an error rate of 29.8%. Prior to retrieval, topic text was processed to locate phrases, which were then searched as phrases. Certain proper nouns were expanded with synonyms. A form of pseudo relevance feedback (known as local context analysis) was used to expand topic texts with additional terms taken from the recognized transcript collection.
6. shef-s1, University of Sheffield with collaborators at Cambridge University (Abberley et al, 1998). Recognition was performed using the Abbot recognizer system with a vocabulary of 65,532 words producing a transcript with a 35.9% WER. Retrieval was performed with a locally built IR system using Okapi-style BM25 weights.
7. cuhkt-s1, University of Cambridge (Johnson et al, 1998). Recognition was performed using the HTK speech toolkit recognizing from 65,000 word vocabulary. The resulting transcript had a 24.8% WER. Retrieval used the Okapi system using BM25 weights. Expansion of selected topic terms with synonyms and with additional terms using pseudo-relevance feedback was used, as was phrase spotting in topic text. Matches on proper nouns and nouns were preferred over adjectives, adverbs and verbs as this strategy was found to bring improvements in retrieval effectiveness.

As can be seen from the descriptions, the seven runs represent a relatively diverse set of retrieval and recognition approaches. The average WER of the transcripts ranged from 24.8% to 66%. Note that two further recognizer transcripts were produced and archived in this year of TREC, nist-b1 and nist-b2 (Garofolo et al, 1999). However, no associated retrieval runs performed on these transcripts were located and so were not used in this experiment.

3.1 The Experiment

For each run, rankings for each of the 23 topics (51-73) were gathered from the TREC web site. NIST's sclite software was used to calculate the WER of each document retrieved in the top 200 rank positions. Since sclite only calculates WER based on speaker id, the original recognized transcripts were modified by replacing speaker ids with document ids so that WER could be measured on each document. After obtaining WER of each story across all systems, the average error at each rank position across the 23 queries was calculated and graphed.

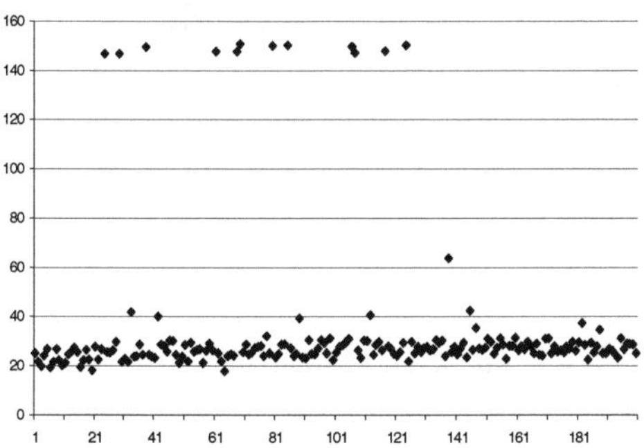

Fig. 1. Document rank (x-axis) vs. word error rate (y-axis) for dragon-s1 system

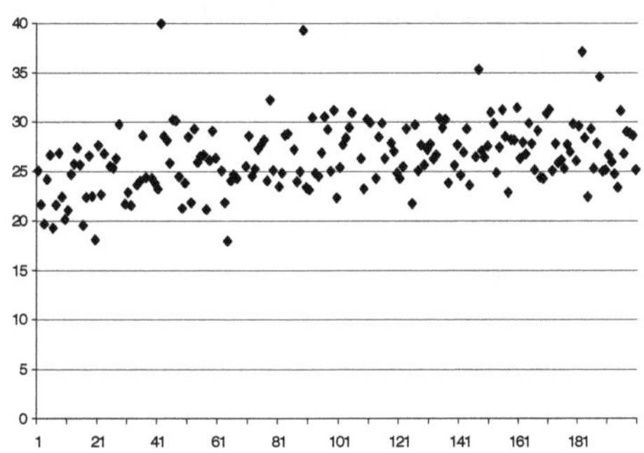

Fig. 2. Graph of Figure 1 with y-axis adjusted to focus on majority of retrieved documents

The graph (in Figure 1) shows a slight increase in error rate for recognized documents at higher ranks. A small set of documents with a very a high error rate across the ranking was observed (the twelve points at the top of the scatter plot). The reason for this effect was investigated and found to be related to mistaken insertions of large amounts of text into short documents by the recognizer (such erroneous documents were found in all six transcripts). Ignoring these few high error rate documents by focusing the scatter plot on the main band of documents reveals the trend of increasing error rate more clearly. It can be seen that top ranked documents (those on the left side of the graph) have a lower word error rate than those ranked further down the ranking. The plot such as that shown in Figure 2 was repeated for all other six runs and is displayed in Figure 3 – 8. Across all runs, the average WER for the very top ranked documents (those in the top 10) is lower than the WER for documents in the wider part of the ranking. Such differences in WER are also shown in Table 1 where the average WER is calculated in the top 10, 50 and 200 rank positions and it can be seen that for all recognizers and runs WER is lower for higher ranked documents.

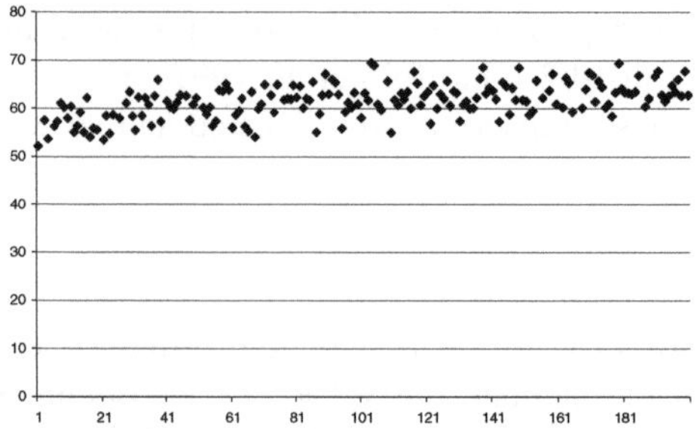

Fig. 3. derasru-s1, rank vs. WER

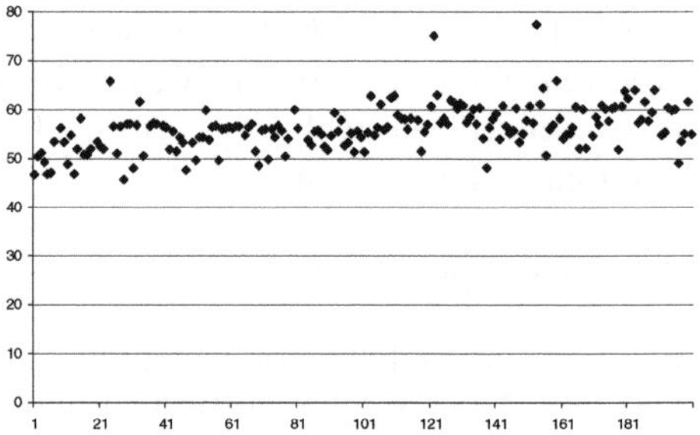

Fig. 4. derasru-s2, rank vs. WER

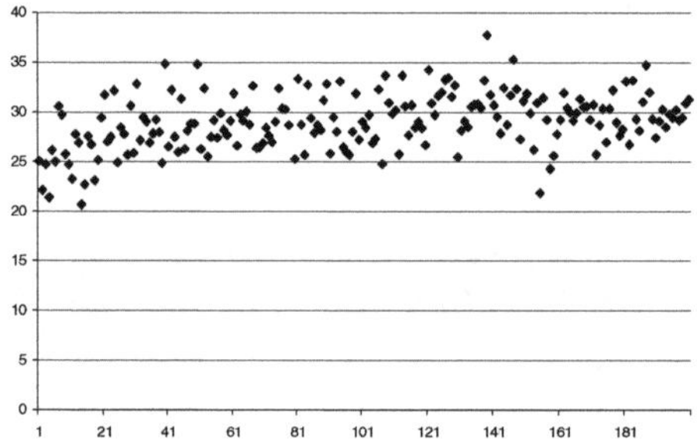

Fig. 5. att-s1, rank vs. WER

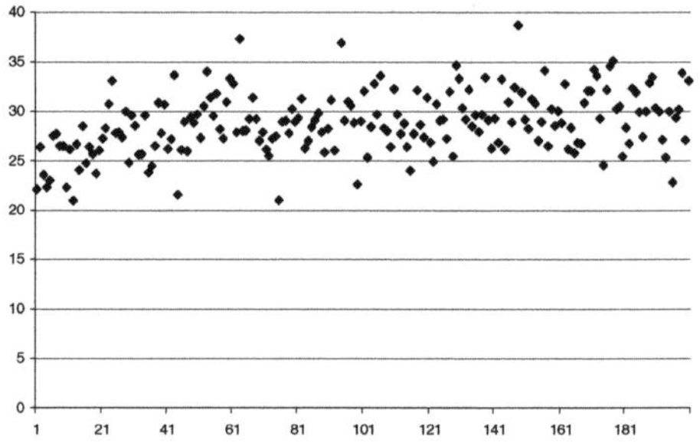

Fig. 6. att-s2, rank vs. WER

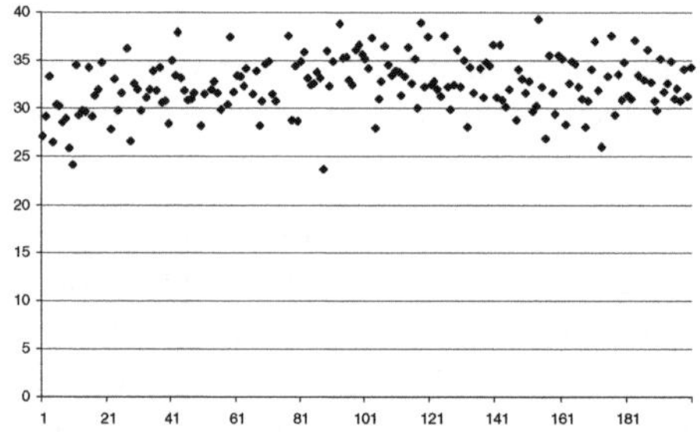

Fig. 7. shef-s1, rank vs. WER

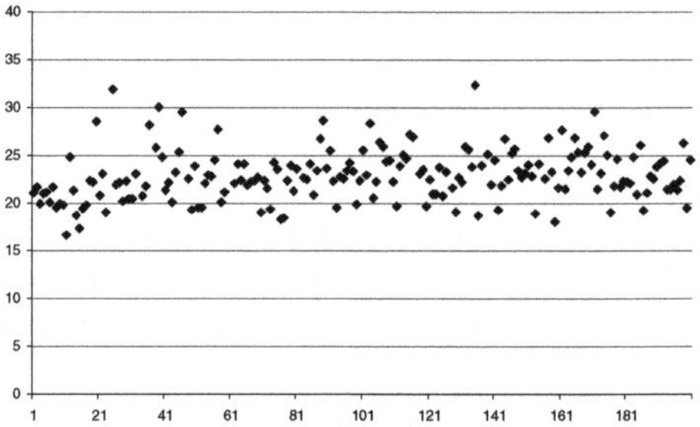

Fig. 8. cuhtk-s1, rank vs. WER

The slight, though consistent trend measured across all data sets provides evidence that when retrieving speech recognized documents, those with lower word error rates tend to be ranked higher. The trend also appears to occur independent of the mix of retrieval strategies used across the runs (e.g. different weighting schemes, use of pseudo-relevance feedback, use of document expansion, etc) and independent of the accuracy of the speech recognizer used.

Table 1. Word Error Rate differences for top 10, 50 and 200 retrieved documents

Run	Average WER in top 10 (%)	Average WER in top 50 (%)	Average WER in top 200 (%)
derasru-s1	57.1	58.8	62.8
derasru-s2	50.5	53.1	56.2
att-s1	25.5	27.4	29.1
att-s2	24.8	26.8	29.0
dragon-s1	22.8	25.1	26.9
shef-s1	28.4	32.7	33.4
cuhtk-s1	20.6	22.2	23.4

Although the trend is consistent across the data sets, it is not immediately clear what the cause of such a trend is: one explanation is that top ranked documents tend to contain a broader range of query words than those documents ranked lower. Another explanation mentioned by Sanderson and Shou (2002) is that transcripts of spoken documents containing query words assigned a high *tf* weight – which tend to be ranked highly by retrieval systems – often have a lower overall WER. Determining which of these possible causes might explain the observed effect was the subject of the next experiment.

4 Determining the Cause of the Effect

As valuable as it can be to examine the search output of other research groups' retrieval systems (as was conducted in Section 3), analyzing the ranked output of a system that one has no access to is often limiting. This is because a common consequence of such analysis is the discovery that new experiments need to be conducted to generate different versions of the data, which requires access to the retrieval system of other research groups, something that is rarely possible. Therefore, in order to conduct more detailed analysis of WER in retrieved documents, the six recognized transcripts used in the experiment of Section 3 along with the two NIST transcripts (nist-b1 and nist-b2) were indexed and searched so that new search output could be created for further experimentation. The aim of the experiments was to examine the relationship between WER, *tf* weights and the number of words in common between a query and a document.

In the experiment, the average WER of top ranked documents retrieved by queries of different length was measured. The TREC-7 SDR collection holds only 23 topics. In order to produce a larger number of topics of different lengths, (non-stop) words were randomly sampled without repeated words from each of the topics. The number of words sampled was varied, producing sets of topics of length 1, 2, 5, 10 and 15. Each of the 23 topics was sampled 1,000 times for each of the five different lengths. The queries were submitted to two versions of the GLASS search engine, an in house IR system that implements Robertson et al's BM25 ranking algorithm (1995) as well as a simple quorum scoring (coordination level matching) algorithm that ranks documents by the number of query words found in a matching document (making no use of *tf*, *idf* weights or of document length normalization). No relevance feedback or other expansion methods were employed in both algorithms. The tables of the results

Table 2. The average WER measured across the ten top ranked documents retrieved by quorum scoring for each of the 1,000 topics randomly sampled

Topic length	cuhtk-s1	dragon98-s1	att-s1	shef-s1	nist-b1	nist-b2	derasru-s2	derasru-s1
1	19.3	22.0	22.2	25.8	26.7	39.5	51.4	52.3
2	18.9	22.7	22.1	25.9	26.4	39.1	50.6	52.7
5	16.7	21.0	21.1	23.9	24.3	38.0	44.6	48.6
10	15.9	19.7	20.4	22.1	23.5	36.9	42.9	47.1
15	15.5	19.6	19.9	21.2	23.0	36.7	42.1	46.4

Table 3. Precision at 10 measured in the retrieved documents shown in Table 2

Topic length	cuhtk-s1	dragon98-s1	att-s1	shef-s1	nist-b1	nist-b2	derasru-s2	derasru-s1
1	0.06	0.07	0.06	0.06	0.05	0.05	0.06	0.06
2	0.09	0.08	0.09	0.09	0.08	0.07	0.07	0.07
5	0.20	0.18	0.19	0.19	0.18	0.15	0.17	0.17
10	0.26	0.23	0.26	0.25	0.24	0.20	0.21	0.23
15	0.28	0.24	0.27	0.27	0.25	0.21	0.21	0.23

Table 4. The average WER measured across the ten top ranked documents retrieved by BM25 for each of the 1,000 topics randomly sampled

Topic length	cuhtk-s1	dragon98-s1	att-s1	shef-s1	nist-b1	nist-b2	derasru-s2	derasru-s1
1	18.1	21.7	22.3	25.3	25.7	39.0	46.9	51.0
2	17.8	22.2	22.4	25.2	25.2	39.2	46.3	50.7
5	17.2	21.1	22.2	24.4	23.9	38.7	44.3	50.6
10	17.2	20.9	21.9	24.1	24.1	38.8	43.0	49.4
15	16.9	20.7	21.9	23.9	24.1	38.4	42.1	48.9

Table 5. Precision at 10 measured in the retrieved documents shown in Table 4

Topic length	cuhtk-s1	dragon98-s1	att-s1	shef-s1	nist-b1	nist-b2	derasru-s2	derasru-s1
1	0.10	0.09	0.09	0.09	0.08	0.08	0.08	0.08
2	0.16	0.15	0.15	0.15	0.15	0.13	0.13	0.13
5	0.30	0.28	0.28	0.28	0.28	0.24	0.25	0.27
10	0.36	0.35	0.35	0.35	0.35	0.29	0.31	0.33
15	0.37	0.36	0.35	0.36	0.36	0.30	0.33	0.33

of this experiment are shown in Table 2 and Table 4, which record average WER and Table 3 and Table 5, which display precision measured at rank ten.

As can be seen, across all eight transcripts for both form of ranking algorithm, as the length of topic increases, the WER measured in the top ranked documents reduces, while precision at 10 increases. This effect is consistent for both forms of ranking algorithm used. From the result with the quorum scoring, it can be concluded that the reduction in WER shown in Table 2 was caused by the change in top ranked documents: as topic length increases the top ranked documents hold more query words. Documents that match on a broader range of query words tend to have a lower WER. While a relationship between the rank position of recognized documents and their WER was observed in the past, to the best of our knowledge a causal effect has not been determined before. From the results in Table 2, we conclude that the process of retrieval itself is locating documents that have a lower WER.

Table 6. Comparison of average word error rate (WER) measured across the eight transcripts shown in Table 2 and in Table 4

Topic length	av. WER BM25	av. WER Quorum	difference	ttest (p)
1	31.3	32.4	-1.2	0.058
2	31.1	32.3	-1.2	0.057
5	30.3	29.8	0.5	0.099
10	29.9	28.6	1.4	**0.001
15	29.6	28.1	1.6	**0.001

The number of words in common between a document and a query is not the full story, however, as it can be seen that for topics of length one, for all but one transcript, WERs are lower using BM25 ranking (Table 4) than when using quorum scoring (Table 2). Here, top ranked documents retrieved by a single word query using BM25 are those documents in the collection that contain the query word repeated the most number of times (normalized by document length). Observing a query word repeated many times in a document would appear to be an indicator that that document was recognized well. The comparison of WERs is summarized in Table 6. The amount of WER reduction is relatively small and for topics of length one or two the difference is not significant. In comparing the error rates across the two ranking algorithms for longer queries (five, ten or fifteen words) the quorum scoring algorithm retrieves documents with lower WERs and for the longest queries lengths, the differences between quorum and BM25 are significant.

However, it must be remembered that quorum scoring though retrieving documents with low WERs is not retrieving the most relevant documents as across the Tables, precision at ten is consistently higher for BM25 ranking. We believe that this effect is due to BM25 top ranked documents matching on fewer query words than the documents top ranked by quorum scoring but with higher *tf*, which means a query word is repeatedly recognized, so BM25 has the effect of ranking higher documents with fewer matching terms.

5 Experiments with Manual Calculation of WER on Top Ranked SpeechBot Snippets

To provide further confirmation of the results in Section 4, measurements were made of the word error rate in the snippets of top ranked transcripts retrieved by a publicly available spoken document retrieval system, SpeechBot (Van Thong et al, 2000; Moreno et al 2000). We would like to test whether the correlation between word error rate and document ranking could be generally applied to other systems using different speech recognition technologies. A white paper published on the engine's Web site (Quinn, 2000) described that the engine indexed streaming spoken audio using a speech recognizer. Several thousand hours of audio data were crawled and stored in a searchable collection composed of mainly US-based radio stations producing predominantly news, current affairs and phone-in shows. The snippets in the result list summary presented by SpeechBot were brief sections of speech transcript that strongly matched a user query; most likely selected by a within document passage ranking approach.

The WER of each retrieved snippet was computed by manually comparing the snippet text with human listening to the corresponding part of the audio recording noting any inserted, deleted or substituted words. The WER was calculated using the total number of errors divided by the total number of words in the returned snippets. This method was consistent with NIST's WER calculation tool sclite which was used in the TREC SDR track. Because the majority of the SpeechBot collection was audio news, 34 current affair queries were created for the experiment. The number of words in the examined snippets ranged from twenty to forty. It was found that audio files were not available with some of the retrieved results (usually occurring with old audio recordings dated before 1999 or with the recordings of certain shows). The authors were made aware that a number of the transcripts used by SpeechBot for certain radio programs like PBS's News Hours are manually written transcripts and not generated by an SR system (Quinn, 2000), such transcripts were also ignored. Therefore, 311 out of a possible 350 snippets were assessed, the average WER measured within the snippets was 19.29%, and the standard deviation was 14.04%. Among the snippets, the maximum calculated WER was 68.75% while the minimum was 0%. The measured rate was substantially lower than the estimated 50% WER reported to exist across the whole SpeechBot collection (Quinn, 2000). This constitutes further evidence of the retrieval process assigning high rank to well recognized documents.

6 Conclusions and Future Work

This paper described experiments that demonstrated that when there is variability in the word error rate across the documents of a speech recognized collection, retrieval systems tend to retrieve highest documents with low word error. This effect was demonstrated through experimentation on an operational spoken document retrieval system as well as a series of analyses across multiple speech recognizers and retrieval algorithms. It was shown that documents holding many query words tend to have low WER.

We plan to extend our investigation to other retrieval research areas where documents containing varying levels of error are retrieved. Research topics such as retrieval of transcripts produced by Optical Character Recognition (OCR) of scanned document images or retrieval of documents translated into a different language may be worthy of further investigation. When retrieving OCR'ed documents, understanding if the top ranked are more readable or are the product of a better scan would be a straightforward experiment to undertake. A potentially more intriguing question is if in the context of cross language information retrieval, if top ranked documents are better translated than those retrieved further down the ranked list. To the best of our knowledge, this question has not been addressed within the cross language research community.

References

Abberley, D., Renals, S. and Cook, G. (1998) Retrieval of broadcast news documents with the THISL system; In Proceeding IEEE ICASSP, 3781-3784

Allan, J., Callan, J. Sanderson, M., Xu, J. (1998) INQUERY and TREC-7 in the proceeding of the 7th Text REtrieval Conference (TREC 7)

Allan, J. (2002) Perspectives on Information Retrieval and Speech, in Information Retrieval Techniques for Speech Applications, Coden, Brown and Srinivasan, editors, Lecture Notes in Computer Science, Volume 2273 1-10.

Garofolo, J.S., Voorhees, E.M., Auzanne, C.G.P., Stanford, M., Lund, B.A. (1999) TREC-7 Spoken Document Retrieval Track Overview and Results, in the Proceedings of the DARPA Broadcast News Workshop

Garofolo J.S., Auzanne, C.G.P., Voorhees, E.M. (2000) The TREC Spoken Document Retrieval Track: A Success Story; Proceeding of RIAO

Johnson, S.E., Jourlin, P., Moore, G.L., Spärck Jones, K., Woodland, P.C. (1998) Spoken Document Retrieval For TREC-7 At Cambridge University, in the proceeding of the 7th Text REtrieval Conference (TREC 7)

Johnson, S.E., Jourlin, P., Spärck Jones, K., Woodland, P.C. (1999): Spoken Document Retrieval for TREC-8 at Cambridge University, in the proceedings of the 8th Text REtrieval Conference (TREC 8)

Jones, G.J.F. and Lam-Adesina, A.M. (2002) An Investigation of Mixed-Media Information Retrieval, in the proceedings of the 6th European Conference on Research and Development for Digital Libraries (ECDL), 463-478

Moreno, P., Van Thong, P.M., Logan, B., Fidler, B., Maffey, K., Moores, M. (2000) SpeechBot: A Content-based Search Index for Multimedia on the Web, in the proceedings of the 1st IEEE Pacific-Rim Conference on Multimedia, (IEEE-PCM 2000)

Nowell, P. (1998) Experiments in Spoken Document Retrieval at DERA-SRU, in the proceeding of the 7th Text REtrieval Conference (TREC 7)

Quinn, E. (2000) SpeechBot: The First Internet Site for Content-Based Indexing of Streaming Spoken Audio Technical Whitepaper, Compaq Computer Corporation, Cambridge, Massachusetts, USA

Robertson, S Walker, S Jones, MM Hancock-Beaulieu (1995) Okapi at TREC-3, in the proceeding of the 3rd Text REtrieval Conference (TREC 3)

Sanderson, M., Crestani, F, (1998) Mixing and Merging for Spoken Document Retrieval, in the proceedings of the 2nd European Conference on Digital Libraries; Heraklion, Greece. Lecture Notes in Computer Science N. 1513, Springer Verlag, 397-407

Sanderson, M., Shou, X.M. (2002) Speech and Hand Transcribed Retrieval; Lecture Notes in Computer Science N.2273, Information Retrieval techniques for Speech Application, Springer, 78-85

Shou, X.M., Sanderson, M., Tuffs, N. (2003) The Relationship of Word Error Rate to Document Ranking, in the proceedings of the AAAI Spring Symposium Intelligent Multimedia Knowledge Management Workshop, Technical Report SS-03-08, ISBN 1-57735-190-8, 28-33.

Siegler, M., Berger, A., Witbrock, M., Hauptmann, A. (1998): Experiments in Spoken Document Retrieval at CMU, In the proceedings of the 7th TREC conference (TREC-7)

Singhal, A., Choi, J., Hindle, D., Lewis, D.D., Pereira, F. (1998) AT&T at TREC-7, in the proceeding of the 7th Text REtrieval Conference (TREC 7)

Van Thong, J.M., Goddeau, D., Litvinova, A., Logan, B., Moreno, P. Swain, M. (2000) SpeechBot: A Speech Recognition based Audio Indexing System for the Web in the proceedings of the International Conference on Computer-Assisted Information Retrieval, Recherche d'Informations Assistee par Ordinateur (RIAO2000) 106-115

Zechner, K., Waibel, A. (2000) Minimizing Word error rate in Textual Summaries of Spoken Language, in the proceedings of NAACL-ANLP-2000, 186-193

Natural Language Processing for Usage Based Indexing of Web Resources

Anne Boyer and Armelle Brun

INRIA Lorraine - BP 239 - 54506 Vandœuvre lès Nancy, France
{boyer, brun}@loria.fr

Abstract. The identification of reliable and interesting items on Internet becomes more and more difficult and time consuming. This paper is a position paper describing our intended work in the framework of multimedia information retrieval by browsing techniques within web navigation. It relies on a usage-based indexing of resources: we ignore the nature, the content and the structure of resources. We describe a new approach taking advantage of the similarity between statistical modeling of language and document retrieval systems. A syntax of usage is computed that designs a Statistical Grammar of Usage (SGU). A SGU enables resources classification to perform a personalized navigation assistant tool. It relies both on collaborative filtering to compute virtual communities of users and classical statistical language models. The resulting SGU is a community dependent SGU.

1 Introduction

The amount of available information has exponentially increased in the last years due to the development of information and communication technologies and the success all over the world of Internet applications. Therefore the identification of reliable and interesting items becomes more and more difficult and time consuming, even for skilled people using dedicated tools, such as powerful search engines. Due to the huge amount of online resources, the major difficulty is nevermore to know if a pertinent document is available but to identify the more reliable and interesting items among the overwhelming stream of available information. A key factor of success in information retrieval and delivery is the development of powerful tools easy-to-use for a large audience.

Different approaches for resources retrieval use to be explored, such as content analysis, keywords indexing and identification, topic detection, etc. [1]. A major difficulty inherent to such approaches is that one keyword may have different meanings, or not, dependent of the user, his/her context and the history of his/her past navigations. Moreover two different keywords may have similar meanings, depending on the context. Expressing a query is a difficult task for many people and a lot of research and industrial projects deal with query assistance. Furthermore automatic indexing of multimedia resources is still a hard research problem. To cope with these difficulties (query expression, multimedia

G. Amati, C. Carpineto, and G. Romano (Eds.): ECIR 2007, LNCS 4425, pp. 517–524, 2007.

indexing, etc.) we decide to investigate another way by ignoring the content, the nature, the format and the structure of resources.

This paper describes our intended work, relying on our past researches both on collaborative filtering [2,3] and statistical language modeling [4,5]. Our objective aims at providing a new web browsing tool based on an analysis of usage. This tool enables multimedia information retrieval by browsing techniques without expressing any query. It means that users are modelized without requiring any preferences elicitation. This approach enables to easily manage heterogeneous items (video, audio, textual, multimedia) with a single treatment, this is an advantage since classical methods require dedicated tools for resource tagging.

We plan to extract frequent patterns of consultations by taking advantage of the analogy between language-based statistical modeling and resource retrieval. These frequent patterns will allow the design of syntax of usage, relying on the hypothesis that there is logic and coherency defining implicit "rules" inside a navigation. The resulting Statistical Grammar of Usage (SGU) enables a classification, clustering and selection of resources to design personalized filtering.

In the next section, the problem of retrieving resources when browsing is stated and our approach based on the use of statistical language models is detailed. The following section presents the most popular statistical language models and their appropriateness to web browsing. Section 4 puts forward the community-based Statistical Grammar of Usage we design. Then, discussion and perspectives conclude the paper.

2 Principle of Our Web Browsing Tool

Our web browsing tool helps users during a navigation process: it suggests the pertinent items to a specific user, given his/her past navigation and his/her context. The aim is to compute the **pertinence** of any resource. The pertinence of a resource is defined as the interest of a user for it and allows to compute **predictions** of resources (the highest the pertinence of a resource is, the highest is its probability to be suggested to the user).

First, we hypothesize an **implicit search**, it means that the active user has no explicit queries to formulate. Secondly, we consider as a **consultation** the sequence of one or more items, dedicated to a given search. A multi-navigation is the mix of different consultations within a single browsing process. A **resource** is any item (textual, audio, video or multimedia document, web page, hyperlink, forum, blog, website, etc.), viewed as an elementary and indivisible entity without any information about its format, its content or any semantic or topic indexing. The only data describing *a priori* a resource is a normalized mark called **identifier**, enabling to identify and to locate it. Our approach relies on an analysis of usage. A **usage** is any data, explicitly or implicitly left by the user during navigation. For example, history of consultation, click-stream or log files are implicit data about the interest of the visited items for the active user. This measure can be either an explicit information as votes, annotations or any estimation computed from implicit data [6].

An advantage of our approach is that it only takes into account a measure of the user's interest for a given resource, which is directly linked to the pertinence criterion: the user's satisfaction. Let us remember that we decide to ignore any structural or thematic information about a resource. Our approach computes a personalized indexing of resources not in terms of its intrinsic nature but in terms of a more subjective but more reliable and pertinent criterion, i.e. the user's context, preferences and habits. It is the reason why this approach manages heterogeneous resources with a single treatment.

The question to solve is the following: how to estimate the *a priori* pertinence of a resource for a given user. The difficulty relies on sparsity of data: we don't have any appreciation of a resource if this user has not seen it and usually many resources have not been seen by this user. To compute the *a priori* pertinence of a resource, we plan to design a grammar of usage. As a **grammar of language** is the set of rules describing the relation between words, a **grammar of usage** is the set of rules describing the relation between resources. A grammar of language estimates if a word is pertinent given the beginning of a sentence. A grammar of usage allows to estimate if a resource is relevant for a specific user given his/her previous consultations. There is no *a priori* grammar of usage, as Internet is a dynamic and moving environment. A means to cope with the difficulty of designing an *a priori* grammar is the use of a statistical approach based on usage analysis. As huge usage corpora are available (log files, clickstream, etc.) it makes it possible to explicit regularities in terms of resource consultations. This statistical approach can be investigated in a similar way to language modeling based on statistical models.

The resulting grammar is called a **Statistical Grammar of Usage** (SGU). It enables the computation of the probability of a resource given the active user and his/her sequence of navigation. This probability measures the pertinence of the resource. A SGU, if trained on the whole usage corpus, is a general grammar since it is learned for all users in all contexts. The accuracy of such a grammar is insufficient and furthermore, the presupposed logic and coherency between users becomes a too strong and unrealistic hypothesis. Given two users, it seems unlikely that they exhibit the same resource consultation behavior: the SGU has to be personalized. Nevertheless, learning a user-specific SGU requires a large amount of data for each user and it is irrealistic to wait for collecting enough data to train it. It is the reason why we will determine groups of users with similar behavior called **communities**. Thus we plan to compute a SGU for each community and design a **community-based SGU**. Users are preclassified into a set of coherent communities, in terms of resource consultation behavior. Collaborative filtering techniques are a means to build coherent communities in terms of usage. This gathering can be compared to topic classification in natural language processing.

The principle of collaborative filtering techniques [7] amounts to identifying the active user to a set of users having the same tastes and, that, based on his/her preferences and his/her past visited resources. This approach relies on a first hypothesis that users who like the same documents have the same topics of interests and on a second hypothesis that people have relatively constant likings. Thus, it is possible to predict resources likely to match user's expectations by taking advantage of experience of his/her community.

A first comment on usual collaborative filtering techniques is that the structure of navigation is ignored. However, this aspect can be crucial in some applications such as web browsing. For example, a user may not like a resource because he/she has not previously read a prerequisite resource. Thus the SGU will submit a resource when it becomes pertinent for a user, for example when he/she has read all prerequisites. As statistical language models emphasize the order of words in sentences, it seems interesting to determine if such models and collaborative filtering can be used together to improve the quality of suggestions.

3 Statistical Language Models

3.1 Overview of Statistical Language Models

The role of a statistical language model (**SLM**) is to assign a likelihood to a given sentence (or sequence of words) in a language [8]. A SLM is defined as a set of probabilities associated to sequences of words. These probabilities reflect the likelihood of those sequences. SLM are widely used in various natural language applications such as automatic translation, automatic speech recognition, etc. Let the word sequence $W = w_1, \ldots, w_S$. The probability of W is computed as the product of the conditional probabilities of each word w_i in the sentence. To estimate these probabilities, a vocabulary $V = \{w_j\}$ is stated. The probability of the sequences of words are trained on a training text corpus.

3.2 How Can Web Browsing Take Advantage of SLM?

Web browsing and statistical language modeling domains seem to be similar in several points. First, statistical language modeling uses a vocabulary made up of words. This set can be viewed as similar to the set of resources R of the web. Then, the text corpus is made up of sentences of words, they can be viewed as similar to the sequences of consultations of the usage corpus. A sequence of S words in a sentence is similar to a sequence of consultation of S resources. Finally, the presence of a word in a sentence mainly depends on its previous words, as the consultation of a resource mainly depends of the preceding consultations.

Given these similarities, we can naturally investigate the exploitation of SLM into a web browsing assistant. As noticed in the previous section, these models have the characteristic that the order of the elements in the history is crucial. This aspect may be important for specific resources in web browsing.

However, we have to notice that web browsing and natural language processing have two major differences. The first one is that it is possible that a user may mix different queries within a single history (we will call this "multi-navigation") but it is unrealistic to mix different sentences when speaking or writing. This first remark brings us to consider a generalization of SLM to integrate "multi-navigation" in the browsing process. The second one is that natural language exhibits strongest constraints: each word in a sentence is important and deleting or adding a word may change the meaning of the sentence. Web browsing is not so sensitive and adding or deleting a specific resource within a navigation may have no impact. Then we will have to consider permissive models, able to take into account less constrained histories such as navigation has.

3.3 n-Grams Language Models

Due to computational constraints and probability reliance, the whole history h_i of w_i cannot be systematically used to compute the probability of W. Classical SLM aim at reducing the size of the history while not decreasing performance.

n-grams models reduce the history of words to their $n-1$ previous words. These models are the most commonly used in most of natural language applications. n-grams model can be directly used in web browsing assistance. In the previous section, we put forward that the quality of the model will be increased if it is dedicated to a community and trained on the corresponding community usage corpus. Thus, the usage corpus is split into community usage corpora and a model is trained on each community corpus.

Let a community c_j and a sequence of consultations of resources $h_j = R_{j1}, \ldots, R_{ji-1}$. For each resource $R_i \in R$, the n-grams model computes the probability $P_n(R_i \mid R_{i-n+1}, \ldots, R_{i-1}, c_j)$. The history h_j is reduced to the $n-1$ last resources consulted, other resources are discarded. Thus, this model assumes that the consultation of a resource R_i does not mainly depend on resources consulted far from R_i.

As previously mentioned, the behavior of users is less constrained than language: adding or deleting a resource in a sequence of consultations has a lower influence on the result of the search than adding or deleting a word in a sentence. This model does not ideally match our retrieval problem since the history considered is the exact sequence of consultations $R_{i-n+1} \ldots R_{i-1}$, that may be too restrictive in the general case. However, this model may be suitable for frequent sequences of consultations, that can be considered as "patterns of consultation". They are assigned a high probability, thus increasing the probability of resources inside such sequences. It should be interessant to take into account, in a more adequate way, such "patterns of consultations".

As n-grams models exhibit strong constraints, we are also interested in more permissive models. Trigger-based language models seem to me more adequate to less constraint histories such as navigation.

3.4 Trigger-Based Language Models

Trigger-based models [9] aim at considering long-time dependence between two words (w_x and w_y for instance). Dependence is measured by Mutual Information (MI) [10]. This measure can easily integrate long-time dependence by using a distance parameter d. d is the maximum number of words occurring between w_x and w_y, a window of d words is thus considered.

A couple (w_x,w_y) with a high MI value means that w_x and w_y are highly correlated and the presence of w_x raises the probability of occurrence of w_y, at a maximal distance of d words. (w_x,w_y) is named a trigger. This model considers only highly correlated pairs of words (corresponding to high MI values), useless pairs are discarded. The resulting set is called S.

Given history $h_j = w_1, \ldots w_{i-1}$, the trigger model computes the probability of w_i as:

$$P_t(w_i \mid h_j) = \frac{\sum_{w_j \in h_j} \delta_{w_j,w_i,h_j,S}}{\sum_{w_j \in h_j} \sum_{w_t \in V} \delta_{w_j,w_t,h_j,S}} \quad (1)$$

$$\text{with } \delta_{\mathrm{w_j,w_i,h_j,S}} = \begin{cases} 1 & (w_j, w_i) \in S \ \ and \ \ d_j(w_j, w_i) \leq d \\ 0 & otherwise \end{cases}$$

where $d_j(w_j, w_i)$ is the distance between w_j and w_i, in terms of words in h_j.

In our web browsing assistant tool, the trigger model is made up of triggers of resources (R_x,R_y). The consultation of R_x triggers the consultation of R_y, at a maximal distance of d resources. As MI measure is not symmetric ($MI(R_x; R_y) \neq MI(R_y; R_x)$), this model integrates order between resources, that may be crucial for specific resources.

The advantage of such a model is the long-time dependence between both resources. In a consultation, two resources can be viewed with various values of distance without changing the meaning of the consultation. Trigger models enable to modelize this kind of influence, when the value of the distance between items is not discriminant but the order of occurrence is meaningful. Such a model is less constrained than n-grams models and seems to be adequate to the navigation problem.

Similarly to n-grams model, a trigger-model is developed for each community c_j. MI values are computed for each couple of resources and for each community. A set of the most related triggers (S_{c_j}) is extracted for each community c_j.

The probability of a resource R_i, given the community c_j, its corresponding set of triggers S_{c_j} and the sequence of consultation of resources $h_j = R_1, \ldots, R_{i-1}$ is:

$$P_t(R_i \mid h_j, c_j) = \frac{\sum_{R_x \in h_j} \delta_{R_x,R_i,h_j,S_{c_j}}}{\sum_{R_x \in h_j} \sum_{R_y \in R} \delta_{R_x,R_y,h_j,S_{c_j}}} \quad (2)$$

$$\text{with } \delta_{\mathrm{R_x,R_i,h_j,S_{c_j}}} = \begin{cases} 1 & (R_x, R_i) \in S_{c_j} \ \ and \ \ d_j(R_x, R_i) \leq d \\ 0 & otherwise \end{cases}$$

where $d_j(R_x, R_i)$ is the distance between R_x and R_i in history h_j.

4 Towards a Community-Based SGU

The SGU we propose in this article has the advantage of considering both the community of the active user and his/her consultation history (sequence of consultation), whereas state of the art models usually exploit the set of consultations. The use of this model relies on two steps:

4.1 Determination of the Community of the Active User

The first objective is to compute a set of user communities based on an analysis of usage. To achieve this goal, we use collaborative filtering techniques. The set of users is split into classes by using a recursive k-means like algorithm [2], the similarity between two users is estimated as the mean of the distance for each commonly voted resource [11]. The whole corpus is then split into community sub-corpora. Each one is made up of usage of any user of the community. A user is then assigned to the closest community using the same similarity measure.

4.2 Computation of the Probability of a Resource

Given the community c_j of user U_j, and his history h_j, the computation of the probability of a resource R_i relies on three sub-models based on language models presented in section 4. The first sub-model computes the probability $P_n(R_i \mid h_j, c_j)$, by exploiting the probabilities of resources sequences of the n-grams model. The second sub-model is the trigger model, it computes the probability $P_t(R_i \mid h_j, c_j)$. The last sub-model is devoted to resources out of the training corpus. A probability *a priori* $P_a(R_i \mid c_j)$ is set to each resource $R_i \in R$. The resulting model, that can be viewed as a community-based SGU, computes the linear combination of the three previously described sub-models.

$$P(R_i \mid h_j, c_j) = \lambda_n P_n(R_i \mid h_j, c_j) + \lambda_t P_t(R_i \mid h_j, c_j) + \lambda_a P_a(R_i \mid c_j) \quad (3)$$

where λ are optimized with EM algorithm [12] on a development corpus.

Thus, given a user U_j and his/her history h_j, we first have to determine the community c_j he/she belongs to. Then, the probability of any available resource is computed given the SGU learned for this community.

Then, the N most likely resources are selected. The systematic selection of resources in the same subset of likely items avoids the introduction of novelty in resources suggestion. To enable novelty in suggestion, we randomly select a subset of unlikely resources (SUR) that is added to the previous subset of N likely resources (SLR) to build the set of candidates (SC). We determine the suggested resources for a specific user using a roulette wheel. We assign a sector of the wheel to any resource in SC; the size of this sector is proportional to the probability of occurrence of this resource as given by the SGU. One or several resources from SC are then drawn independently using this roulette wheel principle and are submitted to user U_j.

5 Discussion and Perspectives

This paper aims at describing a new web browsing assistant, based on usage and natural language processing. This approach exempts the difficult task of content indexing and facilitates heterogeneous resources management. Similarities between SLM and web browsing are put forward, therefore the integration of SLM is investigated. The resulting model is a Statistical Grammar of Usage (SGU).

As a single SGU may be unefficient, it has to be personalized. To tackle sparsity of data, a preclassification of users into communities is performed. Community-based SGU are then proposed. A second contribution consists in the design of community-based SGU, predicting the sequentiality of resources during navigation. Moreover, a community-based SGU builds an *a posteriori* structure of navigation based on the subjective but reliable measure of pertinence of a resource for a user. Consequently it performs a personalized indexing of resources, based on usage analysis.

As collaborative filtering techniques used to build communities and triggers used to suggest resources have both proved their efficiency in their respective domain, a first perspective is the validation of the community-based SGU in terms of quality of predictions in web browsing. A second perspective is the use of the community-based SGU to compute a personalized classification of resources, depending not only on topics but also on user's preferences and context.

References

1. R. Baeza-Yates and B. Ribeiro-Neto, *Modern Information Retrieval*, ACM Press, New York, 1999.
2. S. Castagnos and A. Boyer, "A client/server user-based collaborative filtering algorithm model and implementation," in *Proceedings of the 17th European Conference on Articial Intelligence (ECAI 2006)*, Riva del Garda, Italy, august 2006.
3. S. Castagnos and A. Boyer, "Frac+: A distributed collaborative filtering model for client/server architectures," in *2nd conference on web information systems and technologies (WEBIST 2006)*, Setbal, Portugal, 2006.
4. K. Smaïli, A. Brun, I. Zitouni, and J.P. Haton, "Automatic and manual clustering for large vocabulary speech re cognition: A comparative study," in *European Conference on Speech Communication and Technology*, Budapest, Hungary, 1999.
5. A. Brun, K. Smaïli, and J.P. Haton, "Contribution to topic identification by using word similarity," in *International Conference on Spoken Language Processing (ICSLP2002)*, 2002.
6. P. Chan, "A non-invasive learning approach to building web user profiles," in *5th International Conference on Knowledge Discovery and Data Mining - Workshop on Web Usage Analysis and User Profiling*, San Diego, USA, august 1999.
7. J. Herlocker, J. Konstan, L. Terveen, and J. Riedl, "Evaluating collaborative filtering recommender systems," *ACM Transactions on Information Systems (TOIS)*, vol. 22, no. 1, pp. 5–53, january 2004.
8. R. Rosenfeld, "Two decades of statistical language modeling: Where do we go from here," 2000.
9. R. Rosenfeld, "A maximum entropy approach to adaptative statistical language modeling," *Computer Speech and Language*, vol. 10, pp. 187–228, 1996.
10. N. Abramson, *Information Theory and Coding*, McGraw-Hill, New-York, 1963.
11. U. Shardanand and P. Maes, "Social information filtering: algorithms for automating "word of mouth"," in *Proceedings of the ACM CHI'95 - Conference on Human Factors in Computing Systems*, 1995, vol. 1, pp. 210–217.
12. A.P. Dempster, N.M. Laird, and D.B. Rubin, "Maximum likelihood from incomplete data via the em algorithm," *J. of the Royal Statistical Society*, vol. 39, 1977.

Harnessing Trust in Social Search*

Peter Briggs and Barry Smyth

Adaptive Information Cluster,
School of Computer Science & Informatics,
University College Dublin, Ireland
{firstname.surname}@ucd.ie

Abstract. The social Web emphasises the increased role of millions of users in the creation of a new type of online content, often expressed in the form of opinions or judgements. This has led to some novel approaches to information access that take advantage of user opinions and activities as a way to guide users as they browse or search for information. We describe a social search technique that harnesses the experiences of a network of searchers to generate result recommendations that can complement the search results that are returned by some standard Web search engine.

1 Introduction

The world of the Web has changed in some very important ways recently. As if to emphasise this new "release" we have the all too frequently hyped *Web 2.0* designation, but, more importantly, we can see how these changes are being driven by the (re-)establishment of the end-user as a critical component of the Web experience. The people-power driving this *social Web* extends beyond the content-layer and into the computational-layer that underpins a new breed of information access technologies such as *social search* engines. Put simply, social search techniques attempt to harness the views, opinions, and behaviour of users to improve Web search. This includes the new generation of social tagging and bookmarking services, which allow users to label and share all sorts of content, from pictures (Flickr) to news articles (Digg) and from blogs (Technorati) to Web pages (Del.icio.us). However, the concept also includes a range of fundamentally new search engine technologies, which attempt to use human relevance judgments to replace or augment more traditional forms of algorithmic search.

In this paper we focus on a novel approach to social search, one that is informed by the idea that, when we search, frequently there will have been other people who have previously searched for the same or similar information. As such, these people, or more precisely their search histories, can be used to make useful recommendations for our current search. We describe a model of social search that has evolved from the *Collaborative Web Search* (CWS) approach

* This material is based on works supported by Science Foundation Ireland under Grant No. 03/IN.3/I361.

G. Amati, C. Carpineto, and G. Romano (Eds.): ECIR 2007, LNCS 4425, pp. 525–532, 2007.

described by [7], which harnesses the search patterns of communities of like-minded searchers in order to identify relevant results for promotion. While CWS assumes a centralised community-based search engine, in this paper we describe a distributed model of collaborative Web search in which groups of searchers are organised as a peer-to-peer *search network* that facilitates inter-searcher co-operation at search time. We also describe and evaluate how an explicit model of user-user trust drives the evolution of the search network, enabling the propagation of queries through the network and influencing the rank of recommended results in the final result-list.

2 Social Information Access

2.1 Social Navigation and Recommendation

Human judgement can come in a variety of forms and be harnessed in many different ways. Perhaps the simplest form of human judgement relates to link-following behaviour as users browse: the selection of a particular link to follow can be viewed as an expression of interest in the link by the user. Many researchers have sought to record and leverage this type of information to adapt the structure of a Web site according to a user's navigation preferences. For example, the work of Smyth & Cotter [9] describes such an approach to personalizing the structure of mobile portals. The browsing behaviour of users has also been used to indicate important sections of online documents [3], and to construct navigation trails for users [11,2].

Many recommendation technologies have always relied very directly on the judgments of groups of users as the basis for their recommendations. For example, automated collaborative filtering (ACF) [6,4] is a popular approach to social recommendation in which users are profiled according to the ratings that they attach to particular content items. These ratings can take the form of explicit judgments on the part of users — e.g. the PTV system's [8] profiles comprised manually rated TV programs. Alternatively, ACF systems can infer judgments by monitoring other user activities — e.g. Amazon converts purchasing behaviours into judgments. To generate a set of recommendations for a particular user, an ACF system will identify a set of similar users — users whose ratings are highly correlated with the target user — and suggest items that these similar users have liked in the past, avoiding items that they have disliked. Of particular relevance, [5] discusses the role of trust in such recommender systems.

2.2 Social Search

Human judgments have played a role in Web search for some time. Google's PageRank [1] technique harnesses human judgments through link analysis. More directly, an early search engine called DirectHit ranked results by their selection popularity. This idea that human relevancy judgments may help to improve Web search is a powerful concept, and has led to the development of a number of so-called *social search* initiatives. For example, the work of [10] builds user profiles from the bookmark collections of users in order to implement a collaborative filtering style approach to search result recommendation.

Recent research has sought to take the concept of social search one step further. For example, the work of [7] on *collaborative Web search* (CWS) is especially important in this paper. CWS is a meta-search engine that focuses on a community-based approach to Web search in which each search community is made up of a set of like-minded searchers; for example, a motoring Web site will attract an ad-hoc community of car enthusiasts. CWS stores the search behaviour of a community of like-minded users in a matrix (*hit-matrix*, H_C) such that $H_C(i,j)$ is the number of times that the members of a community C have selected page p_j for query q_i. When a new query q_T is submitted by a community member, a set of result recommendations is generated by selecting results that have been submitted for similar queries; ultimately these recommendations are promoted within the result-list returned by some underlying search engine. Query similarity is typically measured in terms of query-term overlap (see Equation 1) and results are ranked by their weighted relevance. The relevance of p_j to q_i is calculated as the proportion of times that p_j has been selected for q_i (see Equation 2), and these local relevance values are combined across all similar queries and weighted by the similarity of these queries to q_T; see Equation 3.

$$Sim(q,q') = \frac{|q \cap q'|}{|q \cup q'|} \tag{1}$$

$$Relevance(p_j, q_i) = \frac{H_C(i,j)}{\sum_{\forall j} H_C(i,j)} \tag{2}$$

$$WRel(p_j, q_T, q_1, ..., q_n) = \frac{\sum_{i=1...n} Relevance(p_j, q_i) \bullet Sim(q_T, q_i)}{\sum_{i=1...n} Exists(p_j, q_i) \bullet Sim(q_T, q_i)} \tag{3}$$

3 A Distributed Model of Collaborative Web Search

The key contribution of this paper is to recast the centralised model of CWS as a network of search partners connected by trust-weighted linkages that will evolve over time. Accordingly, search communities emerge as a consequence of network evolution, avoiding the need for the pre-defined communities required by the current model of CWS described by [7].

3.1 Building a Trust-Enhanced Search Network

Our new network-based model of CWS associates each user u with a hit-matrix, H_u, which locally stores their personal search history. In turn each user is supported by a CWS agent that works in the usual way (as outlined in the previous section) to generate recommendations from their personal hit-matrix (and an underlying search engine) in response to their queries. Of course, the limitation is that these recommendations are now drawn from an individual user's personal search history, instead of a much richer community search history. The solution is to facilitate the sharing of recommendations between user search agents in a collaborative fashion. Thus we need to connect individual search agents to form a search network. Connected search agents can cooperate directly during future searches as queries and results are propagated across these connections.

3.2 Trust-Based Ranking

By connecting two users u_T and u_S, we can allow them to collaborate during future searches so that when u_T enters a new query, recommendations can come from his/her own hit-matrix but also from the hit-matrix of u_S; we will discuss query propagation through the network in the next section. For now it is important to understand that the strength of the connection between a pair of users is based on the response of one user to recommendations from another.

The *trust score* between users u_T and u_S is determined by the proportion of u_S's recommendations that u_T has selected; see Equation 4, where $SelRecs(u_S, u_T)$ is the number of times that u_T has selected one of u_S's recommended results, and $NumRecs(u_S, u_T)$ is the total number of recommendations made by u_S to u_T; note that $Trust(u_T, u_S) \neq Trust(u_S, u_T)$ so the search network takes the form of a directed graph; see Figure 1.

$$Trust(u_T, u_S) = \frac{SelRecs(u_S, u_T)}{NumRecs(u_S, u_T)} \tag{4}$$

Recommendations that originate from users with high trust scores should be considered more reliable than recommendations that originate from users with lower trust scores. This is captured by the trust-based relevance formula shown in Equation 5; a recommendation from u_S to u_T is weighted by the level of trust that u_T has for the recommendations of u_S based on past experiences.

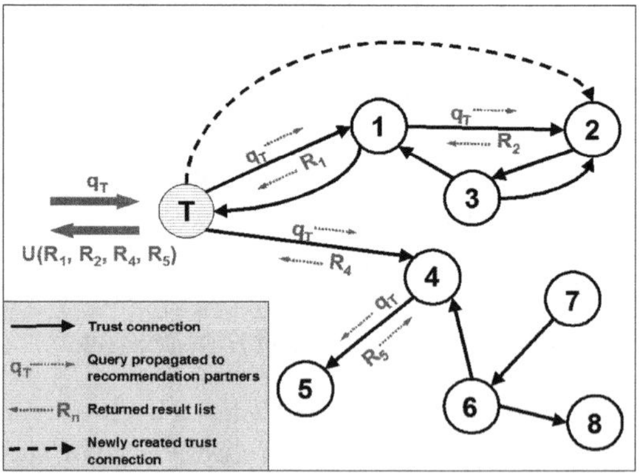

Fig. 1. Example of a search network

The final list of result recommendations can then be ranked according to their trust-weighted relevance scores as in the standard model of CWS.

$$TRel(u_T, u_S, p_j, q_T, q_1, ..., q_n) = Trust(u_T, u_S) \bullet WRel_{u_S}(p_j, q_T, q_1, ..., q_n) \quad (5)$$

3.3 Query Propagation

In our search network, users are not all interconnected; for example, in Figure 1 we see that user u_T is connected to users u_1 and u_4. However, this does not mean that u_T is limited to receiving recommendations from only these two directly connected users. We allow queries to propagate through the network at search time, along chains of connected users, with recommendations propagated back to the target user. In this way, a query q_T submitted by u_T is first propagated to u_1 and u_4 and then on to users u_2, u_3, and u_5, and so on, as shown in Figure 1. If a user has recommendations for q_T then these are returned to the intermediary who passed on the query, and so on back up the chain to u_T. Recommendations that originate from an indirect search partner are weighted with a trust score that is the combination of the trust scores of the sequence of partners along the chain. Thus, as shown in Equation 6, when u_T is connected to u_S by a sequence of user connections, $u_1, ..., u_n$, then the trust score between u_T and u_S (used in the relevance calculation, Equation 5) is given by $Trust'(u_T, u_S)$.

$$Trust'(u_T, u_S) = Trust(u_T, u_1) \bullet ... \bullet Trust(u_n, u_S) \quad (6)$$

3.4 Maintaining the Search Network

Our model allows for the creation of new links if users happen to select results that have been recommended by a distant search partner. For instance, in Figure 1, u_T selects a recommendation from u_2, thereby leading to the creation of a new link between u_T and u_2. Currently we are exploring a variety of heuristics for determining the initial trust scores of such connections. At the end of each search session the selections of the target user, in this case u_T, are used to update the trust scores of all users who contributed a recommendation. If a user contributed a recommendation that has been selected then their trust score will increase, if not it will decrease. Once again, we are currently evaluating a number of different trust models as alternatives to the proportional model described in Equation 4. In addition, we can use a simple threshold model to delete connections between users if their trust-scores fall below a certain level, thus eliminating the influence of low-quality agents within the network. This is an important point because it also provides a useful mechanism for dealing with malicious users and search spam. Such users will suffer from falling trust scores and will eventually become disconnected from the search network as recommendation providers.

4 Evaluation

4.1 Setup

The data for our evaluation comes from the profiles of 50 users of the online social bookmarking site Del.icio.us. For each user we view their bookmarked pages as proxies for search results that they are interested in, and the tags they associate with these pages as the queries that they might have used when searching for these pages. In order to achieve some degree of overlapping user interest, we downloaded the profiles of the top 50 users who tagged *http://www.foaf-project.org* (the home page of the *Friend of a Friend* project for encouraging the use of machine-readable metadata to express online relationships between people). For each user, we retrieved all of their bookmarked URLs and the tag(s) used to label them. On average we downloaded 406 bookmarked pages and tag-sets for each user, with each profile containing an average of 242 unique tags (or queries). We then created a hit-matrix for each user, associating every bookmarked page with its respective queries (tags) to simulate a search history for each individual. Each user hit-matrix and CWS agent serves as a node on our search network and below we describe the initialisation of this network.

For the purposes of this experiment we are primarily interested in the ability of a network of collaborating searchers to generate relevant recommendations from their hit-matrices, and so in what follows we do not rely on an underlying search engine to produce complete result-lists. We adopt a *leave-one-out* approach to the evaluation: for each user we simulate a search for each one of their queries, q_T, in each case temporarily removing that row from their hit-matrix, and comparing the recommended search results from the search network to the known selections that this user has made for this query. A result r_S suggested by some user u_S, for a query q_T submitted by user u_T, is considered relevant if and only if r_S is actually contained within the hit-matrix of u_T as a selection for q_T. In this way we can measure precision and recall for each q_T by comparing the recommendations made to u_T to this user's prior selections for q_T.

This method is repeated for a number of iterations, each iteration or *epoch* replaying the leave-one-out strategy for all users to allow the trust scores, which start out at 0.5 for all connected users, to evolve. In the initial stages of the experiment each network node (user) is connected to 10 other user nodes at random. During the experiment, the calculation of trust scores between users is delayed until such time that at least 10 recommendations have been exchanged between those users, ensuring that there is sufficient recommendation history between them to calculate a reliable trust score.

4.2 The Evolution of Search Performance

Ultimately we are interested in the ability of the search network to deliver improved search performance as it evolves, and so we compute the average precision

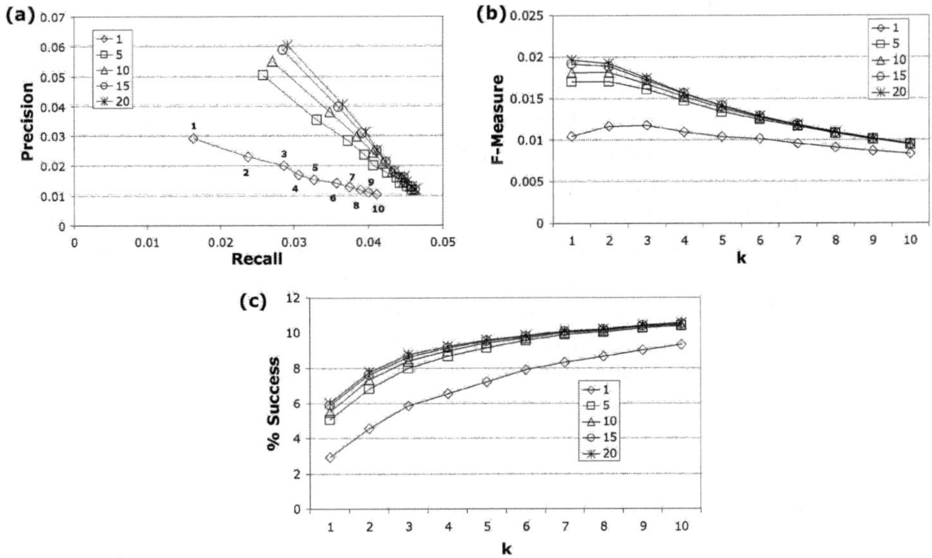

Fig. 2. (a) Precision vs. recall for result-lists sizes (k) from 1 to 10. Numbers alongside epoch-1 nodes indicate the value of k for this precision-recall pair. **(b)** F-measure for $k = 1...10$. **(c)** Percentage of sessions with ≥ 1 relevant top-k result.

and recall scores for the recommended result-lists of varying sizes ($k = 1...10$) at the end of each epoch. We present the results for epochs 1,5,10,15, and 20 as a standard precision and recall graph in Figure 2(a). The first thing to note is that the precision and recall values are very low, but this is more an artifact of the strict test for relevance used in this experiment (it requires that both the result recommender and the target user labelled the recommended URL with the same tag/query) than a reflection of search quality. Nevertheless, we see marked increases in the precision and recall statistics of the search network as it evolves over epochs. Figure 2(b) looks at the F-measure (the harmonic mean of precision and recall) over epochs, and again shows that the quality of the recommendations made by the network improves over time.

In Figure 2(c) we present an alternative performance graph which computes the average percentage of sessions that include at least one relevant result within the top k recommendations in sessions where recommendations are actually made. Once again we see a steady increase as the search network evolves. For example, during epoch 1, successful results are found in the top result-list position about 3% of the time, rising to just over 9% of the time if we consider the top 10 result-list positions. By epoch 20, this success rate has more than doubled for $k = 1$, with a success rate of over 6% at this position, and increased to nearly 11% for the top 10 results.

5 Conclusions

Social search techniques attempt to harness human judgments as a way to guide search. In this paper we have built on the work of [7] and the collaborative Web search technique, one of the shortcomings of which has been the need for predefined communities of searchers. We have introduced a new distributed model of CWS that generates community-focused search result recommendations from a network of trusted search partners. Communities develop over time, and preliminary experimental evidence supports the view that as trust relationships evolve the recommendations become increasingly accurate. A number of issues remain to be resolved. For example we have described one approach to initialising and updating the search network at search time and highlighted a number of options for alternatives. Our future work will look to explore these alternatives and provide a more comprehensive evaluation of different options, and to examine the scalability of our network model as the number of users increases.

References

1. S. Brin and L. Page. The Anatomy of a Large-Scale Web Search Engine. In *Proceedings of the 7th International World-Wide Web Conference*, volume 30, pages 107–117. Networks and ISDN Systems, 1998.
2. R. Farzan and P. Brusilovsky. *Social Navigation Support Through Annotation-Based Group Modeling*, volume 3538. Springer, 2005.
3. W. C. Hill, J. D. Hollan, D. Wroblewski, and T. McCandless. Edit Wear and Read Wear. In *CHI '92: Proceedings of the SIGCHI Conference on Human Factors in Computing Systems*, pages 3–9, New York, NY, USA, 1992. ACM Press.
4. J.A. Konstan, B.N Miller, D. Maltz, J.L. Herlocker, L.R. Gorgan, and J. Riedl. GroupLens: Applying Collaborative Filtering to Usenet News. *Communications of the ACM*, 40(3):77–87, 1997.
5. P. Massa and P. Avesani. Trust-Aware Collaborative Filtering for Recommender Systems. In *CoopIS/DOA/ODBASE (1)*, pages 492–508, 2004.
6. U. Shardanand and P. Maes. Social Information Filtering: Algorithms for Automating "Word of Mouth". In *Proceedings of the Conference on Human Factors in Computing Systems*, pages 210–217. ACM Press, 1995. New York, USA.
7. B. Smyth, E. Balfe, J. Freyne, P. Briggs, M. Coyle, and O. Boydell. Exploiting Query Repetition and Regularity in an Adaptive Community-Based Web Search Engine. *User Modeling and User-Adapted Interaction: The Journal of Personalization Research*, 14(5):383–423, 2004.
8. B. Smyth and P. Cotter. PTV: Intelligent Personalised TV Guides. In *Proceedings of the 12th Conference on Innovative Applications of Artificial Intelligence. (IAAI-2000)*. AAAI Press, 2000.
9. B. Smyth and P. Cotter. Personalized Adaptive Navigation for Mobile Portals. In *ECAI*, pages 608–612, 2002.
10. A. Voss and T. Kreifelts. SOAP: Social Agents Providing People with Useful Information. In *Proceedings of the International ACM SIGGROUP Conference on Supporting Group Work*, pages 291–298. ACM Press, 1997.
11. A. Wexelblat and P. Maes. Footprints: History-Rich Web Browsing. In *Proceedings of the Third International Conference on Computer-Assisted Information Retrieval*, 1997. Montreal, Quebec, Canada.

How to Compare Bilingual to Monolingual Cross-Language Information Retrieval

Franco Crivellari, Giorgio Maria Di Nunzio, and Nicola Ferro

Department of Information Engineering – University of Padua
Via Gradenigo, 6/a – 35131 Padova – Italy
{crive, dinunzio, ferro}@dei.unipd.it

Abstract. The study of cross-lingual *Information Retrieval Systems (IRSs)* and a deep analysis of system performances should provide guidelines, hints, and directions to drive the design and development of the next generation *MultiLingual Information Access (MLIA)* systems. In addition, effective tools for interpreting and comparing the experimental results should be made easily available to the research community. To this end, we propose a twofold methodology for the evaluation of *Cross Language Information Retrieval (CLIR)* systems: statistical analyses to provide MLIA researchers with quantitative and more sophisticated analysis techniques; and graphical tools to allow for a more qualitative comparison and an easier presentation of the results. We provide concrete examples about how the proposed methodology can be applied by studying the monolingual and bilingual tasks of the *Cross-Language Evaluation Forum (CLEF)* 2005 and 2006 campaigns.

1 Introduction

The growing interest in *MultiLingual Information Access (MLIA)* is witnessed by the international activities which promote the access, use, and search of digital contents available in multiple languages and in a distributed setting, that is, digital contents held in different places by different organisations. In particular, the *Cross-Language Evaluation Forum (CLEF)*[1] aims at evaluating MLIA systems which operate on European languages in both monolingual and cross-lingual contexts. The experimental evaluation carried out on MLIA systems takes on a twofold meaning: on the one hand, it should provide guidelines, hints, and directions to drive the design and development of the next generation MLIA systems; on the other hand, effective tools for interpreting and comparing the experimental results should be made easily available to the community.

We focus our attention on the study of cross-lingual IRSs and on a deep analysis of performance comparison between systems which perform monolingual tasks, i.e. querying and finding documents in one language, with respect to those which perform bilingual tasks, i.e. querying in one language and finding documents in another language.

[1] `http://www.clef-campaign.org/`

G. Amati, C. Carpineto, and G. Romano (Eds.): ECIR 2007, LNCS 4425, pp. 533–540, 2007.

The work presented in this paper aims at improving the current way of comparing bilingual and monolingual retrieval, which is basically a comparison of two averages of performance measures, and strives to provide better methods and tools for assessing the performances. A twofold methodology for the evaluation of CLIR systems is proposed: statistical analyses to provide MLIA researchers with quantitative and more sophisticated analysis techniques; graphical tools to allow for a more qualitative comparison and an easier presentation of the results. We discuss an example of application of the proposed methodology by studying the monolingual and bilingual tasks of the CLEF 2005 and 2006 campaigns. Note that these application examples also serve the purpose of validating the proposed methodology in a real setting.

The paper is organized as follows: Section 2 introduces the proposed methodology; Section 3 describes the experimental setting used for applying the proposed methodology, provides the application examples and reports the experimental results; finally, Section 4 draws some conclusions and provides an outlook for future work.

2 Cross-Lingual Comparison Methodology

A common method used to evaluate how good bilingual retrieval systems are is to compare results against monolingual baselines. The *Mean Average Precision (MAP)* is often used as a summary indicator among different performance figures available in Information Retrieval. For example, the overviews of the last CLEF workshops report figures where the MAP of a bilingual IRS is around 80% of the MAP of a monolingual IRS for the main European languages [1,2,3]. Even the recent literature on CLIR evaluation [5] compares performances between a monolingual baseline MAP and a MAP of bilingual approach. However, some hints of a deeper analysis can be found: statistical and query-by-query performance analyses.

In order to go beyond the simple comparison of MAP values between monolingual and bilingual performances, we propose a comparison methodology consisting of two complementary techniques which are both based on a comparison of results on single topics: a deep statistical analysis of both the monolingual and the bilingual tasks, described in Section 2.1; and a graphical comparison of both the monolingual and the bilingual tasks, described in Section 2.2.

2.1 Statistical Analysis Methodology

As pointed out by [4], a statistical methodology for judging whether measured differences between retrieval methods can be considered statistically significant is needed and, in line with this, CLEF usually performs statistical tests on the collected experiments [1,2] to assess their performances. On the other hand, these statistical tests are usually aimed at investigating the differences among the experiments within the same task, e.g. the monolingual French experiments alone or the bilingual French experiments alone, but they do not perform any kind of cross-task analysis, i.e. some kind of direct comparison between monolingual and bilingual tasks.

Given the average performance for each single topic of the monolingual and bilingual task, we want to study the distribution of these performances and employ different statistical tests to verify the following conditions:

1. the distributions of the performances are similar. This suggests that bilingual systems behave in a similar way with respect to monolingual ones, which represent our empirical baseline of the best attainable performances;
2. the variances of the two distributions are similar. This suggests that even though the passing from one language to another causes a decrease in the performances, the effect of the translation does not increase the dispersion of performances, which would add more uncertainty;
3. the mean of the two distributions are different and, in particular, the mean of the monolingual distribution is greater than the mean of the bilingual one. This suggests some loss of performances due to the effect of the translations from one language to another.

Note that we do not aim to demonstrate whether all these conditions simultaneously hold or not. Rather, we want to develop an analysis methodology which allows researchers to gain better insights into these conditions. We can anticipate here from Section 3 that these conditions do not hold simultaneously for all the monolingual and bilingual tasks we have analysed even though the general claim is usually complied with.

In order to verify the first condition, since the distribution of the experiments is unknown, we can adopt a quantile-quantile plot, which allows us to compare the distribution of the monolingual experiments with respect to the distribution of the bilingual experiments. In a quantile-quantile plot the quantiles of the two distributions are increasingly ordered and compared and, if the samples do come from the same distribution, the plot will be linear.

The last two conditions are analyzed and studied by means of statistical tests for the equality of two variances and for the equality of two means; the tests that are used in the paper (the F-test and the t-test, respectively) assume that collected data are normally distributed. Therefore, before proceeding, we need to verify the normality of the involved distributions by using graphical tools for inspection (i.e. the boxplot, or the normality plot) or normality tests (i.e. the Lilliefors test, or the Jarque-Bera test). However, if the normality assumption is violated, a transformation of the data should be performed. The transformation for proportion measures that range from 0 to 1 is the arcsin-root transformation which Tague-Sutcliffe recommends for use with precision/recall measures.

After the check on the normality of data, a test for the equality of variances, the F-test, is carried out to check whether the distributions have the same variance or not, and this step allows us to verify the second condition. Finally, in order to assess whether the mean of the monolingual performances is greater than the bilingual one, a t-test is used. In particular, since we have two paired sets (monolingual and bilingual) of m measured values, where m is the number of topics, the paired t-test is used to determine whether they differ from each other in a significant way under the assumptions that the paired differences are independent and identically normally distributed. This step allows us to verify the third condition reported above.

2.2 Graphical Comparison Methodology

The graphical tool which allows us to easily compare the performances for each topic of the monolingual and bilingual tasks, and to gain a visual explanation of the behavior of the two distributions, first needs a retrieval effectiveness measure to be used as a performance indicator. Then, we compute a descriptive statistic for the selected measure for each topic. In our case, the average precision and the mean were used, respectively. Finally, we increasingly order the monolingual topics by the computed descriptive statistic; the bilingual topics are ordered in the same order as the monolingual ones, because we are performing a topic-by-topic comparison and we want to compare a monolingual topic with the corresponding bilingual one. Note that ordering of the bilingual topics is usually different from what we would obtain if we increasingly ordered the bilingual topics by the computed descriptive statistic. From this ordered data we can produce two plots which provide us with an indicator for summarizing the trend of the monolingual and bilingual distributions. Examples of these plots are shown in Figure 1. For space reasons, the graphical examples are limited to only two plots.

Figure 1a shows the first plot where, for each topic, the monolingual performances on the x-axis, ordered increasingly, are plotted against the corresponding bilingual performances on the y-axis. If monolingual and bilingual behave in a similar way, the points are placed close and around the bisector of the first and third quadrant; on the other hand, if the monolingual performs better than the bilingual, the points are shifted towards the right whereas they are shifted symmetrically if the bilingual performs better than the monolingual. This plot provides us with a qualitative estimate about whether the three conditions introduced in the previous section hold: in that case, the plot would appear roughly linear and the points would be shifted towards the right. It is important to stress that this plot resembles a quantile-quantile plot but with an important difference: in a quantile-quantile plot the two distributions are independently ordered by their increasing quantiles, while in this plot the bilingual distribution is in the same order as the monolingual one.

Figure 1b presents a different view of the same data: for each topic on the x-axis, we plot on the y-axis both the monolingual (circles) and the bilingual performances (squares). This representation allows us to directly inspect the differences of the performances in a topic-by-topic fashion and provides us with hints about which topics require a deeper investigation because, for example, performances are too low or differences in the performances are too great. Moreover, this plot also allows us to qualitatively assess the three conditions reported in the previous section: in that case, the bilingual points would have a trend roughly similar to the monolingual ones and they would be below the monolingual ones.

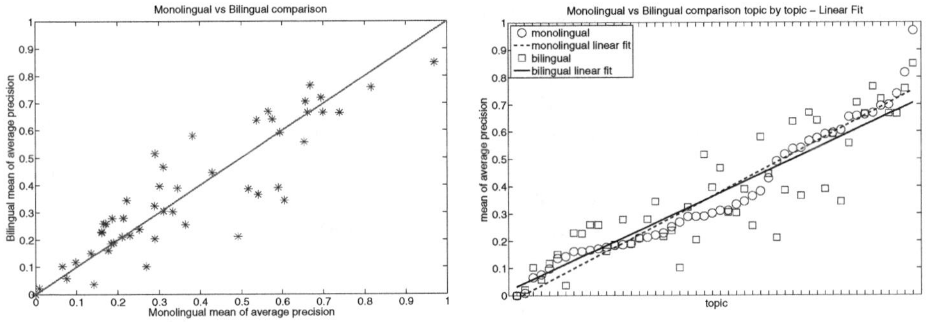

(a) Example of direct comparison. (b) Comparison on French 2006 data.

Fig. 1. Example of monolingual vs bilingual French 2006 comparison plots

Figure 1b also shows an example of linear fit: if the three conditions introduced in the previous section hold, the two straight lines would have roughly the same slope and the bilingual line would be right shifted with respect to the monolingual one.

3 Experimental Setting

The experimental collections and experiments used are fully described in [1,2].

In the CLEF 2005 and 2006 campaigns the languages of the target collection used for the monolingual and bilingual tasks were the same: Bulgarian, English, French, Hungarian, and Portuguese. However, we decided to take only those experiments of a bilingual task that have English as the source language in order to remove outliers from the data (for example extremely low performances due to very difficult languages such as Hindi or Amharic). Moreover, since we needed a sufficient number of experiments for each task to have reliable statistical analyses, we selected the tasks with the most experiments. These constraints led use to choose monolingual French (38 experiments for 2005 and 27 for 2006), monolingual Portuguese (32 experiments for 2005, 37 for 2006), bilingual French (12 experiments for 2005, 8 for 2006), and bilingual Portuguese (19 experiments for 2005, 18 for 2006). Remember that each one of these tasks has 50 topics.

For each task, we built a matrix $n \times m$ of n experiments and m topics where at position (i, j), with $1 \leq i \leq n$ and $1 \leq j \leq m$, we have the average precision (AP) of experiment e_1 on topic t_j. Then, we took the mean of the transformed performances by columns, that is, we took the average performances for each topic. As a result we had a vector for each task, like: $v_{task}^T = [mean_1 \; mean_2 \; \ldots \; mean_m]$, where $mean_1$ is the mean calculated for the first column, that is, the first topic of the task. The aim of the experimental analysis is to study the distribution of the mean of both the monolingual and bilingual tasks and compare them.

Table 1. Variance tests (F-tests) and Two-samples Paired t-test on CLEF 2005 data. Values in bold indicate hypotheses rejected.

		$H_0 : \sigma^2_{mono} = \sigma^2_{bili}$	$H_0 : \sigma^2_{mono} <= \sigma^2_{bili}$	$H_0 : \sigma^2_{mono} => \sigma^2_{bili}$
French	p-value	0.8281	0.5859	0.4141
Portuguese	p-value	0.9661	0.4831	0.5169
		$H_0 : \mu_{mono} = \mu_{bili}$	$H_0 : \mu_{mono} <= \mu_{bili}$	$H_0 : \mu_{mono} => \mu_{bili}$
French	p-value	0.8532	0.4266	0.5734
Portuguese	p-value	**0.0000**	**0.0000**	1.0000

The results presented are divided into years (2005 and 2006) and language (French and Portuguese). First the result of the normality test is presented, then the results of the analysis of variance are shown, and finally the analysis of the mean is discussed. Each calculation was carried out using MATLAB (version 7.2 R2006a) and MATLAB Statistics Toolbox (version 5.2 R2006a).

3.1 Statistical Analysis Methodology

Since the data resulted normal after a normality test, no arcsin-root transformation was adopted. In all the analyses, an alpha level of 5% was used.

CLEF 2005. The first analysis examines the variances of the data of the monolingual and the bilingual tasks. In Table 1, the results for the monolingual French vs bilingual French and the monolingual Portuguese vs bilingual Portuguese are presented. All the hypotheses are shown, starting from the most important one: the variances of the monolingual, σ^2_{mono}, and the bilingual, σ^2_{bili}, are equal. The other two hypotheses are important because the outcome shows that it is better not to reject them instead of accepting the alternative hypothesis which is, in those cases, σ^2_{mono} is either greater or less than σ^2_{bili}.

The second analysis considers the means of the monolingual, μ_{mono}, and bilingual, μ_{bili}, performances. Even though the hypothesis stated in Section 2.1, that is, the mean of the monolingual performances are better than the bilingual ones, is the main one, we believe it is important to consider all the aspects of the analysis. For this reason, we have presented the results for all the hypotheses in Table 1. It is interesting to see the differences between the French tests that result all in favor of the null hypothesis, that is to say it is preferable never to accept the alternative hypotheses that μ_{mono} is either greater or less than μ_{bili}. On the other hand, the analysis of Portuguese tasks shows that with the combination of all the hypotheses there is strong evidence that the mean of the performance of the monolingual Portuguese is greater than the bilingual one.

CLEF 2006. The analyses of the variances of the data of the monolingual and the bilingual tasks are shown in Table 2 for both the monolingual French vs bilingual French and the monolingual Portuguese vs bilingual Portuguese. All the tests confirm the hypothesis that the variances of the monolingual and bilingual tasks are equal.

Table 2. Variance tests (F-tests) and Two-samples Paired t-test on CLEF 2006 data. Values in bold indicate hypotheses rejected.

		$H_0 : \sigma^2_{mono} = \sigma^2_{bili}$	$H_0 : \sigma^2_{mono} <= \sigma^2_{bili}$	$H_0 : \sigma^2_{mono} => \sigma^2_{bili}$
French	p-value	0.8019	0.4009	0.5991
Portuguese	p-value	0.4270	0.7865	0.2135
		$H_0 : \mu_{mono} = \mu_{bili}$	$H_0 : \mu_{mono} <= \mu_{bili}$	$H_0 : \mu_{mono} => \mu_{bili}$
French	p-value	0.6860	0.3430	0.6570
Portuguese	p-value	**0.0001**	**0.0001**	0.9999

The two-samples paired t-test on the mean of the performances, shown in Table 2, confirm the outcome of the CLEF 2005: the tests on the French tasks are all in favor of the null hypothesis, that is to say the means are equal; the tests on the Portuguese tasks confirm that there is strong evidence that the mean of the performance of the monolingual Portuguese is greater than the bilingual one.

3.2 Graphical Comparison Methodology

In addition to the statistical analyses, we also present an effective graphical tool that gives a visual explanation of the behavior of the distributions of the monolingual and bilingual performances. Figures and plots were already shown in Section 2.2 and we cannot report the complete set of plots here for space reasons. On the other hand, we would like to comment on those plots in the light of the statistical analyses carried out in the previous section.

First, testing whether two distributions have similar shape and testing the normality of data can be done by means of standard tools such as the quantile-quantile plot and the normal probability plot. The quantile-quantile plots show that any monolingual-bilingual pair, both for French and Portuguese, has a regular linear trend, that is to say the shapes of the distributions are similar. The normal probability plot also shows the same regularity, which is sometimes violated along the tails of the distributions.

Second, the analysis of the performance on single topics can be appreciated in Figure 1: Figure 1a shows the performances on French 2006 data, ordered in the way explained in Section 2.2. This plot shows that there is a strong correlation between monolingual and bilingual performances; this correlation is above 0.80 for each pair of French, Portuguese tasks. Moreover, the cloud of points is located around the bisector of the first and third quadrant (above and below it) which means that, in this case, it is difficult to decide whether the monolingual is better than the bilingual or not. This confirms the results of the test on the means for French. Instead, the plot for the Portuguese (not shown here) shows a cloud of points which is more below than above the line, confirming that the monolingual Portuguese performs generally better than the bilingual.

In Figure 1b, a linear interpolation of the French 2006 tasks is performed. The two lines are very close and cross themselves; this figure clearly shows that even the linear interpolation of the monolingual and bilingual French data gives a positive response to the question that, in this case, the monolingual and bilingual performances are equal. Notice that we also have an indication of when the monolingual performance is better or worse than the bilingual; for example, for low performances bilingual performs better than monolingual while for high performances monolingual performs better.

4 Conclusions and Future Work

In this paper, we proposed a methodology which exploits both statistical analyses and graphical tools for the evaluation of MLIA systems. The statistical analysis provides MLIA researchers guidelines to drive the design and development of the next generation MLIA systems; the graphical tool provides a means to interpret experimental results and to present the results to other research communities easily. We provided concrete examples about how the proposed methodology can be applied by the analysis of the monolingual and bilingual tasks of the CLEF 2005 and 2006 campaigns.

Acknowledgements

The work reported in this paper has been partially supported by the DELOS Network of Excellence on Digital Libraries, as part of the Information Society Technologies (IST) Program of the European Commission (Contract G038-507618). The authors would like to warmly thank Carol Peters, coordinator of CLEF, for her continuous support and advice.

References

1. G. M. Di Nunzio, N. Ferro, G. J. F. Jones, and C. Peters. CLEF 2005: Ad Hoc Track Overview. In *Accessing Multilingual Information Repositories: Sixth Workshop of the Cross–Language Evaluation Forum (CLEF 2005). Revised Selected Papers*, pages 11–36. LNCS 4022, Springer, Heidelberg, Germany, 2006.
2. G. M. Di Nunzio, N. Ferro, T. Mandl, and C. Peters. CLEF 2006: Ad Hoc Track Overview. In *Working Notes for the CLEF 2006 Workshop*. `http://www.clef-campaign.org/2006/working_notes/workingnotes2006/dinunzioOCLEF2006.pdf`, 2006.
3. J. Gonzalo and C. Peters. The Impact of Evaluation on Multilingual Text Retrieval. In *Proc. 28th Annual International ACM SIGIR Conference on Research and Development in Information Retrieval (SIGIR 2005)*, pages 603–604. ACM Press, New York, USA, 2005.
4. D. Hull. Using Statistical Testing in the Evaluation of Retrieval Experiments. In *Proc. 16th Annual International ACM SIGIR Conference on Research and Development in Information Retrieval (SIGIR 1993)*, pages 329–338. ACM Press, New York, USA, 1993.
5. J. Wang and D.W. Oard. Combining Bidirectional Translation and Synonymy for Cross-Language Information Retrieval. In *Proc. 29th Annual International ACM SIGIR Conference on Research and Development in Information Retrieval (SIGIR 2006)*, pages 202–209. ACM Press, New York, USA, 2006.

Multilingual Text Classification Using Ontologies

Gerard de Melo and Stefan Siersdorfer

Max Planck Institute for Computer Science, Saarbrücken, Germany
{demelo,stesi}@mpi-inf.mpg.de

Abstract. In this paper, we investigate strategies for automatically classifying documents in different languages thematically, geographically or according to other criteria. A novel linguistically motivated text representation scheme is presented that can be used with machine learning algorithms in order to learn classifications from pre-classified examples and then automatically classify documents that might be provided in entirely different languages. Our approach makes use of ontologies and lexical resources but goes beyond a simple mapping from terms to concepts by fully exploiting the external knowledge manifested in such resources and mapping to entire regions of concepts. For this, a graph traversal algorithm is used to explore related concepts that might be relevant. Extensive testing has shown that our methods lead to significant improvements compared to existing approaches.

1 Introduction

Text classification (TC) is the process of associating text documents with the classes considered most appropriate, thereby distinguishing topics such as particle physics from optical physics. Research in this area, despite the considerable amount of work on cross-lingual information retrieval, has almost entirely neglected cases of documents being provided in multiple languages. Apart from truly multilingual environments as in large parts of Africa, people all over the world work with a lingua franca such as English or Spanish in addition to their native languages. Therefore, most applications of TC, e.g. digital libraries, news wire filtering as well as web page and e-mail categorization, also turn out to be interesting applications of *multilingual text classification* (MLTC), where documents given in different languages are to be classified by topic or similar criteria.

In this paper, we provide linguistic arguments against existing approaches and devise a novel solution that exploits background knowledge from ontologies and lexical resources. Section 2 discusses related work in this area, followed by Section 3, which briefly recapitulates fundamental ideas in TC and delivers arguments against existing approaches. Section 4 then presents Ontology Region Mapping as an alternative, which is then evaluated in Section 5, while the concluding section outlines the implications for continued research in this area.

G. Amati, C. Carpineto, and G. Romano (Eds.): ECIR 2007, LNCS 4425, pp. 541–548, 2007.

2 Related Work

There has been research on MLTC in the case of enough training documents being available for every language [1,2], however such scenarios are not particularly interesting as they can be resolved with separate monolingual solutions. A more universal strategy is to use translation to ensure that all documents are available in a single language [3,4,5], which also corresponds to the dominant approach in *cross-lingual information retrieval* (CLIR) [6]. However, our work shows that translations alone entail suboptimal TC results. An alternative approach to monolingual TC [7,8,9] and CLIR [10] related to the path pursued in our work relies on ontologies or thesauri to construct concept-based representations. While some authors pay respect to hypernyms and other directly related concepts [11,12,13], our approach is apparently the first to use an activation spread model in TC or CLIR. Multilingual solutions based on *latent semantic analysis* (LSA) have also been studied [14,15], however LSA differs from our approach in that it does not use formal background knowledge, but rather identifies concepts implicitly present in a set of documents, computed statistically by detecting terms with similar occurrence patterns.

3 Background

A *classification* is an assignment of class labels to objects such as text documents, and automatic text classification is the process of establishing and deploying classifiers that approximate text classifications made by human experts. When a set of pre-classified training documents is available, and an appropriate representation of their contents as numerical *feature vectors* is constructed, one of several learning algorithms can be employed to learn a classification, e.g. the *Support Vector Machine* (SVM) [16], which distinguishes two classes by using the hyperplane that maximizes the distance to the closest positive and negative examples as a binary decision surface. The conventional way of establishing the vector space for text documents involves well-known techniques such as stopword removal and stemming as preprocessing steps to computing TF-IDF values that are used to construct feature vectors based on the bag-of-words model [17].

Machine translation has been proposed as an ad hoc means of making multilingual document sets amenable to such TC processing [3]. However, certain drawbacks of the bag-of-words model then become particularly severe, e.g. when Spanish '*coche*' is generally mapped to '*car*', whereas French '*voiture*' is translated to '*automobile*', the learning algorithm remains unaware of the synonymy. Consider also that AltaVista Babel Fish [18] translates Spanish '*Me duele la cabeza*' to '*It hurts the head to me*', which does not contain the word '*headache*'.

Simple *concept mappings* have been used for alternative text representations in monolingual TC. Rather than using the original terms, one considers the *concepts* associated with their meanings as e.g. captured by WordNet [19]. Although the idea of mapping from several languages to language-neutral concepts seems particularly attractive, it may have a detrimental effect on the efficiency. For instance, lemmatizing inflected words to their base forms means that the distinct base forms of '*protected*' and '*protection*' prevent the two from being identified. Furthermore, WordNet lists many senses of the word '*school*', of which, in TC, at least seven should be seen as a thematic cluster rather than being distinguished, including school as an educational institution, as a building, as the process of being educated, etc. The idiosyncrasies of different languages pose additional problems, e.g. the English term '*woods*' is much narrower than the French '*bois*', so the two might not be mapped to the same concept. In Japanese and Chinese, there are separate words for older and younger sisters. German (as well as several other languages) allows for almost arbitrary compounds such as '*Friedensnobelpreisträgerin*' (woman awarded the Nobel Peace Prize).

4 Ontology Region Mapping

As all of the problems mentioned above involve terms being treated as distinct despite being closely related, we present *Ontology Region Mapping* (ORM) as a novel approach, where ontologies are construed as semantic networks in which entire regions of related concepts are considered relevant, rather than just individual ones. Our approach first maps terms to the concepts they are immediately associated with and then explores further related concepts.

4.1 Ontologies and Ontology Mapping Functions

An *ontology* is a theory of what possesses being in the world or in a domain. For our purposes, the requirements are a set of concept identifiers (*concept terms*) and a function τ providing information about how they are connected. For a concept term c, $\tau(c)$ should deliver a finite set of entries (c_i, r_i, w_i), where r_i indicates the type of relation (hypernymy, antonymy, etc.), c_i is a concept term, and $w_i \in [0, 1]$ is a relation weight specifying to what degree c and c_i are related.

Additionally, we construct *ontology mapping functions* that map document terms t from languages such as Spanish to such concepts, returning a set of pairs (c, w) where c is a concept term that possibly represents t's meaning and w is c's weight, i.e. the degree of relevance of c estimated with respect to the local context in which t occurred (the words surrounding t in the document). In our implementation, the functions look up terms in the English and Spanish WordNet [20], which serve as our ontological resources, using the lemmatized base form when no entry exists for the inflected form. In order to determine to what degree the concepts listed in WordNet for a term t are relevant in a particular context, part-of-speech information determined via morphological analysis is used to eliminate certain candidates. The remaining ones are then distinguished using an existing word sense disambiguation technique [21], where additional context strings are constructed for the candidate concepts by concatenating their human language description provided by WordNet with the respective descriptions of their immediate holonyms, hyponyms, as well as two levels of hypernyms. The similarity of two context strings for a document term t and a concept, respectively, is assessed by creating feature vectors for them using bag-of-words TF-IDF weighting and then applying the cosine measure. Our approach deviates from [21] in that we do not merely select the concept with the highest score because many related senses might be equally relevant when classifying. Instead, all candidate concepts are maintained with their cosine values, normalized with respect to the sum of all values, as their respective weights.

4.2 Weight Propagation

The mapping functions map document terms to the concept terms that immediately represent their respective meanings. ORM, however, not only maps to individual concepts but to entire regions of concepts by propagating a part of a concept's weight to related concepts. For every relation type r, an associated *relation type weight* $\beta_r \in [0, 1)$ is used, e.g. 0.8 for hypernym concept terms and 0.2 for hyponym terms. If a mapping function linked a document term to some concept term c_0 with weight 1.0, the relation type weights mentioned above would provide the direct parent hypernym of c_0 a weight of 0.8, the grandparent would obtain 0.64, and so on, until the values fall below some predetermined threshold. The amount of weight passed on is additionally also governed by the fixed relation weights stored in the ontology (cf. Section 4.1). When multiple paths from a starting concept term c_0 to another term c' exist, the path that maximizes the weight of c' is chosen. For this, Algorithm 4.1, inspired by the A* search algorithm [22], traverses the graph while avoiding cycles and suboptimal paths (see Fig. 1).

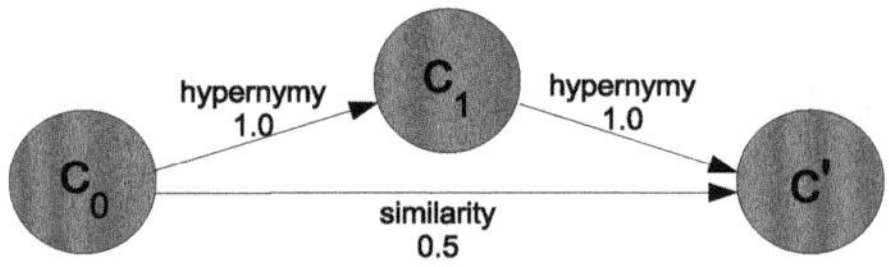

Fig. 1. Suboptimal paths: If c_0 has weight 1.0 and 80% is passed to hypernyms and 40% for similarity, then the direct path from c_0 to c' would only yield a weight of $0.5 \cdot 0.4 = 0.2$ for c', whereas for the indirect path we have $(1.0 \cdot 0.8)^2 = 0.64$

The algorithm's objective is to determine the optimal weights for related concepts and then accordingly update global concept term counts ctc_c that represent the sum of all weights assigned to a concept while processing an entire document. A list of nodes to be visited is maintained, sorted by weight and initially only containing the starting concept c_0 in conjunction with its weight w_{c_0}. The node c with the highest weight is then repeatedly removed from this list and the counter ctc_c is incremented by c's weight. The algorithm evaluates all neighbours of c, computes their weights and adds them to the list, provided the new weight is greater than a pre-determined threshold $w_{\min}$ as well as any previously computed weight for that particular neighbour. A parameter space search heuristic can be used to empirically determine suitable values for $w_{\min}$ and the β_r values. It can be shown that this algorithm always chooses the optimal weight and terminates if the parameter constraints are fulfilled. In order to decrease the runtime, one

Algorithm 4.1. Ontology-relation-based feature weighting

Input: initial concept c_0 with weight w_{c_0} from an ontology with relation function τ, initial term counts ctc_c for concept terms c, a relation type weight $\beta_r < 1$ for every relation type r, weight propagation threshold $w_{\min} > 0$

Objective: update concept term counts ctc_c for all relevant concepts c from ontology

$weight_{c_0} \leftarrow w_{c_0},\ weight_c \leftarrow -\infty$ for all $c \neq c_0$
$open \leftarrow \{c_0\},\ closed \leftarrow \emptyset$
while $open \neq \emptyset$ **do**
  choose concept c with greatest $weight_c$ from $open$
  $open \leftarrow open \setminus \{c\},\ closed \leftarrow closed \cup \{c\}$ ▷ Move c to *closed*
  $ctc_c \leftarrow ctc_c + weight_c$ ▷ increase concept term count
  for each relation entry $(c_i, r_i, w_i) \in \tau(c)$ **do** ▷ visit neighbours c_i of c
    $w \leftarrow \beta_{r_i} \cdot weight_c \cdot w_i$
    if $w \geq w_{\min}$ and $c_i \notin closed$ **then** ▷ proceed only if over threshold
      $open \leftarrow open \cup \{c_i\}$
      $weight_{c_i} \leftarrow \max\{weight_{c_i}, w\}$

may add a $|closed| < k$ condition to the while-loop, causing the algorithm to visit only k highest-ranking concepts.

4.3 General Procedure

Instead of multilingual ontologies, one may also use translated documents. In both cases, each document is tokenized and stop words are removed using a fixed list, resulting in a sequence of terms $d = (d_1, \ldots, d_l)$. For each term, an appropriate mapping function then returns a list of corresponding concepts with associated weights. These are then each submitted as input to Algorithm 4.1 with their respective weight such that the concept term counts ctc_c of any additionally relevant concept terms are updated, too. Despite the non-integral values of these concept term counts, one can proceed to compute concept TF-IDF scores similar to those in conventional TC. While a normalization of the ctc_c to concept term frequencies $ctf(d, c)$ is straightforward, the notion of occurrence required for document frequencies does not emanate from our definition of concept term counts as it does in the case of conventional term counts, for it is unclear whether concept terms with a minuscule weight qualify as occurring in the document. We thus use a threshold $\alpha \in [0, 1]$ and define $ctfidf_\alpha(d, c)$ as $ctf(d, c) \cdot \log \frac{1}{cdf_\alpha(c)}$, where $cdf_\alpha(c)$ returns the fraction of all training documents for which $ctc_c > \alpha$ is obtained. The feature space is then constructed by associating each concept term with a separate dimension, and the respective $ctfidf_\alpha$ values can be used to create feature vectors, which are finally normalized. Though not ordinarily covered by mapping functions, technical terms as well as names of people or organizations, for instance, might be crucial when categorizing a document. Hence, an extended setup may be considered, where the ctc_c are combined with conventional term counts. The $ctfidf_\alpha$ values are then computed globally with respect to all such term counts and the feature space has dimensions for concept terms as well as for stems of original document terms.

5 Evaluation

5.1 Experimental Setup and Tuning

In order to evaluate our solution we performed a large number of cross-lingual tests using the SVMlight implementation [23] of Vapnik's SVM with its default settings. We imported WordNet 2.1 [19] for the English language, and additionally applied mapping tables [24] to the Spanish WordNet [20] to synchronize the two resources. A Japanese version of WordNet does not exist, so only translation-based approaches were tested in that case. All translations were performed by AltaVista's Babel Fish Translation service [18].

Two datasets were extracted from *Reuters Corpus Vol. 1* and *2* (RCV1, RCV2) using English training (RCV1) and Spanish test documents (RCV2): one based on topic codes and another one on geographical codes (industry codes could not be used due to inconsistencies between RCV1 and RCV2). An additional dataset with Japanese test documents was generated from Wikipedia [25]. As virtually all TC problems can be reduced to binary ones [17], we tested 105 binary problems per dataset, resulting from 15 randomly selected classes, with 100 training and 600 test documents (Wikipedia: 300) per setup, also selected randomly, however with equal numbers of positive and negative examples in order to avoid biased error rates. A separate validation set was generated based on the same principles as our Reuters topic dataset and then used to tune the relation type weights for hypernyms, holonyms, derivations, etc., as well as other parameters. We chose a value of 0.5 for the α in our $ctfidf_\alpha$ formula. For each setup, we also determined the most suitable numbers of features for feature selection based on Information Gain, which turned out to be 1000 for Ontology Region Mapping (ORM).

5.2 Results and Discussion

First of all, Table 1 shows that the conventional bag-of-words method without any translation whatsoever (B) worked surprisingly well, probably due to named entities and because of the relatedness of English and Spanish. Nonetheless, the error rates are unsatisfactory for production use and similar results cannot be achieved for arbitrary language pairs. For Japanese, in fact, this method could not be used directly as advanced tokenization heuristics would be required. As expected, the translation approach T leads to significant improvements.

Applying ORM clearly is beneficial to efficiency compared with a simple concept mapping setup without propagation (CM). The error rates depend on the ontology employed. Better results than with the English/Spanish WordNet setup (CM and ORM) may be obtained by using our ORM approach with translations (TORM), even more so by also including the document terms in the final representation (TORM+T). This is a positive result, implying that ORM with the freely available English WordNet as well as translation software, which also tends to be more available than multilingual lexical resources, suffices for MLTC, as in the case of Japanese, for which a WordNet version currently does not exist.

Table 1. Test Results for Reuters English-Spanish and Wikipedia English-Japanese datasets (micro-averaged F_1 scores in %, average error rates in % with 95% confidence intervals) where **B:** conventional bag-of-words method without translations, **CM:** simple concept mapping approach without weight propagation, **ORM:** Ontology Region Mapping, **ORM+B:** Ontology Region Mapping combined with bag-of-words, **T:** bag-of-words from English translations, **TCM/TORM/TORM+T:** same as CM/ORM/ORM+B but with English translations as input

	Reuters Spanish			
	Topics		**Geography**	
	F_1	error rate	F_1	error rate
B	80.97	18.61 ±0.30	81.86	18.12 ±0.30
CM	89.23	10.49 ±0.24	85.74	14.58 ±0.28
ORM	89.53	10.36 ±0.24	87.33	12.97 ±0.26
ORM+B	91.88	8.04 ±0.21	91.92	8.22 ±0.21
T	90.96	8.80 ±0.22	88.76	11.43 ±0.25
TCM	90.75	9.06 ±0.22	91.12	9.16 ±0.23
TORM	91.12	8.74 ±0.22	93.89	6.28 ±0.19
TORM+T	92.46	7.43 ±0.20	94.44	5.68 ±0.18

	Wikipedia Japanese	
	F_1	error rate
T	86.26	14.00 ±0.38
TCM	85.38	15.10 ±0.40
TORM	86.67	13.52 ±0.38
TORM+T	87.29	12.86 ±0.37

For news and encyclopedic articles, outperforming the T method is a rather difficult task in light of the considerable discriminatory power of the terms in the introduction paragraphs. Nonetheless, our methods delivered superior results that are statistically significant. Geographical references, in contrast, are often less explicit, so our methods pay off even more. Given that the relation type weights were tuned with respect to the Reuters topic-based validation set, we may presume that even better results than the ones indicated are achievable.

6 Conclusions and Future Work

In the past, many attempts to use natural language processing for monolingual TC have failed to deliver convincing results [17]. A linguistic analysis led us to a novel approach called Ontology Region Mapping, where related concepts, too, are taken into consideration when mapping from terms to concepts, so additional background knowledge is exploited, which is particularly useful in multilingual settings. Our experimental evaluation confirmed our intuitions.

In the future, we would like to devise strategies for constructing multilingual resources that integrate more background knowledge and better reflect the semantic relatedness of concepts than WordNet. Additionally, a more sophisticated word-to-concept mapping setup could be used that recognizes compounds and disambiguates better. Finally, it could be explored how well ORM performs for multilingual information retrieval and text clustering. Indeed, we believe our feature weighting approach or extensions of it to have a wide range of interesting applications, in multilingual as well as monolingual settings, because it captures the general meaning of a text more adequately than established schemes.

References

1. Bel, N., Koster, C.H.A., Villegas, M.: Cross-lingual text categorization. Proc. ECDL 2003 (2003) 126–139
2. García Adeva, J.J., Calvo, R.A., de Ipiña, D.L.: Multilingual approaches to text categorisation. Europ. J. for the Informatics Professional **VI**(3) (2005) 43 – 51
3. Jalam, R.: Apprentissage automatique et catégorisation de textes multilingues. PhD thesis, Université Lumière Lyon 2, Lyon, France (2003)
4. Olsson, J.S., et al.: Cross-language text classification. Proc. SIGIR 2005 (2005) 645–646
5. Rigutini, L., et al.: An EM based training algorithm for cross-language text categorization. In: Proc. Web Intelligence 2005, Washington, DC, USA (2005) 529–535
6. Oard, D.W., Dorr, B.J.: A survey of multilingual text retrieval. Technical report, University of Maryland at College Park, College Park, MD, USA (1996)
7. de Buenaga Rodríguez, M., et al.: Using WordNet to complement training information in text categorization. Proc. 2nd RANLP (1997)
8. Moschitti, A., Basili, R.: Complex linguistic features for text classification: a comprehensive study. Adv. in IR, Proc. ECIR 2004 (2004)
9. Ifrim, G., Theobald, M., Weikum, G.: Learning word-to-concept mappings for automatic text classification. Proc. 22nd ICML - LWS (2005) 18–26
10. Verdejo, F., Gonzalo, J., Peñas, A., et al.: Evaluating wordnets in cross-language text retrieval. Proceedings LREC 2000 (2000)
11. Scott, S., Matwin, S.: Text classification using WordNet hypernyms. Proc. Worksh. Usage of WordNet in NLP Systems at COLING-98 (1998) 38–44
12. Bloehdorn, S., Hotho, A.: Boosting for text classification with semantic features. Proc. Worksh. on Mining for/from the Semantic Web at KDD 2004 (2004) 70–87
13. Ramakrishnanan, G., et al.: Text representation with WordNet synsets using soft sense disambiguation. Ing. systèmes d'information **8**(3) (2003) 55–70
14. Gliozzo, A.M., et al.: Cross language text categ. by acq. multil. domain models from comp. corpora. Proc. ACL Worksh. Building and Using Parallel Texts (2005)
15. Dumais, S.T., et al.: Automatic cross-language retrieval using latent semantic indexing. AAAI Symposium on CrossLanguage Text and Speech Retrieval (1997)
16. Vapnik, V.N.: The Nature of Statistical Learning Theory. Springer-Verlag New York, Inc., New York, NY, USA (1995)
17. Sebastiani, F.: Machine learning in automated text categorization. ACM Computing Surveys **34**(1) (2002) 1–47
18. AltaVista: Babel fish translation. `http://babelfish.altavista.com/` (2006)
19. Fellbaum, C., ed.: WordNet: An Electronic Lexical Database (Language, Speech, and Communication). The MIT Press (1998)
20. Farreres, X., Rigau, G., Rodríguez, H.: Using WordNet for building WordNets. Proc. Conf. Use of WordNet in NLP Systems (1998) 65–72
21. Theobald, M., Schenkel, R., Weikum, G.: Exploiting structure, annotation, and ontological knowledge for automatic classification of XML data. 6th Intl. Worksh. Web and Databases (2003) 1–6
22. Russell, S., Norvig, P.: Artificial Intelligence: A Modern Approach. Prentice-Hall, Englewood Cliffs, NJ, USA (1995)
23. Joachims, T.: Making large-scale support vector machine learning practical. Advances in Kernel Methods: Support Vector Machines (1999)
24. Daudé, J., et al.: Making Wordnet mappings robust. Proc. Congreso de la Sociedad Española para el Procesamiento del Lenguage Natural (SEPLN) (2003)
25. Wikimedia Foundation: Wikipedia. `http://www.wikipedia.org/` (2006)

Using Visual-Textual Mutual Information and Entropy for Inter-modal Document Indexing

Jean Martinet and Shin'ichi Satoh

National Institute of Informatics
Multimedia Information Research Division
Tokyo, Japan
{jean, satoh}@nii.ac.jp

Abstract. This paper presents a contribution in the domain of automatic visual document indexing based on inter-modal analysis, in the form of a statistical indexing model. The approach is based on inter-modal document analysis, which consists in modeling and learning some relationships between several modalities from a data set of annotated documents in order to extract semantics. When one of the modalities is textual, the learned associations can be used to predict a textual index for visual data from a new document (image or video). More specifically, the presented approach relies on a learning process in which associations between visual and textual information are characterized by the mutual information of the modalities. Besides, the model uses the information entropy of the distribution of the visual modality against the textual modality as a second source to select relevant indexing terms. We have implemented the proposed information theoretic model, and the results of experiments assessing its performance on two collections (image and video) show that information theory is an interesting framework to automatically annotate documents.

Keywords: Indexing model, mutual information, entropy, inter-modal analysis.

1 Introduction

The increasing amounts of multimedia information is almost useless if users have no mean to efficiently access specific information they search. There is a crucial need to index multimedia documents in order to provide efficient access to them. Given this huge quantity of information, the indexing process must be automated. A promising approach to automatically annotate multimedia documents is to exploit the information redundancy across modalities with inter-modal analysis [1]. In human-computer interaction, a modality refers to the type of communication channel used to convey or acquire information. Multimodality is defined in [2] as "the capacity of the system to communicate with a user along different types of communication channels and to extract and convey meaning automatically". In multimedia content analysis, however, a modality rather refers to the nature of a signal. A modality, which is embedded in one

G. Amati, C. Carpineto, and G. Romano (Eds.): ECIR 2007, LNCS 4425, pp. 549–556, 2007.

medium[1], is the particular way in which the information is encoded for presentation to humans. Multi-modal (or more specifically inter-modal) analysis refers to the process of combining several modalities – for instance in a synergistic (collaborative) way, in order to automatically extract semantics. In particular, when one of the modalities is textual, inter-modal analysis can be used to automatically annotate other modalities. For instance, for image and video annotation, the visual modality contains the visual part of image and video documents, and textual modality contains textual any resource that can be used as a description of the content of the document (e.g. document name, annotation, text from a web page containing the document, open and closed captions, etc.) Applications of inter-modal analysis include automatic document – or image region – annotation, organization of a document collection, and text illustration (finding visual documents related to a given text in order to illustrate it).

We introduce in this paper an information theoretic framework for the task of indexing visual data. Our approach uses mutual information as a measure of the strength of the dependency between modalities. Besides, we consider Shannon's information entropy [3] of the distribution of the visual modality over the textual modality as a quality criterion to select annotation keywords. Intuitively, the quality of a keyword refers to its discriminative power in selecting images. The entropy provides a straightforward way to unify two aspects of a distribution, namely the size of its range, and its uniformity. Given some visual data, a textual annotation can be generated according to both the strength of the dependency with the visual data and the quality of the annotation.

2 Related Work

A number of approaches have been proposed to discover relationships between several modalities. In particular, much effort has been made to learn associations between visual features and text from an annotated data set and to apply the learned association to predict annotation keywords for unseen documents. Different models and machine learning techniques have been developed for this purpose. For instance, the co-occurrence model of by Mori et al. [4] counts co-occurrences between annotation keywords and image features in a train set, and generates keywords for test images according to the highest co-occurrence values. Later, the problem has also been examined using statistical machine translation ([5] and [6]) where an analogy is made with the problem of translating words from a source language (visual features) to a target language (annotation keywords). Other models such as Latent Dirichlet Allocation (LDA) [7] and relevance language models [8] have been applied to automatic image annotation. Information theory has been used by Salton [9] to measure the extent to which two index terms i and j deviate from independence, by evaluating the Expected Mutual Information Measure (EMIM). Information theory has also been used to multi-modal signal analysis. In particular, the mutual information measure

[1] Used as the singular of media. Note that a single medium – using one type of communication channel – may contain several modalities.

(while being often used in machine learning for feature selection), has also been applied to multi-modal analysis, namely for the problem of volumetric medical image registration ([10]). This problem consists in aligning data from different sources (magnetic resonance images and tomography images for instance), each taken as a modality. In this case, one wants to maximize mutual information to increase the dependency between features and to minimize the mapping error between modality feature spaces.

In the section 3 and 4, we describe how we use the mutual information and entropy in the proposed annotation model. In the section 6, we show and discuss experimental results, and the section 7 gives a conclusion.

3 Visual-Textual Mutual Information

We choose to demonstrate the proposed approach in the specific context where the modalities are visual and textual. Documents are represented with a pair of low-level visual feature data and textual data, written $\mathcal{D} = (\mathcal{F}, \mathcal{A})$, where $\mathcal{F} = \{\boldsymbol{f_1}, \ldots, \boldsymbol{f_n}\}$ denotes the set of quantized feature vectors of the visual modality and $\mathcal{A} = \{a_1, \ldots, a_m\}$ denotes the set of annotation keywords of the textual modality. Feature vectors are clustered, quantized, and further identified by their closest centroid so that the visual modality can be represented with a reasonably small number of discrete visual descriptors. Considering visual feature vectors and annotation keywords as binary random variables F_i and A_j denoting the occurrence of $\boldsymbol{f_i}$ and a_j in $\mathcal{F}$ and $\mathcal{A}$ respectively, we are interested in finding the dependency between F_i and A_j in order to predict an annotation $\mathcal{A}$ from the feature set $\mathcal{F}$ of a new document. The prediction is generated using two sources: the mutual information of F_i and A_j, and the conditional entropy of the distribution of the feature vectors given a keyword.

The mutual information of F_i and A_j measures the mutual dependency of the two random variables by quantifying the distance between the joint distribution of F_i and A_j and the product of their marginal distributions. Formally, the mutual information $I(F_i, A_j)$ of F_i and A_j is defined as:

$$I(F_i, A_j) = \sum_{k \in \{0,1\}} \sum_{l \in \{0,1\}} p(F_i = k, A_j = l) log \left(\frac{p(F_i = k, A_j = l)}{p(F_i = k)p(A_j = l)} \right) \quad (1)$$

where $p(F_i, A_j)$ is the joint probability distribution of F_i and A_j , and $p(F_i)$ and $p(A_j)$ are the marginal probability distributions of F_i and A_j respectively. The mutual information, which is a non-negative and symmetric measure, equals to 0 when F_i and A_j are independent. A high value indicates a strong dependency. The maximum value is reached when the random variables are identical: F_i and A_j always occur in the same context, which means that all information conveyed by F_i is shared by A_j. The annotation model uses the mutual information to represent the strength of associations between visual feature vectors and keywords.

4 Entropy of Feature Vectors Distribution

An intuitive understanding of the information entropy relates to the amount of uncertainty about an event associated with a given probability distribution. Considering the distribution of the visual feature vectors associated to an annotation keyword, the entropy measures the uncertainty of this association. If we note $F|a_j$ the random variable yielding the conditional distribution of visual feature vectors in the context of a_j, then $p(F = f_i|a_j)$ (or $p(f_i|a_j)$) is the probability of f_i being associated to a_j. The information entropy of the distribution of $F|a_j$ is defined as:

$$H(F|a_j) = -\sum_{i}^{n} p(f_i|a_j) \times log\,(f_i|a_j)) \quad (2)$$

This entropy can be seen as a unified measure of both the number of feature vectors to which the keyword is associated and the uniformity (flatness) of the distribution. A large range together with a flat distribution will result in a high entropy. In this case, the annotation keyword is scattered across visual feature vectors and there is no preferential association with some particular feature vectors. The maximum value corresponds to the case where the keyword is uniformly distributed over feature vectors. Inversely, a small range and a sharp distribution will result in a low entropy. In this case, the annotation keyword is considered accurate and discriminative from an annotation point of view. The minimal value is 0, corresponding to the case where the keyword is associated with one feature vector with probability 1. This value also corresponds to the unlikely case where the keyword is associated with no feature vector. The proposed model uses entropy to select accurate and discriminative annotation keywords, that is to say keywords with low entropy.

5 Annotation Model for an Unknown Image

Both mutual information and entropy contribute to select annotation keywords for a new image. The first source measures the *local* quality $L(\mathcal{F}, a_j)$ of a keyword a_j as a descriptor of the document $\mathcal{D} = (\mathcal{F}, \emptyset)$ to be annotated, for which $\mathcal{A}$ is unknown:

$$L(\mathcal{F}, a_j) = \sum_{i}^{n} I(F_i, A_j) \quad (3)$$

This value is high when a_j is a good descriptor of $\mathcal{F}$. The second source quantifies the *global* quality $G(a_j)$ of a keyword a_j in the context of the whole document collection, regardless of the new specific document to be annotated:

$$G(a_j) = \frac{1}{H(F|a_j) + \epsilon} \quad (4)$$

The value ϵ added under the ratio guaranties that the denominator is never 0. This value is high when a_j is associated with few visual feature vectors, meaning that it is a good discriminator of the images of the collection.

Given a new document $\mathcal{D}$ for which only the set of visual feature vectors $\mathcal{F} = \{\boldsymbol{f_1}, \ldots, \boldsymbol{f_n}\}$ is known, we define the prediction weight $w(\mathcal{F}, a_j)$ of an annotation keyword a_j as follows:

$$w(\mathcal{F}, a_j) = L(\mathcal{F}, a_j) \times G(a_j) \tag{5}$$

Annotation keywords for a new image are generated according to the equation 5.

The justification of using two sources if inspired by the standard *tf.idf* weighting scheme [12] widely used in text retrieval, in the sense that a keyword can be considered a "good" annotation keyword if it both describes well the image and discriminates well documents in the collection.

6 Experimental Results

In this section, we describe and discuss the results of experiments carried out to assess the efficiency of the proposed annotation model. Low-level visual features include standard color and texture descriptors. Vectors of dimension 168 are composed with 128-bin HSV histograms and 40 Gabor wavelet coefficients. Vectors are extracted from 32 × 32 fixed size square patches from images, and clustered and quantized using a variation of ISODATA algorithm. Euclidean distance is used for the clustering algorithm and quantization process.

The performance of the proposed model has been evaluated with two annotated data collections: one containing still images from a home photograph collection and the other containing keyframes extracted from a collection of video news. The still images are drawn from a personal home photographs collection, which have been thoroughly annotated manually. A dedicated ontology describing landscapes elements and general physical objects likely to appear in home photographs has been defined for this collection. This ontology is composed with 196 entries, and the collection contains 800 annotated images, with an average of 6.7 annotation keywords per image. The video collection is a subset of TRECVID 2004 data provided by NIST which is composed with CNN and ABC broadcast news videos. The keyframes provided by CLIPS-IMAG are used for the visual modality. For the textual modality, the experiments are based on keywords derived from automatic speech recognition transcripts provided by LIMSI. Words from raw speech transcripts are tagged with Schmid's probabilistic part-of-speech tagger [13], and only the base form (lemma) of nouns are kept in the annotation vocabulary. In total, the vocabulary is composed with over 4.000 words, and this collection contains about 10.000 annotated keyframes, with an average of 4.63 annotation keywords per image. Each collection is randomly divided in train and test sets with proportion 0.7 and 0.3 respectively.

The proposed approach is compared with two models: a co-occurrence model counting the co-occurrence values between feature vectors and keywords, and a Naive Bayes classifier estimating the conditional posterior probability $p(A_j|\mathcal{F})$ of keywords given a set of feature vectors from a test image. The test documents are annotated with the proposed model (resp. co-occurrence model, Naive Bayes classifier) by generating words with highest prediction weight(resp. co-occurrence values, conditional posterior probabilities). We consider the case where the number of predicted keywords is equal to the number of ground truth annotation keywords, and the case where the number of predicted keywords is fixed. Parameters for the 3 models are estimated on the train set, and the performance is evaluated by comparing image ground truth annotations and predictions of the models in the test set.

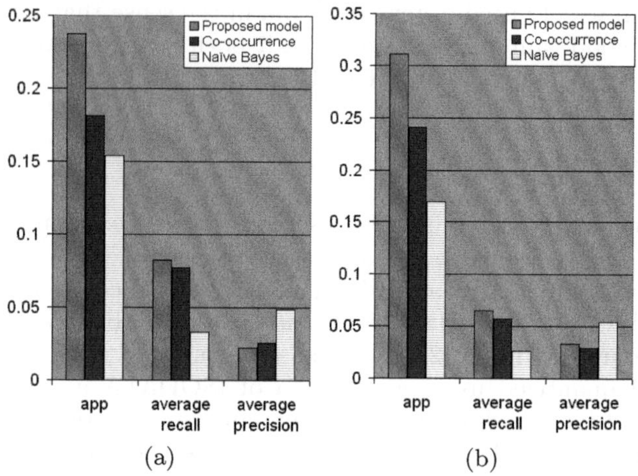

Fig. 1. Evaluation results for the image collection, in the two cases of generated prediction length: (a) variable-length annotation, (b) fixed-length annotation

The figure 1 (resp. 2) shows evaluation results for the image (resp. video) test collection. We used standard evaluation measures to assess the performance of the models : the prediction precision of a document is the number of correct prediction words divided by the number annotation words. The average prediction precision (app) is the prediction precision averaged for all documents in the test set. Average recall and precision values are evaluated for all words in the annotation vocabulary, used as a single term query.

Results show that the information theoretic model is globally more efficient that the two other models in terms of the considered evaluation measures. The model gives the highest app and average recall values for the two collections. However, for the video collection (figure 2), these values are comparable the co-occurrence model.

As for the Naive Bayes classifier, while it achieves highest average precision values for the image collection, the average precision values are comparable to the two other models for the video collection. In general, for all models the recall is higher for the case of fixed-length prediction. This is because the fixed length is set to 10, which is larger than the average ground truth annotation length. Therefore, images are annotated with more keywords and more images are retrieved. In this case, as one can expect, the precision is decreased, particularly for the video collection. We note that all values are lower for the video collection. Indeed, one can expect less evidence in the association between the keyframes and the ASR output than for the manually annotated image collection.

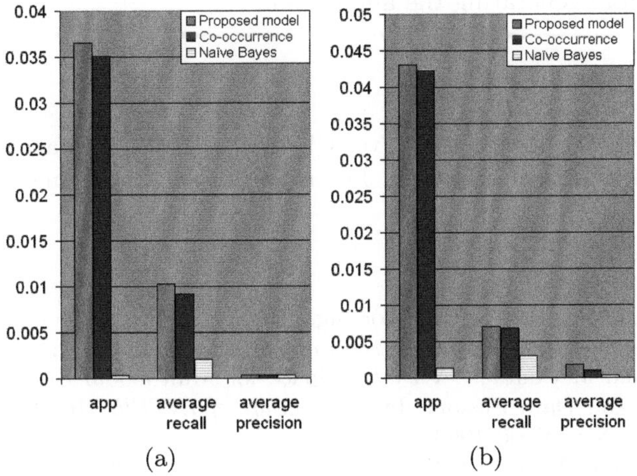

Fig. 2. Evaluation results for the video collection, in the two cases of generated prediction length: (a) variable-length annotation, (b) fixed-length annotation

We selected those two collections are because they are very different in terms of type and quality of image, and annotation process. While the first one is composed with high quality photographs with an accurate manual description of the visual content, the second one is composed with average quality keyframes with an automatic annotation process extracted from the speech, which is more loosely related to the visual content.

We highlight that for the video collection, we do not filter high or low frequency words as the entropy measure selects relevant words according to the visual feature distribution, independently from their frequency. Indeed, a high frequency word can be given the same entropy value as a low frequency word it is always observed in the same context. But in the general case, high frequency words are associated more sparsely with feature vectors, yielding higher entropy.

7 Conclusion

This paper describes a document indexing model based on inter-modal analysis. This model is based on the mutual information of visual and textual modalities and on the entropy of the distribution of visual features for annotation keywords. The model uses these two sources to generate annotation keywords for a new document. The main contribution of this work is to demonstrate the efficiency of a first approach of an information theoretic approach for the problem of automatic indexing.

We believe that information theory is a dedicated framework for semantic extraction from inter-modal analysis, which requires more investigations. Further research includes a refinement of the annotation value definition, in order to integrate a selection of "best" (or more representative) feature vectors from a new image before generating the annotation.

Acknowledgments

The authors wish to thank Prof. Yves Chiaramella for the image test collection. Jean Martinet is supported by the Japan Society for the Promotion of Science.

References

1. Cees G.M. Snoek and Marcel Worring. Multimodal video indexing: A review of the state-of-the-art. *Multimedia Tools and Applications*, 25(1):5–35, 2005.
2. L. Nigay and J. Coutaz. A Design Space for Multimodal Systems: Concurrent Processing and Data Fusion. In *Proceedings of INTERCHI '93*, pages 172–178. ACM Press: New York, 1993.
3. C. E. Shannon. A mathematical theory of communication. *Bell System Technical Journal*, 27:379–423, 623–656, 1948.
4. Y. Mori, H. Takahashi, and R. Oka. Image-to-word transformation based on dividing and vector quantizing images with words. In *In MISRM'99 First International Workshop on Multimedia Intelligent Storage and Retrieval Management*, 1999.
5. Kobus Barnard, Pinar Duygulu, and David Forsyth. Recognition as translating images into text. In *Internet Imaging IX, Electronic Imaging*, 2003.
6. Muhammet Bastan and Pinar Duygulu. Recognizing objects and scenes in news videos. In *Proceedings of International Conference on Image and Video Retrieval (CIVR 2006)*, 2006.
7. David Blei and Michael Jordan. Modeling annotated data. In *26th International Conference on Research and Development in Information Retrieval (SIGIR'03), New York*, 2003.
8. J. Jeon, V. Lavrenko, and R. Manmatha. Automatic image annotation and retrieval using cross-media relevance models. In *Proceedings of the ACM SIGIR Conference on Research and Development in Information Retrieval*, pages 119–126, 2003.
9. G. Salton, C. Buckley, and C. T. Yu. An evaluation of term dependence models in information retrieval. In *SIGIR '82: Proceedings of the 5th annual ACM conference on Research and development in information retrieval*, pages 151–173, New York, NY, USA, 1982. Springer-Verlag New York, Inc.
10. William M. Wells, Paul Viola, Hideki Atsumi, and Shin Naka jima. Multi-modal volume registration by maximization of mutual information. *Medical Image Analysis*, 1(1):35–51, 1996.
11. T. Butz and J.-Ph. Thiran. Multi-modal signal processing: An information theoretical framework. Technical Report 02.01, Signal Processing Institute (ITS), Swiss Federal Institute of Technology (EPFL), 2002.
12. G. Salton. *The SMART Retrieval System*. Prentice Hall, 1971.
13. Helmut Schmid. Probabilistic part-of-speech tagging using decision trees. In *International Conference on New Methods in Language Processing*, Manchester, UK, 1994. unknown.

A Study of Global Inference Algorithms in Multi-document Summarization

Ryan McDonald

Google Research
76 Ninth Avenue, New York, NY 10011
ryanmcd@google.com

Abstract. In this work we study the theoretical and empirical properties of various global inference algorithms for multi-document summarization. We start by defining a general framework for inference in summarization. We then present three algorithms: The first is a greedy approximate method, the second a dynamic programming approach based on solutions to the knapsack problem, and the third is an exact algorithm that uses an Integer Linear Programming formulation of the problem. We empirically evaluate all three algorithms and show that, relative to the exact solution, the dynamic programming algorithm provides near optimal results with preferable scaling properties.

1 Introduction

Automatically producing summaries from large sources of text is one of the oldest studied problems in both IR and NLP [7,13]. The expanding use of mobile devices and the proliferation of information across the electronic medium makes the need for such technology imperative. In this paper we study the specific problem of producing summaries from clusters of related documents – commonly known as multi-document summarization. In particular, we examine a standard paradigm where summaries are built by extracting relevant textual units from the documents [5,9,11,18].

When building summaries from multiple documents, systems generally attempt to optimize three properties,

- **Relevance:** Summaries should contain informative textual units that are relevant to the user.
- **Redundancy:** Summaries should not contain multiple textual units that convey the same information.
- **Length:** Summaries are bounded in length.

Optimizing all three properties jointly is a challenging task and is an example of a *global inference problem*. This is because the inclusion of relevant textual units relies not only on properties of the units themselves, but also properties of every other textual unit in the summary. Unlike single document summarization, redundancy is particularly important since it is likely that textual units from different documents will convey the same information. Forcing summaries to obey a length constraint is a common set-up in summarization as it allows for a fair empirical comparison between different possible

G. Amati, C. Carpineto, and G. Romano (Eds.): ECIR 2007, LNCS 4425, pp. 557–564, 2007.

outputs [1,12]. It also represents an important "real world" scenario where summaries are generated in order to be displayed on small screens, such as mobile devices.

The global inference problem is typically solved in two ways. The first is to optimize relevance and redundancy separately. For example, the work of McKeown et al. [14] presents a two-stage system in which textual units are initially clustered, and then representative units are chosen from each cluster to be included into the final summary. The second approach is to treat the problem truly as one of global inference and optimize all criteria in tandem. Goldstein et al. [9] presented one of the first global models through the use of the maximum marginal relevance (MMR) criteria, which scores sentences under consideration as a weighted combination of relevance plus redundancy with sentences already in the summary. Summaries are then created with an approximate greedy procedure that incrementally includes the sentence that maximizes this criteria. More recently, Filatova and Hatzivassiloglou [8] described a novel global model for their event-based summarization framework and showed that inference within it is equivalent to a known NP-hard problem, which led to a greedy approximate algorithm with proven theoretical guarantees. Daumé et al. [6] formulate the summarization problem in a supervised structured learning setting and present a new learning algorithm that sets model parameters relative to an approximate global inference algorithm.

In this work we start by defining a general summarization framework. We then present and briefly analyze three inference algorithms. The first is a greedy approximate method that is similar in nature to the MMR algorithm of Goldstein et al. [9]. The second algorithm is an approximate dynamic programming approach based on solutions to the knapsack problem. The third algorithm uses an Integer Linear Programming (ILP) formulation that is solved through a standard branch-and-bound algorithm to provide an exact solution. We empirically evaluate all three algorithms and show that, relative to the exact solution, the dynamic programming algorithm provides competitive results with preferable scaling properties.

2 Global Inference

As input we assume a document collection $\boldsymbol{D} = \{D_1, \ldots, D_k\}$. Each document contains a set of textual units $D = \{t_1, \ldots, t_m\}$, which can be words, sentences, paragraphs, etc. For simplicity, we represent the document collection as the set of all textual units from all the documents in the collection, i.e., $\boldsymbol{D} = \{t_1, \ldots, t_n\}$ where $t_i \in \boldsymbol{D}$ iff $\exists\ t_i \in D_j \in \boldsymbol{D}$. We let $S \subseteq \boldsymbol{D}$ be the set of textual units constituting a summary.

We define two primary scoring functions,

1. *Rel*(i): The relevance of textual unit t_i participating in the summary.
2. *Red*(i, j): The redundancy between textual units t_i and t_j. Higher values correspond to higher overlap in content.

These scoring functions are completely arbitrary and should be defined by domain experts. For instance, scores can include a term to indicate similarity to a specific query for query-focused summarization or include terms involving entities, coherence, domain specific features, etc. Scores can also be set by supervised learning algorithms when training data is available [18]. Finally, we will define the function $l(i)$ to indicate the length of textual unit t_i. Length is also arbitrary and can represent characters, words, phrases, etc. As in most summarization studies, we assume that as input an integer K, for which the length of any valid summary cannot exceed.

Formally the multi-document summarization inference problem can be written as:

$$S = \arg\max_{S \subseteq D} s(S) = \arg\max_{S \subseteq D} \sum_{t_i \in S} Rel(i) - \sum_{t_i, t_j \in S,\ i<j} Red(i,j) \quad (1)$$

$$\text{such that} \sum_{t_i \in S} l(i) \leq K$$

We refer to $s(S)$ as the *score* of summary S. We assume that redundancy scores are symmetric and the summation of scores is over $i < j$ to prevent counting each more than once. If desired, we could unevenly weight the relevance and redundancy scores to prefer one at the expense of the other. It is also worth mentioning that the redundancy factors in Equation 1 are pairwise. This is a slight deviation from many systems, in which the redundancy of unit t_i is calculated considering the rest of the summary in its entirety. For now, we have simplified the redundancy factor to a sum of pairwise relationships because it will allow us to define an Integer Linear Programming formulation in Section 2.1. In turn, this will allow us to compare our approximate algorithms to an upper bound in performance.

It can be shown that solving the inference problem in Equation 1 is NP-hard. For space reasons we omit the proof. It is not difficult to show that the major source of intractability are the redundancy terms from. When the redundancy terms are removed, the problem is still NP-hard and can be shown to be equivalent to the *0-1 knapsack problem* [4]. There does exist a $O(Kn)$ algorithm for solving the knapsack problem, but this only makes it pseudo-polynomial, since K is represented as $\log K$ bits in the input. However, for the summarization problem K is typically on the order of hundreds, making such solutions feasible. We will exploit this fact in Section 2.1.

2.1 Global Inference Algorithms

Greedy Algorithm. A simple approximate procedure to optimizing Equation 1 is to begin by including highly relevant textual units, and then to iteratively add new units that maximize the objective. This algorithms is outlined in Figure 1a and is a variant of MMR style algorithms. The advantage of this algorithm is that it is simple and computationally efficient. The runtime of this algorithm is in the worst case $O(n \log n + Kn)$ due to the sorting of n items and because each iteration of the loop takes $O(n)$ and the loop will iterate at most K times. This assumes the unlikely scenario when all sentences have a length of one. In practice, the loop only iterates a small number of times. We also assume that calculating $s(S)$ is $O(1)$ when it is really a function of loop iterations, which again makes it negligible.

It is not difficult to produce examples for which this greedy procedure will fail. In particular, the choice of including the most relevant sentence in the summary

Input: $\boldsymbol{D} = \{t_1, \ldots, t_n\}$, K

(a) Greedy Algorithm

1. sort $\boldsymbol{D}$ so that $Rel(i) > Rel(i+1)\ \forall i$
2. $S = \{t_1\}$
3. while $\sum_{t_i \in S} l(i) < K$
4. $\quad t_j = \arg\max_{t_j \in \boldsymbol{D}-S} s(S \cup \{t_j\})$
5. $\quad S = S \cup \{t_j\}$
6. return S

(b) Knapsack Algorithm

1. $S[i][0] = \{\}\ \forall 1 \leq i \leq n$
2. for i: $1 \ldots n$
3. $\quad$ for k: $1 \ldots K$
4. $\quad\quad S' = S[i-1][k]$
5. $\quad\quad S'' = S[i-1][k-l(i)] \cup \{t_i\}$
6. $\quad\quad$ if $s(S') > s(S'')$ then
7. $\quad\quad\quad S[i][k] = S'$
8. $\quad\quad$ else
9. $\quad\quad\quad S[i][k] = S''$
10. return $\arg\max_{S[n][k],\ k \leq K} s(S[n][k])$

Fig. 1. (a) A greedy approximate algorithm. (b) A dynamic programming algorithm based on a solution to the knapsack problem.

(Figure 1a, line 2) can cause error propagation. Consider the case of a very long and highly relevant sentence. This sentence may contain a lot of relevant information, but it may also contain a lot of noise. Including this sentence in the summary will help maximize relevance at the cost of limiting the amount of remaining space for other sentences.

Dynamic Programming Algorithm. To alleviate this problem we devise a dynamic programming solution. Recall that the input to the problem is a set of textual units, $\boldsymbol{D} = \{t_1, \ldots, t_n\}$, and an integer K. Let $S[i][k]$, where $i \leq n$ and $k \leq K$, be a high scoring summary of exactly length k that can only contain textual units in the set $\{t_1, \ldots, t_i\}$. Figure 1b provides an algorithm for filling in this table. This algorithm is based on a solution to the 0-1 knapsack problem [4]. In that problem the goal is to fill a knapsack of capacity K with a set of items, each having a certain weight and value. The optimal solution maximizes the overall value of selected items without the total weight of these items exceeding K. Clearly if one could ignore the redundancy terms in Equation 1, the summarization problem and knapsack problem would be equivalent, i.e., value equals relevance and weight equals length. Of course, redundancy terms are critical when constructing summaries and we cannot ignore them.

The crux of the algorithm is in lines 4-10. To populate $S[i][k]$ of the table, we consider two possible summaries. The first is $S[i-1][k]$, which is a high scoring summary of length k using textual units $\{t_1, \ldots, t_{i-1}\}$. The second is a high scoring summary of length $k - l(i)$ plus the current unit t_i. $S[i][k]$ is then set to which ever one has highest score. The knapsack problem is structured so that the principle of optimality holds. That is, if for $i' < i$ and $k' \leq k$, if $S[i'][k']$ stores the optimal solution, then $S[i][k]$ will also store the optimal solution. However, the additional redundancy factors in the multi-document summarization problem, which are included in the score calculations of line 6, break this principle making this solution only approximate for our purposes.

The advantage of using a knapsack style algorithm is that it eliminates the errors caused by the greedy algorithm inserting longer sentences and limiting the space for future inclusions.The runtime of this algorithm is $O(Kn)$ if we again assume that $s(S) \in O(1)$. However, this time K is not a worst-case scenario, but a fixed lower-bound on

maximize $\sum_i \alpha_i Rel(i) - \sum_{i<j} \alpha_{ij} Red(i,j)$

such that $\forall i, j$:
(1) $\alpha_i, \alpha_{ij} \in \{0, 1\}$
(2) $\sum_i \alpha_i l(i) \leq K$
(3) $\alpha_{ij} - \alpha_i \leq 0$
(4) $\alpha_{ij} - \alpha_j \leq 0$
(5) $\alpha_i + \alpha_j - \alpha_{ij} \leq 1$

Fig. 2. ILP formulation of global inference

runtime. Even so, most summarization systems typically set K on the order of 100 to 500, making such solution easily computable (see Section 3). Note also that the correctness of the algorithm as given in Figure 1 is based on the assumption that there is a valid summary of every length $k \leq K$. It is not difficult to modify the algorithm and remove this assumption by checking that both S' and S'' truly have a length of k.

One additional augmentation that can be made to both the greedy and knapsack algorithms is the inclusion of a beam during inference. This was implemented but found to have little impact on accuracy unless a beam of substantial size was used.

ILP Formulation. It would be desirable to compare the previous two algorithms with an exact solution to determine how much accuracy is lost due to approximations. Fortunately there is a method to do this in our framework through the use of Integer Linear Programming (ILP). ILP techniques have been used in the past to solve many intractable inference problems in both IR and NLP. This includes applications to relation and entity classification [17], sentence compression [3], temporal link analysis [2], as well as syntactic and semantic parsing [15,16].

An ILP is a constrained optimization problem, where both the cost function and constraints are linear in a set of integer variables. Solving arbitrary ILPs is an NP-hard problem. However, ILPs are a well studied optimization problem with efficient branch and bound algorithms for finding the optimal solution. Modern commercial ILP solvers can typically solve moderately large optimizations in a matter of seconds. We use the GNU Linear Programming kit[1], which is a free optimization package.

The multi-document global inference problem can be formulated as the ILP in Figure 2. In this formulation we include indicator variables α_i and α_{ij}, which are 1 when a textual unit or pairs of textual units are included in a summary. The goal of the ILP is to set these indicator variables to maximize the payoff subject to a set of constraints that guarantee the validity of the solution. The first constraint simply states that the indicator variables are binary. The second constraint states that for all sentences included in the summary, the sum of their lengths must be less than our predefined maximum. Constraints (3) to (5) ensure a valid solution. Constraints (3) and (4) simply state that if the summary includes both the units t_i and t_j then we have to include them individually as well. Constraint (5) is the inverse of (3) and (4).

2.2 Implementation Details

When implementing each algorithm it is important for the scale of the score functions to be comparable. Otherwise, the algorithms will naturally favor either relevancy or

[1] http://www.gnu.org/software/glpk/

redundancy. Furthermore, there are quadratically many redundancy factors in a summary score compared to relevance factors. Depending on the scoring functions this can lead to summaries with a small number of very long sentences or a lot of very short sentences. One way to avoid this is to add new constraints specifying a desired range for sentence lengths. Alternatively, we found that replacing every score with its z-score alleviated many of these problems since that guaranteed both positive and negative values. When scores are predominantly negative, then the algorithms return summaries much shorter than K. This is simply fixed by changing the constraints to force summary lengths to be between K-c and K, where c is some reasonably sized constant.

In the ILP formulation, the number of constraints is quadratic in the total number of textual units. Furthermore, the coefficient matrix of this problem is not unimodular [17]. As a result, the ILP algorithm does not scale well. To alleviate this problem, each algorithm passed through a preprocessing stage that sorted all textual units by relevance. Every textual unit not in the top 100 was discarded as unlikely to be in the final summary. In this way, all algorithms ran under the same conditions.

3 Experiments

In this study we used sentences as textual units. Each textual unit, document and document collection is represented as a bag-of-words vector with $tf*idf$ values. Length bounds are always in terms of words. In addition to the three algorithms described in this paper, we also ran a very simple baseline that is identical to the greedy algorithm, but does not include redundancy when scoring summaries.

We ran two primary sets of experiments, the first is on generic summarization and the second query-focused summarization. Results are reported using the ROUGE evaluation package [12]. ROUGE is a n-gram recall metric for an automated summary relative to a set of valid reference summaries. We report ROUGE-1 and ROUGE-2 scores, which capture unigram and bigram recall.

In the generic setting, a system is given a document collection $\boldsymbol{D}$, and length bound K, and is asked to produce a summary that is most representative of the entire document collection. For these experiments, we used the DUC 2002 data set [10]. This data set contained 59 document collections, each having at least one manually created summary for lengths 50, 100, 200. We define the score functions as follows:

$$\begin{aligned} \mathit{Rel}(i) &= POS(t_i, D)^{-1} + SIM(t_i, \boldsymbol{D}) \quad \text{(where } t_i \in D \text{ and } D \in \boldsymbol{D}\text{)} \\ \mathit{Red}(i,j) &= SIM(t_i, t_j) \end{aligned}$$

where $POS(t, D)$ is the position of textual unit t in document D and $SIM(a, b)$ is the cosine similarity between two vectors. Relevance scores prefer sentences that are near the beginning of documents and are maximally informative about the entire document collection. Again, these score functions are general and we only use these particular scoring criteria because the data is drawn from news sources.

Results are shown in Table 1a. The first thing to note is that incorporating redundancy information does improve scores, verifying previous work [8,9]. Next, we see that scores for the sub-optimal knapsack algorithm are very near scores for the exact ILP algorithm and are even sometimes slightly better. This is due to the fact that redundancy

Table 1. (a) Results for generic summarization experiments using DUC 2002 data set. Each cell contains the ROUGE-1 and 2 scores (R1 / R2). (b) Results for query-focused summarization experiments using DUC 2005 data set.

(a)

	Summary Length		
	50	100	200
Baseline	26.6 / 5.3	33.0 / 6.8	39.4 / 9.6
Greedy	26.8 / 5.1	33.5 / 6.9	40.1 / 9.5
Knapsack	27.9 / 5.9	34.8 / 7.3	41.2 / 10.0
ILP	28.1 / 5.8	34.6 / 7.2	41.5 / 10.3

(b)

Baseline	34.4 / 5.4
Greedy	35.0 / 5.7
Knapsack	35.7 / 6.2
ILP	35.8 / 6.1

scores are more influential in the ILP solution. Highly relevant, but semantically different, sentences will often contain identical terms (i.e., person names or places). These sentences are then forced to compete with one another when constructing the summary, when it may be desirable to include them both. The final point we will make is that the greedy algorithms performance is consistently lower than the knapsack algorithm. An analysis of the resulting summaries suggests that indeed long sentences are getting included early, making it difficult to add relevant sentences later in the procedure.

The query-focused setting requires summaries to be relevant to a particular query that has been supplied by the user. For these experiments, we used the DUC 2005 data sets [5]. This data consists of 50 document collections, each with a corresponding query. For each collection and query, multiple manually constructed summaries of 250 words were provided. The redundancy score of the system remained unchanged from the previous experiment. However, for a document collection $\boldsymbol{D}$ and query Q, the relevance score was changed to the following:

$$Rel(i) = SIM(t_i, Q) + SIM(t_i, \boldsymbol{D})$$

Thus, relevance is an equally weighted combination of similarity to the query and similarity to the entire document collection. Results are shown in Table 1b. Again we see that the knapsack algorithm outperforms the greedy algorithm and has a score comparable to the ILP system.

Another important property of these systems is the efficiency in which they produce a summary relative to a document collection. For document collections with 50 textual units (i.e., $|\boldsymbol{D}| = 50$), the greedy, knapsack and ILP algorithms take 5, 8 and 25 seconds on average to produce a summary. For document collections with 100 textual units, the algorithms take 7, 16 and 282 seconds. This trend for the ILP solution as $|\boldsymbol{D}|$ grows, making it infeasible to use for large real world data sets.

4 Conclusions

In this work we studied three algorithms for global inference in multi-document summarization. We found that a dynamic programming algorithm based on solutions to the knapsack problem provided optimal accuracy and scaling properties, relative to both a greedy algorithm and an exact algorithm that uses Integer Linear Programming.

References

1. Document Understanding Conference (DUC). http://duc.nist.gov
2. P. Bramsen, P. Deshpande, Y.K. Lee, and R. Barzilay. Inducing temporal graphs. In *Proceedings of the Empirical Methods in Natural Language Processing (EMNLP)*, 2006.
3. J. Clarke and M. Lapata. Models for sentence compression: A comparison across domains, training requirements and evaluation measures. In *Proceedings of the Annual Meeting of the Association for Computational Linguistics (ACL)*, 2006.
4. T.H. Cormen, C.E. Leiserson, and R.L. Rivest. *Introduction to Algorithms*. MIT Press/ McGraw-Hill, 1990.
5. H.T. Dang. Overview of duc 2005. In *Proceedings of the Document Understanding Conference (DUC)*, 2005. http://duc.nist.gov
6. Hal Daumé III, John Langford, and Daniel Marcu. Search-based structured prediction. 2006. In Submission.
7. H.P. Edmundson. New methods in automatic extracting. *Journal of the Association for Computing Machinery*, 1(23), 1968.
8. E. Filatova and V. Hatzivassiloglou. A formal model for information selection in multi-sentence text extraction. In *Proceedings of the International Conference on Computational Linguistics (COLING)*, 2004.
9. J. Goldstein, V. Mittal, J. Carbonell, and M. Kantrowitz. Multi-document summarization by sentence extraction. *In Proceedings of the ANLP/NAACL Workshop on Automatic Summarization*, 2000.
10. U. Hahn and D. Harman, editors. *Proceedings of the Document Understanding Conference (DUC)*, 2002. http://duc.nist.gov
11. J. Kupiec, J. Pedersen, and F. Chen. A trainable document summarizer. In *Proceeding of the Annual Conference of the ACM Special Interest Group on Information Retrieval (SIGIR)*, 1995.
12. C.Y. Lin and E. Hovy. Automatic evaluation of summaries using n-gram cooccurrence statistics. In *Proceedings of the Joint Conference on Human Language Technology and North American Chapter of the Association for Computational Linguistics (HLT/NAACL)*, 2003.
13. P.H. Luhn. The automatic creation of literature abstracts. *IBM Journal of Research and Development*, 2(2), 1959.
14. K. McKeown, J. Klavansn, V. Hatzivassiloglou, R. Barzilay, and Eleazar Eskin. Towards multidocument summarization by reformation: Progress and prospects. In *Proceedings of the Annual Conference of the American Association for Artificial Intelligence (AAAI)*, 1999.
15. V. Punyakanok, D. Roth, W. Yih, and D. Zimak. Semantic role labeling via integer linear programming inference. In *Proceedings of the International Conference on Computational Linguistics (COLING)*, 2004.
16. S. Riedel and J. Clarke. Incremental integer linear programming for non-projective dependency parsing. In *Proceedings of the Empirical Methods in Natural Language Processing (EMNLP)*, 2006.
17. D. Roth and W. Yih. A linear programming formulation for global inference in natural language tasks. In *Proceedings of the Conference on Computational Natural Language Learning (CoNLL)*, 2004.
18. S. Teufel and M. Moens. Sentence extraction as a classification task. In *Proceedings of the ACL/EACL Workshop on Intelligent Scalable Text Summarizaion*, 1997.

Document Representation Using Global Association Distance Model

José E. Medina-Pagola, Ansel Y. Rodríguez, Abdel Hechavarría, and José Hernández Palancar

Advanced Technologies Application Center (CENATAV),
7th Avenue # 21812 % 218 and 222, Siboney, Playa, Havana City, Cuba
{jmedina,arodriguez,ahechavarria,palancar}@cenatav.co.cu

Abstract. Text information processing depends critically on the proper representation of documents. Traditional models, like the vector space model, have significant limitations because they do not consider semantic relations amongst terms. In this paper we analyze a document representation using the association graph scheme and present a new approach called Global Association Distance Model (GADM). At the end, we compare GADM using K-NN classifier with the classical vector space model and the association graph model.

1 Introduction

Nowadays, due to the rapid scientific and technological advances, there are great creation, storage and data distribution capacities. This situation has boosted the necessity of new tools to transform this big amount of data into useful information or knowledge for decision makers. This transformation process is known as Knowledge Discovery in Databases (KDD).

Recent studies and analyses have concluded that complex data require a high number of components to be completely described. this data has to be embedded in high-dimensional spaces (from tens to thousands dimensions). Examples are spectrophotometer data, gene expression data, pictures and texts. In this paper, we focus our analysis on textual data and their representation.

The representation model that is used affects critically almost any text processing task; like information retrieval, classification, clustering, summarization, question-answering, etc. The vector space model is the classic one and by far the most used model. Nevertheless, some studies have shown that the weakness of this model is to leave out semantic complexity of the textual data.

As terms appear related to other terms in any document, their meanings strongly depend on the meanings of the surrounding terms; even more, term meanings emerge from mutual sense reinforcement. If we assume that sentences are the main semantic unit in texts, then mutual sense reinforcements or associations amongst their terms should be the strongest possible. Nevertheless, it is well known that these reinforcements or associations are feasible in other contexts as, for instance, paragraphs or groups of them. The association graph scheme is an approach that includes this consideration.

G. Amati, C. Carpineto, and G. Romano (Eds.): ECIR 2007, LNCS 4425, pp. 565–572, 2007.

In this paper we propose a new document representation model using the mutual sense reinforcement among terms, called Global Association Distance Model (GADM). Section 2 is dedicated to related work. In section 3 we analyze the strengths and weaknesses of the Association Graph Model. In section 4 we describe GADM. Experimental results are discussed in the last section.

2 Related Work

Document categorization, clustering and information retrieval tasks are often based on a good data representation. At a high level, the manipulation of textual data can be described as a series of processing steps that transform the original document representation to another one, simpler and more useful to be processed automatically by computers.

This usually involves enriching the document content by adding information, using background knowledge, normalizing terms, etc. At the start of the process, the textual data may exist as a paper, for instance, and the final representation of the preprocessing could be a straight ASCII text enriched with some additional information. This preprocessing final representation is used to represent data in a useful way for computer calculation.

These terms could be organized in different forms but, in general, they are considered as groups or bags of words, usually structured using a vector space model [1]. In this model, the term sequences, or their syntactical relations, are not analyzed; therefore, they are considered as unigrams supposing an independence of their occurrences.

In the vector space model, each document is a vector of terms. The values of these vectors could be assumed as weights, considering the following interpretations [2]:

- Boolean: Each term is associated with a Boolean value representing whether it is present or not in a document.
- Tf (Term Frequency): Each term is associated with a frequency of appearance in a document, absolute or normalized.
- Tf-Idf (Term Frequency - Inverse Document Frequency): The term is associated with its frequency, adjusted by the inverse of the number of documents containing each term.

These vectors of terms are used in a second stage, among other tasks, to analyze the similarities between documents, or groups of them, using different measures as the cosine, applied to the angle between the vectors, defined as [2]:

$$sim(d_i, d_j) = cos(d_i, d_j) = \frac{(d_i \bullet d_j)}{||d_i|| * ||d_j||} = \frac{\sum w_{ir} * w_{jr}}{\sqrt{\sum w_{ir}^2 * \sum w_{jr}^2}}, \tag{1}$$

where d_i, d_j are the vectors of documents i, j, $||di||$, $||dj||$ the norms of the vectors, and w_{ir}, w_{jr} are the term weights in the vectors di, dj, respectively. Other common measures are Dice and Jaccard coefficients.

Alternative approaches to the vector space model are the language models. These consider the probabilities of occurrence of a phrase S in a language M, indicated by $P(S/M)$. However, the phrases are usually reduced to one term, assuming again unigrams and independence among them. An example of this model is the Kullback-Leibler Divergence (a variation of the cross-entropy), defined as:

$$D(d_i||d_j) = \sum P(t/d_i) log \frac{P(t/d_i)}{P(t/d_j)}. \quad (2)$$

This expression could be combined in both directions to obtain a similarity measure, as was pointed out by *Feldman and Dagan* [3].

Other implementation is the proposal of Kou and Gardarin [4]. This proposal is a kind of language model, considering the similarities between two documents as:

$$sim(d_i, d_j) = d_i \bullet d_j = \sum_r w_{ir} w_{jr} + \sum_r \sum_{s \neq r} w_{ir} w_{js} (t_r \bullet t_s), \quad (3)$$

where w_{ir} and w_{js}, using Kou-Gardarin terminology, are the term weights in document vectors d_i, d_j, respectively, and $(t_r \bullet t_s)$ is the a priori correlation between terms t_r and t_s. Actually, the authors included in the first part of the expression the self-correlation in t_r, considering that $(t_r \bullet t_r) = 1$. The authors propose the estimation of the correlation through a training process. As can be noticed, those correlations express the probabilities $P(t_r, t_s/M)$ of phrases containing the terms t_r, t_s in a language M. Besides, that expression could be reduced to the cosine measure (normalized by the length of the vectors) if the term independence is considered and, for that reason, the correlation $(t_r \bullet t_s)$ is zero.

Another vector space model is the Topic-based Vector Space Model (TVSM) [5]. The basic premise of the TVSM is the existence of a space R which only has positive axis intercepts. Each dimension of R represents a fundamental topic. It is assumed that all fundamental topics are independent from the others. In this model, each document is represented as a vector of term-vectors; each term-vector is a set of weights between the term and the fundamental topics.

The approaches mentioned above are variants of the Generalized Vector Space Model proposed by *S.K.M Wong et al.* [6]. In their work, they expressed that there was no satisfactory way of computing term correlations based on automatic indexing scheme.

We believe that up to the present time that limitation has not been solved yet. Although several authors have proposed different methods of recognizing term correlations in the retrieval process, those methods try to model the semantic dimension by a global distribution of terms, but not with a local evaluation of documents.

In general, it could be assumed that the better the semantic representation of the information retrieved and discriminated, the better this information is mined.

3 Association Graph Model

The Association Graph Model assumes that a same term in two documents could designate different concepts [7]. Besides, two terms could have different relations, according to the subject of each document, and those relations could exist only in the context of some documents, forming a specific group, and regardless of the relations in a global dimension or language.

In order to model the relation between two terms in a document, the shortest physical distance between those terms was considered. The basic premise of this model can be expressed as follows: *Two documents should be closer if the number of common terms is greater and the shortest physical distances among them in each document are similar.*

Considering the distance by paragraph, without ignoring the natural co-occurrence when appearing in the same sentence, and considering: (p_r, n_r), (p_s, n_s), the paragraph and sentence numbers of terms t_r and t_s respectively, the physical distance between these terms in a document i is defined as follows:

$$D^i_{rs} = \begin{cases} 1 & (r = s) \vee [(p_r = p_s) \wedge (n_r = n_s)] \\ |p_r - p_s| + 2 & Othercase \end{cases} . \tag{4}$$

Observe that the minimum value of D^i_{rs}, as could be expected, is not zero, but one. This consideration is only a convenient assumption to expressions defined further on.

According to this, a document is modeled by a graph, where the vertices are the distinguished terms and the edges are their relations, weighted by their distances. Notice that this is a full connected graph, where any term has some relation (stronger or not according to the distance) with the others.

In spite of the fact that the physical distance between two terms is the basic notion that supports the association graph scheme, the model proposed included the frequency of the terms.

Therefore, *Association Graph Model* was defined as a graph representation weighed by vertex, considering the weights of the distinguished terms, and by edge, considering the physical distance between the adjacent terms. As the main components of any graph are the edges, and trying to synthesize in a couple of value the strength of the association, the vector A^i_{rs} was proposed as the edge weight of the related terms t_r, t_s in a document i, defined as:

$$A^i_{rs} = \left(\frac{w^i_r}{\sqrt{D^i_{rs}}}, \frac{w^i_s}{\sqrt{D^i_{rs}}} \right) \tag{5}$$

where w^i_r and w^i_s are the weights of the terms t_r and t_s, respectively, in a document i.

Since the weight A^i_{rs} is a two-dimensional vector, the association strength is evaluated as the Euclidean norm $||A^i_{rs}||$. In this case, the strength is greater if the term weights are greater and the distance between the terms is shorter. Besides, the upper value of the edge weight is (w^i_r, w^i_s) and the lower tends to $(0, 0)$ when the distance is very long.

As an association graph does not possess a structural or spatial representation, the similarity between two documents was defined considering each graph as a set of edges. The measure proposed is called Weighted Coverage and it is defined as follows.

$$sim(d_i, d_j) = \frac{1}{2} \frac{\sum\limits_{t_r, t_s \in T_{ij}} S^{ij}_{rs} ||A^i_{rs}||}{\sum\limits_{t_r, t_s \in T_i} ||A^i_{rs}||} + \frac{1}{2} \frac{\sum\limits_{t_r, t_s \in T_{ij}} S^{ij}_{rs} ||A^j_{rs}||}{\sum\limits_{t_r, t_s \in T_j} ||A^j_{rs}||}, \qquad (6)$$

The weight S^{ij}_{rs} represents a similarity measure between A^i_{rs} and A^j_{rs}. This weight is defined as:

$$S^{ij}_{rs} = cos(A^i_{rs}, A^j_{rs}) * (1 - \frac{1}{2}(||A^i_{rs}|| - ||A^j_{rs}||)^2) \qquad (7)$$

where the first part of the expression represents the cosine between those vectors, defined in a similar way as (1).

Although this model shows good performances, at least in classification tasks, it has some drawbacks: this model intrinsically keeps the frequency point of view, conserving the basis of the vector space model; the experimental results of this model are not clearly better than in the vector space model; and the computational, temporal and spatial costs of this model is relatively higher than vector space model. Aiming at solving these drawbacks, other assumptions, also based on the association graph scheme, were analyzed.

4 Global Association Distance Model

In order to evaluate the effect of the frequency over the weighted coverage measure, an experiment was performed, considering and removing the frequency in (5). But, this experiment produced an amazing result: the performance of the association graph model when it includes term frequency was very similar to the one removing the frequency. These results suggested that frequency does not seem to be such an important term feature, and only the physical distance between terms is enough.

The above observation led us to analyze the basic premise of the association graph scheme and to consider other assumptions that simplify the complexity of the association graph model.

We know that in the classical vector space model the term relevance is assumed by the number of occurrences in a document. Nevertheless, if we want to include its semantic context, we should consider not only its occurrence but also its co-occurrences (and association strength) with other terms in sentences, paragraphs and so on. For that reason, we keep the expression (4), taking the formula $1/\sqrt{D_{rs}}$ - we omited superindex i only for simplicity considerations - as a similarity expression for evaluation purposes.

In order to simplify the association graph model, based on a set of weigthed edges, we consider the preliminary ideas of the vector space model proposed by *Salton*, i.e. to consider "a document space consisting of documents ... , each identified by one or more index terms ... weighted according to their importance" [1]. In our case, the importance of a term t_r in a document d is related to its global association strength and can be calculated as follows.

$$g_{t_r} = \sum_{t_s \in d} \frac{1}{\sqrt{D_{rs}}} \tag{8}$$

Therefore, *Global Association Distance Model* can be defined as a vector space model where each term is weighted by their global association strength. So, a document d can be modeled by a vector $\overrightarrow{d}$.

$$\overrightarrow{d} = (g_{t_1}, ..., g_{t_n}) \tag{9}$$

It is not difficult to understand that the similarity measure in this model can be calculated by any of the measures defined for the classic vector space model. Up to the present time, we have only considered the cosine measure (1).

5 Experiments and Analysis

In order to evaluate the proposed model, TREC-5[1] in Spanish and Reuter-21758[2] in English corpora were used, considering topic classification. TREC-5 is a AFP news corpus with 695 news published during 1994-1995 and classified in 22 topics. Reuter 21758 is a Reuters LTD news corpus with 21758 news classified in 119 topics; due to memory restrictions we randomly selected 800 news items. All news were preprocessed, normalizing and lemmatizing their terms with TreeTager [8].

We compared the Global Association Distance Model (GADM) with the standard Vector Space Model (VSM) and the Association Graph Model (AGM). A K-NN classifier was used with $k = 1, 5, 10, 15, 20$. The experimental evaluation was done using k-fold with $k = 10$.

Three evaluation measures of performance are commonly used for a single category or topic [9]: Precision is the quotient of the correctly assigned and the ones assigned to the category, Recall is the quotient of the correctly assigned and the ones belonging to the category; and $F - measure = \frac{2*Recall*Precision}{Recall+Precision}$.

For evaluating the performance average across categories, there are two conventional methods:

- Macro-averaging performance: Scores are computed by a simple average of the performance measures for each category. Macro-averaged performance score gives equal weights to every category or topic, regardless of its frequency.

[1] http://trec.nist.gov/pubs/trec5

[2] ftp://canberra.cs.umass.edu/pub/reuters

– Micro-averaging performance: Scores are computed by first accumulating the corresponding variables in the per-category expressions, and then using those global quantities to compute the scores. Micro-averaged performance score gives equal weights to every document.

Tables 1 and 2, and their corresponding Fig. 1(a) and Fig.1(b), show the experiment results.

As can be noticed 1, GADM mostly outperforms both models: VSM and AGM, for all values of k. This shows that the use of a vector of global association strengths can actually improve the performance of document processing, at least for classification tasks using K-NN. Although these results are only preliminaries,

Table 1. AFP classification results (F-measure)

k	Macro-averaging performance			Micro-averaging performance		
	VSM	AGM	GADM	VSM	AGM	GADM
1	0.618342	0.621376	0.628072	0.852003	0.864963	0.874922
5	0.613379	0.618476	0.619129	0.872003	0.876437	0.873555
10	0.597129	0.576622	0.597338	0.859100	0.856458	0.864963
15	0.574918	0.551977	0.583869	0.844880	0.848993	0.863341
20	0.538896	0.523529	0.551356	0.828865	0.831809	0.846111

Table 2. Reuters classification results (F-measure)

k	Macro-averaging performance			Micro-averaging performance		
	VSM	AGM	GADM	VSM	AGM	GADM
1	0.042262	0.042397	0.045202	0.662500	0.642500	0.682500
5	0.035451	0.034263	0.039288	0.703750	0.686250	0.717500
10	0.031761	0.028553	0.033216	0.710000	0.681250	0.731250
15	0.027341	0.026778	0.029069	0.705000	0.676250	0.716250
20	0.025304	0.023235	0.026642	0.698750	0.672500	0.706250

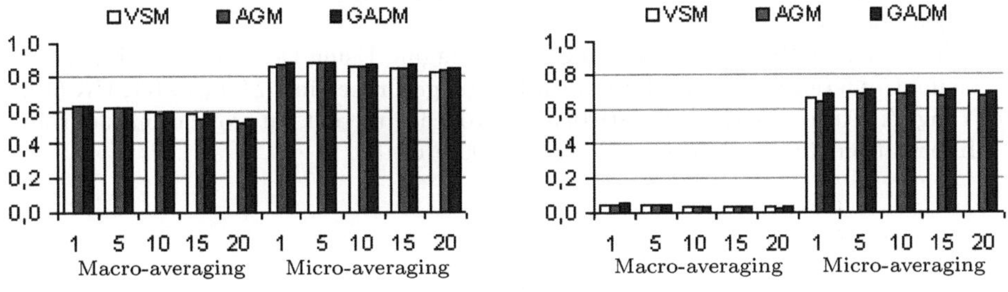

(a) AFP

(b) Reuters

Fig. 1. Classification results (F-measure)

they show that our proposed model presents a good behavior and seems to be a better solution considering a semantic approach.

6 Conclusion

In order to achieve a better document representation, we have proposed a new concept named global association strength, which uses semantic relations amongst terms in documents. Also, we presented a new document representation model named *Global Association Distance Model*, with a complexity no greater than the classic vector space model, and showing better results in documents classifications.

The experiment results have shown interesting improvements. Although other experiments must be done, the proposed model gives promising results and opens new branches of work.

References

1. Salton, G.: The SMART Retrieval System - Experiments in Automatic Document Processing. Prentice-Hall, Englewood Cliffs, New Jersey, (1971).
2. Berry, M.: Survey of Text Mining, Clustering, Classification and Retrieval. Springer, (2004).
3. Feldman, R., Dagan, I.: Knowledge Discovery in Textual Databases (KDT). In Proc. of the first International Conference on Data Mining and Knowledge Discovery, KDD'95, Montreal, (1995) 112–117.
4. Kou, H., Gardarin G.: Similarity Model and Term Association for Document Categorization. NLDB 2002, Lecture Notes in Computer Science, Vol. 2553, Springer-Verlag Berlin Heidelberg New York, (2002) 223–229.
5. Becker, J., Kuropka, D.: Topic-based Vector Space Model. In Proc. of Business Information Systems (BIS) 2003, (2003).
6. Wong, S.K.M, Ziarko W. and Wong, P.C.N.: Generalized Vector Space Model in Information Retrieval. Proc. of the 8th Int. ACM SIGIR Conference on Research and Development in Information Retrieval, New York, ACM 11 (1985).
7. Medina-Pagola, J.E., Guevara-Martinez, E., Hernández-Palancar, J., Hechavarría-Díaz, A., Hernández-León, R.: Similarity Measures in Documents using Association Graphs. In Proc. of CIARP 2005, Lecture Notes in Computer Science, Vol. 3773, (2005) 741–751.
8. Schmid, H.: Probabilistic Part-Of-Speech Tagging Using Decision Tree. In: International Conference on New Methods in Language Processing, Manchester, UK (1994)
9. Yang Y.: An evaluation of statistical approaches to text categorization. Journal of Information Retrieval, Vol. 1, No. 1/2, (1999) 67–88.

Sentence Level Sentiment Analysis in the Presence of Conjuncts Using Linguistic Analysis

Arun Meena and T.V. Prabhakar

Department of Computer Science and Engineering,
Indian Institute of Technology, Kanpur
208016 Kanpur, UP, India
{arunm, tvp}@iitk.ac.in

Abstract. In this paper we present an approach to extract sentiments associated with a phrase or sentence. Sentiment analysis has been attempted mostly for documents typically a review or a news item. Conjunctions have a substantial impact on the overall sentiment of a sentence, so here we present how atomic sentiments of individual phrases combine together in the presence of conjuncts to decide the overall sentiment of a sentence. We used word dependencies and dependency trees to analyze the sentence constructs and were able to get results close to 80%. We have also analyzed the effect of WordNet on the accuracy of the results over General Inquirer.

Keywords: Sentiment analysis, favorability analysis, text mining, information extraction, semantic orientation, text classification.

1 Introduction

In recent years, there has been a rapid growth of web-content, especially on-line discussion groups, review sites and blogs. These are highly personal and typically express opinions. To organize this information, automatic text categorization and identification of sentiment polarity is very useful. Most work done in this field has been focused on topic based categorization, which is sorting the documents according to their subject content.

Sentiment classification is a special case of text categorization problem, where the classification is done on the basis of attitude expressed by the authors. Sentiment analysis requires a deep understanding of the document under analysis because the concern here is how the sentiment is being communicated. Previous attempts in solving this problem, for example [2], [4] focused on the use of machine learning methods (N-gram, etc.), ignoring the importance of language analysis which is being used to communicate sentiments. Therefore, we need to find new methods to improve the sentiment classification exploring the linguistic techniques.

Our work differs from earlier work in four main aspects: (1) our focus is not on classifying each review as a whole but on classifying each sentence in a review. (2) We give more consideration/importance to the language properties of the sentence and in understanding the sentence constructs, for each sentence we recognize the subjects of the feeling and the feature being described. (3) We concentrate on the effects of conjunctions and sentence constructions which have not been researched for

G. Amati, C. Carpineto, and G. Romano (Eds.): ECIR 2007, LNCS 4425, pp. 573–580, 2007.

sentiment analysis. (4) Our method does not need a training set since it depends on linguistic analysis.

2 Previous Work

The cornerstone on sentiment analysis is Pang and Lee's 2002 paper [2]. The authors of that paper compare Naive Bayes, Maximum Entropy, and Support Vector Machine approaches to classify sentiment of movie reviews. They explain the relatively poor performance of the methods as a result of sentiment analysis requiring a deeper understanding of the document under analysis. Document level sentiment classification assumes the whole document to have a single overall sentiment [1], [9], [11], [15].

In [14] the authors assigned sentiment to words, but they relied on quantitative information such as the frequencies of word associations or statistical predictions of favorability. A number of researchers have also explored learning words and phrases with prior positive or negative polarity (another term is semantic orientation) [1], [10]. Although we use a similar technique we don't limit ourselves to any limited word list, instead we use WordNet to find the semantic orientation of the words which are not found in the General Inquirer word list.

An approach similar to ours is taken by Matsumoto, et al. in [3]. The authors of that paper recognize that word order and syntactic relations between words are extremely important in the area of sentiment classification, and therefore it is imperative that they are not discarded. They construct a dependency tree for each sentence and then prune them to create subtree for classification.

Our work is most close to this work but still has a great deal of difference as we are not training on the trees - we use POS-tagging and dependency trees to analyze the sentence constructs. We analyze the effects of conjunctions in detail on the overall semantic orientation of the sentence. Our analysis is not confined to adjectives and verbs as we have also dealt with nouns, adverbs, conjunctions and prepositions which act as feeling words or affect the sentiment of the phrase.

3 Sentiment Analysis

The essential issue in sentiment analysis is to identify how sentiments are expressed in texts and whether the expressions indicate positive (favorable) or negative (unfavorable) opinions towards the subject. Thus, sentiment analysis involves identification of sentiment expressions, polarity and strength of the expressions and their relationship to the subject.

Often times, two words that are syntactically linked in a sentence are separated by several words. In these cases, small N valued N-gram models would fail at extracting a correlation between the two words, but we have used typed dependencies and dependency tree to deal with Non-Local Dependency Problem.

Another problem is The Word Sense Disambiguation Problem, consider two sentences ***I love this story***, ***this is a love story,*** the first sentence is communicating positive sentiment, whereas the second sentence is an objective statement with neutral sentiment. By looking at the typed dependencies of the words *(love, story)* in the first sentence, one can identify that they have a direct object relation ***{dobj(love-2, story-4)}*** which identifies it as a sentence with a sentiment, while in the second sentence ***{nn(story-5, love-4)}*** *love* just acts as an noun modifier to the word *story*, stating that the story is a love story identifying it as an objective sentence.

Role of Conjunctions: A conjunction is a word that links words, phrases, or clauses, and it may be used to indicate the relationship between the ideas expressed in a clause and the ideas expressed in rest of the sentence. They play a vital role in deciding the overall polarity of a sentence. They often change the sentiment into the opposite orientation or add in the strength of the sentiment.

For example, The Pacifica is exceptionally quiet most of the time, but it suffers some engine blare under hard acceleration. If we only consider the word exceptionally, we will mistake the sentiment for positive. However, the word but in the sentence changes its sentiment orientation, actually it is negative. The difficulty with conjunctions is that they can occur almost anywhere in the structure of a sentence and therefore demands a thorough analysis of the sentence construct as we need to find the main clause in a sentence in order to decide the sentence level polarity.

4 Sentiment Classification (Evaluation)

Step1 does the POS-tagging, generates the dependency tree and gives the typed dependencies of the words, for this the *"Stanford Lex-Parser"* [7] is used. We then select the feeling words (a feeling word is anchored by some substantive meaning and describes an author's attitude towards the subject). In Step 2 and 3, we determine the presence of a conjunction and if present we identify all the individual phrases containing sentiments. Step 4 calculates the polarity of individual phrases using the default polarity calculation method with the help of the general inquire word list [13] or WordNet to get the semantic orientation of the words.

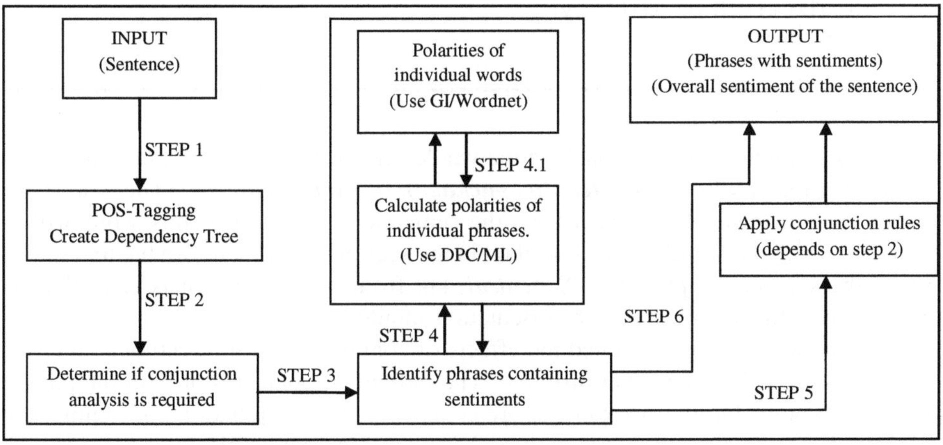

Fig. 1. Steps in the process of sentiment classification using our system

If a conjunction is found, Step 5 applies a rule (which we created with a very comprehensive list of conjunctions) to the sentence depending upon the type of conjuncts. The rules help in deciding the main clause of the sentence, i.e. the phrase that will decide the overall polarity of the sentence. If no conjunction is found, then the default method for finding the polarity of sentiment expression is used. Finally, in Step 6 the total polarity of the sentence is decided.

4.1 Conjunction Analysis

Full syntactic parsing plays an important role to extract sentiments correctly, because the local structures obtained by a shallow parser are not always reliable [8]. We start by passing the current sentence to the lex-parser, the output of the lex-parser is the dependency tree with POS tagging and the typed dependencies of the words.

Example: Following example explains the process of sentiment classification with the help of dependency tree and the typed dependencies of the words in a sentence. The tagset used is the penn-tagset, where JJ=adjective, NN=noun, VP=verb phrase etc.

Example 1. Supergirl is definitely a terrific DVD package, but a very lousy movie

```
(ROOT
 (S
  (NP (NNP Supergirl))
  (VP (VBZ is)
   (ADVP (RB definitely))
   (NP
    (NP (DT a) (JJ terrific)
     (NNP DVD) (NN package))
    (, ,)
    (CC but)
    (NP (DT a) (RB very) (JJ lousy)
     (NN movie))))
  (. .)))
```

```
nsubj(package-7, Supergirl-1)
cop(package-7, is-2)
advmod(package-7, definitely-3)
det(package-7, a-4)
amod(package-7, terrific-5)
nn(package-7, DVD-6)
det(movie-13, a-10)
advmod(movie-13, very-11)
amod(movie-13, lousy-12)
conj_but(package-7, movie-13)
```

In this example, if we just analyze the whole sentence with the default polarity calculator the result will be ***{+1 (terrific, package); -1 (lousy, movie)}***. (The word *lousy* is not present in the GI word list, so the semantic orientation of this word will be searched using the WordNet as described in the section 4.3). Therefore the total polarity of the sentence is ***{+1 -1 = 0} neutral***, but from the sentence it is clear that the author didn't like the movie so the orientation should be negative.

Now, as we have incorporated the effects of conjunction in the sentiment analysis, the analysis will be somewhat different. The individual polarities of both the phrases will be as above, but the two phrases are connected by a conjunction *but* joining *(terrific package, lousy movie*), which are *(NP, NP)*. Comparing the tags and the conjuncts with the rules from the rule file, it's clear that the second phrase is the main clause and therefore it will be used to decide the overall polarity of the sentence which is negative ***(-1)***.

4.2 Conjunction Rules

We have compiled rules to analyze the effects of more than 80 conjunctions (including all types of conjunctions) with 10 rules on average for each conjunction. A sample rule is given below:

Table 1. Rules for usage and effects of the conjunction *but*

```
<conjunction id="but" class="CC" subClass="ADVERSATIVE">
        <rule LC="NN" RC="NN" result="!RC" />
        <rule LC="JJ" RC="JJ" result="RC" />
        <rule LC="S" RC="S" result="RC" />
        ...................
</conjunction>
```

The rule is for the conjunction *but*, it also shows the class and subclass of the conjunction. Each *rule* tag describes a rule for different conjuncts, according to the first rule if the left clause and right clause of the conjunction are NN; for example *everyone/NN but/CC John/NN is/VBZ present/JJ,* the polarity for the right NN will be opposite of the polarity of the left NN. Therefore, as the sentence is positive towards *everyone* so it is negative toward *John*, this is what the rule describes. Similarly we can conclude for other rules.

4.3 Default Polarity Calculation (DPC)

We start the polarity classification by identifying the Positive and Negative words using the General Inquirer (GI) [13]. While determining the orientation of a word, if a word is not found in the GI list, we search that word in the WordNet dictionary and all the synonyms are searched for semantic orientation in the GI word list (as synonyms generally have same semantic orientation) which helps us in determining the polarity of the word. While calculating polarity of a word we have also considered effects of ***negations*** (*good* is positive, while *not good* is negative) [12]. Further effects of words like *very, little, rather* etc. which intensifies or decreases the polarity of a word have been analyzed.

One problem with the method of counting positive and negative terms is that we may need to remove the suffix of a given term in order to see if it exists in our list of terms. To do this we are using the stemming algorithm of *"Stanford Lex-Parser"* [7].

4.4 Overall Sentiment Determination

Once we get the individual polarities of the phrases, we decide the polarity of the sentence as described earlier. The product review analysis is also possible with this Sentiment Analyzer; you can even provide specific subject or subjects to find the polarity.

Example 2. An example explaining how conjunctions change the whole sentiment of the sentence

[INPUT]
The notchy gear box was a worry and needed some time getting used to, but today with a three month old Aveo I can safely say it was no problem.

[OUTPUT]
Finding Polarity:
notchy: -1 {gear box}
worry: -1 {gear box}
Conjunction Found [and]: Current Polarity = -2
Conjunction Found [but]: Current Polarity= 0
problem: -1 (preceding *no* found) : +1 {gear box, worry}
Total Polarity: [-1 and 0 but +1]
and(-1, 0) = -1
but(-1, +1) = +1
Final Polarity: +1

Once each sentence in a given text is evaluated, combining sentence level ratings to a global score is still a tough problem.

Summarization of the results depends on the query type, i.e you may want to have all the phrases containing the sentiments or you may need the polarity of the entire sentence then the polarities of individual phrases will be combined together using conjunction analysis to find the overall sentiment of the sentence.

5 Experimental Results

Our experiments use car reviews as the dataset, compiled from different car review sites like http://www.motortrend.com, http://wardsautoworld.com, etc. The dataset contains more than 10,000 pre-labeled sentences, 5000 positive and 5000 negative.

We will first look at the results of the machine learning algorithms. There were more than 60% (40,000/64,000) sentences in the movies review dataset with one or more than one conjunction and as expected the results were very poor (less than 40%). It is clear from the results that the algorithm trained on the documents can't be used for the sentence level analysis; further the machine learning can't be efficiently used for sentences with conjunction(s).

Table 2. Results of learning algorithm (Naïve Bayes) in presence of one or more conjunctions

Tested on sentences with	Trained with No Conjuncts	Trained with Conjuncts
No Conjunction	70%	73%
Conjunction(s)		
1 Conjunction	51%	58%
More than 1 Conjunctions	40%	55%

Our system gave a poor result with the movie dataset i.e. just 39%, as the reviews were labeled at document level. There were many sentences with overall negative polarity in the positive reviews and vice-versa.

For the evaluations, we check whether the polarity of the sentiment is appropriately assigned to the given subject in each input in terms of the sentiment expression in the output, and calculated the precision and recall.

Table 3. Accuracy of the various classification algorithms considered in this paper

	Machine Learning	DPC		Conjunction Analysis	
		GI	GI with WordNet	GI	GI with WordNet
Sentences with No Conjuncts	72%	54%	66%	...	...
Sentences with Conjuncts	56%	39%	51%	62%	78%

6 Conclusions and Future Work

From Table 3, we can observe that use of WordNet substantially enhances the accuracy of the sentiment analysis. We can also see that conjunction analysis improves the sentiment classification by more than 25% and it is clear from the results that Machine learning algorithm is superior to DPC. Performing *base level* sentiment analysis using a learning algorithm and employing conjunct analysis for combing these phrase level sentiments to sentence level sentiments appears will result in better accuracy.

Our current system requires manual development of sentiment lexicons, and we need to modify and add sentiment terms for new domains, so automated generation of the sentiment lexicons in order to reduce human intervention in dictionary maintenance will also be our priority. This will improve precision and recall for new domains.

We believe that our major challenge is in the conjunction rules; we need to find a way of dealing with situations for which there is no rule specified. In addition we can implement named identity tagging for domain specific information and it should help remove objective sentences to a greater extent.

References

1. Peter Turney. Thumbs Up or Thumbs Down? Semantic Orientation Applied to Unsupervised Classification of Reviews. In Proceedings of the 40th Annual Meeting of the Association for Computational Linguistics (ACL), pages 417-424, 2002.
2. B. Pang, L. Lee, S. Vaithyanathan, "Thumbs up? Sentiment Classification using Machine Learning Techniques," Proceedings of the 2002 Conference on Empirical Methods in Natural Language Processing, 2002.
3. S. Matsumoto, H. Takamura, M. Okumura,"Sentiment Classification using Word Sub-Sequences and Dependency Sub-Tree," Proceedings of PAKDD, 2005.
4. B. Pang, L. Lee,"Seeing stars: Exploiting class relationships for sentiment categorization with respect to rating scales," Proceedings of the ACL, 2005.

5. Yu, Hong and Vasileios Hatzivassiloglou. 2003. Towards answering opinion questions: Separating facts from opinions and identifying the polarity of opinion sentences. In Proceedings of EMNLP.
6. Tetsuya Nasukawa, and Jeonghee Yi, "Sentiment Analysis: Capturing Favorability Using Natural Language Processing" K-CAP'03, October, 2003, pp. 70-77.
7. The Stanford Natural Language Processing Group (http://nlp.stanford.edu/software/lex-parser.shtml)
8. Kanayama Hiroshi, Nasukawa Tetsuya and Watanabe Hideo. Deeper Sentiment Analysis Using Translation Technology, pp. 4-5.
9. Kushal Dave, Steve Lawrence, and David M. Pennock, "Mining the peanut gallery: Opinion extraction and semantic classification of product reviews," in Proceedings of the 12th International World Wide Web Conference (WWW-2003), 2003.
10. Vasileios Hatzivassiloglou and Kathleen R. McKeown, "Predicting the semantic orientation of adjectives," in Proceedings of the 35th Annual Meeting of the Association for Computational Linguistics (ACL-1997), 1997.
11. Bo Pang and Lillian Lee, "A sentimental education: Sentiment analysis using subjectivity summarization based on minimum cuts," in Proceedings of the 42th Annual Meeting of the Association for Computational Linguistics (ACL-2004), 2004.
12. Livia Polanyi and Annie Zaenen. Contextual valence shifters. In Proceedings of the AAAI Symposium on Exploring Attitude and Affect in Text: Theories and Applications (published as AAAI technical report SS-04-07), 2004.
13. Philip J. Stone, Dexter C. Dunphy, Marshall S. Smith, Daniel M. Ogilvie, and associates. The General Inquirer: A Computer Approach to Content Analysis. The MIT Press, 1966.
14. S. Morinaga, K. Yamanishi, K. Teteishi, and T. Fukushima. Mining product reputations on the web. In Proceedings of the ACM SIGKDD Conference, 2002.
15. Beineke, Philip, Trevor Hastie, Christopher Manning, and Shivakumar Vaithyanathan. 2004. Exploring sentiment summarization. In AAAI Spring Symposium on Exploring Attitude and Affect in Text: Theories and Applications (AAAI tech report SS-04-07).

PageRank: When Order Changes

Massimo Melucci and Luca Pretto

University of Padova
Department of Information Engineering
{massimo.melucci,luca.pretto}@dei.unipd.it

Abstract. As PageRank is a ranking algorithm, it is of prime interest to study the *order* induced by its values on webpages. In this paper a thorough mathematical analysis of PageRank-induced order changes when the damping factor varies is provided. Conditions that do not allow variations in the order are studied, and the mechanisms that make the order change are mathematically investigated. Moreover the influence on the order of a truncation in the actual computation of PageRank through a power series is analysed. Experiments carried out on a large Web digraph to integrate the mathematical analysis show that PageRank — while working on a real digraph — tends to hinder variations in the order of large rankings, presenting a high stability in its induced order both in the face of large variations of the damping factor value and in the face of truncations in its computation.

1 Introduction

As PageRank is an algorithm to rank webpages or websites, questions regarding the order induced by PageRank values are in principle of interest in many real-life applications. Our main contribution regards a mathematical analysis of the situations in which a variation of the value of the damping factor α forces a variation in the order induced by PageRank. More precisely, if $p_x(\alpha)$ denotes the PageRank of the page x when the damping factor is α, it is our interest to answer questions such as: If $p_i(\alpha) < p_j(\alpha)$, does α_1 such that $p_i(\alpha_1) > p_j(\alpha_1)$ exist? What are the mechanisms that cause changes in the order, and what are the mechanisms that hinder them? Do digraphs in which a change is impossible exist? Which webpages are likely to be favoured to the detriment of others by a change in α? What is the influence on the order of a truncation in the actual computation of PageRank through a power series? Experiments were carried out on a large Web digraph to integrate the mathematical analysis and to test the hypothesis that the order induced by the PageRank values computed on the whole digraph is significantly different from the order induced by the PageRank values computed on the local, smaller digraph around every webpage.

To study the problem of PageRank-induced order changes with a mathematical perspective is technically much more difficult than studying PageRank value changes. When value changes are studied, in fact, a linear system is under investigation, but when order changes are examined linearity is lost, so that traditional linear algebra tools become useless.

G. Amati, C. Carpineto, and G. Romano (Eds.): ECIR 2007, LNCS 4425, pp. 581–588, 2007.

2 Preliminary Results

The basic facts on PageRank used in this paper can be found in [4].

Let us denote by $\mathbf{W}$ the transition probability matrix of the *Web Markov chain*, i.e. the Markov chain represented by the Web digraph after the dangling nodes problem is addressed. In order to obtain irreducibility and aperiodicity, the Web Markov chain is modified: when in a state i, an existing arc is followed with probability α, while a jump towards a randomly chosen state is performed with probability $1-\alpha$, where $\alpha \in [0,1)$ is the damping factor. This modification of the Web Markov chain will be called *PageRank Markov chain*, and its associated digraph will be called *PageRank digraph.* The PageRank vector is the limit probability vector of the PageRank Markov chain; it can be obtained by solving

$$\mathbf{p}(\alpha) = \frac{(1-\alpha)}{N}\mathbf{1} + \alpha\mathbf{W}^T\mathbf{p}(\alpha) \tag{1}$$

where N is the total number of webpages under consideration and $\mathbf{1}$ is an $N \times 1$ vector of ones. Moreover, each entry of $\mathbf{p}(\alpha)$ is a continuous function in the interval $[0,1)$ [1].

Let us denote by $\mathbf{I}$ the identity matrix. When $\alpha < 1$, $\mathbf{I} - \alpha\mathbf{W}^T$ is non-singular [5, Lemma B.1], and so (1) gives

$$\mathbf{p}(\alpha) = \frac{(1-\alpha)}{N}(\mathbf{I} - \alpha\mathbf{W}^T)^{-1}\mathbf{1} \ . \tag{2}$$

The inverse matrix $(\mathbf{I}-\alpha\mathbf{W}^T)^{-1}$ can be expressed as $(\mathbf{I}-\alpha\mathbf{W}^T)^{-1} = \mathbf{I}+\alpha\mathbf{W}^T + \alpha^2(\mathbf{W}^T)^2 + \cdots = \sum_{n=0}^{+\infty}\alpha^n(\mathbf{W}^T)^n$ and so by considering expression (2) we get:

$$\mathbf{p}(\alpha) = \frac{(1-\alpha)}{N}\left(\sum_{n=0}^{+\infty}\alpha^n(\mathbf{W}^T)^n\right)\mathbf{1} \ . \tag{3}$$

Since the order induced by PageRank is affected neither by the factor $(1-\alpha)$ nor by the first member of the sum, what governs the order is the power series

$$\alpha(\mathbf{W}^T)\frac{1}{N}\mathbf{1} + \alpha^2(\mathbf{W}^T)^2\frac{1}{N}\mathbf{1} + \cdots + \alpha^n(\mathbf{W}^T)^n\frac{1}{N}\mathbf{1} + \cdots \tag{4}$$

Let us denote by $\Pr[A]$ the probability of the event A. It is worth noting that the generic term $(\mathbf{W}^T)^n\frac{1}{N}\mathbf{1}$ gives a vector whose entry j is the probability $\Pr[x(n) = j]$ that the Web Markov chain $x(\cdot)$ is in state j after n steps, when the initial distribution of the chain is the uniform one. Therefore the order induced by PageRank depends on the transitory behaviour of the Web Markov chain. In order to understand the behaviour of PageRank-induced order when α varies, Formula (4) must be considered under different points of view — this is what will be done in the following sections. From Formula (4) it is straightforward to observe that two pages i, j do not change their relative order when the damping factor varies if $\Pr[x(n) = i] \geq \Pr[x(n) = j]$ or $\Pr[x(n) = i] \leq \Pr[x(n) = j]$ for $n = 1, 2, 3, \ldots$, when the initial distribution of the chain is the uniform one.

3 When the Order May Change

3.1 Global Behaviour

Let us focus our attention on the global behaviour of order changes when the damping factor varies from 0 to 1. The changes that occur on the basis of the general, i.e. global, structure of the digraph, independently from its specific, 'local', characteristics will henceforth be known as 'global' changes.

Let us consider first the case $\alpha \approx 0$. To study order problems mathematically in this case, it is convenient to write

$$\mathbf{p}(\alpha) = \frac{(1-\alpha)}{N}(\mathbf{I} - \alpha \mathbf{W}^T)^{-1}\mathbf{1} = \frac{(1-\alpha)}{N}(\mathbf{1} + \alpha \mathbf{W}^T\mathbf{1} + o(\alpha)\mathbf{1}) . \quad (5)$$

When $\alpha \approx 0$, but $\alpha \neq 0$, the pages' relative order is governed by the vector $\mathbf{W}^T\mathbf{1}$, that is, for the generic page i, by the sum $\sum_{h \to i} \frac{1}{d_h}$, where d_h is the out-degree of page h and '$h \to i$' indicates that the sum is extended to all and only the pages which have at least one link to page i. In the rest of this paper this quantity will be called the *weighted in-degree* of page i. PageRank's continuity and Formula (5) allow one to state that if page i has a weighted in-degree greater than page j, then $p_i(\alpha) > p_j(\alpha)$, at least when $\alpha \approx 0$.

Let us consider now the other extreme, that is the case $\alpha \approx 1$. In this case, the study can be carried out by considering the PageRank Markov chain. It is known that the limit $\lim_{\alpha \to 1-} \mathbf{p}(\alpha)$ exists, and that when $\alpha \to 1-$ the PageRank value of all the states that in the Web Markov chain are inessential is 0, that is, all the PageRank values are swallowed up by the closed communicating classes of the Web Markov chain, that is by the rank sinks of the Web Markov chain [1].

These results, showing that a global PageRank shift occurs from inessential towards essential states when α moves from 0 to 1, suggest that this shift can change the PageRank-induced order. Of course, this can happen if in the Web Markov chain inessential states exist. Studies on the Web digraph structure show that this is the case; indeed, most of the states are inessential [2]. In the case of a Web Markov chain with inessential states this shift of PageRank might *not* generate an order change (comparing the two cases $\alpha \approx 0$ and $\alpha \approx 1$) only in rather pathological cases, practically impossible in the Web digraph structure: this could happen, indeed, if the PageRank shift did not influence the PageRank-induced order present when $\alpha \approx 0$. In Figure 1(a) a Web Markov chain with a rank sink is depicted, from which it is easy to obtain: $p_1(\alpha) = p_2(\alpha) = p_3(\alpha) = \frac{1-\alpha}{5}$, $p_4(\alpha) = \frac{1+2\alpha-3\alpha^2}{5}$, $p_5(\alpha) = \frac{1+\alpha+3\alpha^2}{5}$ so that an order change between page 4 and 5 occurs when $\alpha = 1/6$. In Figure 1(b), on the contrary, no order change occurs when α moves from 0 to 1, because the initial order persists even when all the PageRank values are swallowed up by the rank sink.

3.2 Local Behaviour

Hitherto it has been shown that PageRank-induced order is governed by the weighted in-degree of each page when $\alpha \approx 0$, and by the swallowing-up in the

Fig. 1. A Web Markov chain with a rank sink which causes a change in order (a), and a Web Markov chain with a rank sink without changes in order (b)

rank sinks when $\alpha \approx 1$. What happens between these two extremes? More precisely: Are the order changes that may occur when α moves from 0 to 1 only due to a PageRank shift towards the rank sinks (global influence), or might they also be due to a local influence, different from the one seen up to now? The following example shows that global influence on its own cannot explain *all* the PageRank-induced order changes.

Example 1. Let us consider the Web digraph structure in Figure 2. Since this

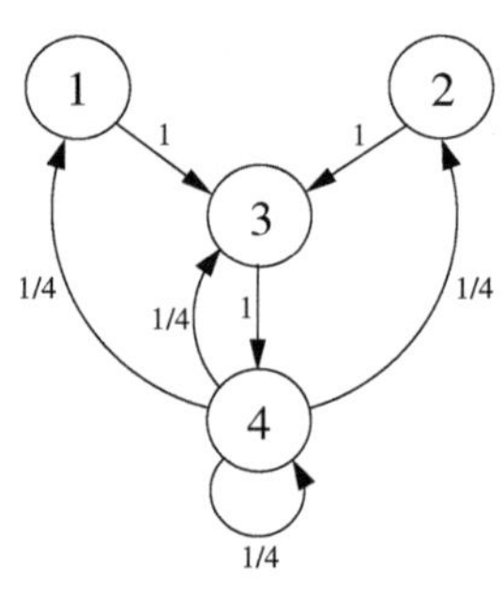

Fig. 2. A strongly connected Web digraph in which the PageRank-induced order changes when the damping factor varies

digraph is strongly connected, that is the Web Markov chain it represents is irreducible, no PageRank shift towards a rank sink can occur. Nevertheless, the PageRank-induced order of the pages of the digraph changes when α varies. In fact, elementary computations give:

$$p_3(\alpha) = p_1(\alpha)(1 + 2\alpha)$$
$$p_4(\alpha) = p_1(\alpha)(1 + \alpha + 2\alpha^2)$$

and so, while $p_1(\alpha)$ is always less than the other PageRank values, $p_3(\alpha) > p_4(\alpha)$ when $\alpha < 1/2$ and $p_3(\alpha) < p_4(\alpha)$ when $\alpha > 1/2$. Note that the symmetry of the digraph gives: $p_1(\alpha) = p_2(\alpha)$.

These order changes can be explained with reference to the local structure of the digraph. For the purposes of the present analysis it should be noted that in (3) the PageRank-induced order is governed by:

$$\alpha(\mathbf{W}^T)\mathbf{1} + \alpha^2(\mathbf{W}^T)^2\mathbf{1} + \cdots + \alpha^n(\mathbf{W}^T)^n\mathbf{1} + \cdots$$

Let us focus our attention on two pages i and j. As regards page i there is:

$$\alpha \sum_{h_1 \to i} \frac{1}{d_{h_1}} + \alpha^2 \sum_{h_2 \to h_1} \sum_{h_1 \to i} \frac{1}{d_{h_2}} \frac{1}{d_{h_1}} + \cdots + \\ + \alpha^n \sum_{h_n \to h_{n-1}} \sum_{h_{n-1} \to h_{n-2}} \cdots \sum_{h_1 \to i} \frac{1}{d_{h_n}} \frac{1}{d_{h_{n-1}}} \cdots \frac{1}{d_{h_1}} + \cdots . \qquad (6)$$

which is a weighted sum; the generic member

$$B(n,i) = \sum_{h_n \to h_{n-1}} \sum_{h_{n-1} \to h_{n-2}} \cdots \sum_{h_1 \to i} \frac{1}{d_{h_n}} \frac{1}{d_{h_{n-1}}} \cdots \frac{1}{d_{h_1}}$$

gives the sum of the probabilities of all and only the paths that end in page i after n steps. Formula (6) shows that the sums of the probabilities of the paths that end in page i — which in their turn depend on the topology of the digraph and on the weights of the links — govern the order induced locally by PageRank. More precisely, when local behaviour is under examination, only that subgraph of the Web digraph which is made up of the pages that lead to page i should be considered. This digraph will henceforth be referred to as the *back digraph* of page i. Now, when α is close to 0 the order is governed by the part of the back digraph near the page, while when α is close to 1 a broader part should be considered.

Definition 1. *The back digraph of page i* expands at step n with regard to page *j if $B(n,i) > B(n,j)$. $B(n,i) - B(n,j)$ is the* size *of the expansion.*

Now, it can be easily proved that the two pages i and j may change their relative order when α varies only if an expansion change between the back digraphs of pages i and j takes place, that is only if n and m exist such that the back digraph of page i expands at step n with regard to page j, and the back digraph of page j expands at step m with regard to page i. This fact gives a *necessary* condition for an order change to take place; note that, for instance, in the Web digraph in Figure 2, the back digraph of page 3 expands with regard to page 4 at step 1, while the back digraph of page 4 expands with regard to page 3 at step 2.

On the other hand, this condition is far from being a *sufficient* condition for a local change in order to take place. This fact has essentially two causes:

1. The size of the expansions should be considered; for example, an expansion of the back digraph of page i with regard to page j with a size s cannot be compensated for by a subsequent expansion of the back digraph of page j with regard to page i with a smaller size. Moreover, as a consequence

of the damping effects of the damping factor, the size of the compensating expansion necessarily gets bigger the further the subsequent expansion is from the first.

2. The more subsequent expansions of the back digraph of page i with regard to page j take place, the harder it becomes for page j to compensate for these expansions, and so allow an order change.

4 Effects of Truncation

The mathematical analysis expounded up to now is also of use in throwing light on another important question: how is the PageRank-induced order affected by a truncation in PageRank computation? The power series (3) computes the PageRank of each webpage after an infinite number of iterations. For obvious reasons, the computation has to be stopped after a finite number n, that is, the power series has to be truncated. Such a truncation may affect the PageRank value of each webpage, although to varying extents. Can this truncation also affect the relative order of the pages?

It should be noted that, from a mathematical point of view, when local behaviour is studied, our attention is focused on the first few terms of (3) — a truncation, in a sense — while when global behaviour is under examination, our attention ranges over the whole series, i.e. the terms with a very high value of n are also considered. So when a truncation at step n occurs, only those local effects involving expansion that take place no later than at step n are taken into consideration, while global effects are wholly ignored. While the effects of partly ignoring the expansions depend on the local structure of the Web digraph, and so cannot be completely foreseen, the mathematical analysis on global behaviour shows that, when α is near to 1, the webpages belonging to a rank sink are liable to be the most damaged, that is, their PageRank-induced ranking positions tend to be lower than they would be if no truncation occurred.

5 Experiments

The experiments fundamentally aimed at investigating: (1) The correlations between the various PageRank-induced rankings computed for different values of α, and (2) the correlation between the PageRank-induced ranking and the ranking induced by a truncated computation of PageRank.

In every experiment the correlations between rankings were assessed by Kendall's τ [3]. The experiments were carried out on the 10GB Web Track (WT10g) test collection of the Text Retrieval Conference (TREC).

The results were computed as follows. First, the value of Expression (4) was computed for $\alpha = 0.1, 0.5, 0.8$ and for $n = 50$ — for $n = 50$ it was observed that a ranking no longer changes, i.e., the value of the sum can vary but the relative order does not. Then the value of Expression (4) was computed for the same α's and for $n = 1, 2, 5, 10, 50$. τ was then computed for each α between each pair of rankings, that is, the ranking after $n = 50$ and the ranking after a different n.

Table 1. Correlation between 10-page subrankings with $\alpha = 0.1, 0.5$ and $n = 1, 2, 5, 10$

From	To	$n=1$	$n=2$	$n=5$	$n=10$
1	N	0.999	1.000	1.000	1.000
1	10	0.956	0.956	0.956	0.956
11	20	0.733	0.956	1.000	1.000
21	30	0.556	0.911	0.956	0.956
31	40	0.600	0.911	1.000	1.000
41	50	0.733	0.911	1.000	1.000
51	60	0.200	1.000	1.000	1.000
61	70	0.600	0.822	1.000	1.000
71	80	0.022	0.911	1.000	1.000
81	90	0.289	0.822	0.956	0.956
91	100	0.556	0.956	1.000	1.000

$\alpha = 0.1$

From	To	$n=1$	$n=2$	$n=5$	$n=10$
1	N	0.997	0.998	0.998	1.000
1	10	0.867	0.956	0.956	0.956
11	20	0.644	0.778	0.911	1.000
21	30	-0.644	-0.467	0.822	0.956
31	40	0.289	0.244	0.644	0.956
41	50	-0.022	0.022	0.467	0.911
51	60	-0.022	-0.022	0.778	0.956
61	70	0.378	0.511	0.733	1.000
71	80	0.111	0.111	0.511	0.956
81	90	0.156	0.244	0.778	1.000
91	100	0.333	0.289	0.511	1.000

$\alpha = 0.5$

As regards the complete N-webpage rankings, the results are reported in the first line of Tables 1 and on the left of Table 2 for $\alpha = 0.1, 0.5, 0.8$, respectively — in the first line, N means that all the webpages were involved in computing the correlation.

Another analysis was conducted by investigating the behaviour of the *sub*-rankings computed at every 10 webpages of the ranking sorted by decreasing final PageRank values p_i's. At this aim the correlation with the ranking induced by $p_i^{(n)}$'s — PageRank values after n steps — was computed. The results are reported in Table 1 for $\alpha = 0.1$ and $\alpha = 0.5$ and on the left of Table 2 for $\alpha = 0.8$. The correlations increased as n increased, as one would expect.

The influence of α on the variation of the correlations observed at different subrankings was less strong, as shown in Table 2 which reports the correlations between the subrankings at $\alpha = 0.1$ after $n = 50$ steps and the ones at

Table 2.

From	To	$n=1$	$n=2$	$n=5$	$n=10$
1	N	0.998	0.998	0.997	0.996
1	10	0.244	0.289	0.200	0.333
11	20	0.511	0.511	0.600	0.911
21	30	0.244	0.200	0.333	0.778
31	40	0.022	0.067	0.022	0.511
41	50	-0.244	-0.467	-0.422	0.378
51	60	-0.156	0.067	0.289	0.733
61	70	-0.111	-0.111	-0.022	0.333
71	80	0.200	0.200	0.422	0.022
81	90	-0.111	-0.111	-0.289	0.200
91	100	0.289	0.156	0.156	0.156

Correlation between 10-page subrankings with $\alpha = 0.8$ and $n = 1, 2, 5, 10$.

From	To	$\alpha = 0.5$	$\alpha = 0.8$
1	N	0.999	0.998
1	10	0.911	0.778
11	20	0.289	0.111
21	30	0.689	0.600
31	40	-0.022	-0.200
41	50	0.689	0.556
51	60	0.778	0.733
61	70	0.244	0.244
71	80	0.511	0.467
81	90	0.333	0.200
91	100	-0.200	-0.289

Correlation between 10-page subrankings with $n = 50$, $\alpha = 0.1$ and subrankings with $\alpha = 0.5, 0.8$.

$\alpha = 0.5, 0.8$ after $n = 50$ steps; for example the correlations between the ranking after $n = 50$ steps with $\alpha = 0.1$ and the ranking after $n = 50$ steps with $\alpha = 0.5$ is reported in the third column of Table 2. In particular, if $\alpha = 0.8$, the convergence of τ to the state of perfect correlation is quite slow.

To sum up, the experiments showed that PageRank-induced order is stable enough if the damping factor is close to 0 and when a truncation in the computation is effected after n steps. As far as concerns the top ranked webpages, some instability can be observed if α tends to 1.

To test if the damping factor affects the final ranking, the correlation was also computed between the rankings at $n = 50$ and at different α's independently of n. It was found that the correlations were very high if all the N webpages were involved in the computation, whereas the correlations varied if the subrankings were instead considered.

When all the N webpages were considered, the experiments showed that PageRank-induced order is considerably stable in large digraphs, both when variations in the damping factor values take place and when a truncation in the computation is effected. The experiments showed that the effects of PageRank shift towards rank sinks are negligible in large digraphs, at least if α is not close to 1.

The whole body of these considerations suggests that the PageRank definition has a mathematical structure that naturally tends to *hinder* order changes in practice. Moreover, the influence of the back digraph near each page is predominant, especially when α is not too large.

References

1. P. Boldi, M. Santini, and S. Vigna. PageRank as a function of the damping factor. In *Proc. of WWW*, pages 557–566. ACM, May 2005.
2. A. Broder, R. Kumar, F. Maghoul, P. Raghavan, S. Rajagopalan, R. Stata, A. Tomkins, and J. Wiener. Graph structure in the web. In *Proc. of WWW*, 2000.
3. M. Kendall. *Rank Correlation Methods.* Charles Griffin & Co. Ltd., fourth edition, 1975.
4. A. N. Langville and C. D. Meyer. *Google's PageRank and Beyond: The Science of Search Engine Rankings.* Princeton University Press, Princeton, 2006.
5. E. Seneta. *Non-negative Matrices and Markov Chains.* Springer, New York, second edition, 1981.

Model Tree Learning for Query Term Weighting in Question Answering

Christof Monz

Department of Computer Science
Queen Mary, University of London
Mile End Road, London E1 4NS, United Kingdom
christof@dcs.qmul.ac.uk
www.dcs.qmul.ac.uk/~christof

Abstract. Question answering systems rely on retrieval components to identify documents that contain an answer to a user's question. The formulation of queries that are used for retrieving those documents has a strong impact on the effectiveness of the retrieval component. Here, we focus on predicting the importance of terms from the original question. We use model tree machine learning techniques in order to assign weights to query terms according to their usefulness for identifying documents that contain an answer. Incorporating the learned weights into a state-of-the-art retrieval system results in statistically significant improvements.

1 Introduction

Current question answering systems rely on document retrieval as a means of identifying documents which are likely to contain an answer to a user's question. The documents returned by the retrieval engine are then further analyzed by computationally more expensive techniques to identify an answer to a given question.

The effectiveness of the retrieval component is critical for the performance of a question answering system: If the retrieval system fails to find any relevant documents for a question, further processing steps to find an answer will inevitably fail as well.

In particular, the way queries are formulated has a strong impact on retrieval effectiveness. Boolean retrieval is especially sensitive to query formulation. Using all content words from the question for the query can steer the retrieval process in a wrong direction. For example, consider the question *What is the abbreviation for the London stock exchange?*. It seems natural to include the word *abbreviation* in a boolean query like:

(1) `abbreviation AND london AND stock AND exchange`

However, most documents that contain an answer to this question express it in the form of '...London Stock Exchange (LSE)...,' not using the term *abbreviation* or one of its morphological variants at all. Hence, in boolean retrieval, a query such as (1) might be too strict and retrieve no documents at all.

Vector-space retrieval is less strict with respect to presence or absence of query terms in documents, but a similar problem arises. Even vector-space retrieval approaches will prefer documents containing the term *abbreviation* over those that do not contain it,

G. Amati, C. Carpineto, and G. Romano (Eds.): ECIR 2007, LNCS 4425, pp. 589–596, 2007.

although, as discussed above, most documents providing an answer to the question above do not contain it. The reason is that the term *abbreviation* receives a high term weight, because it is much less frequent in the collection than the other terms and is therefore considered a well-discriminating term for this query.

Our approach tries to predict the importance of a term for a given question by applying machine learning techniques. The learned importance weights are then used to improve the retrieval engine. Learning query term weights is appealing in the context of question answering because a user's information need is expressed as a well-formed sentence as opposed to sets of keywords, used in regular information retrieval.

2 Related Work

Previous work on query formulation for question answering has mainly been done for Web question answering. Brill et al. [2] focus on formulating query strings that approximate the way an answer is likely to be expressed. This involves automatically mapping the syntax of an interrogative to the syntax of a declarative sentence. Documents are required to either match one of the strings or the boolean query. However, they do not address the issue of term weighting.

Paşca [11] does address the issue of term selection and term relevance. His work is closely related to the work presented in this paper. For query formulation, he distinguishes between three types of terms: high-relevance, medium-relevance, and low-relevance query terms. Deciding which class a given term belongs to is based on a number of rules, some of which are also integrated in our approach.

Although machine learning techniques have been used before to find answer extraction patterns, see, e.g., [8,14], they have not been applied before to query formulation in the context of question answering. On the other hand, machine learning has been applied to query formulation in the context of ad hoc retrieval. Cooper et al. [4] use logistic regression to assign weights to matching clues, such as the number of times a query term occurs in the query, the number of times a query term occurs in a document, the idf score of a matching term, and the number of distinct terms common to both query and document. [3] did apply machine learning techniques for selecting query terms, but it was done in the context of relevance feedback retrieval.

The work by Mayfield and McNamee [9] and Agichtein et al. [1] is to some extent complementary to our work as it expands retrieval queries with terms or phrases that are likely to be found in the context of phrases which are of the expected answer type. While the approach in [9] is based on simple co-occurrence statistics, [1] also incorporate the BM25 retrieval weighting scheme to assign weights to the expansion phrases.

3 Optimal Query Term Selection

In this section we estimate the effect query formulation, in the form of term selection, can have on retrieval performance, by using the TREC-9, TREC-10, and TREC-11 data sets consisting of 500 different questions each and the AQUAINT document collection.

For evaluating the retrieval effectiveness, we used NIST's judgment files. Each file indicates for each submitted answer-document-id pair, whether the answer is correct.

Questions with no correct answer were disregarded. The documents that are known to contain a correct answer form the gold standard for our evaluation, which contains 480, 433, and 455 questions for TREC-9, TREC-10, and TREC-11, respectively.[1]

In order to compute the optimal term selection for each question, we compare all possible ways of selecting terms from a question. That is, given a question q in which the set of terms T occurs, we consider all possible subsets of T, and evaluate the respective performances. More formally, the set of term selection variants (*tsv*) is defined as $tsv(q) = POW(T) - \{\emptyset\}$. For each question in the three data sets, we determined the query variant with the highest average precision. Table 1 shows the performance gains that can be achieved by using an oracle to pick the optimal query variant.

Table 1. Comparison of the a@n scores of optimal retrieval queries to baseline runs

	TREC-9		TREC-10		TREC-11	
a@n	Lnu.ltc	opt	Lnu.ltc	opt	Lnu.ltc	opt
a@5	0.700	0.823 (+17.6%)▲	0.649	0.749 (+15.4%)▲	0.523	0.690 (+31.9%)▲
a@10	0.785	0.890 (+13.4%)▲	0.734	0.815 (+11.0%)▲	0.626	0.767 (+22.5%)▲
a@20	0.845	0.921 (+9.0%)▲	0.801	0.887 (+10.7%)▲	0.705	0.824 (+16.9%)▲
a@50	0.914	0.956 (+4.6%)▲	0.875	0.924 (+5.6%)▲	0.795	0.881 (+10.8%)▲

In the context of question answering, it is common to measure the effectiveness of a retrieval system in terms of answer-at-n (a@n) which is the percentage of questions for which the system returned at least one document containing an answer in the top-n ranked documents. For our baseline we did not use blind relevance feedback, as Monz [10] has shown that simple Lnu.ltc weighting with stemming outperforms approaches using blind feedback for question answering.

As one could expect, query formulation has a significant impact on the overall performance of a retrieval system, even if query formulation is just based on term selection without expanding the queries with semantically related terms. This comparison shows that much can be gained from better query formulation, but, of course the problem of identifying an optimal query without having any relevance assessments remains open. In the remainder we explore ways to solve this issue.

Two retrieval methods a and b are compared by one-tailed statistical significance testing, using the bootstrap method [5]. Improvements at a confidence level of 95% are marked with "△" and at a confidence level of 99% with "▲".

4 Computing Query Term Weights

Our approach is to use the different query variants of a question to distinguish between terms that help retrieve relevant documents, and terms that harm the retrieval effectiveness for that particular question.

In the previous section, we considered only one single best-performing query variant, but often there are several almost equally well-performing query variants. Looking at

[1] The original TREC-9 data set contains 243 questions that are re-formulations of the questions in the main TREC-9 set. For our experiments, these variants were disregarded.

the ranked query variants, reveals that some terms occur more frequently in higher-ranked variants than other terms.

An analysis of the distribution of query terms over the ranked query variants allows one to assign a weight to each query term: If a term occurs mainly in query variants that have a high average precision it should receive a high weight, whereas a term that occurs mainly in query variants that have a low average precision should receive a low weight. Thus, the weight of a query term depends on two factors: The average precisions of the query variants in which the term occurs (its presence weight: $w_{+}(t)$), and the average precisions of the query variants in which the term does not occur (its absence weight: $w_{-}(t)$). Presence and absence weights are normalized by the sum of the average precisions of all query variants, so the weights will range between 0 and 1.

Given a question q and all its query variants $tsv(q)$, the *presence weight* of term t ($w_{+}(t)$) is computed as:

$$w_{+}(t) = \frac{\sum_{q' \in tsv(q) \wedge t \in q'} \text{avg\_prec}(q')}{\sum_{q' \in tsv(q)} \text{avg\_prec}(q')} \tag{2}$$

The *absence weight* of term t ($w_{-}(t)$) is computed as:

$$w_{-}(t) = \frac{\sum_{q' \in tsv(q) \wedge t \notin q'} \text{avg\_prec}(q')}{\sum_{q' \in tsv(q)} \text{avg\_prec}(q')} \tag{3}$$

The presence and absence weights of a term t, can be combined into a single weight by subtracting the absence weight from the presence weight, which we call the *gain* of term t: $gain(t) = w_{+}(t) - w_{-}(t)$. If a query term has a positive gain it should be included in the query, but excluded if its gain is negative.

This approach of computing term weights assumes that terms occur independently of each other. This assumption does not hold in practice, but it is commonly used in information retrieval and allows us to simplify the computation of term weights.

5 Representing Terms by Feature Sets

In the previous section, the computation of the term weights was based on the distribution of the terms themselves over the query variants. This is problematic for two reasons. First, the same term can have a high gain in one query, and a low gain in another. Second, if the learning algorithm is based on the surface terms themselves, it cannot assign weights to terms that did not occur in the training data. The first problem is a direct consequence of the term independence assumption. It could be solved by conditioning the weight of a term on a number of terms that also occur in the question, but then data sparseness becomes even more of an issue.

One way to address both problems is to represent terms and their contexts in a more abstract manner. Here, we use a set of features that represent certain characteristics of a term and its role in a question. The list of features contains information about the term's part-of-speech, whether it semantically includes other terms in the question, the type of question it occurs in, etc. As mentioned above, some of the features capture aspects inherent to a term, such as part-of-speech, while others capture contextual aspects, such as semantic inclusion. Table 2 lists all features. The features *question focus*, *superlative*, *quoted*, *number of leaves*, *modified noun*, and *person name* are based on [11].

Table 2. List of features for question words

Feature	Value	Feature	Value
part-of-speech	Penn Treebank part-of-speech tag	location	Whether the word is part of a location name
question focus	Whether the word is part of the question focus	abbreviation	Whether the word is an abbreviation
superlative	Whether the question contains a superlative adjective	upper case	Whether the word starts with an uppercase letter
question class	A fixed list of question classes	classif. word	Whether the word was used to classify the question
multpl. occurr.	Whether the word occurs more than one in the question	person name	What part of a person's name is the word, if applicable
quoted	Whether the word occurs between quotation marks	honorific	Whether the word is a honorific term (e.g., Dr.)
modified noun	Whether the word is a noun that is preceded (modified) by another noun	no. edges	The number of edges pointing to a word in the dependency parse graph of the question
term ratio	$1/m$, where m is the number of unique terms in the question	hypernym	Whether the word is a hypernym of another question word
no. leaves	The number of hyponyms in WordNet that do not have any further hyponyms	relative idf	The relative idf compared to the other words in the question

Most of the features in Table 2 are fairly general and self-explanatory, but some are more specific and do require some further explanation.

The *focus* of a question is a phrase describing a semantic type of which the answer is an instance. For example, in the question *In what country did croquet originate?*, the focus is *country*. The answer to this question, which is *France*, is an instance of *country*. Whether a word is part of the question focus affects query formulation, because many documents containing an answer to the question do not explicate the instance relation.

Classifying words help determine the type of a question. E.g., in the question *What province is Calgary located in?*, the word *located* indicates that the question is a location question. However, words that are good indicators for question classification, are infrequent in answers as expressed in documents. For instance, it is rather unlikely that the word *located* is used in a declarative sentence answering this question.

The *number of incoming edges* refers to the dependency parse graph of the question which is generated by MINIPAR [7]. If a word has a larger number of incoming edges, several words in the question are in a modifier or argument relationship with this word, and therefore it is more likely to play a central role in the question. For each word in any question, the features listed in Table 2 are extracted automatically, by using off-the-shelf tools, including a POS tagger, dependency parser, and named-entity recognizer.

6 Learning Term Weights

Instead of using some heuristics for predicting the query term importance we apply machine learning techniques to assign a weight to each term in the question, where the actual query that is used for retrieval will include these weights. The input for the learning algorithm is the set of feature vectors as described in the previous section, and the classes are the terms' gains as described in Section 4.

For the purpose of learning term weights, the machine learning algorithm should learn to predict the degree of the query term's usefulness for query formulation. Decision trees, naive Bayes, and linear regression, all allow for interval classification and generate transparent classification rules.

Naive Bayes classification is known to be well-performing for nominal classification, but performs badly for interval classification [6].

The best-known algorithm for decision tree learning is Quinlan's C4.5 [13], but C4.5 cannot deal with cases where the class to be learned is not a label, but a real number. M5 [12], on the other hand, which is an extension of C4.5, does allow for this type of continuous classification, also referred to as regression.

The M5 algorithm builds model trees combining conventional decision tree learning with the possibility of linear regression models at the leaves of the tree. The resulting representation is transparent because the decision structure is clear and the regression models are normally easily interpretable. The idea of model trees is largely based on the concept of regression trees. The advantage of M5 is that model trees are generally much smaller than regression trees and have proved to be more accurate in a number of tasks; see [12]. M5 is suited for learning query term weights because it combines decision tree learning with linear regression, which allows it to consider dependencies between features. The learning algorithm used here, is M5′ [15], which is a reconstruction of Quinlan's M5 algorithm. M5′ is part of the WEKA ML package [16].

7 Results

For our experiments, the weight of a query term depends on two factors: The frequency of a term in a document, and the collection frequency, i.e., the number of documents containing that term. If we want to integrate the learned term weights, as described above, the computation of the retrieval status value (*RSV*) has to be adapted appropriately. We use the learned query term weights in combination with the original retrieval status value that resulted from computing the similarity between a query q and a document d according to the Lnu.ltc weighting scheme, which results in the new retrieval status value: RSV_L, which is defined as follows:

$$RSV_L(q,d) = \sum_{t \in q \cap d} RSV(q,d) \cdot \mathrm{weight}(fr(t,q)) \cdot \mathrm{idf}(t)$$

Here, $fr(t,q)$ is the feature representation of term t in query q, and weight($fr(t,q)$) is the learned weight, which results from applying the M5′ model tree to t's feature vector. $RSV(q,d)$ is the document similarity according to the Lnu.ltc weighting scheme, and idf(t) is the idf value of term t, i.e., $\log_2(N/df_t)$, where N is the collection size and df_t is the number of documents containing t.

For the evaluation three different model trees were generated, one for each of the TREC data sets. The model tree for assigning weights to terms in the TREC-9 data set was trained on feature representations of words from TREC-10 and TREC-11 (2854 instances), the model tree for the TREC-10 data set used feature representations from TREC-9 and TREC-11 (3167 instances), and the model tree for the TREC-11 data set used feature representations from TREC-9 and TREC-10 (2769 instances).

First, we considered the performance with respect to the answer-at-n (a@n) measure. Table 3 shows the results of using learned query terms weights in contrast to the Lnu.ltc baseline. The improvements are rather modest, although still statistically significant in some cases.

Table 3. Comparison of the a@n scores of learned-weights retrieval (LWR) to the baseline

	TREC-9		TREC-10		TREC-11	
a@n	Lnu.ltc	LWR	Lnu.ltc	LWR	Lnu.ltc	LWR
a@5	0.700	0.727 (+3.7%)$^{\triangle}$	0.649	0.654 (+0.1%)	0.523	0.547 (+4.6%)$^{\triangle}$
a@10	0.785	0.806 (+2.7%)$^{\triangle}$	0.734	0.730 (-0.1%)	0.626	0.637 (+1.8%)
a@20	0.845	0.863 (+2.1%)	0.801	0.804 (±0.0%)	0.705	0.732 (+3.8%)$^{\triangle}$
a@50	0.914	0.908 (-0.1%)	0.875	0.859 (-1.8%)	0.795	0.815 (+2.5%)

8 Conclusions

In this paper we investigated to what extent it is possible to learn query term weights for better query formulation. As we have seen in Section 3, keyword selection has a strong impact on the performance of the retrieval component. In order to learn query term weights, we considered all possible ways of selecting terms from the original question for query formulation, and used the performance results of each possible formulation in order to determine individual query term weights.

Query terms are represented as sets of features on which the M5′ model tree learning algorithm is trained. The resulting model trees confirm some of the heuristics and intuitions for keyword selection than can be found in the literature; see, e.g., [11]. The improvements are modest for a@n, yet statistically significant in some cases, and stayed behind the potential improvements optimal query selection can yield. On the other hand, our term weight learning approach yields significantly better results than the baseline for mean average precision. Hence question answering systems that are more sensitive to the rank and number of a retrieved document can benefit from using our approach.

In some cases the issue of whether a term is helpful for retrieving answer documents simply depends on idiosyncrasies of the documents that contain an answer, but we do not believe that this had a noticable impact on our results as we used large and varied training data to generalize properly.

References

1. Eugene Agichtein, Steve Lawrence, and Luis Gravano. Learning to find answers to questions on the web. *ACM Transactions on Internet Technology*, 4(2):129–162, 2004.
2. Eric Brill, Susan Dumais, and Michelle Banko. An analysis of the AskMSR question-answering system. In *Proceedings of Emperical Methods in Natural Language Processing (EMNLP 2002)*, pages 257–264, 2002.
3. Hsinchun Chen, Ganesan Shankaranarayanan, Linlin She, and Anand Iyer. A machine learning approach to inductive query by examples: An experiment using relevance feedback, ID3, genetic algorithms, and simulated annealing. *Journal of the American Society for Information Science*, 49(8):693–705, 1998.
4. William Cooper, Aitao Chen, and Frederic Gey. Full text retrieval based on probalistic equations with coefficients fitted by logistic regression. In *Proc. of the 2nd Text REtrieval Conference*, pages 57–66, 1993.
5. Brad Efron. Bootstrap methods: Another look at the jackknife. *Annals of Statistics*, 7(1):1–26, 1979.
6. Eibe Frank, Leonard Trigg, Geoffrey Holmes, and Ian H. Witten. Naive bayes for regression. *Machine Learning*, 41(1):5–25, 2000.
7. Dekang Lin. Dependency-based evaluation of minipar. In *Proceedings of the Workshop on the Evaluation of Parsing Systems*, 1998.
8. Lucian Vlad Lita and Jaime Carbonell. Unsupervised question answering data aquisition from local corpora. In *Proceedings of the Thirteenth Conference on Information and Knowledge Management (CIKM 2004)*, pages 607–614, 2004.
9. James Mayfield and Paul McNamee. JHU/APL at TREC 2005: QA retrieval and robust tracks. In Ellen M. Voorhees and Lori P. Buckland, editors, *Proceedings of the Fourteenth Text REtrieval Conference (TREC 2005)*, 2005. NIST Special Publication: SP 500-266.
10. Christof Monz. Document retrieval in the context of question answering. In Fabrizio Sebastiani, editor, *Proceedings of the 25th European Conference on Information Retrieval Research (ECIR-03)*, LNCS 2633, pages 571–579. Springer, 2003.
11. Marius Paşca. *High-Performance Open-Domain Question Answering from Large Text Collections*. PhD thesis, Southern Methodist University, 2001.
12. John R. Quinlan. Learning with continuous classes. In *Proceedings of the 5th Australian Joint Conference on Artificial Intelligence*, pages 343–348, 1992.
13. John R. Quinlan. *C4.5: Programs for Machine Learning*. Morgan Kaufmann, 1993.
14. Deepak Ravichandran and Eduard Hovy. Learning surface text patterns for a question answering system. In *Proceedings of the 40th Annual Meeting of the Association for Computational Linguistics*, pages 41–47, 2002.
15. Yong Wang and Ian H. Witten. Induction of model trees for predicting continuous classes. In *Proceedings of the Poster Papers of the European Conference on Machine Learning (ECML)*, pages 128–137, 1997.
16. Ian H. Witten and Eibe Frank. *Data Mining: Practical Machine Learning Tools and Techniques with Java Implementations*. Morgan Kaufmann, 1999.

Examining Repetition in User Search Behavior

Mark Sanderson[1] and Susan Dumais[2]

[1] Department of Information Studies, University of Sheffield, Sheffield, S1 4DP, UK
[2] Microsoft Research, Redmond, Redmond, WA, USA
m.sanderson@shef.ac.uk, sdumais@microsoft.com

Abstract. This paper describes analyses of the repeated use of search engines. It is shown that users commonly re-issue queries, either to examine search results deeply or simply to query again, often days or weeks later. Hourly and weekly periodicities in behavior are observed for both queries and clicks. Navigational queries were found to be repeated differently from others.

1 Introduction

With the advent of large scale logging of user's activities on search engines, analysis of those logs has produced much valuable information. Early examples of such work come from Broder (2002), who highlighted the differing forms of queries that users of Web search engines issue. He classified queries into *navigational queries* (where the goal is to find a particular web site); *informational queries* (where the user is seeking information on a particular topic) and *transactional queries* (where the user is looking to find sites, which themselves have to be searched to locate required information).

In almost a decade of research in web search engine query logs much of the published work has analyzed small samples of logs often from a single day. The many works of Jansen & Spink (summarized in their 2006 paper) cover no more than a sample of a days worth of activity from a particular search engine. Although such analyses provide insights about short-term interactions with search engines, they do not shed light on longer-term patterns, which are of interest in this paper.

An early log study by Silverstein et al. (1998) summarized characteristics of almost a billion queries collected over a 43 day period of time. However, the authors did not specifically look at temporal effects or individual usage over time. A temporal analysis of queries covering a week's worth of data was recently reported by Beitzel et al. (2004). Their analyses focused on daily periodicities for queries in different topical categories. More recently, Teevan et al. (2006) examined the search behaviors of 114 anonymized users over the course of one year. In this work, users' repetition of queries and items clicked on in search result lists were of interest. Teevan et al. found that across the year, 33% of user queries were repetitions of queries previously issued by the same user. Repetition across users was lower at around 18%. Teevan et al. also separated out navigational queries, which they defined as queries issued at least twice and where the same URL was clicked in the result list for each query. They found that 71% of repeated queries were navigational; if duplicate repeat queries were eliminated, this number fell to 47%. They also examined *clicks*: the item a user chose in the search result list. They found that 29% of clicks recorded in the logs had been clicked on by the same user before. Teevan et al. identified different patterns of repetition and described types of user search behavior that fitted the patterns.

G. Amati, C. Carpineto, and G. Romano (Eds.): ECIR 2007, LNCS 4425, pp. 597 – 604, 2007.

From this study, it was clear that search repetition is common. A more detailed study of repetition was undertaken in this paper to better understand search behavior over time with the goal of improving the search experience. We started with simple measurements of the repetition of queries and clicks, but it became clear that many forms of repetition in user behavior exist. Therefore the analysis was broadened to explore repetition and periodicities in general.

2 Data Set Examined

The query log analyzed was gathered from a major Web search engine for a 3 month period, from 9th Jan. – 13th Apr. 2006, consisting of approximately 3.3 million queries and 7.7 million search result clicks gathered from 324,000 unique users. Users were selected from a voluntary opt-in feedback system. The query text and the date/time when the query was received by the search engine were recorded. In addition, the URL of items in the search result list that were clicked on, the query that generated that result list, and the rank position of the clicked item were recorded. When users retrieved additional search results for a query, such events were treated as a separate query. This differs from Teevan et al.'s work where all identical queries issued by a user less than thirty minutes apart were treated as the same query. Their approach attempted to capture the notion of a search session. Any such approximation of sessions is prone to error, so we choose to examine the query data in its rawer form.

Users were identified by an anonymised ID associated with a user account on a particular PC. As is the case with most log analyses, if a user has more than one computer each with the opt-in feedback system working, they have multiple IDs. Conversely, if more than one person used the same account on a PC, they were amalgamated into a single user.

3 Initial Analysis of Query and Click Repetitions

The first analysis conducted on this data replicated the initial analyses of Teevan et al. determining the level of repetition in queries and in clicked results. Of the 3.3 million queries submitted, 1.62 million were unique to a single user (although many queries were repeated across users); the rest (1.68 million) were submitted more than once by a user. Repeat queries represented a little over 50% of all the queries submitted. This compares to the 33% observed by Teevan et al. We speculate that the different proportion of repeat queries is due to the difference in definition of what counts as a repeat query. The repeated queries were examined to determine how many were navigational queries. Based on Teevan et al.'s definition (same query and same URL click), around 80% of the 1.68 million repeat queries were navigational queries. This compares with the 71% observed by Teevan et al. From this analysis and those published before, it is clear that users repeat queries often on a search engine.

The search result clicks of users were also examined. Of the 7.6 million clicks recorded in the three month period, 1.3 million (17.5%) were found to be clicks accessed more than once by individual users. Within that group of repeat clicks, 83%

were from the same query, and 17% were from different queries. A similar ratio of repeat clicks from same or different queries was found by Teevan et al.

We next examined temporal differences in click patterns over time, which Teevan and colleagues did not analyze. Specifically, we examined the number of repeat clicks over time as well as whether repeat clicks were more or less likely to come from the same query as the time between repeat clicks increased.

3.1 Change in Repetition for Varying Differences in Time

Figure 1 shows a histogram of the counts of repeated clicks as a function of the number of days between the two clicks. The numbers of same click pairs steadily declines as the difference in time between the click events grows – searchers are more likely to click on the same URL in close temporal proximity. The curve drops off smoothly for several months and then more suddenly around 90 days due to *windowing effects* in the query logs which cover only 94 days.

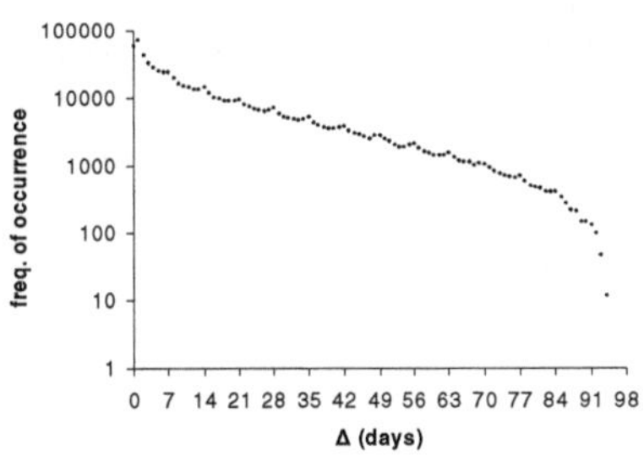

Fig. 1. Histogram of counts of same click events in the query log binned by the Δ in days between the two events. Note in all graphs in the paper, the number of paired events = the number of events-1.

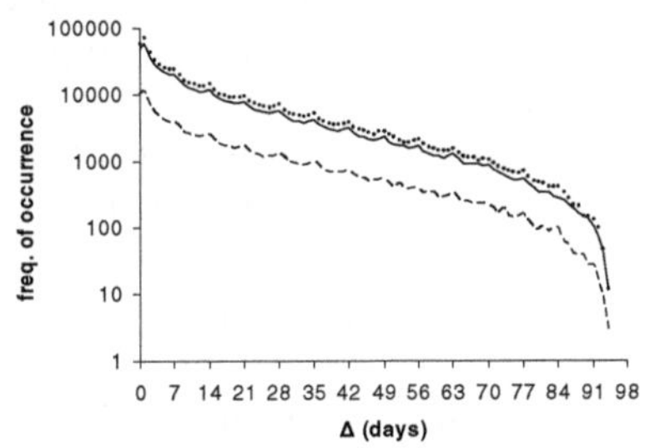

Fig. 2. As Figure 1, with the addition of: same click pairs resulting from the same query (middle curve); same click pairs resulting from different queries (lower curve)

The repeat click data was sub-divided into two sets, click pairs resulting from the same query and click pairs resulting from different queries. Figure 2 shows this breakdown. The upper of the two new lines shows click pairs resulting from the same query; the lower line shows clicks resulting from different queries. As can been seen in the graph, the number of repeat clicks from different queries is substantially lower than from the same query. The gap between the two lines decreases slightly as the difference in time between clicks grows. Table 1 charts the proportional difference between the two curves, calculated by the following formula

$$\frac{(\text{Same Queries} - \text{Different Queries})}{\text{Total Queries}}$$

As the Δ between the paired events of a user clicking on a particular search result URL grows, the user is more likely to reach that URL via a different search query.

Table 1. Table of the percentage relative difference between the same query and different query lines in Figure 2

Δ (days)	0	14	28	42	56	70
Relative difference	80%	79%	77%	76%	77%	74%

3.2 Periodicities in Repetitions

It can also be seen in the histograms in Figures 1 and 2 that there is a seven day periodicity in the data. From this data we can infer that if a user uses a search engine on a particular day of the week, they are more likely to re-use the engine on the same day in the following weeks. The weekly periodicity is observable for pairs of click events that occur months apart. A more detailed analysis of the log data revealed that the periodicity was due to a *weekend effect*. Users who access the search engine on a weekend are more likely to use the engine again on a weekend than on a weekday. It was found that if one observes a user event on a weekend, the probability of that user's next event also happening on a weekend was 55% (by chance, the probability was 28.6%, $^{2}/_{7}$). For weekdays, the probability of the next event also occurring on a weekday was 81%, by chance it was 71% ($^{5}/_{7}$). The frequency of occurrence of search events on the engine was also different for each day (see Table 2). This combination of factors leads to the observed periodicity.

Table 2. Distribution of queries by days of the week

Sun	Mon	Tues	Wed	Thurs	Fri	Sat
14%	16%	15%	15%	14%	13%	13%

Recalling the definition of user set out in Section 2 (an ID associated with a user account on a particular PC), we conclude that accounts on a PC used for searching on a weekday tend to be used more for searching on other weekdays and that accounts used on a weekend tend to be used more on other weekends.

4 Further Temporal Analysis

From the histogram in Figure 1, it was clear that a number of factors were influencing the shape of the graph: the 94 day windowing effect (which caused the slope and sharp tail off) and the weekend effect (which caused the 7 day periodicity). Therefore, we further analyzed the data using a series of normalizations to remove such effects.

4.1 Normalizing the Data

In order to examine just the windowing effect in the data (independent of any repetition), query events in the log were randomly paired ignoring which user or query they came from. A histogram of the events binned by the number of days between the two events is shown in Figure 3. The windowing effect is clear. We can now use this curve to remove the windowing artifact from other analyses. Queries issued by the same user were randomly paired. A histogram of this plot is shown in

Figure 4: both the windowing and weekend effects are present. To remove the windowing effect, the data in Figure 4 was normalized using the randomly paired data in Figure 3 producing the graph in Figure 5. The normalization formula is shown below: the count c at a certain Δ_i (expressed as a fraction of the total counts) is divided by a similarly calculated fraction from the normalizing data.

$$\left(c(\Delta_i) \Big/ \sum_{t=0}^{94} c(\Delta_t) \right) \Big/ \left(cn(\Delta_i) \Big/ \sum_{t=0}^{94} cn(\Delta_t) \right)$$

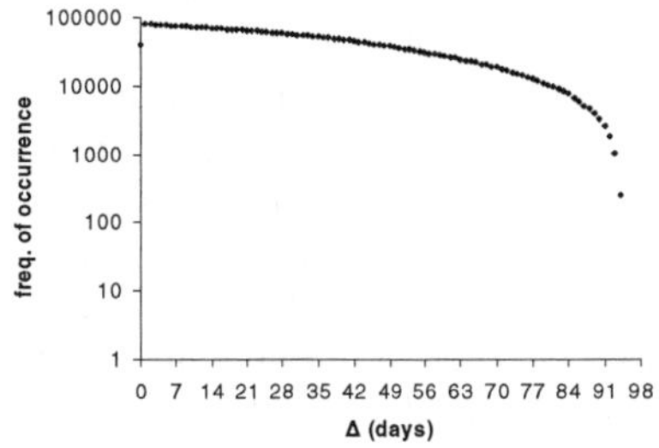

Fig. 3. Histogram of randomly paired events

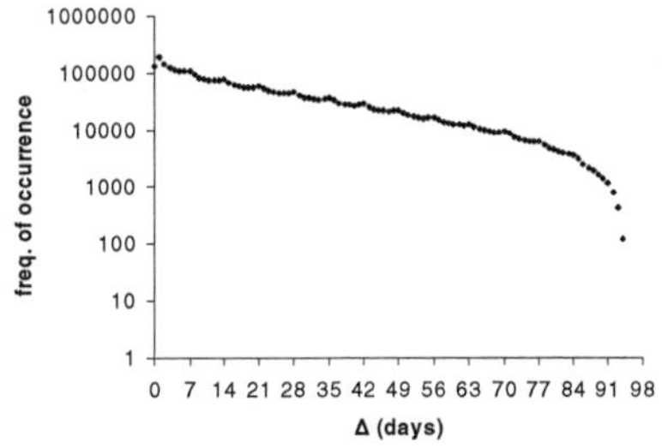

Fig. 4. Histogram of randomly paired events from the same user

In this normalized view, the horizontal line in Figure 5 crossing the y-axis at 1 is where the data points in Figure 3 would be plotted. Anything appearing above or below the line constitutes a deviation from the norm. Plots above the line are events occurring more often than found in the normalizing data; in contrast, anything plotted below occurs less often.

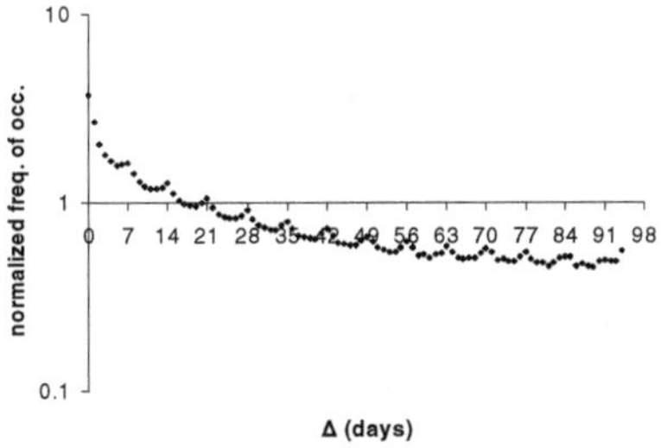

Fig. 5. Normalized histogram of events from the same user (Figure 4 normalized by data in Figure 3). Y-axis is a normalized count of the number of events in each bin.

From these graphs it can be seen that if a user issues a query to a search engine, the chances of them issuing another query on the same day (Δ=0) is 3-4 times more likely than would be expected by chance. Users are more likely than chance to re-use the search engine after issuing a query for a period of up to 20-21 days. Two search events by the same user with a difference greater than 21 days are less likely to occur than would be expected by chance. This graph shows that on average, users' search engine use tends to be *bursty*. If we observe users searching, we are likely to observe them searching again relatively soon, probably within the next three weeks; beyond that time, however, there is an increasing chance they may not be observed again.

4.2 Analyzing Query Repetition

Replicating the methodology used above, we examined user behavior with repeated queries. Events in the logs from the same user issuing the same query were randomly paired and a histogram of those events binned by the time difference in days was plotted. The graph produced was normalized by the data in Figure 4, so as to eliminate the windowing & weekend effects as well as the burstiness of user search behavior. The results are shown in Figure 6, where it can be seen that users re-issuing of queries to a search engine is more bursty than user search engine re-use. If a user issues a particular query, they are likely to re-issue that query again within the following 7 days. After that, however, the chance of observing the same query from the same user reduces. Users appear to have a limited interest in pursuing a particular query. Whether this is because the user's information need was satisfied or because the user gave up is left for study in future work.

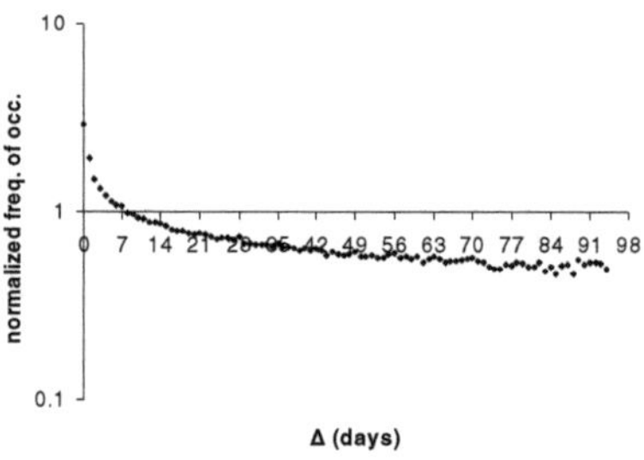

Fig. 6. Normalized histogram of counts of the same user issuing the same query

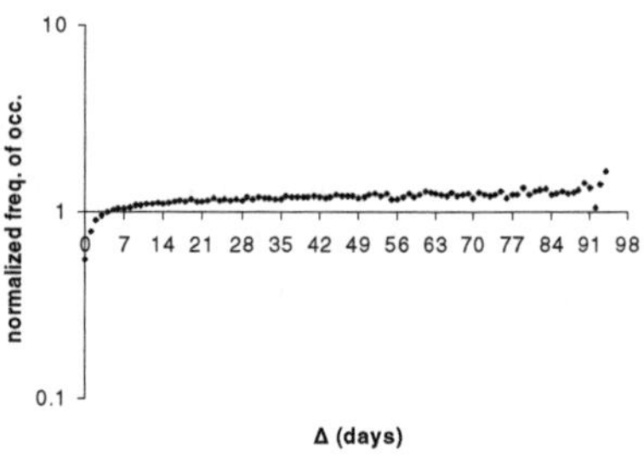

Fig. 7. Normalized histogram of repeat navigational queries issued by the same user

4.2.1 Examining Query Types

We examined whether the pattern of query repetition observed so far varied for different query types. Based on heuristics like those identified by Teevan et al. and by Lee et al. (2005), we generated a list of navigational queries. Queries in the logs that matched items in this list were randomly paired and normalized by the data used to generate the graph in Figure 6. The results are shown in Figure 7. As can be seen, if a user issues a navigational query, we are less likely to observe the same query being issued again within a few days than would be expected from general repeat query behavior by users (Figure 6). Thus the burst of repeat queries observed in Figure 6 appears due to non-navigational queries, which tend to be more information seeking focused. From this examination of the data sets, we conclude that repeat query behavior is different depending on the nature of the users' query. Navigational queries are less likely to be repeated by users within a few days than queries with a more information seeking focus, and navigational queries are more likely to be repeated at later points in time.

A final aspect of repeat searching behavior was examined: queries repeatedly submitted to a search engine by different users. For this analysis, search requests from different users were examined. It was found that certain such queries occurred in bursts. Within this set, queries that were topical for the time period covered by the query logs such as "April fools day" and "spring cleaning tips" were found as well as novelty or news queries such as "33-pound cat", "Grammy music awards", "testing of a scramjet engine", etc. It was found that such queries had a higher than expected frequency of occurrence for around 1-2 weeks.

5 Hourly Analysis of User Queries

The presence of hourly periodicities in user behavior was also examined. The data used to produce Figure 6 was re-binned to hourly differences between events to produce the results in Figure 8. As can be seen, users' use of search engines follows a strong 24 hour periodicity. Users who query at a particular time on one day are likely to query at that same time on a different day. The data in Figure 6, which shows repeated queries from the same user, was similarly re-binned (shown in Figure 9). Remembering that the data in Figure 6 was normalized to remove windowing & weekend effects as well as user search periodicities, it is striking that users issuing the same query seem to show a stronger 24 hour periodicity in their behavior than is observed in general user search behavior.

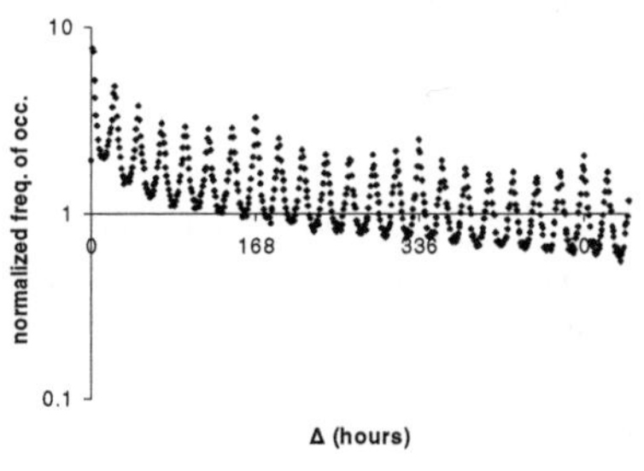

Fig. 8. Histogram of the first 22 days of randomly paired query log events from the same user. Note points on the x axis mark out weeks.

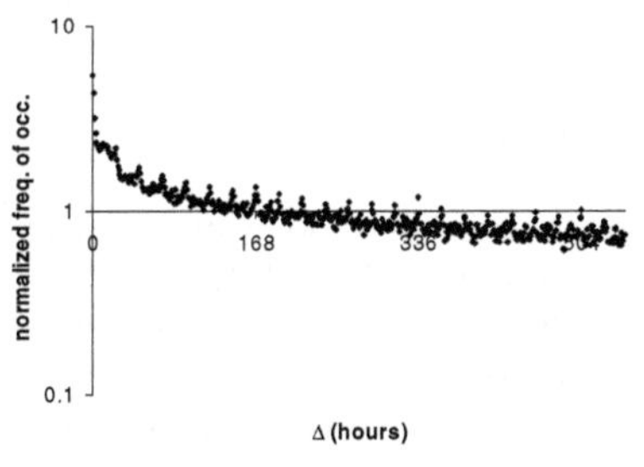

Fig. 9. Histogram of the first 22 days of randomly paired events from the same user issuing the same query

It is not entirely clear why users repeatedly searching with the same query would be more likely to do so at the same time of day. A preliminary examination of these regular queries revealed that some queries seemed to be associated with a particular time of day, such as queries related to a TV show (e.g. "deal or no deal", "american idol"). Others appeared to be queries that a user issued regularly to monitor a particular event or topic (e.g. queries for lotteries; see also Kellar et al., 2006 for a description of monitoring queries). Although there was no quality to these queries that in themselves would indicate they should be issued at a common hour, there was an indication in the data that the peaks were more likely to occur on weekdays than on weekends. One might speculate that during the week, times when search engines are used are regulated by the structure of people's work and school lives. The exact reason for this periodicity is left to future work.

6 Conclusions and Future Work

This paper presented an analysis of repetitions in user search behavior. Many queries and URL clicks are repeated over time. Users show both seven day and 24 hour periodicities in their use of search engines and these periods repeat and stay consistent over many weeks. Use of search engines is also bursty with users tending to repeatedly use search engines within a short period of time (e.g. a few weeks). Users re-issuing queries displayed an even stronger burstiness, although for navigational queries, the opposite was observed with users unlikely to re-issue navigational queries within a few days of first issuing the query. Queries that are repeatedly issued by different users were also examined and found to be related to temporally varying events or news (see also Vlachos et al., 2004).

The work in this paper constitutes a preliminary analysis of the topic of repetitions in user interactions with search engines. All analyses presented in this paper, described the general behavior of a large user population. We have not yet examined the variation within the averages and the degree to which individuals deviate from the norm. It is also unclear to what extent the periodicities that we have observed are related uniquely to search engine use or are a reflection of general use of the Web or even of general computer use. An examination of such behavior would be one avenue of future work.

References

Andrei Broder. A taxonomy of web search. *SIGIR Forum 36(2), Fall 2002*, 3-10, 2002.

Steven M. Beitzel, Eric C. Jensen, Abdur Chowdhury, David A. Grossman and Ophir Frieder: Hourly analysis of a very large topically categorized web query log. ACM *SIGIR 2004*, 321-328, 2004.

Bernard J. Jansen, Amanda Spink: How are we searching the World Wide Web? A comparison of nine search engine transaction logs. *Information Processing and Management, 42(1)*, 248-263, 2006.

Melanie Kellar, Carolyn Watters and Michael Shepherd. A goal-based classification of web information tasks. *ASIST*, 2292-2315, 2006.

Uichin Lee, Zhenuy Lui and Junghoo Cho. Automatic identification of user goals in web search. *WWW 2005*, 391-400, 2005.

Craig Silverstein, Monica Henzinger, Hannes Marais and Michael Moricz. Analysis of a very large web search engine query log. *DEC SRC Technical Note 1998-014*, 1998.

Jaime Teevan, Eytan Adar, Rosie Jones and Michael Pott. History repeats itself: Repeat queries in Yahoo's logs. ACM *SIGIR 2006*, 703-704, 2006.

Michail Vlachos, Christopher Meek, Zografoula Vagena, Dimitris Gunopulos. Identifying similarities, periodicities and bursts for online search queries. ACM *SIGMOD 2004*, 131-142, 2004.

Popularity Weighted Ranking for Academic Digital Libraries

Yang Sun and C. Lee Giles

Information Sciences and Technology
The Pennsylvania State University
University Park, PA, 16801, USA

Abstract. We propose a popularity weighted ranking algorithm for academic digital libraries that uses the popularity factor of a publication venue overcoming the limitations of impact factors. We compare our method with the naive PageRank, citation counts and HITS algorithm, three popular measures currently used to rank papers beyond lexical similarity. The ranking results are evaluated by discounted cumulative gain(DCG) method using four human evaluators. We show that our proposed ranking algorithm improves the DCG performance by 8.5% on average compared to naive PageRank, 16.3% compared to citation count and 23.2% compared to HITS. The algorithm is also evaluated by click through data from CiteSeer usage log.

Keywords: weighted ranking, citation analysis, digital library.

1 Introduction

Effectively indexing and retrieving information from large document databases continues to be a challenging task. Automated digital libraries make it easier for users to access and use these databases. In Web search, PageRank [14] and HITS [9] algorithms created to measure importance or authority as a ranking factor showed a great success compared to lexical similarity measures. Citation count is also widely used in evaluating the importance of a paper. However, unweighted citation counting often does not accurately describe the impact of papers [11].

The publication process of academic papers makes the citation graph much different from the Web graph. First, publication date and content of papers usually do not change over time whereas those of the Web pages can. Second, the typical citation graph of academic papers is an acyclic digraph without loops (there are rare exceptions to this). It is also common that a paper does not cite future papers. (Except in unusual cases where papers can cite unpublished work that is published later. In that case these papers can be treated as multiple versions.) Thus, the interpretation of the naive PageRank algorithm would be problematic[13,17]. We introduce a popularity factor weighted ranking algorithm based on PageRank with significantly improved performance for ranking academic papers. Our contributions are as follows:

G. Amati, C. Carpineto, and G. Romano (Eds.): ECIR 2007, LNCS 4425, pp. 605–612, 2007.

- We define a new popularity factor that reflects the influence of publication venues and overcomes the limitations of a venue's impact factor.
- A popularity factor weighted ranking score of a paper in the proposed ranking method is defined by the weighted citations from other papers and the popularity factor of its publication venue and is implemented on the CiteSeer metadata.
- A user study with four evaluators shows the improved performance of this new algorithm. We also use clickthrough data from CiteSeer usage log to validate the results.

2 Weighted Ranking

According to information foraging theory[15], users of information retrieval systems will evaluate the value of documents by information cues (such as title, author, venue, citation count, publication date of a paper in academic digital libraries) and follow the most valuable document. The more cues they encounter, the better they can evaluate the value. Lexical similarity only shows a limited information cue about a document. Citation count as an information cue is usually considered to be strongly correlated to academic document impact [12]. Although it is widely used in academic evaluation, citation count has limitations which make it less precise. The citation count of an individual paper by itself does not reflect the different citation structure in each discipline[19]. The papers with high impact and the ones with low impact are treated the same in citation count [4].

2.1 Weighted Ranking Method

In our research users are modeled as optimal information foragers who evaluate the cost of pursuing a document by information cues and follow the most valuable document [15]. According to this user model, the reference in a high impact paper will have a high probability to be followed by users. The citations of a paper should be viewed as weighted links. Not only the count of citations but also the impact of citations matters in this sense. Furthermore, the quality of the publication venue where a document is published is also an important information cue for users to evaluate the value of a document.

Popularity Factor. All serious research publication venues have a peer review process for publishing papers. It is fair to consider that the impact of a paper is partially reflected by where the paper is published. Impact factors of journals are widely used in evaluating the quality of publication venues[3]. There are limitations about the definition and the usage of impact factors. The impact factor is not normalized across research areas[6]. The calculation of an impact factor only considers a 3 year period. But important papers may receive many citations after 3 years[18]. Conferences are not considered in the calculation. Conferences often play very important roles in computer and information science research because of their timeliness and popularity.

We introduce the popularity factor to consider venue as an information cue and to reflect the influence of a publication venue. The popularity factor is defined based on citation analysis of publication venues. Note that the popularity factor does not distinguish journals from conference or workshop proceedings. The popularity factor of a publication venue v in a given year is defined by Equation 1:

$$PF(v,t) = \frac{n_v}{N} * \sum_{i \in P} \frac{PF(i,t) \times w(i)}{N(i)}. \quad (1)$$

where *PF(v,t)* is the popularity factor of publication venue v in a given year t, P is the set of publication venues i which cite v in that year, and n_v is the number of papers published in venue v in that year. If N is the total number of papers published in that year in a particular discipline, n_v/N represents the probability of a reader having access to a paper published in the publication venue v. Let *w(i)* be the weight which represents the frequency that venue i cites venue v. *N(i)* is the total number of references generated by venue i, and *PF(v,t)* is normalized so their squares sum to 1 for reasons of convergence: $\sum_v (PF(v,t))^2 = 1$ and has a range from 0 to 1 with larger values ranked higher. In our definition, a discipline is considered as a collection of related publication venues. The number of total papers in a discipline can be obtained by counting all the papers in all venues of the discipline. Multiple popularity factor values for one venue may occur in different disciplines.

According to our definition, this popularity factor differs from the impact factor by considering the impact of all publication venues, recent to long ago papers, and the probability of reader access. These differences overcome several shortcomings of the impact factor and provide a robust and reliable measure for publication venues. The popularity factor is computed with a simple iterative algorithm and achieve convergence point after 18 iterations [9,14]. Top 5 venues ranked by popularity factors in 2004 based on our database (includes primarily Computer Science and Engineering papers) is listed in Table 1.

Table 1. Popularity factors for computer science venues in 2004. Conferences and journals are both included.

Popularity factor	Name
0.05868	INFOCOM
0.04277	ACM SIGPLAN
0.04027	ACM SIGCOMM
0.02731	Human factors in computing systems
0.02622	Mobile computing and networking

Ranking Score. A paper will nearly always be cited after it is published. A citation relationship in a published document should not change over time (revisions to technical reports may be an exception). Figure 1 illustrates the temporal effect of citation graphs.

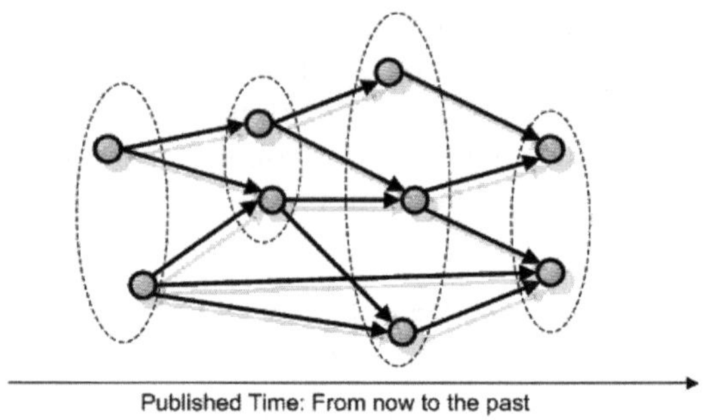

Fig. 1. A schematic illustration of a citation graph. Each circle represents a paper. Arrows represent citations. It can be seen that citation graphs typically do not have backward citations. They are acyclic digraphs.

With the popularity factor and temporal effect, we define the ranking score $R(d_T)$ of an academic paper d at a previous time T in Equation 2 as

$$R(d_T) = PF(v_{d_T}) + \sum_{t>T, d_t \in D} \frac{R(d_t)}{N(d_t)}. \tag{2}$$

where $R(d_t)$ is the ranking score of a paper dt which is published at time t and cite paper d_T. D is the set of papers which cite d_T. $N(d_t)$ is the number of references in paper d_t. $PF(v_{d_T})$ is the popularity factor of the publication venue v where paper d_T is published. The ranking score has a range of 0 to very large numbers. The vector PF is considered as an initial score of a paper when there is no citation record for this paper. This ranking assumes that the ranking score of a previously published paper will not have any impact on later published ones. Our algorithm does not permit bidirectional citations. If a paper is cited before it is published (no a common case), the paper will be considered to have two versions and the two versions will have separate rankings. The citation graph can be constructed as a function of publication time because of the temporal property of papers. The adjacency matrix of the graph can be then sorted and form a strict upper triangular matrix. Then, the equation of ranking scores can be written to a system of n linear equations with n unknowns, which has a single unique solution. As such there are no convergence issues. Notice that the ranking function has a computational complexity of $O(nm)$ where n is the number of papers in the database and m is the average citations to a paper. m typically ranges from 0 to less than 50000 with a power law distribution for academic papers [16], making the ranking algorithm scalable for large digital libraries. The evaluation method and the results are discussed in the next section in order to demonstrate how our algorithm improves the ranking performance.

3 Evaluation

Evaluating the quality of ranking algorithms used in an information retrieval system is a nontrivial problem[5,7]. Most evaluation methods involve human evaluators judging the relevance of the retrieved documents to specific queries. In this research we compare our ranking method to naive PageRank, citation count and HITS using discounted cumulative gain method with four human evaluators. Two of the evaluators were graduate students from computer science department whose research interest is in data mining and Web search. One is a research programmer who has experience on machine learning program development. The other evaluator is a software engineer with Master's degree in computer science. The four ranking algorithms are implemented in a basic information retrieval system which indexes 536,724 papers' metadata from the CiteSeer database[1] using Lucene[2] in which *tf-idf* is used as the basic ranking function. The evaluation is expected to compare the performance of the re-ranking functions beyond lexical similarity. Top 5 ranked papers by our algorithm with comparison to naive PageRank, citation count and authority scores of HITS are shown in Table 2.

Table 2. Ranking scores of the top 5 papers in computer science

Title	Weighted	PageRank	Citation	HITS
The Anatomy of a Large-Scale Hypertextual Web Search Engine - Brin (1998)	0.13504	0.11924	521	0.89659
Boosting a Weak Learning Algorithm By Majority - Freund (1995)	0.07568	0.06363	174	0.27771
A Tutorial on Learning With Bayesian Networks - Heckerman (1996)	0.04321	0.04048	203	0.20106
Irrelevant Features and the Subset Selection Problem - John (1994)	0.04097	0.05434	290	0.14429
Dynamic Itemset Counting and Implication Rules for Market Basket Data - Motwani (1997)	0.03944	0.04392	232	0.43546

3.1 Evaluation Method

Discounted cumulative gain (DCG) [8,20] measure considers the ranking of papers in addition to the relevant score from evaluators. The assumption in DCG is the lower the ranked position of a paper the less valuable it is for the user because the less likely it is going to be examined by users. Human evaluators are required to rate the ranked papers with a score from 0 to 2, where 0 represents non relevance, 1 represents marginal relevance, and 2 represents high relevance. The DCG value is computed separately at each relevance level. The agreement among evaluators is examined by kappa statistics to show the confidence level by using the evaluation results. The kappa statistic is a standard measure of agreement with categorical data to access the degree to which two or more raters agree in assigning data to categories [10].

Fifty queries were selected from the CiteSeer query log. Papers are ranked by the four ranking algorithms separately in addition to the *tf-idf* lexical similarity measure. Top twenty papers are mixed and presented to four human evaluators (using four evaluators is considered enough for DCG evaluation methods [8,20]).

3.2 Results

Kappa Measure. The agreement among the four evaluators was examined for each of the 50 queries. The average kappa agreement among the four evaluators is 53% which is considered to be in the level of moderate agreements [10].

Precision Recall. The precision-recall curves at different relevance levels for the four algorithms using standard methods [8] are presented in Figure 2.

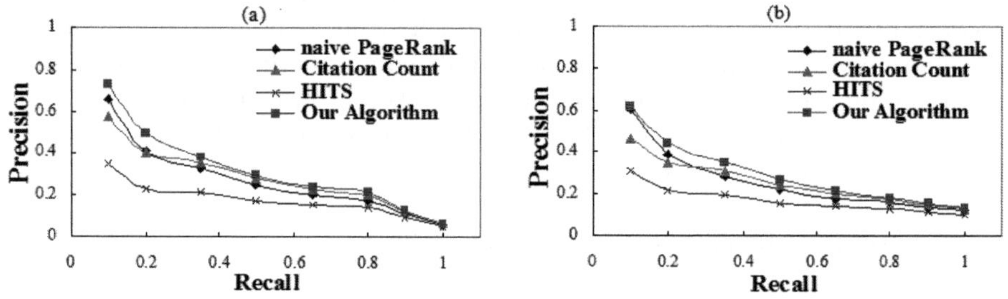

Fig. 2. Precision-Recall of the four ranking algorithms at relevance level 1(a) and 2(b)

DCG. The DCG vector curves for ranks 1-20 is shown in Figure 3. The DCG scores are shown in Table 3. Both the curve and statistics show that our algorithm significantly outperforms the other three algorithms for documents ranked after rank 10.

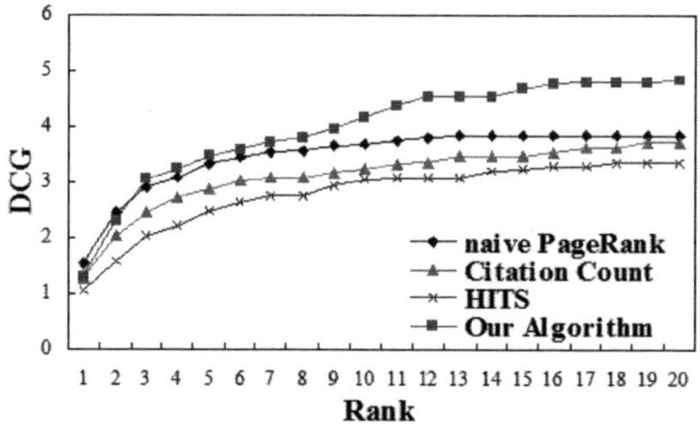

Fig. 3. DCG at various document rankings

3.3 Validity of Results

To validate the results, a pairwise measure of each ranking algorithm is calculated using the clickthrough data extracted from CiteSeer usage log. For any pair of papers, if the clickthrough rate of the high-ranked paper is larger than the low-ranked paper, we consider this pair is correctly ranked. The average pairwise accuracy of each algorithm is listed in Table 4.

Table 3. The DCG score for the four ranking algorithms

	Our Algorithm	naive PageRank	Citation Count	HITS
DCG @ rank 1	1.3	1.54	1.28	1.05
DCG @ rank 5	3.49	3.33	2.87	2.49
DCG @ rank 10	4.18	3.71	3.24	3.06
DCG @ rank 15	4.69	3.86	3.49	3.25
DCG @ rank 20	4.86	3.86	3.71	3.36
Ave. DCG(1-10)	3.28±0.87	3.12±0.68	2.69±0.62	2.35±0.64
Ave. DCG(10-20)	4.68±0.16	3.85±0.04	3.54±0.14	3.25±0.11

Table 4. Average pairwise accuracy based on Clickthrough data

	Our Algorithm	naive PageRank	Citation Count	HITS
Ave. accuracy	74.18%	70.29%	67.1%	67.13%

4 Discussion and Conclusions

A new weighted ranking score of a paper was defined by the weighted citations from other papers and the popularity factor of its publication venue. A ranking system based on the Lucene index and Citeseer metadata was built and was used to evaluate our algorithm with comparison to other popular ranking algorithms.

The algorithm is evaluated by DCG method using human evaluators and we compare its results to the ranking of naive PageRank, citation counts and HITS. The comparison results show that the weighted ranking algorithm improves the DCG ranking performance by 8.5% compared to naive PageRank, 16.3% compared to citation count and 23.2% compared to HITS. We also use the clickthrough data from CiteSeer usage log to validate the results. This leads us to believe that our weighted ranking algorithm is more accurate than those currently being used.

Our evaluation experiment shows that the user agreement on paper rankings is not very high. Effective personalized ranking algorithms would most likely satisfy the diversity of most user information needs.

References

1. Citeseer, http://citeseer.ist.psu.edu.
2. D. Cutting, "The Lucene Search Engine," http://lucene.apache.org/, 2006.
3. E. Garfield, "The impact factor," Current Contents, 25, 3-7, 1994.
4. S. Harnard, "The New World of Webmetric Performance Indicators: Mandating, Monitoring, Measuring and Maximising Research Impact in the Open Access Age," Proc. of the 1st ECSP in Biomedicine and Medicine, 2006.
5. D. Hawking, N. Craswell, P. Thistlewaite, and D. Harman, "Results and challenges in Web search evaluation," Proc. of the 8th International World Wide Web Conference, 1321-1330, 1999.

6. F. Hecht, B. Hecht, and A. Sandberg, "The journal "impact factor": a misnamed, misleading, misused measure," Cancer Genet Cytogenet, 104(2), 77-81, 1998.
7. D. Hull, "Using statistical testing in the evaluation of retrieval experiments," Proceedings of the 16th annual international ACM SIGIR Conference, 329-228, 1993.
8. K. Jarvelin and J. Kekalainen, "IR evaluation methods for retrieving highly relevant documents," Proc. of the 23rd SIGIR conference, 41-48, 2000.
9. J. M. Kleinberg, "Authoritative sources in a hyperlinked environment," Journal of ACM, 48, 604-632, 1999.
10. R. J. Landis and G. G. Koch, "The measurement of observer agreement for categorical data," Biometrics, 33, 159-174, 1977.
11. S. Lehmann, B. Lautrup, and A. D. Jackson, "Citation networks in high energy physics," Physical Reivew E68, 026113 (2003).
12. F. Narin, "Evaluative bibliometrics: The use of publication and citation analysis in the evaluation of scientific activity," Cherryhill, N.J.: Computer Horizons, 1976.
13. Z. Nie, Y. Zhang, J. Wen, and W. Ma, "Object-Level Ranking: Bringing Order to Web Objects," Proc. of the 14th International World Wide Web Conference, 2005.
14. L. Page and S. Brin, "The PageRank citation ranking: bringing order to the web," tech. report SIDL-WP-1999-0120, Stanford University, Nov. 1999.
15. P. Pirolli and S. Card, "Information foraging in information access environments," Proc. of the SIGCHI conference, 51 - 58, 1995.
16. S. Redner, "How Popular is Your Paper? An Empirical Study of the Citation Distribution," European Physical Journal B, 4, 131-134, 1998.
17. M. Richardson, A. Prakash, and E. Brill, "Beyond PageRank: Machine Learning for Static Ranking," Proc. of the 15th International World Wide Web Conference, 2006.
18. P. Seglen, "Why the impact factor of journals should not be used for evaluating research," British medical journal, 314(7079), 498-502, 1997.
19. Thomson and Corporation, "In Cites: Analysis Of" http://www.in-cites.com/analysis/, 2005.
20. E. Voorhees, "Evaluation by Highly Relevant Documents," Proc. of the 24th annual international ACM SIGIR conference, 74-82, 2001.

Naming Functions for the Vector Space Model

Yannis Tzitzikas and Yannis Theoharis

Computer Science Department, University of Crete, Greece, and
Institute of Computer Science, FORTH-ICS, Greece
{tzitzik, theohari}@ics.forth.gr

Abstract. The Vector Space Model (VSM) is probably the most widely used model for retrieving information from text collections (and recently from over other kinds of corpi). Assuming this model, we study the problem of finding the best query that "names" (or describes) a given (unordered or ordered) set of objects. We formulate several variations of this problem and we provide methods and algorithms for solving them.

1 Introduction

This paper elaborates on the "naming problem" i.e. the problem of finding a query (or the best query) that describes (names) a given (unordered or ordered) set of objects. The motivation for studying this problem is that several variations of it are useful in a number of tasks (e.g. for relevance feedback, document representation learning). In addition, naming functions can be exploited for providing flexible information access services, for instance, they can be exploited for realizing the context-based interaction scheme described in [9].

More precisely, naming functions can be exploited for supporting an alternative query formulation process. The user selects a number of objects (i.e. a set A') and asks the system to formulate the query that "describes" these objects. The system returns a query q' and subsequently the user can change it in a way that reflects his/her information need. Roughly this resembles the Query By Example (QBE) process in relational databases. It also resembles the relevance feedback mechanisms in IR systems. For example, consider a user who has formulated a query q and has been given an answer A. She can select a subset A' of the current answer A consisting of those elements of A which she finds relevant to her information need. Subsequently, the system has to change appropriately the query. Notice that here the user is not directly driven to the answer of the new query (as it is done in classical relevance feedback mechanisms). Instead, she is given the new query, so she has the opportunity to modify it before submitting it for evaluation. Probably this (intermediate step) could alleviate the problem of unexpected results that frequently occur after a feedback cycle. Notice that if $A' \subseteq A$ then it is like giving explicit negative relevance feedback (to the elements of $A - A'$), and "tacit" positive relevance feedback (to the elements of A'). However, this interaction scheme is more general than the classical feedback mechanisms of IR as it allows giving positive feedback to documents not already in the current answer. In addition, naming functions can help users to be attuned

G. Amati, C. Carpineto, and G. Romano (Eds.): ECIR 2007, LNCS 4425, pp. 613–620, 2007.

with the indexing and query language of the system. It is not hard to foresee additional applications of naming functions, in building mediators (metasearchers) and in cross-language retrieval. For these reasons, we believe that the naming problem deserves separate attention and analytical elaboration.

The rest of this paper is organized as follows. Section 2 states the naming problem and Section 3 provides solutions assuming the vector space model. Section 4 reports some experimental results. Finally, Section 5 concludes the paper and identifies issues for further research.

2 The Naming Problem

An IR system S, hereafter source, can be viewed as a function $S : Q \rightarrow \mathcal{A}$ where Q is the set of all queries that S can answer, and $\mathcal{A}$ is the set of all answers to those queries, i.e. $\mathcal{A}=\{ S(q) \mid q \in Q\}$. As the focus is given on retrieval queries, it is assumed that $\mathcal{A}$ is a subset of $\mathcal{P}(O)$, the powerset of O, where O is the set of all objects stored at the source. In order to capture sources that return sets of ranked objects (based on best-match retrieval models), we shall use B's to denote ordered answers (and A's for unordered ones). Given an ordered set B, we use $B(k)$ to denote the first k elements of B and $B\{k\}$ to denote the *set* of elements that appear in $B(k)$. We shall also use the notation $\{B\}$ to denote the *set* of elements that appear in B (in other words $\{B\} = B\{|B|\}$). So if $B = \langle o_1, o_2, o_3\rangle$ and $B' = \langle o_3, o_2, o_1\rangle$, then we can write $\{B\} = \{B'\}$. Furthermore, we use $B_{|A}$ to denote the restriction of B on the set A, i.e. the ordered set obtained if we exclude from B those elements that do not belong to A. Similarly, $\{B\}_{|A}$ denotes the restriction of $\{B\}$ on the set A, i.e. the unordered set obtained if we exclude from $\{B\}$ those elements that do not belong to A.

2.1 The Naming Problem for Unordered Sets

Def. 1. *A query q is an* exact name *of A and we write $q=n(A)$, if $S(q)\{|A|\}=A$.*

This means that the first $|A|$ elements of $S(q)$ are occupied by elements of A.

Def. 2. *A query q is the best* upper name *of A and we write $q = n^+(A)$, if there exists $m \geq |A|$, such that $S(q)\{m\}_{|A} = A$, where m is the minimum possible.*

This means that the restriction of the set of the first m ($m \geq |A|$) elements of $S(q)$ over A is A.

Def. 3. *A query q is the best* lower name *of A and we write $q = n^-(A)$, if there exists $0 < m \leq |A|$, such that $S(q)\{m\} \subseteq A$, where m is the maximum possible.*

This means that the first m ($m \leq |A|$) elements of $S(q)$ are occupied by elements of A. However, more often than not, non of these three names exists. For this reason below we define a more relaxed definition, which generalizes both *upper* and *lower* name definitions.

Def. 4. *A query q is a* relaxed name *of A and we write $q = \widetilde{n}(A)$, if $|S(q)\{m\} \cap A| = j$, where $m \geq j > 0$.*

If $m = j = |A|$ then q is an exact name. If $m > j = |A|$, then q is an upper name (the best one, if m is the least possible). If $m = j < |A|$, then q is a lower name (the best one, if m is the greatest possible).

2.2 The Naming Problem for Ordered Sets

Here we elaborate on the naming problem for ordered sets. Specifically, we consider weakly ordered sets as this captures both sets and linearly ordered sets. We use $\langle o_1, o_2, o_3 \rangle$ to denote a linearly ordered answer where o_1 is the most relevant object, and $\langle o_1, \{o_2, o_3\}, o_4 \rangle$ to denote a weakly ordered answer where o_2 and o_3 are equally relevant (e.g. they have the same degree of relevance).

Def. 5. *Let B be a weakly ordered set. A query q is:*

- *an* exact name *of B (and we write $q = n(B)$), if $S(q)(|\{B\}|) = B$.*
- *the best* upper name *of B (and we write $q = n^+(B)$), if $S(q)(m)_{|\{B\}} = B$, where m is the minimum possible.*
- *the best* lower name *of B (and we write $q = n^-(B)$), if $S(q)(m) = B(m)$, where m is the maximum possible (and $m > 0$).*
- *a* relaxed name *of B (and we write $q = \widetilde{n}(B)$), if $S(q)(j)_{|\{B(m)\}} = B(m)$, where m is the maximum and j is the minimum possible (and $m > 0$).*

3 Naming Functions for the Vector Space Model

The Vector Space Model (VSM) is one of the classical models for information retrieval over text collections and has been recently used for information retrieval over XML documents [2] and ontology-based sources [3]. Let $T = \{t_1, \ldots, t_k\}$ be the set of different words that appear in the documents of the collection O (we ignore stemming, stopwords). According to the VSM, the degree of relevance equals the cosine of the angle between the vectors $\boldsymbol{q}$ and $\boldsymbol{o_j}$.

3.1 Naming Functions for Unordered Sets

As one can imagine, it is very difficult to find the upper name for a given set of objects assuming the classical index structures that IR systems and Web search engines currently use (i.e. inverted files). However, several methods could be used to find an approximate upper name, the most well known example being the best query of the Rocchio method [6]. To be more specific, below we list a number of conditions that a candidate query q could satisfy:

(a) minimizes the maximum distance between q and an element of A,
(b) minimizes the average distance between q and the elements of A,
(c) is the result of subtracting from a query q_1 that satisfies (b) a query q_2 that minimizes the average distance between itself and all those elements not in A.

A query that satisfies (b) is $q = \frac{1}{|A|} \sum_{d \in A} d$. A query that satisfies (c) is $q = \frac{1}{|A|} \sum_{d \in A} d - \frac{1}{|Obj \setminus A|} \sum_{d \notin A} d$, what is called "best query" on which the standard Rocchio method for relevance feedback is based on. A query that satisfies (a) is $q = \frac{1}{2}(a_1 + a_2)$, where a_1, a_2 are the most distant (i.e., less similar) documents of A.

Note that none of these methods guarantees that the resulting query is the best upper name (as defined in Def. 2). In particular, (a) can be better than (b), as depicted in Figure 1 (left). In this case, A comprises 3 elements, i.e., a_1, a_2 and a_3. y is an element that does not belong to A. o is the query vector that method (b) yields, while u is the one that method (a) yields. Obviously, o will rank documents in the rank $\langle a_2, a_1, y, a_3\rangle$, while u will rank them as follows, $\langle a_2, \{a_1, a_3\}, y\rangle$. Also, (a) can be better than (c), as depicted in Figure 1 (right). In that case y_1, y_2 are documents not in A, o is the query vector that method (c) yields, while u is the one that method (a) yields. Note that the vector $q_2 = y_1 + y_2$ (i.e., $\sum_{d \notin A} d$) has the same direction as $q_1 = \frac{1}{3}(a_1 + a_2 + a_3)$ (i.e., $\sum_{d \in A} d$) and thus the resulting query o will rank the documents in the rank $\langle a_2, a_1, \{y_1, y_2\}, a_3\rangle$, while u will rank them as follows, $\langle y_2, a_2, \{a_1, a_3\}, y_1\rangle$. The latter is a better upper name than the former according to Def 2, because it has m=4, while the former has m=5.

We should stress that the query yielded from method (a) can be evaluated more efficiently than the queries yielded by the other two methods, since it only comprises the words of the two most distant documents.

Methods $(b), (c)$ yield a relaxed name. Specifically, and with respect to Def. 4, we can say that method (b) attempts to maximize j, irrespective of the value of m. On the other hand, method (c) attempts to maximize j and minimize $m - j$ simultaneously. Hence, method (c) yields a better approximation of the best lower name than (b).

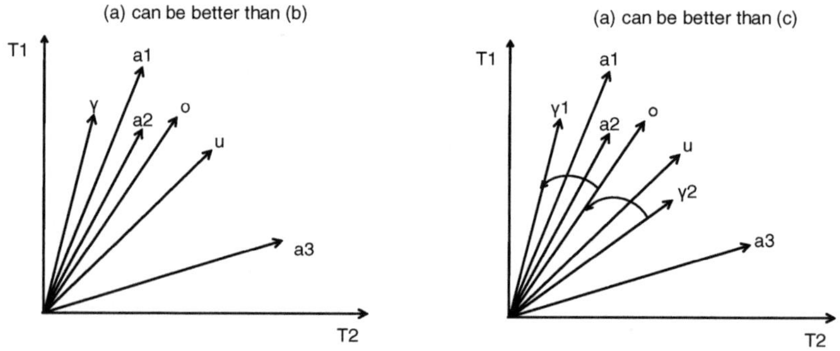

Fig. 1. Different methods to approximate the upper name

Bellow, we elaborate on the method (a) to approximate the upper name. Let a_1 and a_2 be the most distant (less similar) elements of A, i.e. $\cos(\boldsymbol{a_1}, \boldsymbol{a_2}) = \min(\{cos(\boldsymbol{a}, \boldsymbol{b}) \mid a, b \in A\})$. Since the measure of similarity equals the cosine between the corresponding vectors and the maximum angle between two document vectors equals $\frac{\pi}{2}$ (because the vector coordinates are positive) and cosine is monotonically decreasing (with maximum value at 0 and minimum at $\frac{\pi}{2}$), it follows that the most distant documents will be those that form the bigger angle. If there exist more than one such pairs a_1, a_2, then we choose arbitrarily one of them. The two less similar vectors of A define the $area(\boldsymbol{a_1}, \boldsymbol{a_2})$, which is the geometric space of all vectors of $[0,1]^{|T|}$ such that $\forall b \in area(\boldsymbol{a_1}, \boldsymbol{a_2})$, $|(\widehat{\boldsymbol{b}, \boldsymbol{a_1}})| < |(\widehat{\boldsymbol{a_1}, \boldsymbol{a_2}})|$ and $|(\widehat{\boldsymbol{b}, \boldsymbol{a_2}})| < |(\widehat{\boldsymbol{a_1}, \boldsymbol{a_2}})|$. As one can easily see, every query $q \in area(\boldsymbol{a_1}, \boldsymbol{a_2})$ ranks each document of $area(\boldsymbol{a_1}, \boldsymbol{a_2})$ higher than either a_1 or a_2. The query that minimizes the maximum distance between q and an element of A is the vector $\boldsymbol{q_A} = \frac{1}{2}(\boldsymbol{a_1} + \boldsymbol{a_2})$ whose direction is that of the vector that corresponds to the composed force of $\boldsymbol{a_1}$ and $\boldsymbol{a_2}$. So we can decide whether there is an exact name in two steps:

(1) We find the vectors $\boldsymbol{a_1}$ and $\boldsymbol{a_2}$, i.e.:

$$(\boldsymbol{a_1}, \boldsymbol{a_2}) = \arg_{\boldsymbol{a},\boldsymbol{a'}} min(\{cos(\boldsymbol{a}, \boldsymbol{a'}) \mid a, a' \in A\})$$

(2) we check if there are documents in $area(\boldsymbol{a_1}, \boldsymbol{a_2})$ that do not belong to A and how many they are. Let denote with n the number of these documents. If $n = 0$, then the exact name exists and it equals to q_A, i.e. $n(A) = q_A$. Otherwise, i.e. if $n > 0$, then there is not an exact name, because, since $q_A \in area(\boldsymbol{a_1}, \boldsymbol{a_2})$, it ranks each of these n documents of $O \setminus A$ higher than either a_1 or a_2.

Clearly, the cost of Step (1) is $\mathcal{O}(|A|^2)$, while the second step of the algorithm costs $\mathcal{O}(|O|)$ time, since we need to iterate over the whole collection in the worst case. However, in most cases (and with the availability of an index) we can just evaluate the query q_A and see if the first $|A|$ elements of the answer is A.

Let's now consider the case where there is not an exact name. In that case we would like to find an upper name and a lower name (ideally the best ones). Note that q_A (as defined earlier) will not rank first the elements of A, however it will rank the elements of A quite high. Specifically, it will be $S(q_A)\{|A| + n\}_{|A} = A$, i.e. q_A is an upper name of A. One can easily see that q_A is definitely the best upper name because for any other query q, if $S(q)\{j\}_{|A} = A$ then certainly $j \geq |A| + n$.

Finding the best lower name means finding a query q that ranks first the biggest number of elements of A, i.e. finding the maximum m ($m \leq |A|$) such that $S(q)\{m\} \subseteq A$. One expensive solution that would certainly yield the lower name would be to find the biggest subset of A that has an exact name (using the method presented earlier). The time complexity of this method is exponential with respect to $|A|$.

Another method is presented next. For each pair $b, b' \in (O \setminus A) \bigcap area(\boldsymbol{a_1}, \boldsymbol{a_2})$ we first investigate the $area(\boldsymbol{b}, \boldsymbol{b'})$: we want only elements of A to be there and if this is the case we compute their number. At the end we select the pair that gave the bigger number and consider as lower name the "composed force" of that pair[1]. The time complexity of this method is $\mathcal{O}(|O|^3)$, since all possible pairs of

[1] If there were an element not in A, then the result of this query would contain it.

(b, b') are $|O|^2$ and for such pair the method iterates over the whole collection in the worst case. Again a multidimensional index would greatly reduce the cost of computing $(O \setminus A) \bigcap area(\boldsymbol{a_1}, \boldsymbol{a_2})$ as well as the cost for computing $area(\boldsymbol{b}, \boldsymbol{b'})$.

3.2 Naming Functions for Ordered Sets

A common subproblem implied by all variations of the naming problem is to investigate how many documents of B are topologically located in the "right" order. For instance, if $B = \langle a_1, a_2, a_3 \rangle$ and a_3 rests in $area(\boldsymbol{a_1}, \boldsymbol{a_2})$, then there doesn't exist an exact name for B. Let a_i be the i-th in order document of B (a_1 stands for the most relevant and $a_{|\{B\}|}$ for the less relevant document of B), i.e. $B = \langle a_1, ..., a_{|B|} \rangle$. It is evident that a relaxed name exists, if $\exists i$ s.t. $2 \leq i \leq |B|$ and $sim(\boldsymbol{a_1}, \boldsymbol{a_2}) \geq ... \geq sim(\boldsymbol{a_1}, \boldsymbol{a_i})$. In that case a_1 is the relaxed name of B. If $i = |B|$ then a_1 is an upper or an exact name of B. Specifically, it is an exact name of B, if $S(a_1)(|B|) = B$, otherwise it is an upper name of it. Finally, if $i < |B|$ and $S(a_1)(i) = B(i)$, then a_1 is a lower name of B. The algorithm *LinearlySorted*, shown below, returns the maximum i.

The complexity of Alg. *LinearlySorted* is $\mathcal{O}(|B|)$. To check whether a_1 is a lower or upper or exact name, a query evaluation step is needed. If B is not a linear ordered but a weakly ordered set, e.g. $\langle \{a_1, a_2\}, a_3 \rangle$, then we treat the class of the most highly ranked elements in B (in our example the set $\{a_1, a_2\}$) as we did in Section 3.1, i.e. we try to find an exact name of it. This results in a query $q_{\{a_1,a_2\}}$ that is either an exact or an upper name. Then, we replace in B the set $\{a_1, a_2\}$ with $q_{\{a_1,a_2\}}$ and use the algorithm *LinearlySorted* with some slight modifications. Specifically, we view B as a linear order of sets, i.e. $B = \langle c_1, ..., c_j \rangle$, where $c_i \subseteq O$ and let $|B| = j$. The variation of the algorithm *LinearlySorted* for this case is the algorithm *WeaklySorted* shown below.

Alg. *LinearlySorted*	Alg. *WeaklySorted*
Input: B	Input: $B = \langle c_1, ..., c_j \rangle$
(1) $M := 1$	(1) $M := 1$
(2) for $i = 2$ to $\lvert\{B\}\rvert$ do	(2) for $i = 2$ to $\lvert B\rvert$ do
(3) if $sim(\boldsymbol{a_1}, \boldsymbol{a_i}) > M$	(3) $X := max(\{sim(n(c_1), a_k) \mid a_k \in c_i\})$
(4) then return i-1	(4) if $X > M$ then return i-1
(5) else $M := sim(\boldsymbol{a_1}, \boldsymbol{a_i})$	(5) else $M := X$
(6) return i	(6) return i

4 Experimental Evaluation

We have implemented naming functions on top of the experimental Web Search Engine, *GRoogle* (http://groogle.csd.uoc.gr:8080/groogle/), in order to investigate whether our approach can be applied on large collections. Experiments were carried out on a "normal" PC (Pentium IV 3.2GHz, 512MB RAM, Suse Linux v9.1). We used a collection crawled from *www.csd.uoc.gr* and *www.ics.forth.gr*. We considered two subsets of it, one consisting of 1,000 documents and 40,000 terms and another consisting of 5,000 documents and 250,000 terms. In order to select the documents that play the role of the answer set A (or B for ordered sets), we formulated a random query and defined A and B using the first 10/100 documents. We repeated our experiment 10 times and computed the average times. Recall that the computation of naming functions comprises two steps: a) find the name query, b) find what kind of name (upper/exact) this query is. The second step requires evaluating the name query, and clearly this depends on the underlying IR system. Since the name computed at the first step is certainly an upper name for the case of unordered sets (and relaxed name for the case of ordered sets), it is reasonable to measure separately the first step. Also, note that for ordered sets we always consider the first document of B as name and thus, the cost of the first step is negligible. Table 1 reports the execution times (t_a denotes the time to find the query, t_b the time to evaluate it and t_c the time to identify what kind of name the query is). Recall that T denotes the vocabulary of the collection. The main conclusion drawn is that the most costly task is to evaluate the name query in order to decide whether it is an upper/exact name. For this reason and in order to show the effect of the length of the computed name, we evaluated it once by considering only the 5 better (i.e., more weighted) terms and once more with the 10 better terms. Issues for reducing the query evaluation time are beyond the scope of this paper. From the results of the evaluation we can say that our method is efficient. Specifically, the cost of the first step depends only on the size of A and not on the collection. Concerning the effect of the vocabulary size, the cost of the first step is independent of it, more precisely it depends on the vocabulary of the documents in A (for the computation of the inner product of two vectors we only consider the non-zero coordinates of the vectors, which are usually much less than the vocabulary size).

Table 1. Execution times for computing names of unordered and ordered sets

Collection			Naming Functions (in sec)					
			Unordered			Ordered		
$\|O\|$	$\|T\|$	$\|\{A\}\|$	t_a	$t_b(query\ terms)$	t_c	t_a	$t_b(query\ terms)$	t_c
1K	40K	10	0.015	1.566 (5) - 3.174 (10)	0.001	0	1.878 (5) - 3.575 (10)	0.008
1K	40K	100	0.328	1.56 (5) - 3.176 (10)	0.005	0	1.624 (5) - 3.192 (10)	0.004
5K	255K	10	0.015	112.1 (5) - 262.5 (10)	0.001	0	131.2 (5) - 264.7 (10)	0.048
5K	255K	100	0.328	116.0 (5) - 251.1 (10)	0.006	0	153.4 (5) - 271.1 (10)	0.152

5 Concluding Remarks

Naming functions for taxonomy-based sources were originally proposed in [8] and were subsequently exploited for defining a flexible interaction scheme for information sources [9]. However, they were defined only for sources that return sets of objects. In this paper we formulated the naming problem for sources that return ordered sets of objects. Subsequently, we provided algorithms for computing naming functions appropriate for systems that are based on the vector space model. Somehow related work includes [4,7,1].

Although we provided polynomial algorithms for the naming function problem, they are by no means optimal and a lot of work has to be devoted for finding more efficient algorithms and appropriate indexing structures that would allow the application of this interaction scheme on large corpi of documents. For instance, indexes like [5] could be exploited for speeding up the computation of naming functions.

Another issue for further research and experimental investigation is how we could shorten the usually long queries that are computed by the naming functions. For instance, a user might prefer to receive only short queries, e.g. queries that contain at most k terms. One method to address this requirement would be to reformulate the naming problem by introducing one additional parameter (i.e. the maximum number of terms desired). An alternative approach (followed in the Section 4) would be to present to the user only the best k terms of the queries that the current naming functions compute, i.e. the k terms having the highest weights. The relative evaluation of these two approaches is an open issue.

Acknowledgement. This work was partially supported by the EU project CASPAR (FP6-2005-IST-033572).

References

1. R. S. Bot and Y. B. Wu. "Improving Document Representations Using Relevance Feedback: The RFA Algorithm". In *Procs of the 13th ACM intern. Conf. on Information and Knowledge Management*, Washington, USA, 2004.
2. Vinay Kakade and Prabhakar Raghavan. "Encoding XML in Vector Spaces". In *Proceedings of the 27th European Conference on Information Retrieval*, Santiago de Compostela, Spain, 2005.
3. Latifur Khan, Dennis McLeod, and Eduard Hovy. "Retrieval effectiveness of an ontology-based model for information selection". *The International Journal on Very Large Data Bases*, 13(1):71–85, January 2004.
4. Massimo Melucci. "Context Modeling and Discovery Using Vector Space Bases". In *Proceedings of the 14th ACM international Conference on Information and Knowledge Management*, Bremen, Germany, 2005.
5. G. Qian, S. Sural, Y. Gu, and S. Pramanik. Similarity between Euclidean and cosine angle distance for nearest neighbor queries. *Proceedings of the 2004 ACM symposium on Applied computing*, pages 1232–1237, 2004.
6. J.J. Rocchio. "Relevance Feedback in Information Retrieval". In G. Salton, editor, *The SMART Retrieval System*. Prentice Hall, Englewood Cliffs, NJ, 1971.
7. Xuehua Shen, Bin Tan, and ChengXiang Zhai. "Implicit User Modeling for Personalized Search". In *Proceedings of the 14th ACM international Conference on Information and Knowledge Management*, Bremen, Germany, 2005.
8. Y. Tzitzikas and C. Meghini. "Ostensive Automatic Schema Mapping for Taxonomy-based Peer-to-Peer Systems". In *7th Intern. Workshop on Cooperative Information Agents, CIA-2003*, pages 78–92, Helsinki, Finland, August 2003.
9. Y. Tzitzikas, C. Meghini, and N. Spyratos. "A Unified Interaction Scheme for Information Sources". *J. of Intelligent Information Systems*, 26(1):75–93, Jan. 2006.

Effective Use of Semantic Structure in XML Retrieval

Roelof van Zwol[1] and Tim van Loosbroek[2]

[1] Yahoo! Research, Barcelona, Spain
[2] Utrecht University, Department of Computer Science, Utrecht, The Netherlands

Abstract. The objective of XML retrieval is to return relevant XML document fragments that answer a given user information need, by exploiting the document structure. The focus in this article is on automatically deriving and using semantic XML structure to enhance the retrieval performance of XML retrieval systems. Based on a naive approach for named entity detection, we discuss how the structure of an XML document can be enriched using the Reuters 21587 news collection.

Based on a retrieval performance experiment, we study the effect of the additional semantic structure on the retrieval performance of our XSee search engine for XML documents. The experiment provides some initial evidence that an XML retrieval system significantly benefits from having meaningful XML structure.

1 Introduction

With XML becoming the de-facto standard for tagging content and exchanging information over the Internet, it becomes possible to exploit the structure of a document when searching for information. However, the mainstream of the XML document collections available today are based on a mixture of HTML-based presentation tags and logical tags, such as chapter, section, and image. These types of tags have little to no semantic meaning to the users.

The success of XML search engines not only depends on efficient and effective retrieval strategies, but also on the availability of meaningful structure in the document collections being searched, and the capability of the user to use this structure as part of his information need [11]. The focus in this article is on (1) automatically deriving semantic ($\sim$ meaningful) structure based on named entity detection, and (2) evaluating the impact of the additional semantic structure on the retrieval performance of XML retrieval systems, and in particular on our XSee search engine [12].

INEX, the Initiative for the evaluation of XML retrieval [4] provides an international forum for the evaluation of XML element retrieval systems. The objective is to exploit the available structural information in documents, to implement a more focused retrieval strategy and to return document components, the so called XML elements - instead of complete documents - in response to a user query. This has its impact on all facets of the retrieval process: query formulation, retrieval strategy, and result presentation.

G. Amati, C. Carpineto, and G. Romano (Eds.): ECIR 2007, LNCS 4425, pp. 621–628, 2007.

Within INEX, the user information need is expressed in different formats, using content-only or content-and-structure approaches. The content-only (CO) approach reflects the keyword based approach commonly used for information retrieval on the Internet. The content-and-structure (CAS) approach uses a combination of textual and structural clues, expressed in NEXI [8]. NEXI is the query language adopted by INEX, and is heavily based on XPath. If structural clues are provided in the information need, it is open to the system to use these clues in a *strict* (SCAS) or *vague* (VCAS) fashion.

The effect of having semantic structure in both the information need and document structure is studied in a small-scale retrieval performance experiment, of which the results are presented here.

Organization

Section 2 elaborates on the motivation for the research presented in this article and formulates the two hypotheses that will be put to the test. In Section 3 our approach for named entity detection and semantic tagging of the Reuters 21587 news collection is presented. The retrieval model of our XSee search engine is presented in Section 4, to provide some insights in the underlying theories used for XML retrieval. Section 5 describes the setup and results of the experiment, while the conclusions are presented in Section 6.

2 Motivation and Hypotheses

The underlying motivation for the research described in this article is that using the structure of an XML document allows the system to more effectively retrieve relevant fragments of information. As in passage retrieval, returning only those pieces of information to the user for a given information need allows the user to work more efficiently. One advantage of XML retrieval over passage retrieval should be that the available document structure improves answering the user's information need. Recent work by [7] on using XML document structure reveals that using structural clues (hints) using a vague interpretation is not by definition aiding the retrieval performance, as the user is bad at guessing where in the document the information should be found. As a consequence, we believe that the underlying document structure should be meaningful to the user, to allow him to effectively use the structure of a document in the information need. Unfortunately, this is not the case for the majority of the available XML document collections.

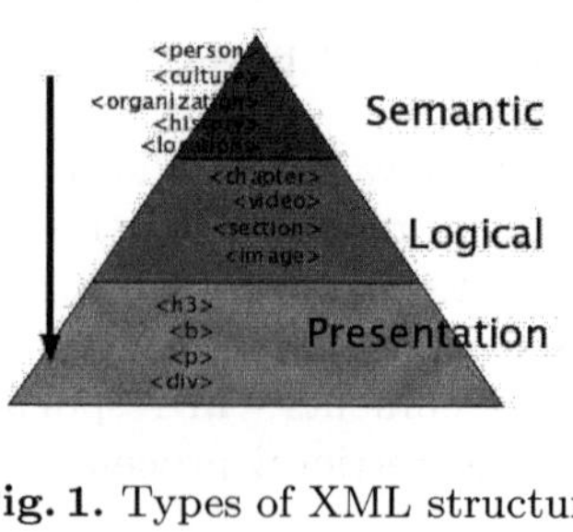

Fig. 1. Types of XML structure

In Figure 1 a classification of XML element types is given in three categories based on the nature of the encapsulated content. Traditionally HTML is used to present the contents of a document, allowing the author to make appealling presentations for the reader.

Typical examples of such elements are `h2`, `b`, and `i` tags, which allow the author to emphasize on particular phrases of text. In XML collections these tags are still used, together with a set of logical elements such as `section`, `image`, and `title`. Logical elements are frequently used in XML standards such as docbook, and newsML to define the *role* of a text fragment. However, logical elements still miss the expressive power of what is defined here as semantic structure. Intuitively, semantic elements label the contents of a text. This can be a large XML fragment, such as a `cultural` description of a `destination`, or a textual phrase that identifies a `person` or `location`. The latter are typical examples of semantic elements, which can automatically be derived using named-entity detection.

Semantic tagging of large XML fragments is a complex task, which is hard to derive automatically. A number of initiatives employed in the Semantic Web movement address this issue, but in general it is assumed to be a task for the author.

2.1 Hypotheses

The main objective of this article is to find evidence that having a richer semantic document structure contributes to the retrieval performance of XML retrieval systems. Therefore the following two hypotheses have been defined.

***Hypothesis 1*: XML retrieval systems can effectively exploit structure to increase retrieval performance.**

The first hypothesis states that an XML retrieval system, in this case XSee, benefits from additional structure that is present in a document, even if the user information request does not contain structural clues. The underlying assumption is that the XML retrieval systems use a ranking mechanism that prefer small elements containing relevant information, over larger elements with the same amount of relevant information. In the case of named entities, the XML retrieval systems should be able to judge such elements to be too small to be relevant units of information to be returned to the users, since they do not provide the contextual information needed to answer a user information need [6].

***Hypothesis 2*: XML retrieval systems can effectively exploit semantic structure to increase retrieval performance, when this structure is used by the user.**

Hypothesis 2 extends Hypothesis 1, by focusing on those cases where structural clues form part of the information request.

3 Named Entity Detection and Semantic Tagging of the Reuters News Collection

In this section we'll give a summary of our naive approach for named entity detection. A more extensive description and evaluation of the approach can be found in [9]. Furthermore, we discuss the effect of the semantic enrichment on the structure of the Reuters news collection.

Automated detection of named entities in text has been practised for a long time and many different and sophisticated approaches [3,1] are available today. Using linguistic, syntactic, or document patterns a set of machine learned extraction rules is normally derived that allow for the detection of named entities in a set of documents. In these cases, an annotated training set is needed to bootstrap the system.

The approach used for named entity detection here, is based on a combination of heuristically derived rules and pre-defined thesauri. The first step focuses on the detection of possible candidates using regular expressions that aim at finding sequences of words starting with a capital. The second step involves eliminating potential false-positives, e.g. words found at the beginning of a sentence and are unlikely to be part of a named entity. Next, we use dictionaries of person, company, and organization names to categorize the named entities, as well as gazetteers for the identification of place names. In the final step of the named entity detection phase, contextual clues are used to categorize the remaining (un-identified) candidates. Examples of such contextual clues are: "... minister", " ... organization ...", "... inc.", etc. Those candidates that were not successfully recognized are abandoned.

Based on the derived set of named-entities, the Reuters collection is extended by tagging the matching text fragments with the corresponding named-entity tags. Table 1 provides some statistics, which show how the structure and content of the Reuters 21587 collection is affected by our approach. In total four new element types have been added to the collection, using the approach for named-entity detection as described above, e.g. `person`, `company`, `organization`, `location`. From Table 1 it can be seen that it duplicates the number of XML elements in the collection, and drastically reduces the average (text) size of the leaf nodes. In [9] the quality of the named-entity detection is evaluated. It is clear that this approach is far from optimal, and fails to classify a huge number of named entities, which results in a low recall. However, it proved to be an accurate approach for maintaining a high precision in classifying named entities, e.g. avoiding false-positives. We believe that the latter is crucial to the success of effective use of semantic ($\sim$ meaningful) structure in XML retrieval.

Table 1. Collection statistics for Reuters 21587

	Original	**Semantic**
documents	20841	20841
XML elements	209135	427854
avg. leaf size	23.6	10.6
max. node depth	3	4
avg node depth	2.6	3.42
unique elements	6	11

4 The XSee XML Search Engine

The XSee XML SEarch Engine [12] is built for participation in INEX 2006. In this section, a brief overview of the internals of the XSee system is given to provide insight into the particulars of an XML retrieval system in general, but also as a reference for the explanation of the results of the experiment described in Section 5.

Data Structure. During the indexing phase, an inverted file is derived from the XML collection containing the triples (file, path, term). This inverted file is processed into the following data structure:

```
NodeTermWeight(node id, term weight)
TermIndex(term, stemmed term, start, end)
Node(id, parent, name, file, path, depth, size)
```

`NodeTermWeight` is a large table, that contains an entry for each term that is contained by a node. The entry stores the unique id of that node, together with the termweight for that term/node combination, the table is sorted on term. The `TermIndex` contains an entry for each unique term, its stemmed variant, and a start-/end- position that specificies a continuous slice of the `NodeTermWeight` table, which contains the selected node/termweights for that term. The `Node` table contains information about each node, such as its id, parent node id, element name, file, the unique path to that node, the path depth, and number of terms contained by that node, respectively.

Retrieval Model. At indexing time, a termweight for a term q and a node n is calculated using:

$$termweight(n,q) = \begin{cases} \frac{tf_{n,q}}{TF_q}, if is\_leaf(n), \\ \frac{tf_{n,q}}{TF_q} + \sum_{c \in children(n)} D_{node} \cdot termweight(c,q) \end{cases} \quad (1)$$

For a leaf node, the weight of a term is determined using the term frequency of that term within the node divided by the number of times the term occurs in the document collection. If the node is not a leaf, the termweight also depends (recursively) on the termweight of its children. The influence of a term occuring in one of its children is reduced via the node decay D_{node}. Typical values for D_{node} are in the range [0.09..0.39], as is heuristically determined.

At query time, the relevance score of a node for a given query Q is calculated as:

$$rsv(n,Q) = F_{term}^{|\{q|q \in Q \ \wedge q \in contains(n,q)\}|} \cdot \sum_{q \in Q} termweight(n,q) \quad (2)$$

For each queryterm q in the query, it is determined whether this term is contained in the node. The sum over the corresponding termweights is taken and multiplied by the termfactor F_{term} to the power of the number of unique query terms occuring in node n. This assures that nodes containing all the query terms will be pushed to the top of the ranking.

This simple model proves to be effective in terms of retrieval performance [12], but also allows for better scalability compared to the GPX and B3-SDR models [5,11], which are based on the same principles.

5 Experiment

In this section the setup and results of the retrieval performance experiment are described.

5.1 Setup

For the experiment we have used two versions of the Reuters 21587 collection of news items. One version without the semantic markup (Original), and one version with the semantic markup (Semantic), as discussed in Section 3. Furthermore we have created a set of 15 topics that describe information needs containing named entities. For each topic a description, narrative, and three different title versions are available, as shown in Example 5.1.

Example 1. Topic 14

Description: Find out what the connection is between Intel and IBM.
Narrative: IBM decided to use Intel processor chips inside their PCs.
CO-title: `IBM Intel`
CAS-title: `//par[about(., IBM) and about(., Intel)]`
Sem. CAS-title: `//par[about(.//company, IBM) and about(.//company, Intel)]`

The CO-title only contains keywords, while the CAS-title specifies the desired element of retrieval. Finally, the Semantic CAS-title specifies that both IBM and Intel are companies.

The methodology used to evaluate our ideas on effective use of semantic tagging for XML retrieval is based on the TREC evaluation procedure. We will therefore use binary relevance judgements, which are determined at XML element level, and have been gathered with TERS [10], a general purpose testbed for the evaluation of retrieval systems, which implements the blind pooling review method for collecting the judgements. If an XML element is considered relevant, all its ancestors are also by definition relevant.

Instead of evaluating multiple search engines, we'll only discuss the results of our XSee XML retrieval engine. For creating the assessment pools we have used an additional XML retrieval engine and selected the top 100 results of each topic into the assessment pool. This will not guarantee that the relevance judgements are complete. Therefore, and also based on the small number of topics used, we'll report the binary preference (bpref) besides the commonly accepted mean average precision (MAP) and recall-precision curves. In [2] it is discussed that the binary preference is more stable in such situations, and therefore more reliable.

5.2 Results

Table 2.a shows the MAP and bpref for the CO-title based runs on both collections. For both measures, the results on the Semantic collection are better than the Original collection. In fact we found that the differences are significant using a T-test (MAP: $t[15] = -3.090$, $p < 0.01$). This means that we have found evidence to support our first hypothesis, which stated that the XSee system can effectively exploit structure to improve on retrieval performance. In more detail, the precision at document cutoff graph of Figure 2.a shows how this impacts the precision of the ranking. With the exception of the first item in the ranking, the results on the Semantic collection are always better. At current we have no plausible explanation for the dip at the top of the ranking.

Investigation of the performance on the (strict) CAS-based runs is split in three categories. One CAS-title run on the Original collection, and two CAS-title runs on the Semantic collection. Of the two runs, one uses the semantic clues as part of the information request.

The results of Table 2.b show that best results are obtained with the *semantic* CAS-title run on the Semantic collection. Second best is the CAS-title run on the Semantic collection. In both cases we find that the results are significantly different (MAP: $t[15] = -2.817$, $p < 0.015$ and $t[15] = -2.475$, $p < 0.030$). The comparison of the CAS-title runs on both collections provides supporting evidence for our first hypothesis, as was the case for the CO-based runs. While based on the comparison of the two different CAS-title runs on the Semantic collection we can conclude that our XSee system can effectively exploit semantic structure to increase retrieval performance, when this structure is used by the user (Hypothesis 2). Inspecting the graph that plots the precision at document cutoff values for the CAS-title runs (Figure 2.b) shows that use of semantic clues in the information need on the Semantic collection is has a particularly positive influence on the precision at the top of the ranking.

Table 2. Summary

(a) CO titles

	Original	**Semantic**
MAP	0.3418	**0.3867**
bpref	0.5558	**0.5846**

(b)CAS titles

	CAS title on Original	**CAS title on Semantic**	**Semantic CAS title on Semantic**
MAP	0.4624	0.4963	**0.6378**
bpref	0.687	0.7072	**0.7182**

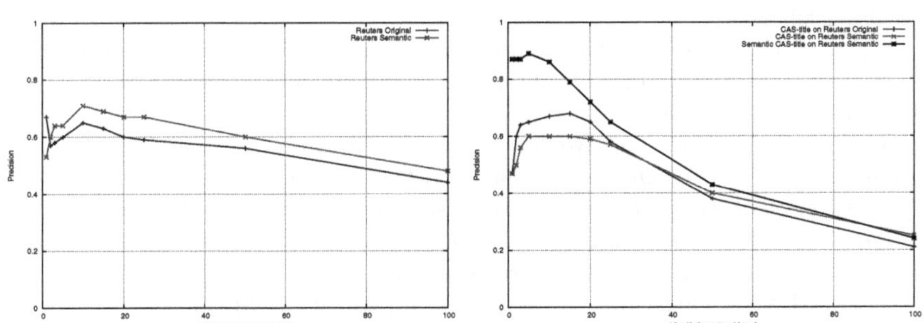

(a) CO-titles (b) CAS-titles

Fig. 2. Precision at document cutoff values

6 Conclusions

In this article we have shown in which cases our XSee search engine for XML retrieval can effectively exploit the available structure. Furthermore, we have pleaded for more meaningful and semantic structure in XML documents, and have indicated how this can be achieved using named entity recognition. For this approach to be effective, we believe that aiming at a high precision in the tagging process is more crucial to the effectiveness of XML retrieval, than having a high recall in terms of full named entity detection coverage.

Based on a small-scale retrieval performance experiment we have found evidence that XML retrieval systems, and XSee in particular, can effectively exploit the additional structure (Hypothese 1) as it allows for identifying and retrieving more relevant XML elements in the top of the ranking. This is especially true, when the users specify semantically structured clues in the information request on the semantic Reuters collection, as stated in Hypothesis 2.

References

1. Alias-i. Lingpipe - `http://www.alias-i.com/lingpipe/`, 2006.
2. Chris Buckley and Ellen M. Voorhees. Retrieval evaluation with incomplete information. In *SIGIR'04*, pages 25 – 32. ACM Press, 2004.
3. M. Ciaramita and Y. Altun. Named-Entity Recognition in Novel Domains with External Lexical Knowledge. In *Workshop on Advances in Structured Learning for Text and Speech Processing (NIPS 2005)*, 2005.
4. N. Fuhr, M. Lalmas, S. Malik, and G. Kazai. *Advances in XML Information Retrieval and Evaluation*, volume 3977 of *LNCS*. Springer-Verlag, 2006.
5. Shlomo Geva. GPX : Gardens point XML information retrieval at INEX 2005. In *Advances in XML Information Retrieval*, volume 3977 of *LNCS*, 2006.
6. Georgina Ramirez, Thijs Westerveld, and Arjen P. de Vries. Using small XML elements to support relevance. In *SIGIR'06*, pages 693–694. ACM Press, 2006.
7. Andrew Trotman and Mounia Lalmas. Why Structural Hints in Queries do not Help XML-Retrieval. In *SIGIR'06*, Seatle, Washington, USA, August 2006. ACM Press.
8. Andrew Trotman and Börkur Sigurbjörnsson. Narrowed Extended XPath I (NEXI). In *Advances in XML Information Retrieval*, number 3493 in LNCS, pages 16–40, 2005.
9. Tim M. van Loosbroek. An ad hoc approach for creating a semantic enhanced document collection. Master's thesis, Department of Computer Sciences, Utrecht University., April 2006.
10. Herre van Oostendorp and Roelof van Zwol. Google's "im feeling lucky", truly a gamble? In *Web Information Systems - WISE'04*, pages 378–390, November 2004.
11. Roelof van Zwol. B3-SDR and effective use of structural hints. In *Advances in XML Information Retrieval*, volume 3977 of *LNCS*, Dagstuhl, Germany, 2006. Springer.
12. Roelof van Zwol. XSee: Structure Xposed. In *Advances in XML Information Retrieval*, LNCS, Dagstuhl, Germany. To appear, 2007. Springer.

Searching Documents Based on Relevance and Type

Jun Xu[1], Yunbo Cao[1], Hang Li[1], Nick Craswell[2], and Yalou Huang[3]

[1] Microsoft Research Asia, No. 49 Zhichun Road, Beijing, China
[2] Microsoft Research Cambridge, UK
[3] Nankai University, No. 94 Weijin Road, Tianjin, China
{junxu, Yunbo.Cao, hangli, nickcr}@microsoft.com,
yellow@nankai.edu.cn

Abstract. This paper extends previous work on document retrieval and document type classification, addressing the problem of 'typed search'. Specifically, given a query and a designated document type, the search system retrieves and ranks documents not only based on the relevance to the query, but also based on the likelihood of being the designated document type. The paper formalizes the problem in a general framework consisting of 'relevance model' and 'type model'. The relevance model indicates whether or not a document is relevant to a query. The type model indicates whether or not a document belongs to the designated document type. We consider three methods for combing the models: linear combination of scores, thresholding on the type score, and a hybrid of the previous two methods. We take course page search and instruction document search as examples and have conducted a series of experiments. Experimental results show our proposed approaches can significantly outperform the baseline methods.

1 Introduction

Traditionally, the document search problem can be described as follows. The user submits a query to the search system and the search system attempts to return documents that the user will find relevant. In many cases, the user not only has an idea of what 'document content' they are looking for, but also what 'type of document'. For example, sometimes users know that they want to search for information from technical papers, homepages, or instruction documents.

In this paper we consider a setting for search, which we refer to as 'typed search'. In typed search, we ask the user to enter a query as usual and at the same time allow them to designate the document type which they want. Then the system returns documents that are not only relevant to the query, but also likely to be of the designated type. Assuming the user indeed will be more satisfied by documents of the requested type, a typed search system is potentially more effective than a system where the user can not specify a document type.

Web search engines like Google, MSN Search, and Yahoo! already provide search by type features, but usually in cases where a perfect distinction can be made between types. For example, by selecting scopes such as Image search or News search the user can specify the type of result that they require. Similarly there may be operators such as 'filetype:pdf'. Not all document types are easily specified, and in some cases it

G. Amati, C. Carpineto, and G. Romano (Eds.): ECIR 2007, LNCS 4425, pp. 629 – 636, 2007.

may even be difficult for humans. In this paper we consider useful document types where it is not easy to make a perfect type classifier.

A number of papers considered the need for document genre classification and the development of type classifiers[14],[15]. This paper considers the problem of typed search, assuming that a type has already been identified that is helpful to end-users and a classifier can be developed. For this specific search problem past research was limited to search on special types of documents such as homepage [1],[2],[7],[13].

This paper aims at being a thorough investigation of this search problem. We choose two document types as examples in our experiments, namely course page search (search for course web pages of colleges) and instruction document search (search for instruction documents). We try to answer the following three questions which we think are crucial for constructing typed search systems. (1) Is it possible to develop a general framework for typed search? (2) What is the best strategy for combining the relevance information and type information? (3) Is it possible to construct typed search systems that are easily to be extended to different types and are easily adapted to different domains?

For our general framework, we propose the use of two probabilistic models for typed search: relevance model and type model. The former represents the relevance of documents to queries, and the latter represents the likelihood of documents of being the designated type. In our experiments we use BM25 as relevance model and Logistic Regression as type model.

We propose three methods for combing the relevance and type models: linear combination, thresholding, and a hybrid method using both thresholding and linear combination. We found that linear combination and thresholding can work well with the default parameter settings. Hybrid can take advantage of the other two strategies and works best, but it needs parameter tuning.

Our methods outperform the baselines on both instruction document search and course page search. It is also possible to conduct domain adaptation, applying a type model trained on one corpus to a separate corpus. Therefore, it is feasible to create generic typed search systems.

2 Related Work

Our work on document types is related to work on identifying and classifying documents by genre, such as [6],[10][14],[15]. We use the term 'type' to indicate maximum generality. Our framework can be applied for any given type of document, even if not everyone would agree that it constitutes a distinct 'genre'. Some authors use 'genre' and 'type' interchangeably, defining genre as: a document type based on similarity of form and purpose [4].

Homepage search can be regarded as a specific typed search. Much research work was conducted on that issue. For instance, TREC had a task called 'home/named page finding'[2]. Many systems were developed for the task [1],[7],[13]. In homepage search, both relevance information and type information are needed in web pages ranking. For example, information about the URL can be used to indicate type [13], because homepages tend to have shallow URLs ending in '/'. In [16], users have the option of specifying some concepts (e.g., a catalog, a call for paper, etc.) of interest when submitting a query.

Table 1. Two views of documents

	Relevant	Irrelevant
Designated type	**A**	B
Not designated type	C	D

Our experiments apply BM25[11] and Logistic Regression[5]. BM25 was introduced as part of Okapi, a system for document retrieval with a probabilistic approach. Specifically, BM25 attempts to rank documents in order of decreasing probability of relevance. Logistic Regression (LR) is one model for classification. LR outputs probability values rather than confidence scores in classification.

3 Problem Description

The search system has a mechanism allowing users to designate the types of documents which users can search. The type of a document (or web page) represents the genre or the functional category of the document. Users can use a menu or a special search operator to designate document types.

If the user knows what type of document they wish to find, they can select that type. They then type a search query as usual. The search system receives the query and the document type. It automatically retrieves and ranks documents on the basis of not only relevance but also likelihood of being the specified type.

Typed search is useful for helping users to find information. Traditional information retrieval conducts search on the basis of relevance of documents to the query [8],[9][12]. Similarly, typed search needs to assure that the retrieved documents are relevant to the query. However, typed search also needs to assure that the retrieved documents belong to the designated document type. Table 1 shows two views of documents. From Table 1, we see that **A** is the set of documents that we want to collect in typed search. By introducing types into search, one can drastically reduce the numbers of documents returned to users.

Various document types can be considered such as resume, blog, homepage, email, etc. For a recent example of a detailed study of document types, see [4]. Here, we assume that the search provider (e.g., librarian, search engine designer) can identify typed searches that are valuable to users, and apply our typed search approach.

4 Our Approach

4.1 General Framework

We propose a general framework for ranking in typed search. Given a query *q* and a document *d*, we rank the documents with the conditional probability $\Pr(r,t \mid q,d)$, where *r* and *t* take 1 or 0 as values and they denote 'relevant or not' and 'in the same type or not'. In instruction document search, for example, $t = 1$ means that a document is an instruction document. In typed search, we rank documents using the probability scores of documents.

Here, we assume that *r* and *t* are conditionally independent given *q* and *d*. We further assume that *t* is only dependent on *d*, not on *q*. Hence we have,

$$\Pr(r,t \mid q,d) \approx \Pr(r \mid q,d) \cdot \Pr(t \mid q,d) \approx \Pr(r \mid q,d) \cdot \Pr(t \mid d) \tag{1}$$

We take Equation (1) as a general model for typed search. We call the two sub-models $\Pr(r \mid q,d)$ and $\Pr(t \mid d)$ 'relevance model' and 'type model', respectively. The relevance model judges whether or not a document is relevant to the query. The type model judges whether or not a document is in the designated document type.

4.2 Relevance Model and Type Model

Given a query and a document, the relevance model outputs a relevance score. In typed search, for a given query, we create a list of *<document, relevance_score>* pairs using the relevance model.

In this paper, we employ BM25[11] as the relevance model. In practice, for indexing, we index the title and the body of a document separately, calculate a BM25 score for each, then linearly combine the scores. We view this combination of scores for the title and the body as the *relevance_score*.

Given a document, the type model outputs a type score. In typed search, we create a list of *<document, type_score>* pairs using the type model.

We take a statistical machine learning approach to constructing a type model. More specifically, given a training data set $D = \{x_i, y_i\}_1^n$, we construct a model $\Pr(y \mid x)$ that can minimize the number of errors when predicting y given x (generalization error). Here $x_i \in X$ and $y_i \in \{1,-1\}$. x represents a document and y represents whether or not a document is a document in the designated type. When applied to a new document x, the model predicts the corresponding y and outputs the score of the prediction. In this paper, we adopt Logistic Regression[5] as our type model. Logistic Regression calculates the 'type probability' of $\Pr(y=1 \mid x)$ a document.

In our approach, we actually use the *type_score* of a document:

$$type\_score = \log \frac{\Pr(y=1 \mid x)}{1-\Pr(y=1 \mid x)} \tag{2}$$

4.3 Combining Strategy

We propose three strategies for combing the scores calculated by the relevance and type models. They are linear combination, thresholding, and hybrid respectively. We rank documents using the combined scores in typed search.

In linear combination, we calculate *ranking_score* by linearly interpolating *relevance_score* and *type_score*.

$$ranking\_score = \lambda \cdot type\_score + (1-\lambda) \cdot relevance\_score\,, \tag{3}$$

where $\lambda \in [0,1]$ is a parameter. In thresholding, we calculate *ranking_score* by descretizing *type_score* to 1 or 0 based on a predetermined threshold.

$$ranking\_score = \begin{cases} relevance\_score & \text{if } Pr(y=1|x) > \theta \\ 0 & \text{otherwise} \end{cases}. \tag{4}$$

Here $\theta \in [0, 1]$ is threshold. In cases where we are certain of our *type_score* we can apply a strict threshold, as is the case for News search on the Web (all pages that are not news are strictly filtered out). If we are less confident about our *type_score* we can apply a lower threshold or try the linear combination strategy.

In hybrid method, we calculate *ranking_score* using both linearly combination and thresholding. Here $\lambda \in [0,1]$ and $\theta \in [0,1]$ are parameters.

$$ranking\_score = \begin{cases} \lambda \cdot type\_score + (1-\lambda) \cdot relevance\_score \text{ if } \Pr(y=1 \mid x) > \theta \\ 0 \text{ otherwise} \end{cases} \quad (5)$$

How to determine the parameter values (λ and θ) is an issue we need to consider. In linear combination, by default can be set as 0.5, since Equation (3) is equivalent to Equation (1), when λ=0.5. In thresholding, θ can also be set as 0.5 by default, since the document is likely to belong to the type when $Pr(y=1|x)>0.5$. In our experiments, we found that nearly best search performances can be achieved when λ and θ are 0.5. In the hybrid method, however, λ and θ have to be tuned empirically.

5 Experiments

In our first experiment, the document type is 'course page', a page describing a course, as would be available on a university website. In the second experiment, the document type is 'instruction document', for example an online manual.

As the first baseline method, we solely use the scores of BM25 to conduct ranking. This can also be seen as an extreme case of our proposed approach in which the parameter λ in Equation (3) equals 0.

As the second baseline method, we solely use the scores of Logistic Regression in ranking, as explained in Section 4.2. This baseline is the other extreme case of our proposed approach, when the parameter λ in Equation (3) equals 1.

As the third baseline method, we add keywords to queries and employ BM25. For course search, we combine the original query words with the keyword 'course' to generate a new query. (We tried to use other keywords, but they did not work well). For instruction document search, we combine the query with 'howto' and 'how to'.

As the fourth baseline method, we use BM25 to conduct ranking, then employ rules to filter the results. As rules, we implement the major features in the type model.

The third and fourth baselines are the simplest ways of combining type information and relevance information in typed search ranking. Hereafter, we denote the four baseline methods as 'BM25', 'Logistic Regression', 'BM25 + Keyword', and 'BM25 + Heuristics'. (We denote our methods as 'Combined (linear)', 'Combined (thresholding)', and 'Combined (hybrid)'.)

We make use of MAP (Mean Average Precision) and *MRR* (Mean Reciprocal Rank) for evaluation of typed search.

5.1 Course Page Search

Data Set

We used 'Four Universities Data Set'[1] in the experiment. The 8,282 WWW-pages in the data set were assigned to six category labels (Course, Student, Staff, Department, Project, and Other). In our case, course pages are positive examples.

---

[1] http://www.cs.cmu.edu/afs/cs.cmu.edu/project/theo-20/www/data/

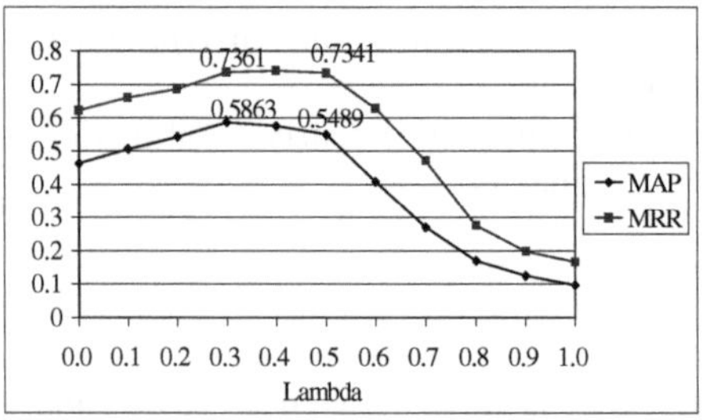

Fig. 1. Performance linear combination w.r.t. λ for course page search.

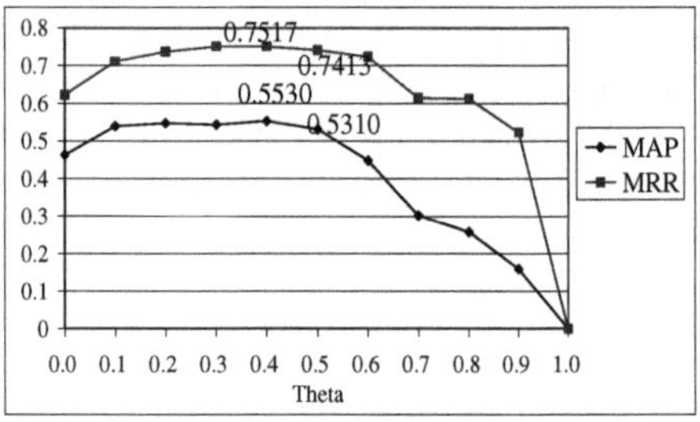

Fig. 2. Performance of thresholding w.r.t. θ for course page search.

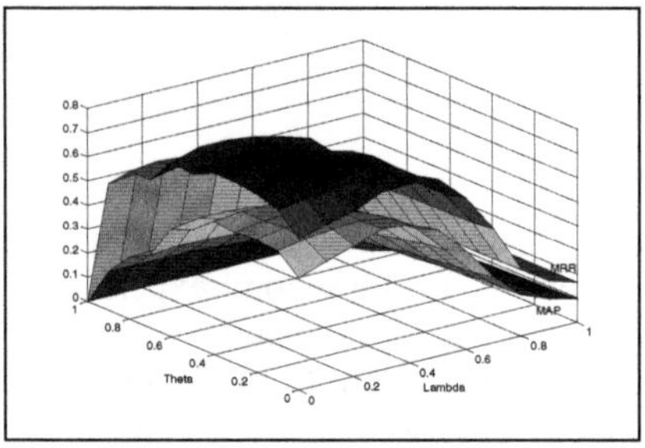

Fig. 3. Performance of hybrid w.r.t. λ and θ for course page search

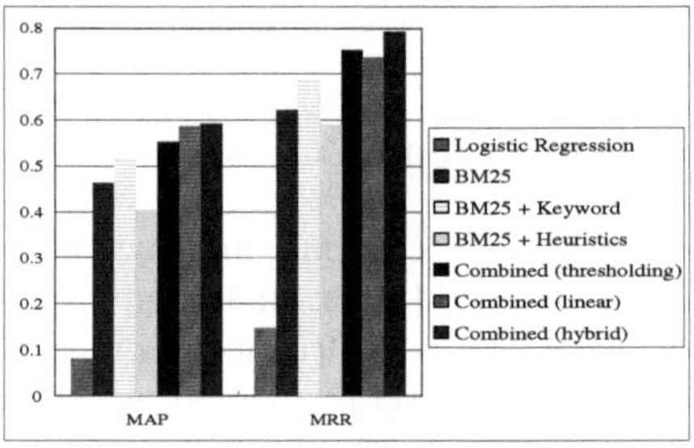

Fig. 4. Performance of course page search

As search queries, we collected the course names from the web sites of the Computer Science department of CMU[2] and MIT Open Courseware, Electrical Engineering and Computer Science[3].For each query, we retrieved at most top 100 course web pages. The retrieved web pages were judged manually by human annotators whether they are really relevant to the query one by one. In this way, all the retrieved documents for each query got the labels **A**, **B**, or **C-D** as in Section 3. Our final query set contains 52 queries. On average, a query has 7.8 relevant web documents.

Experiment on Typed Search

We compared the performances of the typed search ranking methods. We randomly divided queries into four even subsets and carried out 4-fold cross-validation. The result reported below are thus those averaged over the four trials.

We tried various values for the parameter λ and θ in our methods of linear combination, thresholding and hybrid. Fig. 1, Fig. 2, and Fig. 3 show the performance curves when the parameters changes. The best result is accomplished by hybrid, if we happen to know the parameter values.

Fig. 4 shows the results of our proposed methods (linear combination, thresholding, and hybrid method) together with baseline methods. Our methods perform much

---

[2] https://acis.as.cmu.edu/gale2/open/Schedule/SOCServlet?Formname=ByDept

[3] http://ocw.mit.edu/OcwWeb/Electrical-Engineering-and-Computer-Science/index.htm

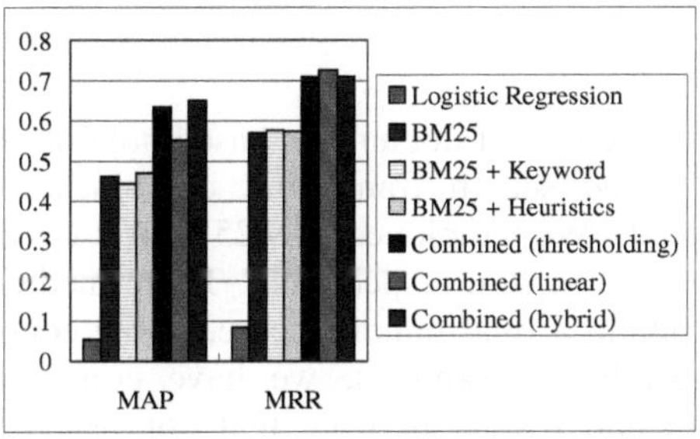

Fig. 5. Performance of instruction search

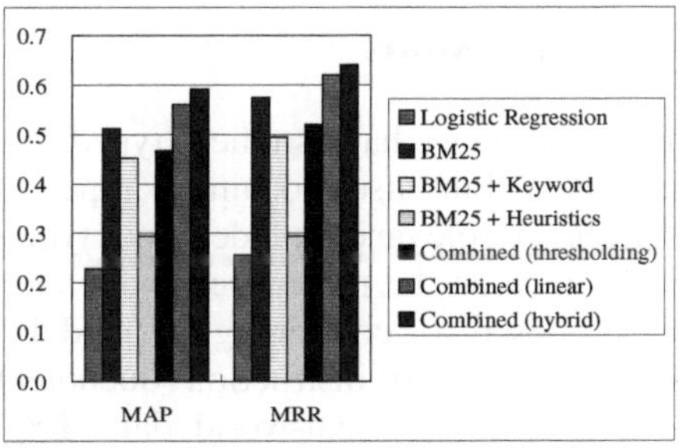

Fig. 6. Performance of domain adaptation.

better than baselines. The results indicate that the traditional information retrieval approach cannot solve the problem of 'typed search' well. Our approach of combing type information and relevance information is effective.

It is not surprising to see that the baseline 'BM25' cannot work well for the task, because it is designed for search of relevant documents. As we have discussed in Section 3, typed search needs consider both relevance and document type. For a similar reason, it is also not surprising to observe that the baseline 'Logistic Regression' cannot achieve good result.

For 'BM25+Keyword', it is hard to construct a query that can filter out documents that not belong to the desired type. For 'BM25+Heuristics', it is hard to make a combination between BM25 and rules. Thus, employing probabilistic models for both relevance and type ranking, as in our approach, appears to be a reasonable choice.

5.2 Instruction Document Search

In the experiment, we investigated typed search applied to instruction document. We created a document set by crawling from the intranet of an international company. We also collected all the real queries about instruction document search from the query log of a search engine on the intranet.

We created a dataset which contains 50 queries, similar to Section 5.1. 61 documents are labeled with **A**, 352 with **B,** and the others with **C-D**. We conducted experiment with 5-fold cross-validation. The results are reported in Fig. 5. Our methods outperform baseline methods for instruction document search.

5.3 Domain Adaptation for Instruction Document Search

In the experiment, we tested whether a generic model can be constructed for typed search. Specifically, we investigated whether a type model trained on intranet can still work well on TREC W3C corpus [3].

We created a data set which contains 50 queries, similar to Section 5.1. Among the documents, 74 are labeled with **A**, 361 with B and others with **C-D**. Fig. 6 shows the results. The experimental results show that our method achieves good result on the TREC W3C corpus, although the type model is trained on a different domain.

6 Conclusions

In the paper, we have studied 'typed search' – search of documents based not only on relevance, but also document type. Our typed search framework combines two models: a relevance model and type model. We employed BM25 and Logistic Regression as the relevance model and the type model, respectively. Three methods are proposed for combining the models to obtain a final ranking score. Using course page search and instruction document search as examples, we have conducted experiments with real-world data. Experimental results indicate that our proposed approaches are effective for typed search and perform significantly better than the baselines. Experimental results also indicate that our proposed approach can perform consistently well across different domains.

References

1. Craswell, N., Hawking, D., and Robertson S.E.: Effective site finding using link anchor information. In: Proc. of the 24th SIGIR conference, New Orleans, USA. (2001), 250-257
2. Craswell, N. and Hawking, D.: Overview of the TREC-2004 Web Track. In: NIST Special Publication: 500-261, the Thirteen Text REtrieval Conference. (2004)
3. Craswell, N., Vries, A., and Soboroff, I.: Overview of the TREC-2005 Enterprise Track. The Fourteenth Text Retrieval Conference. (2005)
4. Freund, L., Toms, E.G., and Clarke, C.L.: Modeling task-genre relationships for IR in the workplace. In: Proc. of the 28th ACM SIGIR Conference, Salvador, Brazil. (2005)
5. Hastie, T., Tibshirani, R., and Friedman, J.: The Elements of Statistical Learning. Springer, New York (2001)
6. Kessler, B., Nunberg G., and Schutze, H.: Automatic Detection of Text Genre. In: Proc. of the 35th Association for Computational Linguistics, Madrid, Spain. (1997), 32–38
7. Kraajj, W., Westerveld, T., and Hiemstra, D.: The Importance of Prior Probabilities for Entry Page Search. In: Proc. of the 25th ACM SIGIR conference. (2002)
8. Mizzaro, S: Relevance: The whole history. Journal of the American Society for Information Science 48(9): (1997), 810-832
9. Mizzaro, S: How many relevancies in information retrieval? Interacting With Computers, 10(3):305-322, Vol. 10. (1998) 321-351.
10. Rauber, A. and Müller-Kögler, A.: Integrating automatic genre analysis into digital libraries. In: Proc. of the 1st ACM/IEEE Joint Conf. on Digital Libraries. Virginia, USA. (2001), 1-10.
11. Robertson, S.E., Walker, S., Beaulieu, M.M., Gatford, M., and Payne, A.: Okapi at TREC-4. In: Proc. of the 4th Text REtrieval Conference. (1996), 73-96
12. Salton, G., McGill, M.: Introduction to Modern Information Retrieval. McGraw-Hill (1983).
13. Song, R., Wen, J.R., Shi, S., Xin, G., Liu, T.Y., Qin, T., Zheng, X., Zhang, J., Xue, G., and Ma, W.Y.: Microsoft Research Asia at Web Track and Terabyte Track of TREC 2004. In: NIST Special Publication: 500-261, the 13th Text REtrieval Conference. (2004)
14. Voorhees, E.: Overview of the TREC 2003 Question Answering Track. In: Proc. of the 12th Annual Text Retrieval Conference. (2003).
15. Zaragoza, H., Craswell, N., Taylor, M., Saria, S., and Robertson, S.: Microsoft Cambridge at TREC 13: Web and Hard Tracks. In the 13th Text REtrieval Conference. (2004)
16. Matsuda, K. and Fukushima, T.: Task Oriented World Wide Web Retrieval by Document Type Classification. In: Proc. of the 8th CIKM. Kansas, USA. (1999)

Investigation of the Effectiveness of Cross-Media Indexing

Murat Yakıcı and Fabio Crestani

i-lab group
Department of Computer and Information Sciences
University of Strathclyde
26 Richmond Street, Livingstone Tower, G1 1XH, Glasgow, UK
{murat.yakici, fabio.crestani}@cis.strath.ac.uk

Abstract. Cross-media analysis and indexing leverage the individual potential of each indexing information provided by different modalities, such as speech, text and image, to improve the effectiveness of information retrieval and filtering in later stages. The process does not only constitute generating a merged representation of the digital content, such as MPEG-7, but also enriching it in order to help remedy the imprecision and noise introduced during the low-level analysis phases. It has been hypothesized that a system that combines different media descriptions of the same multi-modal audio-visual segment in a semantic space will perform better at retrieval and filtering time. In order to validate this hypothesis, we have developed a cross-media indexing system which utilises the *Multiple Evidence* approach by establishing links among the modality specific textual descriptions in order to depict topical similarity.

1 Introduction

A major challenge lies in developing representations suitable for crossing media and languages in the processes of *retrieval*, *filtering*, *categorization* and *summarisation*. The process of cross-media indexing consists of building relationships among concepts extracted from different modalities such as image, speech and text while reducing their weaknesses. A robust indexing model could reconcile and recognize the relationships among concepts identified in these individual modalities and build not only a unified representation but also enrich the descriptions. Therefore, a cross-media analysis system should aim at minimising uncertainty, imprecision and inconsistency across the indexing performed by the single modalities. The aim of the EU-IST Reveal-This (R-T) project is to design, develop, test a complete umbrella infrastructure that will integrate a whole range of information access technologies across media and languages. A critical objective of the project is to benefit from multi-modal analysis and indexing techniques that can extend the possibilities of content push and pull technologies and increase the quality of content provided. In order to address the challenges mentioned previously and leverage the potential of different modalities, we have developed a prototype cross-media indexing system as part of the project. The

G. Amati, C. Carpineto, and G. Romano (Eds.): ECIR 2007, LNCS 4425, pp. 637–644, 2007.

prototype implements a standard way of identifying, accessing, manipulating, storing and retrieving semantic links between modalities. The prototype system constitutes the last layer of the chain of processing in the Cross-Media Analysis and Indexing subsystem of the R-T system. It utilises the *Multiple Evidence* approach by establishing links among the modal descriptions in order to depict topical similarity in the semantic textual space. The prototype's input is a merged representation of the high-level features that are extracted by the speech processing component (speaker turns, transcriptions, persons, topics), the text-processing component (named entities, terms, facts), the image processing component (faces, key-frames, visual features), the image and the text categorizers. The media files were automatically segmented to form stories. After this stage, the descriptions are merged. This representation is further processed by the our prototype to produce a unified view of the content and enrichment.

In this study, we want to know whether cross-media indexing using *Multi Evidence* approach is effective or not. We admit that, neither cross-media indexing is an easy task nor such a question can be answered quickly. Therefore, we break down the question into several testable hypotheses and investigate them one by one.

We state our *generic* hypothesis as whether *Multi Evidence* approach performs better over a baseline system given the processed test collection, queries and graded relevance judgements. We also state *specific* hypotheses as follows:

1. No modality is dominant over other modalities, which means every modality has a significant contribution for effective retrieval.
2. Imprecision in the speech modality (recognition errors) can be remedied by other modalities for effective retrieval.
3. Document expansion may remedy the imprecision of speech modality.

We now gave an overview of the cross-media indexing task, the project and the prototype system as well as our research questions. In the rest of the paper, we describe the indexing models, the evaluation strategy that is followed for our hypotheses and the prototype and discuss our initial results. The last section summarises key points of the paper and explores future directions.

2 Indexing Model

The prototype utilises Dempster-Shafer's Theory of Evidence [1] approach for establishing links among the modal descriptions in order to depict *topical similarity* in the textual space. The theory has been extensively studied in image retrieval [2] and structured document retrieval [3], but has never been applied in such a context.

Dempster's work lays the foundations of a non-Bayesian theory of probability which was then extended by Shafer. The theory combines two or more bodies of evidence defined with in the same *frame of discernment* T into one body of evidence. We used the work of Lalmas [3] in a slightly different way. In our application of the model, we consider different modalities as sources of evidences.

A document d containing a term t's existence in a modality m is counted as an *evidence* to support the topical similarity hypothesis. Therefore, each modality is treated as a probability density function also called as *Base Probability Assignment (BPA)*. For the details of our approach and a short review of related work, please refer to our previous works [4,5].

3 Evaluation

Evaluating the accuracy and robustness of individual processing modalities should be done in their own domain and with specially crafted test beds. However, when it comes to evaluating the performance of the merged and enriched representation of these processors, one needs to find a solution. In an attempt to validate our arguments and thus our prototype system, we considered construction of a test collection using expert users. For building such a collection, one would need to have a set of "documents" (these are video/radio segments which we call stories), a set of queries and a set of documents that have been found to be relevant to the queries (having exhaustively gone through the collection and found that these were all relevant documents contained in the collection).

3.1 Building of the Test Collection

For the purposes of the project, we collected TV and radio broadcast parallel content in English and Greek languages, approximity 30+30 hours that covers news(40%), politics(21%) and travel(35%) domains. The collection contains multi-modal documents of different genre that guarantees a significant variety in modality-specific characteristics.

We reserved a subset of four-hours English version of the content for generating cross-media representations, queries and relevance judgements to be used in testing specifically our cross-media indexing prototype (see table 1). Despite being relatively small, our intention was to have a data collection completely annotated with graded expert relevance judgements. Admittedly, manual annotation of every single segment is a labour and time intensive task especially when using non-binary relevance judgements. This is one of the trade-offs of building a test collection one would confront. Due to the same reasons, we were not able to manually create transcriptions and annotate feature in different modalities. Contribution of the manual transcripts and feature annotation are valuable to make a data set complete. These would provide us an established data set for comparison with other work for improving and evaluating individual and corporate performances. The lack of these led us to consider using other well known collections that can be acquired to test some of our other hypotheses.

Nevertheless, considering the characteristics of the modalities and the content of our collection, we created a set of 87 queries. Three experts were involved in the annotation task. They were asked to go through the collection exhaustively for each query and manually annotate categories and provide relevance judgements. The experts were also asked to justify their judgements when considering stories

Table 1. Description of the collection used for testing Cross-Media Indexing prototype

The subset of the test collection			
Domain	*Medium*	*Duration*	*Stories*
Politics (EU Plenary sessions and Press Conferences)	Radio Video	- 01:09:58	- 31
News	Radio Video	- 00:44:26	- 33
Travel	Radio Video	00:28:12 03:01:44	24 188
Video sub-total		03:46:10	252
Radio sub-total		00:28:12	24
Grand total		04:14:23	276

as "partially relevant" to indicate how strict/lenient they had been in their judgements. The segments that were not mentioned at all were assumed to be judged as irrelevant by the annotators.

For the analysis of the media files (subset collection), we have defined a process work-flow so that every time one of the processors is improved, a new version of the collection could be regenerated. This approach enabled us to resolve dependencies amongst the processing modules and more importantly test the effects of relative improvements of individual processors and their contribution to the overall enriched representation.

3.2 Retrieval Experiments

In order to assess the performance of our model, we prepared a baseline system which utilised the Vector Space Model and *tf-idf* weighting scheme. Both systems were run using using the same query set over the processed content on every single modality and multiple modalities (see section 1 for extracted features). Figure 1 depicts the precision vs. recall values from our experiments utilising multiple modalities.

The results suggest that our Dempster-Shafer (DS) approach is performing significantly worse than the baseline system at the high recall end. At the high precision end, the results seem to be identical. Before drawing any quick conclusions and rejecting our general hypothesis, we wanted to make sure whether root cause of the problem was our proposed model. Despite being better than DS model, the precision and recall values for the baseline system does not seem very high either. This may also be the result of factors such as the size of the collection or the imprecision of individual modalities. Further analysis of our empirical result also showed that there is no positive performance contribution of modalities that are other than speech by which we reject our first and second hypothesis. Although not significantly, in some cases it was observed that modalities other than speech altogether had decreased the performance in both the baseline and the DS system. In other words, running both systems using speech only content yielded better performance. Given our set-up and experiments, we conclude that

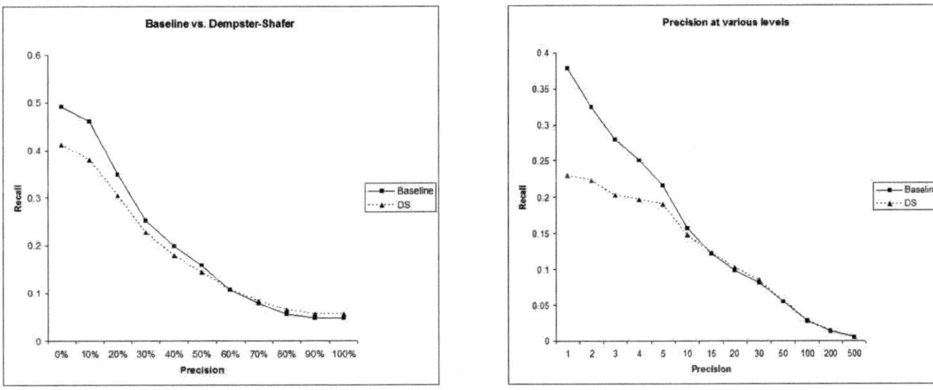

Fig. 1. Precision vs. Recall values for the Baseline and the Dempster-Shafer systems

speech modality seems to be dominant over other modalities. This might be due to three reasons in addition to the above mentioned ones:1) Test queries might be biased towards speech 2) Size of the vocabulary is larger in this modality 3) Reducing every single modality to textual space may not be appropriately capture relations.

In order to understand the effects of queries (at least partially), we analysed the Mean Average Precision (MAP) values of both systems to see for which queries the performance of the prototype decreases or increases. The queries were grouped according to their domain. We took the difference between the baseline system's MAP values and DS's. The positive values showed that baseline system's performance is better than DS for the specific queries. This is particularly the case for the queries belonging to the travel domain. Although not conspicuous at first sight, DS performs slightly better for politics domain. From our prior knowledge we know that, acoustics of politics domain is totally different from other domains. We also know that the ASR module is well trained in this domain and not very well in others. Therefore, it is not surprising to see the speech processor doing better job in politics than news and travel. However, DS seems to be competitive in news domain as well. This diversity may well be one of the reasons for not catching up with baseline.

3.3 Discussion

Given that speech is dominant over all other modalities and there is no significant contribution from them, we returned to our third specific hypothesis. Admittedly, ASR (D_{asr}) is less accurate than manual transcripts (D_{mt}). As mentioned before, document expansion techniques can be employed to reasonably repair the ASR where the expanded representation(D_{asr_exp}) is closer to the original speech(D_{org}) representation (bearing in mind that manual transcripts may be slightly different than original speech). Therefore, ideally a retrieval model would return the same document if we have an ASR representation which is closer to original document. We formalise this assumption as $D_{org} \approx D_{mt} > D_{asr_exp} > D_{asr}$. However, we also acknowledge that document expansion models should be used carefully as they may insert terms to the document which might move up non-relevant documents higher in the ranking.

3.4 Document Expansion from Web

Singhal et.al [6] makes several assumptions for using document expansion for Spoken Document Retrieval and advices that using a parallel corpus is beneficial for obtaining good results. Given the purposes and dimensions of our project, acquiring a parallel corpus which has the same period of time and diversity of domains is an unrealistic assumption. However, in order apply document expansion with some corpus that might show some resemblance in topicality, we decided to use WWW.

Initially, Google search engine results were utilised to expand each story in the collection. Unlike running ASR documents as queries themselves [6], we selected top 10 representative and discriminative terms to be used as queries from the speech modality of the stories. The selection of terms was done by a slightly different algorithm using Kullback-Leibler Divergence measure applied in [7].

KL is typically used for measuring the difference between two probability distributions [8]. When applied to the problem of measuring the distance between two term distributions as in Language Modelling [9], KL estimates the relative entropy between the probability of a term t occurring in the actual collection Θ_c (i.e. $p(t|\Theta_c)$), and the probability of the term t occurring in the estimated Topic Language Model Θ_d (i.e. $p(t|\Theta_d)$). KL is defined as,

$$KL(\Theta_d||\Theta_c) = \sum_{t \in V} p(t|\Theta_d) log \frac{p(t|\Theta_d)}{p(t|\Theta_c)} \quad (1)$$

where,

$$p(t|\Theta_c) = \frac{n(t, \Theta_c)}{\sum_{t \in \Theta_c} n(t, \Theta_c)} \quad (2)$$

and,

$$p(t|\Theta_d) = \frac{n(t, d) + \alpha}{(\sum_{t \in d} n(t, d)) + \alpha} \quad (3)$$

The expression $n(t, d)$ is the number of times term t occurs in a document d. The sparsity problem within the Θ_d is handled by Laplace smoothing. A non-zero constant α is introduced to alleviate the zero probability [9]. The smaller the KL divergence the closer the document is to the actual collection. A zero KL score indicates two identical distributions.

In our case, we are interested in each term t's contribution to the KL score for a document d, instead of determining the difference between two term distributions. The greater the contribution to the document model the higher the KL score will be. Therefore, for each term t's the contribution is calculated as in the following:

$$KL_d(t) = p(t|\Theta_d) log \frac{p(t|\Theta_d)}{p(t|\Theta_c)} \quad (4)$$

The top 20 ranked terms in a document model are then ranked further according to each term's representative (generality) and discriminative (specificity) properties. The first $p(t|\Theta_d)$ part in the equations 4 determines the representativeness and the later *log* component determines the disciminativeness. This approach makes it possible to measure the effectiveness of the two categories of terms in addition to our main objectives. Top ten qualifying terms are then used as queries to be submitted to the search engines. The contents of the first ten links were gathered and each story's speech content was expanded using the formula suggested by [10] and applied in [6] without additional weights. We continued our experiments using the base index (BI), base index extended by representative queries (IRQ) and base index extended by discriminative queries (IRD). Figure 2 depicts the precision and recall values for baseline and DS over BI, IRQ and IDQ indexes. Although not significantly different than the baseline, it is observed that DS system shows some improvement over the IRQ index on the left hand side graph.

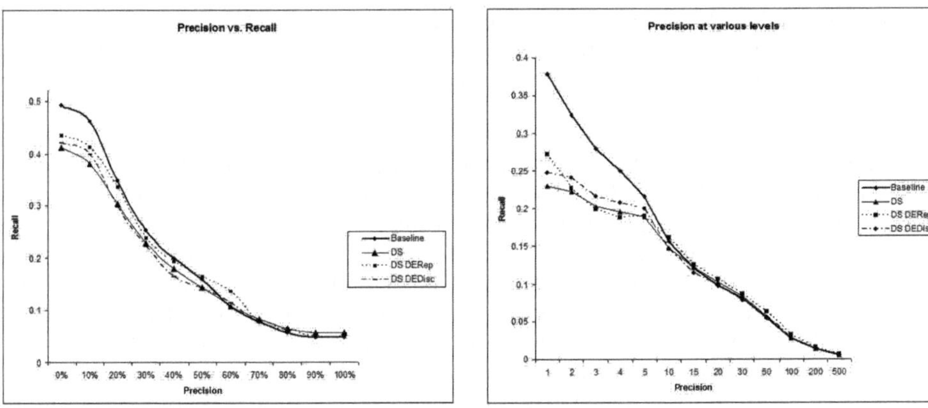

Fig. 2. The graph on the left illustrates the precision and recall values for Baseline, DS and DS expanded with Representative and Descrimitive queries. The graph on the right hand side illustrates the precision values in various levels.

In addition to these experiments, we had also indexed ASR only and its expanded versions using Terrier system to run other 8 retrieval models [11]. We found out that there is no significant performance improvement for each retrieval model over these three indexes (F=1.694, Significance=0.184). The results also suggest that none of the retrieval models perform significantly better than the other ones (F=0.80, Significance=0.999).

4 Conclusions and Future Work

In this paper, our approach to cross-media indexing and our initial experiments were presented. Given the available parameter space it is quite early to say that Dempster-Shafer approach is not working. The hypothesis of cross-media indexing and the models in use are still an open research area. There are many issues, variables and problems yet to be tackled for a reliable system performance. There might be other cross-media models that can combine different modalities in their own feature space or equalize them in one modality such as text.

We will be working on various ways to improve the processes such as "cross-media" query generation and document expansion models provided in this study. In order to further investigate the effects of document expansion models using web, we acquired TDT-2 and AQUAINT collections.

References

1. Shafer, G.: A Mathematical Theory of Evidence. Princeton University Press (1976)
2. Jose, J.M., Harper, D.J.: A retrieval mechanism for semi-structured photographic collections. In: Proceedings of the DEXA, Springer-Verlag (1997)
3. Lalmas, M., Moutogianni, E.: A Dempster-Shafer indexing for the focussed retrieval of a hierarchically structured document space: Implementation and experiments on a web museum collection. In: Proceedings of RIAO, Paris, France (2000)
4. Yakici, M., Crestani, F.: Design of a cross-media indexing system. In: Proceedings of ECDL. (2006) 477–480
5. Yakici, M., Crestani, F.: Design and implementation of a cross-media indexing system for the reveal-this system. In: Proceedings of Axmedis Conference. (2006)
6. Singhal, A., Pereira, F.: Document expansion for speech retrieval. In: Proceedings of ACM SIGIR'99, Berkeley, California, United States, ACM Press (1999)
7. Baillie, M., Elsweiler, D., Nicol, E., Ruthven, I., Sweeney, S., Yakici, M., Crestani, F., Landoni, M.: University of Strathclyde: the i-labs first big day out at TREC HARD. In Voorhees, E.M., Buckland, L.M., eds.: Proceedings of the Fourteenth Text Retrieval Conference (TREC-14), NIST Special Publication (2005)
8. Kullback, S.: Information theory and statistics. Wiley, New York (1959)
9. Xu, J., Croft, W.B.: Cluster-based language models for distributed retrieval. In: Proceedings of ACM SIGIR'99, New York, NY, USA, ACM Press (1999) 254–261
10. Rocchio, J.J.: Relevance feedback in information retrieval. In Salton, G., ed.: The SMART retrieval system: experiments in automatic document processing, Englewood Cliffs, NJ, USA, Prentice-Hall (1971) 313 – 323
11. Ounis, I., Amati, G., Plachouras, V., He, B., Macdonald, C., Lioma, C.: Terrier: A High Performance and Scalable Information Retrieval Platform. In: Proceedings of ACM SIGIR'06, Workshop on Open Source Information Retrieval (OSIR 2006). (2006)

Improve Ranking by Using Image Information

Qing Yu, Shuming Shi, Zhiwei Li, Ji-Rong Wen, and Wei-Ying Ma

Microsoft Research Asia
{qingyu, shumings, zli, jrwen, wyma}@microsoft.com

Abstract. This paper explores the feasibility of including image information embedded in Web pages in relevance computation to improve search performance. In determining the ranking of Web pages against a given query, most (if not all) modern Web search engines consider two kinds of factors: text information (including title, URL, body text, anchor text, etc) and static ranking (e.g. PageRank [1]). Although images have been widely used to help represent Web pages and carry valuable information, little work has been done to take advantage of them in computing the relevance score of a Web page given a query. We propose, in this paper, a framework to contain image information in ranking functions. Preliminary experimental results show that, when image information is used properly, ranking results can be improved.

Keywords: Web search, image information, image importance, relevance.

1 Introduction

Web search engines commonly evaluate Web pages based on two factors [1]: text relevance scores, and static ranks. The former describes how relevance a page is according to the text contents of a page; and static rank is a query-independent page quality measurement. Multimedia information (especially images) carries valuable information and plays an important role in representing a Web page to end users, just like the saying goes "a picture is worth a thousand words." However, for some reason, current Web search engines are prone to discard multimedia information within Web pages in computing relevance scores, and therefore lose some important pieces of information of the original Web pages.

In this paper, we study the problem of how to leverage image information within Web pages to improve Web search rankings from two aspects. On one side, the relevance between an image and a text query is evaluated based on the automatically extracted image annotation. On the other side, we deduce an image's importance for a page by analyzing the image's content. Finally the two aspects of an image are combined by a multi-evidence combination method.

2 Our Approach

In this section, we provide a detailed description of our approach. First, because images cannot be used directly for indexing and retrieval, it is required to annotate each image by assigning some pieces of text to them. Then, for each image, its relevance score to a query can be computed on its text description by adopting

G. Amati, C. Carpineto, and G. Romano (Eds.): ECIR 2007, LNCS 4425, pp. 645 – 652, 2007.

existing document retrieval techniques. Third, we use a machine learning approach to compute each image's importance in its Web page, as different images in a Web page play different roles and therefore have different levels of importance. In the fourth step, we compute an overall image score for a Web page based on all images' information in the page. And this score is then combined, in the final step, with other evidence of the page to get the final score, which is ultimately used for ranking. In the following sections, we will describe the steps in a little more detail.

2.1 Image Annotation and Relevance Score Computation

Ideally, in order to use image information for text search, we must acquire precise equivalent text information represented by an image and then index these text segments. However, it can be difficult (if not impossible) to automatically get the precise text representation of an image. Fortunately, some research work [3][8] has shown that ALT text, URL and surrounding text from an image can be treated as approximate text presentation of the image, as shown in Fig.1. We have chosen to employ these text segments as the approximate image annotation. With this kind of annotation, each image can be transformed into a small structured document with multiple text fields. The precision of approximate annotation is a key factor in the final performance of our approach. If some incorrect text descriptions are added to an image, the calculated image relevance score would have bias. In paper [2], the author discusses some kinds of surrounding text extraction methods. In this paper, we adopt a rule-based DOM-tree parsing method, which focuses on dealing with table elements. Extensive engineering efforts were made, which are ignored here due to space limitation. Although not perfect, we found that the extracted surrounding text is commonly acceptable image description.

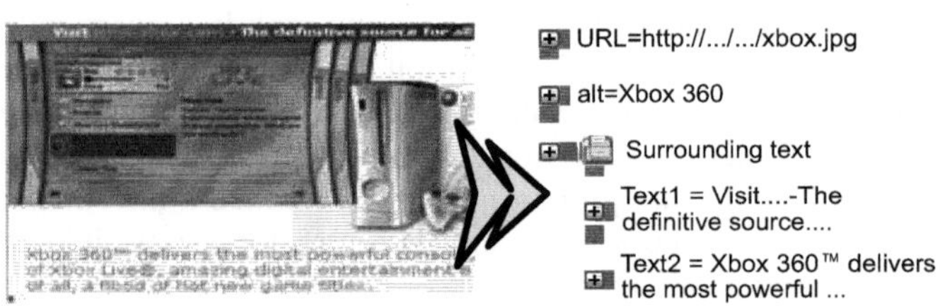

Fig. 1. Image text fields and annotations

In the previous step, each image has been represented by a structured document with multiple text fields. Therefore, existing IR techniques for structured document retrieval can be employed to compute a relevance score for each image. A simple approach is to separately calculate a relevance score for each text field separately and then linearly combine the scores. Further, assuming the text fields of an image are F_i (i=0, 1, …, m), the relevance score for the image can be computed as $R = \sum_{i=1}^{m} w_i \cdot R_i$,where w_i is the weight of field F_i, and R_i is the relevance score of field

F_i with respect to a query. For each field, the relevance score F_i can be computed by adopting some existing IR models. We use the BM25 formula [9],

$$R_i = \sum_{t \in Q} \frac{(k_1 + 1) \cdot tf(t, F_i)}{k_1 \cdot ((1-b) + b \cdot \frac{fl_i}{avfl_i}) + tf(t, F_i)} \cdot W(t) \tag{1}$$

where tf is the occurrence frequency of the term t within a specific field F_i, fl_i and $avfl_i$ are field length and average filed length of F_i, and k_1 and b are two parameters. $W(t)$ is the term weight, which can be computed by $\log \frac{N - df(t) + 0.5}{df(t)}$,where N is the total fields number of all contained images in a collection and $df(t)$ is the number of structured documents containing term t.

2.2 Image Importance Computation

Images within a Web page tend to be highly heterogeneous in terms of their functionality and importance [4]. Types of images include topical images (images which help to clarify the main topic of a Web page), advertising images, banners, logos, formatting images, and others. Some of them carry more detailed information than others and are therefore judged more important to their contained page. By contrast, some images are inserted into a page only for decoration, advertisement or navigation, and therefore are relatively less important, or even harmful when used for ranking. To make best use of image information in general Web search, it is required to differentiate between important and less important images. We use a real-valued score, called *image importance score*, to depict the importance of an image within a Web page. We adopt a machine learning approach to automatically learn the importance score based on three kinds of features: image level, page level, and site level, as listed in Table 1.

Table 1. Features used for computing image importance scores

Feature kind	Feature name
Image level	size, width/height ratio, blurriness, contrast, colorfulness, photo vs. graphic
Page level	relative pos, relative size, relative width/height ratio
Site level	IsInnerSite, FrequencyInSite

Image level features are extracted from the image content. Features contained in this category include size, width/height ratio, blurriness, contrast, colorfulness, and others. These features are mainly used to represent the quality of an image[1]. Page level features are used to describe the relationship between an image and the page containing it, or the relationship between different images within the same page. Feature *RelativePos* is used to describe the relative horizontal and vertical position of an image in a Web page. Relative positions are used to achieve comparable results among different Web pages. Feature *Relative size* represents the percentage of the

[1] We ignore image low-level features such as color histogram in this paper for two reasons: First, it is relatively expensive to include them in the feature set, as they are typically high-dimensional data. Second, they are relatively less informational to our scenario.

image area covering the Web page. And the *relative width/height ratio* feature describes the relationship between the image shape and the page shape. Site level features are mainly used to represent the relationship between an image and the Web site containing it. The *IsInnerSite* presents whether the URL of the image is on the same site as its hosting page. Many advertising images and other non-important images are located on different sites from the page on which they are displayed. The *FrequencyInSite* represents the number of times the image appears in (the different Web pages of) the given site. Its functionality is just like DF (document drequency: the number of documents contain a term within a collection) in ranking functions in information retrieval.

We use a machine learning method to deduce the image importance. And a model is trained from pre-labeled training sample set T. Each labeled sample (image) can be expressed as $(\mathbf{x}_{i,j}, y_{i,j})$, where $\mathbf{x}_{i,j}$ is the extracted features vector of the image i in the page j and $y_{i,j}$ is its labeled importance (degree). We crawled 2,000 Web pages and the images contained within them (total 6,152 images). Each image is assigned an important level, namely 0 (useless), 1(important) and 2 (highly important). To train the model, we need to find a function f such that $\sum_{(\mathbf{x},y)\in T}\left|f(\mathbf{x}_{i,j})-y_{i,j}\right|^2$ is minimized. We use linear Ranking-SVM [7] to solve this problem.

2.3 Page's Overall Image Score

After getting an image's relevance score R and importance weight I, we can naturally compute the image's overall score by $S = R \cdot I$. However a Web page commonly contains multiple images, so a function f is needed to combine all images together to get the page's *overall image score*. The most straightforward way of computing the overall image score is summing all image scores. Another simple way is using the maximal image score as the overall image score. These two straightforward methods both have their apparent disadvantages. Web pages typically have various numbers of images. If we sum scores directly the resulting overall image score would be unreasonably higher for pages that contain many images. Considering page A having an image of score 1.0 and page B having 20 images of score 0.1 each, it is apparently not feasible to assign the overall image score for page B as 2.0 (=0.1*20, twice of page A's score). The disadvantage of the maximum method is that it loses a lot of information by only considering the most important image in each Web page.

In this paper, we utilize the homogeneous evidence combination technique [10] to combine the image relevance scores. In a little more detail, the following formula is adopted,

$$S_{image} = \sum_{i} discount(i) \cdot S_i \tag{2}$$

where $discount(i) = \frac{1}{(1+c \cdot i)^2}$, and c is a parameter controlling the degree of discount.

After computing a page's overall image score S_{image}, we can simply combine it linearly with the page's text relevance score S_{text} and static rank S_{static}: $S_{total} = \lambda_1 \cdot S_{img} + \lambda_2 \cdot S_{text} + \lambda_3 \cdot S_{static}$, where $\lambda_1 + \lambda_2 + \lambda_3 = 1$.

Most image related applications require much computation. However our approach can be implemented efficiently. Among all steps of our approach, only the image relevance score computation is done online. Web pages typically have lots of images, and the processing for them may be costly. So we first remove noisy images before online computation. This pre-processing step increases efficiency greatly with little impact on search quality.

3 Experiments

Experimental Setup. In order to examine our method, we crawled Web pages and their images from a large website[2]. This website contains huge number of Web pages that cover plenty of information and various Web page types and styles. We totally crawled about 7M Web pages, with about 25% pages containing images (after removing noisy images). We randomly selected 83 queries from the query log of the site search engine. These queries were manually classified into 13 categories. For each query, we labeled the top 30 results returned by each ranking algorithm, which is tested in follow experiments, by assigning each page a relevance value from 1 (meaning 'poor match') to 5 (meaning 'perfect match'). We adopted the nDCG metric [6] to measure the performance. In our experiments, the discount factor b was fixed to 2 and the gain values for levels (from 1 to 5) were 0, 1, 3, 7 and 15, respectively.

Overall Results. The baseline retrieval system adopts the BM25 formula to compute relevance scores on the body, title, anchor, and URL fields. And the resultant text score is linearly combined with the PageRank [1] score to get the baseline score. For our approach, the overall image score is computed and is linearly combined with the baseline score. All the algorithms are tested under the optimal parameters. To effectively find approximate optimal parameter settings in the algorithm, we take an iterative approach. For each iteration step, we free two parameters for tuning (by grid search) and fix other parameters. The Table 2 shows the preliminary comparison of the baseline method and our method. The search performance is improved when image information is considered. Such preliminary experimental results verify that image information could be useful for improving web search performance.

From Figure 2, we can further observe that image information is especially useful for queries on which the baseline system does not perform well. We divide all queries into five ranges based on the performance of the baseline system. We can clearly see that the performance improvement is significant on those queries whose baseline performance is very low. On the contrast, for queries over which the baseline system already behaves well, little performance improvement can be achieved (we can see that the performance even slightly decreased on queries in the [0.8, 1] range). And for queries in most queries ranges, the performance is improved moderately by including image information.

---

[2] http://www.msn.com. We didn't use TREC dataset in our experiments because they only contain the plain text of Web pages (while images were not downloaded). Even the latest dataset was generated two years ago, and many pages and their images may have been deleted, so we can't re-crawl the required images.

Query Categories. Queries can be divided into some categories according to certain criteria. It is not difficult to imagine that different categories of queries may see different degrees of performance improvement when considering image information. To achieve a better understanding of it, we manually classified the queries into 13 categories[3]. The categories and the performance improvements (measured by nDCG@3) for each category are shown in Table 4. We can see the improvements are relatively high for some categories such as information resources, services, people, and others. While the performance on some categories remains unchanged or becomes slightly worse. If the queries can be automatically classified into some categories (we must admit that it is hard to do) and be assigned different combination weights of image scores and baseline scores, the overall performance would be further improved.

Table 2. General Performance Comparison

nDCG	Baseline	With Image	%
@1	0.616	0.638	3.56%
@3	0.604	0.641	6.08%
@5	0.588	0.622	5.70%
@10	0.535	0.548	2.45%

Table 3. Stat. of avg importance for different labeling levels (nDCG@3)

Labeled Relevance Level	Total importance
5 (Best)	1.022
4	0.547
3	0.454
2	0.832
1	0.397

Table 4. Performance improvements for different query categories (nDCG@3)

Category	Query #	Improve	%
Information resources	6	0.166	35.40%
Services	6	0.166	35.40%
People	10	0.057	9.75%
Products	9	0.047	12.44%
Knowledge	7	0.042	6.63%
Geography	6	0.031	6.17%
Entertainment	14	0.028	4.45%
Companies	6	0.003	0.47%
Hobby	5	0.000	0.00%
Holiday /Events	5	0.000	0.00%
Navigational	5	-0.021	-2.43%
Others	13	0.000	0.00%

Alternative Techniques. As we have seen from Section 2, our approach is comprised of several steps. Although straightforward methodologies are used in some steps, technologies used in some other steps are crucial to achieve such performance gains. Firstly, we adopted a machine learning approach to automatically compute image importance from a Web page. Fig.3 illustrates the effect of it by comparing an alternative way which simply uses uniform image importance. We can see that, although the "uniform importance" approach performs slightly better than the baseline, differentiating image importance can bring great performance improvement. In the overall image score computation step, we use the homogeneous evidence combination technique to compute a page's overall image score based on all image information on the page. Fig. 4 shows the performance of various methods described in section 2.3. As

[3] Some queries belong to more than one category.

expected, our approach behaves the best among them. In our approach, we compute an overall image score against a given query for each page. That is, we adopt a query-dependent approach. A simpler and attractive way is to use image information as query-independent page quality evidence [5]. A primary precondition to do so is that good search results should be more likely to contain high-quality images. Does this precondition actually hold? To answer this question, we provide statistics on labeled search results data. Table 3 summarizes the mean of a page's total importance for each relevance level, which is the sum of the importance of all images within a page and can indicate whether the page contains high-quality images. We can observe from the table that, although the best search results (with label 5) are commonly more likely to contain high-quality images than other levels, this assumption does not hold for some other levels (e.g. the quality of images in level-2 pages is on average higher than the quality of images in level-3 and 4). Fig.5 shows the experimental results of the above query-independent image usage. It is clear from the figure that this kind of usage is quite ineffective relative to our query-dependent approach.

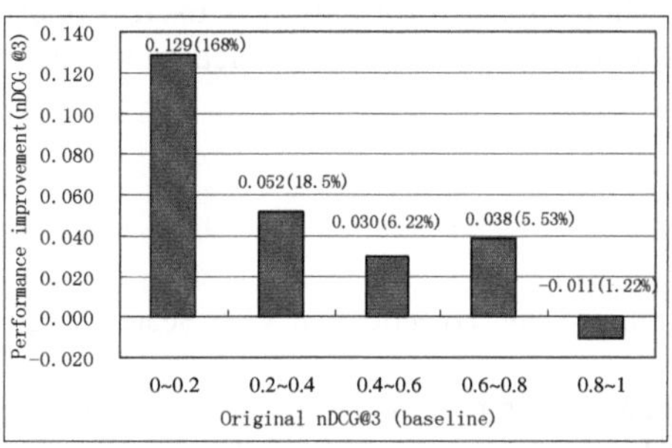

Fig. 2. Performance improvement on queries with different baseline performance ranges

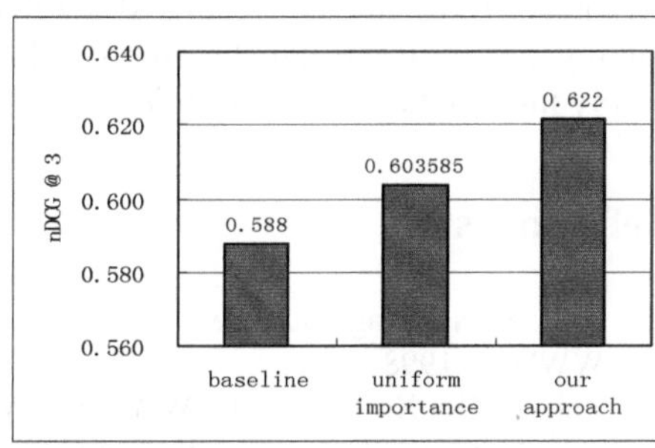

Fig. 3. Effect of differentiating image importance

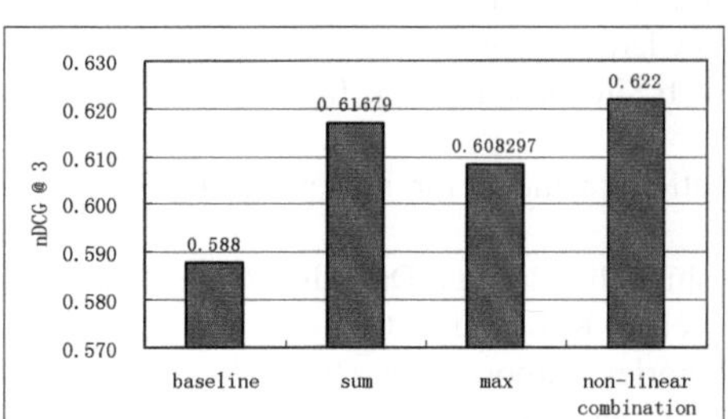

Fig. 4. Performance comparison among various image score combination techniques

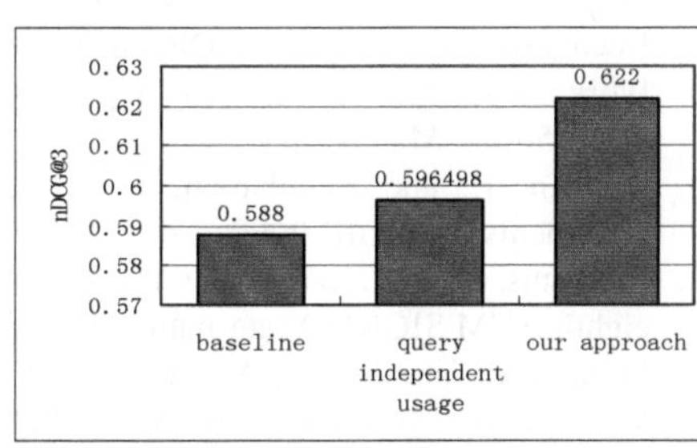

Fig. 5. The performance of using image information as query-independent evidence

4 Related Work

In the image retrieval field, some work has been done on keyword-based image indexing and search [2][3][8]. These systems extract text information related to an image as its annotation from its host web page. Then, images are indexed by using their text information. In content-based image retrieval [11], the contents of images are used to calculate the similarity between a query image and some candidate images.

Some methods are proposed, which uses images as a query independent evidence of page quality, such as [5]. Differently, our method is to explore the semantic information of images and to give their information an importance score. Some works give different weights for different parts of text on a page based on visual information such as font size, color, bold attributes or Web blocks [12]. However, these methods give different weights to text, but they don't use extra information besides text.

5 Conclusion

The important role of images in helping representing Web pages motivates us to utilize image information for relevance ranking. Computing a query-dependent overall image score for a Web page based on all images embedded in it has distinct advantage. Experimental results show that when image information is used properly, ranking performance can be improved. And our approach is practical and can be simply integrated into existing Web search engines.

References

1. Brin, S., and Page, L. The Anatomy of a Large-Scale Hypertextual Web Search Engine. WWW-7, 1998
2. Cai, D., He, X., Li, Z., Ma, W.Y., and Wen, J.R. Hierarchical Clustering of WWW Image Search Results Using Visual, Textual and Link Information. In Proc. Of the 12th ACM international conference on Multimedia, 2004.
3. Frankel, C., Swain, M., and Athitsos, V. Webseer: an Image Search Engine for the World Wide Web. University of Chicago Technical Report TR96-14, 1996.
4. Hu, J., and Bagga, A. Categorizing Images in Web Documents. SPIE Document Recognition and Retrieval (DR&R X), Santa Clara, January 2003.
5. Ivory, M.Y., and Hearst, M.A. Improving Web Site Design. IEEE Internet Computing, v.6 n.2, p.56-63, March 2002
6. Jarvelin, K. and Kekalainen, J. IR evaluation methods for retrieving highly relevant documents. In SIGIR 2000.
7. Joachims, T. Optimizing Search Engines Using Clickthrough Data. In: Proceedings of the eighth ACM SIGKDD international conference on Knowledge discovery and data mining.
8. Lempel, R., and Soffer, A. PicASHOW: Pictorial Authority Search by Hyperlinks on the Web. In Proc. 10th Int. World Wide Web Conf., pp. 438-448, Hong Kong, China, 2001.
9. Robertson, S.E., Walker, S., and Beaulieu, M. Okapi at TREC–7: Automatic Ad-hoc, Filtering, VLC and Filtering Tracks. In Proceedings of TREC'99.
10. Shi, S., Song, R., and Wen, J.R. Latent Additivity: Combining Homogeneous Evidence. Technique report, MSR-TR-2006-110, Microsoft Research, August 2006.
11. Smith, J.R. and Chang, S. VisualSeek: A Fully Automated Content-Based Image Query System. In Proc. ACM Conf. Multimedia, ACM Press, New York, 1996.
12. Song, R., Liu, H., Wen, J.R., and Ma, W.Y. Learning Block Importance Models for Web Pages. WWW'2004.

N-Step PageRank for Web Search

Li Zhang[1], Tao Qin[2], Tie-Yan Liu[3], Ying Bao[4], and Hang Li[3]

[1] Department of Mathematics, Beijing Jiaotong University,
Beijing, 100044, P.R. China
li.zhang1982@gmail.com
[2] MSPLAB, Dept. Electronic Engineering, Tsinghua University
Beijing, 100084, P.R. China
tsintao@gmail.com
[3] Microsoft Research Asia, No. 49, Zhichun Road, Haidian District,
Beijing, 100080, P.R. China
{tyliu,hangli}@microsoft.com
[4] Academy of Mathematics and Systems Science, Chinese Academy of Sciences
Beijing, 100080, P.R. China
ybao@amss.ac.cn

Abstract. PageRank has been widely used to measure the importance of web pages based on their interconnections in the web graph. Mathematically speaking, PageRank can be explained using a Markov random walk model, in which only the direct outlinks of a page contribute to its transition probability. In this paper, we propose improving the PageRank algorithm by looking *N*-step ahead when constructing the transition probability matrix. The motivation comes from the similar "looking *N*-step ahead" strategy that is successfully used in computer chess. Specifically, we assume that if the random surfer knows the *N*-step outlinks of each web page, he/she can make a better decision on choosing which page to navigate for the next time. It is clear that the classical PageRank algorithm is a special case of our proposed *N*-step PageRank method. Experimental results on the dataset of TREC Web track show that our proposed algorithm can boost the search accuracy of classical PageRank by more than 15% in terms of mean average precision.

1 Introduction

PageRank [6] is one of the most successful link analysis algorithms for Web search. PageRank simulates a random walk on the web graph (nodes in the graph represent web pages, and edges represent hyperlinks), and uses the stationary probability of visiting each webpage to represent the importance of that page. Consider a random surfer that is visiting web page *a* at present. At each of successive steps, the surfer will proceed from page *a* to a web page randomly chosen from all the pages that *a* links to. Take Fig.1 for instance. There are three hyperlinks starting from page *a* to pages *b*, *c* and *d* respectively. The surfer will then visit each of these three pages with a probability of 1/3. In other words, the transition probability of this Markov random walk only depends on the information of the page that is currently being visited.

To our knowledge, however, such a Markov model is not always the best choice in many real-world applications. Take computer chess game for example. The key to

G. Amati, C. Carpineto, and G. Romano (Eds.): ECIR 2007, LNCS 4425, pp. 653 – 660, 2007.

the winning of computer "Deep Blue" [2] over human is that it can predict all the situations within much more steps than a human being can do at the same time. That is, if one knows more information, he/she may have more opportunities to make the right decision. Similarly, we argue that if the surfer in the PageRank model can have more information than the direct outlinks, (for example, how many pages one can find by *N*-step jumps after choosing one of the direct outlinks), he/she may choose his/her next step with quite different probabilities in order to maximize his/her information gain. This is just the motivation of our proposed *N*-step PageRank method. It is clear that this *N*-step PageRank is a generalized version of classical PageRank.

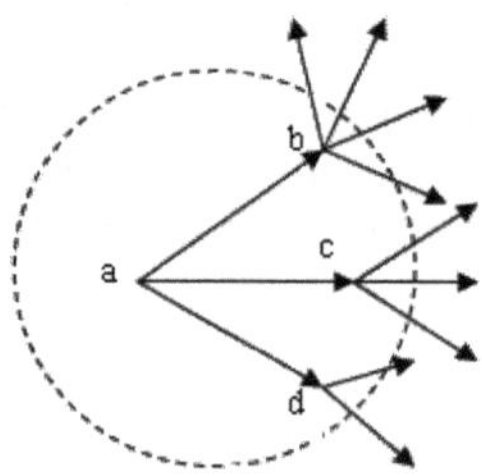

Fig. 1. Illustration of *N*-step

To better elaborate on this new method, in Section 2, we will briefly introduce the classical PageRank algorithm. Then in Section 3, we will give the mathematical formulation of the *N*-step PageRank and prove how it can be effectively computed. Experimental results are reported in Section 4. Finally, we give the conclusions and future work in the last section.

2 PageRank Review

Since our *N*-step PageRank is based on PageRank, we shall introduce PageRank in this section.

The directed link graph of the Web is usually modeled as $G=<V,E>$, where V={1,2, …, *n*} is the set of vertices, the elements of which correspond to all the pages on the Web; $E=\{<i,j>| i,j\in V\}$ is the set of edges, the elements of which correspond to the links between web pages (from page *i* to page *j*).

Based on this modeling, we can further define the adjacency matrix *A* of the link graph as follows,

$$a_{ij} := \begin{cases} 1, if <i,j>\in E \\ 0, \quad otherwise \end{cases} \tag{1}$$

That is, if there is a hyperlink from page *i* to page *j*, $a_{ij}=1$; otherwise, $a_{ij}=0$.

Most link analysis algorithms are based on this adjacency matrix, including PageRank. PageRank simulates a random walk process on the Web to calculate the importance of each web page. Suppose there is a surfer in an arbitrary page. At each step, he/she can transit to one of the destination pages of the hyperlinks on the current page with probability α or to another page randomly in the graph with probability 1-α. Normalize each non-zero row of the adjacency matrix A with its sum, we will get an initial transition matrix *P*. Dealing with zero rows[1], we will get a probability transition matrix $\overline{P}$. Then the above random walk behavior can be modeled as

$$\overline{\overline{P}} = \alpha\overline{P} + (1-\alpha)U \tag{2}$$

Where U is a uniform probability transition matrix, all elements of which equal to $1/n$ (n is the dimension of U).

If we use $\pi = (\pi_1, \pi_2, \cdots, \pi_n)^T$ to denote the stationary distribution of the probability transition matrix $\overline{\overline{P}}$, by the ergodic theory [3], we have:

$$\lim_{m\to\infty} E\left\{\frac{1}{m}\sum_{k=0}^{m-1} I_{\{\text{visiting } i \text{ at the } k\text{-th step}\}}\right\} = \lim_{m\to\infty}\frac{1}{m}\sum_{k=0}^{m-1}\overline{\overline{p}}_{ji}^{(k)} := \pi_i,\ a.s. \tag{3}$$

With the above interpretations, it is reasonable to assume that the more clicks on a webpage, the more important it is. Then the clicking ratio can be interpreted as a measurement of the relative importance of web pages. Thus the stationary distribution $\pi = (\pi_1, \pi_2, \cdots, \pi_n)^T$ can be regarded as the rank scores of the web pages. In fact, π can be computed through the following iterative process.

$$\pi(t+1) = \overline{\overline{P}}^T \pi(t) \tag{4}$$

Furthermore, it has been proved that the stationary value of π corresponds to the principal eigenvector of $\overline{\overline{P}}^T$ when $\overline{\overline{P}}^T$ has its unique principal eigenvalue [5].

For clarity, we will denote the above PageRank algorithm [6] as classical PageRank in the following discussions.

3 *N*-Step Pagerank

In classical PageRank, when the surfer chooses the next webpage, he/she uses only the information of direct outlinks of the current page, i.e., chooses one of the outlink pages with equal probability. In other words, classical PageRank assumes that outlinks are non-distinguishable to the surfer. In this paper, we argue that outlinks can actually be distinguished from many aspects. For example, the surfer may find more useful information or more hyperlinks to new pages after clicking one outlink than the

[1] If the sum of a row is 0, the elements in this row are all given value 1/n after normalization [5].

other. Inspired by the "look N-step ahead" strategy in a computer chess, we propose using the information contained in the next N-step surfing to represent the information capacity of an outlink, and thus distinguish different outlinks. To make it clearer, we take Fig.1 for example.

Suppose a user is browsing webpage a. According to the classical PageRank algorithm, when he selects the next webpage, he can only choose from b, c, and d with equal probability. In our N-step PageRank algorithm, the user has more information to decide which page to select. That is, he also knows outlinks of page b, c, and d. For the case of looking 2-step ahead, the probabilities that he selects b, c, and d can be defined as proportional to the outlink numbers of b, c, and d. That is, p_{ab}=4/9, p_{ac}=3/9, and p_{ad}=2/9. For the case of looking 3-step ahead, the probabilities that he selects b, c, and d will be proportional to the number of web pages he can reach after the next two steps. Recursively, we can come to the conclusion that for the case of looking N-step ahead, the probabilities that the surfer selects b, c, and d are proportional to the number of web pages he can reach after the next (N-1) steps.

3.1 Transition Matrix $P^{(n)}$

After the above intuitive explanation of N-step PageRank method, we will give the expression of the corresponding transition probability matrix. Actually, for two arbitrary vertexes i and j, we have $p_{ij}^{(N)} = \frac{d_j^{(N-1)} 1_{\{(i,j)\in e(G)\}}}{\sum_{(i,k)\in e(G)} d_k^{(N-1)}}$, where $d_j^{(N)}$ is the vertex number after vertex j jump N steps, and $d^{(0)} = (1,1,\ldots,1)_n^T$.

Theorem 1. *The transition matrix of N-step PageRank algorithm* $P^{(N)}$ *can be computed as following:*

$$P^{(N)} = (D^{(N)})^{-1} A D^{(N-1)} \tag{5}$$

where A is the adjacent matrix of the directed graph G, $D^{(N)}$ *is a diagonal matrix generated by the vector* $d^{(N)}$, $d^{(N)} = Ad^{(N-1)} = A^N d^{(0)}$, *the elements in* $d^{(N)}$ *is the vertex number after each vertex jumps N steps*[2]

Proof: Considering

$$AD^{(N-1)} = \begin{bmatrix} a_{11}\times d_1^{(N-1)}, a_{12}\times d_2^{(N-1)}, \cdots, a_{1n}\times d_n^{(N-1)} \\ a_{21}\times d_1^{(N-1)}, a_{22}\times d_2^{(N-1)}, \cdots, a_{2n}\times d_n^{(N-1)} \\ \cdots \\ a_{n1}\times d_1^{(N-1)}, a_{n2}\times d_2^{(N-1)}, \cdots, a_{nn}\times d_n^{(N-1)} \end{bmatrix},$$

---

[2] $(D^{(N)})^{-1}$ is the extended inverse matrix, and if a certain diagonal element in $D^{(N)}$ is 0, the corresponding element in $(D^{(N)})^{-1}$ is 0.

we have

$$(D^{(N)})^{-1}AD^{(N-1)} = \begin{bmatrix} \frac{a_{11}\times d_1^{(N-1)}}{d_1^{(N)}}, \frac{a_{12}\times d_2^{(N-1)}}{d_1^{(N)}}, \dots, \frac{a_{1n}\times d_n^{(N-1)}}{d_1^{(N)}} \\ \frac{a_{21}\times d_1^{(N-1)}}{d_2^{(N)}}, \frac{a_{22}\times d_2^{(N-1)}}{d_2^{(N)}}, \dots, \frac{a_{2n}\times d_n^{(N-1)}}{d_2^{(N)}} \\ \dots \\ \frac{a_{n1}\times d_1^{(N-1)}}{d_n^{(N)}}, \frac{a_{n2}\times d_2^{(N-1)}}{d_n^{(N)}}, \dots, \frac{a_{nn}\times d_n^{(N-1)}}{d_n^{(N)}} \end{bmatrix}$$

Note that $d^{(N)} = Ad^{(N)}$, and $d_i^{(N)} = \sum_{j=1}^{n} a_{ij} d_j^{(N-1)}$, then it is easy to verify that

$$p_{ij}^{(N)} = \frac{d_j^{N-1} 1_{\{(i,j)\in e(G)\}}}{\sum_{(i,k)\in e(G)} d_k^{N-1}} = \left[(D^{(N)})^{-1} AD^{(N-1)}\right]_{ij}$$

The above theorem shows that we can easily get the transition matrix $P^{(N)}$ of N-step PageRank from the adjacent matrix A. Since D is a diagonal matrix, and some of its diagonal elements may be zeros, the matrix $P^{(N)}$ is more sparse than the initial matrix P in classical PageRank which can be used to speed the computation of the stationary distribution.

3.2 Convergence Rate

Similar to the classical PageRank algorithm, we can obtain irreducible random matrix $\bar{\bar{P}}^{(N)}$ from $P^{(N)}$. According to the ergodic theory [3], the corresponding Markov chain has a unique stationary distribution $\pi^{(N)}$, and $\pi^{(N)} = (\bar{\bar{P}}^{(N)})^T \pi^{(N)}$. We still use $\pi^{(N)}$ to measure the importance of web pages. We can use power iteration method to compute the final distribution $\pi^{(N)}$, just like equation (5). In this sub section, we will discuss the convergence rate of the corresponding power iteration method.

Theorem 2. *The stationary distribution is*

$$\pi^{(N)} = \alpha(P^{(N)})^T \pi^{(N)} + (\alpha(\sum_{i\in D} r_i \pi_i^{(N)}) + \tfrac{1}{n}(1-\alpha))e_n,$$

where $\pi^{(N)}(0)$ *denotes the initial vector of the iteration,* $\pi^{(N)}(t)$ *denotes the vector after t steps of iteration, and D is a collection of nodes,*

$$D = \{i \mid \sum_{i=1}^{n} p_{ij}^{(N)} = 0, i \in V\}$$

And the convergence rate is

$$\left\|\pi^{(N)} - \pi^{(N)}(t)\right\|_1 \le \alpha^t \left\|\pi^{(N)} - \pi^{(N)}(0)\right\|_1 \le 2\alpha^t$$

Proof: Omitted since the proof process is very similar to that of classical PageRank.

4 Experimental Results

4.1 Settings and Results

To compare our *N*-step PageRank algorithms with classical PageRank (PR) [6], we chose the topic distillation task in Web track of TREC 2003 as the benchmark. For each query, we first use BM25 [7] to get a list relevance pages. We then choose the top 2000 pages from this list, and combine the relevance score with importance score as follow:

$$score_{combination} = (1-\alpha)\times score_{importance} + \alpha\times score_{relevance}$$

To evaluate the search accuracy, we adopted two widely-used criteria in our experiments: mean precision at *n* (P@n) [8], and mean average precision (MAP) [8].

The MAPs of two PageRank algorithms under investigation are shown in *Fig.2*. We can see that our 2-step PageRank almost uniformly outperforms classical PageRank, which shows the effectiveness of considering multi-step hyperlink information.

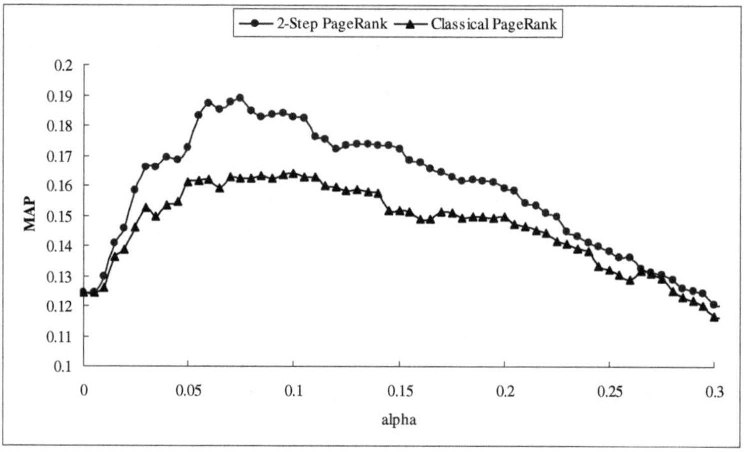

Fig. 2. Search accuracy comparison

We list the best performance of these two algorithms in Table 1 for further comparison. From this table, the best performance of 2-step PageRank is much better than classical PageRank. The MAP of 2-step PageRank is more than 2.5 points higher than that of classical PageRank, which corresponds to over 15% relative improvement. And 2-step PageRank gets more than 6% relative improvement over classical PageRank in terms of *P*@10. Note that the combination of importance score and relevance score overcomes the relevance score only, which implies the value of link analysis.

Table 1. Comparison of different ranking algorithm

Ranking methods	Best MAP	Best P@10
Relevance only	0.1243	0.104
2-step PageRank + relevance	0.1891	0.134
PageRank + relevance	0.1639	0.126
Best Result on TREC2003 [1]	0.1543	0.128

4.2 The Influence of Loops

Real-world Web graph may contain loops. One may argue that these loops will influence the result of *N*-step PageRank algorithm, and the hyperlink structure is too complex to quantify this influence. For example, in Fig.3, the loop *b*→*u*→*v*→*b* will make web page *b* double counted when computing the outdegree of web page *b* in 3-step PageRank.

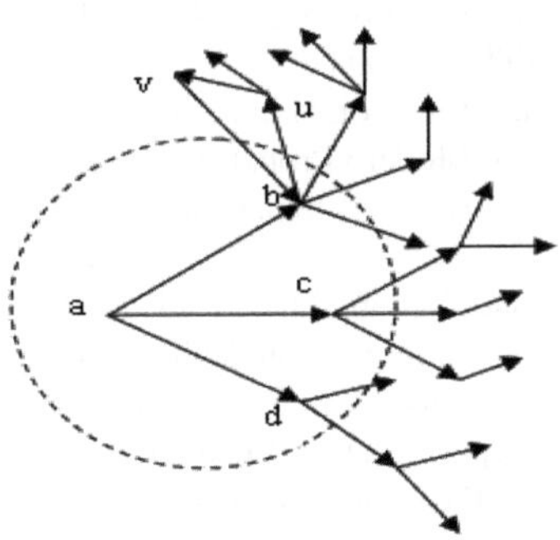

Fig. 3. Illustration of circle in 3-step outlinks

To investigate the influence of the loop, we conducted a small experiment on the case of N=2. After eliminating loops while counting the 2-step outlinks of a webpage, we find the ranking result is almost the same as the previous result. We will investigate more complicated cases in our future work.

5 Conclusions and Future Work

Inspired by the computer chess, in this paper, we pointed out that the random walk model in classical PageRank algorithm could be improved by considering more information. Specifically, we modified the transition matrix of classical PageRank algorithm by using multi-step outlink information, and proposed the *N*-step PageRank algorithm. Experiments on the topic distillation task of TREC2003 showed that the new algorithm outperforms the classical PageRank algorithm.

In the future work, we plan to investigate the following problems:

(1) We have not given an explicit relationship between the stationary distribution of *N*-step PageRank and that of classical PageRank in this paper. Though the transition matrices of the two algorithms seem not very complex, they cannot represent each other by elementary transformation. As a result, it is not clear what will happen to the corresponding eigenvectors.

(2) As aforementioned, the "look *N*-step ahead" model is to leverage more information when computing page importance. Actually the proposed *N*-step PageRank algorithm is just a simple implementation. We will study how to make use of other information, such as website structure.

Acknowledgments

We would like to thank anonymous reviewers for their hard work.

References

1. Craswell, N., Hawking, D. (2003). Overview of the TREC 2003 Web Track, in the twelfth Text Retrieval Conference (TREC 2003).
2. Hsu, F.-h. Behind Deep Blue, Princeton University Press, Princeton, NJ, 2002.
3. Kallenberg, O. Foundations of Modern Probability, Page152.
4. Kleinberg, J. Authoritative sources in a hyperlinked environment, Journal of the ACM, Vol. 46, No. 5, pp. 604-622, 1999.
5. Ng, A. Y., Zheng, A. X., and Jordan, M. I. Link analysis, eigenvectors, and stability. In Proc. 17th International Joint Conference on Artificial Intelligence, 2001.
6. Page, L., Brin, S., Motwani, R., and Winograd, T. The PageRank citation ranking: Bringing order to the web, Technical report, Stanford University, Stanford, CA, 1998.
7. Robertson, S. E. Overview of the okapi projects, Journal of Documentation, Vol. 53, No. 1, 1997, pp. 3-7.
8. Salton, G. and McGill, M. J. Introduction to Modern Information Retrieval. McGraw-Hill, 1983.

Authorship Attribution Via Combination of Evidence

Ying Zhao and Phil Vines

School of Computer Science and Information Technology, RMIT University
GPO Box 2476V, Melbourne, Australia
{yizhao,phil}@cs.rmit.edu.au

Abstract. Authorship attribution is a process of determining who wrote a particular document. We have found that different systems work well for particular sets of authors but not others. In this paper, we propose three authorship attribution systems, based on different ways of combining existing methodologies. All systems show better effectiveness than the state-of-art methods.

1 Introduction

Authorship attribution systems use stylistic elements to classify documents according to authorship. Apart from solving literary disputes, it is being used in forensics, to identifying authorship of emails, postings on newsgroup, and can be used to provide legal support [1,4]. In this paper we have studied the effect of combining different attribution systems in various ways. Rather than simply providing a marginal performance improvement, we have found a principled way of combining methods that leads to a statistically significant improvement in classification effectiveness. Additionally, when a number of different models all attribute a document to the same author, we can have a higher degree of confidence in regard to the correctness of the attribution. This is likely to be important in legal situations.

In our previous work the results showed that different methods seem to provide more effective discrimination for some author pairs but not others [12]. We conjecture that some authors have stylistic habits that result in the use of one form of style markers in a manner that is more distinctive than others. In this paper we propose three systems, all based on the principle of using Kullback-Leibler diverence, can be combined [12].

2 Background

There are two main steps involved in any authorship attribution system: feature extraction, and classification based on the extracted features. The simplest test is two-class authorship attribution, where a document is classified according to whether it is more similar to one of the two pre-defined authors. An alternative evaluation is the somewhat harder multi-class classification problem where a classification decision is made amongst n authors, where $n > 2$. In this paper we evaluated the proposed systems on both types of attribution tasks.

The extraction of stylometric features is the key aspect of effective authorship attribution. Both lexical markers, such as function words and punctuation symbols [2,5,11],

G. Amati, C. Carpineto, and G. Romano (Eds.): ECIR 2007, LNCS 4425, pp. 661–669, 2007.

and more advanced grammatical markers obtained using natural language processing (NLP) [3,7,12] have been proposed. On the other hand a variety of different classification methods have been used, including *support vectors machines* (SVMs) [5], *Bayesian networks* [9,11], and *principal component analysis* (PCA) [3,6]. In previous work we have proposed an entropy and uni-gram language model based technique for authorship attribution. The method has outperformed the state-of-art techniques, including SVM and bayesian networks. The core algorithm of computation is reviewed below, followed by the description of 3 new proposed systems. All systems are an improvement of the baseline results[1] achieved previously.

The technique we have used involves building language models for each set of training documents and for the test document, based on a set of pre-defined features.In this work, we have experimented with four different types of features. *Function words (**FW**)*, *Part-of-speech tags (**POS**)*,[2] *Function words with the corresponding POS tags (**F/P**)*,and *Function words in a first order Markov model (**2Gram**)*.

Once the language models are constructed, we use Kullback Lieblier divergence to measure the difference in the models for the training documents and the individual test documents to be attributed. Kullback-Leibler divergence is used to calculate the relative entropy thus quantifying the difference between two distributions. We incorporate language models for approximation of feature distributions into the computation of Kullback-Leibler divergence. Formally, the divergence-based model is given by:

$$KLD\,(p_d \| p_C) = \sum_{c \in C \cup d} \left[\left(\frac{f_{c,d}}{\mu + |d|} + \frac{\mu}{\mu + |d|} p_B(c) \right) \times \log_2 \frac{\frac{f_{c,d}}{\mu+|d|} + \frac{\mu}{\mu+|d|} p_B(c)}{\frac{f_{c,C}}{\mu+|C|} + \frac{\mu}{\mu+|C|} p_B(c)} \right] \qquad (1)$$

where p_C is the model built from a group of training texts that are of a particular author category and p_d is the single document model, with which the document is to be identified. Since a given document will not normally contain all the features of each type that we pre-defined, a smoothing technique is used to estimate the likelihood of missing features in the documents. Dirichlet Prior, also known as Bayesian Smoothing [8], is one of the most effective smoothing approaches in IR [10]. It re-allocates the probabilities using the collection model $p_B(c)$. μ is a smoothing parameter to adjust the significance of feature c from $p_B(c)$. In general for short documents, the background probabilities dominate, on the principle that the in-document authorial evidence is weak; as the document length grows, however, the influence of the background model diminishes. Choice of an appropriate value for μ is a tuning stage in the use of this model. The document with unknown authorship is assigned to the author model that has the smallest divergence calcuated amongst all available author models.

3 Methodology

We have previously experimented with systems that build single models using multiple feature types, but with little success [12]. It seems that the distinguishing features

[1] The baseline results related to this work has been re-presented in Table 1.

[2] We use NLTK (a Natural Language ToolKit) package that is available from `http://nltk.sourceforge.net/index.html`

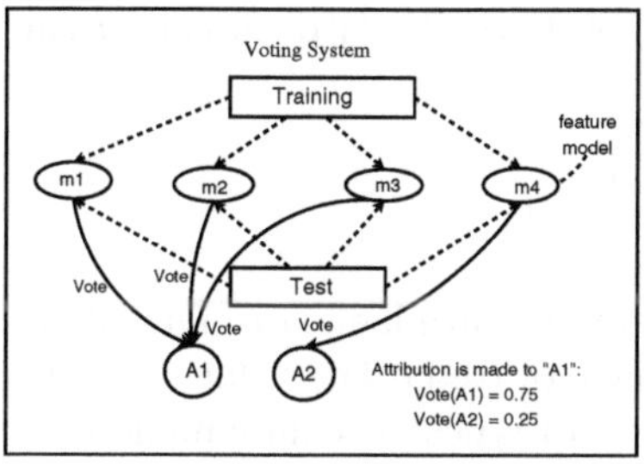

Fig. 1. The voting system

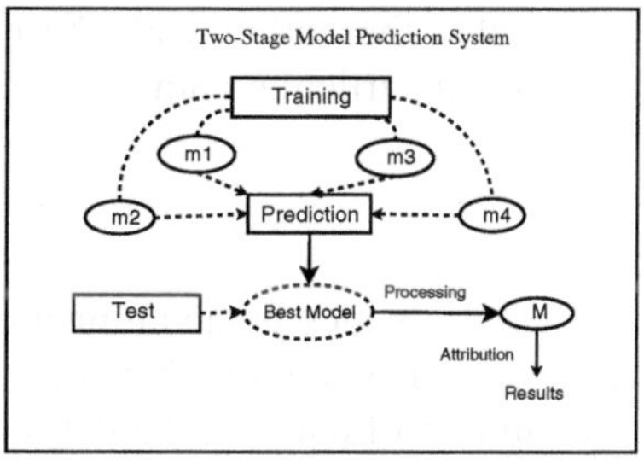

Fig. 2. The prediction system

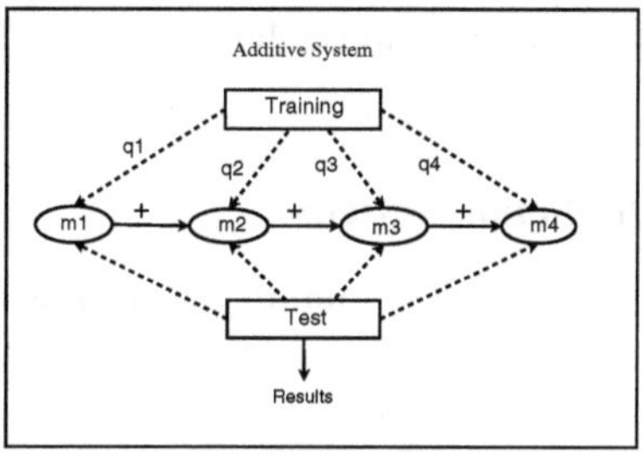

Fig. 3. The additive system

peculiar to given author combinations have a greater tendency to be overwhelmed by other features. In this paper we propose three authorship attribution systems, all based on different ways of combining the evidence resulting from models that use a single feature type for classification.

3.1 Voting System

The idea of the voting system is simple: several different methods are used to attribute a document with unknown authorship to an author. The author that gets the most votes wins. In this system, each model is built based on a different type of style markers. The framework of the voting system is depicted in Figure 1.

Given several types of style markers $\{m_1 \ldots m_j\}$, we need to build and train j feature models for each author candidate A_i. The divergence is then measured between a document model and an author model that consists of j different feature models:

$$Vote(A_i) = \forall_j \left(\forall_i \; Argmin_i \; KLD\left(\Theta_j(d) \parallel \Theta_j(A_i) \right) \right) \tag{2}$$

According to Equation 2, a document with unknown authorship d gets one vote for author A_i from feature model Θ_j, if the divergence between d and A_i for feature model Θ_j is the smallest amongst all author candidates. Therefore each author A_i can receive at most j votes. $Vote(A_i)$ is the number of votes that are received by a particular author A_i ($Vote(A_i) \leqslant j$).

Informally, the author with the most votes wins. Formally, a document d can be attributed to a particular author A_i:

$$d \rightarrow A_i \ \text{ if } \ \frac{Vote\,(A_i)}{|\Theta|} \geqslant t \tag{3}$$

where $|\Theta| = j$, which is the number of the feature models available. t is a threshold for making the attribution that has the value in the range of $(0, 1]$. The value of t depends on the number of author candidates and the number of types of feature models. Additionally t determines the strictness of the voting system. The bigger the value of t, the more strict the system is. Using binary authorship attribution as an example, a typical range of values for t is: $t > \lceil|\Theta|/2\rceil$ if $|\Theta|$ is an even number, otherwise $t \geqslant \lceil|\Theta|/2\rceil$.

The methodology of this approach also provides a principled way of merging various approaches with little modification while enhancing the attribution accuracy and the reliability of the results.

3.2 Two-Stage Model Prediction System

In the previous voting system, both the training and testing samples are modeled based on individual types of style markers for each individual author. In contrast we propose a two-stage model prediction system, which is less expensive in terms of computational cost compared to that of the voting system. The outline of the two-stage model prediction system is shown in Figure 2.

Based on the observation that different types of style markers are more effective for different authors [12]. The idea of this model is to dynamically predict the best feature model for each attribution task individually, and then use predicted feature model for attribution. Given j different types of style markers available for the system, feature models $\Theta_1 \ldots \Theta_j$ are built. The training documents for each author are then split into two groups. The first group is used to train each of the feature models. The second group is used as test data to determine which model has the most effective discrimination ability amongst a pair, or group of authors. That feature model is then used to attribute the documents with unknown authorship.

3.3 Additive System

The additive system uses the same modeling process as the voting system. The outline of the additive system is shown in the Figure 3. For both training and test samples, full modeling is required for each author and each individual type of style markers.

However the additive system is different from the voting system in that different rules are used for the final attribution. Rather than giving a unit vote by each feature model, we calculate the "size" of the vote in accordance with the magnitude of divergence that each feature model produces. For example, three models might slightly favor author A, whereas one model may favor author B. In the simple voting model, this would select author A whereas in this model, author B may be selected if the divergence measured by one feature model is small enough. In our experiments we have effectively summed the divergences produced by each feature model (by setting $q_j = 1/j$), however it is possible that in a more sophisticated system we may also assign a different weight

q_j to each of the models. Formally, this model measures divergence, and attributes a document as follows:

$$KLD(d \parallel A_i) = \sum_j q_j \left(KLD\left(\Theta_j(d) \parallel \Theta_j(A_i)\right)\right) \quad \text{where} \quad \sum_j q_j = 1 \tag{4}$$

From which the document d is attributed to author A_i, if $i = Argmin_i\ KLD(d \parallel A_i)$.

4 Experiments and Results

The experimental setup is consistent with our previous work, [12,11]. Seven newswire authors are selected. Each author has contributed over 800 documents and therefore is considered to have reasonable style in writing. The average document length is 724 words. All experiments in this paper are based on this collection of AP newswire articles. Our baseline results were shown in Table 1.

In this paper we have applied all 4 types of style markers to the three proposed systems. Therefore $j = 4$, and the value i refers to the number of authors that are to be differentiated in the experiments.

Voting system. We implemented the voting system as described in Figure 1. We have experimented with binary authorship attribution using both three and four different types of feature models. Thus, the threshold t is set at $t > 0.5$ ($t = 1$ or $t = 0.75$ when voting with 4 feature models; $t = 1$ or $t = 0.67$ when voting with 3 feature models). Results are shown in Table 2, given different number of sample documents for training.

In Table 2, Rec_{voting} shows the percentage of total number of test documents that were attributed with correct authorship, and is somewhat akin to a recall measure. Since this system requires a threshold to be met before a document can be attributed, some documents remain unattributed if they do not meet the threshold. Pre_{voting} shows the number of documents that were correctly attributed with authorship as a percentage of those for which attribution decisions were made (i.e. that meet the threshold, and is somewhat akin to a measure of precision). It can be seen that when the best 3 feature models are used and a lower threshold $t = 0.67$ is set, the system performs better than the baseline, in terms of percentage of documents correctly attributed. On the other hand, if the highest threshold $t = 1$ is used, we can have a higher degree of confidence that the attributed documents have been done so correctly — up to 98.7%—using 4 feature models with sufficient training samples.

Table 1. The effectiveness (the percentage of test documents attributed with correct authorship) for 2, 3, 4, and 5-class attribution. The data is extracted from the AP collection [11], with function words as features, using Dirichlet smoothing.

	25 training docs				100 training docs				400 training docs			
N	FW	POS	F/P	2Gram	FW	POS	F/P	2Gram	FW	POS	F/P	2Gram
2	**88.2**	87.9	87.3	82.0	**91.7**	89.5	90.7	87.1	**92.8**	89.6	92.1	92.6
3	81.8	**82.1**	81.1	72.0	**86.6**	83.7	85.9	79.5	**88.3**	84.1	88.1	88.1
4	77.1	**78.0**	76.7	64.9	**82.7**	79.8	82.2	74.3	84.9	80.4	**85.2**	84.8
5	73.5	**74.8**	73.2	59.6	79.7	76.9	79.2	70.5	82.2	77.5	**82.9**	82.1

Table 2. The effectiveness of the voting system on binary authorship attribution. Results are compared to the best results of using individual feature.

	25 Training docs				100 Training docs				400 Training docs			
	4/1	4/.75	3/1	3/.67	4/1	4/.75	3/1	3/.67	4/1	4/.75	3/1	3/.67
Rec_{voting}	70.9	86.0	79.6	89.2	75.7	90.2	83.6	92.2	81.1	92.1	83.6	92.9
Pre_{voting}	**95.4**	91.5	93.9	89.2	**97.9**	94.7	96.0	92.2	**98.7**	95.2	95.7	92.9

Table 3. The effectiveness of using the two-stage model prediction system for 2, 3, 4, and 5-class authorship attribution tasks

	M_c (# of combinations)						
N Class	FW	POS	FW/POS	2GramFW	M_c(pre)	Acc.	Acc.(Pre)
2 (/21)	6	4	4	7	**9**	92.3	**93.5**
3 (/35)	10	4	9	12	**21**	87.5	**88.8**
4 (/35)	11	3	16	9	**19**	83.8	**85.2**
5 (/21)	5	1	**12**	3	11	80.9	**82.4**
overall (/112)	33	12	40	32	**60**		

Finally, instead of implementing a new model voting system from scratch, the methodology provides a principled way of integrating existing approaches with little modification of the approach itself. Researchers have proposed many different ways to perform authorship attribution [3,5,6,9]. However the open challenge is the lack of benchmarks and consistent experimental setup, which has led difficulties to compare these approaches. The proposed voting system is a plausible alternative to make use of existing approaches as far as possible. There is no need to modify the approach itself, instead the output can be used as the input to the the voting system.

Two-stage model prediction system. We implemented the two-stage model prediction system described in Section 3.2. 400 training samples are used in total so as to be comparable with other experiments, however, these are divided into two sets. As shown in Figure 2, the first set is used to train each of the feature models and the second set is used to predict the best feature model to use for that author combination. The system is evaluated with both binary and multi-class authorship attribution, results are presented in Table 3.

As shown in Table 3, 9 correct predictions are made out of 21 attribution tasks in total for binary classification. In cases where the best system was not chosen, the second best system was usually chosen and in these instances there was usually not a lot if difference between the top systems. Importantly the prediction system avoids choosing a system that is particularly bad for a given author combination.

We have experimented with authorship attribution tasks from 2-class up to 5-class authorship attribution across all possible author combinations. Thus we have in total 112 attribution tasks, given 7 authors in total ($C_7^2 + C_7^3 + C_7^4 + C_7^5$). The likelihood for an individual type of style markers to be the best choice of the task is 35.7% (40/112) in the experiment. In contrast, by model prediction the likelihood can be increased to

Table 4. The comparison between results from additive system and the best baseline results

	25 training docs				100 training docs				400 training docs			
$M_{selected}$	2	3	4	5	2	3	4	5	2	3	4	5
FW+POS	**90.3**	**85.4**	**81.7**	**78.6**	92.6	88.3	85.3	**82.9**	92.9	88.6	85.0	82.9
FW+F/P	88.4	82.2	77.8	74.3	91.6	86.5	82.8	79.8	92.7	88.3	85.1	82.6
FW+2Gram	87.3	80.0	74.6	70.2	92.2	86.9	82.8	79.5	95.0	91.9	89.5	87.6
POS+F/P	89.5	84.2	80.4	77.3	92.5	88.4	**85.4**	**82.9**	92.9	89.0	86.2	83.9
POS+2Gram	86.2	78.6	73.0	68.5	91.3	85.4	81.2	77.8	95.1	92.1	89.7	89.6
F/P+2Gram	87.3	80.6	75.6	71.6	91.8	86.5	82.7	79.7	95.2	92.4	90.2	88.4
-FW	88.8	83.2	79.0	75.7	92.7	88.1	84.7	81.9	95.5	92.7	90.7	89.0
-POS	88.8	82.8	78.4	74.8	92.6	87.8	84.3	81.5	95.5	92.8	90.8	89.3
-FW/POS	89.1	83.3	78.9	75.3	93.0	88.2	84.5	81.5	95.7	93.2	91.2	89.6
-2Gram	89.4	84.0	80.0	76.7	92.5	**88.8**	85.2	82.8	93.1	89.0	86.0	83.5
all	89.4	83.9	79.8	76.5	**93.1**	88.7	**85.4**	82.7	**95.7**	**93.2**	**91.5**	**90.0**

53.6% (60/112). Approximately 18% improvement is achieved in terms of the probability of picking up the best feature model for a particular attribution task.

On the other hand, the overall attribution accuracies are also improved by the two-stage model prediction system. On average around 1.5% improvement is achieved in terms of the overall accuracy, shown as the last column in Table 3.

Additive system. Finally we implemented additive system. We evaluated the system by varying the number of feature models and the number of available training samples that are provided to the system. Given 4 types of features in total, we provided all possible combinations of features, the numbers of combinations are, C_4^2, C_4^3, and C_4^4. Results of the experiments are presented in Table 4, given different number of training samples. In the table a "+" indicates the combination of individual types of style markers, while "-" indicates the exclusion of that type of style markers.

It can be seen that the additive system gives the best effectiveness amongst all three proposed systems. The best result is achieved by modeling all 4 types of features with 400 documents per author as the training data. 90% effectiveness is achieved for the most difficult 5 class attribution test. This is an 8% improvement on the best result achieved using any of the individual models with the same experimental setup as shown in Table 1. In addition, as mentioned earlier, these results are also far superior that that which can be obtained simply by a combination of different types of style markers into one model [12].

In addition the choice of style markers affects the effectiveness of the additive system. Given sufficient 400 training samples for instance, we observed that the combination of "FW/POS" and "2GramFW" gives the best results amongst all feature model pairs. While using all types of features but "FW/POS" gives the best results of any three types of feature models. Table 5 shows the results of significance tests between the performance of the best baseline system from Table 1, and the best additive system using all feature models. The additive system performs significantly better, especially with the harder multi-class attribution tasks, with which the p-values are extremely small.

Table 5. The significance test between the best baseline results and the best additive modeling results across all attribution tasks. (400 documents for training).

	2 class	3 class	4 class	5 class
p-value	0.018 Y	$1.291e^{-7}$ Y	$2.053e^{-13}$ Y	$5.177e^{-13}$ Y

5 Conclusions

We have shown in this paper that there is no one type of style markers that always works the best for all attribution tasks. The suitability of the types of style markers is author dependent and task dependent. Importantly, it is shown from previous work that simply adding more features into a given model does not guarantee better results.

We propose three systems in this paper that effectively make use of the advantages of different types of style markers. The results achieved by these systems, especially the additive system, are shown to be significantly more effective than any previous method based on an individual type of style markers. In addition, the voting system provides a principled way of integrating existing approaches with little modification while enhancing the attribution performance. Finally the two-stage model prediction system is much less expensive in terms of computational cost, while providing with not only higher attribution accuracy but also better choices of style markers in relation to a particular attribution task.

References

1. S. Argamon, M. Saric, and S. S. Stein. Style mining of electronic messages for multiple authorship discrimination: First results. In *Proc. 9th ACM SIGKDD Int. Conf. on Knowledge Discovery and Data Mining*, pages 475–480. ACM Press, 2003.
2. H. Baayen, H. V. Halteren, A. Neijt, and F. Tweedie. An experiment in authorship attribution. *6th JADT*, 2002.
3. H. Baayen, H. V. Halteren, and F. Tweedie. Outside the cave of shadows: Using syntactic annotation to enhance authorship attribution. *Literary and Linguistic Computing*, 11(3):121–132, 1996.
4. C. E. Chaski. Computational stylistics in forensic author identification. In *SIGIR workshop: Stylistic Analysis of Text For Information Access*, August 2005.
5. J. Diederich, J. Kindermann, E. Leopold, and G. Paass. Authorship attribution with support vector machines. *Applied Intelligence*, 19(1-2):109–123, 2003.
6. D. I. Holmes, M. Robertson, and R. Paez. Stephen Crane and the New York Tribune: A case study in traditional and non-traditional authorship attribution. *Computers and the Humanities*, 35(3):315–331, 2001.
7. A. Kaster, S. Siersdorfer, and G. Weikum. Combining text and linguistic doument representations for authorship attribution. In *SIGIR workshop: Stylistic Analysis of Text For Information Access*, August 2005.
8. D. Mackay and L. Peto. A hierarchical dirichlet language model. *Nat. Lang. Eng.*, 1(3):289–307, 1995.

9. A. Sarkar, A. D. Roeck, and P. H. Garthwaite. Term re-occurrence measures for analyzing style. In *SIGIR workshop: Stylistic Analysis of Text For Information Access*, August 2005.
10. C. X. Zhai and J. Lafferty. A study of smoothing methods for language models applied to information retrieval. *ACM Transaction on Information System*, 22(2):179–214, 2004.
11. Y. Zhao and J. Zobel. Effective authorship attribution using function words. In *Proc. 2nd AIRS Asian Information Retrieval Symposium*, pages 174–190. Springer, 2005.
12. Y. Zhao, J. Zobel, and P. Vines. Using relative entropy for authorship attribution. In *Proc. 3rd AIRS Asian Information Retrieval Symposium*. Springer, 2006. 92-105.

Cross-Document Entity Tracking

Roxana Angheluta and Marie-Francine Moens

Katholieke Universiteit Leuven
roxana.angheluta@student.kuleuven.be,marie-france.moens@law.kuleuven.be

1 Introduction

The main focus of current work is to analyze useful features for linking and disambiguating person entities across documents. The more general problem of linking and disambiguating any kind of entity is known as entity detection and tracking (EDT) or noun phrase coreference resolution. EDT has applications in many important areas of information retrieval: clustering results in search engines when looking for a particular person; possibility to answer questions such as "Who was Woodward's source in the Plame scandal?" with "senior administration official" or "Richard Armitage" and information fusion from multiple documents. In current work person entities are limited to names and nominal entities. We emphasize the linguistic aspect of cross-document EDT: testing novel features useful in EDT across documents, such as the syntactic and semantic characteristics of the entities. The most important class of new features are contextual features, at varying levels of detail: events, related named-entities, and local context. The validity of the features is evaluated on a corpus annotated for cross-document coreference resolution of person names and nominals, and also on a corpus annotated only for names.

2 Corpora

In order to build a dedicated corpus for the task, we downloaded and annotated articles linked by GoogleNews (`http://news.google.com/`) - further called GoogleNews corpus. After removing erroneous articles and cleaning the HTML source code, the final corpus contained 103 articles (statistics in table 1, second column). Documents were parsed, parse trees were manually corrected and annotated for coreference resolution of person names both within- and across-documents.[1] We tested also some of the features on the corpus mentioned in [2][2] - further called Named-entity corpus (statistics in table 1, third column).

3 Features

Each mention is characterized by a number of features. Some of them (called indirect features) are not used directly in coreference resolution, but rather in the computation of other features. Most of the indirect features were computed

[1] For within-document coreference annotation we used WordFreak [6] and followed the guidelines defined in the Message Understanding Conference MUC7 [3].

[2] Available from: *http://l2r.cs.uiuc.edu/~cogcomp/Data/MIRROR/*

G. Amati, C. Carpineto, and G. Romano (Eds.): ECIR 2007, LNCS 4425, pp. 670–673, 2007.

Table 1. Corpora used in the experiments

	GoogleNews	Named-entity
Total number of documents	103	294
Average document length (words)	516	882
Total number of person mentions	3169	4237
Names	1621	4237
Nominals	1548	-
Average number of person mentions per document	30.76	14.41
Total number of entities	607	1323

using OpenNlp package [4]. They include: head noun phrase (head NP), head noun, part of speech (POS) tags, position and type of mention: NE or nominal. Other features are used directly in cross-document coreference resolution. They are grouped in different classes as follows:

Compatibility features: In order for two mentions to match, they have to agree in gender and number. The gender is computed heuristically using lists of first names and titles and also WordNet [7] synsets. The number is detected from the POS tag of the head noun.

String matching features: We have used a normalized substring matching between the head NPs of the two mentions.

Contextual features: Various contextual features that are based on the redundancy of the data signal coreference relations. They are the focus of current work and they are valid only across documents.

1. **Events:** If two documents speak about the same event[3], entities appearing in them have higher chances to corefer. In current experiments, for the Named-entity corpus, documents are clustered into events using a similarity function based on the vocabulary: cosine between vectors of tf*idf weighted terms. For the GoogleNews corpus, we use the events detected automatically by Google.
2. **Local context:** Another contextual feature is the local context of mentions (as in [1]). In current experiments we have used the sentence to which the mention belongs. The cosine between vectors of terms weighted with tf*idf is used as the similarity measure, binned in the interval [0-1].
3. **Related named-entities (NE):** At a more fined-grained level than events and local contexts, one can look at the named-entities related to the current mention. Often, a person can be fully described by the entities related to it, such as other persons, locations, organizations. We look at named-entities that are present inside the same sentence as the mention. The similarity between two mentions is computed as the normalized number of common related named-entities.

[3] Defined in the Topic Detection and Tracking challenge [5]: the same topic happening in the same time in the same place.

4 Similarity Function and Algorithm

The similarity function between mentions works as follows: compatibility features are first used as hard constraints for non-coreference and then string matching and contextual features are linearly combined into a final similarity metric, normalized in [0,1]. The similarity function is extended to a proximity measure between coreference chains: hard constraints for mentions remain hard-constraints for the chains, based on at least one pair of mentions. Additionally, 2 more constraints are imposed: 1) name compatibility i.e. the most representative name from each chain, if it exists, should be the same and 2) head compatibility i.e. the chains should have at least one pair of mentions with the same head noun. The final metric for compatible chains is computed as a linear combination of the maximum similarity between pairs of mentions for strings and related named-entities, and the similarity between the concatenation of the local contexts.

The clustering algorithm used is a variant of single-pass to solve the problem of dependency on the order of the input.[4] For GoogleNews corpus, we assume that within-document coreference resolution has been solved, we cluster correct coreference chains inside documents. For the Named-entity corpus we cluster mentions. We compare our results with a baseline: cluster within-document coreference chains/mentions using only local context.[5]

5 Evaluation and Results

Similarly with [2], we evaluated with the F-measure applied to pairs of objects which corefer: $F = \frac{2PR}{P+R}$, where P=Precision=percentage of correctly predicted pairs and R=Recall=percentage of correct pairs that have been predicted.

The best result on GoogleNews corpus has been obtained using string-matching, events and local context: 67.72% F-measure. The baseline got a very low result (9.52%), signaling that local context alone is not a very discriminative feature for person entity disambiguation and linking. Relative contribution of each of the non-compatibility features is represented in table 2[6]. For each combination of features we show the best run in terms of feature weights. We begin with all features and iteratively remove them one by one so that we get the best performance. The result obtained using all features is 66.37% while the result obtained using only compatibility features is 40.67%.

In the Named-entity corpus we cluster mentions. We tested 2 combinations of features: string and string + events. String-matching features alone are already responsible for 78.64% of the F-measure. When adding events the results increase up to 90.38% F-measure, which is comparable with the figures reported in [2]: for persons, they obtained 87.4 %, 92.7% and 90.5% F-measure, depending on the

[4] The threshold has been tuned empirically for the run involving only string-matching and the same value was kept for the other experiments. Feature weights have been varied, multiple experiments have been run for the same combination of features.

[5] No efforts have been made to tune the threshold specifically for the baseline.

[6] Compatibility features have been used in all runs except the baseline.

Table 2. F-measure obtained by clustering coreference chains on the GoogleNews corpus. Different combinations of features are considered, eliminating at each step the feature whose removal leads to best performance.

	String-matching	Local context	Events	Related NE	All
	53.69	64.22	64.51	**67.72**	66.37
	53.46	64.21	**66.69**		
	46.79	**63.00**			
None	40.67				

model used. Our results confirm that names are easy to link and to disambiguate, based only on string-matching features. Although the corpus is not really suited for event detection (documents are far away in time), preclustering documents proved to be a useful feature. Also in this corpus, the baseline obtained a low score: 16.53% F-measure.

6 Conclusions

In this paper, we focused on cross-document entity tracking or noun phrase coreferent resolution with regard to person entities, where we restricted the problem to proper name and nominal description disambiguation and linking. We have suggested a number of features useful for linking coreference chains across documents, some of which are novel, i.e., clustering documents according to events and related named-entities. We have evaluated their usefulness on two corpora. From the results, we concluded that string-matching features are the most relevant for detecting cross-document links, followed by local context, events and related named-entities. Names are the easiest to be resolved and string-matching features alone yield high results. Local contexts alone are not discriminative.

References

1. Bagga, A., Baldwin, B.: Entity-Based Cross-Document Coreferencing Using the Vector Space Model In Proceedings of the 36th Annual Meeting of the Association for Computational Linguistics and the 17th International Conference on Computational Linguistics (COLING-ACL'98), pp. 79-85, August 1998.
2. Li, X., Morie, P., Roth, D.: Robust Reading: Identification and Tracing of Ambiguous Names: Discriminative and Generative Approaches In Proceedings of the Annual Meeting of the North American Association of Computational Linguistics (NAACL), pp. 17-24, 2004.
3. MUC-7 Coreference Task Definition (version 3.0) Message Understanding Conference Proceedings, 1998. http://www.itl.nist.gov/iaui/894.02/related_projects/muc/proceedings/muc_7_toc.html
4. OpenNlp http://opennlp.sourceforge.net/
5. Topic Detection and Tracking http://www.nist.gov/speech/tests/tdt/
6. WordFreak http://wordfreak.sourceforge.net/
7. Felbaum, C. (ed.) WordNet: An Electronic Lexical Database ISBN-10:0-262-06197-X

Enterprise People and Skill Discovery Using Tolerant Retrieval and Visualization

Jan Brunnert, Omar Alonso, and Dirk Riehle

Hasso Plattner Institut, Potsdam, Germany
jan@brunnert.de
SAP Research, Palo Alto, USA
{omar.alonso, dirk.riehle}@sap.com

Abstract. Understanding an enterprise's workforce and skill-set can be seen as the key to understanding an organization's capabilities. In today's large organizations it has become increasingly difficult to find people that have specific skills or expertise or to explore and understand the overall picture of an organization's portfolio of topic expertise. This article presents a case study of analyzing and visualizing such expertise with the goal of enabling human users to assess and quickly find people with a desired skill set. Our approach is based on techniques like n-grams, clustering, and visualization for improving the user search experience for people and skills.

1 Introduction

Expertise identification requires data which is usually scattered among different enterprise systems, such as groupware, address books or human resources systems. Accessing this data is often difficult and not aimed at exploring organizational capabilities by topic. Search functionality is often string-based and many searches have to be submitted until the user can put together a mental picture of how different topics relate and who the relevant employees are. Enterprise people search is of critical importance: when decisions during a business workflow require an expert, when a co-worker needs help or when assisting customers it is critical to quickly find people that have certain expertise or interests. Cohen et al. [1] pointed out that it is crucial to understand where in the company expertise resides in order to establish efficient communication. By having a visual overview of the expertise map, the knowledge quality in an organization can be assessed. As outlined by Ashrafi et al. [2] understanding the quality of knowledge in an organization can be used to sense opportunities, develop strategies and implement them effectively and efficiently.

The goal in our case study was to implement a search and visualization application designed to be powerful yet easy to use. Search and retrieval capabilities have been identified by Stein and Zwass [3] as being of importance to the success of an enterprise knowledge management. Because the spelling of names is not always apparent in large international organizations, we had to apply forgiving search strategies. The application provides a novel interface that improves the user experience. Skills can be browsed using visual interaction components like a tag cloud and a graph visualizing connections between tags. A secondary objective of the project was to evaluate how quickly one can bootstrap such an application using only open source components.

G. Amati, C. Carpineto, and G. Romano (Eds.): ECIR 2007, LNCS 4425, pp. 674–677, 2007.

2 Tolerant Retrieval and Cluster Clouds

Users frequently enter imprecise queries that contain spelling errors or, in the case of names, phonetic variations. By using n-grams the application returns good search results even when terms or names were misspelled or not clearly known. Providing a result set with similar terms, users can quickly identify the correct term even when they are uncertain what they are looking for and initiate browsing from this starting point. Employees usually provide very coarse-grained, generic keywords describing their area of expertise, while users will search for very fine-grained, specific skills [4]. Also, in international organizations like SAP, the spelling of names is not often clear. A meaningful tag cloud of skills is automatically generated as proposed by John and Seligman in [5].

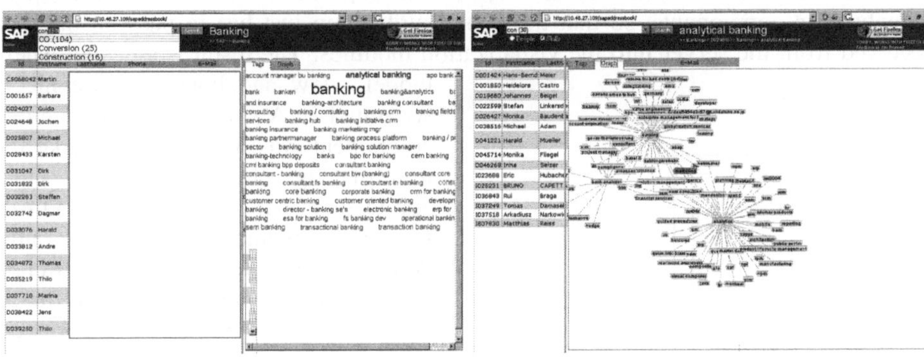

Fig. 1. AJAX auto complete-search, user list, tag clouds, and topic-graph visualization

By clustering the tags around the most frequent terms into sets of related terms the total number of visible keywords is reduced into fewer, more meaningful entities. These skill-clusters are automatically discovered from the supplied set of keywords. For each keyword the total number of elements in a cluster determines the size of the cluster. This clustered set can now be displayed in a tag cloud that projects the size to the keywords' font-size, giving an intuitive view on the importance of a given skill. When drilling down, the application will shows terms similar to the selected tag, in effect showing the keywords that are contained in the cluster and relevant users having these skills are shown. When selecting users, their skills are shown in the visualization pane for further exploration. To support the user on their mental journey, the user's navigation path is always visible and the breadcrumbs can be used to return to previous steps.

The clustering algorithm starts by picking the most frequently used keyword and then searches for terms having a similarity value (as determined by the trigram module) within a certain range. The total number of occurrences of these words is then calculated and saved with the picked keyword. All words identified in this step are removed from the set of available words. This process iterates until no more words can be found. The total number is cluster generated from the data set is 2880.

The application also puts terms in relation to each other in order to visualize them in a graph. We decided to generate the graph based on keyword co-occurrence as suggested by Widdows et al [6] with the edges being weighted by number of co-occurrences. Two keywords are connected when one or more person has specified both keywords in their profile (6355 pairs were identified in the production set).

3 Architecture and Implementation

The browser client follows a two-view paradigm: for every action the user takes, relevant data is displayed in all view modules. For example, when selecting a term, matching persons are shown and the tag cloud is repopulated with related tags. Likewise, selecting a person shows the person's keywords for further drilling down. The entire interface is built on the Dojo toolkit [7]. All rendering is done in these visualization components, so that the client application just has to pass on the data received from the server into the visualization modules. The result is a design where model, view and controller are nicely separated. A custom tag cloud component was developed to make the use of tag clouds easier in other applications. It automatically adjusts font-size within a given range depending on a tag's associated weight and features a fade-over effect for an increased visual browsing experience. The graph connecting keywords based on their co-occurrence in persons' profiles is displayed using the prefuse toolkit, which uses a force-based algorithm to align the graph [8].

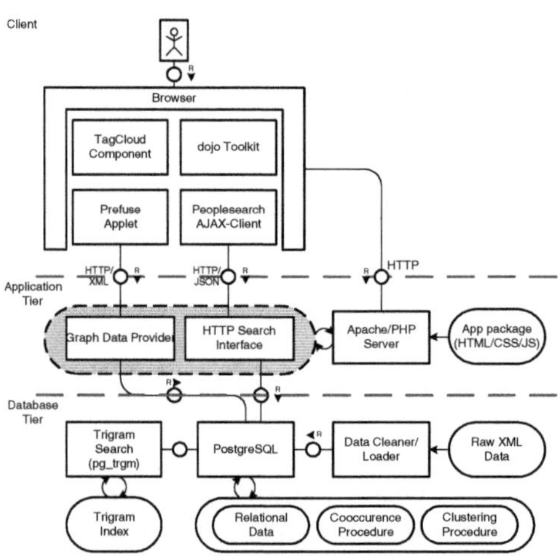

Fig. 2. Architectural Overview

A back-end had to be built to accept data from multiple data sources that are loaded into a single database. In the prototype, keywords are extracted from free-text fields in an XML dump of the corporate address book. The data is then loaded into a PostgreSQL relational database. The database schema contains trigram indexes on the person names and keywords, which is made possible by the pg_trgm [9] module. The database currently contains all of SAP's employees world-wide, around 41000 keywords (15000 distinct) showing the power of the setup on a real-life production dataset. The backend provides facilities to cluster similar keywords and determine the importance of the cluster in the organization. This is accomplished by having a stored procedure in the database. Co-occurrence is also done close to the database using a PL/pgSQL procedure.

Search queries coming from the AJAX-interface are passed on to the PostgreSQL backend via the web server application layer. The returned JSON data can be used to drive multiple visualization modules. Data is served to the prefuse applet in a similar way, providing XML data of edges and nodes in the graph, derived from the co-occurrence data.

4 User Evaluation

A preliminary evaluation survey was conducted among users to get feedback on the prototype. From the ten users that responded to the questionnaire, all reported that the tool was useful compared to the existing system. In a scale 1 to 5 (1=bad, 5=excellent), the people search feature averaged 3.4, the skill search 3.8, the cluster cloud 3.87, and the graph visualization 3.4. For the short development lifecycle of the prototype, the results are encouraging.

5 Future Work

The next step is the integration of additional data sources like documents collections with rich meta-data [10], e-mail archives [11] and publications [12]. Temporal and spatial search mechanisms are also planned.

References

[1] Cohen, W.M., and Levinthal, D.A. "Absorptive Capacity: A New Perspective on Learning and Innovation" *Administrative Science Quarterly* (35:1), 1990, 128-152
[2] Ashrafi, N. *et al.* "Boosting Enterprise Agility via IT Knowledge Management Capabilities" Proceedings of the 39th Hawaii International Conference on System Sciences, 2006
[3] Stein, E.W. and Zwass, V. "Actualizing Organizational Memory with Information Systems" Information Systems Research (6:2), 1995, 85-117
[4] Balog, K. *et al.* "Formal Models for Expert Finding in Enterprise Corpora" SIGIR'06, Seattle, WA, USA.
[5] John, A. and Seligmann, D. "Collaborative Tagging and Expertise in the Enterprise" WWW 2006, Edinburgh, UK.
[6] Widdows, D. *et al.* "Visualisation Techniques for Analysing Meaning" Fifth International Conference on Text, Speech and Dialogue, Brno, Czech Republic, 2002, 107-115.
[7] http://dojotoolkit.org/
[8] http://www.prefuse.org/
[9] http://www.sai.msu.su/~megera/postgres/gist/
[10] Reichling, T. *et al.* "Matching Human Actors based on their Texts: Design and Evaluation of an Instance of the ExpertFinding Framework" GROUP'05, Sanibel Island, FL, USA.
[11] Campbell, C.S. *et al.* "Expertise Identification using Email Communications" CKIM'03, New Orleans, LA, USA.
[12] Tho, Q.T. *et al.* "A Web Mining Approach for Finding Expertise in Research Areas" Proceedings of the 2003 International Conference on Cyberworlds (CW'03).

Experimental Results of the Signal Processing Approach to Distributional Clustering of Terms on *Reuters-21578* Collection

Marta Capdevila Dalmau and Oscar W. Márquez Flórez

University of Vigo, Telecommunication Engineering School,
Signal and Communications Processing Dpt.
Rúa Maxwell s/n, Campus Universitario, 36310 Vigo, Spain
{martacap, omarquez}@gts.tsc.uvigo.es
http://www.gts.tsc.uvigo.es

Abstract. *Distributional Clustering* has showed to be an effective and powerful approach to supervised term extraction aimed at reducing the original indexing space dimensionality for *Automatic Text Categorization* [2]. In a recent paper [1] we introduced a new *Signal Processing* approach to *Distributional Clustering* which reached categorization results on *20 Newsgroups*[1] dataset similar to those obtained by other information-theoretic approaches [3][4][5]. Here we re-validate our method by showing that the 90-categories *Reuters-21578*[2] benchmark collection can be indexed with a minimum loss of categorization accuracy (around 2% with *Naïve Bayes* categorizer) with only 50 clusters.

Keywords: Automatic text categorization, Distributional clustering, Signal processing, Variance, Correlation coefficient.

1 Introduction

In the field of *Automatic Text Categorization*, the high dimensionality and data-sparseness of the indexing term space is problematic for most of the categorizers commonly used [2]. *Distributional Clustering* [3][4][5] is a supervised clustering technique that has shown to be very effective at reducing the document indexing space with residual loss in categorization accuracy.

We recently introduced [1] a new approach to *Distributional Clustering* based on techniques commonly used in *Signal Processing* [6] in *Communications Engineering* and validate its performance towards *20 Newgroups* dataset, obtaining similar results as the ones published by other information-theoretic approaches [3][4][5].

Here we re-confirm the potential of our *Signal Processing* approach to Distributional Clustering of terms on *Reuters-21578* benchmark collection, which differs from the *20 Newsgroups* dataset in its uneven distribution over categories.

[1] http://people.csail.mit.edu/jrennie/20Newsgroups/

[2] http://kdd.ics.uci.edu/databases/reuters21578/reuters21578.html

G. Amati, C. Carpineto, and G. Romano (Eds.): ECIR 2007, LNCS 4425, pp. 678–681, 2007.

2 Method

In text categorization, we are given a pre-labelled training set of documents[3]. In base of this information, any term t_k can then be characterized by its probability distribution function over the discrete variable category C defined in $\{c_1, ..., c_{|C|}\}$, $|C|$ being the number of distinct categories, that is

$$f_k : c_i \mapsto f_k(c_i) = P(c_i|t_k) \text{ with } \sum_{i=1}^{|C|} f_k(c_i) = 1 \tag{1}$$

In our *Signal Processing* approach, we interpret t_k as being a probabilistic signal that encloses the random variable category. At a first stage, all signals that present a very flat distribution, and thus have a low *variance*[4], are not informative of the variable category and will be considered as noisy signals that can be eliminated with virtually no loss of information. Subsequently, signals with similar probability distribution can be clustered together since they contain the same category information. The degree of similarity between signals is measured by the *correlation coefficient* which estimates the interdependence between both signals (equals 0 if they are statically independent and, in the other extreme, +-1 if they are linearly dependent).

$$\text{Correlation coefficient:}^5\ \rho_{f_j f_k} \triangleq \frac{1}{\sigma_{f_j}\sigma_{f_k}} E[(f_j - m_{f_j})(f_k - m_{f_k})] \tag{2}$$

Initially we have adopted an agglomerative hard clustering[6] algorithm [1] inspired by the method proposed in [3]. We have implemented three further variants of this algorithm. The static window of dimension M (M being the final number of clusters) used in the original algorithm has been dynamically expanded in order to avoid the merging of poorly correlated clusters in case no other merging could be done. This implies the implementation of a dynamic window expansion/compression approach. Furthermore, soft clustering[7] has been implemented in both static and dynamic window approaches.

3 Experimental Scenario

We have used a common subset of the *Reuters-21578 ModApté* training/test partition which only considers the set of 90 categories with at least one positive training example and one positive test example. The distribution of categories in

[3] In the following, we are using the classic *bag-of-words* indexing [2].

[4] The *variance* provides a measure of the *spread* of the values of f_k relative to its mean m_{f_k} and is defined as $\sigma^2_{f_k} \triangleq E[(f_k - m_{f_k})^2] = E[{f_k}^2] - {m_{f_k}}^2$.

[5] $E[\,]$ denotes the expectation operation and $m_{f_k} = E[f_k]$ the *mean* of f_k.

[6] Hard clustering allocates each term to a single cluster.

[7] In a soft clustering approach a single term can belong to more than one cluster (i.e. clusters may overlap).

this corpus is highly skewed, with 36.7% of the documents in the most common category (i.e. label 'earn'), and overlapped (i.e. a document may have a multi-category tag).

After removing *stopwords* (using the list provided by *Weka*[8] software) and *non-alphabetical* words, the collection training set is indexed by 32,539 words. Upon this, we further eliminated terms occurring in less than 4 documents or appearing less than 4 times in the whole dataset. The *variance* threshold has been empirically set to 0.002, which ends in 10,859 informative words (7,483 if we use the *sample variance* introduced in next paragraph), resulting after the elimination of noisy words.

We applied our four clustering variants on the former *Reuters-21578 ModApté* pre-processed training set. We have used both the *variance* and *correlation coefficient* as defined in the probabilistic approach of section 2 and also under an statistical interpretation that assumes equi-probable categories[9] (i.e. $P(c_i) = \frac{1}{|C|}$). The initial threshold for the *correlation coefficient* was set to its maximum value 1 and its decremental step in the contraction phase of the dynamic window approaches was empirically set to 0.00001. Finally, the whole collection of documents has been indexed in the resulting space of clusters. The quality of the clustering has then been evaluated quantitatively by means of the categorization accuracy of the well-known *multinomial Naïve Bayes* classifier [2] over the test set. The results obtained have been referenced to analogous categorization accuracy of the same pre-processed collection (removal of *stopwords* and *non-alphabetical* words) indexed in the feature space resulting from classic *Information Gain* and *Chi-square* term selection functions[10].

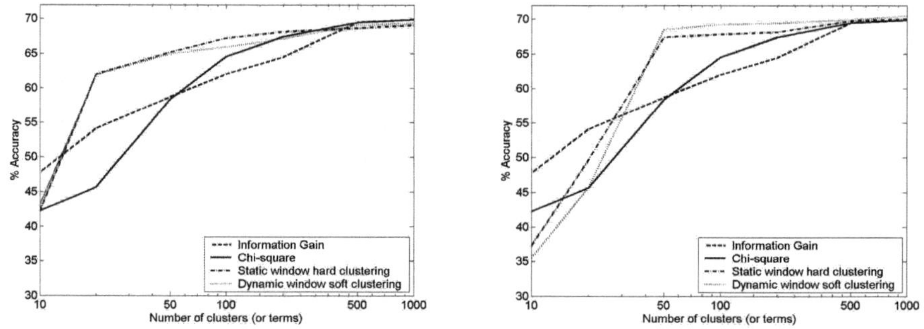

Fig. 1. Categorization accuracy of *Distributional Clustering* vs. *Information Gain* and *Chi-square* with *Naïve Bayes* categorizer. Dispersion and similarity measures used: (left) *variance* and *correlation coefficient.* (right) *sample variance* and *sample correlation coefficient,* i.e. assuming equi-probable categories.

[8] Our clustering implementation is based in the *Weka 3 Data Mining Software* [7].

[9] Statiscal measures are referred as *sample variance* and *sample correlation coefficient.*

[10] For graphical simplicity only the curves corresponding to *Static window hard clustering* and *Dynamic window soft clustering* algorithms are represented in figure 1.

4 Discussion of Results and Conclusions

Figure 1 shows similar results to the ones obtained by the previous researches on *Distributional clustering* [1][3][4][5] using *20 Newsgroups* dataset (i.e. the 20-categories *Newsgroups* dataset can be indexed with only 20 clusters). In particular, when using the statistical measures (right side of figure 1), the curves corresponding to our *Distributional clustering* algorithms present an abrupt initial increase up to 50 clusters (accuracy 67-68%) and from there they asymptotically get to an accuracy of around 70%. In other words, the indexing space can be reduced from the original 32,539 words to 50 clusters (i.e. three orders of magnitude) with a residual loss of categorization accuracy of 2%. The 90-categories *Reuters-21578* collection can thus be painlessly indexed with only 50 clusters.

The curves showed in the left side of figure 1 correspond to the proper probabilistic measures and show a softer evolution: with small number of clusters (less than 50) the clustering works better than with the statistical assumption, while the other way round occurs from this point on. The interpretation of these results is heavily concerned by the fact that *Reuters-21578* is such an unevenly distributed collection: categorization errors in most common categories have a strong impact on overall categorization accuracy.

We are currently deepening our research on this aspect, introducing other measures of similarity. We are also working on the tuning of the clustering parameters such as the *variance* threshold and the dynamic window contraction step in order to optimize the overall categorization results. Finally, we are planning the design of a full categorizer based on our *Signal Processing* approach.

Acknowledgments. For this research Marta Capdevila was supported in part by a predoctoral grant from the *Xunta de Galicia* regional government (Spain).

References

1. Marta Capdevila, Oscar W. Márquez: A signal processing approach to distributional clustering of terms in automatic text categorization. Proceedings of INSCIT2006, I Int. Conf. on Multidisciplinary Information Sci. and Tech., 2006.
2. Fabrizio Sebastiani: Machine Learning in Automated Text Categorization. ACM Computing Surveys, Vol. 34, No. 1, pp. 1-47, March 2002.
3. L.Douglas Baker, Andrew Kachites McCallum: Distributional Clustering of Words for Text Classification. Proceedings of SIGIR-98, 21st ACM International Conference on Research and Development in Information Retrieval, 1998
4. Noam Slonim, Naftali Tishby: The Power of Word Clusters for Text Classification. 23rd European Colloquium on Information Retrieval Research, 2001
5. Inderjit S. Dhillon, Subramanyam Mallela, Rahul Kumar: A Divisive Information-Theoretic Feature Clustering Algorithm for Text Classification. Journal of Machine Learning Research 3 1265-1287, 2003
6. A. Bruce Carlson: Communication Systems an Introduction to Signals and Noise in Electrical Communications. McGraw-Hill Book Company, third edition, 1986
7. Ian H. Witten and Eibe Frank: Data Mining: Practical machine learning tools and techniques. 2nd Edition, Morgan Kaufmann, San Francisco, 2005.

Overall Comparison at the Standard Levels of Recall of Multiple Retrieval Methods with the Friedman Test

José M. Casanova[1], Manuel A. Presedo Quindimil[2], and Álvaro Barreiro[1]

[1] IRLab, Department of Computer Science
[2] Department of Mathematics,
University of A Coruña,
Campus de Elviña s/n, 15071, A Coruña, Spain
{jcasanova,mpresedo,barreiro}@udc.es
http://www.dc.fi.udc.es/irlab/

Abstract. We propose a new application of the Friedman statistical test of significance to compare multiple retrieval methods. After measuring the average precision at the eleven standard levels of recall, our application of the Friedman test provides a global comparison of the methods. In some experiments this test provides additional and useful information to decide if methods are different.

1 Introduction

Evaluation is a basic need in Information Retrieval (IR). In order to assess whether a retrieval method performs better than others or not, it is necessary to apply a test of significance, because metrics comparisons are strictly valid for the same collection and queries. The tests of significance determine if the difference is not caused by chance based on statistical evidence. Different applications of these tests have been described [1] and widely used in IR. Evaluation is still an open issue that affects the basis of IR. Recent work assesses that "significance substantially increases the reliability of retrieval effectiveness" [2].

The Friedman test is a non parametric statistical significance test that can be employed with ordinal data [3]. In the ordinary use of this test, it is applied to the Mean Average Precision (MAP) or other metrics using the query as the block variable [4,5]. We propose to use the Friedman test to compare multiple retrieval methods using the eleven standard levels of recall as the block variable. This new application of the Friedman test provides a global comparison through the levels of recall.

2 The Friedman Test

The Friedman significance test [6] allows the comparison of multiple methods, in situations where the random variable is ordinal (rank-order) and the block variables are mutually independent.

G. Amati, C. Carpineto, and G. Romano (Eds.): ECIR 2007, LNCS 4425, pp. 682–685, 2007.

Let b be the number of blocks, k the number of methods to be compared and X the random variable. The function $R(X_{ij})$ returns the rank of method j in the i-th block. In case of tied values, the final rank is the average of the corresponding tied ranking scores. Let $R_j = \sum_{i=1}^{b} R(X_{ij})$ be the sum of ranks for a method. Then the following values A and B are calculated as:

$$A = \sum_{i=1}^{b}\sum_{j=1}^{k} R(X_{ij})^2 \qquad B = \frac{1}{b}\sum_{j=1}^{k} R_j^2 \tag{1}$$

The statistic T is defined as:

$$T = \frac{(b-1)[B - bk(k+1)^2/4]}{A - B} \tag{2}$$

The null hypothesis states that the methods are the same and it is rejected at an α level of significance if the quantile $1-\alpha$ of the F distribution (Fisher-Snedecor distribution) with $(k-1)$ and $(b-1)(k-1)$ degrees of freedom is greater than T.

Paired comparisons among the methods are done when the null hypothesis is rejected. If methods i and j are significantly different the following inequality is satisfied:

$$|R_j - R_i| > t_{1-\alpha/2}\left[\frac{2b(A-B)}{(b-1)(k-1)}\right]^{\frac{1}{2}}, \tag{3}$$

where $t_{1-\alpha/2}$ is the $1-\alpha/2$ quantile of the t-distribution with $(b-1)(k-1)$ degrees of freedom.

3 The Friedman Test with the Standard Levels of Recall as the Block Variable

The Friedman test has been used to determine if differences in MAP and other precision metrics are significant, always using the query as the block variable. We propose a novel application of the Friedman test using as the random variable (X_{ij}) the interpolated average precision at eleven standard levels of recall, i.e. the level of recall acts as the block variable. Intuitively, using the Friedman test as described allows a global comparison of the common precision/recall figures, but with the support of a test of significance. The results obtained with this new test can be different than those obtained when MAP acts as the random and the query as the block variable.

We motivate our evaluation method in the need of an analytical method to compare precision/recall curves. Our approach uses only the information of the ranking position for each method at the 11 standard levels of recall.

The independence assumption regarding the precision at the 11 standard levels of recall may not always hold. If values were dependent, an statistical method for the analysis of repeated measurements [7] could be applied instead of the Friedman test.

4 Experimental Results

To illustrate the use of this test we compared the behaviour of several smoothing methods for language models in the ad-hoc retrieval task. We tried to assess if the following methods described in [8] are different: Jelinek-Mercer (`jm`), Dirichlet (`dir`) and absolute discount (`ad`). The implementation was developed using the retrieval framework Lemur Toolkit[1].

After setting up the best smoothing parameters, we ran the different retrieval methods. Then we computed the MAP and the interpolated average precision at every standard level of recall to evaluate the results. In order to analyse if there was a significant difference among the three smoothing methods we apply the Friedman test to the MAP values and for each of the precisions values at the eleven levels of recall using the query as the block variable. The level of significance of the tests was fixed at $\alpha = 0.05$.

For example, for WEB8 collection (TREC small web) using the titles from topics 401-450 with the smoothing parameters `jm` $\lambda = 0.01$, `ad` $\rho = 0.80$ and `dir` $\mu = 2200$, the Friedman test using the query as the block variable shows that MAP values are significantly different for the three methods, and average interpolated precision values are also significant for the eleven standard levels of recall and the three methods. The Friedman test using the level of recall as the block variable also shows that there are significant differences for the three methods. After doing paired comparisons, the test of MAP values reveals that Dirichlet is better than the other two methods. The tests applied to the precisions at the levels of recall indicate that Dirichlet is significantly better than absolute discounting for all levels of recall and that is better than Jelinek-Mercer for six of the eleven levels. The Friedman test using the recall level as the block variable confirms that Dirichlet is significantly better than the other methods.

Now we describe another scenario where the proposal presented in this paper complements the information provided by previous tests. For the FBIS collection (TREC disk 5) using only the titles from topics 351-400 with the smoothing parameters `jm` $\lambda = 0.05$, `ad` $\rho = 0.75$ and `dir` $\mu = 4000$, the Friedman test using the query as the block variable only finds significant differences among the three methods for precision values at levels of recall from 0.40 to 0.80 and at level 1.00, but it does not find significantly different MAP values. Paired comparisons support that Dirichlet performs better than the other two methods. The Friedman test using the level of recall as the block variable indicates that the three methods are significantly different, and paired comparisons also show that Dirichlet performs better than the others. Therefore, in this experiment the use of the recall level as the block variable for the Friedman test reinforces that Dirichlet is the smoothing algorithm that outperforms all others.

Another interesting experiment is the following. For the LATIMES collection (TREC disk 5) using the title, description and narrative from topics 351-400 with the smoothing parameters `jm` $\lambda = 0.80$, `ad` $\rho = 0.75$ and `dir` $\mu = 3000$, the Friedman test using the query as the block variable shows that MAP values

[1] http://www.lemurproject.org

are no significantly different for the three methods, and average interpolated precision values are only significant at the levels of recall of 0.30, 0.70 and 0.80. The Friedman test using the level of recall as the block variable also shows that there are not statistically significant differences among the three methods.

5 Conclusions and Further Work

We proposed a new application of the Friedman significance test using the levels of recall as the block variable. The use of this variant gives additional information to previous applications of the Friedman test that used the query as the block variable. This way the test provides an approximation to the comparison of the precision/recall curves. We illustrated the method, giving evidence that in fact in some experiments the proposed test helps to decide whether the methods are different or not. We plan to apply our method using the collection as the block variable and the MAP or other single measure as the random variable because this ensures the independence assumption. An alternative to our method for the cases where independence may not hold could be the application of statistical methods for the analysis of repeated measurements. Future work will also address the problem of determining the utility of the test of significance following the methodology introduced by Zobel in [9].

Acknowledgements. The work reported here was cofunded by the SEUI and FEDER funds under the project MEC TIN2005-08521-C02-02 and "Xunta de Galicia" under project PGIDIT06PXIC10501PN. José M. Casanova is supported by a grant of DXID of the "Xunta de Galicia". We also thank the support of the "Galician Network of NLP & IR" Ref. 2006/03.

References

1. Hull, D.: Using statistical testing in the evaluation of retrieval experiments. In: Proc. of ACM SIGIR '93. (1993) 329–338
2. Sanderson, M., Zobel, J.: Information retrieval system evaluation: effort, sensitivity, and reliability. In: Proc. of ACM SIGIR '05. (2005) 162–169
3. Sheskin, D.: Handbook of parametric and nonparametric statistical procedures. Chapman & Hall/CRC (2000)
4. Hull, D.: Stemming algorithms: A case study for detailed evaluation. JASIS **47**(1) (1996) 70–84
5. Kekäläinen, J., Järvelin, K.: The impact of query structure and query expansion on retrieval performance. In: Proc. of ACM SIGIR '98. (1998) 130–137
6. Conover, W.: Practical Nonparametric Statistics. 2ed edn. John Wiley & Sons, Inc (1980)
7. Davis, C.: Statistical Methods for Analysis of Repeated Measurements. Springer-Verlag NY, Inc (2002)
8. Zhai, C., Lafferty, J.: A study of smoothing methods for language models applied to information retrieval. ACM Trans. Inf. Syst. **22**(2) (2004) 179–214
9. Zobel, J.: How reliable are the results of large-scale information retrieval experiments? In: Proc. of ACM SIGIR '98. (1998) 307–314

Building a Desktop Search Test-Bed

Sergey Chernov[1], Pavel Serdyukov[2],
Paul-Alexandru Chirita[1], Gianluca Demartini[1], and Wolfgang Nejdl[1]

[1] L3S / University of Hannover, Appelstr. 9a D-30167 Hannover, Germany
[2] Database Group, University of Twente, P.O. Box 217, 7500 AE Enschede, The Netherlands
{chernov, chirita, demartini, nejdl}@l3s.de,
serdyukovpv@cs.utwente.nl

Abstract. In the last years several top-quality papers utilized temporary Desktop data and/or browsing activity logs for experimental evaluation. Building a common testbed for the Personal Information Management community is thus becoming an indispensable task. In this paper we present a possible dataset design and discuss the means to create it.

1 Introduction

In the last years several top-quality papers utilized Desktop data and / or activity logs for experimental evaluation. For example, in [4], the authors used indexed Desktop resources (i.e., files, etc.) from 15 Microsoft employees of various professions with about 80 queries selected from their previous searches. In [3] Google search sessions of 10 computer science researchers have been logged for 6 months to gather a set of realistic search queries. Similarly, several papers from Yahoo [2], Microsoft [1] and Google [5] presented approaches to mining their search engine logs for personalization. We want to provide a common public Desktop specific dataset for this research community.

The most related dataset creation effort is the TREC-2006 Enterprise Track [1]. Enterprise search considers a user who searches the data of an organisation in order to complete some task. The most relevant analogy between the Enterprise search and Desktop search is the variety of items which compose the collection (e.g., in the TREC-2006 Enterprise Track collection e-mails, cvs logs, web pages, wiki pages, and personal home pages are available). The biggest difference between the two collections is the presence of *personal documents* and especially *activity logs* (e.g., resource read / write time stamps, etc.) within the Desktop dataset.

In this paper we present an approach we envision for generating such a Desktop dataset. We plan our new dataset to include *activity logs* containing the history of each file, email or clipboard usage. This dataset will bring a basis for designing and evaluating of special-purpose retrieval algorithms for different Desktop search tasks.

2 Dataset Design

File Formats and Metadata. The data for the Desktop dataset will be collected among the participanting research groups. We are going to store several file formats: TXT,

[1] `http://www.ins.cwi.nl/projects/trec-ent/`

G. Amati, C. Carpineto, and G. Romano (Eds.): ECIR 2007, LNCS 4425, pp. 686–690, 2007.

Table 1. Logged Application and File Formats

Application	File Format
Acrobat Reader	PDF files
MS Word	DOC, TXT, RTF
MS Excel	XSL
MS Powerpoint	PPT
MS Internet Explorer	HTML
MS Outlook	PST
Mozilla Firefox	HTML
Mozilla Thunderbird	MSF and empty extension, mbox format

Table 2. Timeline and Permanent Logged Information

Permanent Information	**Applied to**
URL	HTML
Author	All files
Recipients	Email messages
Metadata tags	MP3
Has/is attachment	Emails and attachments
Saved picture's URL and saving time	Graphic files
Timeline information	
Time of being in focus	All files
Time of being opened	All files
Being edited	All files
History of moving/renaming	All files
Request type: bookmark, clicked link, typed URL	HTML
Adding/editing an entry in calendar and tasks	Outlook Journal
Being printed	All files
Search queries in Google/MSN Search/Yahoo!/etc.	Search fields in Internet browsers
Clicked links	URL
Text selections from the clipboard	Text pieces within a file and the filename
Bookmarking time	Bookmarks in Internet browsers
Chat client properties	Status, contact's statuses, sent filenames and links
Running applications	Task queue
IP address	User's address and addresses user connects to
Email status	Change between "received" and "read" status

HTML, PDF, DOC, XLS, PPT, MP3 (tags only), JPG, GIF, and BMP. Then, each group willing to test its system would submit 1-2 Desktop dumps, using logging tools for a number of applications listed in the Table 1.

The set of logged applications can be extended in the future. Loggers save the information which we describe in Table 2.

Data Gathering. As the privacy issue is very important here, we propose two options for possible information gathering.

1. Optimistic approach. We assume there are volunteers ready to contribute some of their personal information to the community, given that this information would be redistributed only among a restricted group of participants. As a test case, we gave two laptops to students for half a year. They were able to use them for free, but the condition was that all the information on these laptops will be available for future research. They were also warned not to store highly private information like passwords or credit card numbers. As this approach worked well, we expect that all participating groups will find similar reasonable incentives to attract more volunteers.

2. Pessimistic approach. While some people are ready to share information with their close friends and colleagues, they do not like to disclose it to outsiders. For this case, there is a way to keep information available only for a small number of people: Personal data is collected from participating groups by some coordinators and preprocessed into the publicly available uniform XML format. Every group can adapt its search prototypes to this format and submit binary files to the coordinators. Runs are then produced locally by a coordinator and results are sent back to the participants. This way, only trusted coordinators have access to the actual documents, while it is possible for all participants to evaluate their results. Similar schemes has been tested in TREC Spam Track [2], and it might be a necessary solution for future TREC tracks as well, whenever they involve highly private data (i.e. medical, legal, etc.).

3 Relevance Assessments and Evaluation

As we are aiming at real world tasks and data, we want to reuse real queries from Desktop users. Since every Desktop is a unique set of information, its user should be involved in both query development and relevance assessment. Thus, Desktop contributors should be ready to give 10 queries selected from their everyday tasks. This also solves the problem of subjective query evaluation, since users know best their information needs.

In this setting queries are designed for the collection of a single user, but some more general scenarios can be designed as well, for example finding relevant documents in every considered Desktop. It is thus possible to see the test collection as partitioned in sub-collections that represent single Desktops with their own queries and relevance assessments. This solution would be very related to the MrX collection used in the TREC SPAM Track, which is formed by a set of emails of an unknown person.

The query can have the following format:

- <num> KIS01 </num>
- <query> Eleonet project deliverable June</query>
- <metadataquery>date:June topic:Eleonet project type:deliverable </metadataquery>
- <taskdescription>I am combining new deliverable for the Eleonet project. </taskdescription>
- <narrative> I am looking for the Eleonet project deliverable, I remember that the main contribution to this document has been done in June. </narrative>

We include the <metadataquery> field so that one could specify semi-structured parameters like metadata field names, in order to narrow down the query. The set of possible metadata fields would be defined after collecting the Desktop data.

The Desktop contributors must be able to assess pooled documents 6 months after they contributed the Desktop. Moreover, each query will be supplemented with the description of context (e.g., clicked / opened documents in the respective query session), so that users could provide relevance judgments according to the actual context of the

[2] `http://plg.uwaterloo.ca/~gvcormac/spam/`

query. As users know their documents very well, the assessment phase should go faster than normal TREC assessments. For the task of known-item search, the assessments are quite easy, since only one (at most several duplicates) document is considered relevant. For the adhoc search task we expect users to spend about 3-4 hours to do relevance assessment per query.

4 Proposed Tasks

1. AdHoc Retrieval Task. Ad hoc search is the classic type of text retrieval when the user believes she has relevant information somewhere. Several documents can contain pieces of necessary data, but she does not remember whether or where she stored them, and she is not sure which keywords are best to find them.

2. Known-Item Retrieval Task. Targeted or known-item search task is the most common for the Desktop environment. Here the user wants to find a specific document on the Desktop, but does not know where it is stored or what is its exact title. This document can be an email, a working paper, etc. The task considers that the user has some knowledge about the context in which the document has been used before. Possible additional query fields are: time period, location, topical description of the task in which scope the document had been used, etc.

3. Folder Retrieval Task. It is very popular among users to have their personal items topically organized in folders. Later they may search not for a specific document, but for a group of documents in order to use it later as a whole - browse them manually, reorganize or send to a colleague. The retrieval system should be able to estimate the relevance of folders and sub-folders using simple keyword queries.

5 Conclusion

Building a Desktop IR testbed seems to be more challenging than creating a Web Search or an XML Retrieval dataset. In this paper we presented the concrete parameters for defining the features of such a Desktop Dataset and discussed the possible means for creating it, as well as utilizing it for algorithm assessments.

References

1. E. Agichtein, E. Brill, S. Dumais, and R. Ragno. Learning user interaction models for predicting web search result preferences. In *SIGIR '06: Proceedings of the 29th annual international ACM SIGIR conference on Research and development in information retrieval*, pages 3–10, New York, NY, USA, 2006. ACM Press.
2. R. Kraft, C. C. Chang, F. Maghoul, and R. Kumar. Searching with context. In *WWW '06: Proceedings of the 15th international conference on World Wide Web*, pages 477–486, New York, NY, USA, 2006. ACM Press.
3. F. Qiu and J. Cho. Automatic identification of user interest for personalized search. In *WWW '06: Proceedings of the 15th international conference on World Wide Web*, pages 727–736, New York, NY, USA, 2006. ACM Press.

4. J. Teevan, S. T. Dumais, and E. Horvitz. Personalizing search via automated analysis of interests and activities. In *SIGIR '05: Proceedings of the 28th annual international ACM SIGIR conference on Research and development in information retrieval*, pages 449–456, New York, NY, USA, 2005. ACM Press.
5. B. Yang and G. Jeh. Retroactive answering of search queries. In *WWW '06: Proceedings of the 15th international conference on World Wide Web*, pages 457–466, New York, NY, USA, 2006. ACM Press.

Hierarchical Browsing of Video Key Frames

Gianluigi Ciocca and Raimondo Schettini

DISCo, Università degli Studi di Milano-Bicocca
Via Bicocca degli Arcimboldi 8, 20126, Milano, Italy
{ciocca,schettini}@disco.unimib.it

Abstract. We propose an innovative, general purpose, method to the selection and hierarchical representation of key frames of a video sequence for video summarization. It is able to create a hierarchical storyboard that the user may easily browse. The method is composed by three different steps. The first removes meaningless key frames, using supervised classification performed by a neural network on the basis of pictorial features and a visual attention model algorithm. The second step provides for the grouping of the key frames into clusters to allow multilevel summary using both low level and high level features. The third step identifies the default summary level that is shown to the users: starting from this set of key frames, the users can then browse the video content at different levels of detail.

1 Introduction

The growing interest of consumers in the acquisition of and access to visual information has created a demand for new technologies to represent, model, index and retrieve multimedia data. Very large databases of images and videos require efficient algorithms that enable fast browsing and access to the information pursued [1]. We focus our attention here on the creation of a visual summary using still images, called key frames, extracted from the video. Key frame based video representation views video abstraction as a problem of mapping an entire segment to some small number of representative images. The extraction of key frames must be automatic and content based so that they maintain the salient content of the video while avoiding all redundancy. Regardless of the key frame extraction strategy adopted (e.g. [2], [3], [4]), the visual summaries obtained may contain uninteresting and meaningless key frames, overexposed and underexposed key frames; close-ups with very few details, etc. Moreover, successive very similar key frames can be extracted disturbing the visual summary since they convey the same information. Finally, summaries formed by hundreds or even thousands of key frames are dispersive since they require intensive browsing. We propose an innovative method for the selection and hierarchical representation of key frames of a video sequence for video summarization.

2 Proposed Method

The method presented in this paper makes it possible to create exhaustive and not redundant visual video summaries. The method comprises three steps (Figure 1): the

G. Amati, C. Carpineto, and G. Romano (Eds.): ECIR 2007, LNCS 4425, pp. 691 – 694, 2007.

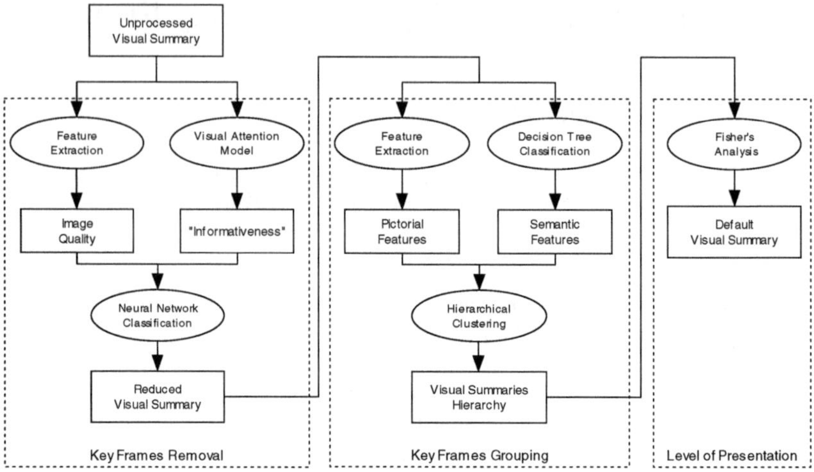

Fig. 1. The pipeline of the proposed method

first removes the meaningless key frames, the second groups the key frames into clusters, and the third identifies the default summary level.

2.1 Detection of Meaningless Key Frames

To remove meaningless key frames, we extract a set of features that describe them in terms of quality and information supplied. To assess the image quality we have chosen some of the features used in assessing the quality of images taken by digital cameras, such as the percentage of dark and bight pixels, for identifying overexposed and underexposed images; the dynamic range of the image, for detecting flat-looking images; and the color balance. The amounts of information that key frames convey, is assessed via a visual attention algorithm to detect the Regions of Interest (ROI) from a saliency map [5]. Only portions of the input image associated with relevant information are detected. Intuitively if the percentage of ROIs is low or the ROIs are distributed outside the central part of the image, the corresponding key frame should be discarded. Actually, a trained neural network classifier makes the decision for discarding a key frame. The neural network has eight input nodes that correspond to the computed features, two hidden layers with eight nodes each, and a single final node whose output could be interpreted as the probability that the processed key frame should be discarded. The learning process is based on the back-propagation algorithm with bias terms and momentum [6].

2.2 Hierarchical Key Frames Representation

Similar key frames are grouped by a hierarchical clustering algorithm. Initially each key frame corresponds to a cluster and represents it. The clustering is performed analyzing only pairs of consecutive clusters in order to retain their temporal ordering. At each step, the two clusters with the smallest maximum pair-wise distance (complete link strategy) are merged. The representative key frame of the cluster resulting by the merging of two clusters is selected between the two representative frames and corresponds to the key frame with higher percentage of ROIs. Two clusters are compared using both low-level and high-level features. Low-level features are used to describe the pictorial content of the key frames while high-level features describe their semantic content. The low level features used are the color histogram, the wavelet statistics and the edge's direction histogram [4]. The high level features are obtained applying the classification strategies proposed in [7] where the images are classified as indoor, outdoor or close-up images. Instead of assigning an image to a single class we have chosen to use the probabilities that a given image belong to each of the three classes as a semantic histogram signature. The similarity degree between two clusters is the combination of the normalized similarity degrees computed on both the low level and high level features. The representative key frames are used to visually represent the clusters in the hierarchical visual summary. The clusters are arranged according to the chronology of the representative key frames.

2.3 Optimal Video Summary

The idea underlying the strategy for detecting the optimal summary level is that the frame differences used in merging the clusters in the previous step be considered merging costs, that is, the costs of reducing the summary by one key frame. In the initial phase of the clustering process, the costs will be relatively small, that is, the merged clusters will have small merging costs because the representative key frames are similar, and their merging will not significantly reduce the summary information contents. As clustering continues, the merging costs will increase, meaning that the representative key frames will be increasingly dissimilar, and their merging will result in a summary with fewer information contents. We select the level at which the merging costs rise significantly as the level of the optimal summary. The summaries at lower levels contain more key frames and redundant information. The summaries at higher levels contain few key frames. We apply the Fisher's discriminant analysis [8], to the sequence of increasing merging costs that minimizing the intra-group differences while maximizing their inter-group separation, makes it possible to define low and high merging costs. The global maximum in the Fisher's analysis curve corresponds to the default clustering level in the summary presentation. If the maximum is very close to the initial or ending clustering levels, we assume that, for that summary, the default presentation summary do not differ from the original summary.

3 Results and Conclusions

Since all the summary's post processing steps (removal of meaningless key frames and classification) rely on semantic information which no objective quality measure can effectively incorporate, the post-processing algorithm was heuristically tested by domain experts on a set of videos belonging to [9]. These experts manage video footages on a daily basis, manually extracting relevant information the videos to use for content cataloging and publication through distributed channels. The test set was composed of 14 non-professional videos, about 4 hours of footage. The experts evaluated the processed summary in terms of compactness, and information contents as well as the effectiveness of the multilevel summary. They judged the results positive, thus our post-processing pipeline has been used in the AESS archive. Some results can be found at http://www.ivl.disco.unimib.it/Activities/VideoIndexing.html. As the post-processing algorithm does not use previous knowledge about the video contents, nor is any assumption made about the input data, it can be used in different domains as a general purpose algorithm. Nevertheless, some improvements can be made. The key frame removal stage could be extended with more pictorial quality features in order to better cover the many factors that can cause a user to reject a frame (e.g. wrong skin tone, half faces, etc…). The hierarchical key frame grouping stage could also be extended. We have introduced three generic classes for the classification of the key frames but more classes can be added in the decision trees to enlarge the semantic dictionary.

References

1. Dimitrova, N., Zhang, H., Shahraray, B., Sezan, M., Huang T., Zakhor, A.: Applications of video-content analysis and retrieval. IEEE MultiMedia, 9(3), (2002), 44-55.
2. Hanjalic A., Lagendijk R. L., Biemond J.: A new Method for Key Frame Based Video Content Representation. Image Databases and Multimedia Search, World Scientific, 1998.
3. Han S., Yoon K., and Kweon I.: A new Technique for Shot Detection and Key Frames Selection in Histogram Space. Proc. 12th W. on Im. Pr. and Im. Und., (2000), 475-479.
4. Ciocca, G., Schettini, R. Dynamic key-frame extraction for video summarization. Journal of Real-Time Image Processing, 1(1), (2006) 69-88.
5. Corchs, S., Ciocca, G., Schettini, R., Deco, G.: Video Summarization Using a Neurodynamical Model of Visual Attention. IEEE Int. W. MM. Sig. Proc., (2004), 71-74.
6. Rumelhart, D.E., Hinton, G.E., Williams, R.J.: Learning Internal Representations by Error Propagation. Parallel Distributed Processing, MIT Press, Volume 1, (1986), 318-362.
7. Schettini, R., Brambilla, C., Cusano, C., Ciocca, G.: Automatic classification of digital photographs based on decision forests. Int. J. Patt. Rec. and A. I., 18(5), (2004), 819-846.
8. Fisher, R. A.: The use of multiple measures in taxonomic problems. Ann. Eugenics, Volume 7, (1936), 179-188.
9. AESS, Archivio di Etnografia e Storia Sociale: http://aess.itc.cnr.it/index.htm.

Active Learning with History-Based Query Selection for Text Categorisation

Michael Davy* and Saturnino Luz

Department of Computer Science,
Trinity College Dublin
{Michael.Davy, Saturnino.Luz}@cs.tcd.ie

Abstract. Automated text categorisation systems learn a generalised hypothesis from large numbers of labelled examples. However, in many domains labelled data is scarce and expensive to obtain. Active learning is a technique that has shown to reduce the amount of training data required to produce an accurate hypothesis. This paper proposes a novel method of incorporating predictions made in previous iterations of active learning into the selection of informative unlabelled examples. We show empirically how this method can lead to increased classification accuracy compared to alternative techniques.

1 Introduction

Automated text categorisation [1] systems are designed to label unseen documents into a set of predefined categories. They use inductive learning algorithms to construct a hypothesis of the target concept from a given set of labelled examples (training data) which can then be used to classify unseen data. The more labelled training data available to the learner the greater the chances of inducing an accurate hypothesis (PAC learnability [2]). However, there are many domains where labelled data is difficult to obtain while unlabelled data is abundant.

Active Learning [3] has shown to significantly reduce the number of examples required to induce an accurate hypothesis. In this paper we examine the query selection function of active learning, which is responsible for selecting informative unlabelled examples. This paper proposes a method of incorporating classifier predictions from previous iterations of active learning into the query selection function. We show how this method can lead to increased accuracy over alternative techniques in text classification problems.

2 Active Learning

Traditional learning techniques are *passive* since they have no control over the training data - it is simply supplied to them. Passive learners require large amounts of training data to ensure there are enough informative examples for

* Supported by the Irish Research Council for Science Engineering and Technology.

G. Amati, C. Carpineto, and G. Romano (Eds.): ECIR 2007, LNCS 4425, pp. 695–698, 2007.

the induction of an accurate hypothesis. Active learning gives the learner control over its training data. An active learner is supplied with a small amount of initial training data but can supplement this by selecting informative unlabelled examples and requesting their correct labels from an oracle. Labels are obtained only for the selected examples which equates to labelling only the informative examples, thereby reducing the number of labelled examples required.

Algorithm 1. Generic Active Learner

Input: tr - training data
Input: ul - unlabelled examples

for $i = 0$ **to** *stopping criteria met* **do**
- $\Phi_i =$ `Induce(`tr`)` // Induce classifier
- $s =$ `QuerySelect(`ul, Φ_i`)` // select query example
- $l =$ `Oracle(`s`)` // Obtain query's label
- $ul \leftarrow ul \setminus \{s\}$ // remove example from ul
- $tr \leftarrow tr \cup \{(s, l)\}$ // update training data

Output: $C =$ `Induce(`tr`)`

Algorithm 1 outlines a generic active learner. In each iteration a probabilistic classifier (Φ_i) is induced from all the known labelled data. The classifier is used to give predictions on the unlabelled data (ul). A query example (s) is selected, removed from the unlabelled data[1] and true label (l) requested from the oracle. The training data is updated with the newly labelled example and the next iteration begins. The final output is a classifier trained on all labelled data.

3 History Based Query Selection

The success of active learning relies on its ability to detect informative unlabelled examples. Uncertainty Sampling (US) [3] is perhaps the best known method for query selection. The query is selected as the unlabelled example for which the current classifier is most uncertainty about ($s = \arg\min_{x \in ul} |\Phi_i(x) - 0.5|$).

Each iteration of active learning produces predictions for the unlabelled examples. These predictions are ephemeral, that is, after the query example has been selected they are discarded. We propose a novel method which stores and utilises the historical predictions in order to find informative unlabelled examples. Predictions for a particular unlabelled example can be retrieved for all iterations of active learning to date allowing us to study their variance. Below we propose two methods that incorporate historical predictions into the query selection function.

[1] We use Pool-Based active learning.

History Uncertainty Sampling. (HUS) is an extension to US where we redefine uncertainty as the sum of the uncertainty of the last k predictions, as shown in Equation 1. Unlabelled examples are ranked according to their HUS value and the query example is selected as the one which has the lowest value.

$$s = \arg\min_{x \in ul} \sum_{j=1}^{k} (|\Phi_{i-j}(x) - 0.5|) \tag{1}$$

History KLD. The past k predictions can be thought of as the output of a committee of size k. In this context we can measure uncertainty as the disagreement among committee members using *Kullback-Leibler divergence to the mean* [4]. The KL divergence of a particular example (x) for a committee (of k members) is calculated as in Equation 2, where the class distribution of an individual committee member is $P_m(C|x)$ and the mean distribution of the committee is given by $P_{avg}(C|x) = (\sum_m P_m(C|x))/k$. The KL divergence between two class distributions $P_1(C)$ and $P_2(C)$ with categories $C = \{c_1, \ldots, c_{|C|}\}$ is given in Equation 3.

$$KLD(x) = \frac{1}{k} \sum_{m=1}^{k} D(P_m(C|x)||P_{avg}(C|x)) \tag{2}$$

$$D(P_1(C)||P_2(C)) = \sum_{j=1}^{|C|} P_1(c_j) \log \frac{P_1(c_j)}{P_2(c_j)} \tag{3}$$

$$s = \arg\max_{x \in ul} KLD(x) \tag{4}$$

We rank the unlabelled examples according to Equation 4 and select the highest value example. History KLD will select those examples which have erratic label assignments.

4 Experimental Results

We used the four *one-v-one* binary problems of the 20 Newsgroup dataset as given in [5]. Documents were tokenised on non-word characters, stopwords removed and stemming applied using the Porter algorithm. Feature selection was not performed. In each iteration of active learning we selected just one query example. The classifier used was a 2-Nearest Neighbour using a normalised linear kernel. Initial training data consisted of 4 positive and 4 negative examples. The history depth (k) was tested for $k = \{2, 3, 5, 7\}$.

Figure 1 shows the *Macro*-averaged accuracy from 10 trials of 10-Fold cross validation where active learning is halted after 100 iterations. In all four problems accuracy was increased when using HKLD. Accuracy increases over US using HKLD were (A) 2.178%, (B) 0.946%, (C) 2.027% and (D) 3.887%. All increases are statistically significant using a paired t-test ($\alpha = 0.05$). We believe HUS does not improve accuracy as it fails to capture the variation in class prediction.

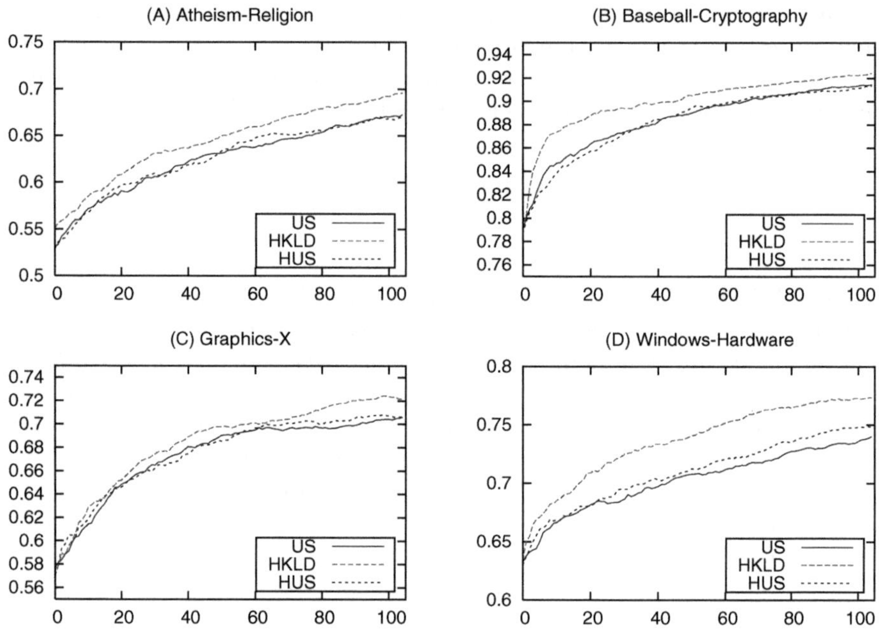

Fig. 1. 20NG results with history depth ($k = 3$). Accuracy is given on the Y axis and the iterations of active learning is given on the X axis. Active learning was stopped after 100 iterations.

5 Conclusions and Future Work

Incorporating history information using HKLD can lead to increased accuracy when compared to US. The depth of history used can have an impact on the improvement obtained when using the history methods. We wish to explore ways to select an optimal history depth in addition to early stopping mechanisms for the active learning process.

References

1. Sebastiani, F.: Machine Learning in Automated Text Categorization. ACM Computing Surveys (CSUR) **34**(1) (2002) 1–47
2. Mitchell, T.: Machine Learning. McGraw-Hill Higher Education (1997)
3. Lewis, D., Gale, W.: A sequential algorithm for training text classifiers. Proceedings of the 17th International ACM SIGIR (1994) 3–12
4. McCallum, A., Nigam, K.: Employing EM in Pool-Based Active :earning for Text Classification. Proceedings of the 15th International Conference on Machine Learning (1998) 350–358
5. Schohn, G., Cohn, D.: Less is more: Active learning with Support Vector Machines. Proceedings of the 17th International Conference on Machine Learning (2000) 285–286

Fighting Link Spam with a Two-Stage Ranking Strategy

Guang-Gang Geng, Chun-Heng Wang, Qiu-Dan Li, and Yuan-Ping Zhu

Key Laboratory of Complex System and Intelligent Science, Institute of Automation
Chinese Academy of Sciences, Beijing 100080, P.R. China
{guanggang.geng,chunheng.wang,qiudan.li,yuanping.zhu}@ia.ac.cn

Abstract. Most of the existing combating web spam techniques focus on the spam detection itself, which are separated from the ranking process. In this paper, we propose a two-stage ranking strategy, which makes good use of hyperlink information among Websites and Website's intra structure information.The proposed method incorporates web spam detection into the ranking process and penalizes the ranking score of potential spam pages, instead of removing them arbitrarily. Preliminary experimental results show that our method is feasible and effective.

1 Introduction

Link spam is something trying to unfairly gain a high ranking on a search engine for a web page without improving the user experience, by mean of trickily manipulating the link graph. The issue of link spam is important not only because it can render significant gains in the rankings of target pages, but also because many instances of it are very hard to detect [5].

So far, most of the techniques for combating link spam, especially classification based techniques, focus on the spam detection itself [1][3][6][7]. In these literatures, web spam identification is regarded as the preprocessing step of ranking, which increases the CPU burden. Additionally, with removing the "spam", the left "cleaning data" holds more dangling pages and no inlink pages; furthermore, because of no definite criteria to judge the web spam, arbitrarily removing detected "spam pages" may lead to some potential useful pages never be retrieved.

According to the fact that users generally only explore the first few search engine results, by analyzing the characteristics of link spam, we propose a two-stage ranking strategy, which makes full use of inter-sites hyperlinks information and intra-site structure information.

2 Two-Stage Ranking Strategy

Several link spams are depicted in Fig. 1, where a website is denoted by a dash rectangular; the solid rectangular represent webpages, and the line with arrow denotes hyperlink.

G. Amati, C. Carpineto, and G. Romano (Eds.): ECIR 2007, LNCS 4425, pp. 699–702, 2007.

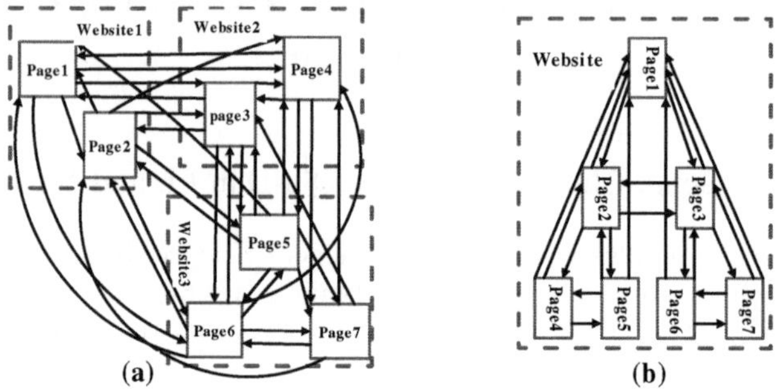

Fig. 1. Typical Link Spams. (a). A Link Alliance and (b). A Link Farm.

By exploiting the link structures of link spam, we find that pages in the spam farm are densely connected, and many common pages exist in both the inlinks and outlinks of these pages. The site level close connection relationship is also ubiquitous in link alliance, such as Fig. 1.(a). On the other hand, supporters of a reputable page should not overly depend on one another, i.e. they should be spread across different quality. One possible way to combat web spams is to decrease the influence of intra-site links and penalize the overly dependent relation in the web graph. This is just the motivation of our paper. In this section, a two-stage ranking strategy is proposed.

Site Level Web Model and Page Level Web Model. The site level Web model is defined as a directed graph: $G_S = (V_S, E_S)$, where V_S represents the set of websites; and $E_S = \{< i, j, k > | i, j \in V_S, k \in N\}$ is the collection hyperlinks between websites, k is the number of hyperlink.

For penalizing the densely connection relationship between Websites, we define the adjacent matrix A of the site level web graph as follows:

$$A_{ij} = \begin{cases} \log_m (k + t), & \text{if } < i, j, k > \in E_S \\ 0, & \text{otherwise} \end{cases} \tag{1}$$

where $m \in N$, t is a regulated parameter.

Correspondingly, the page level Web model is defined as: $G_P = (V_P, E_P)$, where $V_P = \{1, 2, \cdots, n\}$ is the collection of pages; and $E_P = \{< i, j > | i, j \in V_P\}$ is the set of edges, each of which represents a hyperlink. Different from traditional Web model, the model doesn't take the intra-site hyperlinks into account. For convenience, we denote the website as $Site(P_i)$, where page i lies.

Site Level Link Analysis. Site Level Link Analysis is similar to the original PageRank algorithm in spirit. We first construct a probability transition matrix M by renormalizing each row of A. The importance of website S_i is denoted as $SR(S_i)$, which can be computed as follows:

$$SR(S_i) = \varepsilon_1/N + (1 - \varepsilon_1) \cdot \Sigma_{<j,i,k> \in E_S}(SR_j \cdot M_{ji}) \tag{2}$$

where ε_1 is a dampening factor; N is the number of nodes in G_S.

Page Importance. G. R. Xue et al. [2] proposed a Dissipative Heat Conductance model to compute the importance of the pages in a hierarchical structure. However, they treated all hyperlinks identically. We argue that different hyperlinks should have different weights in the voting process.

The weight of the hyperlink from page r to page l is denoted as W_{rl}, which is computed with the formula:

$$W_{rl} = \begin{cases} 1/(1+\ln(L_r \cdot L_l)), & \text{if } <r,l> \in E_P \\ 0, & \text{otherwise} \end{cases} \tag{3}$$

where L_r and L_l is the level of page r and page l in respective website. If page r is on the highest level, let $L_r = 1$. And L_r increases by one when r goes down to the next level of the website tree.

According to the weighted adjacent matrix, firstly normalize the weights of outlinks from the same website with their sum and get a new weight matrix $\overline{W}$. Secondly, use the formula (4) to penalize the hyperlinks from one site to a single page.

$$\overline{W}_{rl} = \overline{W}_{rl} \cdot \log_m (k+m-1)/k, \tag{4}$$

where k is the link number of $Site(P_r)$ to page l. Then the page importance can be computed as follows:

$$PR_l = \varepsilon_2/N + (1-\varepsilon_2) \cdot \Sigma_{<r,l>\in E_P}(\overline{W}_{rl} \cdot SR(Site(P_r))), \tag{5}$$

Compared with the iterative PageRank [4] algorithm, once gets the SR value, the page importance can be computed in one step with formula (5).

3 Experiments

The collection of experiments is a set of 16,367,815 pages from the .cn domain, downloaded in 2005. Some obvious content spams have been removed with some heuristic rules provided by [6] in preprocessing stage. Given the large size of the collection, instead of classifying all the pages into spam and normal, we firstly return the ranking results, then with 11 postgraduates inspected a few pages of the websites manually to classify the result page. They also found the most relevant documents from the returned results as relevance pages, with 50 queries from the daily life. The average relevant documents of every query is 13.75.

In experiment, BM2500 [8] was used as the relevance weighting function. The top 2000 results according to the relevance score were chosen, and then the relevance(BM) and importance(PR) were combined as: $\mu \cdot BM_i + (1-\mu) \cdot PR_i$.

Fig. 2 shows the comparison of the average spam number in the top n ranked results, using our algorithm and PageRank. We can see that the proposed two-stage ranking method can effectively penalize the web spam, especially in the top 20 ranked documents. According to the fact that users generally only explore the first few search engine items, the results is inspiring, in spite of the performance dropping when more top ranked results are considered.

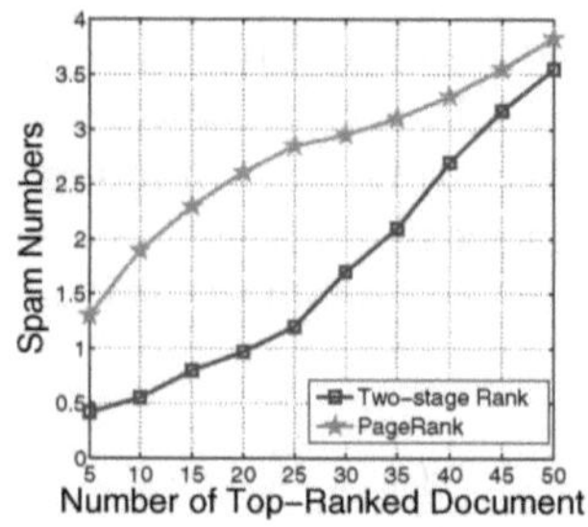

Fig. 2. Comparison of the Average Link Spam Number in the Top n Ranked Results

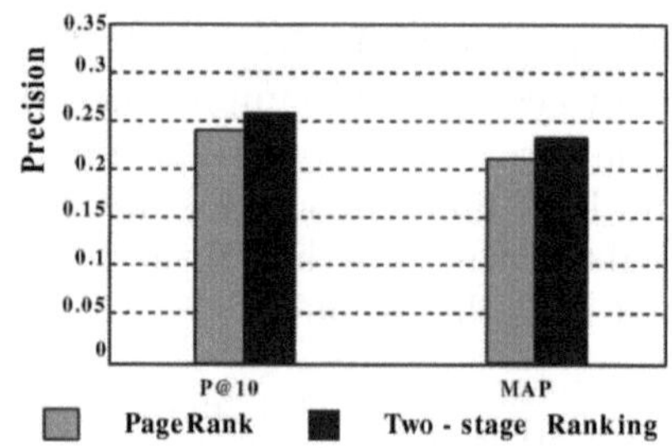

Fig. 3. Comparison of IR Precision

As we can see in Fig. 3, the proposed algorithm significantly outperforms PageRank on average precision and P@10 by 10.7% and 8.1% respectively. All the data in Fig. 2 and Fig. 3 are obtained with the same parameters, i.e. $\varepsilon_1 = \varepsilon_2 = 0.15$, $m = 2.71828$, $t = m - 1$ and $\mu = 0.85$.

4 Conclusions

In this paper, a two-stage ranking strategy was proposed, which penalized link spams with inter-sites hyperlinks information and intra-site structure information. Experimental results showed that the proposed method punished the spam pages well and improved the retrieval precision effectively.

References

1. A. A. Benczúr, K. Csalogány, T. Sarlós, and M. Uher. Spamrank: Fully Automatic Link Spam Detection. In *Proc. of AIRWeb'05*, Chiba, Japan, May 2005.
2. G. R. Xue, Q. Yang, H. J. Zeng, Y. Yu, and Z. Chen. Exploiting the Hierarchical Structure for Link Analysis, In *Proc. of SIGIR'05*, Salvador,Brazil, Auguest 2005.
3. A. L. C. Carvalho, P.A. Chirita, E.S. Moura, P. Calado, and W. Nejdl. Site Level Noise Removal for Search Engines. In *Proc. of the World Wide Web conference*, May 2006.
4. L. Page, S. Brin, R. Motwani, and T. Winograd. The PageRank Citation Ranking: Bringing Order to the Web. *Stanford Digital Library Technologies Project*, 1998.
5. Z. Gyöngyi, and H. G. Molina. Link Spam Alliances. *Technical Report*, September 2005.
6. A. Ntoulas, M. Najork, M. Manasse, and D. Fetterly. Detecting Spam Web Pages through Content Analysis. In *Proc. of the World Wide Web conference*, May 2006.
7. L. Becchetti, C. Castillo1 D. Donato1, S. Leonardi, and R. Baeza-Yates. Using Rank Propagation and Probabilistic Counting for Link Based Spam Detection. In *Proc. of WebKDD'06*, August 2006.
8. Robertson, S. E. Overview of the Okapi Projects. *Journal of Documentation*, Vol. 53, No. 1, 3-7, 1997.

Improving Naive Bayes Text Classifier Using Smoothing Methods

Feng He and Xiaoqing Ding

Dept. of Electronic Engineering, Tsinghua University, 100084, Beijing, China
State Key Laboratory of Intelligent Technology and Systems

Abstract. The performance of naive Bayes text classifier is greatly influenced by parameter estimation, while the large vocabulary and scarce labeled training set bring difficulty in parameter estimation. In this paper, several smoothing methods are introduced to estimate parameters in naive Bayes text classifier. The proposed approaches can achieve better and more stable performance than Laplace smoothing.

1 Introduction

Text categorization is the task of automatically assigning one or more predefined class labels to new text document. Many machine learning approaches have been introduced for text categorization in the literature [1]. Naive Bayes (NB) classifier is a kind of simple Bayesian classifier. It assumes that the occurrence of each word in a document is independent of the position and neighbors of that word.

Two types of NB text classifiers have been extensively compared in [2]. Since the multinomial model is almost uniformly better than the Bernoulli model, we will focus on the multinomial model text classifier. The multinomial model is a generative model. It uses labeled document sets to estimate the parameters in training. New text document is labeled with the class which is most likely to generate the document according to the learned model.

In real life tasks, the vocabulary often contains tens of thousands of words. Since making labels for training documents by human labor is tedious and time consuming, usually, it's hard to collect enough labeled document sets for training. This data sparseness problem brings difficulty in parameter estimation for NB text classifier. Various feature selection methods have been proposed for text categorization [3], by which, only a reduced size of vocabulary are used for model training and testing. These feature selection methods have demonstrated their effectiveness in text categorization. However, in real life application, often it's hard to determine the suitable feature size for text classifier.

Actually, the multinomial model is a unigram language model [2]. In language modeling, n-gram model is also subjected to the data sparseness problem. Several smoothing methods have been proposed to tackle this [4]. The smoothing techniques not only avoid zero probability for unseen words, but also achieve an improved estimation for the whole model. Thus, in this paper, these smoothing methods are introduced from language modeling area to estimate the parameters in NB text classifier.

G. Amati, C. Carpineto, and G. Romano (Eds.): ECIR 2007, LNCS 4425, pp. 703 – 707, 2007.

2 Multinomial Naive Bayesian Model

In multinomial model, regardless of the words' position and context in documents, each document is expressed as a vector of word weights with dimension |V|, where V denotes the vocabulary. We denote a document as d_i, i = 1, 2, …D, where D is the total number of training documents, denote a word as w_t, t = 1, 2, ...|V|, denote the class labels as c_j, j = 1, 2,…C, where C is the total number of classes, and denote the number of times word w_t occurring in document d_i as $n(w_t, d_i)$. According to the independence assumption, document d_i is represented as $(n(w_1, d_i), n(w_2, d_i), \ldots n(w_{|V|}, d_i))$. The class conditional probability of the document is given by the multinomial distribution, as follows:

$$P(d_i \mid c_j) = P(|d_i|) \frac{|d_i|!}{\prod_{t=1}^{|V|} n(w_t, d_i)!} \prod_{t=1}^{|V|} P(w_t \mid c_j)^{n(w_t, d_i)},$$

wherein, $P(w_t|c_j)$ is the probability of each word in each class. To avoid zero probability, conventionally, Laplace smoothing is applied to estimate $P(w_t|c_j)$:

$$P(w_t \mid c_j) = \frac{1 + \sum_{d_i \in c_i} n(w_t, d_i)}{|V| + \sum_{t=1}^{|V|} \sum_{d_i \in c_j} n(w_t, d_i)}.$$

Based on the learned multinomial model, new document is labeled as the class which has the highest posterior probability, by applying Bayes rule, we have:

$$C(d_i) = \arg\max_{c_j} (\log P(c_j) + \sum_{t=1}^{|V|} P(w_t \mid d_i) \log P(w_t \mid c_j)) \cdot$$

3 Smoothing Algorithms

These smoothing methods aim to redistribute the probability mass by discounting the probabilities of words seen and assigning the extra probability mass to words unseen. These smoothing algorithms include absolute smoothing, Good-Turing smoothing, linear smoothing and Witten-Bell smoothing, listing below:

Absolute Smoothing

$$P_{ab}(w_k \mid c_j) = \begin{cases} \frac{r - \delta}{T}, & r > 0 \\ \frac{\delta \cdot \sum_{i>0} N_i}{T \cdot N_0}, & r = 0 \end{cases}$$

Good-Turing Smoothing

$$P_{gt}(w_k \mid c_j) = \begin{cases} \frac{r^*}{T}, & 0 < r \le K \\ \frac{r}{T}, & r > K \\ \frac{N_1}{N_0 T}, & r = 0 \end{cases}$$

The small constant δ in absolute smoothing has already been determined in previous study [5], arriving at: $\delta = N_1 / (N_1 + 2N_2)$.

Linear Smoothing

$$P_{\ln}(w_k \mid c_j) = \begin{cases} (1 - \frac{N_1}{V}) \cdot \frac{r}{T}, & r > 0 \\ \frac{N_1}{V \cdot (V - \sum_{i>0} N_i)}, & r = 0 \end{cases}$$

Witten-Bell Smoothing

$$P_{wt}(w_k \mid c_j) = \begin{cases} \frac{r}{T + V\text{-}N_0}, & r > 0 \\ \frac{V\text{-}N_0}{N_0 \cdot (T + V\text{-}N_0)}, & r = 0 \end{cases}$$

Wherein, $r = \sum_{d \in c_j} n(w_k, d)$; $T = \sum_{k=1}^{|V|} \sum_{d \in c_j} n(w_k, d)$;

$$N_i = \sum_{k=1}^{|V|} I(\sum_{d \in c_j} n(w_k, d) = i), \quad I(x) = \begin{cases} 1, & x \text{ is true} \\ 0, & x \text{ is false} \end{cases}.$$

4 Experimental Results

This section validates the performance of the newly introduced smoothing algorithms by empirically comparing them with Laplace smoothing. The experiments are conducted on Chinese text documents.

Data set and Performance measure

The data set comprises the documents of 10 topic classes from the corpus of China Encyclopedia. It contains 18,478 text documents. Each document is assigned a single topic label. Among the 10 topic classes, the largest one is military affairs, which consists of 2683 documents, while the smallest one is sports, which consists of 700 documents. Since there is no natural delimiter between Chinese words, using a dictionary of 119,817 vocabularies, the popular maximum matching method is performed to segment Chinese texts into word strings. Then, stop-words and words that occur only once are removed. After this tokenization, the data set has a vocabulary size of 138,743. For this multi-label single-class categorization task, we adopt classification accuracy as our performance measure.

Experiments and Results

NB text classifier is performed with randomly selected train-test splits. To focus on the issue of parameter estimation, in particular, the effect of size of train set and size of feature on probability estimation in NB text classifier, gradually, we change the size of training set by varying the ratio of train set to the whole data set from 1% to 90%, and, change the number of features obtained by information gain method [3] from 500 words to full vocabulary. We run 10 training and testing trials under each feature size and training set size, and take the average accuracy as result reported here.

Figure 1 shows results in the condition that the ratio of training set to whole data set is 90%. Here, these smoothing methods have similar performance, except that linear smoothing is slightly inferior. Figure 2 and 3 show results for the training ratio of 2% and 1% respectively. We can see, as the feature size increases, the accuracy obtained from Laplace smoothing and linear smoothing decrease evidently, while for other methods, the performances are stable, relatively immune to the change of feature size. Even with full vocabulary, the other three methods can achieve results comparable to the Laplace smoothing with optimally chosen feature size. We also conduct experiments on other training set ratio, limited by the length, they are not shown here.

Figure 4 shows results for different training ratio, with feature size optimally selected under each training ratio. The traditional Laplace method is slightly but consistently inferior to the proposed smoothing methods, except the linear smoothing method.

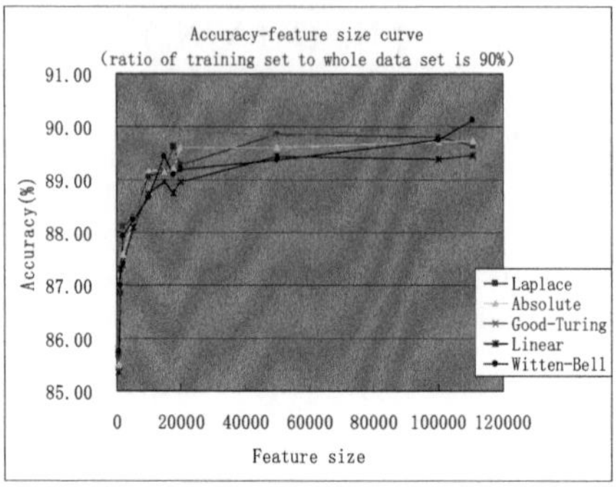

Fig. 1. A comparison of smoothing methods for different feature size, while the ratio of training set to whole data set is 90%

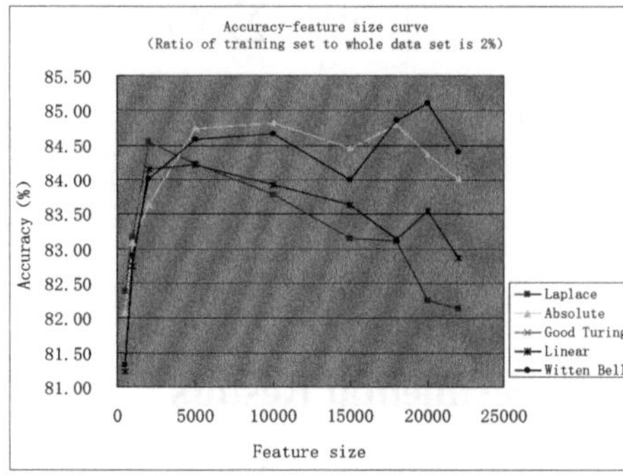

Fig. 2. A comparison of smoothing methods for different feature size, while the ratio of training set to whole data set is 2%

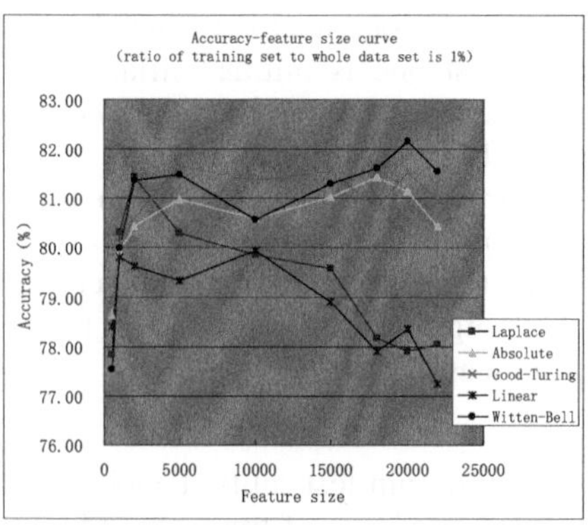

Fig. 3. A comparison of smoothing methods for different feature size, while the ratio of training set to whole data set is 1%

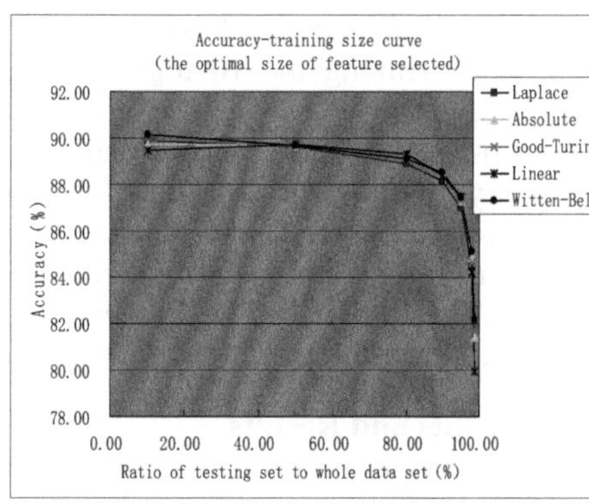

Fig. 4. A comparison of smoothing methods for different training set size, while the feature sizes are optimally selected under each given size of training set

5 Conclusion and Future Works

In this paper, several smoothing algorithms popularly applied in language modeling are introduced for parameter estimation in NB text classifier. The effectiveness of these smoothing methods is demonstrated through empirical study. The proposed approaches can avoid the burden of repeating feature selection procedure to determine the suitable feature size, especially in case of small training set, thus, achieving substantial improvement over the standard NB text classifier. Even without the computation intensive feature selection process, the proposed smoothing methods can obtain experimental results comparable to the Laplace smoothing with optimally chosen feature size. Future works include apply these smoothing algorithms to data sets other than Chinese corpus. Further, we will continue to study the scarce labeled training set issue in text categorization.

References

[1] F.Sebastiani, Machine learning in automated text categorization, *ACM Computing Surveys*, 34(1), pp. 1–47, 2002.

[2] A.McCallum and K.Nigam, A comparison of event models for naive Bayes text classification, In *AAAI/ICML-98 Workshop on Learning for Text Categorization*, pp. 41–48.

[3] Y.Yang and J.O.Pedersen, A comparative study on feature selection in text categorization, In D.H.Fisher, editor, *Proceedings of ICML'97,* pp. 412–420, Nashville, US, 1997.

[4] S.F.Chen, J.Goodman, An empirical study of smoothing techniques for language modeling, In Joshi, A., Palmer, M., eds. *Proceedings of Annual Meeting of the ACL'96*, pp. 310–318, 1996.

[5] Ney, Hermann, Ute Essen, and Reinhard Kneser, On structuring probabilistic dependences in stochastic language modeling. *Computer, Speech, and Language*, pp.1-38, 1994.

Term Selection and Query Operations for Video Retrieval

Bouke Huurnink and Maarten de Rijke

ISLA, University of Amsterdam,
Kruislaan 403, 1098 SJ Amsterdam, The Netherlands
{bhuurnin,mdr}@science.uva.nl

Abstract. We investigate the influence of term selection and query operations on the text retrieval component of video search. Our main finding is that the greatest gain is to be found in the combination of character n-grams, stemmed text, and proximity terms.

1 Introduction

Widespread availability of digital video technology has led to increasing amounts of digital video data. Evidently, we need retrieval systems and algorithms to help us search through it. Automated Speech Recognition (ASR) transcripts are often the only text information source available for video search; the challenge is that this text is subject to speech recognition errors and that the spoken language used is substantially different from that in written query text. Additionally, written queries may request information contained in the visual modality.

Past experiments have shown that ASR transcriptions of video are a valuable source of information [8]. Despite this fact, most research in video retrieval centers around multimodal analysis. Typically, only standard text retrieval methods are used for search of speech extracted from video. As part of an agenda aimed at optimizing the textual components of video retrieval, we report on experiments with different types of text representation for video retrieval. Specifically, we consider character n-grams (i.e., sub-word units), stemming, and proximity terms (i.e., multi-word units) and determine their impact on text retrieval effectiveness for video search. Our main finding is that video retrieval performance can be significantly improved through combination of term selection strategies.

2 Experimental Setup

For evaluation purposes we use the TREC Video Retrieval Evaluation (TRECVID) [8] datasets from 2003–2006. The combined dataset yields over 300 hours of news broadcast video which has been automatically segmented into over 190,000 shots—the basic units of retrieval. It is accompanied by 95 topics, each consisting of a short natural language description of the visual content that is desired from relevant shots. For each topic a ground truth has been manually created. Other components of the TRECVID dataset utilised for our retrieval experiments

G. Amati, C. Carpineto, and G. Romano (Eds.): ECIR 2007, LNCS 4425, pp. 708–711, 2007.

include ASR transcripts, machine translation text, and news story boundary annotations, all of which have been generated automatically.

Each shot is associated with the (English language) transcription of words spoken during that shot, as well as the transcription of the news story in which it occurs. By incorporating the surrounding news story we compensate for the temporal mismatch between the occurrence of an entity in speech and its occurrence in the associated video. We focus on the effects of textual term selection for video retrieval. Therefore we choose to use an existing, well researched, vector space model as our retrieval mechanism, rather than tweaking parameters for this specific task. Indexing and retrieval is done using Lucene [4]. Evaluation of the ranked shots is done using Mean Average Precision (MAP) [8].

3 Experiments

We ran three sets of experiments, one on character n-gram tokenization, one on stemming, and one on the use of proximity terms. For each set of experiments we determine the winning algorithm and perform a comparison to the text baseline. Our baseline representation is an index containing words as they occur in the ASR with stopwords removed. To see to what extent the representation methods just listed have complementary effects, we ran several combination experiments.

Character N-Gram Tokenization. Character n-gram tokenization has been shown to boost retrieval in certain situations [5]. E.g., the use of character 4-grams has proved useful for retrieval from English newspapers [2]. We investigate the effects of n-gramming at different sizes of n (and in different combinations). We follow the tokenization strategy used in [5], creating overlapping, cross-word character n-grams, using values for n from 3 to 7.

Stemming. For our stemming experiments we used the Porter stemming algorithm [7] to normalise morphological variants of words.

Proximity Terms. Our third set of experiments follows [6], who found the use of proximity to be beneficial in web queries, prioritising results in which query words occur close together. Here we experiment with word n-grams, varying the magnitude of n and the proximity required between query words. We investigate the effects of retrieving consecutive sequences of 1 to 7 query terms, and combinations thereof. We also explore the effects of proximity terms—where terms are required to occur in a window of n words—and vary the window size.

Run Combinations. Here, we evaluate all linear combinations of runs produced using the optimal term selection strategies determined in the previous three experiments [3].

4 Results and Analysis

Table 1 provides an overview of the best performing settings for each of the methods identified in the previous section, and for their combinations. Due to

Table 1. Best performing settings per method and for combinations; Δ indicates the percentage change compared to the baseline. Significant changes are indicated with * and ** (two-tailed Wilcoxon Matched-pair Signed-Ranks Test, improvements at the 0.05 and 0.01 level, respectively).

Individual			Combinations		
Method	MAP	Δ	Methods	MAP	Δ
Baseline	0.0609	–	Char. n-grams	0.0596	−2.1
Char. n-grams	0.0574	−5.7	Char. n-grams + stem.	0.0647	+6.3
Stemming	0.0647	+6.2	Char. n-grams + Prox. terms	0.0616	+1.2
Prox. terms	0.0627*	+3.0	Prox. terms + stem.	0.0658	+8.2
			Char. n-grams + stem. + Prox. terms	0.0691**	+13.5

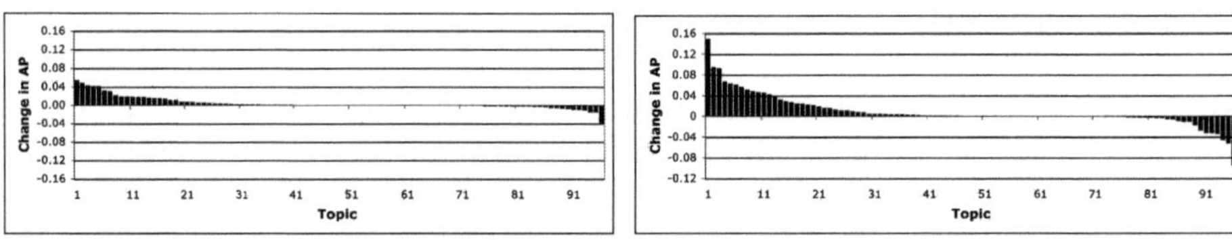

Fig. 1. (Left): Per-topic change in average precision compared to the baseline, using stemming. (Right): Per-topic change in average precision compared to the baseline, using the grand combination strategy. Topics sorted in descending order by change over the baseline.

space limitations we cannot provide detailed tables per method; instead, we briefly discuss the findings per group of experiments.

Character N-Gram Tokenization. The best performing method here is 5-grams (which differs from the best settings reported in [2, 5], who found 4-grams to be optimal). It performs below the baseline, but not significantly. The best performing combination of character n-gram tokenizations combines 4, 6, and 7-grams and performs somewhat better, but still below the baseline.

Stemming. Retrieval on stemmed ASR text outperformed all single n-gram techniques, with a MAP of 0.0647; the difference with the baseline was not significant, though: as Figure 1(Left) suggests, all gains were offset by losses of practically the same size.

Proximity Terms. Proximity terms (allowing up to 10 non-query term to occur between query terms) outperformed the baseline, and did so significantly, as they did in [6]. The best results were achieved by using 2 word sequences for the proximity query, and combining these with the baseline run (i.e., "1 word sequences").

Combinations. Turning to combinations of different methods, we observe that the following best combinations all outperformed the baseline (but not significantly): character n-grams plus stemming, character n-grams plus proximity terms, and proximity terms plus stemming. The grand combination that combines runs created using the best settings for each of character n-gramming, stemming, and proximity terms results in a run that significantly outperforms the baseline (at the 0.01 level). Figure 1(Right) shows the change in topic average precision values using the final grand strategy. Space does not allow an analysis of the effects on individual topics, but it is evident that the majority of topics show an increase in performance.

5 Conclusion

We examined whether term selection alone can be used to significantly improve video retrieval performance. We see that the textual component of video retrieval is similar to other forms of retrieval in that the use of proximity term pairs significantly improves retrieval effectiveness. Stemming also improves performance, while character n-gramming proves not to be directly useful for video retrieval. However, a combination of character n-grams, stemmed terms, and proximity terms results in the best performance.

In future work we plan to explore further avenues for improving the effectiveness of textual search for video retrieval. These include query and document expansion techniques [9] and the linking of textual topics to visual detectors [1].

Acknowledgments

Both authors were supported by the Netherlands Organization for Scientific Research (NWO) MUNCH project under project number 640.002.501. Maarten de Rijke was also supported by NWO under project numbers 017.001.190, 220-80-001, 264-70-050, 354-20-005, 600.065.120, 612-13-001, 612.000.106, 612.066.-302, 612.069.006, 640.001.501, and by the E.U. IST programme of the 6th FP for RTD under project MultiMATCH contract IST-033104.

References

1. C.G.M. Snoek, B. Huurnink, L. Hollink, M. de Rijke, G. Schreiber and M. Worring Adding semantics to detectors for video retrieval. *Journal paper* (submitted).
2. V. Hollink, J. Kamps, C. Monz, and M. de Rijke. Monolingual document retrieval for European languages. *Inf. Retr.*, 7(1-2):33–52, 2004.
3. J. Kamps and M. de Rijke. The effectiveness of combining information retrieval strategies for European languages. In *Proc. 19th Annual ACM Symp. Applied Computing*, pages 1073–1077, 2004.
4. Lucene. The Lucene search engine, 2006. `http://lucene.apache.org/`.
5. P. McNamee and J. Mayfield. Character n-gram tokenization for European language text retrieval. *Inf. Retr.*, 7(1-2):73–97, 2004.
6. G. Mishne and M. de Rijke. Boosting web retrieval through query operations. In *Proc. ECIR 2005*, LNCS 3408, pages 502–516. 2005.
7. M. F. Porter. An algorithm for suffix stripping. In *Readings in Information Retrieval*, pages 313–316. 1997.
8. A.F. Smeaton, P. Over, and W. Kraaij. TRECVID: Evaluating the effectiveness of information retrieval tasks on digital video. In *ACM Multimedia*, 2004.
9. E. Voorhees and J. Garofolo. Retrieving noisy text. In E. Voorhees and D. Harman, editors, *TREC: Experiment and Evaluation in Information Retrieval.* 2005.

An Effective Threshold-Based Neighbor Selection in Collaborative Filtering

Taek-Hun Kim and Sung-Bong Yang

Dept. of Computer Science, Yonsei University
Seoul, 120-749, Korea
{kimthun,yang}@cs.yonsei.ac.kr

Abstract. In this paper we present a recommender system using an effective threshold-based neighbor selection in collaborative filtering. The proposed method uses the substitute neighbors of the test customer who may have an unusual preferences or who are the first rater. The experimental results show that the recommender systems using the proposed method find the proper neighbors and give a good prediction quality.

1 Introduction

A recommender system utilizes in general an information filtering technique called collaborative filtering. Since collaborative filtering is based on the ratings of the neighbors who have similar preferences, it is very important to select the neighbors properly to improve prediction quality.

Collaborative filtering is widely used for the recommender systems, however, many issues in collaborative filtering remain unresolved such that privacy, user's trust, attribute, scalability, sparsity, so called the cold start problem and the first rater problem[5].

In this paper we show that the recommender systems using an effective threshold-based neighbor selection. The proposed method consists of two stages for selecting the neighbors. In the first stage we find the best n substitute neighbors according to the test customer. And then we find the proper neighbors using the threshold which is the correlation of similarities and dissimilarities based on the substitute neighbors.

The experimental results show that the proposed recommender system selects meaningful neighbors for the high prediction quality. Therefore the proposed system can be a choice to solve the first rater problem in collaborative filtering.

2 An Effective Threshold-Based Neighbor Selection

The threshold-based neighbor selection method selects the neighbors who belong to a certain range with respect to the similarities of the preferences. The number of neighbors selected by this method varies because it selects neighbors according to a certain threshold value δ.

G. Amati, C. Carpineto, and G. Romano (Eds.): ECIR 2007, LNCS 4425, pp. 712–715, 2007.

We propose an effective threshold-based neighbor selection method in collaborative filtering and we call it *SNS*. It is designed for solving the first rater problem in collaborative filtering and can be applied the clustering-based recommender systems. The proposed method utilizes the substitute neighbors of the test customer in the process of finding the neighbors for high prediction quality.

SNS finds the best n substitute neighbors. It then searches the proper neighbors who have similarities either larger than δ_H or smaller than δ_L, where δ_H and δ_L are some threshold values for the Pearson correlation coefficients. Note that as the threshold values changes, so does the size of the neighbors. The pseudo code for *SNS* is shown in Algorithm 1.

Algorithm 1. A pseudo code for *SNS*

```
Unmark all the customers in the input dataset S;
n = N; {n is for counting of the number of substitute neighbors N}
while (n is not 0) do
    Find the substitute neighbor s who has the best similarity with the test customer
    t, in the dataset S;
    Mark s;
    Insert s into Q;
    {Q holds temporarily the neighbors for a threshold-based search}
    Add s to Neighbors;
    Subtract 1 from n;
end while
while (Q is not empty) do
    Delete w from Q;
    for each unmarked neighbor x of w do
        if the weight similarity between x and w is either greater than δH or less than
        δL then
            Mark x;
            Add x to Neighbors;
        end if
    end for
end while
return Neighbors;
```

3 Experimental Results

In the experiments we used the MovieLens dataset of the GroupLens Research Center[3]. We used two types of evaluation metrics which are prediction accuracy metrics and recommendation list accuracy metrics. One of the statistical prediction accuracy metrics is MAE. MAE is the mean of the errors of the actual customer ratings against the predicted ratings in an individual prediction [2]. *Precision* and *recall* are also used for evaluating recommendation list in the *information retrieval* community[1]. And the standard *F-measure* is used in order to evaluate the quality as a single measure[4]. It is given by Equation (1).

$$F\!-\!measure = \frac{2 \cdot Precision \cdot Recall}{Precision + Recall} \ . \qquad (1)$$

For the experiment, we have chosen randomly 10% of customers out of all the customers in the dataset as the test customers. The rest of the customers are regarded as the training customers. For each test customer, we chose ten different movies randomly that are actually rated by the test customer as the test movies. The final experimental results are averaged over the results of ten different test sets for a statistical significance.

For comparing the recommender systems, we have implemented three recommender systems for the experiments. The first system is collaborative filtering with the threshold-based neighbor selection, called TNS. The second one is collaborative filtering with the clustering-based neighbor selection, called CNS. The third one is collaborative filtering with the proposed refined threshold-based neighbor selection, called SNS.

For TNS and SNS, three different systems have been implemented. The first system is TNS(or SNS) with positive neighbors ($TNS_p(SNS_p)$). The second one is TNS(or SNS) with negative neighbors ($TNS_n(SNS_n)$). And the last one is TNS(or SNS) with both positive and negative neighbors ($TNS_a(SNS_a)$). For SNS, we have implemented n different settings according to the best n substitute neighbors (SNS^n). For CNS, we have used the k-means clustering method for the clustering.

The experimental results are shown in Table 1 and Table 2. We determined the parameters which gave us the smallest MAEs and the largest precision, recall, and F-measure through various experiments.

The experimental results show us that SNS have an approximate performance to TNS. For both TNS and SNS, $TNS_a(SNS_a)$ is outperform other systems for both the prediction accuracy and the recommendation list accuracy. For SNS_a, SNS_a^1 is the better than SNS_a^n for the prediction accuracy but not for recommendation list accuracy.

For the clustering-based recommender systems, the clustering-based systems are worse prediction qualities than the threshold-based systems. However TNS and SNS are approximate performance just like non-clustering systems show.

These fact means that the proposed neighbor selection is very useful to the recommender systems because it can be substituted for the threshold-based

Table 1. The experimental results

	MAE	$Precision$	$Recall$	$F-measure$
TNS_p	0.739936	0.739142	0.773789	0.755284
TNS_n	0.814013	0.688163	0.726987	0.706330
TNS_a	0.739440	0.740468	0.776128	0.756053
SNS_p^1	0.739697	0.739770	0.773321	0.755274
SNS_n^1	0.848319	0.685816	0.721465	0.702517
SNS_a^1	0.739494	0.740010	0.777172	0.756606
SNS_a^n	0.739503	0.741065	0.773923	0.756849

Table 2. The experimental results based on the clustering

	MAE	$Precision$	$Recall$	$F-measure$
CNS	0.750977	0.733624	0.769937	0.751082
TNS_a	0.750465	0.733748	0.770103	0.751082
SNS_a^1	0.750563	0.733922	0.770679	0.751355
SNS_a^n	0.750787	0.734199	0.770657	0.751665

neighbor selection. Therefore, it could be used the recommender systems to solve the first rater problem.

4 Conclusions

Although collaborative filtering can be regarded as a good choice for a recommender system, there is still much more room for improvement in prediction quality. In this paper we proposed an effective threshold-based neighbor selection that uses substitute neighbors to find the proper neighbors. The substitute neighbors are the neighbors of the test customer who may have an unusual preferences or who is the first rater.

The experimental results show that the proposed recommender system finds meaningful neighbors using substitute neighbors and it is also useful for clustering-based recommender systems. Therefore the clustering-based collaborative filtering using the proposed method could be a choice to the very large scale dataset problem and also the first rater problem.

Acknowledgements

We thank the GroupLens Research Center for permitting us to use the MovieLens dataset. This work has been supported by the BK21 Research Center for Intelligent Mobile Software at Yonsei University in Korea.

References

1. H. Nguyen, P. Haddawy: The Decision-Theoretic Interactive Video Advisor. Proceedings of the Conference on UAI. (1999)
2. J.S. Breese, D. Heckerman, and C. Kadie: Empirical Analysis of Predictive Algorithms for Collaborative Filtering. Proceedings of the Conference on UAI. (1998)
3. MovieLens dataset, GroupLens Research Center, url: http://www.grouplens.org/.
4. R.J. Mooney, L. Roy: Content-Based Book Recommending Using Learning for Text Categorization. Proceedings of the ACM Conference on Digital Libraries. (2000)
5. N.Yamamoto, M. Saito, M. Miyazaki, H.Koike. Recommendation Algorithm Focused on Individual Viewpoints. Proceedings of the Conference on CCNC. (2005)

Combining Multiple Sources of Evidence in XML Multimedia Documents: An Inference Network Incorporating Element Language Models

Zhigang Kong and Mounia Lalmas

Department of Computer Science, Queen Mary, University of London
{cskzg, mounia}@dcs.qmul.ac.uk

Abstract. This work makes use of the semantic structure and logical structure in XML documents, and their combination to represent and retrieve XML multimedia content. We develop a Bayesian network incorporating element language models for the retrieval of a mixture of text and image. In addition, an element-based collection language model is used in the element language model smoothing. The proposed approach was successfully evaluated on the INEX 2005 multimedia data set.

1 Introduction

We believe that structure can play an essential role in the retrieval of multimedia content in XML multimedia documents. The proposed approach makes use of the semantic structure and logical structure in XML documents, and their combination for representing and retrieving XML multimedia document content.

This work develops a general framework for combining multiple sources of evidence from various structured elements in XML multimedia documents. The multimedia content here refers to any type of multimedia data or a mixture of text and multimedia data. An element language model is applied upon each XML element. The framework combines the language models associated with the elements used to perform the retrieval of the multimedia content, using the Inference network model. An element-based smoothing method for the element language model is proposed.

The general framework has been applied in the context of a retrieval task based on a mixture of text and image retrieval on the INEX 2005 Multimedia collection [4].

2 The Proposed Approach

XML documents are composed of structured elements that are nested within each other in a hierarchical tree. There are *logical* relationships between these nested elements. The elements logically surrounding a given element can be used to provide additional sources of evidence for representing this given element. In addition, the elements closer to it in the document hierarchy could provide more accurate representation than those further away.

XML document contents are surrounded by text markups. These, which here refer to the element names, can provide meaningful semantics that can be viewed as metadata describing the nature of the element. This semantics can also be used for the

G. Amati, C. Carpineto, and G. Romano (Eds.): ECIR 2007, LNCS 4425, pp. 716 – 719, 2007.

representation and retrieval of elements. We call the structure based on the meaningful markup *semantic structure*.

The two types of structures are based on different characteristics. The former is identified according to document logical hierarchy and the latter is classified according to the semantics of the markups (the name of the elements). They work in different ways to represent multimedia contents in XML documents. It would be expected that the combination of these two types of representations could lead to better effective retrieval of multimedia content than that based on only one of them.

We can directly query the semantically structured elements as well as query the multimedia data. This can be viewed as a retrieval of a mixture of text and multimedia data. Each semantically structured element (or the multimedia data) is represented by its logically surrounding elements (as shown in the left network of figure 1).

Furthermore, there could be other semantic structures (not the queried structures) in the XML documents. We can further structure the elements logically surrounding a queried element into semantic ones and non-semantic ones (as presented in the right network of figure 1). This could improve the representation based on the surrounding elements.

We use the first method (the left network of the figure 1) in this work as the test collection is not suitable for evaluating the second method. Due to the nested tree structure of XML documents, those structured elements are disjoint from each other so that they have non-overlapping content.

3 An Inference Network Incorporating Element Language Models

The inference network framework was explicitly designed for combing multiple representations and retrieval algorithms [1]. However, the heuristic estimation formulas (such as tf-idf) used in [3] do not correspond well to real probabilities. To address this [1, 2] incorporate the language model in the Bayesian network. We follow this idea but apply it to XML multimedia retrieval.

Figure 1 shows the Bayesian network combining the structured elements. In the networks, the node *D* models a document. The one or more nodes *S* model the queried elements, the element containing multimedia data and/or the semantically structured elements. The nodes between *D* and *S* model the elements logically surrounding the queried elements (the *S* nodes), where *own* is the queried element itself, 1st is its parent element, and so on. The *Q* node models a query and the *I* node models the information need.

The probability of a structured element can be estimated as the probability of it generating the query. In this work, the Dirichlet prior smoothing is used for the smoothing:

$$P(q \mid e)=\prod_{t\in q}(\lambda_1 P(t \mid e)+\lambda_2 P(t \mid C)) \quad (1); \quad P(t \mid e)=\frac{f(t,e)}{|e|} \quad (2); \quad \lambda_1=\frac{|e|}{|e|+\mu} \quad \lambda_2=\frac{\mu}{|e|+\mu} \quad (3)$$

where μ is a parameter set to the average length of the same type of structured elements in the collection; for example, it is the average length of <history> elements in the collection when querying //history. $|e|$ is the length of the element. $f(t,e)$ is the occurrences of a query term in the element.

$P(t|C)$ is a collection language model used for smoothing. However, we focus on estimating the structured elements instead of the entire document. Therefore, we compute the $P(t|C)$ as the probability of observing the term in the collection of the same type of queried elements. For our previous example, it is the probability of observing the term in all <history> elements in this collection. This is called an element-based collection language model.

We use the #WAND operator in the combination as using #WSUM can result in the smoothing component having no influence on the ranking [2].

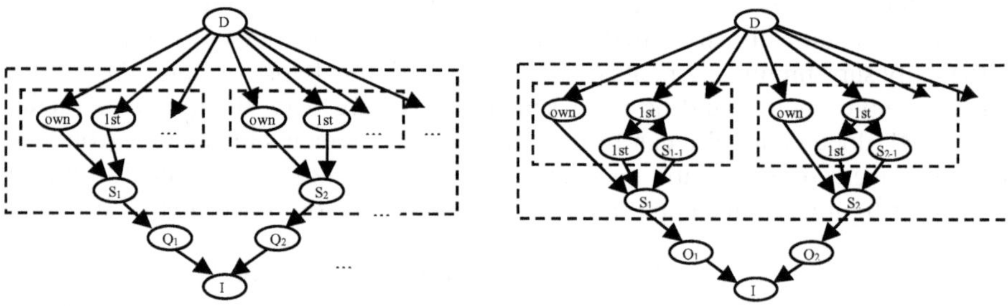

Fig. 1. The Bayesian networks for combining structured elements

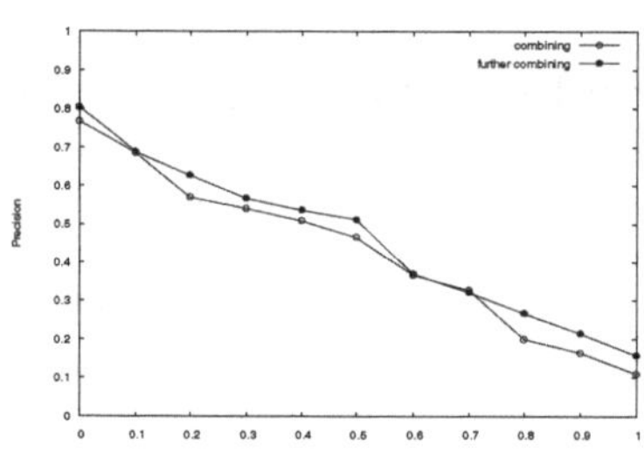

Fig. 2. Combining structured elements

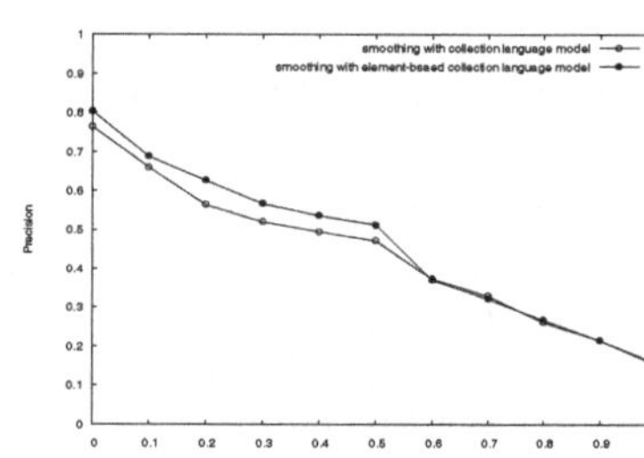

Fig. 3. Using different reference models

4 Retrieving a Mixture of Text and Image

We evaluate our approach using the INEX 2005 multimedia collection, which consists of 2633 images in 462 XML documents. It contains INEX CAS topics made of content and structural constraints. This work directly queries the image element and the structural constraints (viewing them as semantic structures).

Figure 2 shows the results of our approach. At first, we combine the semantically structured elements using their own contents. Then we further combine the logically surrounding elements to estimate their probabilities. The MAP and precision@10 are 0.4079 and 0.4105 in the former and 0.4408 and 0.4158 in the latter. Compared with the best official submission of the INEX 2005 multimedia track, the latter increases by 58.62% (MAP) and 33.91% (precision@10).

Further combining the logically surrounding elements of each semantically structured element achieves 8.07% improvement. We perform experiments to study each semantically structured element (as a query contains several structural constraints). The elements are grouped according to their depth. The results show that combining surrounding elements increases by 56.38% (MAP) and 31.86% (precision@10) for 2 depth elements, and 16.46% (MAP) and 30.73% (precision@10) for 3 depth elements.

The overall improvement of combining the logically surrounding elements is much lower than improvement of each structured element. This is due to most (12 out of 19) of the topics query for the document root element (/destination), which can not be improved. For those not querying for root element, the overall improvement is 16.00%, which increases 98.27% over that (8.07%) of all topics. Therefore, we can expect further improvement in the overall performance as there are only 7 topics not querying the document root element.

Figure 3 shows the results using different reference models in the smoothing. The first uses the standard collection language model and the second uses our element-based collection language model. The MAP of the former is 0.4184 and that of the latter is 0.4408. The latter increases 5.35% over the former.

As discussed above, most topics query for a document root element. In this situation, the element-based collection language model is the same as the standard collection language model. When restricted to topics not querying for a root element, the results show that the MAP of using element-based collection language model increases by 16.10% over that of using standard collection language model. Therefore, we expect the element-based collection language model to lead to improved effectiveness.

5 Conclusions and Future Work

This work makes use of the combination of semantic structure and logical structure in XML documents to represent and retrieve XML multimedia content. An inference network incorporating element language models was developed. In addition, we used an element-based collection language model. The experiments performed on the INEX 2005 multimedia collection showed promising results. Future work needs to be carried out into the use of the framework within a larger XML multimedia document collection.

References

1. Croft , W. B. (2000). Combining approaches to information retrieval. In W. B. Croft, editor, Advances in Information Retrieval, pages 1--36. Kluwer Academic Publishers.
2. Metzler, D. and Croft, W. B. (2004). Combining the language model and inference network approaches to retrieval, IP&M, 40(5):735-750.
3. Turtle, H. and Croft, W. (1991). Evaluation of an inference network-based retrieval model. ACM TOIS, 9(3):187–222.
4. van Zwol, R., Kazai, G, and Lalmas, M. (2006). INEX 2005 multimedia track. Advances in XML Information Retrieval and Evaluation, INEX 2005 Workshop.

Language Model Based Query Classification

Andreas Merkel and Dietrich Klakow

Spoken Language Systems
Saarland University
D-66123 Saarbrücken, Germany

Abstract. In this paper we propose a new way of using language models in query classification for question answering systems. We used a Bayes classifier as classification paradigm. Experimental results show that our approach outperforms current classification methods like Naive Bayes and SVM.

1 Introduction

Unlike most current information retrieval systems, which just return documents to keyword queries, a question answering (QA) system tries to return an accurate answers to natural language questions. So, the step of retrieving relevant documents is just a part of a complex QA system. In order to simply find the one correct answer such a system has to "understand" the meaning of a question. For example, if the question is "*In what country is Luxor?*", the answer should be a country name and not a date. That means the question has to be analyzed in a separate step. There, the question is classified into several semantic categories. This categorization not only helps to find and verify the answer by reducing the search space but also may determine the search strategies for further QA modules ([1]).

Normally, most QA systems just use no more than 20 coarse classes for classification, but in this paper we decided to use the taxonomy proposed by [1] which takes 6 coarse and 50 fine grained classes into account We used the fine grained classification in our experiments because they showed that they are more useful to locate and verify answers. Based on this categorization we used a Bayes classifier with language models as classification paradigm. We show that this approach outperforms systems in current literature.

2 Methods

2.1 Language Models for Classification

Next we have to introduce a suitable classification framework. Ponte and Croft suggested in [3] language models to information retrieval from text collections and showed that they can outperform more traditional methods. We want to propose to use this technique also in the classification of questions.

G. Amati, C. Carpineto, and G. Romano (Eds.): ECIR 2007, LNCS 4425, pp. 720–723, 2007.

As already mentioned we used the Bayes classifier as categorization paradigm. The Bayes classifier is defined by

$$\hat{c} = \text{argmax}_c P(Q|c)P(c) \tag{1}$$

which provides minimum error rate if all probabilities are exactly known. Here, $P(Q|c)$ is the probability of a question Q and a given semantic class c. $P(c)$ is the prior for that class. The probability of $P(Q|c)$ is a language models trained on the class c. In case of unigram language models $P(Q|c)$ is calculated as the product of $P(w|c)$ for all w in Q. The major advantage of the language modeling (LM) approach is that a huge amount of techniques are available to estimate and smooth probabilities even if there is just little training data available. On average, there are only about 100 training questions per question type.

Unlike in previous work ([5]) we do not assume a uniform prior. $P(c)$ can be considered as a unigram language model on semantic classes. As all classes are seen at least four times and therefore sufficiently often, there is no need for smoothing at all and relative frequencies can be used to estimate the language model.

Next we will describe how to estimate the probability $P(w|c)$. It is essential to avoid zero probabilities because that would exclude specific terms from the classification. Hence language model smoothing methods come into play.

2.2 Absolute Discounting

Absolute discounting and its variants are the most popular smoothing techniques in speech recognition. It is defined by

$$P_\delta(w|c) = \frac{\max(N(w,c) - \delta, 0)}{\sum_w N(w,c)} + \frac{\delta B}{\sum_w N(w,c)} P_{BG}(w|c) \tag{2}$$

where $N(w,c)$ is the frequency of observations of the term w together with class c. $P_{BG}(w|c)$ is a background model used for smoothing, δ is the smoothing parameter and B denotes how often $N(w,c)$ is larger than δ.

2.3 Dirichlet Prior

Using a Dirichlet prior results in

$$P_\mu(w|c) = \frac{N(w,c) + \mu P_{BG}(w|c)}{\sum_w N(w,c) + \mu} \tag{3}$$

where μ is a smoothing parameter to be determined on the development data.

2.4 Linear Interpolation

Linear interpolation was first introduced by Jelinek and Mercer [2] and hence some people refer to it also as Jelinek-Mercer smoothing. It is defined by

$$P_\lambda(w|c) = (1-\lambda)\frac{N(w,c)}{\sum_w N(w,c)} + \lambda P_{BG}(w|c) \tag{4}$$

where $N(w, c)$ are frequencies on the training data, λ is a smoothing parameter to be tuned on the development data.

3 Experiments

3.1 Data

For classification we used 6 coarse and 50 fine grained classes as defined in [1]. So, for example, the coarse class *LOCATION* contains the fine classes *city, country, mountain, other and state* whereas *HUMAN* consists of *group, individual, title and description* and so on. As training data for our experiments we used the 5,500 questions provided by the Cognitive Computing Group at University of Illinois at Urbana Champaign[1]. For the evaluation task we used the TREC 10[2] dataset consisting of 500 questions. Both training and test sets are labeled with the corresponding coarse and fine classes.

3.2 Results

As some examples for our experiments Fig. 1 shows the results for the classification with linear interpolation and absolute discounting as smoothing methods. The x-axis shows the interpolation weight and on the y-axis the accuracy is printed.

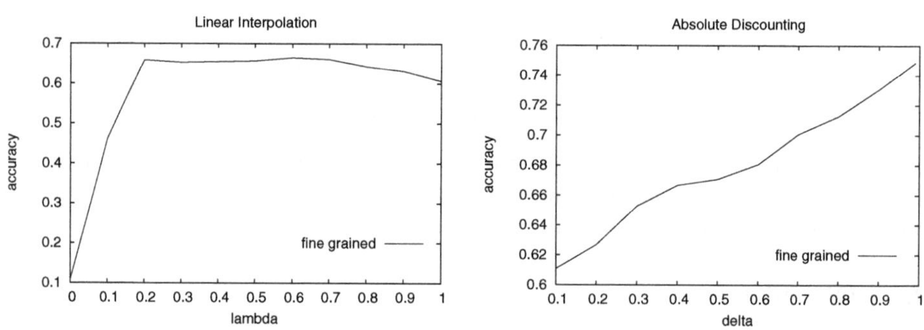

Fig. 1. Accuracy vs. interpolation weight for linear interpolation (a) and absolute discounting (b)

The plot on the left hand side (a) shows the linear interpolation smoothing method. It has a maximum near $\lambda = 0.2$ and is relatively independent to the interpolation weight. The graph on the right hand side shows the absolute discounting method. In contrast to the linear interpolation it strongly depends on the discounting parameter and has it maximum at $\delta = 1$.

[1] http://l2r.cs.uiuc.edu/~cogcomp/Data/QA/QC/

[2] http://trec.nist.gov/

Table 1 compares results from [4] with the proposed LM approach for various machine learning algorithms. They used two different feature sets (bag-of-words and bigram features), so we always print the better result. Our LM approach utilizes optimized bigram features. In particular, we used a log-linear interpolation between a bigram and a unigram distribution.

Table 1. Comparison of various algorithms (Naive Bayes, ... SVM) investigated in [4] with the proposed language model based approach, denoted by LM in the table

Algorithm	Accuracy	Error Rate
Naive Bayes	67.8%	±2.5%
Neural Network	68.8%	±2.5%
Decision Tree	77.0%	±2.1%
SVM	80.2%	±2.0%
LM	80.8%	±2.0%

The table shows that our approach is much better than Naive Bayes. This difference is probably due to the used smoothing techniques. The SVM is the best algorithm from [4] however it is outperformed by the LM approach by a small margin. But in terms of error rate it is not significantly better than SVM. The exact error rates are shown in the last column of the table.

4 Conclusion

In this paper we showed a language modeling approach for query classification based on a Bayes classifier. We experimented with different smoothing methods and various unigram and bigram models. As result we showed that our proposed approach outperforms current categorization methods. So it is significantly better than a classification with Decision Trees and as good as SVM.

References

1. Xin Li and Dan Roth. Learning Question Classifiers. In *Proceedings of the 19th International Conference on Computational Linguistics (2002).*
2. Hermann Ney, Ute Essen and Reinhard Kneser On Structuring Probabilistic Dependencies in Stochastic Language Modeling. In *Computer Speech and Language 8 (1994) 1-38.*
3. Jay M. Ponte and Bruce Croft A Language Modeling Approach to Information Retrieval. In *Proceedings SIGIR (1998).*
4. Dell Zhang and Wee Sun Lee Question Classification using Support Vector Machines. In *Proceedings SIGIR (2003).*
5. Chengxiang Zhai and John Lafferty A Study of Smoothing Methods for Language Models Applied to Ad Hoc Information Retrieval. In *Proceedings SIGIR (2001).*

Integration of Text and Audio Features for Genre Classification in Music Information Retrieval

Robert Neumayer and Andreas Rauber

Vienna University of Technology
Institute for Software Technology and Interactive Systems
{neumayer,rauber}@ifs.tuwien.ac.at

Abstract. Multimedia content can be described in versatile ways as its essence is not limited to one view. For music data these multiple views could be a song's audio features as well as its lyrics. Both of these modalities have their advantages as text may be easier to search in and could cover more of the 'content semantics' of a song, while omitting other types of semantic categorisation. (Psycho)acoustic feature sets, on the other hand, provide the means to identify tracks that 'sound similar' while less supporting other kinds of semantic categorisation. Those discerning characteristics of different feature sets meet users' differing information needs. We will explain the nature of text and audio feature sets which describe the same audio tracks. Moreover, we will propose the use of textual data on top of low level audio features for music genre classification. Further, we will show the impact of different combinations of audio features and textual features based on content words.

1 Introduction

The large-scale adoption of new business models for digital audio material is happening already. What many content providers and online music vendors are still missing are appropriate means of presenting their media to their users according to different information needs. Amazon[1] or last.fm[2] have shown the potential of recommendation engines based on the mining of transactional data.

It further is an intrinsic need for every Music Information Retrieval system to include not only recommendation or playlist generation engines, but also possibilities for searching and browsing. Music Information Retrieval has made huge progress in terms of devising sophisticated descriptors for the acoustic content of music. Research in this direction facilitates numerous content-based search scenarios, such as query by humming, or organisation tasks, such as genre classification, playlist generation, and browsing access by perceived sound similarity.

Song lyrics cover semantic information about a song's contents on a level that could never be covered by audio features only. Many users may rather

[1] http://www.amazon.com
[2] http://www.last.fm

G. Amati, C. Carpineto, and G. Romano (Eds.): ECIR 2007, LNCS 4425, pp. 724–727, 2007.

be interested in songs that cover similar topics than sound alike. Some musical genres, such as e.g. Christmas candles, can only be detected by including textual features, as they occur across many different musical genres and the definition of the genre itself is rather done on a topic level.

We thus combine both textual as well as audio information for music genre classification, i.e. automatically assigning musical genres to tracks based on audio features as well as content words in song lyrics.

The remainder of this paper is organised as follows. Section 2 provides a brief overview of related work. This is followed by a description of our classification experiments in Section 3, as well as outlook on current work in Section 4.

2 Related Work

The area of Music Information Retrieval has been heavily researched, particularly focusing on audio feature extraction. First experiments with content-based Music Information Retrieval were reported in [1] as well as [6], the focus being on automatic genre classification of music. In this paper we use a modified version of the Rhythm Patterns features, previously used within the SOMeJB system [5]. Based on that feature set, [2] shows that the Statistical Spectrum Descriptors yield relatively good results at a manageable dimensionality.

A sophisticated semantic and structural analysis including language identification of songs based on lyrics is conducted in [4]. Artist similarity based on song lyrics is performed in [3], pointing out that similarity retrieval using lyrics is inferior to acoustic similarity. It is also suggested that a combination of lyrics and acoustic similarity could improve results, further motivating the research reported in this paper.

3 Experiments

Due to the lack of public benchmark corpora, we created a parallel corpus of audio and song lyrics files of a music collection of 9.758 titles organised into 41 genres. Class sizes ranged from only a few songs for the 'Classical' genre to about 1.900 songs for 'Punk Rock'. In order to utilise the information contained in music for genre classification, we describe sets of audio features derived from the waveform of audio tracks as well as the bag-of-word features for song lyrics.

3.1 Audio Features

Three features were computed from audio tracks in standard PCM format with 44.1 kHz sampling frequency (e.g. decoded MP3 files). *Rythm Patterns* (RP) [5], also called Fluctuation Patterns, denote a matrix representation of fluctuations on critical bands (parts of it describe rhythm in the narrow sense), resulting in a 1.440 dimensional feature space. *Statistical Spectrum Descriptors* (SSDs, 168 dimensions) are statistical moments derived from a psycho-acoustically transformed spectrogram [2]. *Rhythm Histograms* (RH, 60 dimensions) are calculated as the sums of the magnitudes of each modulation frequency bin of all 24 critical bands.

Table 1. Classification accuracies for different types and combinations of audio features and features based on lyrics. The experiments A1 - A3 denote audio-only, L1 - L4 lyrics-only, and C1 - C3 features combined from audio and lyrics feature sets. The type column shows the types of feature sets used, dimensionality notes the resulting dimensionality of the data.

Name	Type	Dimensionality	Classification Accuracy
A1	RH.	60	.3100
A2	SSD.	168	.4168
A3	RP.	1440	.4128
L1	LYRICS	60	.2451
L2	LYRICS	168	.3204
L3	LYRICS	1440	.4445
L4	LYRICS	3000	.4708
C1	LYRICS + RH	120	.3268
C2	LYRICS + SSD	336	.4817
C3	LYRICS + RP	2880	.4841

3.2 Lyrics Features

For every piece of music, three lyrics portals were accessed, using artist name and track title as queryies. If the results from *lyrc.com.ar* were of reasonable size, these lyrics were assigned to the track. If *lyrc.com.ar* fails, *sing365lyrics.com* will be checked for validity, then *oldielyrics.com.*

All lyrics were processed using the bag-of-words model and weighted by *tfidf* information. Feature selection was done via document frequency thresholding, i.e. the omittance of terms that occur in a very high or very low number of documents. For the matrices used for the experiments terms occurring in more than half of the documents were omitted, the lower threshold was then adjusted to meet the desired dimensionality. Downscaling was performed to different dimensionalities matching the dimensionalities of the audio feature spaces.

3.3 Classification Results

Table 1 shows classification accuracies for a set of experiments based on audio and lyrics features as well as combinations thereof. Experiments were performed by Weka's implementation of Support Vector Machines for ten-fold stratified cross validation. Results shown are the macro averaged classification accuracies.

Results show that a combination of lyrics and audio features improves overall classification performance. The best results were achieved by the 'LYRICS + RP' setting (C3), closely followed by the 'LYRICS + SSD' experiment (C2). The higher-dimensional the data for the lyrics experiments is, the higher is its classification accuracy, implying that there is even more discriminating information contained in lyrics (see experiments L1 - L4), which is not covered in this context because of the limitations of the simple concatenation approach. For combination experiments (C1 - C3) we use balanced combinations of features, i.e. the dimensionality of the lyrics component always equals the dimensionality of the audio feature component.

For statistical significance testing we used a paired T-test for a significance level of $\alpha = .05$. Results showed that A2 performs better than A1 ($p = .0143$), but there is no significant difference between A2 and A3 ($p = .9353$). Further, it is shown that C3 performed better than both A2 ($p = .1934$) and L3 ($p = .0129$). However, the results are not significantly different from experiment L4 ($p = .1755$), leading to the conclusion that high-dimensional lyrics data only is a strong basis for a classifier. Hence a classifier based on differing numbers of lyrics than audio features, e.g. more dimensions in the lyrics than in the audio space, might further improve classification accuracy. Yet, by combining lyrics and audio (C2) the same performance was achieved at a fraction of the dimensionality.

4 Conclusions and Future Work

We showed that the combination of multi-modal features for information retrieval increases classification accuracy. Future work will deal with better means of combining classification results. Ensemble methods might prove useful, overcoming the limitation of implicit feature weighting encountered in the current setting. Additionally, stylistic features for text genre classification are currently being integrated.

References

1. Jonathan Foote. An overview of audio information retrieval. *Multimedia Systems*, 7(1):2–10, 1999.
2. Thomas Lidy and Andreas Rauber. Evaluation of feature extractors and psycho-acoustic transformations for music genre classification. In *Proceedings of the Sixth International Conference on Music Information Retrieval (ISMIR 2005)*, pages 34–41, London, UK, September 11-15 2005.
3. Beth Logan, Andrew Kositsky, and Pedro Moreno. Semantic analysis of song lyrics. In *Proceedings of the 2004 IEEE International Conference on Multimedia and Expo, ICME 2004, 27-30 June 2004, Taipei, Taiwan.* IEEE, 2004.
4. Jose P. G. Mahedero, Álvaro Martínez, Pedro Cano, Markus Koppenberger, and Fabien Gouyon. Natural language processing of lyrics. In *MULTIMEDIA '05: Proceedings of the 13th annual ACM international conference on Multimedia*, pages 475–478, New York, NY, USA, 2005. ACM Press.
5. Andreas Rauber, Elias Pampalk, and Dieter Merkl. Using psycho-acoustic models and self-organizing maps to create a hierarchical structuring of music by musical styles. In *Proceedings of the 3rd International Symposium on Music Information Retrieval*, pages 71–80, Paris, France, October 13-17 2002.
6. George Tzanetakis and Perry Cook. Marsyas: A framework for audio analysis. *Organized Sound*, 4(30), 2000.

Retrieval Method for Video Content in Different Format Based on Spatiotemporal Features

Xuefeng Pan[1,2], Jintao Li[1], Yongdong Zhang[1], Sheng Tang[1], and Juan Cao[1,2]

[1] Institute of Computing Technology, Chinese Academy of Sciences, Beijing 100080, China
{xfpan,jtli,zhyd,ts,caojuan}@ict.ac.cn
[2] Graduate School of Chinese Academy of Sciences, Beijing 100080, China

Abstract. In this paper a robust video content retrieval method based on spatiotemporal features is proposed. To date, most video retrieval methods are using the character of video key frames. This kind of frame based methods is not robust enough for different video format. With our method, the temporal variation of visual information is presented using spatiotemporal slice. Then the DCT is used to extract feature of slice. With this kind of feature, a robust video content retrieval algorithm is developed. The experiment results show that the proposed feature is robust for variant video format.

1 Introduction

With the advances in multimedia and Internet applications, techniques for video retrieval are in increasing demand. The existing approaches in clip-based retrieval with visual cues are mainly based on image retrieval techniques. The most common of these are (1) to use color histogram of key frames (2) to match key frames using color, texture combined with motion information and (3) to use correlation between frames using ordinal measure [1]. As pointed out in [2], ordinal measures based techniques give better performance in comparing with color or motion based methods. But the number of partitions is critical when there existing display format changed contents [3].

It is known that two video clips with same content but compressed in different formats may have distinct color or texture characteristics. The color or texture based features used in image matching are no longer fit for retrieval the same video content in different format. Due to this consideration, we present a retrieval method based on spatiotemporal slice for matching video clips with same content but in different format in this paper. A spatiotemporal slice is a collection of scans in the same position of every frame which indicates the coherency of the video [5]. Ngo used the color and texture of spatiotemporal slice for video clustering and retrieval in [4]. But the retrieval for video content in different format was not discussed in Ngo's paper. In this paper, we will develop a method using the low-frequency AC DCT coefficients of spatiotemporal slice blocks to match video clips with the same content but compressed in different formats.

G. Amati, C. Carpineto, and G. Romano (Eds.): ECIR 2007, LNCS 4425, pp. 728–731, 2007.

2 Clips Retrieval

The horizontal slice which is the collection of scans at the middle row of every frame is used in our method. The slice is converted into gray image and the height of slice is resized to $N \times N$. Then the resized gray slice is segmented into N blocks. By analyzing these blocks with discrete cosine transform (DCT), the variation information of video clip is generated. A sample of grayed slice and blocks are shown in Figure 1. In this way we can transform clip retrieval problem into block

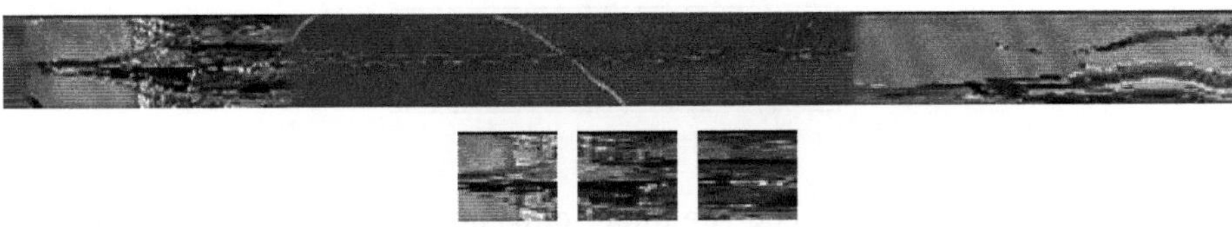

Fig. 1. Grayed slice and blocks from a video clip (n=32)

sequence matching problem. Because most energy of DCT is compacted on low-frequency AC DCT coefficients, the first P low-frequency AC DCT coefficients of each block are gathered to form a feature vector for the blocks and video content. Let $C_Q = (Q_1, Q_2, \cdots, Q_m)$ and $C_T = (T_1, T_2, \cdots, T_n)(m \leq n)$ denote feature vector group of the query clip Q and the target clip T. $Q_i =< Q_i[0], ..., Q_i[P-1] >$denotes the feature vector of the i_{th} block. We use Video Feature Vector Similarity ($VFVS$) to locate Q in T. The matching process is as the following steps.

(1) Shift C_Q along C_T from the beginning of C_T. Let $C_T^i = (Y_i, Y_{i+1}, \cdots, Y_{i+m-1})$ denote some part of C_T that has the same block number as C_Q starting at block $i(i \leq n - m + 1)$.
(2) Compute the $VFVS$ between C_Q and C_T^i using:

$$VFVS(C_T^i, C_Q) = 1 - \frac{\sum_{i=1}^{L} d(C_T^i, C_Q)}{\sum_{i=1}^{L} (abs(C_T^i) + abs(C_Q))} \tag{1}$$

Where $d(\cdot)$ is the distance metric defined on feature vector (here L_1 distance is applied), and $abs(\cdot)$ is the sum of the absolute value of elements of feature vector.
(3) The local maximums of the $VFVS(C_Q, C_T^i)(i = 0, \cdots, n - m + 1)$ values which above a certain threshold are taken as matches.

3 Experiments and Discussion

In this part we present the effectiveness of the proposed algorithm with experimental results.

Firstly, a 30 minutes MPEG-1 video is divided into 24 equal clips and the boundaries of each query clip are the multiple of block size N. Then each clip is taken as query clip Q. The algorithm proposed in section 2 is used to locate Q in original video T. In the experiments, all clips are correctly located. The similarity between Q and sub sequences of T is presented in Table 1.

Table 1. Query clips locating with aligned block boundary

Similarity	Mean	Stdev.
S_{same}	1.0	0.0
S_{diff}	0.232157	0.041292

Note: S_{same}: similarity between the query clips and the located clips having the same content; S_{diff}: similarity between the query clips and other clips having different content; Mean is the average value of the similarity; Standard Deviation. Stdev. is the standard deviation of the similarity.

Table 2. Query clips locating with unaligned block boundary

Similarity	Mean	Stdev.
S_{simi}	0.573746	0.169228
S_{diff}	0.226036	0.033999

Note: S_{simi}: the larger one of similarity value between the query clips and clips start at frame 32m and 32(m+1); S_{diff}: similarity between the query clips and other clips.

Because the feature used in this paper is slice block based, in practice, the block boundaries of query clips may not aligned with the block boundaries of target video. In experiments, the block size is set as $N = 32$. We chose the sub sequences of target clip T starting at frame $32m + i$, $(i = 0, 1, 2, ..., 31)$as query clips Q. So the block boundary of query clips Q and target clip T are not aligned. The similarity between Q and sub sequences of T is presented in Table 2. Further more, the minimum of S_{same} is 0.40695. This is larger than Mean+4×Stdev of S_{diff} in Table 1. The similarity between Q and sub sequences of T is assumed to obey Gaussian distribution. According to the properties of Gaussian distribution, the proposed method can tell the same content as Q from the different content of Q with the probability larger than 0.9999. Four query clips Q are transcoded into 3 different formats. The average similarity between Q and reformatted clips Q' are presented in Table 3. The threshold of similarity for clip matching is set at 0.4 to locate the position of query clip Q in target clip T. In experiment we locate all the 12 reformatted clips in T with no false positive.

Table 3. Similarity between original clips and reformatted clips

Reformatted clip	Similarity
352 × 288 in AVI	0.88623
320 × 180 in AVI	0.87302
320 × 180 in MPEG1	0.90397

4 Conclusion and Future Work

A robust video content retrieval method is proposed in this paper. The proposed method adopts a novel feature extraction scheme for video content representation. It differs from most existing video retrieval methods in that it is no longer using the character of every single frame. The spatiotemporal features extracted with slice DCT are sensitive to clip content variation and robust to video format changing. With this kind of feature, we develop a robust video clip retrieval algorithm. The experiment results show that the proposed feature is robust for variant video format.

There are still things to do with our method. For example, the feature can be more compact, the feature vector matching method can be more efficient. The next target is to revise the method in this paper for large scale video data.

Acknowledgements. This work is supported by the Key Project of Beijing Natural Science Foundation (4051004), and Beijing Science and Technology Planning Program of China (D0106008040291, Z0004024040231).

References

1. Xian-Sheng Hua, Xian Chen, Hong-Jiung Zhnng: Robust Video Signature Based on Ordinal Measure, International Conference on Image Processing (2004). Page(s):685-688
2. Arun Hampapur, Ki-Ho Hyun, Ruud Bolle: Comparison of Sequence Matching Techniques for Video Copy Detection. Proc. Storage and Retrieval for Media Databases, Jan. 2002, Page(s): 194-201
3. Changick Kim, Bhaskaran Vasudev: Spatiotemporal Sequence Matching for Efficient Video Copy Detection, IEEE Transactions on Circuits and Systems for Video Technology, Vol. 15, No. 1, January 2005, Page(s):127-132
4. Chong-Wah Ngo, Ting-Chuen Pong, Hong-Jiang Zhang: On Clustering and Retrieval of Video Shots through Temporal Slices Analysis, IEEE Transactions on Multimedia, Vol. 4, No. 4, December 2002, Page(s):446-458
5. Peng. S. L, Medioni. G, Interpretation of image sequences by spatio-temporal analysis, Workshop on Visual Motion, March 1989. Page(s):344-351

Combination of Document Priors in Web Information Retrieval

Jie Peng and Iadh Ounis

Department of Computing Science, University of Glasgow, United Kingdom
{pj, ounis}@dcs.gla.ac.uk

Abstract. Query independent features (also called document priors), such as the number of incoming links to a document, its PageRank, or the length of its associated URL, have been explored to boost the retrieval effectiveness of Web Information Retrieval (IR) systems. The combination of such query independent features could further enhance the retrieval performance. However, most current combination approaches are based on heuristics, which ignore the possible dependence between the document priors. In this paper, we present a novel and robust method for combining document priors in a principled way. We use a conditional probability rule, which is derived from Kolmogorov's axioms. In particular, we investigate the retrieval performance attainable by our combination of priors method, in comparison to the use of single priors and a heuristic prior combination method. Furthermore, we examine when and how document priors should be combined.

1 Introduction

In Information Retrieval (IR), a document can have query-dependent and query-independent features. Query-dependent features relate to the characteristics of the document, which are specific to the queries and cannot be used before we receive the queries (e.g. the relevance of the document content to a given query). Query-independent features, also referred to as document priors, are features that do not depend on the queries. These document priors can be used to enhance the retrieval performance of a Web IR system, regardless of the query. For example, the number of incoming links to a document (Inlinks), its PageRank, or the length of its associated URL have been shown to be useful in some Web search tasks, such as the Homepage finding and Named Page finding tasks [2] [3].

The language modelling approach to IR provides an elegant framework to integrate single document priors into the retrieval process [3]. However, it is not clear how several document priors should be combined in a principled way. Indeed, most previous work considered either combining document prior probabilities in a heuristic way, usually assuming that document priors are independent from each other [3], or handtuning a linear combination of the priors [4]. Indeed, documents with a high PageRank score usually have a high number of incoming links, suggesting that the PageRank and Inlinks priors are often correlated. In addition, handtuning a linear combination of prior scores is heuristic, and not

G. Amati, C. Carpineto, and G. Romano (Eds.): ECIR 2007, LNCS 4425, pp. 732–736, 2007.

very practical in a realistic setting, where relevance judgements are not always available. In this paper, we present a novel and robust method for combining document priors in a principled way. We use a conditional probability rule, which is derived from Kolmogorov's axioms. The objective of the paper is two-fold: Firstly, we examine how effective our proposed method for the combination of priors is compared with the usually adopted heuristic approach. Secondly, we investigate whether the combination of document priors leads to a better retrieval performance compared to a baseline with and without the use of single priors. In particular, we examine when document priors should be combined.

2 Use of Single Priors in Language Modelling

In language modelling, the probability $P(D|Q)$ of a document D being generated by a query Q is estimated as follows [1]:

$$P(D|Q) = \frac{P(D) \cdot P(Q|D)}{P(Q)} \tag{1}$$

$P(D)$ is a document prior probability. $P(Q)$ can be ignored since it does not depend on the documents and, therefore, does not affect the ranking of documents. $P(Q|D)$ is given by [1]: $P(Q|D) = \sum_{i=1}^{n} \log(1 + \frac{\lambda \cdot tf(t_i,d)(\sum_t cf(t))}{(1-\lambda)cf(t_i)(\sum_t tf(t,d))})$, where λ is a constant given between 0 and 1. n is the number of query terms. $tf(t_i, d)$ is the term frequency of query term t_i in a document d; $\sum_t tf(t, d)$ is the length of document d, i.e. the number of tokens in the document; $cf(t_i)$ is the term frequency of query term t_i in the collection, and $\sum_t cf(t)$ is the total number of tokens in the collection.

In the above language modelling approach, the document prior $P(D)$ usually refers to a single document prior probability. This prior can be omitted from Equation (1), if all documents have a uniform prior. However, it is possible to consider multiple priors for each given document. In this case, it is important to combine the document prior probabilities in a principled way, taking into account the possible dependence between the considered document priors. In the next section, we propose a novel method for appropriately combining document priors.

3 Combination of Multiple Document Priors

For combining document priors, most of the current approaches either assume that the document priors are independent [3], or handtune a linear combination of the priors [4]. In the case the priors are assumed to be independent, the following formula is often used to combine two document priors p_1 and p_2:

$$P(D)_{p_1 \oplus p_2} = P(D)_{p_1} \cdot P(D)_{p_2} \tag{2}$$

where $P(D)_{p_1}$ is the document prior probability related to prior p_1; $P(D)_{p_2}$ is the document prior probability related to prior p_2; $P(D)_{p_1 \oplus p_2}$ is the document prior probability referring to the combination of both priors p_1 and p_2.

However, as mentioned in Section 1, the document priors are not necessarily independent. Therefore, we propose a different approach for combining the prior probabilities. We use a conditional probability rule that is based on Kolmogorov's axioms, given as follows:

$$P(D)_{p_1 \oplus p_2} = P(P(D)_{p_2}|P(D)_{p_1}) \cdot P(D)_{p_1} \tag{3}$$

where $P(D)_{p_1}$ is the document prior probability related to prior p_1, called the *base prior* probability; $P(P(D)_{p_2}|P(D)_{p_1})$ is the conditional probability related to prior p_2, given the prior p_1; $P(D)_{p_1 \oplus p_2}$ is the joint probability of both priors p_1 and p_2 occurring.

Note that the above conditional probability rule can be easily extended to more than two priors. $P(P(D)_{p_2}|P(D)_{p_1})$ can be estimated from a set of relevance judgements as follows: Firstly, we divide the prior probability $P(D)_{p_1}$ into several equal size bins on a logscale. Secondly, inside each bin, we divide the prior probability $P(D)_{p_2}$ into several equal size subset bins, again on a logscale. Finally, the conditional probability of $P(P(D)_{p_2}|P(D)_{p_1})$ in each subset bin is the number of target documents divided by the number of documents in that subset bin.

4 Experiments and Analysis

In this paper, we consider four well-established document priors, namely Page-Rank (PR), information-to-noise ratio (ITN) [5], document length (DL), and the document URL score [3]. We use the standard .GOV Web test collection, and its corresponding TREC 2003 and TREC 2004 Homepage and Named Page finding topic and relevance assessment sets. The official evaluation measure for both tasks is the Mean Reciprocal Rank (MRR).

Firstly, we assess the performance of each of the four single priors (see Table 1). Our baseline (BL) is a language modelling approach, where all documents have a uniform prior probability. From Table 1, we can see that, in general, the single document priors can improve the retrieval performance on the used tasks. The only exception is the ITN prior, which leads to a degradation of the retrieval performance in most cases.

Secondly, we investigate the combination of every pair of priors using our proposed combination approach, and compare it to the performance of the corresponding single priors. Note that the used base prior probability is important in Equation (3). From Table 1, we observe that several combinations of document priors lead to an enhanced MRR score, when we use an effective document prior as base. In particular, combining the best single priors usually leads to an enhanced retrieval performance, compared to their single use.

Finally, we compare our proposed method to a heuristic combination approach, where the priors are assumed to be independent. From Table 1, we observe that our combination way generally outperforms the heuristic method,

Table 1. MRR for the Named Page and Homepage tasks. We use $\lambda = 0.9$ in all experiments. The best retrieval performance is highlighted in bold, and the base prior probabilities are highlighted in italic. Runs statistically different from the best run (Wilcoxon Matched-Pairs Signed-Ranks Test, $p < 0.05$) are underlined. Note that for lack of space, only the most commonly used priors for each task are combined.

Named Page Finding			Homepage Finding		
	MRR			MRR	
	TREC 2003	TREC 2004		TREC 2003	TREC 2004
BL	0.4366	0.3533	BL	0.2363	0.1200
BL+PR	0.4539	0.3588	BL+PR	0.4339	0.3558
BL+DL	0.4546	0.4116	BL+URL	0.4738	0.3976
BL+ITN	0.4186	0.3583	BL+ITN	0.1980	0.0980
Our Proposed Method					
BL+*PR*+DL	0.3730	0.3117	BL+*PR*+URL	0.5247	0.4062
BL+*DL*+PR	0.4732	**0.4365**	BL+*URL*+PR	**0.5424**	**0.4446**
BL+*PR*+ITN	0.3755	0.3098	BL+*PR*+ITN	0.4059	0.3385
BL+*ITN*+PR	**0.4894**	0.4021	BL+*ITN*+PR	0.3889	0.3696
BL+*DL*+ITN	0.4787	0.4130	BL+*URL*+ITN	0.4729	0.4133
BL+*ITN*+DL	0.4377	0.3495	BL+*ITN*+URL	0.4615	0.3508
Priors Independence Assumption Method					
BL+PR +DL	0.4674	0.4065	BL+PR+URL	0.5409	0.4110
BL+PR+ITN	0.4815	0.3867	BL+PR+ITN	0.3526	0.3166
BL+DL+ITN	0.4232	0.3704	BL+URL+ITN	0.4551	0.3470

when the best of the two combined document priors is used as the base prior. The only exception is related to the ITN prior, when it is used as a base prior to combine with the PageRank or URL prior. This combination seems to work very well. Further investigation is required to understand the behaviour of ITN. Overall, our proposed technique can always outperform the heuristic method.

The above results are consistent across both used retrieval tasks. In addition, we observe that, excepting for the TREC 2003 Named Page finding task, using the two best single priors leads to the best overall MRR performance.

5 Conclusion

We have investigated the retrieval performance attainable with query-independent features, in the form of document prior probabilities on two Web search tasks, using a standard Web test collection. We showed that our proposed conditional combination method increases the retrieval performance over the respective single priors, when we use the two best-performing single priors. In addition, we observed that our technique can always outperform a heuristic method, which assumes the independence of priors.

References

1. Hiemstra, D.: Using Language Models for Information Retrieval. PhD thesis. (2001)
2. Kamps, J., Mishne, G., de Rijke, M.: Language Models for Searching in Web Corpora. In Proc. of TREC 2004, Gaithersburg, MD, (2004)
3. Kraaij, W., Westerveld,T., Hiemstra, D.: The Importance of Prior Probabilities for Entry Page Search. In Proc. of SIGIR 2002, Finland, (2002)
4. Metzler, D., Strohman, T., Zhou, Y., Croft, W.B.: Indri at TREC 2005: Terabyte Track. In Proc. of TREC 2005, Gaithersburg, MD, (2005)
5. Zhu X.L., Gauch, S.: Incorporating Quality Metrics in Centralized / Distributed Information Retrieval on the WWW. In Proc. of SIGIR 2000, Athens, (2000)

Enhancing Expert Search Through Query Modeling

Pavel Serdyukov[1], Sergey Chernov[2], and Wolfgang Nejdl[2]

[1] Database Group, University of Twente, P.O. Box 217, 7500 AE Enschede, The Netherlands
[2] L3S / University of Hannover, Appelstr. 9a D-30167 Hannover, Germany
serdyukovpv@cs.utwente.nl, {chernov, nejdl}@l3s.de

Abstract. An expert finding is a very common task among enterprise search activities, while its usual retrieval performance is far from the quality of the Web search. Query modeling helps to improve traditional document retrieval, so we propose to apply it in a new setting. We adopt a general framework of language modeling for expert finding. We show how expert language models can be used for advanced query modeling. A preliminary experimental evaluation on TREC Enterprise Track 2006 collection shows that our method improves the retrieval precision on the expert finding task.

1 The Expert Finding Task

New challenges for the information retrieval research community are posed by the emerging field of Enterprise Search [2]. The diversity of complex information needs within a typical enterprise together with heterogeneity of Intranet data make it difficult to improve the quality of search in general. Instead, researchers concentrate on several important search tasks. One important example of such a task is finding a relevant expert within an organization. This problem implies that user needs to find the most knowledgeable expert to answer her query personally. User submits several keywords to a local Intranet search engine and receives a set of experts, ranked by their likelihood to be an expert for the query. The current developments in expert search are driven by the Expert Finding task within the TREC 2006 Enterprise Track initiative [1]. So far, one of the most comprehensive descriptions of the problem and possible solutions using language modeling approach is presented in [1]. We also adopt a theoretically-sound language modeling method, while using different techniques for the model estimation and ranking.

Numerous ad-hoc query expansion and language model based query modeling methods operate on the top-k ranked documents. At the same time, nobody applied these methods in the scope of expert finding task, what appears to be an omission in our opinion. Our algorithm allows performing a query modeling which consists of pseudo-relevance feedback and query expansion. To the best of our knowledge, this is the first study of query modeling applied to the expert search task. The preliminary evaluation on the official TREC Enterprise Track 2006 test collection shows that our method improves the retrieval performance.

[1] http://www.ins.cwi.nl/projects/trec-ent/wiki/index.php

G. Amati, C. Carpineto, and G. Romano (Eds.): ECIR 2007, LNCS 4425, pp. 737–740, 2007.

2 Expert Finding as a 2-Step Process

A comprehensive description of a language modeling approach to expert finding task is presented in [1]. We adopt the notation from this work and omit some details of model estimation; an interested reader can refer to the original paper. The Step-1 of our expert finding method is similar to Model 1 approach from [1], while The Step-2 contains the actual refinement and is essentially the core of our proposal.

2.1 Step 1: Using Language Model for Expert Ranking

The basic idea of language modeling is to estimate a language model for each expert, and then to rank experts by cross-entropy of estimated query model w.r.t. expert language model [3]. In our setup, each document d in the collection is associated with a candidate ca, the association is defined as $a(d, ca)$. Expert finding problem according to a probability ranking principle in IR is rephrased as: "What is the probability of a candidate ca to be an expert given the query q?" Each candidate ca is represented by a multinomial probability distribution $p(t|ca)$ over a term vocabulary. Expert language model θ_{ca} is computed as the maximum likelihood estimate of a term generation probability, smoothed by the background language model. The query q is also represented by the probability distribution $p(t|q)$, and a query language model is denoted as θ_q. So, the system output should contain the ranking of candidates in descending order of cross-entropy between language models θ_q and θ_{ca}. A cross-entropy of query w.r.t. expert models is computed as shown in Eq.1:

$$ExpertScore_{ca}(q) = -\sum_{t \in q} p(t|\theta_q) \log p(t|\theta_{ca}) \tag{1}$$

The top-k experts with the highest scores are returned to the system (not to the user) as a result of a Step 1, where k is set empirically. So far we described the state-of-art approach, while Step 2 contains our enhancement for the expert search.

2.2 Step 2: Expert Ranking Refinement Using Query Modeling

In order to model a user query more precisely we need a source of additional knowledge about her information need. Traditionally, top-k documents for the query served in IR as such a source and were used to build an expanded and detailed query model. Expert search is a task which differs noticeably from a standard document retrieval. Users search not for the specific pieces of information, but for people who are actually generators and/or collectors of the information. It means that despite the query can be very specific, the experts in this topic can have an expertise in related topics too. Moreover, the broader their expertise, the higher are chances that they can consult on a more specific question. Therefore, we need to utilize two evidences about user information need in the context of expert finding task:

1. The top-k documents retrieved from the whole collection (using classic LM approach to document retrieval)
2. The top-k persons which we could consider relevant experts (retrieved on a Step 1).

The first source enriches our knowledge about the initial user information need. Whereas second one makes it less specific and relaxes a query towards a broader topic. So, as a new query model we use a mixture of two query models: document-based ($DocumentBasedNew\theta_q$) built on top-k documents and expert-based ($ExpertBased New\theta_q$) built on top-k experts:

$$p(t|New\theta_q) = \lambda p(t|DocumentBasedNew\theta_q) + (1-\lambda)p(t|ExpertBasedNew\theta_q) \quad (2)$$

For the both query models estimation, instead of the methods proposed in [1], we use a principled and theoretically-sound method by Zhai and Lafferty from [3], which in our previous experiments for distributed IR outperformed other similar algorithms.

Once it is computed, we mix the new query model with an initial query to prevent a topic drift. As a result, we build a new expert ranking using expanded query and term generation probabilities. In Eq.3 we again measure a cross-entropy, but using a new query model:

$$NewExpertScore_{ca}(q) = -\sum_{t\in q} p(t|New\theta_q) \log p(t|\theta_{ca}) \quad (3)$$

3 Preliminary Results and Discussion

In our experiments we used the W3C collection, provided by the TREC 2006 Enterprise Track, and the Lucene[2] open source information retrieval library. We indexed the mailing lists of W3C dataset[3] and searched for the Title query part of the official topics of the Expert Finding task 2006. The comparison between precision at first 10 results (P@10) of baseline method (Step 1 only) and our method (Step 1 and Step 2) is presented on the Fig. 1 and Fig. 2.

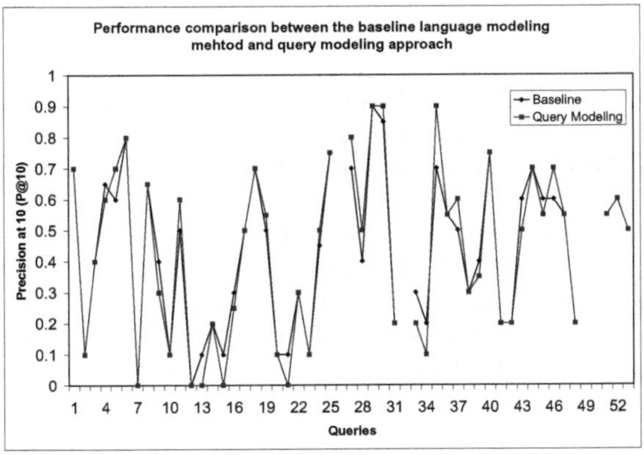

Fig. 1. Performance of the baseline language modeling ranking and query modeling approach

[2] http://lucene.apache.org/

[3] For a rapid experimental setup we used only the mailing list part, while we are planning to evaluate our method on the whole collection later.

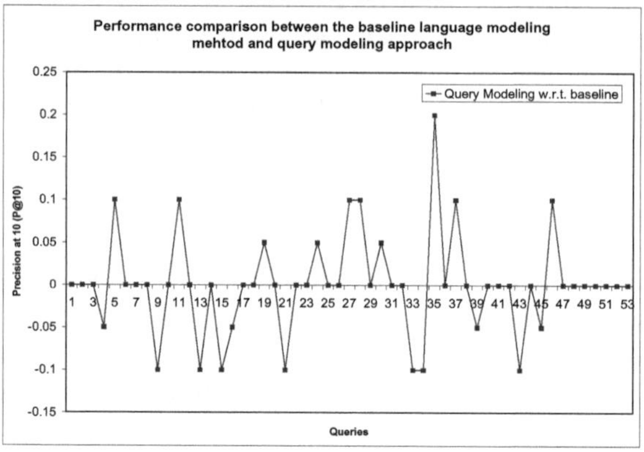

Fig. 2. Difference in performance of the baseline language modeling ranking and query modeling approach

We observe that the improvement of our method is promising, while not significant in the current experiment. Our method is effective when an average precision is high already at the step 1, and fails where average precision is below median. This is explainable since our method uses best top-k experts and documents from the Step 1 for the following query modeling. If the initial ranking is poor, the query modeling is poor too. But the precision for the best queries was improved by 10-20%, so this method is suitable to apply on top of already effective retrieval systems. It appears that a prediction of query performance could be crucial for a query modeling. The further study of the expert-search-specific query modeling and predicting of a query performance is the main focus of our future research.

References

1. K. Balog, L. Azzopardi, and M. de Rijke. Formal models for expert finding in enterprise corpora. In *SIGIR '06: Proceedings of the 29th Annual International ACM SIGIR Conference on Research and Development in Information Retrieval. Seattle, USA*, pages 43–50. ACM Press, 2006.
2. D. Hawking. Challenges in enterprise search. In *Proceedings of the Australasian Database Conference ADC2004, pages 15–26, Dunedin, New Zealand*, 2004.
3. C. Zhai and J. D. Lafferty. Model-based feedback in the language modeling approach to information retrieval. In *CIKM'01: Proceedings of the 2001 ACM CIKM International Conference on Information and Knowledge Management, Atlanta, Georgia, USA, November 5-10, 2001*, pages 403–410, 2001.

A Hierarchical Consensus Architecture for Robust Document Clustering

Xavier Sevillano, Germán Cobo, Francesc Alías, and Joan Claudi Socoró

Department of Communications and Signal Theory
Enginyeria i Arquitectura La Salle. Ramon Llull University. Barcelona (Spain)
{xavis,gcobo,falias,jclaudi}@salle.url.edu

Abstract. A major problem encountered by text clustering practitioners is the difficulty of determining *a priori* which is the optimal text representation and clustering technique for a given clustering problem. As a step towards building robust document partitioning systems, we present a strategy based on a hierarchical consensus clustering architecture that operates on a wide diversity of document representations and partitions. The conducted experiments show that the proposed method is capable of yielding a consensus clustering that is comparable to the best individual clustering available even in the presence of a large number of poor individual labelings, outperforming classic non-hierarchical consensus approaches in terms of performance and computational cost.

1 Introduction

The low availability of labeled document collections has made document clustering techniques become a necessary tool to organize unlabeled corpora according to their thematic contents. However, finding a thematically meaningful partition of an unlabeled document collection is not a straightforward task. This is mainly due to the difficulty of blindly choosing the optimal document representation and clustering method that, for a given a clustering problem, ensure the best match between the true labeling of the documents and the partitioning results.

That is, the performance of document clustering systems requires finding representations of documents that reflect their contents to a maximum extent. Unfortunately, it is difficult to determine *a priori* the optimal type of representation and its dimensionality given a particular document clustering problem. We call this situation the *data representation dependence* effect. Moreover, the application of different clustering methods on the same data often yields different partitions. This gives rise to what we call the *algorithm dependence* effect.

Due to both effects, obtaining the optimal partition of an unlabeled corpus on a single run of a clustering algorithm fed by a specific text representation is a rather challenging aim. Allowing for these circumstances, in this work we define a generic framework for deriving a robust partition of a document collection by building a cluster ensemble upon a wide range of text representations and partitions, following a hierarchical consensus strategy. This proposal is a step towards setting text clustering practitioners free from the obligation of blindly choosing a single document representation and clustering technique.

G. Amati, C. Carpineto, and G. Romano (Eds.): ECIR 2007, LNCS 4425, pp. 741–744, 2007.

2 Hierarchical Consensus Clustering Architecture

The hierarchical consensus clustering architecture is a modular and flexible proposal that allows to obtain a robust partition of the document collection subject to clustering. In this paper, it is assumed that the only knowledge available is the number of clusters we want to group the documents in (i.e. the expected number of thematic categories). Figure 1 depicts the specific implementation of the hierarchical consensus clustering architecture employed in this work.

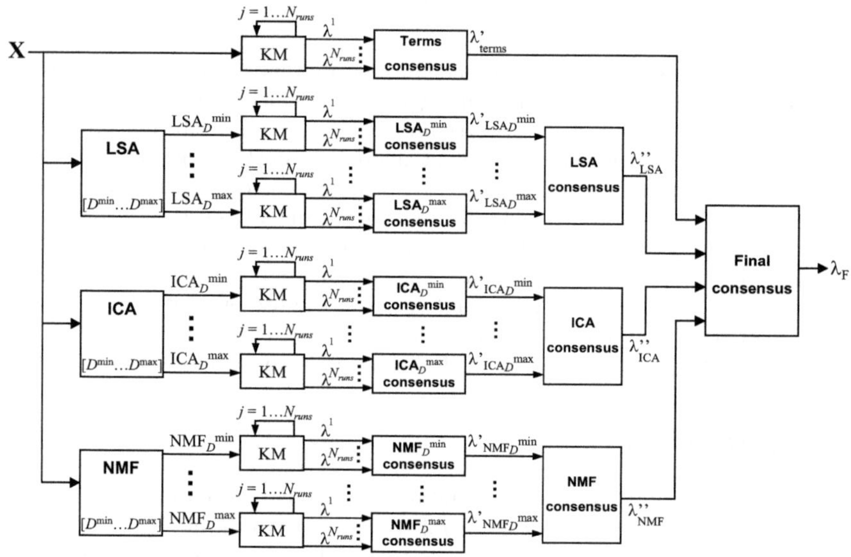

Fig. 1. Hierarchical consensus clustering architecture using k-means (KM) clustering and document representations based on terms, LSA, ICA and NMF

In this paper, the document corpus is initially represented in a term-based vector space (term-by-document matrix $\mathbf{X}$), using the *tfidf* weighting scheme [4]. However, it is a commonplace that the efficiency of clustering systems can be improved by the use of document representations based on feature extraction techniques [5]. Therefore, alternative document representations are derived by applying Latent Semantic Analysis (LSA) [1], Independent Component Analysis (ICA) [2] and Non-negative Matrix Factorization (NMF) [3], with dimensionalities D ranging from $D^{\min}$ to $D^{\max}$. The interval $[D^{\min}, D^{\max}]$ should be wide enough so that the optimal dimensionality of each representation is included in it, but need not be equal for all the representations.

Subsequently, diverse partitions are created by running the k-means (KM) algorithm N_{runs} times with random centroid initialization on each distinct representation[1], yielding what we call *individual clusterings*, denoted as $\lambda^1, \ldots, \lambda^{N_{runs}}$

[1] A more generic strategy for introducing algorithm diversity at this stage would consist in applying distinct clustering algorithms on the same data.

in figure 1. These clusterings form a cluster ensemble which is fed into a first consensus stage, which possibly allows us to overcome the algorithm dependence effect. As a result, a consensus labeling for each distinct document representation is obtained (e.g. for LSA, $\{\lambda'_{\mathrm{LSA}_{D\mathrm{min}}}, \ldots, \lambda'_{\mathrm{LSA}_{D\mathrm{max}}}\}$, or $\lambda'_{\mathrm{terms}}$ for terms).

Next, the labelings corresponding to each feature extraction based document representation method are fed into a second consensus stage, which yields a single labeling per representation scheme (i.e. λ''_{LSA}, λ''_{ICA} and λ''_{NMF} in figure 1). Finally, these labelings (together with $\lambda'_{\mathrm{terms}}$) are used for building the final consensus clustering λ_{F}. Hopefully, λ_{F} will be very similar to the best individual clustering available, thus overcoming the data representation dependence effect.

3 Experiments and Discussion

Experiments have been conducted on the miniNewsgroups corpus, a subset of the 20 Newsgroups document collection that contains 100 documents from each newsgroup. More specifically, six categories of the miniNewsgroups corpus have been used in the following experiments (`comp.graphics`, `rec.autos`, `sci.crypt`, `misc.forsale`, `talk.politics.misc` and `talk.religion.misc`).

In order to build the consensus clusterings, we have compared the following consensus functions: Cluster-Similarity Partitioning Algorithm (CSPA), Hyper-Graph Partitioning Algorithm (HGPA) and Meta-Clustering Algorithm (MCLA) [6]. The tunable parameters of the hierarchical architecture depicted in figure 1 have been set to $D^{\mathrm{min}} = 2$, $D^{\mathrm{max}} = 50$ and $N_{runs} = 30$, hence giving rise to a total of 4440 individual clusterings. Clustering results are evaluated in terms of the F1 measure [4] with respect the true category membership of each document.

Figure 2a presents the F1 measure of the individual and consensus clusterings at the first stage of the hierarchy on a particular case (LSA with $D = 20$). It can be observed that fairly diverse F1 measure values are achieved across the 30 KM algorithm runs, which somehow emulates the algorithm dependence effect. However, the MCLA and CSPA consensus functions are capable of keeping close to the best individual clusterings, while HGPA does not.

In figure 2b, the results of building a LSA consensus clustering on the labelings output by the previous consensus stage are presented, illustrating the influence of the document representation dimensionality. Again, the MCLA and CSPA consensus functions yield λ''_{LSA} clusterings which are very close or even slightly better than the best clustering input to the consensus function.

Finally, the consensus λ_{F} is evaluated by means of the F1 measure histogram depicted in figure 2c, thus comparing the final labeling output by each consensus function to each one of the 4440 individual clusterings. Once more, the MCLA and CSPA are the best performing consensus functions. But more important, their ability to keep track of the best input clusterings is notable, as there are only 11 (out of 4440) individual clusterings superior to the λ_{F} consensus clustering derived through MCLA (19 for CSPA). In other words, the proposed hierarchical consensus clustering architecture has yielded (using the MCLA or CSPA consensus functions) a final clustering that is better than the 99.5% of the individual clusterings, achieving a F1 measure that, in relative terms, is

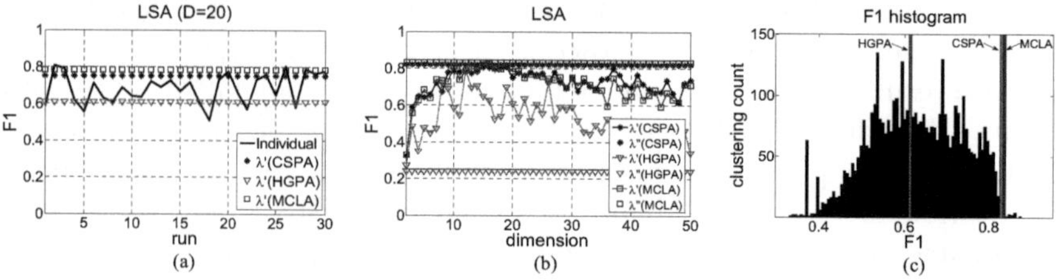

Fig. 2. F1 measure of consensus clusterings: (a) across 30 runs of the KM algorithm, (b) across the $[D^{\min} = 2, D^{\max} = 50]$ dimensionality range, and (c) F1 histogram comparison between the final consensus and individual clusterings

only a 5.5% worse than the best individual clustering. These results reflect high robustness against the data representation and algorithm dependence effects.

Deeper evaluation requires comparing the proposed hierarchical architecture with the classic flat consensus approach (i.e. operating on the 4440 individual clusterings at once). The execution of flat consensus via the MCLA function was halted due to huge memory requirements, whereas the computation of its hierarchical counterpart took 72 seconds of CPU time under Matlab R2006a on a PC P4 3GHz/1GB RAM. As regards hierarchical CSPA and HGPA consensus, their computation was 3 to 13 times faster than flat consensus. Moreover, our hierarchical proposal obtains higher F1 scores (relative increases as high as 38%).

To conclude, this paper has presented a novel strategy for robust document clustering by means of an open hierarchical consensus clustering architecture that operates on highly diverse document representations and partitions. As a result, a final consensus labeling comparable to the best input clustering can be obtained, even in the presence of many poor clusterings. The proposed hierarchical architecture outperforms classic non-hierarchical consensus approaches in terms of both performance and computational cost.

References

[1] Deerwester, S., Dumais, S.-T., Furnas, G.-W., Landauer, T.-K. and Harshman, R.: Indexing by Latent Semantic Analysis. Journal American Society Information Science, Vol. 6, Nr. 41 (1990) 391–407

[2] Kolenda, T., Hansen, L.K. and Sigurdsson, S.: Independent Components in Text. In: Girolami, M. (ed.): Advances in Independent Component Analysis. Springer-Verlag, Berlin Heidelberg New York (2000) 241–262

[3] Lee, D.D. and Seung, H.S.: Learning the Parts of Objects by Non-Negative Matrix Factorization. Nature, 401, pp. 788–791 (1999)

[4] Sebastiani, F.: Machine Learning in Automated Text Categorization. ACM Computing Surveys 34(1), 1-47 (2002)

[5] Shafiei, M., Wang, S., Zhang, R., Milios, E., Tang, B., Tougas, J. and Spiteri, R.: A Systematic Study of Document Representation and Dimension Reduction for Text Clustering. Technical Report CS-2006-05. Dalhousie University (2006)

[6] Strehl, A. and Ghosh, J.: Cluster Ensembles – A Knowledge Reuse Framework for Combining Multiple Partitions. JMLR, Vol. 3, (2002) 583–617

Summarisation and Novelty: An Experimental Investigation

Simon Sweeney[1], Fabio Crestani[1], and David E. Losada[2]

[1] Dept. Computer and Information Sciences
University of Strathclyde, Glasgow, Scotland, UK
{simon, fabioc}@cis.strath.ac.uk
[2] Depto. de Electrónica y Computacion
Universidad de Santiago de Compostela, Spain
dlosada@usc.es

1 Generating Novel Summaries

The continued development of mobile device technologies, their supporting infrastructures and associated services is important to meet the anytime, anywhere information access demands of today's users. The growing need to deliver information on request, in a form that can be readily and easily digested on the move, continues to be a challenge.

Automatic text summarisation is a potential solution to achieving device-friendly content for devices that have limited display screens. An effective way to produce a short summary maybe to include only novel information. However, producing a summary that only contains novel sentences (assuming we employ sentence extraction to build summaries) might imply a loss of context. In this paper we considered summarisation with novelty detection, where information is not only condensed but also attempt is made to remove redundancy. We adopted two strategies to produce summaries that incorporate novelty in different ways; an incremental summary ($SumN_i$) and a constant length summary ($SumN_c$). We compared the performance of groups of users with each of the test systems. The aim was to establish whether a summary that contains only novel sentences provides sufficient basis to determine relevance of a document, or do we need to include additional sentences in the summary to provide context?

Key decisions made at the outset, which influence the production of summaries, relate to the number of summary levels and the length of summaries. We restrict the number of summary levels to 3, primarily to avoid overburdening users in the experimental tasks. In terms of summary length, for each document a number of sentences equal to 7% of its length (with a minimum of 2 sentences and maximum of 6 sentences) were used [2]. Finally, we make use of a similar approach to *NewWords* in [1] as our first attempt to take account of novelty when building summaries. For a full description of the algorithm used to produce the experimental summaries please refer to an extended version of this paper to appear.

The starting point for generating our novel summaries is an initial seed summary, Sum_1, which is a query-biased summary. The query-biased summarisation

G. Amati, C. Carpineto, and G. Romano (Eds.): ECIR 2007, LNCS 4425, pp. 745–748, 2007.

system used to produce the summaries for the experiment was the same as that described in [3]. The length of this query-biased summary, l_1, is determined as a percentage of the original document length. Given a ranked set of sentences, $s_{r_1}, s_{r_2}, ..., s_{r_n}$ (relevance-based ranking), Sum_1 is composed of the top l_1 sentences ordered as they appear in the original document.

Subsequent summaries are generated to include only novel information, and reflect previously seen summary content. To avoid the presentation of material that the user has already seen the focus is on the sentences which, in the original (relevance-based) rank, were ranked right after the ones selected for Sum_1. There are two different ways to produce the next summaries. The first method increases length (N_i) and increments the size of the next summary to be $l_2, = K * l_1$, where $K = 2$, for example, as is the case reported here. This method produces a new summary where all of the material which appeared in Sum_1 is also present in $SumN_{i_2}$. The second method maintains a constant length (N_c) and takes a very different approach producing a new summary, $SumN_{c_2}$, whose size l_2 is equal to l_1. The idea here is to avoid the presentation of material that the user has already seen, and instead focus on the sentences which, in the original (relevance-based) rank, were ranked right after the ones selected for Sum_1. That is, $SumN_{c_2}$ will be composed of sentences selected from $s_{r_{l_1+1}}, s_{r_{l_1+2}}, ..., s_{r_n}$. In contrast, the increasing length method includes both the new sentences and the material already seen, which we consider as the context.

To estimate how novel the candidate sentences are, a history log, composed of previously seen sentences is formed. Each candidate sentence has a relevance score greater than zero. Sentences with a zero relevance score are not included to remove those sentences considered 'not relevant' which, may be novel but off-topic with respect to the query. Next, a WordsSeen list is generated from the history log. The novelty score is based on the proportion of new words with respect to the WordsSeen and compared to all words in the sentence. We compute this as the count of the number of new words divided by the sentence size, including only those words in the sentence that have been stopped and stemmed. To combine the novelty score with the relevance-based score we apply weighting to the novelty score to emphasise novelty scoring over the previous scoring matrix for a sentence. The final score for a candidate sentence is then, the sum of the novelty score with the existing relevance score. Candidate sentences are then ranked according to the combined score.

On the basis of the score ranking and on the required size, a summary is produced. The top scoring candidate sentences form the final summary. The final stage of the process involves reordering summary sentences according to their ordinal position as they occurred in the original document.

2 Experimental Investigation

The documents used in the experiment were taken from the AQUAINT collection and consisted of newswire stories. A total of 5 randomly selected TREC queries and for each query, the 10 top-ranking documents were used as an input for summary generation. The experimental measures to assess the effectiveness of user relevance judgements were the *time to complete the task*, *precision* (P), *recall* (R) and *decision-correctness* (DC). We define DC as the sum of the number of documents marked correctly as relevant, plus the number of documents correctly marked as non-relevant out of the total number of documents marked for that query.

We recruited 20 users to form four experimental groups for the user study. Participants were recruited from members of staff and postgraduate students of the Department of CIS at the University of Strathclyde. For the experiment, each user was given 5 queries, and for each query, the top 10 retrieved documents. These 10 documents were represented as 5 documents summarised using a technique which included novelty, $SumN$, and 5 summarised using a baseline technique that did not use novelty detection, $SumB$, which were query-biased summaries. For each document there are three levels of summary, Sum_1, Sum_2, and Sum_3.

The experimental procedure can be described as follows. Following an initial briefing users were presented with a list of 5 queries. The title and the description of each query (i.e., the 'title' and 'description' fields of the respective TREC topic[1]) provided the necessary background to their 'information need' to allow users to make relevance judgements. For each query, an initial period was allowed to read and digest the query details. Following this, the first of the 10 documents were presented to users, and timing for that specific document started. Users were shown documents from the list where the content for a document consisted of the level 1, 2 and 3 summaries (e.g. $SumN_{c1}$, $SumN_{c2}$, and $SumN_{c3}$). Having seen summary $SumN_{c3}$ users' were required to make a decision as to whether to mark the document as relevant, or non-relevant. After indicating their decision users were presented with the first summary of the next document. The process was repeated until all queries had been evaluated. Once all query tasks were completed a simple online questionnaire was given to the users. The key quantitative data of interest, user decisions and the individual summary timing data, were recorded in log files.

We now report results from the experiment. Due to restrictions of space we are unable to present a full analysis of all the data produced during the experimentation. Table 1 provides a view of the results in the context of the experimental methodology, depicting the allocation of users to groups and associated summary

Table 1. Average performance across all queries for the different summary types

Group	Type	DC	P	R	Time (secs)
1 & 4	$SumB_i$	0.764	0.822	0.845	66
2 & 3	$SumB_c$	0.768	0.850	0.798	53
2 & 3	$SumN_i$	0.776	0.809	0.852	64
1 & 4	$SumN_c$	0.760	0.803	0.752	63

[1] Examples of TREC topics are available at http://trec.nist.gov/data/testq_eng.html

types. The results show a slight increase in DC and R performance with summaries that provide novelty with additional context, $SumN_i$. For P, the baseline summary with a constant length, $SumB_c$, performs best. However, the margins of improvement are somewhat minimal. Appropriate statistical tests found no significance difference in the overall results for the different approaches.

Interestingly, the margin of difference in the time spent on $SumN_i$ compared to $SumN_c$ does not agree with what we might normally expect. A possible reason to explain the similarity in viewing times could be that users may skim the longer summaries, glancing over familiar parts, content already seen, and instead focusing on the new parts. The baseline summaries follow a more expected pattern, though again the margin of difference is small.

A further observation from the table is the similarity in time spent viewing summaries between $SumN_i$ and $SumN_c$, compared to the greater level of separation observed between $SumB_i$ and $SumB_c$. It could be argued then, when we increase the size of the summary, using the baseline approach the user takes more time to digest it whereas, the increasing length summary reads better if it was constructed using novelty.

3 Conclusions and Future Work

In conclusion, findings from the user study suggest that there is little difference in performance (DC, P and R) between novel summaries that include context ($SumN_i$) and those that contain only novel information ($SumN_c$). Therefore, for mobile information access where issues of bandwidth and screen size are paramount then we can conclude that an effective way to produce a short summary is to build one that includes only novel information. However, the lack of improvement over the baseline does place doubt over the merit of building novel summaries and will require more investigation.

Extensions to the work we have presented include investigating the performance of a more refined approach to novelty detection beyond a simple count of new words. In addition, a further point of interest being to study the effects of permitting users to make decisions at any levels; to investigate summary level preference and if there is a corresponding impact on accuracy.

References

1. J. Allan, C. Wade, and A. Bolivar. Retrieval and novelty detection at the sentence level. In *Proceedings of ACM SIGIR'03*, pages 314–321, Toronto, Canada, July 2003.
2. S. Sweeney and F. Crestani. Effective search results summary size and device screen size: Is there a relationship? *Information Processing and Management*, 42(4):1056–1074, 2006.
3. S. Sweeney, F. Crestani, and A. Tombros. Mobile Delivery of News using Hierarchically Query-Biased Summaries. In *Proceedings of ACM SAC'02*, pages 634–639, Madrid, Spain, March 2002.

A Layered Approach to Context-Dependent User Modelling

Elena Vildjiounaite and Sanna Kallio

VTT Technical Research Centre of Finland, Kaytovayla 1, 90580, Oulu, Finland
{FirstName.LastName}@vtt.fi

Abstract. This works presents a method for explicit acquisition of context-dependent user preferences (preferences which change depending on a user situation, e.g., higher interest in outdoor activities if it is sunny than if it is raining) for Smart Home – intelligent environment, which recognises contexts of its inhabitants (such as presence of people, activities, events, weather etc) via home and mobile devices and provides personalized proactive support to the users. Since a set of personally important situations, which affect user preferences, is user-dependent, and since many situations can be described only in fuzzy terms, we provide users with an easy way to develop personal context ontology and to map it fuzzily into common ontology via GUI. Backward mapping, by estimating the probability of occurrence of a user-defined situation, allows retrieval of preferences from all components of the user model.

Keywords: User Model, Context Awareness, Smart Home.

1 Introduction

Context dependency of user preferences in Information and Multimedia Retrieval was first explored by location-based services, and after that other context types, such as time and social context (people co-located together), were shown to be useful for TV [1] and movie [2] recommendations and for tourist guides [3]. However, these works do not aim at explicit acquisition of long-term context-dependent user preferences due to large diversity of interests and situations, which affect interests. Nevertheless methods of explicit acquisition of user preferences, including context-dependent ones, are needed because users want to stay in control of the system and to configure the most important settings. (Users expressed the requirement to stay in control during user study [4] in Amigo project [5], developing user services for home environment.)

Currently explicit acquisition of long-term context-dependent preferences is being developed only for certain applications, such as personalising behaviour of mobile phones [6] and configuring, how Smart Home should record users' conversations and events [7]. Configuring these applications usually requires users to look through many contexts, predefined by an application designer, and to find values for each aspect (when, where, who etc) of each situation. Consequently, each situation is described as a chain of fairly low-level context types and values, such as in the rule "if <location is *home*> AND <device charger is *charging*, then…" [6]. User studies have shown that users are not willing to build long chains even in mobile phone applications, based on a few context types only [6], and thus this approach is not suitable for future smart

G. Amati, C. Carpineto, and G. Romano (Eds.): ECIR 2007, LNCS 4425, pp. 749–752, 2007.

environments, capable of recognition of many context types. Furthermore, this approach makes difficult to use as a context source user calendar entries, expressed in custom terms. This work suggests a method to create context – dependent user profiles via GUI, which provides a convenient way to define personal fuzzy concepts of context ontology and to map them into predefined ontology and sensor data.

2 Context-Dependent User Model and Context Reasoning

An intelligent information retrieval agent aims at providing users with answers to their queries in such a way that these answers are on top of the results' list (it is especially important for small screens of mobile devices), and even at anticipating the users' needs by collecting and presenting information/ multimedia, which users need in a current situation, proactively, e.g., by showing appropriate advertisements or news. In order to do it, an agent needs user model in a form "context" – "preferences for this context". The Amigo system can acquire user preferences in different ways:

- *explicit modelling*: users can set preferences via GUI: e.g., high interest in comedies and no interest in thrillers if a user is accompanied with kids (as opposite to his usual high interest in thrillers); high interest in ICT topics if a user writes a document in a work office; high interest in recipes for open fire cooking if a user is hosting a barbecue party in near future.
- *stereotypes-based modelling*: preference values for common situations can be set by application designers, such as high interest in near future buses from an airport if a user is in the airport and searches for a bus schedule via his phone.
- *dynamic modelling*: learning preferences from observations of user actions, e.g., that a user is more interested in inexpensive hotels in rural area when he searches from home before holidays, and that the same person prefers hotels in city centre when searching for hotels from work location for a business trip.

In order to react to users information needs, the system should recognize users' current and future contexts via physical (e.g., location) and logical (e.g., entries in a user calendar) sensors and to find corresponding preferences. The problem is, that context recognized by the physical sensors is usually expressed as a set of designer-defined descriptors (e.g., numerical values of time, street address, IDs of devices around a user etc), which is inconvenient for users, because they think in higher-level personal semantic descriptors. On the other hand, logical sensors often provide data in such personal terms (e.g., in one user's calendar a "Masks" entry denotes her child's birthday masquerade party; and before and during "Masks" event this user needs information about new food recipes, games and rental of costumes). Thus, we suggest to allow users to add own (custom) terms to ontology of context descriptors, provided by application designers, because it helps in mapping between different sources of information (physical sensors and stereotypes provide raw data or data in designer-defined terms, whereas logical sensors – in personal terms, see Fig.1), and because it saves the users from the need to build long chains of context descriptors manually as many times as they set different preference values for some situation. Instead, users need to edit a set of designer-defined context descriptors, built by an application for them, only when they add a new (custom) term.

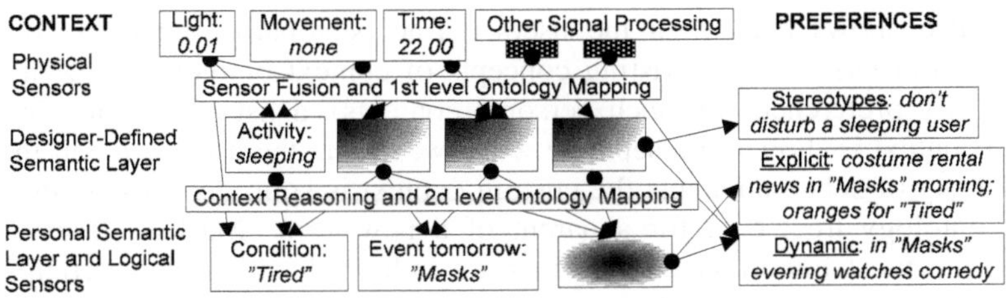

Fig. 1. Layers of Context-Dependent User Model

Since as a regular (annual) event, a "Masks" party can be described only fuzzily (a party at home, with 8-12 guests about 4-12 years old, on Saturday or Sunday 2-10 days after the child's birthday), we developed a GUI for explicit acquisition of user preferences, which allows users to define such personal fuzzy contexts and to set preferences for any personal or predefined context term, as well as for combination of contexts. GUI presents to the users a tree of preference keys and a context ontology tree; and sets a default value 0 for all keys initially. The users can set as many different values for the same key in different contexts as they want by attaching one or several contexts to each value. Users can also set context-independent preferences: the value with no contexts attached is treated as valid always, except for the contexts which degree of similarity with the attached context exceeds certain threshold. Fig.2 shows how the user sets context-dependent and independent values for different information topics and specifies when he needs information on these topics.

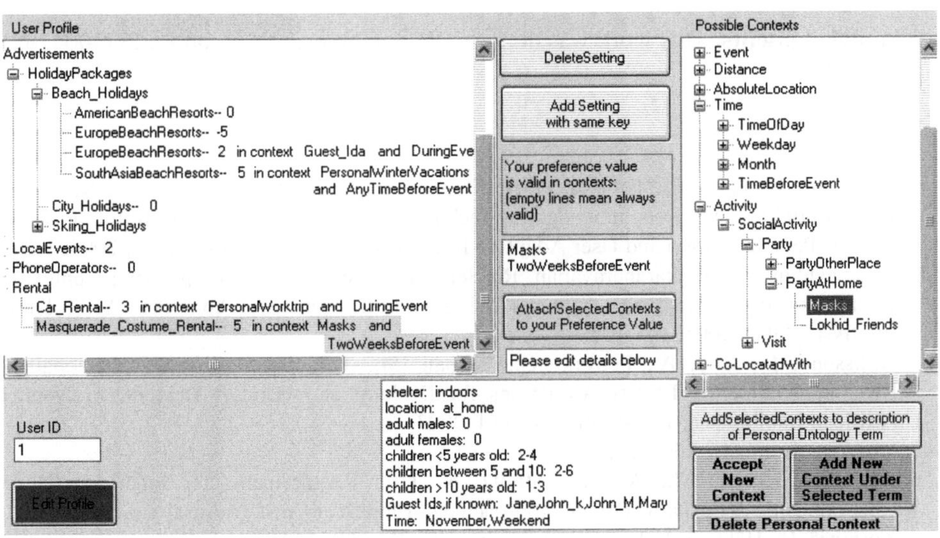

Fig. 2. GUI for acquisition of user preferences, which allows users to express context-dependency of their preferences and to create personal fuzzy concepts of context ontology

Fig.2 also shows that when the users define personally important contexts by adding custom terms under existing concepts of context ontology (as well as when users specify preferences for such personal contexts), a list of subcomponents of a personal context is shown in a separate window, so that the users can edit the values. To which set of subcomponents the custom context is split, depends on its parent in context ontology tree: e.g., for a parent of a new term "Masks", "PartyAtHome" concept, the most important components are indoors location and participants. For "Masks" event also month and approximate day of week (weekend) are known, but for many other personal events they take a default value "any time".

Mapping from sensor data into custom contexts (for retrieval of user preferences) is done by Context Reasoning module, which takes as an input current context and interaction history, both expressed in designer-defined terms (context comes as a set of triples: context value - time stamp - confidence) and in terms of logical sensors, and estimates the probability that a certain situation occurs now, or in a near future.

3 Conclusions and Future Work

This work presented the layers of user model for context-aware applications, aiming at delivering desired information/ multimedia to users in different situations. Since users want to control such applications, we propose a method to set context-dependent preferences via GUI, which serves two goals: first, provides users with a convenient way to express preferences for personally important situations in custom terms; and second, helps in context recognition when custom terms are encountered e.g. in a user calendar. Next, we will collect more data of using it in different applications.

Acknowledgments. This work was done in EU Amigo project (contract IST-004182).

References

1. Ardissono, L., Gena, C., Torasso, P., Bellifemine, F., Chiarotto, A., Difino, A., Negro, B., User Modeling and Recommendation Techniques for Personalized electronic Program Guides, Personalization and User Adaptive Interaction in Digital Television, 6, 2004, 3-26.
2. Adomavicius, G., Sankaranarayanan, R., Sen, Sh., Tughilin, A., Incorporating Contextual Information in Recommender Systems Using a Multidimensional Approach, ACM Trans. Inf. Syst., 23(1), 2005, 103-145.
3. Ardissono, L., Goy, A., Petrone, G., Segnan, M., Torasso, P., Intrigue: Personalized Recommendation of Tourist Attractions for Desktop and Handset Devices, in Applied Artificial Intelligence, 17(8-9) pp. 687-714, 2003
4. http://www.amigo-project.org
5. Röcker, C., Janse, M., Portolan, Streitz, N., User Requirements for Intelligent Home Environments: A Scenario-Driven Approach and Empirical Cross-Cultural Study, 2005.
6. Korpipää, P., Häkkilä, J., Kela, J., Ronkainen, S., Känsälä, I., Utilising Context Ontology in Mobile Device Application Personalisation, Proc. MuM 2004, 133-140.
7. Truong, Kh., Huang, E., Abowd, G., CAMP: A Magnetic Poetry Interface for End-User Programming of Capture Applications for the Home, Proc. Ubicomp 2004, pp. 143-160

A Bayesian Approach for Learning Document Type Relevance

Peter C.K. Yeung, Stefan Büttcher, Charles L.A. Clarke, and Maheedhar Kolla

University of Waterloo, Waterloo ON, Canada
{p2yeung, sbuettch, claclark, mkolla}@plg.uwaterloo.ca

Abstract. Retrieval accuracy can be improved by considering which document type should be filtered out and which should be ranked higher in the result list. Hence, document type can be used as a key factor for building a re-ranking retrieval model. We take a simple approach for considering document type in the retrieval process. We adapt the BM25 scoring function to weight term frequency based on the document type and take the Bayesian approach to estimate the appropriate weight for each type. Experimental results show that our approach improves on search precision by as much as 19%.

1 Introduction

Okapi BM25 [2] is a popular choice for scoring document relevance based on term frequency, document length, and other collection statistics. Recently, Robertson et al. [3] introduced a modified version of BM25 for incorporating *weights* into different fields of a structured document. The intuition for this approach is to consider structured documents and rank them according to the importance of each structure. These structures include document title, author, abstract, content, etc. For our purpose, we study the importance of a particular document structure—document type—and an approach to estimate the appropriate *weight* for each document type.

In order to estimate the weight of each document type, we take the Bayesian approach to calculate the likelihood that documents from a given type are relevant. This probability can then be scaled and used as the weight for the given type. In our experiments, we use the set of document judgments to sample and compute all prior and posterior probabilities in the Bayes' theorem. Results show that our method improves search precision by as much as 19% at 5 documents.

2 Weighting Document Types

2.1 BM25 Retrieval Model

For query terms $T_1, T_2, ..., T_n$, the BM25 relevance score of a document D is

$$S_{BM25}^{(D)} = \sum_{i=1}^{n} w_{T_i} * \frac{(k_1 + 1) * f_{D,T_i}}{f_{D,T_i} + k_1 * ((1 - b) + b * \frac{|D|}{avgdl})} \tag{1}$$

G. Amati, C. Carpineto, and G. Romano (Eds.): ECIR 2007, LNCS 4425, pp. 753–756, 2007.

where f_{D,T_i} is the frequency of T_i in document D, $|D|$ is the length of document D, and *avgdl* is the average document length in the collection. k_1 and b are free parameters, which we set to $k_1 = 1.2$ and $b = 0.75$. w_{T_i} is the *inverse document frequency* weight, which is the significance of T_i in determining document relevance.

$$w_{T_i} = \log\left(\frac{\#\text{ documents}}{\#\text{ documents containing } T_i}\right) \tag{2}$$

Robertson et al. [3] modified BM25 to consider structured documents by computing a linear combination of term frequencies. For each term in a query, its frequency in document D is treated as a combination of its unweighted frequency f_{D,T_i} and the corresponding weight w_j.

$$f'_{D,T_i} = \sum_{j=1}^{N} w_j * f_{D,T_i} \tag{3}$$

For our purpose, *document type* is the only structure that needs to be weighted. The weight for each document type depends on the relevance of its documents. For example, if type A contains more relevant documents than type B, then its weight is larger than type B's. In essence, terms appearing in type A documents are weighted heavily and cause BM25 to produce higher relevance scores for these documents.

2.2 Estimating Weights for Document Types

Another challenge in our approach is to determine the appropriate weight for each document type. The weight influences BM25 on scoring a document from its corresponding type with respect to other types. Hence, weights should reflect the relative difference in document relevance between all document types.

A trivial way to estimate the weight for a document type is by considering the likelihood that its documents are relevant. We apply Bayes' theorem to calculate the probability that given a type, D_T, its documents are relevant, $Pr(\text{Rel}|D_T)$.

$$Pr(\text{Rel}|D_T) = \frac{Pr(\text{Rel}) * Pr(D_T|\text{Rel})}{Pr(D_T)} \tag{4}$$

where $Pr(D_T|\text{Rel})$ is the posterior probability that for a given relevant document, it belongs to type D_T. $Pr(\text{Rel})$ is the prior probability that a document is relevant. $Pr(D_T)$ is the prior probability that documents from type D_T is retrieved, which is also the normalization constant.

2.3 Normalization

We use the set of document judgments to calculate the probabilities (or to train our model). However, since the set of document judgments used in our experiments is incomplete, $Pr(D_T)$ is not an appropriate normalization constant. Therefore, we need to introduce a normalization constant, α.

$$w_j = \alpha * Pr(\text{Rel}) * Pr(D_{T_j}|\text{Rel}) \tag{5}$$

Consider the situation where there is no difference between the relevance of document types, then the weight should have no effect on the document score. This results in reverting our method back to the unstructured case where $w_j = 1$ for all j. Thus, α is used to scale the weight of each document type so that the sum of all w_j equals the number of document types defined, N.

$$N = \alpha * Pr(\text{Rel}) * \sum_{j=1}^{N} Pr(D_{T_j}|\text{Rel}) \tag{6}$$

3 Experimental Setup and Results

3.1 The Corpus

For our experiments, we employ the W3C collection used in the TREC 2006 Enterprise track [1]. The W3C collection contains 331,037 documents with a total uncompressed size of 5.7 gigabytes. These documents are categorized into six different types: mailing lists (lists), public CVS repository (dev), public pages (www), wiki pages (esw), personal pages (people), and other pages (other). Table 1 shows the number of documents in each type and their average document sizes.

Table 1. W3C collection

Scope	Corpus Size	# Docs	Avg Doc Size
www	1.043 (gigs)	45,975	23.8 (kbs)
lists	1.855	198,394	9.8
dev	2.578	62,509	43.2
people	0.003	1,016	3.6
other	0.047	3,538	14.1
esw	0.181	19,605	9.7
all	**5.7**	**331,037**	**18.1**

Our experiments are limited by the nature of the *expert search* task in the TREC Enterprise track. The topics used by this task are limited in a way that the set of document judgments is incomplete. The reason is that some documents might be relevant to a topic but they do not provide the name of an expert. As a result, although these documents contain relevant information, they are not identified in the set of document judgments. On the other hand, any supporting document identified for an expert can be treated as relevant to the topic.

3.2 Results

For the *expert search* task, there were 54 different topics relating to the W3C collection. We separate this set of topics into *training* and *testing* sets: 50% of the topics, along with their relevance judgments, form the *training set* and the other 50% of the topics form the *testing set*. The training set is used to calculate the weight for each document type while the testing set is used to evaluate the performance of our method. Experiments were carried out with each half of the topics taking turn being the training set and the testing set.

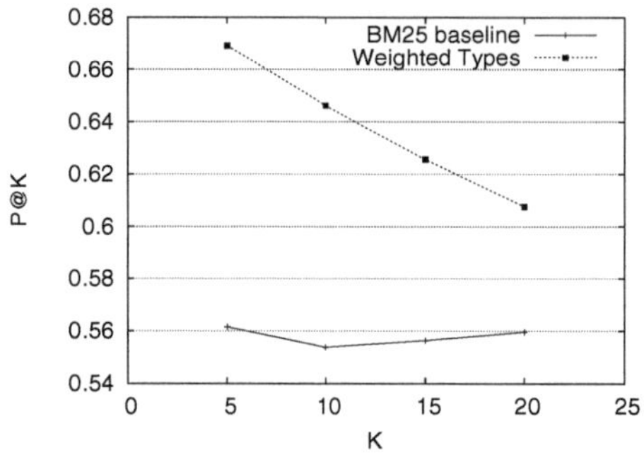

Fig. 1. Precisions for topics 52-78, using topics 79-105 for training

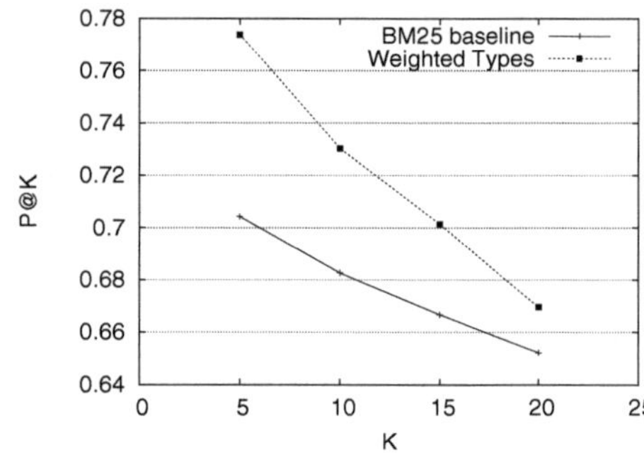

Fig. 2. Precisions for topics 79-105, using topics 52-78 for training

Using topics 79-105 as the training set, Figure 1 shows that $P@5$ improves from 0.5615 to 0.6692 for an 19% increase over the BM25 baseline model. Using topics 52-78 as the training set, Figure 2 shows that $P@5$ improves from 0.7043 to 0.7739 for a 10% increase over BM25.

4 Conclusion

In this poster, we presented a method for weighting document types and took the Bayes approach to estimate the appropriate weight for each type. Our approach is simple, which adapts document type in the retrieval process. Experimental results show that our approach improves search precisions for the TREC W3C collection.

References

1. N. Craswell, I. Soboroff, and A. de Vries. Overview of the trec-2006 enterprise track. to be published in 2006.
2. S. E. Robertson, S. Walker, M. Hancock-Beaulieu, A. Gull, and M. Lau. Okapi at trec-3. In *Proceedings of Text REtrieval Conference*, November 1994.
3. Stephen Robertson, Hugo Zaragoza, and Michael Taylor. Simple bm25 extension to multiple weighted fields. In *CIKM '04: Proceedings of the thirteenth ACM international conference on Information and knowledge management*, pages 42–49, New York, NY, USA, 2004. ACM Press.

Author Index

Lecture Notes in Computer Science

For information about Vols. 1–4328

please contact your bookseller or Springer

Vol. 4376: E. Frachtenberg, U. Schwiegelshohn (Eds.), Job Scheduling Strategies for Parallel Processing. VII, 257 pages. 2007.

Vol. 4374: J.F. Peters, A. Skowron, I. Düntsch, J. Grzymała-Busse, E. Orłowska, L. Polkowski (Eds.), Transactions on Rough Sets VI, Part I. XII, 499 pages. 2007.

Vol. 4373: K. Langendoen, T. Voigt (Eds.), Wireless Sensor Networks. XIII, 358 pages. 2007.

Vol. 4372: M. Kaufmann, D. Wagner (Eds.), Graph Drawing. XIV, 454 pages. 2007.

Vol. 4371: K. Inoue, K. Satoh, F. Toni (Eds.), Computational Logic in Multi-Agent Systems. X, 315 pages. 2007. (Sublibrary LNAI).

Vol. 4370: P.P Lévy, B. Le Grand, F. Poulet, M. Soto, L. Darago, L. Toubiana, J.-F. Vibert (Eds.), Pixelization Paradigm. XV, 279 pages. 2007.

Vol. 4369: M. Umeda, A. Wolf, O. Bartenstein, U. Geske, D. Seipel, O. Takata (Eds.), Declarative Programming for Knowledge Management. X, 229 pages. 2006. (Sublibrary LNAI).

Vol. 4368: T. Erlebach, C. Kaklamanis (Eds.), Approximation and Online Algorithms. X, 345 pages. 2007.

Vol. 4367: K. De Bosschere, D. Kaeli, P. Stenström, D. Whalley, T. Ungerer (Eds.), High Performance Embedded Architectures and Compilers. XI, 307 pages. 2007.

Vol. 4366: K. Tuyls, R. Westra, Y. Saeys, A. Nowé (Eds.), Knowledge Discovery and Emergent Complexity in Bioinformatics. IX, 183 pages. 2007. (Sublibrary LNBI).

Vol. 4364: T. Kühne (Ed.), Models in Software Engineering. XI, 332 pages. 2007.

Vol. 4362: J. van Leeuwen, G.F. Italiano, W. van der Hoek, C. Meinel, H. Sack, F. Plášil (Eds.), SOFSEM 2007: Theory and Practice of Computer Science. XXI, 937 pages. 2007.

Vol. 4361: H.J. Hoogeboom, G. Păun, G. Rozenberg, A. Salomaa (Eds.), Membrane Computing. IX, 555 pages. 2006.

Vol. 4360: W. Dubitzky, A. Schuster, P.M.A. Sloot, M. Schroeder, M. Romberg (Eds.), Distributed, High-Performance and Grid Computing in Computational Biology. X, 192 pages. 2007. (Sublibrary LNBI).

Vol. 4358: R. Vidal, A. Heyden, Y. Ma (Eds.), Dynamical Vision. IX, 329 pages. 2007.

Vol. 4357: L. Buttyán, V. Gligor, D. Westhoff (Eds.), Security and Privacy in Ad-Hoc and Sensor Networks. X, 193 pages. 2006.

Vol. 4355: J. Julliand, O. Kouchnarenko (Eds.), B 2007: Formal Specification and Development in B. XIII, 293 pages. 2006.

Vol. 4354: M. Hanus (Ed.), Practical Aspects of Declarative Languages. X, 335 pages. 2006.

Vol. 4353: T. Schwentick, D. Suciu (Eds.), Database Theory – ICDT 2007. XI, 419 pages. 2006.

Vol. 4352: T.-J. Cham, J. Cai, C. Dorai, D. Rajan, T.-S. Chua, L.-T. Chia (Eds.), Advances in Multimedia Modeling, Part II. XVIII, 743 pages. 2006.

Vol. 4351: T.-J. Cham, J. Cai, C. Dorai, D. Rajan, T.-S. Chua, L.-T. Chia (Eds.), Advances in Multimedia Modeling, Part I. XIX, 797 pages. 2006.

Vol. 4349: B. Cook, A. Podelski (Eds.), Verification, Model Checking, and Abstract Interpretation. XI, 395 pages. 2007.

Vol. 4348: S.T. Taft, R.A. Duff, R.L. Brukardt, E. Ploedereder, P. Leroy (Eds.), Ada 2005 Reference Manual. XXII, 765 pages. 2006.

Vol. 4347: J. Lopez (Ed.), Critical Information Infrastructures Security. X, 286 pages. 2006.

Vol. 4346: L. Brim, B. Haverkort, M. Leucker, J. van de Pol (Eds.), Formal Methods: Applications and Technology. X, 363 pages. 2007.

Vol. 4345: N. Maglaveras, I. Chouvarda, V. Koutkias, R. Brause (Eds.), Biological and Medical Data Analysis. XIII, 496 pages. 2006. (Sublibrary LNBI).

Vol. 4344: V. Gruhn, F. Oquendo (Eds.), Software Architecture. X, 245 pages. 2006.

Vol. 4342: H. de Swart, E. Orłowska, G. Schmidt, M. Roubens (Eds.), Theory and Applications of Relational Structures as Knowledge Instruments II. X, 373 pages. 2006. (Sublibrary LNAI).

Vol. 4341: P.Q. Nguyen (Ed.), Progress in Cryptology - VIETCRYPT 2006. XI, 385 pages. 2006.

Vol. 4340: R. Prodan, T. Fahringer, Grid Computing. XXIII, 317 pages. 2007.

Vol. 4339: E. Ayguadé, G. Baumgartner, J. Ramanujam, P. Sadayappan (Eds.), Languages and Compilers for Parallel Computing. XI, 476 pages. 2006.

Vol. 4338: P. Kalra, S. Peleg (Eds.), Computer Vision, Graphics and Image Processing. XV, 965 pages. 2006.

Vol. 4337: S. Arun-Kumar, N. Garg (Eds.), FSTTCS 2006: Foundations of Software Technology and Theoretical Computer Science. XIII, 430 pages. 2006.

Vol. 4336: V.R. Basili, D. Rombach, K. Schneider, B. Kitchenham, D. Pfahl, R.W. Selby, Empirical Software Engineering Issues. XVII, 193 pages. 2007.

Vol. 4335: S.A. Brueckner, S. Hassas, M. Jelasity, D. Yamins (Eds.), Engineering Self-Organising Systems. XII, 212 pages. 2007. (Sublibrary LNAI).

Vol. 4334: B. Beckert, R. Hähnle, P.H. Schmitt (Eds.), Verification of Object-Oriented Software. XXIX, 658 pages. 2007. (Sublibrary LNAI).

Vol. 4333: U. Reimer, D. Karagiannis (Eds.), Practical Aspects of Knowledge Management. XII, 338 pages. 2006. (Sublibrary LNAI).

Vol. 4332: A. Bagchi, V. Atluri (Eds.), Information Systems Security. XV, 382 pages. 2006.

Vol. 4331: G. Min, B. Di Martino, L.T. Yang, M. Guo, G. Ruenger (Eds.), Frontiers of High Performance Computing and Networking – ISPA 2006 Workshops. XXXVII, 1141 pages. 2006.

Vol. 4330: M. Guo, L.T. Yang, B. Di Martino, H.P. Zima, J. Dongarra, F. Tang (Eds.), Parallel and Distributed Processing and Applications. XVIII, 953 pages. 2006.

Vol. 4329: R. Barua, T. Lange (Eds.), Progress in Cryptology - INDOCRYPT 2006. X, 454 pages. 2006.

MIX
Papier aus verantwortungsvollen Quellen
Paper from responsible sources
FSC® C105338

If you have any concerns about our products,
you can contact us on
ProductSafety@springernature.com

In case Publisher is established outside the EU,
the EU authorized representative is:
Springer Nature Customer Service Center GmbH
Europaplatz 3, 69115 Heidelberg, Germany

Printed by Libri Plureos GmbH
in Hamburg, Germany